谨以此书

献给吕春绪教授七十华诞

吕春绪文选

《吕春绪文选》编辑组◎编

化学工业出版社
·北京·

本书是为庆祝吕春绪教授七十华诞而整理的，主要包括吕先生本人及其指导下弟子发表的科研论文。由于文章数量较多，按硝化反应与硝酰阳离子理论、膨化硝酸铵与膨化硝铵炸药自敏化理论、炸药合成及应用、药物中间体及表面活性剂等的合成及工艺四部分选取了86篇代表性论文进行全文收录。

图书在版编目（CIP）数据

吕春绪文选/《吕春绪文选》编辑组编. —北京：化学工业出版社，2012.11
ISBN 978-7-122-15369-2

Ⅰ.①吕…　Ⅱ.①吕…　Ⅲ.①炸药-文集　Ⅳ.①TQ564-53

中国版本图书馆CIP数据核字（2012）第221964号

责任编辑：刘　军　徐世峰　　　装帧设计：关　飞
责任校对：边　涛

出版发行：化学工业出版社（北京市东城区青年湖南街13号　邮政编码100011）
印　　刷：北京永鑫印刷有限责任公司
装　　订：三河市万隆印装有限公司
889mm×1194mm　1/16　印张33½　插页11　字数986千字　2013年1月北京第1版第1次印刷

购书咨询：010-64518888（传真：010-64519686）　售后服务：010-64518899
网　　址：http://www.cip.com.cn
凡购买本书，如有缺损质量问题，本社销售中心负责调换。

定　　价：268.00元

南京理工大学　吕春绪　教授

吕春绪教授，男，河北乐亭人，生于1943年3月12日。我国炸药领域著名学者，国家有突出贡献专家，国务院特殊津贴获得者。南京理工大学化学工程与技术一级学科带头人、应用化学国家级重点学科带头人。先后获国家科技进步二、三等奖、国家发明三等奖、部科技进步特等奖、民爆行业科技进步特别奖共16项。为我国含能材料赶超世界先进水平与持续发展起到重大促进作用。

1. 幸福家庭

与夫人在南京理工大学

与夫人参观上海世博会

与夫人及大孙女

与夫人及小孙女

全 家 福

2. 留学与访问

1994年留学瑞典斯德哥尔摩大学

与导师Garegg教授

在阿仑尼乌斯实验室

1990年访问日本东京大学

1996年陪同兵器工业部于桂臣副部长（左五）访问瑞典、挪威和英国

1997年率团访问美国盐湖城DYNO公司

会见美国Austing副总裁萨姆

2004年德国化学工艺研究院（ICT）
第35届年会与大会主席Fred Volk教授（左五）

1999年应邀访问印度SUA公司

吕春绪教授访问新加坡义安理工学院

3. 国际交流与合作

1999年主持第二十六届国际烟火会议。国际烟火协会主席N. Davies教授（左三）、
兵总火炸药局金长荣局长（左四）、南京理工大学校长李鸿志院士（左五）、
日本工业火药学会会长吉田忠雄教授（左六）等出席会议

2001年主持第一届国际民用爆破器材学术研讨会，国防科工委副主任张维民（左六）、
张嘉浩局长（左五）、吴燕萍董事长（右三）、曹志超主席（左七）、
国际烟火协会副主席Fred Volk教授（左四）、
波兰装备技术研究院院长Ryzad Kostrow教授（右四）等出席会议

1997年，代表南京理工大学与挪威DYNO公司签订技术合作及DYNO奖学金协议

2009年12月28日，会见美国Austing
国际公司总裁、炸药公司副总裁Mike Gleason（前左四）

2011年10月8日，会见国际爆炸协会主席、美国含能材料杂志（Journal of Energetic Materials）主编James Short教授（右三）

4. 桃李满天下

2003年9月，华东理工大学田禾院士应邀回母校参加50周年校庆

2009年与弟子中北大学化工学院院长曹端林教授一起

1990年在东京大学与我校在日留学生座谈

1997年在美国Austing公司与学生蒋云峰

2005年在日本与弟子权衡道、吴建洲

2001年与缅甸研究生高棉、金松

2006年与巴基斯坦研究生Masler Ayaza

与研究生一起

2005年在从教40周年学术研讨会上做
“硝酰阳离子理论”学术报告

会上弟子赠送对联，上书：
三尺讲台精析世界四十载
六百弟子深探乾坤八万里

与夫人（右四）接受弟子花篮，上书：
书硝化创膨化满园桃李
教硕士育博士四十春秋

在从教四十周年会上接受弟子祝福

5. 硝酰阳离子理论的提出及发展

在硝化理论的基础上，提出硝酰阳离子理论并撰写成书

1990年在日本东京大学做“硝酰阳离子反应”演讲

1999年做“甲苯选择性硝化”专题报告

2004年在广州中国绿色高新精细化工论坛做“绿色硝化反应及其应用”的主题报告

2007年应邀在中国工程物理研究院903所做“硝酰阳离子理论”专题报告

2009年应邀在上海有机所做“炸药合成反应及制造技术”专题报告

2009年应邀在北京理工大学做“炸药研究及发展”学术报告

2010年应邀在银光集团做 “炸药合成反应新发展”主题报告

2010年应邀在成都做 “对炸药研究与发展的几点建议”专题报告

6. 膨化硝铵炸药自敏化理论的形成及应用

提出膨化硝铵炸药自敏化理论并撰写成书

2010年“自敏化理论”获中国爆破器材行业协会科学技术特别奖。颁奖会上接受中国民爆行业协会张维民理事长颁奖，工信部吴风来司长出席会议

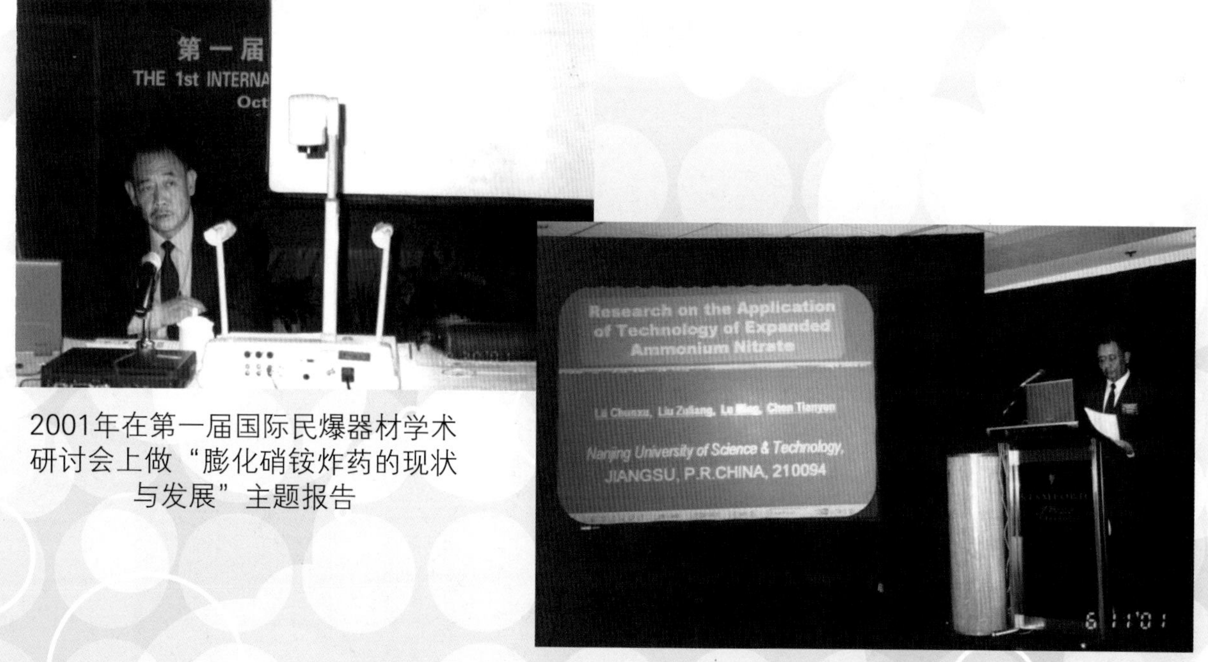

2001年在第一届国际民爆器材学术研讨会上做“膨化硝铵炸药的现状与发展”主题报告

2001年率团参加澳大利亚第28届国际烟火会议并做“膨化硝铵炸药自敏化理论及其应用研究”主题报告

2002年应邀在日本首届含能材料国际会议上做“硝酸铵膨化技术及应用”大会报告

2005年在第8届含能材料发展方向研讨会（捷克）上做“表面活性理论在含能材料应用研究”主题报告

2005年应邀参加日本第二届含能材料国际会议，并做“膨化硝铵炸药理论研究”的大会主题演讲

2005年应邀在武汉科技大学做“表面活性理论”专题报告

2006年应邀在中北大学做“膨化硝铵炸药自敏化理论”专题报告

2009年应云南民爆集团邀请在曲靖做专题讲座

2010年应中国科学院青岛能源研究所邀请做“表面活性剂技术及发展”学术报告

7.社会兼职

2007年在国防科工委民爆行业专家委员会成立大会上，吕春绪教授（右三）任副主任委员

2002年作为火炸药专业委员会副主任委员在火炸药年会上做“火炸药技术新发展”专题报告

2008年作为兵总火炸药专家委员会副主任委员在三亚钝感炸药及973项目学术研讨会上做大会报告

2011年作为中国民爆学会会刊《爆破器材》编委会主任参加中国民爆学会工作研讨会，工信部安全生产司金鑫副司长参加会议，并作重要讲话

2000年在江苏省药物中间体工程技术研究中心成立大会上做报告，任中心主任

2000年在江苏省表面活性剂及助剂工程研究中心成立大会上做报告，任中心主任

主持青海海西东诺董事会会议

作为湖北凯龙化工独立董事与荆门市张之嶙市长一起参观生产线

2011年主持南京东诺工业炸药高科技
有限公司董事会

2004年作为南岭民爆副董事长
与南岭民爆陈光正董事长（左三）、 明
景谷监事会主席（左一）、李铁良总经
理（左四）在一起

2012年8月作为海西东诺董事长
热情接待工信部安全生产司吴风来司长一行来公司指导工作

2012年在广州作为民用爆破器材专业委员会主任委员参加会议

8. 领导及专家参观指导

2000年兵器工业部民爆局张嘉浩局长
来中心参观指导

2000年中组部杨志海副部长（右二）
考察中心

2000年国防科工委吴伟仁司长（左一）来中心
考察、指导

2000年与唐明述院士

2000年与袁渭康院士

2003年江西省委副书记彭明松（右一）回母校参加50周年校庆

2003年日本工业火药学会会长吉田忠雄教授（左二）、日本东京大学田村昌三教授（左一）、国际烟火协会副主席久保田浪之介教授（左三）应邀来南京理工大学参加50周年校庆

2004年江苏省委书记陈焕友（左一）来中心参观指导

2006年，国防科工委张维民副主任（右二）在南京理工大学徐复铭校长（右一）陪同下指导工作

2006年，兵总火炸药局王成国局长参观指导

2007年，兵总火炸药局林菊生局长、金鑫副局长在南京理工大学校长王晓锋教授（左一）陪同下参观指导

2008年，南京工业大学副校长乔旭教授（前排左五）、江苏省化学化工学会秘书长赵伟建教授（前排左三）等专家领导来中心参观指导

2011年中国兵器工业集团副总杨卓教授（右二）参观指导

编辑组成员（按姓名汉语拼音为序）

序 一

吕春绪教授是我国含能材料学科领域著名的学者、专家，我校化学工程与技术一级学科带头人及国家级重点应用化学博士点学科带头人。在近50年的教学与科研生涯中，他辛勤耕耘，取得了丰硕成果。

吕春绪教授著作等身、硕果累累。本文选涵盖了吕春绪教授的四个主要研究领域：硝化反应与硝酰阳离子理论、膨化硝酸铵与膨化硝铵炸药自敏化理论、炸药合成及应用、药物中间体及表面活性剂等的合成及工艺，内容丰富，涉及面广。因篇幅所限，文选只收录了吕春绪教授不同时期发表的80余篇学术论文以及其他一些研究成果。文选阐述严谨，数据翔实，理论与研究开发、工程实践紧密结合，既具有重要的学术价值，又对理解吕春绪教授学术思想的形成具有参考意义。文选反映出吕春绪教授取得的两个突出科研成果：一个是硝酰阳离子（NO_2^+）理论，另一个是硝酸铵自敏化理论。

吕春绪教授特别注重选择性及绿色硝化等有机合成方法的研究。他深入研究 NO_2^+ 反应，以原子经济性为指导，重点研究 NO_2^+ 的选择性硝化与绿色硝化反应机理与技术，并将固载体成功地用于绿色硝化反应中；吕春绪教授结合其 NO_2^+ 定向硝化的科研成果，撰写了国内第一部关于 NO_2^+ 的科技专著——《硝酰阳离子理论》。该专著的出版对硝化反应理论研究及实际应用都有重要指导意义。

吕春绪教授利用表面活性理论，提出了硝酸铵自敏化理论，该理论是膨化硝酸铵发明点的核心。根据自敏化理论制备的膨化硝铵炸药系列彻底革除梯恩梯，使炸药成本大幅度降低，同时使工人免受梯恩梯毒害，消除环境污染，具有十分显著的经济效益和社会效益。膨化硝铵炸药技术被誉为“粉状工业炸药的一次革命”。至2010年底全国各地的近90家工厂已使用该技术，新增产值50多亿元，新增利税近15亿元。该技术的发明为我国粉状工业炸药创新与换代，为我国工业炸药赶超世界先进水平做出突出贡献。

吕春绪教授就这样孜孜不倦地奉献着、追求着，他说自己不过是无数献身祖国科技事业的知识分子中的普通一员。是啊，正是有了无数像吕春绪教授这样知难而进的攀登者忘我无私的付出，才有了伟大祖国科学技术蓬勃发展的美好前景。

中国工程院院士、南京理工大学原校长 李鸿志

2012年5月8日

序　二

本文选是吕春绪老师的学生们送给他的一份生日礼物。其中汇总了吕老师学习、工作及生活中的一些照片，具有代表性的研究论文及著作、发明专利，以及其他一些研究成果。

吕老师任教 40 多年来，在含能材料与应用化学领域从事教学与科学研究，成绩卓著。他长期担任本科生、研究生的专业基础，专业课程的教学，指导硕士生、博士 120 多名。

在教学上，吕老师认真贯彻教育方针，严谨治学。

他为人师表、教书育人，注重学生综合素质的培养。对军工、火化工领域有一定危险性的实验，他都认真摸索，并到现场指导。他爱生、重教、重师德的高尚品德，深受学生们的尊敬与爱戴。凡是听过吕老师授课的老师和学生都会深深体会到：吕老师的课生动实在、开放活跃。高深的理论在精辟的论述和形象的比喻中变得那么简单易懂，繁杂的公式经过具体的数据和图文的分析深刻地印在每个人的脑海中，启发式教学和创新思维不断激起大家的学习热情、打开人们心灵的窗口，很多学生受益匪浅。

在科研上，吕老师知难而进，锲而不舍，成果卓越。

随着研究的不断深入，吕春绪教授以独树一帜的硝化理论在学术界声名显赫，入选并载入《国际名人传记词典》。他曾两次赴日本东京大学就硝化理论进行讲学，得到国外专家的赞许与好评。后来，瑞典 STOCKHOLM 大学邀请他前去就定位反应进行合作研究，可优厚的工作、生活待遇并没有让他留恋，研究期满后，吕老师就回到了我们身边。他提出的“硝酰阳离子理论”曾被西安 204 所、中国工程物理研究院 903 所及中科院上海有机所邀请作专题讲学。

新型混合炸药配方设计与性能研究是吕老师又一个取得重要成果的研究领域。吕老师经过多年探索，独辟蹊径地提出硝酸铵“自敏化”的总体设计，并发展为自敏化理论：利用表面活性技术和控制析晶技术，成功制备了膨化硝酸铵及其炸药，充分改善了产品的物化性能。使得工业炸药无须添加梯恩梯也能达到良好的性能指标。有关成果经专家鉴定认为属国内外首创，并达到国际先进水平。该技术被列入“国家级科技成果重点推广计划”。目前已在全国 90 多个工厂推广使用，让一线工人彻底免受梯恩梯的毒害，同时还大幅度降低了炸药成本。被誉为“粉状工业炸药的一次革命”。

吕老师严谨的治学态度、丰富的科研经验以及敏锐的思维方式是我们学习的楷模。在吕老师七十大寿之际，我谨代表吕老师的学生们祝老师：身体健康，万事如意！

中国科学院院士，学生代表

2012 年 6 月 5 日

前　言

本书是为庆祝吕春绪教授七十华诞而整理的，主要包括吕先生本人及在吕先生指导下弟子公开发表的科研论文。由于文章数量较多，本书从中选取学术论文 80 多篇，汇编成本文选，以赐读者。

吕春绪，男，祖籍河北乐亭，1943 年 3 月 12 日生于吉林桦甸。1965 年毕业于中国人民解放军炮兵工程学院炸药专业，毕业后留校任教。吕春绪教授 1992 年开始享受国务院政府特殊津贴，1994 年以高级访问学者身份留学瑞典斯德哥尔摩大学，就定向反应进行合作研究，1997 年被评为江苏省“333”工程第一层次专家，1999 年被评为国家有突出贡献的中青年专家，1996—2003 年任南京理工大学副校长，同时兼任国防科工委民爆器材研究所所长、国家民爆质检中心主任；2003 至今任江苏省药物中间体工程技术研究中心主任。

吕春绪教授长期从事炸药设计、合成与工艺及应用领域教学和科研工作，是学校含能材料（炸药）及应用化学学科带头人，在炸药领域，尤其在硝化理论、耐热炸药合成、液体炸药配方设计以及膨化硝铵技术等方面有重要发明创造和科技成果。主要成就和贡献为：

1. 提出并发展了硝酰阳离子理论

硝化反应是炸药合成中的主反应，硝化理论是硝化反应的核心，硝酰阳离子（NO_2^+）是硝化理论的精华，一直是炸药工作者关注的热点。

在深入研究硝化反应基础上，以《硝化通论》（1975 出版）及《硝化理论》（1993 出版）为蓝本，结合十多年科研及硕士、博士学位论文和所获得的科研成果中硝化反应部分，吕春绪教授提出了“硝酰阳离子理论”。此理论包括：硝酰阳离子结构及光谱、载体特征及生成反应、区域选择性硝化、N_2O_5 硝化以及绿色硝化。硝酰阳离子与硝酸合氢离子和有机物结合过程及动力学，特别是与胺类及醇类的反应具有独特见解及创新性。这些内容反映了当前该领域的新成就及新发展，有其新颖性及先进性。根据这些成果，2006 年吕春绪教授撰写《硝酰阳离子理论》一书。该书是硝化反应领域的一部科技专著，在国际上尚属首次出版。其对硝化反应理论研究及实际应用具有重要指导意义。尤其是 N_2O_5、离子液体等中性硝化，站在节能减排、绿色环保角度上，被誉为炸药制造史上的重大革命。

2. 二步法合成耐热炸药六硝基芪（HNS）

他主持二步法合成 HNS 新工艺研究，以 TNT 为起始原料，经中间体六硝基联苄（HNBB），二步合成 HNS 新方法。二步法革新了传统工艺，使生产更加安全、平稳。他提出的稀硝氧化及用晶癖改良剂可同时制得 HNS-Ⅰ及 HNS-Ⅱ两种产品，母液套用及直接湿转工艺，使其产率及熔点均超过美国 Shipp 法。该方法具有新颖性、独创性，达到国际先进水平。该成果获得国家发明三等奖。

结合多年对耐热炸药合成及工艺研究的成果，1999 年撰写了耐热炸药领域的专著——《耐热硝基芳烃化学》，并被列入普通高等教育“九五”国家级重点教材。

3. 设计成功 SJY 高爆速液体炸药并填补国内空白

该液体炸药是由硝酸肼、水合肼及添加剂组成，具有高爆速、高猛度的特点。其关键技术是硝酸肼的制备。他突破国外采用的惰性介质甲醇（有毒、价昂）减缓反应的难题，创新性地提出制备硝酸肼的直接硝化和减压浓缩工艺，使硝酸肼产率及纯度大幅度提高，成本显著降低。该成果获国家科技进步三等奖。

4. 提出膨化硝铵炸药自敏化理论及其应用

以表面活性剂理论及表面化学为基础，吕春绪教授提出了膨化硝铵自敏化理论。采用特殊技术途径将微气泡引入硝酸铵晶体内并连同晶体表面歧形化成为热点载体，使其起爆感度发生质的突跃，这就是自敏化理论设计。该理论具有我国自主知识产权，是膨化硝铵炸药主要发明点的核心。

硝酸铵自敏化设计是对国内外传统方法的突破，是国内外炸药工作者未曾想到的创新思路，对膨化硝铵炸药的诞生起到关键作用。膨化硝铵炸药成果获兵器工业部特等奖和国家科技进步二等奖，该理论及其发明的技术为我国工业炸药创新及换代、赶超世界先进水平做出突出贡献。2008年，吕春绪教授撰写《膨化硝铵炸药自敏化理论》一书。

本文选涵盖了硝化反应及硝酰阳离子理论、膨化硝酸铵及膨化硝铵炸药自敏化理论、炸药合成及应用、药物中间体及表面活性剂等的合成及工艺等含能材料、工业炸药、精细化工品等相关学科及专业，对理解吕春绪教授学术思想具有重要参考意义。

因时间比较仓促，文选中不当及疏漏在所难免，敬请批评指正。

编辑组

2012年6月26日

目　录

第一部分　硝化反应与硝酰阳离子理论/1

第二部分　膨化硝酸铵与膨化硝铵炸药自敏化理论/87

第三部分 炸药合成及应用/233

第四部分 药物中间体及表面活性剂等的合成及工艺/319

第一部分

硝化反应与硝酰阳离子理论

金属硝酸盐-醋酐硝化剂的研究

吕春绪，蔡春
（南京理工大学）

摘要： 用拉曼光谱研究了金属硝酸盐-醋酐硝化剂中的活化硝化成分，认为该体系中的活化硝化剂是硝鎓离子。利用气相色谱建立了该硝化剂在有机溶剂中进行硝化反应的Hammett方程，求得反应常数值$\rho=-6.24$。一定条件下该硝化剂具有较好的反应位置选择性，这一是由于反应介质低的介电常数，二是多孔载体也具有明显的作用。

关键词： 硝化反应，拉曼光谱，Hammett方程，位置选择性

当有醋酐存在时，金属硝酸盐可在比较温和的条件下对含有邻对位定位基团的一元取代苯进行硝化反应。由于该反应体系中低的介电常数，使提高反应的位置选择性变得较为容易。例如，将金属硝酸盐吸附在k10蒙脱土上对甲苯、氯苯进行一段硝化时，可明显提高一硝产物中对位异构体的生成比例[1-2]。一般认为该反应体系中的活化硝化成分是硝酸乙酰酯[3]。对于在甲苯硝化反应中得到的低的邻对异构体比例，有人认为是因金属离子与醋酐的作用而使该条件下的反应处于轨道控制[4]。

1 实验部分

1.1 金属硝酸盐一醋酐硝化剂中活化硝化成分的拉曼光谱研究

拉曼光谱是研究活化硝化剂硝鎓离子NO_2^+的有效手段之一，NO_2^+和CO_2具有相似的结构，在1400cm^{-1}附近有其特征谱线[5]。实验中采用法国Jobin Yvon公司HRD 2型双单色仪和Coherent公司Innova 70型Ar^+激光器，激光功率是150mW，波长519.5nm。

为了避免待测定体系的颜色吸收激光束，采用硝酸锌为研究对象。室温下将20g六水硝酸锌溶解于50ml醋酐中，两者摩尔数之比约为1∶8，取该样品溶液在900-1600cm^{-1}范围内进行扫描，记录所得的图线即为该体系的拉曼光谱。

1.2 用竞争法测定定各一元取代苯相对于苯的硝化反应速度

实验中采用的一元取代苯是苯甲醚、甲苯、乙基苯、氯代苯和溴代苯，这些化合物及其一硝产物的异构体都可由气相色谱分离并加以定量。用竞争法测定各一元取代苯在四氯化碳溶液中相对于苯的硝化反应速度，所用金属硝酸盐是三水硝酸铜，醋酐用量是所用硝酸铜摩尔数的5倍。将硝酸铜在丙酮溶液中吸附于一定数量的载体硅藻土上，这样可增大固液反应的相间接触面积，使反应平稳进行。反应于室温下在普通的玻璃仪器中进行，反应结束后将有机相处理至中性，用气相色谱分析其中各组分的含量。

气相色谱分析用美国Varian公司SP 3700型气相色谱仪。色谱柱OV-101毛细管柱，25m×0.2mm；柱前压82.7kPa；气体流量：空气270cm^3/min，氮气27cm^3/min，氢气24cm^3/min；温度：

柱温 175℃，汽化室 210℃，检测器 200℃。

2 结果和讨论

2.1 活化硝化成分的确定

由实验得到的拉曼光谱见图 2.1。从图中可以看出，在 $1380cm^{-1}$ 处有一微弱但却较明显的散射峰，这就是活化硝化剂硝鎓离子 NO_2^+ 的特征谱线，说明该体系中存在 NO_2^+，并可用光谱方法测定。以硝酸铜为例，金属硝酸盐与醋酐间存在下列平衡：

$$5(CH_3CO)_2O+Cu(NO_3)_2 \cdot 3H_2O \rightleftharpoons Cu(CH_3CO)_2+6CH_3COOH+2CH_3COONO_2 \quad (2.1)$$

硝酸乙酰酯经下面两个过程产生 NO_2^+：

$$2CH_3COONO_2 \rightleftharpoons (CH_3CO)_2O+N_2O_5 \quad (2.2)$$

$$N_2O_5 \rightleftharpoons NO_2^+ +NO_3^- \quad (2.3)$$

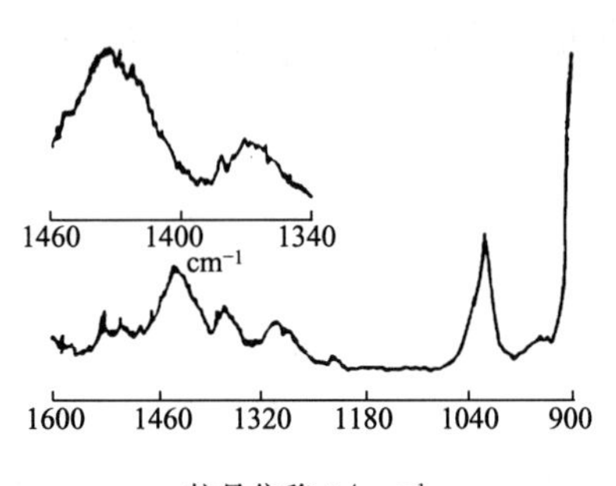

图 2.1 硝酸锌-醋酐体系的拉曼光谱

Fig. 2.1 Raman spectrograph of the system Zn $(NO_3)_2 \cdot H_2O$-Ar_2

拉曼光谱图中同样可看到 $1020cm^{-1}$ 处 NO_3^- 的散射峰，该峰较强，这主要是因硝酸乙酰酯能吸收空气中的水，也产生 NO_3^-。硝化反应发生时，体系中可观察到少量硝烟，硝烟是由 N_2O_5 在空气中分解所产生：

$$N_2O_5 \rightleftharpoons N_2O_4+1/2O_2 \quad (2.4)$$

将反应控制在适当的条件下，可使硝烟尽可能少地产生，这也是该类硝化剂在实际应用中具备的优点之一。

尽管硝酸乙酰酯也可能作为硝化剂进行硝化反应，但由以上分析可知，该反应条件下的活化硝化成分仍是通常的硝鎓离子。

2.2 Hammett 方程的建立

将等摩尔的一元取代苯和苯溶于有机溶剂中，用不足量的硝化剂进行硝化，可求得相对反应速度：

$$\frac{\kappa}{\kappa_0}=\frac{\text{硝基芳烃(不含硝基苯)摩尔数}}{\text{硝基苯摩尔数}}\times\frac{6}{m} \quad (2.5)$$

m 表示待测定化合物苯环上可取代位置的数目，对于一元取代苯 $m=5$。

文中研究的是含有邻对位定位基团的一元取代苯对位的硝化反应，可得 Hammett 方程的具体表达式：

$$\lg f_P=\rho\sigma P^+ \quad (2.6)$$

式中，f_P 为对位反应的分速度因子，$f_P=6\times\frac{k}{k_0}\times P$-%，$P$-%为对位产物分数。由气相色谱分析所得数据进行处理得到的 P-%，k/k_0 和 f_P 见表 2.1。

表 2.1 竞争法硝化得到的实验结果

Tab. 2.1 Results from competitive nitration

一元取代苯	甲苯	氯代苯	溴代苯	苯甲醚	乙苯
P-%	46.1	80.6	84.1	69.4	54.1
k/k_0	24.7	0.0407	0.0151	13416	21.0
f_P	68.32	0.1968	0.07629	55846	68.17

$\lg f_P$ 与 σ_P^+ 之间呈很好的直线关系，见图 2.2。以数理统计方法求得该直线的斜率为－6.24，可得该条件下硝化反应的 Hammett 方程为：

$$\lg f_P = -6.24\sigma_P^+$$

相关系数 $\tau=-0.9996$，$\lg f_P$ 与 σ_P^+ 完全负相关。

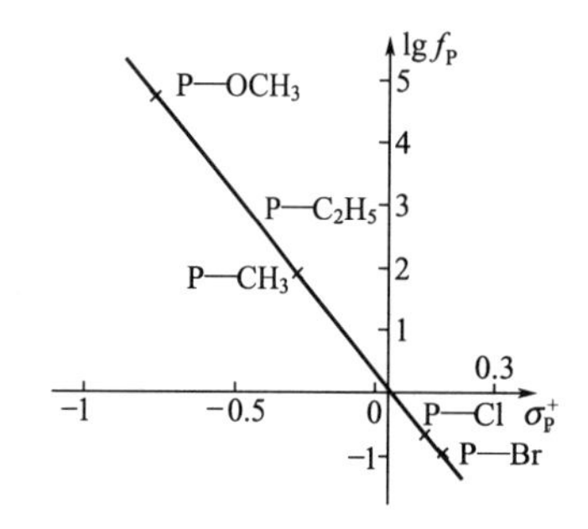

图 2.2 一元取代苯硝化反应 $\lg f_P$ 与 σ_P^+ 的关系

Fig. 2.2 Relationship between $\lg f_P$ and σ_P^+ in the nitration of monosubstituted benzene

反应常数 ρ 反映了与标准反应相比，反应对取代基效应的敏感程度，很明显，ρ 的绝对值愈大，效应愈敏感。根据过渡状态理论，反应速度与过渡状态的结构有关，所以 ρ 反映了与基态相比过渡状态电荷产生和消失的情况，以及对取代基电子效应的要求。ρ 值只与反应进行的条件如温度、溶剂等有关，某一在规定条件下进行的反应，ρ 值是不变的。

在不同条件下得到硝化反应的 ρ 值为：硝酸在硝基甲烷或醋酐中硝化 $\rho=-6.53$[6]；硝硫混酸硝化 $\rho=-6.4$；硝酸乙酰酯在醋酐中硝化 $\rho=-6.0$。

本文实验得到硝酸铜-醋酐在四氯化碳中进行硝化反应时 $\rho=-6.24$，该值与以上三个文献报导值很接近，说明该条件下硝化反应的历程与过渡状态结构和通常的硝化反应是相似的，仍为一般的电荷控制反应。金属离子对该反应并没有起到某种特殊的作用，事实上，铝、铁、锌、铜等金属离子对甲苯一硝产物中异构体的比例没有明显影响。ρ 值为负时，增加反应中心的电荷密度可以加速反应。ρ 标度的数值范围约 0±4，而求得硝化反应的 ρ 值在－6～－7 之间，说明硝鎓离子进攻苯环形成正离子中间体芳鎓离子通常是慢的限速步骤，该过程的速度为供电子基团所促进，因为这类基团有助于使导致生成中间体过渡态上正电荷的离域。另外，ρ 绝对值大，也说明硝化反应过渡态具有较明显的正离子特征。

2.3 位置选择性的解释

作为电荷控制反应，反应介质的介电常数对硝化反应的产物异构体比例有较明显的影响。对甲苯而言，反应介质的介电常数降低，不利于硝鎓离子的稳定，使得反应速度降低，但由于甲基邻位的净电荷大于对位的净电荷，因此反应介质的介电常数降低，生成邻硝基甲苯的速度要降低更多一些，其结果就使得一硝产物中邻对异构体比例下降。

另一方面，在有多孔物质，如酸性黏土、硅胶、活性炭、硅藻土等存在的情况下硝化甲苯，可明显降低一硝产物的邻对比。表 2.2 中列出了将硝酸铜吸附在不同的载体上，在四氯化碳溶液中进行硝化反应所得的产物异构体比例。在通常条件下进行硝化所得的产物邻对异构体比在 1.41～1.62 之间[7]，可以很明显看出，在相同反应条件下，不同类型多孔载体的存在都能有效地改变反应的位置选择性。这主要是由于载体的多孔性质决定的，因为载体的存在会阻碍亲电试剂对邻位的进攻。同时，研究还发现，载体表面的状态和酸性等都明显影响反应的选择性，如在载体表面引入路易斯酸，会明显提高甲苯硝化反应的对位位置选择性，但其作用机制目前尚不清楚。由 ESCA 结果可知，载体并不影响硝酸铜的结构，因此这方面的影响是可以排除的。

表 2.2 载体对产物异构体比例的影响

Tab. 2.2 Effect of various supports on the isomer ratio

载　体	O/P 变化范围	载　体	O/P 变化范围	载　体	O/P 变化范围
活性炭	0.79～0.89	高岭土	1.01～1.08	薄层硅胶	0.97～1.09
硅藻土	1.01～1.06	54＃树脂	0.93～1.09	硅藻土/$FeCl_3$	0.89～0.94
氧化铝	0.96～1.06				

3 结论

在金属硝酸盐—醋酐硝化剂体系中，活化硝化剂 NO_2^+ 的浓度是可测的，该硝化剂进行的硝化反应仍是由 NO_2^+ 直接进行的电荷控制反应，降低反应介质的介电常数或在反应体系中加入多孔载体，都能有效地降低甲苯一段硝化产物的邻对异构体比例。

参考文献

[1] Cornelis A，et al.. Tetrahedron Lett.，1988，29 (49)：5657.
[2] Laszlo P，Pennetreau P. J. Org. Chem.，1987，52：2907.
[3] 福水公寿，木村允．日本化学会志，1973，7：1306.
[4] Laszlo P，Vandormeal J. Chem. Lett.，1988：1893.
[5] 周发歧．炸药合成化学．北京：国防工业出版社，1989：269.
[6] Schofield K，Aromatic Nitration，Cambridge University Press，London. 1980：311.
[7] 奥尔布赖特 L F，汉森 C. 工业与实验室硝化．北京：化学工业出版社，1989：339.

Investigation on a Nitration Agent Consisting of Metallic Nitrates and Acetic Anhydride

Lv Chunxu，Cai Chun
(Nanjing University of Science & Technology)

Abstract：With Laser Raman spectroscopy，the active nitrating agent in a system of metallic nitrates and acetic anhydride has been shown to be nitronium. The Hammett equation for the nitration with this nitrating agent in an organic solvent has been established with the aid of gas chromatography，and the reaction constant has been obtained $\rho = -6.24$. In this system good regioselectivity can be achieved mainly because of the existence of a porous support and the low dielectric constant of the reaction media.

Key words：Nitration，Raman spectroscopy，Hammett equation，regioselectivity

（注：此文原载于　兵工学报，1994，(2)：42-45）

Research on the Selective Nitration of Acetanilide and its Kinetics

Lv Chunxu, Gu Jianliang, Peng Xinhua
(Nanjing University of Science & Technology, Nanjing 210094, China)
EI: 96043144880; ISTP: A1995BF21H00045

Abstract: This paper deals with selective nitration of acetanilide. The best technological conditions and method are obtained, and selective nitration reaction kinetics of acetanilide with mixed acid is studied with the aid of UV-Vis spectrum. It is inferred that nitration of acetanilide with mixed acid is second order reaction and its rate constant is 3.96 L • mol^{-1} • s^{-1} in 81% sulfuric acid.

1. Introduction

P-nitroacetanilide is an important organic intermediate. It is obtained from acetanilide through selective nitration reaction. If nitrating agent is different, result of selective nitration will be different very much[1-6]. For example:

Table 1 Effect of different nitrating agent on results of selective nitration products ratio

Reactant	Tem. of reaction ℃	Nitrating agent	Proportion of ismoer of nitration products % O	M	P
$NHCOCH_3$ (benzene ring)	20	$HNO_3+H_2SO_4$	19.4	2.1	78.5
	20	HNO_3+AC_2O	67.8	2.5	29.7

The result of selective nitration will obviously be different, even though we use the same kind of nitration agent with different composition[7,8]. For example:

Table 2 Effect of composition of mixed acid on results of selective nitration products ration

Reactant	Tem. of reaction ℃	Compositions of mixed acid	Proportion of isomer of nitration products % O	M	P
$NHCOCH_3$ (benzene ring)	25	HNO_3-98% H_2SO_4	5	2	93
	25	HNO_3-80.9% H_2SO_4	17	2	81
	25	HNO_3-77.4% H_2SO_4	23	2	75
	25	HNO_3-69.9% H_2SO_4	37	2	61

The best technological method and conditions of selective nitration with mixed acid are obtained through optimum seeking different reaction methods. Parameter test and orthogonal test. Nitration reaction kinetics is studied.

2. Results and discussions

2.1 Determination of the best technological conditions of selective nitration of acetanilide with mixed acid

2.1.1 Research on parameter test of selective nitration of acetanilide with mixed acid

Table 3　Parameter test of selective nitration of acetanilide

	Factor							Target					
	Concern of NA %	ρ %	V_{NA} ml	Φ	V_{SA} ml	T ℃	t h	Y_1 %	MP_1 ℃	O/P	1/2 O/P	Y_2 %	MP_2 ℃
	98	0	8.6		30	25	3	—	—	—	—	—	—①
Concern. of NA	76.6	0	11.4		30	25	3	91.5	165-92	0.5/1.5	0.167	68.6	211-5②
	65-68	0	13.7		30	25	3	81.1	173-81	0.5/1.5	0.167	60.8	201-5
	65-68	5	14.4		30	25	3	90.8	136-76	0.5/1.5	0.167	60.6	205-15
ρ	65-68	10	15.0		30	25	3	93.1	145-65	0.5/1.5	0.167	69.7	200-9
	65-68	15	15.7		30	25	3	95.0	137-67	0.55/1.45	0.190	68.9	201-9
	65-68	10	15.0	71	16.5	25	3	—	—	—	—	—	—③
Φ(V_{SA})	65-68	10	15.0	81	30	25	3	93.1	145-65	0.5/1.5	0.167	69.7	200-9
	65-68	10	15.0	85	45	25	3	92.2	140-73	0.4/1.6	0.125	73.9	209-13
T	65-68	10	15.0	85	45	15	3	96.7	170-93	0.45/1.55	0.125	77.2	207-11
	65-68	10	15.0	85	45	15	3	98.3	161-90	0.3/1.7	0.145	76.1	207-10
	65-68	10	15.0	85	45	15	1	93.1	156-89	0.3/1.7	0.088	79.2	211-5
	65-68	10	15.0	85	45	15	2	94.4	159-90	0.2/1.8	0.056	85.0	209-13
t	65-68	10	15.0	85	45	15	3	96.7	170-93	0.4/1.6	0.125	77.2	207-11
	65-68	10	15.0	85	45	15	5	90.8	158-90	0.2/1.8	0.056	81.7	206-9
	65-68	10	15.0	85	45	15	7	94.2	166-91	0.25/1.75	0.071	82.5	207-12

① When NA is added dropwise, masses are immediately spurted

② When NA is added dropwise, temperature is not easily controlled

③ Reaction is not completely finished

④ Y_1 is the yield of nitration

⑤ Y_2 is the yield of para-products

When Y_2 is considered as target of judge, better parameters are obtained (they are expressed as* in Tab. 3): concentration of NA 65-68%, 15ml, ρ10%; concentration of SA 95-98%, 45ml, Φ85; T 15℃; t 2h.

2.1.2 Orthogonal experiments of nitration acetanilide (65-68% NA, 95-98% SA) with mixed acid

(1) Table 4 and Table 5 show respectively better datum and mathematics analysis of orthogonal test

Table 4 Orthogonal experiments of nitration of acetanilide (65-68%NA, 95-98%SA)

No.	A V_{NA} ml	A ρ %	B V_{SA} ml	C T ℃	D t h	Target Y_1 %	MP_1 ℃	O/P	1/2 O/P	Y_2 %	MP_2 ℃
1	1(14.4	5)	1(35)	1(10)	1(1.5)	90.8	161-89	0.35/1.65	0.106	75.0	209-13
2	1		2(45)	2(15)	2(2)	94.4	159-86	0.3/1.7	0.088	80.3	210-4
3	1		3(55)	3(20)	3(2.5)	95.3	161-81	0.2/1.8	0.056	85.8	209-13
4	2(15.0	10)	1	2	3	91.9	162-91	0.4/1.6	0.125	73.6	213-6
5	2		2	3	1	93.6	147-91	0.35/1.65	0.106	77.2	208-13
6	2		3	1	2	93.1	154-92	0.2/1.8	0.056	83.9	208-12
7	3(15.7	15)	1	3	2	93.3	161-91	0.35/1.65	0.106	76.9	209-14
8	3		2	1	3	93.1	159-91	0.35/1.65	0.106	76.7	209-13
9	3		3	2	1	94.2	157-97	0.2/1.8	0.056	84.7	210-4

Table 5 Mathematical analysis of orthogonal test of nitration acetanilide (65-68%NA, 95-98%SA)

		A	B	C	D	Better
	K_1	93.5	92.0	92.3	92.9	
	K_2	92.9	93.7	93.5	93.6	
Y_1	K_3	93.5	94.2	94.1	93.4	$B_3C_3D_2A_{1,3}$
	R	0.6	2.2	1.8	0.7	
	K_1	0.083	0.112	0.089	0.089	
1/2 O/P	K_2	0.096	0.100	0.090	0.083	$B_3A_1D_2C_{1,3}$
	K_3	0.089	0.056	0.089	0.096	
	R	0.013	0.056	0.001	0.013	
	K_1	80.4	75.2	78.5	79.0	
Y_2	K_2	78.2	78.1	79.5	80.4	$B_3A_1D_2C_3$
	K_3	79.4	85.8	80.0	78.7	
	R	2.2	9.6	1.5	1.7	
The best combination of level						$B_3A_1D_2C_3$

The best combination of level is : NA 65-68%, 14.4ml, ρ5%; SA 95-98%, 55ml, Φ87; T 20℃, t 2h.

(2) Table 6 shows check, parallel and enlargement test of the best combination of level

Table 6 Check, parallel and enlargement test of selective nitration of acetanilide (65-68%NA, 95-98%SA)

Test	A_1 ρ %	V_{NA} ml	Φ	B_3 V_{SA} ml	C_3 T ℃	D_2 t h	Target Y_1 %	MP_1 ℃	O/P	1/2 O/P	Y_2 %	MP_2 ℃
Check test	5	14.4	87	55	20	2	97.8	161-91	0.25/1.75	0.071	85.6	211-4
Parallel test	5	14.4	87	55	20	2	95.0	165-92	0.25/1.75	0.071	83.1	211-5
Enlargement tes(3 times)	5	43.2	87	165	20	2	100.0	159-90	0.2/0.8	0.056	90.0	211-4

2.2 Measurement of reaction rate constant of selective nitration of acetanilide with mixed acid

2.2.1 Work curve

0.2250g $O_2NC_6H_4NHAc$ (O/P=1/4) is accurately weighted and added into measuring flask. It is exactly diluted to 50ml with 95% alcohol. 2.5×10^{-2} mol (4500ppm) solution is obtained. Then 10ml of above-mentioned solution is added into measuring flask. It is exactly diluted to 100ml with 95% alcohol again. 2.5×10^{-3} mol (450ppm) solution is obtained. 0.0, 0.2, 0.3, ··· 1ml of above-mentioned solution is added into measuring flask, 2.5×10^{-5} mol (0.00, 0.90, 1.35··· 4.50ppm) solution is obtained, and absorbance A is measured at wave length 320nm. The results are given in Tab. 7[10-12].

According to Table 7, Fig. 1 is drew to give A～C work curve.

Table 7 The absorbance A of $O_2NC_6H_4NHAc$ (O/P=1/4) at wave length 320nm in 95% alcohol solution

	ppm	0.00	0.90	1.35	1.80	2.35	2.70	3.15	3.60	4.05	4.50
C	10^{-5}mol	0.00	0.50	0.75	1.00	1.25	1.50	1.75	2.00	2.25	2.50
	A①	0.000	0.114	0.177	0.241	0.305	0.369	0.408	0.480	0.544	0.613

① Instrument: 7520 spectrometer.

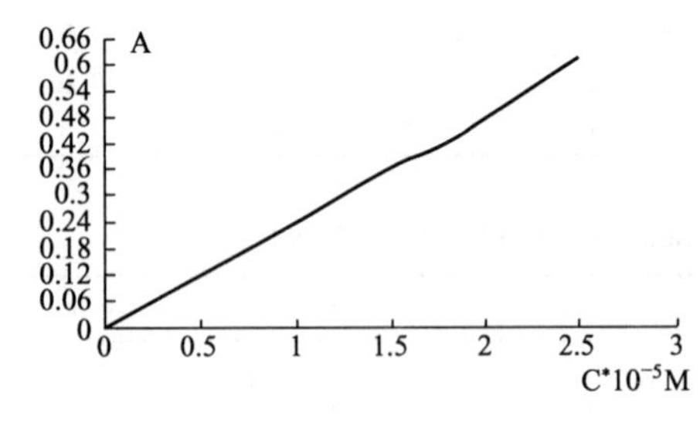

Fig. 1 work curve

The results of calculation with least square method[13] can be expressed as follow:

$$A=0.2437C-0.0038$$

Where C-concentration 10^{-5}mol

R-interrelated coefficient 0.999

2.2.2 Determination of rate constant

(1) Method of test

When acetanilide is nitrated in 81% SA at 25℃, O/P ratio of products is about 1/4[6]. As nitration is an exothermic reaction, low concentration of reactant must be used, so that reaction runs smoothly and reaction time is easily measured. A suitable amount of acetanilide is dissolved in 81 %SA and 65% NA is dissolved in 50ml 81% SA at 25℃. The two solutions are then quickly mixed under stirred. Reckoning of time is begun. At a certain time interval fixed amounts of reaction is taken out, and is quickly added into measuring flask and diluted with 95% alcohol to ceased the reaction[9,10]. After it is uniformly mixed, it is exactly diluted to 100ml with 95% alcohol. The absorbance is measured at wave length 320nm.

(2) Datum and calculation of test

Table 8 No. 1 test: SA 81% 100ml, NA 65% 0.0242g, acetanilide 0.0338g, T 25℃, C_0 2.5×10^{-3}mol

No.	t min	Amount of reactant ml	A	C① 10^{-3}mol	$[1/(C_0-C)]-1/C_0$	$\ln\{[1/(C_0-C)]-1/C_0\}$	lnt②
1	0.1	1	0.242	1.008	270	5.60	3.40
2	1	1	0.318	1.325	451	6.11	4.09
3	2	1	0.438	1.825	1081	6.99	4.79
4	5	1	0.526	2.192	2847	7.95	5.70
5	12	1	0.563	2.346	6094	8.72	6.58
6	30	1	0.588	2.450	19600	9.88	7.50
7	90	1	0.594	2.475	39600	10.59	8.59

① C means concentration of $O_2NC_6H_4NHAc$ in the reaction solution at certain time.

② ln { [1/ (C_0-C)] -1/C_0} Unit of it is second.

According to Tab. 8, a plot of $\ln\{[1/(C_0-C)]-1/C_0\}$ against lnt (Fig. 2) shows a straight line. From this. it is inferred that nitration of acetanilide with mixed acid is second order with $k_2=2.10$ and $k_3=8.17L\cdot mol^{-1}\cdot s^{-1}$.

The parameters of No. 2 test are SA 81% 100ml, NA 65% 0.0485g, acetanilide 0.0676g, T 25℃, $C_0=5\times10^{-3}$mol. The datum of experiment are obtained. According to above-mentioned datum, a plot of $\ln\{[1/(C_0-C)]-1/C_0\}$ against lnt shows

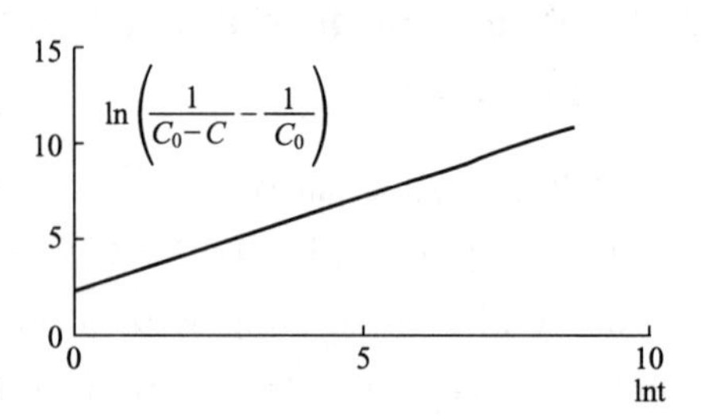

Fig. 2 No. 1 test, $\ln\{[1/(C_0-C)]-1/C_0\}$～lnt

a straight line. From this, it is inferred that nitration of acetanilide with mixed acid is second order, with k_2=2.25 and k_3=9.49 L·mol^{-1}·s^{-1}.

Average value of k_2 of No. 1 and No. 2 test is 8.83L·mol^{-1}·s^{-1}

Tab. 9 shows results of caculation with least square method.

Table 9 Results of caculation with least square method

	No. 1 Test				No. 2 Test			
	T s	A	C 10^{-3}mol	$[1/(C_0-C)]-1/C_0$	T s	A	C 10^{-3}mol	$[1/(C_0-C)]-1/C_0$
1	30	0.242	1.009	271	30	0.352	2.921	281
2	60	0.318	1.321	448	60	0.437	3.618	524
3	120	0.438	1.813	1056	90	0.496	4.103	915
4	300	0.526	2.175	2677	120	0.507	4.196	1039
5	720	0.563	2.326	5347	180	0.530	4.382	1418
6	1800	0.588	2.429	13685	360	0.565	4.669	2821
7	5400	0.594	2.454	21339	900	0.589	4.866	7263
8					3600	0.597	4.932	14506
Regression equation	$\ln\{[1/(C_0-C)]-1/C_0\}=1.628+3.965t$				$\ln\{[1/(C_0-C)]-1/C_0\}=0.958+3.951t$			
k_2	3.97				3.95			
R	0.957				0.971			

Average value of k_2 of No. 1 and No. 2 test is 3.96L·mol^{-1}·s^{-1}.

3. Conclusions

(1) Mixed acid is the best nitrating agent for selectively nitration of acetanilide into P-nitroacetanilide.

(2) Technological enlargement test shows: the best suitable technological conditions are as follows: 27g (0.2mol) acetanilide; 65-68% NA, 14.4ml, ρ5%, 95-98%SA, 55ml, Φ87; the best reaction temperature 20℃, reaction time 2 h. These technological conditions are extremely more advantageous than currently used technology.

The conditions of currently used technology by factories are : reaction temperature 5℃, reaction time 20h.

We enhance reaction temperature to 20℃ and shorten reaction time to 2h, and yield of nitration is more than 93% with 87.5-90% para-products[14]. Quality targets required by factory are satisfied. A vast amount of cost of cooling are saved, capability of production are raised and production cost are decreased.

(3) Selective nitration reaction kinetics of acetanilide with mixed acid is studied with the aid of Uv-Vis spectrum. It is inferred that nitration reaction kinetics of acetanilide with mixed acid is second order, its rate constant being 3.96L·mol^{-1}·s^{-1} in81%sulfuric acid.

Reference

[1] Shofield K. Aromatic Nitration. Cambridge University Press, 1980.

[2] Lv Chunxu. Nitration Theory. Nanjing: Jiangsu Science and Technology Press, 1993.

[3] Brian M. Can. J. Chem. 1986 (46): 1141-1148.

[4] Cook M. USP, 3981933. 1976.

[5] Schossler C. USP，4600797. 1986.
[6] Tomasz A. Can. J. Chem.，1976 (54) 560-564.
[7] Hartshorn S. R. J. Chem. Soc.，1971 (B)：2454.
[8] Moodic R. J. Chem. Soc.，1977 (Perkin Ⅰ)：1693-1705.
[9] Wen Jinquan. Research of Reaction Mechnism of Org. Chem. Nanjing：East China Institute of Technology Press，1986.
[10] Wang Qi. Introduction to Chemical Dynamics，Jiling：Jinling People Press，1982.
[11] Fu Xiancai. Physical Chemistry (No. Ⅰ). Beijing：Peoples Education Press，1990.
[12] Chen Jie，Song Qize. Spectrum Analysis. Nanjing：East China Institute of Technology Press，1990.
[13] Liang Ziyu. Probablity and Statistics. Beijing：Advanced Education Press，1990.
[14] Gu Jianliang. The Studies of Selective Nitration of Acetanilide and 4-Methoxyacetanilide (Thesis). Nanjing：East China Institute of Technology Press，1992.

（注：此文原载于 Proceedings of the Third International Symposium on Pyrotechnics and Explosives. 1995：225-230）

苯酚两相硝化反应及其机理研究

吕春绪，蔡春
（南京理工大学，南京，210094）
EI：98084353667

摘要： 研究了利用乙醚和硝酸钠-硫酸水溶液对苯酚进行两相硝化反应，并确定了反应最佳工艺条件。利用HPLC技术等研究了苯酚两相硝化反应机理，结果认为：反应产物的异构体比例不随反应条件而变，实验测得值与半经验分子轨道的计算结果相符较好，有机相溶剂的性质对产物异构体比例无明显影响，并证实反应所经历的是自由基历程。

关键词： 苯酚，两相硝化，异构体，自由基

硝基苯酚是有机合成的重要中间体。苯酚与稀硝酸在室温条件下很容易反应生成邻硝基苯酚（*o*-MNP）和对硝基苯酚（*p*-MNP）。但因酚羟基和苯环易被氧化，产物的得率很低。若能在减少氧化反应的条件下采用直接法硝化得到硝基苯酚，就能使原料和设备成本大大降低，具有明显的经济效益。近年来人们非常关心简单酚类的两相硝化，结果能使简单酚类直接硝化的得率明显提高[1,2]。结合两相硝化反应特征，我们比较深入地研究了苯酚两相硝化反应影响因素和可能的反应机理。

1 苯酚两相硝化反应的影响因素

1.1 实验

10g苯酚溶解在一定体积的乙醚中，将所得溶液加入到1.1g硝酸钠与60ml一定浓度硫酸组成的溶液中，体系呈两相。另将一定量的亚硝酸钠溶于10ml水中，在磁力搅拌条件下慢慢滴加到反应体系中，控制温度在指定的值，使反应维持一定的时间。而后分离出乙醚相，往其中加入30ml水，在50℃水浴之下蒸除乙醚，乙醚蒸完后有黑色油状物沉于水的底部。用水蒸气蒸馏的方法分离出*o*-MNP，以乙醇-水重结晶，得嫩黄色棱状晶体，熔点44～45℃。向水蒸气蒸馏的残液中加入3g活性炭，搅拌，趁热过滤，将所得滤液静置过夜，有淡黄色针状晶体析出，过滤即得产物*p*-MNP，熔点113～119℃。

1.2 苯酚两相硝化反应影响因素分析

1.2.1 亚硝酸钠用量对反应的影响

乙醚100ml，反应温度20℃，反应时间1h，结果见表1.1。亚硝酸钠的作用是诱导硝化反应发生。简单酚类的两相硝化存在诱导期，它的长短主要取决于酚的活性、水相中酸的浓度和反应进行时的温度。实验结果表明，当亚硝酸钠的用量为0.5g时，反应有较好的结果；用量过少，由于反应速度较慢，在一定反应时间内产物得率较低；用量过多时，由于自催化作用会导致一定程度的氧化反

应，也会使产物的得率下降。

表 1.1 亚硝酸钠用量对反应的影响

Tab. 1.1 Effects of the amount of $NaNO_2$ on the reaction

$NaNO_2$ 用量/g	*o*-MNP 生成量/g	*p*-MNP 生成量/g	MNP 得率%	乙醚回收率%
0.1	5.8	3.4	62.2	76
0.5	8.1	4.9	87.8	78
1.0	7.6	4.4	81.1	81
2.0	7.4	4.5	80.4	79

1.2.2 反应温度的影响

亚硝酸钠 0.5g，乙醚 150ml，反应时间 1h，结果如表 1.2 所示。温度明显影响反应速度。由于苯酚较活泼，因此 20℃以上温度的影响就不明显了。而且反应温度高，无疑会导致氧化反应，但在本实验的浓度范围内，尚未产生明显的氧化反应。

表 1.2 温度对反应的影响

Tab. 1.2 Effects of temperature on the reaction

反应温度/℃	*o*-MNP 生成量/g	*p*-MNP 生成量/g	MNP 得率%	乙醚回收率%
15	5.7	3.0	58.8	86
20	7.2	3.5	72.3	83
25	6.8	4.1	73.6	82
30	6.9	4.3	75.7	79

1.2.3 反应时间对结果的影响

亚硝酸钠 0.5g，乙醚 150ml，反应温度 20℃，结果见表 1.3。显然，延长反应时间可以提高产物得率。

表 1.3 反应时间对反应的影响

Tab. 1.3 Effects of time on the reaction

反应时间	*o*-MNP 生成量/g	*p*-MNP 生成量/g	MNP 得率%	乙醚回收率%
0.5	6.0	3.9	66.9	81
1.0	6.7	4.3	74.3	80
1.5	7.8	4.5	83.1	76
2.0	8.0	4.6	85.1	78

1.2.4 影响因素的正交实验研究

选择反应温度、乙醚相体积、硫酸浓度、亚硝酸钠用量为 4 个可控因素。其它固定因素，水相由 60ml 硫酸溶液与 10ml 亚硝酸钠溶液组成，苯酚 10g，硝酸钠 11g，反应时间 1h。

由各单因素实验结果可以看出，硝化产物的异构体比例不受诸因素明显影响，因此在正交实验中仅以产物得率为评价指标。对正交实验结果进行极差分析，得反应的优化条件为：反应温度 25℃；乙醚相体积 70ml，主要因素；硫酸浓度 5M；亚硝酸钠用量 0.1g。

在此优化条件下其反应结果为：*o*-MNP 8.5g；*p*-MNP 4.9g，MNP 总得率 90.5%。

2 苯酚两相硝化反应可能的机理

对于在亚硝酸钠诱导下进行的苯酚两相硝化反应，提出了反应历程：

$$ArOH + NO^+ \longrightarrow ArOH^+\cdot + NO\cdot \qquad NO\cdot + NO_3^- \longrightarrow NO_2\cdot + NO_2^-$$

$$ArOH^+\cdot \longrightarrow ArO\cdot + H^+ \qquad ArO\cdot + NO_2\cdot \longrightarrow O_2NC_6H_5O\text{(二烯酮)}$$

$$O_2NC_6H_5O \longrightarrow O_2C_6H_4OH(\text{互变体})$$

该历程可以由半经验分子轨道计算的结果加以证实。关键步骤是以 NO^+ 作为电子转移试剂与苯酚形成正离子自由基的碰撞对（ET），反应的诱导期可看成是 NO^+ 的形成过程，其后的步骤都是容易进行的。

2.1 实验

实验在配有 JB60-D 型增力电动搅拌机的玻璃仪器中进行。10g 苯酚溶解于 100ml 有机溶剂中形成有机相，12g 硝酸钠溶于 80ml 4M 硫酸中形成酸性水相，反应温度控制在 15±1℃，搅拌条件下加入 1ml 10%的亚硝酸钠溶液诱导反应。从加完亚硝酸钠的时刻起计时，在一定的时间间隔中从反应体系的有机相中移取 2ml 反应液，用 25ml 相同的有机溶剂稀释使反应中止随后以 HPLC 分析其中 *o*-MNP 和 *p*-MNP 的含量，由于间硝基苯酚在反应中的生成量很少，故分析过程中其量忽略不计。

分析用 HPLC 为日立 653 型高效液相色谱仪。测试条件：紫外检测器，检测波长 314nm，流动相为二氯甲烷，流速 $3cm^3/min$，色谱柱为 3040 硅胶填充柱，柱压 1.5×10^7Pa，*o*-MNP 和 *p*-MNP 标样自制。

2.2 苯酚两相硝化反应可能的机理讨论

（1）以乙醚为有机相时，在反应的不同阶段 *o*-MNP 和 *p*-MNP 的生成量之比（o/p）见表 2.1。在反应过程中，o/p 值随时间无明显的变化规律，其值在 1.13～1.23 之间波动。

表 2.1 不同反应阶段产物的异构体比例

Tab. 2.1 The isomer ratio in the production of reaction at different time

时间/min	1	4	9	19	39	69
邻对比(o/p)	1.13	1.14	1.19	1.23	1.19	1.17

实验结束后分离出的酸性水相以 100ml 乙醚分两次萃取，萃取液以 HPLC 分析，发现其中不含硝基苯酚，因此反应过程中生成的硝基苯酚在水相中分配的量可予以忽略。

（2）不与水混溶的有机溶剂都可作为该条件下两相硝化反应的有机相，且都能在不同程度上降低氧化反应的发生。不同的有机相对反应的影响主要表现在反应速度上，而对产物异构体比例无明显影响。以往人们曾认为有亚硝酸存在时硝化苯酚，经历的是亚硝化-氧化的反应途径，该结论遭到否定的主要原因是亚硝化主要产生对亚硝基化合物，进而可能得到很高比例的对硝基化合物，而实际产物为 *o*-MNP 和 *p*-MNP 的混合物，且 *o*-MNP 所占比例要高[3,4]。近年来有人研究了电荷转移硝化与亲电硝化的区别[5]，由静电学知识也很容易得知，硝化产物的异构体比例较大程度上受苯环的各个可取代位电荷密度大小的影响，苯酚分子中苯环上各位置的 π 电子云密度如图所示[6]。

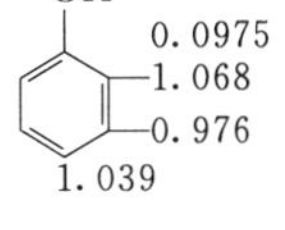

OH
0.0975
1.068
0.976
1.039

显然，若亚硝酸存在时硝化苯酚经历的是亲电反应历程，在忽略立体效应的前提下，o/p 的理论值约为 2，且也应有一定比例的间位取代产物生成。

自由基反应无论在气相或溶液中进行都是十分相似的，酸碱的存在或溶剂极性的改变对自由基反应都没有什么大的影响，而且很少受溶剂的离子溶剂化力影响[7]。计算得到的酚氧自由基中未偶电子在各位置的自旋密度见表 2.2。可以看出，C-4 位具有最高的自旋密度，其次是 C-2、C-6 位，苯环各位置上的相对自旋密度可用来决定在不同位置上取代数量的多少。

表 2.2 利用 AMISCF（UHF）方法计算得到的酚氧自由基未偶电子在各位置的自旋密度

Tab. 2.2 Unpaired electron spin density of the phenoxy radical determined using the AMISCF（UHF）method

位置	C_1	C_2	C_3	C_4	C_5	C_6	O
自旋密度	0.0322	0.2199	0.0043	0.3847	0.0042	0.2204	0.1338

根据计算结果可以推测，所得硝化产物的邻对异构体比例为

$$o/p=\frac{C_2+C_6}{C_4}=\frac{0.2199+0.2204}{0.3847}=1.15$$

表 2.1 所列的实验结果能与之较好吻合，同时也可估计到间位取代产物的生成比例低于 1%。离子反应通常无诱导，由于溶剂效应，反应速度要受到相当大的影响[8]。一元取代苯进行硝化反应生成不同的产物异构体可以看成是一组平行反应，显然溶剂效应也会影响生成不同产物异构体的分速度，宏观上表现为产物异构体比例的改变。以上研究结果表明，该条件下所进行的苯酚两相硝化反应，既不是由 NO_2^+ 直接进行的硝化反应，也不是由 NO^+ 先进行亚硝化，再经过氧化得到硝化产物，由此可证实反应所经历的是自由基历程。酚氧自由基与 $NO_2\cdot$ 在不同位置上结合决定了产物的异构体比例。

3 结论

（1）以乙醚为苯酚溶剂和硝酸钠-硫酸水溶液对苯酚进行两相硝化，工艺要求简单，硝基苯酚得率可达 90%，氧化副反应很少，产物的分离并不复杂，乙醚的回收率在 75%以上，若装置设计合理，该值还可得到提高。活性炭可以重复使用，因此生产成本低，具有工业应用前景。

（2）在优化条件下苯酚的两相硝化有较高得率，产物异构体比例不随反应条件变化 HPLC 研究结果证实了反应的自由基历程。

参考文献

[1] Thompson M J，et al，Tetrahedron Lett.，1988，29（20）：2471.

[2] Thompson M J，et al，Tetrahedron，1989，45（1）：191.

[3] Schofield K. Aromatic Nitration，New York：Cambridge University Press，1980：79.

[4] Usama A O. J. Chem. Soc. Peerkin Trasn. Ⅱ，1985：467.

[5] Sankararaman S，et al. J. Chem. Soc. Perkin Trans. Ⅱ，1991：1.

[6] Carcy F A，et al. Advanced Organic Chemistry，2nd ed.，New York：1984：495.

[7] 穆光照. 自由基反应. 北京：高等教育出版社，1958：38.

[8] 金长振. 理论有机化学. 北京：中国友谊出版公司，1990：217.

Investigation on the Two Phase Nitration of Phenol and its Mechanism

Lv Chunxu, Cai Chun

(Nanjing University of Science & Technology, Nanjing, 210094)

Abstract: The two-phase nitration of phenol with ether and aquatic solution of $NaNO_2^-$ H_2SO_4 is studied. The best conditions of processing are concluded at. With the help of the HPLC technique the mechanism of the two-phase nitration reaction is studied. It is shown that the isomer ratio in the product does not change with variations in reaction conditions, that the result is in good agreement with that of semiempirical orbital calculations, and that the organic solvent has no obvious effect on the isomer ratio, so the free radical mechanism in this reaction can be proved.

Key words: phenol, two-phase nitration, isomers, radical

（注：此文原载于　兵工学报，1996，17（4）：312-315）

乙酰苯胺选择性硝化反应速度研究

吕春绪，顾建良
（南京理工大学，江苏南京，210094）
EI：98054183172

摘要： 采用紫外-可见（Uv-Vis）光谱研究了乙酰苯胺混酸选择性硝化反应速度，推断乙酰苯胺混酸硝化是二级反应，测的在81%硫酸中硝化的二级反应常数为3.96L/(mol·s)。
关键词： 硝化，选择性硝化，反应常数

对硝基乙酰苯胺是重要的有机中间体，它是由乙酰苯胺经选择性硝化反应而获得的，不同的反应路线，选择性硝化结果大不相同。即使同一硝化，不同组成，其选择性硝化结果也有明显差别。特别是混酸硝化，在工业上虽被普遍采用，但是存在着对邻比低，反应时间长，能耗高，生产效率低等缺点。近年来，我们对乙酰苯胺的混酸选择性硝化除了欲改进其工艺外，还比较深入地研究了硝化反应影响因素及其反应动力学。

1 乙酰苯胺混酸选择性硝化最佳实验结果

通过不同反应路线的优化选择，单因素，多因素正交实验，确定出其混酸选择性硝化最佳工艺路线和最佳工艺条件，其最优水平组合为：硝酸（NA）浓度65～68%，用量14.4ml；硝酸过用量（ρ值）为5；硫酸（SA）浓度95～98%，用量为55ml，硫酸有效浓度（Φ值）为87；硝化温度为20℃，硝化反应时间为2h。

表1.1 乙酰苯胺硝化验证，平行，放大试验（65～68%NA，95～98%SA，2g乙酰苯胺）

Tab.1.1 Checking，parallel and enlargement tests in the selective nitration of acetanilide (65～68%NA，95～98%SA，2g acetanilide)

	ρ %	V_{NA} /ml	V_{SA} /ml	T /℃	t/h	y1%	$mp1$ /℃	o/p	$\frac{1}{2}o/p$	y2%	$mp2$ /℃
验证试验	5	14.4	87	55	20	97.8	161-91	0.25/1.75	0.071	85.6	211-4
平行实验	5	14.4	87	55	20	95	165-92	0.25/1.75	0.071	83.1	211-5
放大3倍实验	5	43.2	87	165	20	100	159-90	0.2/1.8	0.056	90	211-4

本工艺与工厂现行工艺比较要优越得多。生产的产品质量达到或超过要求的技术指标，且由于反应温度提高和保温时间缩短，节省大量冷却费用，提高生产能力，降低了成本。与现行工艺比较数据如表1.2。

在研究最佳工艺路线及反应条件的同时，我们还研究了其反应动力学，特别是其反应速度常数。

表 1.2 乙酰苯胺硝化工艺参数比较

Tab. 1.2 Comparision of parameters in the nitration of acetanilide

	每吨乙酰苯胺需酸量 95～98% SA/kg	每吨乙酰苯胺需酸量 65～68% NA/kg	P %	反应温度 /℃	保温时间 /h	硝化得率 %	P/(O+P)%	对硝基乙酰苯胺得率%	硝化产物熔点/℃
工厂现行工艺	3000	735	3	7	20	85～90	80～85	85	210～212
本研究工艺	3750	750	5	20	2	95～100	87.5～90	83～90	>210

2 乙酰苯胺混酸选择性硝化速度常数的测定

假设乙酰苯胺以混酸（MA）硝化是二级反应

$$C_6H_5NHAc+HNO_3 \xrightarrow{H_2SO_4} 6O_2NC_6H_4NHAc+H_2O$$

$$-\frac{d[C_6H_5NHAc]}{dt}=\frac{d[O_2NC_6HNHAc]}{dt}=k_2[C_6H_5NHAc][HNO_3] \tag{2.1}$$

式中，k_2 为乙酰苯胺 MA 硝化的二级反应速度常数。假定反应刚开始时（$t=0$）的 $[C_6H_5NHAc]=[HNO_3]=C_0$，反 t 时$[O_2NC_6H_5NHAc]=C$，式（2.1）可化为：

$$\frac{dC}{dt}=k_2(C_0-C)^2 \tag{2.2}$$

积分得：

$$\frac{1}{C_0-C}-\frac{1}{C_0}=k_2t$$

或

$$\ln(\frac{1}{C_0-C}-\frac{1}{C_0})=k_2t=\ln t+\ln k_2 \tag{2.3}$$

只要测定不同时间 t 生成的 $O_2NC_6H_4Ac$ 的浓度 C，再依式（2.3）作 ln｛$[1/(C_0-C)]-(1/C_0)$｝-ln$t$ 图，如果得一直线，则可能推断乙酰苯胺以 MA 硝化是二级反应的假设正确[5~7]，再由截距 lnk_2 计算得 k_2。

用紫外-可见（Uv-Vis）光谱可以区别不同物质的测定浓度[8]。为了选择适当的吸收波长，分别作了 C_6H_5NHAc，$O_2NC_6H_4Ac$（$o/p=1/4$）95%乙醇水溶液的 Uv-Vis 光谱（图 2.1），图中曲线 1 为乙酰苯胺 2×10^{-5}mol（2.7ppm），曲线 2 为硝基乙酰苯胺 1.5×10^{-5}mol（2.7ppm）。$ON_2C_6H_4NHAc$ 在 320nm 处有最大吸收，且吸收是符合比尔定律的；C_6H_5NHAc 无吸收，

而 NA，SA 的 95%乙醇水溶液在 320nm 处有最大吸收。

$$A=\varepsilon LC \tag{2.4}$$

式中，A，ε 分别为 $O_2NC_6NH_4HAc$（$o/p=1/4$）在 $\lambda_{max}=320$nm 处吸收度和摩尔吸光度系数；L 为样品池厚度，1cm；C 为 $O_2NC_6H_4NHAc$（$o/p=1/4$）的浓度，mol。测定不同时 t 的 A，求得 C，从而由式（2.3）作图，若假设式（2.1）立，即可由截距求得 k_2。

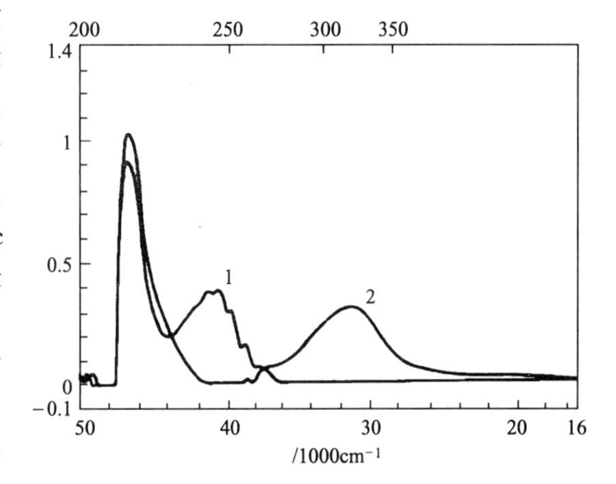

图 2.1 乙酰苯胺，硝基乙酰苯胺（$o/p=1/4$）95%乙醇水溶液的 Uv-Vis 光谱

Fig. 2.1 Uv-Vis spectrum of acetanilide and nitroacetanilide（$o/p=1/4$）in 95%alcoholic solution

2.1 绘制工作曲线

准确称取 0.225g $O_2NC_6H_4NAc$（$o/p=1/4$）置于 50ml 容量瓶中，加 95%乙醇至刻度，得 2×

10^{-2}mol（4500ppm）溶液；取此溶液10ml置于100ml容量瓶中，加95%乙醇至刻度，得2×10^{-3} mol（4500ppm）溶液；取0，0.2，0.3，…，1ml 2×10^{-3}mol的溶液分别置于100ml容量瓶中，加95%乙醇至刻度，得0，0.5，0.75，…，2×10^{-5} mol（0，0.9，1.35，…，4.5ppm）的溶液，在320nm处测其吸光度，列于表2.1中。依表2.1数据绘A-C图（图2.2），即为工作曲线。

表2.1 $O_2NC_6H_4NHAc$（$o/p=1/4$）95%乙醇溶液在320nm处的吸光度

Tab. 2.1 Absorbance A of $O_2NC_6H_4NHAc$（$o/p=1/4$）at 320nm in 95%alcoholic solution

C	/ppm	0	0.9	1.35	1.8	2.35	2.7	3.15	3.6	4.05	4.5
	$/10^{-5}$mol	0	0.5	0.75	1	1.25	1.5	1.75	2	2.25	2.5
A①		0	0.114	0.117	0.241	0.305	0.369	0.408	0.48	0.544	0.613

① 仪器：7520分光光度计

最小二乘法［9］计算的结果是

$$A=0.2437C-0.00388 \tag{2.5}$$

式中，C以10^{-5}mol为单位，相关系数$R=0.999$。

2.2 反应速度常数测定

（1）实验方法 25C时81%的SA中硝化乙酰苯胺产物o/p值约1/4[4]。硝化反应是放热的，因此使用低浓度的反应物使反应平稳放热，便于计算反应时间。具体方法是：将称好的乙酰苯胺溶于50ml 81%SA中，65%NA溶于50ml 81%SA中，恒温至25℃，再在搅拌下迅速混合，开始计时，然后每隔若干时间取出定量的反应液，迅速投入放有95ml 95%乙醇的100ml容量瓶中，稀释，使反应终止[5,6]，混匀，以95%乙醇加满至刻度，在320nm处测定吸光度。

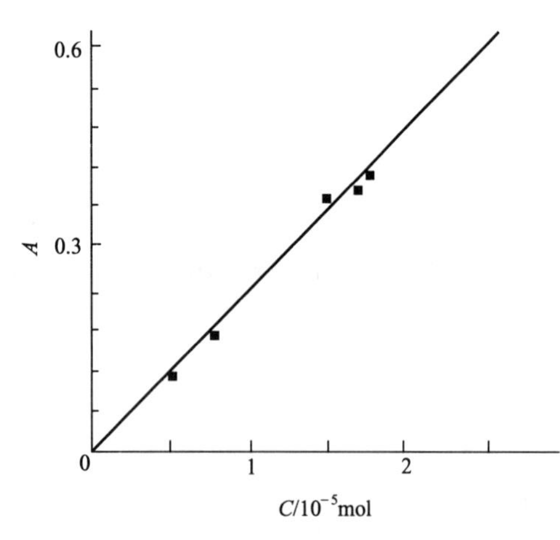

图2.2 工作曲线

Fig. 2.2 Working curve

（2）数据处理 以表2.2中ln｛［$1/(C_0-C)$］−（$1/C_0$）｝对lnt作图（图2.3），得一直线，由此推断乙酰苯胺MA硝化是二级反应，截距$\ln k_2=2.1$，$k_2=8.17$L/(mol·s)。以表2.3中ln｛［$1/(C_0-C)$］−（$1/C_0$）｝对lnt作图（图2.4），得一直线，由此推断乙酰苯胺MA硝化是二级反应，截距$\ln k_2=2.25$，$k_2=9.49$L/(mol·s)。实验一，二的平均值$k_2=8.83$L/(mol·s)。表2.4列出了最小二乘法的处理结果。实验一，二的平均值$k_2=3.96$L/(mol·s)。

表2.2 实验一：SA81%100ml，NA65%0.024 2g，乙酰苯胺0.0338g，$T=25$℃，$C_0=2.5\times10^{-3}$mol

Tab. 2.2 Test No. 1（SA81%100ml，NA65%0.024 2g，acetanilide 0.003g，$T=25$℃，$C_0=2.5\times10^{-3}$mol）

序号	t/min	取样量/ml	A	C① $/10^{-3}$mol	［$1/(C_0-C)$］−（$1/C_0$）	ln{［$1/(C_0-C)$］−（$1/C_0$）}	ln$t$②
1	0.5	1	0.242	1.008	270	5.60	3.40
2	1	1	0.318	1.325	451	6.11	4.09
3	2	1	0.438	1.825	1081	6.99	4.79
4	5	1	0.536	2.192	2847	7.59	5.70
5	12	1	0.563	2.346	6094	8.72	6.58
6	30	1	0.588	2.450	19600	9.88	7.50
7	90	1	0.594	2.475	39600	10.59	8.59

① C是反应液中$O_2NC_6H_4NHAc$的浓度；

② t的单位是秒。

表 2.3 实验二：SA81%100ml，NA65%0.0485g，乙酰苯胺 0.0676g，$T=25℃$，$C_0=2.5\times10^{-3}$mol

Tab. 2.3 Test No. 2 (SA 81%100ml, NA 65% 0.0485g, acetanilide 0.0676g, $T=25℃$, $C_0=2.5\times10^{-3}$mol)

序号	t/min	取样量/ml	A	C① /10^{-3}mol	$[1/(C_0-C)]-(1/C_0)$	$\ln\{[1/(C_0-C)]-(1/C_0)\}$	$\ln t$②
1	0.5	0.5	0.353	2.933	284	5.60	3.40
2	1	0.5	0.437	3.642	536	6.28	4.09
3	1.5	0.5	0.496	4.133	953	6.86	4.50
4	2	0.5	0.507	4.225	1090	6.99	4.79
5	3	0.5	0.530	4.417	1515	7.32	5.19
6	6	0.5	0.565	4.708	3225	8.08	5.89
7	15	0.5	0.589	4.908	10670	9.28	6.80
8	60	0.5	0.597	4.975	39800	10.59	8.59

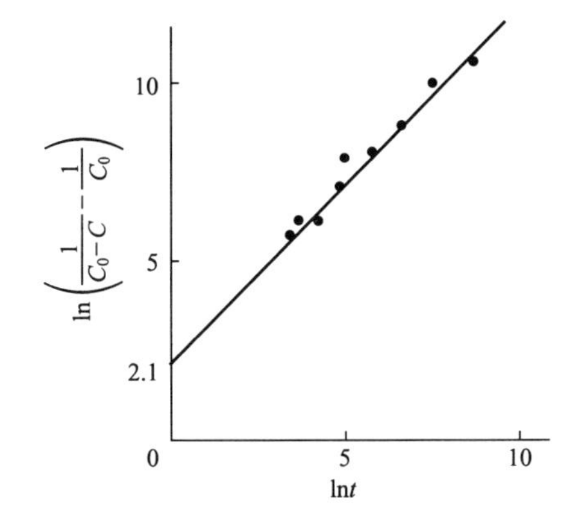

图 2.3 实验一 $\ln\{[1/(C_0-C)]-(1/C_0)\}$-ln$t$ 图

Fig. 2.3 Test No. 1 $\ln\{[1/(C_0-C)]-(1/C_0)\}$ against ln$t$

图 2.4 实验一 $\ln\{[1/(C_0-C)]-(1/C_0)\}$ -ln$t$ 图

Fig. 2.4 Test No. 1 $\ln\{[1/(C_0-C)]-(1/C_0)\}$ against ln$t$

表 2.4 最小二乘法处理结果

Tab. 2.4 Result of cacaulation with method of least square

序号	实验一 t/s	实验一 A	实验一 C/10^{-3}mol	实验一 $[1/(C_0-C)]-(1/C_0)$	实验二 t/s	实验二 A	实验二 C/10^{-3}mol	实验二 $[1/(C_0-C)]-(1/C_0)$
1	30	0.242	1.009	271	30	0.352	2.921	281
2	60	0.318	1.321	448	60	0.437	3.618	524
3	120	0.438	1.813	1056	90	0.496	4.103	915
4	300	0.526	2.175	2677	120	0.507	4.193	1039
5	720	0.563	2.326	5347	180	0.530	4.382	1418
6	1800	0.588	2.429	13685	360	0.565	4.669	2821
7	5400	0.594	2.454	21339	900	0.598	4.866	7263
8					3600	0.597	4.932	14506
回归方程	$\ln\{[1/(C_0-C)]-(1/C_0)\}=1.628+3.965t$				$\ln\{[1/(C_0-C)]-(1/C_0)\}=0.958+3.951t$			
k_2	3.97				3.95			
R	0.957				0.971			

3 结论

(1) 基于紫外-可见（Uv-Vis）光谱可以区别不同物质和测定浓度，从而研究了乙酰苯胺混酸选择性硝化反应动力学。

(2) 从 $\ln[1/(C_0-C)]-(1/C_0)$对 ln$t$ 作图，得一直线，由此推断乙酰苯胺混酸硝化是二级反应。

(3) 用最小二乘法计算乙酰苯胺在 81%硝酸中硝化的二级反应速度常数为 3.96L/(mol·s)。

参考文献

[1] Schofied K. Aromatic nitration. Cambridge University Press, 1980.
[2] 吕春绪. 硝化理论. 南京: 江苏科技出版社, 1993.
[3] Hartshorn S R. J. Chem. Soc., 1971 (B): 2454.
[4] Moodie R. J Chem., 1977 (Perkin Ⅱ): 1693-1705.
[5] 温敬铨. 有机化学反应机理的研究 (上册). 南京: 华东工学院, 1986.
[6] 王琪, 化学动力学导论. 吉林: 吉林人民出版社, 1982.
[7] 傅献彩, 物理化学 (下册). 北京: 人民教育出版社, 1986.
[8] 陈洁, 宋启泽. 波谱分析. 南京: 华东工学院, 1990.
[9] 梁之舜. 概率论及数据统计. 北京: 高等教育出版社, 1990.

On the Reaction Rate in the Selective Nitration of Acetaniline

Lv Chunxu, Gu Jianliang

Abstract: The rate of selective nitration of acetanilide with mixed acid is studied with the help of Uv-Vis spectra. It is inferred that the nitration of acetanilide with mixed acid is a second order reaction and its rate constant is 3.96L/(mol·s) in the presence of 81% sulfuric acid.

Key words: nitration, selective nitration, reaction rate

(注: 此文原载于 兵工学报, 1998, 19 (1): 28-31)

Regioseletivities of Toluene Nitration on Betonite Catalysts by Using of Alkyl Nitrate

Lv Chunxu, Peng Xinhua

(Nanjing University of Science and Technology, Nanjing, 210094, P. R. Chna)

Abstract: One of the main goals of modern nitration methodology is to increase its regioselectivty in order to meet desirable environments. Regioselectivty of toluene nitration were investigated on the acid-treated bentonites catalysts using alkyl nitrates. The reaction was highly para-selective, composition (ortho/para=1.63) of nirtro isomers of toluene nitration with sulfonitric acid, up to ortho-to-para tatios of 0.52、0.47、0.41、0.41 respectively when using n-propyl nitrate、isopropyl nitrate、isooctyl nitrate on bentonite of sulfuric acid treatment time for 1 hour. A radical cationic component was proposed to elucidate nitration mechanism according to the calculated and experimental results.

1. Introduction

Electrophilic nitration of aromatic hydrocarbons has studied extensively. The nitrating agents commonly used are sulfonitric acid, nitri acid-acetic anhydride and nitronium salts ect.[1]. The methods show low regioselectivity. Also, many practical problems remain unsolved. For example, the commercial production of para-nitrotoluene accompanied by great large amounts of less desirable ortho-nitro isomer is highly desirable. Until quite recently the ortho isomer represented a valueless waste product and affected environment seriously. Clearly, there is a need for more regioselective control of electrophlic aromatic substitution. Catalysis of inorganic reactions by inorganic solid[2] is an important new dimension. Olah[3] has surveyed the Lewis acid aluminum chloride catalyzed nitration of aromatics and improved regioselectivity. However, use of standard Lewis acid catalysts is fraught with problems such as their handing, the necessity of setup and workup, solid catalysts have excited much interest. They include resins such as Nafion-H, modified alumina as well as a number of allays. They latter heterogeneous catalysts are inexpensive by comparison with the Nafion-H polymer. They offer a wide range of active sites for catalysts (e. g. , acid site) and most often can regenerated if deactivated during a reaction. Furthermore, they bring other assets. Constraining an organic molecule onto a solid surface has interesting consequences. The adsorbate undergoes reduction in dimensionality, form a three-dimensional reaction to a surface. Thereby the frequency of diffusional encounters increases. The absorbates also migrate to catalytic sites with an attendant electronic redistribution, hence creating a reducing in activation energy. The geometric consequence of anchoring onto a solid is to restrict the angles of attack, i. e., to enhance selectivity. For these reasons, we have sought to devise an improved heterogeneous bentonite catalyst for selective nitration reactions. We now report the successful development of such a reaction, utilizing alkyl nitrates as the reagent and acid-treated bentonite as the catalyst.

2. Results and discussion

Nitration experiments

The reaction procedure was applied to toluene, using alkyl nitrates such as n-propyl nitrate, isopropyl nitrate, cyclohexyl nitrate and isooctyl nitrate on acid-treated bentonite (Table 1). Acidic-bentonite was prepared by 30% sulfuric acid treating sodium bentonite. The results of Table 1 clearly indicate that Para-selective nitration toluene was heightened, although the reaction leads to lower yield.

Table 1 Regioselectivities of Toluene Nitration with Alkyl Nitrates on Acidic Bentonite

Alkyl nitrate	Nitrotoluene(%)			Ortho/para	Yield(%)
	Ortho	Meta	Para		
n-propy lnitrate	34	1	65	0. 52	22
isopropyl nitrate	32	Trace	68	0. 47	14
cyclohexylnitrate	29	Trace	71	0. 41	13
isooctyl nitrate	29	1	70	0. 41	13

For the chosen alkyl nitrates, para-selectivities are in the order cyclo-$C_6H_{11}ONO_2$ > i-$C_8H_{17}ONO_2$ > i-$C_3H_7ONO_2$ > n-$C_3H_7ONO_2$. The ortho/para ratios follow the expected trend from 0. 41 to 0. 52, in accord with decreasing steric hindrance. With respect to the nitration of toluene, Table 2 shows a comparison of its various reagents. Para-selectivity is rather low. In the present study we also found that alkyl nitrates did not act as nitrating agents in the absence of acidic bentonite catalyst. This apparent paradox may be explained by this catalyst offering the best compromise between reactivity and selectivity. We have not striven to optimize the procedure as to yields or selectivities. Undoubtedly, these can be and will be further improved.

PM 3 calculations carried out for the toluene[4], toluene radical cation, alkyl nitrates and active nitrating agents. Electronic structure parameters were listed in Fig. 1 and Table 3.

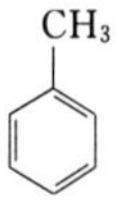

	q	f		q	f
	−0. 2599	0. 0417		−0. 1397	0. 1205
CH_3	−0. 0843	0. 6125	CH_3^+	−0. 1806	0. 5534
	−0. 3855	0. 3944		−0. 1197	0. 3501
	−0. 3717	0. 2539		−0. 1298	0. 1574
	−0. 1986	0. 5965		−0. 1195	0. 6505

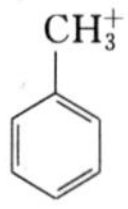

Fig. 1 atomic charges (q) and frontier reaction factors (f) calculated by PM 3 method for the toluene and toluene radical cation.

Table 2 Nitration of Toluene with Various Reagents

Alkyl nitrate	Nitrotoluene(%)			ortho/para
	Ortho	Meta	Para	
HNO_3 - Ac_2O	63	3	34	1. 85
$NO_2^+BF_4^-$	65. 4	2. 8	31. 8	2. 06
HNO_3 - H_2SO_4	59. 5	4	36. 5	1. 63
$AgNO_3$ - BF_3	59	4	37	1. 59
CH_3COONO_2	61	2	37	1. 65
$C_6H_5COONO_2$	59. 6	4. 3	36. 1	1. 65

Nitration mechanism

Fig. 1 revealed that the frontier electron density C4 (para-carbon) involved in a radical substitution is highest in toluene radical cation form. This indicates pare position as the favored site of attack by the $NO_2\cdot$ radical. Thus, one way of achieving regioselectivity is to put the reaction under orbital control with the predominant product arising from interaction of the frontier MO's. Table 3 show that toluene are rather soft Lewis bases. Since orbital takes over, with respect to the competing charge control, in soft-soft sit uation and ELUMO (NO^{2+}) <EHOMO (Ar), it is possible to for electron-transfer from toluene to NO^{2+}. And then there are some active centers to provide strong driving force for electron transfer, such as metal elements and surface radical sites of the $O_3SiO\cdot$ silyloxy type resulting from dehydrative activation of a clay[5]. Indeed, experiment results bear out the explanation of a high para preference for reactions in which the radical cation pathway, reported by Perrin in a thought-provoking article[6] in 1977, is expected to predominate.

Table 3 Frontier Orbital Energy Calculated by PM 3 Method for the Related Substrates

Substrate	ENLUMO/ev	ELUMO/ev	EH(S)OMO/ev	ENHOMO/ev
$NO_2\cdot$	3. 6899	−0. 03692	−10. 3219	−12. 9385
NO_2^+	−10. 2400	−11. 9850	−20. 8480	−22. 5893
$C_6H_5CH_3$	0. 4454	0. 3763	−9. 4419	−9. 7220
$C_6H_5CH_3+\cdot$	−5. 0492	−6. 3391	−15. 2882	−15. 5184

A plausible nitration reaction mechanism on clay catalyst with alkyl nitrates (e. g. n-$PrONO_2$) in the following:

H–O(Si, Al, Si) + $PrONO_2$ ⟶ [Pr, H–O–NO_2, :A⁺ / Si–O–Al(:A⁻)–O–Si] ⇌ [Pr, H–O–NO_2^+ / Si–O–Al⁻–O–Si] ⟶ [NO_2^+, ArH] Si–O–Al⁻–O–Si

$\xrightarrow{M^{n+}\text{etc.}}$ [$NO_2^{\cdot}$, ArH^+] Si–O–Al⁻–O–Si ⟶ H_3C–(C₆H₅⁺·)(H)(NO_2) Si–O–Al⁻–O–Si ⟶ H–O(Si, Al, Si) + $ArNO_2$

Observations consistent with the proposed mechanism. Impressive regioselectively of toluene nitration can be achieved.

3 Experimental section

Procedure for nitration

5. 0g alkyl nitrate and 8. 0g acidic bentonite were mixed with 200ml toluene. The reaction mixture was vigorously stirred and refluxed for 16h under a dry nitrogen atmosphere. At the end of the reaction, the mixture was cooled to room temperature and filtered. The filtrate was analyzed by gas-liquid chromatography.

Preparation of acidic bentonite

10g Na-Bent was added to 800ml of a 30% sulfuric acid aqueous solution. The resulting clay suspension was vigorously stirred under reflux for 1h. then the acid-treated bentonite was washed thoroughly with deionated water until without SO_4^{2-}, and dried at 100℃ in air. Acidic bentonite was calcined in air at 300℃ for at least 12h prior to use.

Acknowledgments

We think financial supports of this work by the Nature Science Foundation of Jiangsu Province (BK97133) and Ph. D. research fellowship from the ministry of Education.

Reference

[1] Lu C. Nitration Theory. Nanjing,: Science and technology of Jiangsu, CH., 1993.
[2] Laszlo P.. Preparative Chemistry Using Supported Reagents, San Diego: Academic Press, 1987.
[3] Olah G. A., Ramaiah P., Sandfrod G., et al. Synthesis, 1994, 468.
[4] Peng X.. Ph. D. Thesis, Nanjing University of Science and technology, Ch., 1997.
[5] Laszlo P.. Pure& Appl. Chem., 1990, 62 (10): 2027.
[6] Perrin C. L.. J. Am. Chem. Soc., 1997, 99 (16): 5516.

(注：此文原载于 Proceedings of the 26th International Pyrotechnics Seminar, 1999: 353-356)

N_2O_5 绿色硝化研究及其新进展

吕春绪
（南京理工大学化工学院　南京　210094）

摘要： 对绿色五氧化二氮在芳烃和氮杂环化合物的硝化反应中的应用进行了详述。五氧化二氮/硝酸体系的硝化能力强，但硝化选择性差，故常用于钝化底物的硝化。五氧化二氮/有机溶剂体系硝化温和，选择性好，主要用于酸敏性和水敏性及含多官能团物质的选择性硝化。同时，对五氧化二氮的合成及放大方法进行了概括。

关键词： 有机化学，五氧化二氮，硝化，绿色

1　引言

绿色化学，在化学品的设计、制造和应用时，与传统的污染末端治理方法不同。它是基于化学原理，通过改变化学品或化学反应过程的内在本质，从源头上减少或消除有害物质的使用或产生，最大限度地保护资源、环境和人类健康[1,2]。

绿色化学是以“原子经济性”为基本原则，并在获取新物质的化学反应中充分利用参与反应的每个原子，寻找在各个环节都洁净和无污染的反应途径和工艺，实现“零排放”，不产生污染；使用无毒无害的溶剂、助剂和催化剂生产有利于生态环境和人类健康的环境友好化学品。

对生产过程来说，绿色化学品包括节约原材料和能源，淘汰有毒原料，降减生产过程废物的数量和毒性。对产品来说，绿色化学旨在减少从原料的加工到产品的最终处置的全生命周期的不利影响。绿色化学是通过改变化学品或生产过程的内在本质来减少或消除有害物质的使用或产生，是化学工业可持续发展的基础。

绿色精细化工，是运用绿色化学的原理和技术，尽可能选用无毒无害的原料，开发绿色合成工艺和环境友好的化工过程，生产对人类健康和环境无害的精细化学品。是要努力实现化工原料的绿色化，合成技术和生产工艺的绿色化，精细化工产品的绿色化，使精细化工成为绿色生态工业[3-5]。

精细化工产品的绿色化，就是要根据绿色化学的新概念、新技术和新方法，研究和开发无公害的传统化学用品的替代品，设计和合成更安全的化学品，采用环境友好的生态材料，实现人类和自然环境的和谐与协调。

硝化反应是经常遇到的有机单元反应。目前，工业上广泛使用的硝化剂有硝硫混酸、硝酸酯、金属硝酸盐等。这些工艺不仅原子经济性不高，而且在生产过程中会产生大量的废酸及有机酸性废水，导致严重的环境污染；混酸的强腐蚀性还会严重损坏设备；反应过程中发生的多硝化、氧化等副反应也会产生严重的安全隐患。

绿色硝化，是以提高反应转化率和选择性，从源头上减少有毒有害副产物的产生，以达到清洁生产的目的。当前国内外研究开发的各种清洁硝化反应，都杜绝了酸的使用，彻底消除了废酸污染。其中最具代表性的新型硝化技术是采用 N_2O_5 作硝化剂的新工艺。

2 N_2O_5 绿色硝化研究

炸药制造通常作用的是硝酸-硫酸混酸、硝酸-硝酸铵、硝酸-醋酐等硝化剂[6]。

N_2O_5 作硝化剂与工业上最常用的上述硝化剂相比具有如下优点：① 温度易控制，反应基本不放热；② 无需废酸处理；③ 产物分离简单，通常蒸出溶剂即可；④ 对多官能团反应物，硝化选择性高；⑤ 不会发生氧化等副反应；⑥产物得率很高（80%～90%）。

此外，当用 N_2O_5-硝酸硝化时，硝化无选择性；而当用 N_2O_5-有机溶剂（尤其是氯代烃）硝化时，反应选择性极强。因此，N_2O_5-硝酸体系和 N_2O_5-有机溶剂体系是两种互补的硝化体系，N_2O_5 这种独特的性质使它在硝化反应中有着广泛的应用前景。

鉴于 N_2O_5 在硝化领域的重要性，越来越多的科研工作者研究了 N_2O_5 的硝化特性及其与芳烃、胺类等的反应。

2.1 N_2O_5 对芳烃的硝化技术研究[7～14]

我们比较深入的研究了 N_2O_4-O_3 体系、N_2O_5-有机溶剂体系及 N_2O_5-HNO_3 体系等对芳烃的硝化。研究了氯苯、甲苯、N_2O_5 在催化剂作用下的硝化。研究了 N_2O_5-CH_2Cl_2 体系对硝基甲苯的催化选择性硝化。

N_2O_5-CH_2Cl_2 运用在芳烃的硝化反应中，最主要的原因就是 N_2O_5-CH_2Cl_2 硝化过程中极强的位置选择性。但由于该硝化体系相对温和，对芳烃的硝化能力很弱，因此需要特殊的催化剂。

以邻硝基甲苯的硝化为例，0℃下反应的结果如下表 1。

表 1 不同固体催化剂对 N_2O_5 硝化的影响

Table 1 Effects of solid catalyst on nitration

催化剂种类	N_2O_5/mol·l^{-1}	Time/min	Yield/%	*o*/*p*
无	0.28	180	6	0.53
HZSM-5	0.17	90	7	0.50
H-丝光沸石	0.17	15	85	0.30
H-八面沸石 780	0.18	3	88	0.28
H-八面沸石 720	0.17	3	92	0.23
Na-八面沸石	0.17	60	16	0.33

由表 1 可见，N_2O_5-CH_2Cl_2 对芳烃的硝化能力很弱，在适当的催化剂下，得率可提高到 90%以上。

由于 N_2O_5-CH_2Cl_2 温和的硝化性能，反应的副反应很少，几乎无氧化、多硝化等副反应发生，杂质含量大大低于用其它方法制备时产生的杂质，具有很好的原子经济性、很高的原料利用率。

2.2 N_2O_5-HNO_3 对胺类的硝化（硝解）技术研究[15～18]

N_2O_5-HNO_3 是中等强度的硝化酸，该体系是一种强有力的、最有应用前景的硝化体系，其硝化能力接近于硝硫混酸的硝化能力，不仅消除了废酸的污染，而且反应可在低温下进行。其温和可操作性使其应用范围愈加广泛，以其作为硝化剂的硝化技术显示了具大的优越性。由于上述特点，主要用于胺类的硝化（硝解），它有望用于 RDX、HMX 及 CL-20 等的制造中。

以 N_2O_5-HNO_3 为硝化剂，硝解乌洛托品与脲，成功制得 1,3,5-三硝基-1,3,5-三氮杂环己酮-2（RDX 酮），当 n（HNO_3）/n（N_2O_5）=4，n（N_2O_5）/n（乌洛托品）=1.5，T=5℃～20℃，t=40min 时，产品得率 120%（按乌洛托品计），熔点 183℃～184℃（文献值为 184℃）。

以 N_2O_5-HNO_3 为硝化剂，也可以制备 HMX。硝解底物有两种，分别是 DADN（1,5-二乙酰基-

3,7-二硝基四氮杂环辛烷）和 TAT（1,3,5,7-四乙酰基-1,3,5,7-四氮杂环辛烷）。

单就硝化剂而言，有许多合适的硝化剂。以 DADN 的硝解为例，将各方法的优缺点比较如表 2。

表 2　N_2O_5-HNO_3 与其它硝化体系的比较

Table 2　The comparison of N_2O_5-HNO_3 with other nitration systems

硝化体系	收率/%	优　点	缺　点
HNO_3	40	步骤简单	得率低
H_2SO_4、NH_4NO_3、HNO_3	80～85	工艺成熟	废酸量大，难处理
$(CF_3CO)_2O$-HNO_3	82～91	工艺成熟	试剂价格昂贵，废液难处理
聚磷酸-HNO_3	86～99	得率高，纯度好(100%)	聚磷酸和 HNO_3 需大大过量，腐蚀性大
P_2O_5-HNO_3	99	得率高，纯度好(100%)	P_2O_5 和 HNO_3 需大大过量，腐蚀性大
SO_3-HNO_3	60	硝化能力强	得率低，纯度低
N_2O_5-有机溶剂	65	无需废酸处理	得率低，溶剂易挥发
N_2O_5-HNO_3	94	得率高，废酸少，硝化剂的过量比小	

由表 2 可知，将 DADN 硝解成为 HMX 的硝化剂中，SO_3、P_2O_5、$(CF_3CO)_2O$ 和 HNO_3 的混合物都能产生 N_2O_5，因此反应的活性硝化剂是 N_2O_5。其中聚磷酸-HNO_3 和 P_2O_5-HNO_3 的产率和纯度最好，但原料耗费量太大，按硝解剂中 N_2O_5 与被硝解底物的摩尔比看，聚磷酸/HNO_3 法是 9.1 倍，P_2O_5-HNO_3 法是 25.8 倍，而 N_2O_5-HNO_3 法只有 3 倍。在保持较高的得率和纯度下，N_2O_5/HNO_3 法不仅节约了原材料，还减少了环境污染。因此，N_2O_5-HNO_3 制备 HMX 是一种得率高，环境友好的硝化方法。

作为一种新型硝化剂，以 N_2O_5 为硝化剂的硝化技术显示了优越性。由于反应可以在无硫酸条件下进行，后处理比较方便，不仅节约了能源，而且基本上做到了无污染。随着人们环保理念的增强，作为绿色硝化剂的 N_2O_5 在有机合成方面将有更加广阔的应用前景。

3　N_2O_5 绿色硝化的新进展

离子液体（ionic liquid）可以定义为完全由正负离子组成的在较低温度下为液体的盐，由于阴阳离子数目相等，因此整体上显电中性。一般来说，我们所指的离子液体是在室温或接近室温（低于 100℃）下为液体状态的通常完全由体积相对较大、不对称的有机阳离子和体积相对较小的无机阴离子组合而成的物质。通常也称为室温离子液体（room temperature ionic liquids）。1948 年以 Lewis 酸和有机阳离子卤化物组成的离子液体为第一代离子液体。最近，酸性离子液体广泛应用于芳香化合物的硝化反应中[19～27]，而且我们同时还研究了它在胺类硝解中的应用[28～40]。

在上述研究工作的基础上，结合某些科研项目任务，我们的重点是在 N_2O_5 制备工程化及 HMX、RDX、CL-20 及 TATB 绿色制备的研究上，并结合离子液体的作用，使 HMX、RDX、CL-20 及 TATB 等产率得到新突破！

3.1　N_2O_5 与离子液体用于 HMX 合成

钱华等[28]研究了 DAPT 在离子液体中进行的硝解反应。以 N_2O_5/HNO_3 为硝解体系，常规反应产率仅为 9.6%，加上超声波后产率提高到 34.5%，离子液体为反应溶剂时，硝解反应产率可以提高到 66.8%，而且二氯甲烷为溶剂时产率也仅为 31.2%，如图 1。原因认为是由于离子液体无蒸气压，能大大改变体系在超声波体系中空穴的特性，有利于空化气泡的产生，同时其具有强烈地吸收超声波的能力，因此可以提高超声波的利用率。而传统的分子溶剂在超声波条件下，空化泡破裂时易挥发而不好回收，同时还对超声的吸收率低。由于反应体系存在固体 DAPT、不互溶的离子液体和硝化剂，因此为非均相反应，超声波的加入可以有效促进相转移进行。

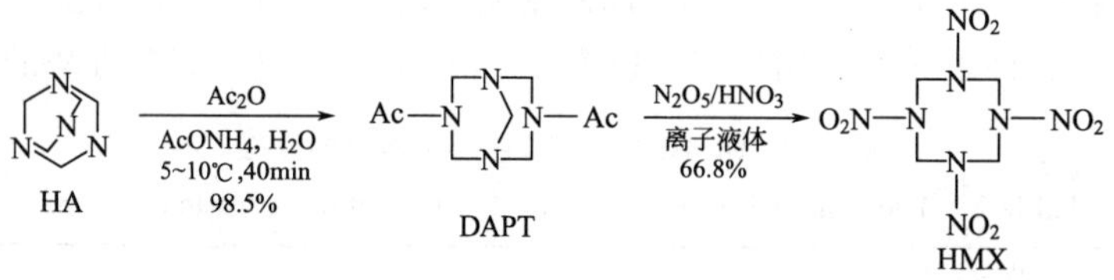

图 1 在超声波条件下，DPT 在离子液体中硝解制备 HMX 路线

Scheme 1 Nitrolysis of DPT to prepare HMX in ionic liquid under ultrasonic

职慧珍等研究了一种新型酸性离子液体[29]，并将之用于 DPT 的硝解反应中，可以有效提高 HMX 的产率[30]；何志勇等研究了 N_2O_5 在 DPT 硝解反应中的应用[31]，同时也研究了酸性离子液体对 DPT 硝解的影响，发现它可以提高 HMX 的产率。

可用于催化硝解反应的酸性离子液体的合成与常用酸性离子液体相似。叔胺可以直接与无机酸（主要是硫酸、硝酸和三氟甲酸等）反应生成的 H 型离子液体。

内酰胺磺酸盐离子液体［Caprolactam］X 也被合成出来并用于硝化反应中[32]。

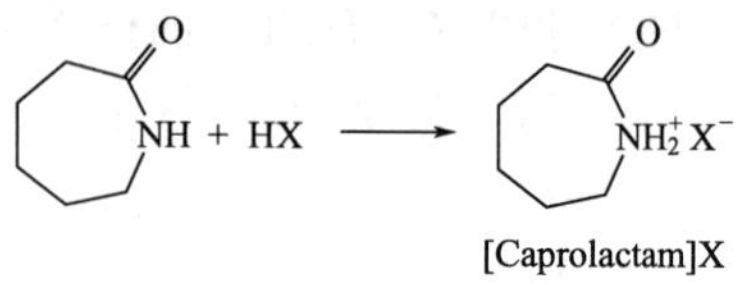

图 2 ［Caprolactam］X 合成路线

Scheme 2 Synthetic route of ［Caprolactam］X

最近一种基于聚乙二醇桥链的新型酸性离子液体被合成出来。离子液体 PEG_{200}-DAIL[29] 和 PEG_{1000}-DAIL[33] 是典型代表。

采用离子液体的用量摩尔分数为底物的 4.0%，硝酸（95%）与 DPT 的摩尔比为 36∶1，考察了 18 种离子液体对 DPT 硝解反应的影响。

研究表明：所用离子液体对 DPT 的硝解反应都有一定的催化作用。7 种类型的离子液体中阴离子不同对 DPT 硝解反应的影响未呈现出一定的规律性，说明离子液体的阴离子对 DPT 硝解反应的影响不大。总体来看，离子液体的阳离子类型对硝解反应的影响较明显，尤其是胺类离子液体，这可能与阳离子的酸性强弱有一定关系。

我们最近的研究结果表明，酸性离子液体和 N_2O_5 均可不同程度的提高 DPT 硝解反应的产率；其中，采用 N_2O_5 与 PEG_{200}-DAIL 结合的新方法可以制得产率为 64% 的 HMX；作为催化剂的酸性离子液体可以多次回收重复使用。

3.2 N_2O_5 与离子液体用于黑索今合成

将酸性离子液体应用于芳香化合物硝化反应的研究，结果表明其对芳香族化合物的硝化有明显的催化作用，可以使硝化产物收率和反应选择性都有一定程度的提高[34~38]，我们还将 Brønsted 酸性离子液体应用于发烟硝酸（质量分数 95%）硝解乌洛托品的反应研究[39,40]，结果表明其对 HA 的硝解反应也有显著的催化作用，黑索金的收率较未加离子液体时显著提高。离子液体作为绿色溶剂、反应试剂和催化剂等应用于有机合成和催化反应中，反应条件温和、选择性好、产率明显提高、易于与产物分离、反应后可回收利用，与传统硝化催化剂相比表现出明显优势和发展前景。

以甲基咪唑 N-质子化离子型酸性离子液体，用于乌洛托品硝解反应体系中，研究了它对 HA 硝解反应制取 RDX 的催化活性和最佳的硝解条件，取得了较满意的结果。

在研究了离子液体的用量、硝酸用量、反应时间、离子液体种类的基础上，对离子液体催化下 HA 硝解生成黑索金的硝解反应进行了正交实验研究。在上述研究工作基础上，作了大量优化组合条件下的验证实验。

在以上正交实验所得出的各个因素中较优水平下，即：离子液体∶硝酸用量：12∶1；IL 用量：1.5%，进行平行实验，并且与不加 IL 条件下的结果进行对比，验证加入 IL 的最终效果。

表 3 较优水平下 HA 硝解反应平行及对比实验

Table 3 Repeated and contrast experiments of the nitrolysis of HA under the best level[a]

Entry	Melting of RDX/℃		RDX/g		Yield/%[b]	
	IL	None	IL	None	IL	None
1	203.8-204.2	203.5-203.8	3.3	2.98	75.9	68.5
2	204.4-204.7	203.3-203.5	3.29	2.94	75.7	67.6
3	203.9-204.0	203.5-203.7	3.31	2.98	76.1	68.5
4	203.8-204.1	203.9-204.0	3.31	2.95	76.1	67.8
5	204.2-204.4	203.8-204.4	3.29	2.98	75.7	68.5
average	204.0-204.3	203.6-203.9	3.3	2.97	75.9	68.3

注：[a]硝解 HA 的反应时间是 90min，HA 的用量为 2.8g，HNO_3（95%）的用量为 33.6g，$n(IL)/n(HA)=1.5\%$。[b]RDX 的得率 $Yield/\%=1.021\times m(RDX)\times 140/[m(HA)\times 222]\times 100\%$。

从表 3 可以看出在较优水平条件下，不加 IL 下 RDX 的得率为 68.3%，加入与 HA 摩尔比为 1.5%的离子液体后得率为 75.9%。加入 IL 后 RDX 的得率明显比不加 IL 的要高，这说明加入 IL 对 HA 硝解反应制取 RDX 有显著的催化作用。离子液体的加入给反应体系提供了离子氛围，从而提高了活性硝解剂 $H_2NO_3^+$ 和少量的 NO_2^+ 的生成速率，加快了 RDX 母体被硝化成 RDX 的速率。但是有关 IL 加入后 HA 硝解生成 RDX 的机理还有待于进一步研究。

3.3 N_2O_5 与离子液体用于 CL-20 合成

鉴于 N_2O_5/HNO_3 在胺类硝解中巨大的应用前景，钱华[15~18]以 N_2O_5/HNO_3 为硝解剂，以 TADNSIW 和 TAIW 为底物，分析工艺条件对收率和纯度的影响，在温和条件下高收率地制备 CL-20。具体实验路线如图 3。

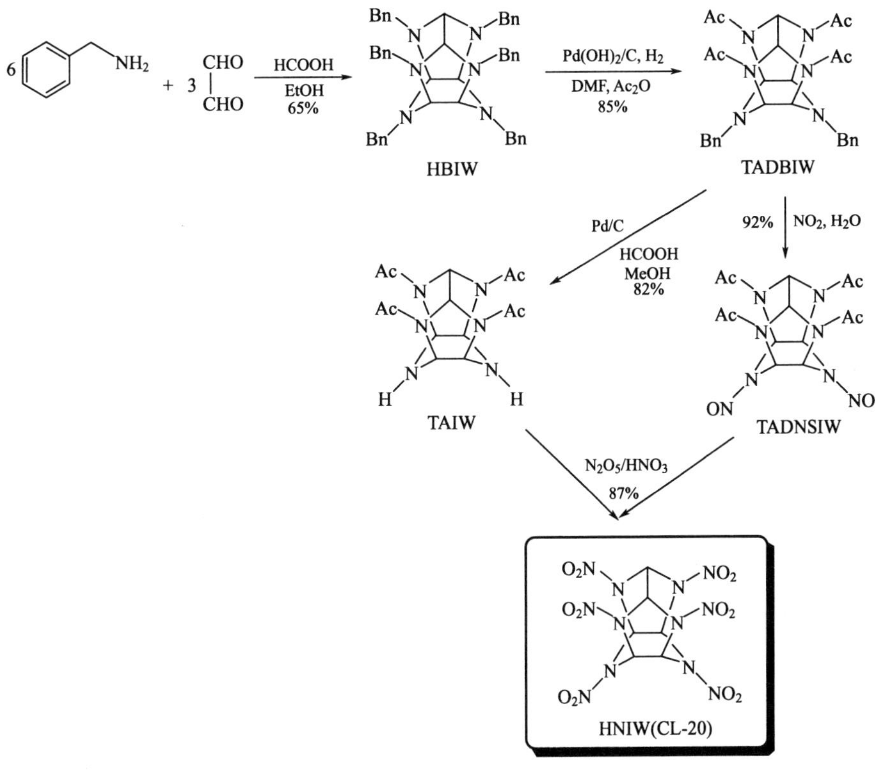

6 NH2 + 3 CHO/CHO —HCOOH, EtOH, 65%→ HBIW —Pd(OH)2/C, H2, DMF, Ac2O, 85%→ TADBIW

TADBIW —Pd/C, HCOOH, MeOH, 82%→ TAIW

TADBIW —NO2, H2O, 92%→ TADNSIW

TAIW, TADNSIW —N2O5/HNO3, 87%→ HNIW(CL-20)

图 3 CL-20 的合成

Scheme 3 Synthesis of CL-20.

3.4 N_2O_5 与离子液体用于 TATB 合成[41]

TATB 是最早的耐热、钝感炸药之一，黄色粉状结晶，在太阳光或紫外线照射下变为绿色。不吸湿，室温下不挥发，高温时升华，除能溶于浓硫酸外，几乎不溶于有机溶剂，高温下略溶于二甲基酰胺和二甲基亚砜。TATB 对枪击碰撞摩擦等意外刺激非常钝感，是美国能源部批准的唯一单质钝感炸药，常被用作衡量钝感炸药的标准物质。美国绝大部分核航弹及核弹头使用了以 TATB 为基的高聚物粘结炸药，其配方 PBX-9502，LX-17 从 1979 年就开始使用。此外，TATB 现还被用作活性钝感剂对 HMX、CL-20 等高能炸药进行钝化处理。在民用方面，TATB 可用于深井射孔弹和制作液晶材料及电磁材料的原材料。

TATB 的含氯量一直是人们关注的问题，含氯量对 TATB 的热安定性有明显影响，其中杂质 NH_4Cl 的分解产物对接触材料有较强的腐蚀作用，对药柱成型、药柱强度、金属弹体等也有不良作用。因此，研究无氯 TATB 合成新方法具有重要意义。从 20 世纪 70 年代末期开始，国内外开始了无氯 TATB 的合成研究。Estes 以三硝基三丙氧基苯为中间体合成了无氯 TATB。Atkins 和 Nielson 以 TNT 为原料，在二氧六环中与 H_2S 反应产生 4-氨基-2,6-二硝基甲苯，然后用硝硫混酸硝化制得五硝基苯胺，再在苯、氯化甲撑或其他合适的溶剂中与 NH_3 反应得到 TATB。魏运洋以廉价的苯甲酸为原料经硝化、Schmidt 反应、硝化和氨化 4 步反应成功合成出 TATB，该路线选择性好，产率高。Anthony J. Bellamy 等人以 1,3,5-三羟基苯为起始反应物通过三步法合成无卤 TATB，避免了卤化物试剂的使用，是比较新颖的无氯 TATB 合成方法。

TATB 的合成：实验以间苯三酚为原料，经过五氧化二氮（N_2O_5）硝化、甲基化和氨气氨化得到 TATB，合成路线如下：

N_2O_5/溶剂 → TNPG → Alkylation → NH_3/Toluene, 0.4～0.6MPa → TATB

采用 N_2O_5/溶剂硝化间苯三酚可获得高纯度的 1,3,5-三羟基-2,4,6-三硝基苯（TNPG），平均收率可达 97%。经过烷基化、氨气氨化可得到无氯 TATB，反应总收率可达 92%，TATB 热分解峰值温度 385℃，达到国军标 GJB 3292—98 要求。

3.5 N_2O_5 合成的工程化技术研究[42～44]

N_2O_5 的合成主要有 P_2O_5 脱水法、氮氧化物氧化法、电解法等，其中 P_2O_5 脱水法仅适于实验室小规模合成；N_2O_4 臭氧氧化法虽然早就出现了中试放大生产，但由于大规模臭氧制备的能耗很高，而且效率不高，因此并没有进一步应用；从工艺、成本、能耗等考虑，HNO_3 电解脱水法是目前最有应用前景的方法[43]，国外称已可达到1000 kg/天规模，最近我们与天津大学建立了合作关系，共同承担某些科研项目，他们以 N_2O_5 工程化为主，我们以 N_2O_5 应用在 RDX、HMX、CL-20 等合成上为主。天津大学可以在实验室达到 1 kg/批规模[44]，而且电解装置体积小，并可进行并联，这为实现硝化场所 N_2O_5 的现场制备打下了良好的工程化基础。因此，N_2O_5 的工程化应用可以方便实现，而且现场制备还可以避免稳定性不高的 N_2O_5 的储存、运输及相应的安全性问题，是以后工业 N_2O_5 清洁硝化的发展方向。

4 结束语

以 N_2O_5 为硝化剂的绿色硝化技术国外研究开发较早。作为一种新型硝化剂，以 N_2O_5 为硝化剂的硝化技术显示了具大的优越性。在有机合成中，使用 N_2O_5/硝酸体系和 N_2O_5/有机溶剂体系，很容易得到理想结构的化合物；对含能材料如 HMX、硝化棉及对酸、水敏感的不稳定化合物的硝化，有其独特的优势。该工艺需解决的主要问题是 N_2O_5 的来源。目前英、美等已开发成功生产能力为 5000kg/h 的电解槽，因此有望实现方便、廉价地生产 N_2O_5。由于反应可以在无硫酸条件下进行，后处理比较方便，不仅节约了能源，而且基本上做到了无污染。随着人们环保理念的增强，作为绿色硝化剂的 N_2O_5 在有机合成方面将有更加广阔的应用前景。

参考资料

[1] 沈玉龙. 绿色化学 [M]. 北京：中国环境出版社，2004：7-16.

[2] 朱宪. 绿色化学工艺 [M]. 北京：化学工业出版社，2007：2-7.

[3] 贡长生. 绿色精细化工导论 [M]. 北京：化学工业出版社，2005：1-8.

[4] 李德华. 绿色化学化工导论 [M]. 北京：北京出版社，2005：10-17.

[5] 韩明汉. 绿色工程原理与应用 [M]. 北京：清华大学出版社，2005：5-12. 2005：10-17.

[6] 吕春绪. 硝酰阳离子理论 [M]. 北京：兵器工业出版社，2006：12-18.

[7] 吕早生. $N2O_4$-O_3 绿色硝化芳烃的宏观动力学及机理研究 [D]. 南京：南京理工大学，2001.

[8] 吕早生，吕春绪. 一种新的绿色硝化技术. 火炸药学报 [J]，2000，23 (4)：9-13.

[9] 吕早生，吕春绪. 绿色硝化技术在硝基氯苯合成中的应用 [J]. 火炸药学报，2000，23 (4)：29-30.

[10] 吕早生，王晓燕，吕春绪. N_2O_4-O_3 硝化芳烃反应的宏观动力学研究 [J]. 火炸药学报，2004，27 (3)：66-69.

[11] 钱华，叶志文，吕春绪. HZSM-5 催化下 N_2O_5 对氯苯的选择性硝化研究 [J]. 含能材料，2007，15 (1)：56-59.

[12] 钱华，吕春绪，叶志文. HZSM-5 催化下 N_2O_5 对甲苯的选择性硝化研究 [J]. 火炸药学报，2006，29 (5)：9-11.

[13] 钱华，叶志文，吕春绪. 绿色硝化剂 N_2O_5 对甲苯的硝化研究 [J]. 化学世界，2006，47 (12)：717-719.

[14] Qian H，Lv C-X，Ye Z-W. HZSM-5 Assisted Nitration of Toluene with Dinitrogen Pentoxide [A]. Preceedings of thr 2nd International Seminar on Industrial Explosive [C]. Beijing：The Publishing House of Ordnance Industry，2006，50-53.

[15] 钱华. 五氧化二氮在硝化反应中的应用研究 [D]. 南京：南京理工大学，2008.

[16] 钱华，吕春绪，叶志文. 绿色硝解合成六硝基六氮杂异伍兹烷 [J]. 火炸药学报，2006，29 (3)：52-53.

[17] 钱华，叶志文，吕春绪. N_2O_5/HNO_3 硝化硝解 TAIW 制备 CL-20 [J]. 应用化学，2008，25 (3)：424-426.

[18] 钱华，吕春绪，叶志文. 绿色硝化剂五氧化二氮及其在硝化中的应用 [J]. 精细化工，2006，23 (6)：620-624.

[19] 齐秀芳，程广斌，段雪蕾，吕春绪. Brønsted 酸性功能离子液体存在下甲苯的硝化反应 [J]. 火炸药学报，2007，30 (5)：12-15.

[20] 岳彩波，魏运洋，吕敏杰. Bronsted 酸性离子液体中芳烃硝化反应的研究 [J]. 含能材料，2007，15 (2)：118-127.

[21] 方东，施群荣，巩凯. 离子液体催化甲苯绿色硝化反应研究 [J]. 含能材料，2007，15 (2)：122-124.

[22] Fang D，Luo J，Zhou X L，et al. Mannich reaction in water using acidic ionic liquid as recoverable and reusable catalyst [J]. *Catal. Lett.*，2007，116 (1-2)：76-80.

[23] Fang D，Luo J，Zhou X-L，*et al*. One-pot green procedure for Biginelli reaction catalyzed by novel task-specific room-temperature ionic liquids [J]. *J. Mol. Catal. A：Chem.*，2007，274：208-211.

[24] Laali K K，Gettwert V J. Electrophilic nitration of aromatics in ionic liquid solvents [J]. *J. Org. Chem.*，2001，66 (1)：35-40.

[25] Qiao K，Hagiwara H，Yokoyama C. Acidic ionic liquid modified silica gel as novel solid catalysts for esterification and nitration reactions [J]. *J. Mol. Catal. A：Chem.*，2006，246：65-69.

[26] Lancaster N L，Llopis-Mestre V. Aromatic nitrations in ionic liquids：the importance of cation choice [J]. *Chem. Commun.*，2003，22：2812-2813.

[27] Smith K，Liu S，El-Hiti G A. Regioselective mononitration of simp learomatic compounds under mild conditions in ionic liquids [J]. *Ind. Eng. Chem. Res.*，2005，44：8611-8615.

[28] Qian H, Ye Z W, Lv C X. Ultrasonically promoted nitrolysis of DAPT to HMX in ionic liquid [J]. *Ultrasonics Sonochem.*, 2008, 15 (4): 326-329.

[29] 职慧珍，罗军，马伟，吕春绪. PEG 型酸性温控离子液体中芳香酸和醇的酯化反应 [J]. 高等学校化学学报，2008，29 (4)：772-774.

[30] Zhi H Z, Luo J, Lü C X. The application of acidic ionic liquids and dinitrogen pentoxide in nitrolysis of DPT. *Chin Chem Lett*, 2009, 20: 379-382.

[31] 何志勇，罗军，吕春绪. N_2O_5 硝解 DPT 制备 HMX [J]. 火炸药学报，2010，33 (2)：36-38.

[32] Qi X F, Cheng G B, Lu C X, et al. Nitration of toluene and chlorobenzene with HNO_3/Ac_2O catalyzed by caprolactam-based Brønsted acidic ionic liquids [J]. *Central European Journal of Energetic Materials*, 2007, 4 (3): 105-113.

[33] Zhi H Z, Lü C X, Zhang Q, Luo J. A new PEG-1000-based dicationic ionic liquid exhibiting temperature-dependent phase behavior with toluene and its application in one-pot synthesis of benzopyrans [J]. *Chem Commun*, 2009, 20: 2878-2880.

[34] Cheng G, Duan X, Qi X, Lu C. Nitration of Aromatics Compounds with NO_2/Air Catalyzed by Sulfonic acid-fundtionalized Ionic Liquids [J]. *Catalysis Communications*, 2008, 10 (2): 201-204.

[35] 程广斌，钱德胜，齐秀芳，吕春绪. 己内酰胺对甲基苯磺酸离子液体中甲苯的选择性硝化反应 [J]. 应用化学，2007，24 (11)：1255-1258.

[36] 齐秀芳，程广斌，吕春绪. 酸性离子液体存在下甲苯的硝酸硝化（Ⅱ）[J]. 含能材料，2008，16 (4)：398-400.

[37] Qi X, Cheng G, Lu C, Qian D. Nitration of toluene and chlorobenzene with HNO_3/Ac_2O Catalyzed by caprolactam-based Brønsted acidic ionic liquids [J]. *Central European Journal of Energetic Materials*, 2007, 4 (3): 105-110

[38] Qi X, Cheng G, Lu C, Qian D. Nitration of simple aromatics with NO_2 under air atmosphere in the presence of novel Brønsted acidic ionic liquids [J]. *Synth. Commun*, 2008, 38 (4): 537-541.

[39] Cheng Guangbin, Li Xia, Qi Xiufang, Lu Chunxu. Synthesis of RDX by nitrolysis of hexamethylenetetramine in the presence of Brønsted acidic ionic liquids [M]. Theory and practice of energetic materials (VOL. Ⅷ), Science Press, 2009, 8: 48-51.

[40] 程广斌，李霞，齐秀芳，吕春绪. 离子液体在乌洛托品硝解反应中的应用 [J]. 2008 年火炸药学术研讨会论文集，2008，7-10.

[41] 马晓明，李斌栋，吕春绪，等. 无氯 TATB 的合成及其热分解动力学 [J]. 火炸药学报，2009，32 (6)：24-27.

[42] 何志勇，罗军，吕春绪. 绿色硝化剂 N_2O_5 的制备方法及其应用进展 [J]. 火炸药学报，2010，33 (1)：1-4.

[43] 张香文，王庆法. 电化学制备五氧化二氮装置及方法 [P]. CN 1746335A，2006-03-15.

[44] 苏敏，王庆法，张香文. 新型绿色硝化剂 N_2O_5 的电化学合成研究进展 [J]. 含能材料，2006，14 (1)：66-67.

Clean Nitrating Agent Dinitrogen Pentoxide and its Application in Nitration

Lv Chunxu

(School of Chemical and Engineering, Nanjing University of Science and Technology, Nanjing 210094)

Abstract: Nitration of aromatic hydrocarbon and nitrogen heterocyclic by N_2O_5 was summarized. N_2O_5/nitric acid system is a very powerful nitration reagent and particularly valuable for the nitration of highly deactivated aromatics. N_2O_5/organic solvent system is a much milder and more selective nitration reagent, which is suitable for the cases that the substrates or products are sensitive to acid or water, and for region-selective O-nitration. Furthermore, the synthesis and pilot-plant-scale process of N_2O_5 were also summarized.

Key words: organic chemistry, dinitrogen pentoxide, nitration, clean

（注：此文原载于 含能材料，2010，18 (6)：611-617）

绿色硝化研究进展

吕春绪
（南京理工大学化工学院，南京　210094）

摘要： 叙述了绿色硝化目前现状，指出新技术如氟两相、微波及超声波应用在硝化反应中的重要作用，特别指出 N_2O_5-离子液体体系在硝化反应中的应用，尤其是在胺类硝化中，对促进硝化反应绿色化有重要意义。附参考文献 71 篇。

关键词： 有机合成；绿色硝化；N_2O_5；离子液体；火炸药

引言

炸药制造过程是精细化工单元反应的集合，其工艺技术正在大幅度向绿色化发展。这是当前炸药研究的重要课题，也是节能减排、零排放、低污染大方向的要求。其废酸及废水的处理研究工作也正在广泛而深入地展开，特别是生物化工技术的应用，为其绿色化工艺奠定了基础。为了加速反应速度，硝化反应以及其他相关反应如缩合、氧化、氟化等都采用了催化技术，其中控制氧化剂微波催化具有显著特征[1-3]。

硝化反应一方面是炸药合成中最重要的反应，军用单体炸药、耐热炸药、高能量密度炸药等新型火炸药的制备生产过程，都需要使用硝化反应技术，引入多个硝基；另一方面，硝化反应生成的有机硝基化合物也是精细化工生产的基本原材料。工业硝化应用硝-硫混酸工艺，此技术已经应用 160 多年，无法满足环境经济发展需求，目前已成为全球最大环境公害之一。

因此，研究绿色硝化技术，寻找安全、环保、节能、无毒、经济的新方法和新技术，是炸药科研与生产发展不容忽视的重要方向，极具现实性和前瞻性。

1　NO_2-O_3 硝化（Kyodai 硝化）

NO_2 本身对芳烃几乎没有硝化能力，通常需要在一定的体系中被活化后，才具有硝化能力。对氮氧化物活化体系的研究已有相当一段时间，早期的研究主要是酸，如 H_2SO_4、BF_3 等，但近期对氮氧化物活化的研究却主要集中在 O_3 上，NO_2-O_3 体系对芳烃的硝化，通常称为 Kyodai 硝化[4,5]。

20 世纪 70 年代前苏联开始研究 NO_2 的无酸硝化，之后美欧国家也相继开展研究。自 1980 年以来，日本京都大学及其合作者发表了许多关于臭氧介质中 NO_2 硝化芳烃的论文，日本将其命名为 Kyodai 硝化[4]。Hitomi Suzuki 教授发现 NO、N_2O_3、NO_2 低氧氮化物在 O_3 存在下能被活化，顺利地与许多芳烃反应，产生相应的硝基化合物。经过 10 多年的探索，研究出臭氧介质二氧化氮（NO_2-O_3）硝化体系及其硝化体系动力学特征与机理，确立了 NO_2-O_3 硝化体系的特点；并在苯、烃基苯、卤苯、胺类及醇类等化合物上获得了应用。

南京理工大学自 1990 年开始研究绿色硝化，内容包括选择性硝化、Kyodai 硝化、氟两相硝化、微波硝化、超声波硝化、N_2O_5 硝化以及离子液体介质中的硝化等，并形成硝化反应的硝酰阳离子理论[6]。以 NO_2 为硝化剂，在特定活化剂作用下硝基化合物的制备，消除了传统硝硫混酸硝化工艺所引起的环境污染，同时可以通过控制反应条件提高硝化反应的区域选择性，以实施定向硝化。以外，由于直接利用 NO_2，省去了 NO_2 制备硝酸的多步能耗步骤，大大降低了硝基化合物制备过程的运行成本。因此，NO_2 硝化技术的研究实施，对其他基本有机单元反应的绿色化研究也必将产生积极影响。

大量小试试验证明[7-11]，NO_2-O_3 绿色硝化技术对芳烃、胺类还是醇类都是可行的，其特点也十分明显。但是，如何实现工程化，也是十分困难的问题。

2 绿色硝化的新技术

2.1 氟两相体系中的硝化

氟两相催化（Fluorous Biphasic Catalysis，FBC）是指在氟两相体系（FBS，Fluorous Biphasic System）中进行的催化反应过程，是近年来发展起来的一种新型均相催化剂固定化（多相化）和相分离技术，1994 年由 Horvath 首次使用[12]。氟两相催化既保持均相催化反应活性高、选择性高的特点，又具有负载催化剂及水两相体系的催化剂易分离、回收的优点。

将催化剂固定在氟相，反应物溶于有机相，在合适的全氟溶剂/有机溶剂体系中，加热（高于该全氟溶剂/有机溶剂对的临界温度 Tc）使两相体系变成均相，从而使反应在均相中进行。反应完成后，温度降低，又分成两相，通过简单的相分离就能方便地分出产物（有机相）和回收催化剂（氟相），不需进一步处理就可将含催化剂的氟相用于新的反应循环。

根据硝化反应的特点，南京理工大学将氟两相体系用于芳烃及胺类的硝化中，较好地提高了选择性及转化率[13-18]。

2.2 微波技术在硝化反应中的应用

甲苯在微波作用下硝化反应，其对邻位比及产率都有很大变化，反应速度可提高十倍以上，异构体比例等数据见表 1。

表 1 甲苯在微波作用下的硝化结果

Table 1 Nitration result of toluene under microwave effect

反应条件	t/min	w(一硝化异构体含量)/%			Δ	产率/%
		o^-	m^-	p^-		
a. t.	30	0	0	0		0
a. t.	60	47.6	2.1	50.3	1.06	93
a. t.	90	46.9	2.5	50.6	1.08	100
MW	4	44.1	2.2	53.7	1.22	54.7
MW	6	43.0	2.1	54.9	1.28	89.0
MW	8	39.7	1.8	58.5	1.47	100

表 1 结果表明，在微波作用下，极短的时间内达到同样的产率，且对邻比明显提高。

2.3 超声波技术在硝化反应中的应用

超声波对化学反应的作用是源于超声波在溶液中产生的空化现象和自由基反应机理。空化现象是指液体中的微小气泡核在超声波作用下被激化，表现为泡核的震荡、生长、收缩及崩溃等一系列力学过程。在空化泡崩溃的极短时间内，空化泡及其周围极小空间范围内出现热点，产生极大的能量，在这些极端条件下，进入空化核内的物质在高温和高压下发生超临界反应。声化学反应的声空化机制和声致自由基的生成机制，集超临界点湿式催化和光催化的优点，从而使超声波催化成为绿色、高效的化学反应的一条新途径。超声波用于硝化反应，是由于其它加快反应速率，提高反应收率。

当用 TRAT 合成 RDX，在 95%（质量分数，下同）HNO_3、100%HNO_3、HNO_3/H_2SO_4 体系作用下产率极低，P_2O_5/HNO_3 体系 RDX 产率仅为 40%，但在超声波作用条件下硫酸硝化产率可提高到 57.3%。

超声波作用下用 DAPT 合成 HMX 产率可由 9.6%提高到 66.8%[19]。

3 N_2O_5 硝化

N_2O_5 在低温下与大多数有机化合物化学反应剧烈，因此被推荐为硝化剂[20]。X 射线研究的结果表明[21]，N_2O_5 分子存在两种不同的结构：固态 N_2O_5 是硝酰阳离子和硝酸根离子（$NO_2^+NO_3^-$）型结构，反应中易发生离子化而成为有效的活化硝化剂 NO_2^+。但在极低的温度下，或在非极性溶剂中则为共价化合物

Olah 等[22]对 N_2O_5 硝化剂的硝化能力进行比较全面的阐述。根据产生的 NO_2^+ 的有效浓度，常用硝化剂硝化能力的排序为：

$$NO_2^+BF_4^- > HNO_3/H_2SO_4 > N_2O_5/HNO_3 > N_2O_5/\text{卤代烃} > HNO_3/Ac_2O > HNO_3$$

由此可见，N_2O_5 硝化剂的硝化能力仅次于硝硫混酸，与硝酸或硝硫混酸硝化剂相比具有很多突出的优点，具有较好的工业应用前景。尤其是近年来，随着环境污染问题日益受到各国关注，对环境污染少的硝化剂颇受青睐。

N_2O_5 硝化反应体系可分为两大类：（1）N_2O_5-硝酸体系，具有高酸度和高活性，适用面广，无选择性，尤其适用高度钝化的硝基芳烃和硝胺类化合物合成；（2）N_2O_5-有机溶剂体系，具有活性不高但反应条件更温和、选择性也更高。尤其适用于含氧或氮的张力环的开环硝化。这两种硝化体系可以优势互补，广泛应用于芳烃、胺类、醇类等的反应。

3.1 N_2O_5 对芳烃的硝化技术研究

不同固体催化剂对 N_2O_5 硝化 *o*-MNT 的影响见表 2。

表 2 不同固体催化剂中 N_2O_5 对邻硝基苯选择性硝化结果

Table 2 Effect of N_2O_5 on the results of selectinally nitrating o-dinitrobenzene in different solid catalysts

催化剂	$C(N_2O_5)/mol\cdot L^{-1}$	t/min	产率/%	p/o
无	0.28	180	6	1.89
HSM-5	0.17	90	7	2.00
H-丝光沸石	0.17	15	85	3.33
H-八面沸石 780	0.18	3	88	3.57
H-八面沸石 720	0.17	3	92	4.35
Na-八面沸石	0.17	60	16	3.03

表 2 结果表明，H-八面沸石 720 具有较高产率，具有较好的原子经济性及很高的原料利用率[23-25]。

3.2 N_2O_5-HNO_3 对胺类的硝化（硝解）技术

N_2O_5-HNO_3 的硝化活性中等，接近于硝硫混酸，是一种最有应用前景的硝化体系。不仅消除了废酸的污染，而且反应可在低温下进行。其温和可操作性使其应用范围愈加广泛[26]。该体系主要用于 RDX、HMX 及 CL-20 等的制造。N_2O_5/HNO_3 体系硝化 TRAT，产率为 86.4%～87.2%，纯度为 98.2%。

南京理工大学研究了用 DADN、TAT、DPT 及 DAPT 四种原料在 N_2O_5-HNO_3 体系中绿色硝解

制备 HMX。制备的 β-HMX 的单晶结构和晶胞结构分别如图 1 所示。

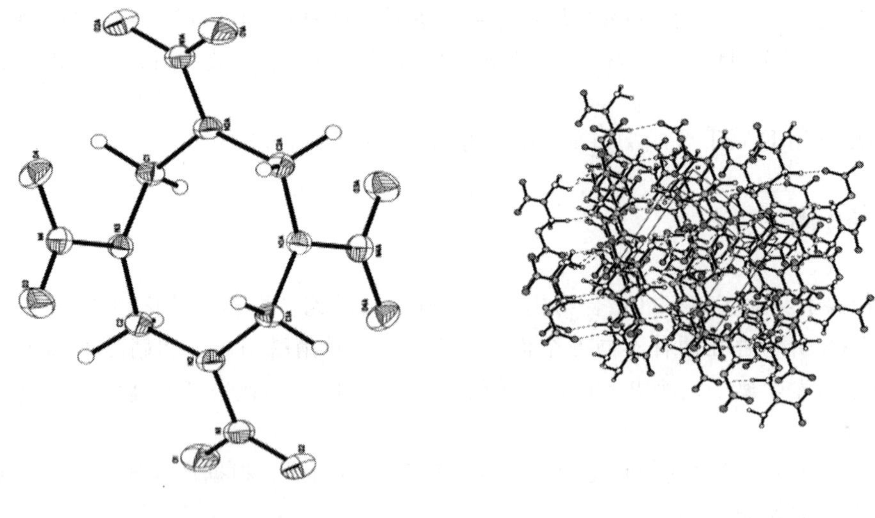

(a) β-HMX 的单晶结构 (b) β-HMX 的晶胞结构

图 1 β-HMX 的单晶和晶胞结构

Fig. 1 Single crystal and crystal lattice structure of β-HMX

以 DADN 为原料，不同硝解方法的优缺点比较见表 3。从表 3 数据看出，N_2O_5/HNO_3 具有最突出的优点。

表 3 以 DADN 为原料不同硝化体系硝化结果比较

Table 3 Comprison of the results of nitrating DADN as starting material with different nitration system

硝化体系	收率/%	优点	缺点
HNO_3	40	步骤简单	得率低
H_2SO_4、NH_4NO_3、HNO_3	80～85	工艺成熟	废酸量大，难处理
$(CF_3CO)_2O$-HNO_3	82～91	工艺成熟	试剂价格昂贵，废液难处理
聚磷酸-HNO_3	86～99	得率高，纯度好(100%)	聚磷酸和 HNO_3 需大大过量，腐蚀性大
P_2O_5-HNO_3	99	得率高，纯度好(100%)	P_2O_5 和 HNO_3 需大大过量，腐蚀性大
SO_3-HNO_3	60	硝化能力强	得率低，纯度低
N_2O_5-有机溶剂	65	无需废酸处理	得率低，溶剂易挥发
N_2O_5-HNO_3	94	得率高，废酸少，硝化剂的过量比小	

以 DPT 为原料合成 HMX，无 N_2O_5 时 25℃下硝酸硝化 30min，产率为 48%；有 N_2O_5 时在 25℃下反应 30 min，产率为 58%，纯度达到 99%[27]。

4 N_2O_5-离子液体中的绿色硝化

4.1 离子液体及其在绿色硝化中的应用

离子液体（Ionic Liquid，IL）作为一种环境友好型溶剂和催化剂，近几年来已经成为研究的热点。南京理工大学合成了 [$(CH_2)_4SO_3HPy$] X 等 50 多种新型结构的离子液体，并应用于芳烃硝化[28-45]和胺类硝解[46-49]中。氯苯在离子液体存在下选择性硝化产率为 55%，对邻位比（p/o）可达 4.0。溴苯选择性硝化产率为 64%，对邻位比高达 14。此外还研究了其它芳烃如甲苯的选择性硝化。HA 在硝酸体系中使用我们自己合成的离子液体使 RDX 产率可达 72.6%，无离子液体时 RDX 产率仅为 59.5%。

今后研究的重点是将 N_2O_5 与离子液体组成的体系用于 HMX、RDX、CL-20 及 TATB 绿色制备。

4.2 N_2O_5-离子液体用于 RDX 合成

纯硝酸 13.5g，乌洛托品 1.5g，五氧化二氮 0.75g，离子液体［Bmim］BF_4（1-丁基-3-甲基咪唑四氟硼酸盐），0.122g（用量为 HA 的物质量的 5%）[50]，在最佳工艺条件下，RDX 的产率为 84%。反应比较平稳，温度波动较小，离子液体可反复使用。

4.3 N_2O_5-IL 离子液体用于 HMX 合成

研究还表明[51]离子液体使 HMX 产率提高到 64%，所用离子液体对 DPT 的硝解反应都有一定的催化作用。7 种不同类型的离子液体中，不同阴离子对 DPT 硝解反应的影响未呈现出一定的规律性，说明离子液体的阴离子对 DPT 硝解反应的影响不大。总体来看，离子液体的阳离子对硝解反应的影响较明显，尤其是胺类离子液体，这可能与阳离子的酸性强弱有一定关系。

4.4 N_2O_5-离子液体用于 CL-20 合成

在 CL-20 合成中，TAIW 或 TADNSIW 硝化成目标化合物 HNIW（CL-20）时可以用各种硝化剂[19,52-54]。研究表明，与混酸硝化相比，用 N_2O_5-IL 体系进行硝化反应，CL-20 的得率可达 87%，污染小，对环境友好，是一条值得深入研究的反应路线。

4.5 N_2O_5-离子液体用于 TATB 合成

Estes 以三硝基三丙氧基苯为中间体合成了无氯 TATB[55,56]。

Atkins 和 Nielson[57]以 TNT 为原料，在二氧六环中与 H_2S 反应产生 4-氨基-2,6-二硝基甲苯，然后用硝硫混酸硝化制得五硝基苯胺，再在苯、氯化甲撑或其他合适的溶剂中与 NH_3 反应得到 TATB。魏运洋[58,59]以廉价的苯甲酸为原料经硝化、Schmidt 反应、硝化和氨化 4 步反应成功合成出 TATB，该路线选择性好，产率高。

Anthony J. Bellamy 等人[60,61]以 1,3,5-三羟基苯为起始反应物通过三步法合成无卤 TATB，避免了卤化物试剂的使用，是比较新颖的无氯 TATB 合成方法。

马晓明[62]以葡萄糖为原料采用生物法高效合成均三苯酚，用 N_2O_5-离子液体绿色硝化，再经烷基化、氨解等反应制得 TATB，是一条无氯工艺路线。N_2O_5-离子液体硝化平均收率为 97%。以均苯三酚计总收率可达 92%。TATB 热分解峰值温度 385℃，产品总体性能达到国军标 GJB 3292-98 要求。

5 N_2O_5 合成的工程化技术研究

N_2O_5 的合成主要有 P_2O_5 脱水法、氢氧化物氧化法、电解法等，其中 P_2O_5 脱水法仅用于实验室合成，从工艺、成本、能耗等考虑，电解法是目前最有应用前景的方法[63]。

电解法合成 N_2O_5 采用隔膜电解槽进行，早期的研究多采用 Pt、Pb 等贵金属的电极、玻璃或陶瓷为膜进行 N_2O_5 的合成，使得电解过程的电流效率低（小于 35%）；20 世纪 80 年代开始，劳伦斯利弗莫尔国家实验室采用电位控制系统和阴极 N_2O_4 循环模式进行电化学制备 N_2O_5，得到 N_2O_5 28.5%（质量分数）的阳极液。电解得到的 N_2O_5/HNO_3 溶液不经分离直接用于有机化合物的硝化，如制备 HMX。

2005 年天津大学开始制备 N_2O_5 的研究工作，取得了较好的研究成果[65-67]，自主设计的电化学制 N_2O_5 装置已获权国家发明专利。此外，在绿色合成工艺改进、电解工艺、电极设计、N_2O_4 回收、N_2O_5 结晶析出等方面开展了相关工作[68-71]。

6 结论

（1）当前国内外研究开发的各种清洁硝化反应，都从根本上杜绝了酸的使用，彻底消除了废酸污染。

（2）作为一种新型硝化剂，以 N_2O_5 为硝化剂的硝化技术显示了具大的优越性。在有机合成中，使用 N_2O_5/硝酸体系和 N_2O_5/有机溶剂体系，很容易得到理想结构的化合物。

（3）N_2O_5 对 TNT、TATB、HMX、RDX 以及 CL-20 等笼型化合物等的硝化，在其选择性及转化率方面呈现出独特的优势，由于可以采用低酸或无酸的中性硝化，基本做到低污染或无污染。

（4）把一些环境友好的催化剂如离子液体，应用于芳烃绿色硝化，乌洛托品（HA）硝解成 RDX 及 DPT 硝解成 HMX 等方面开展的离子液体合成及应用，特别是与 N_2O_5 共同使用，使 RDX 及 HMX 得率较大幅度提高。

参考文献

[1] 吕春绪等著．炸药的绿色制造 [M]．北京：国防工业出版社，2010.

[2] 李德华．绿色化学化工导航 [M]．北京：科学出版社，2005.

[3] 朱宪．绿色化学工艺 [M]．北京：化学工业出版社，2007.

[4] Mori T，Suzuki H. Oone-mediated nitration of aromatic compounds with lower oxides of nitrogen. Synlett. 1995，5：383-392.

[5] Suzuki H，Mori T，Maeda K. Reversal of the ortho-para isomer ratios by altering the initial concentration of substrate in the oone-mediated nitration of chloro-and bromo-benens. J. Chem. Soc.，Chem. Commun，1993，17：1335-1337.

[6] 吕春绪．硝酰阳离子理论 [M]．北京：兵器工业出版社，2006：12-18.

[7] 吕早生．N_2O_4-O_3 绿色硝化芳烃的宏观动力学及机理研究 [D]．南京：南京理工大学，2001.

[8] 吕早生，吕春绪．一种新的绿色硝化技术．火炸药学报 [J]，2000，23（4）：9-13.

[9] 吕早生，吕春绪．绿色硝化技术在硝基氯苯合成中的应用 [J]．火炸药学报，2000，23（4）：29-30.

[10] 吕早生，吕春绪．一种新的绿色硝化技术，火炸药学报 [J]，2000，23（4）：9-12.

[11] 吕早生，王晓燕，吕春绪．N_2O_4-O_3 硝化芳烃反应的宏观动力学研究 [J]．火炸药学报，2004，27（3）：66-69.

[12] Horvath T H，Rabai J. Facile catalyst separation without water：pluorous biphase hydroformylation of olefins [J]. Science，1994，266（1）：72-75.

[13] 易文斌，蔡春．甲苯的氟两相硝化反应研究（Ⅰ）[J]．含能材料．2005，13（1）：52-54.

[14] 易文斌，蔡春．甲苯的氟两相硝化反应研究（Ⅱ）[J]．含能材料．2006，14（1）：29-31.

[15] YI Wen-bin，CAI Chun. Highly efficient dinitration of aromatic compounds in fluorous media using ytterbium perfluorooctanesulfonate and perfluorooctanesulfonic acid as catalysts [J]．Synth Commun，2006，36：2957-2961.

[16] YI Wen-bin，CAI Chun. A novel and highly efficient catalytic system for trinitration of aromatic compounds：ytterbium perfluorooctanesulfonate and perfluorooctanesulfonic acid in fluorous solvents [J]．J Energy Mater，2007，25：129-139.

[17] YI Wen-bin，CAI Chun. Synthesis of RDX by nitrolysis of hexamethylenetetramine in fluorous media [J]．Journal of Hazardous Materials，2008：839-842.

[18] YI Wen-bin，CAI Chun. Preparation of HMX by perfluorooctanesulfonic acid catalyzed nitrolysis of hexamethylenetetramine in fluorous media [J]．Prop，Expl，Pyro，2009：161-165.

[19] Hua Qian，hi-wen Ye，Chun-xu Lv，Ultrasonically Promoted Nitrolysis of DAPT to HMX in ionic liquid [J]. Ultrasonics Sonochemistry，2008，15（4）：326-329.

[20] 黄卫华．硝化剂五氧化二氮的合成与放大 [J]．上海航天，2003（2）：56-59.

[21] McClelland B W，Hedberg，L. Molecular Structure of N_2O_5 in the gas phase. large amplitude motion in a system of coupled rotors [J]．J Am Chem Soc，1983，105：3789-3793.

[22] Olah G A，Malhotra R，Narang S C. Nitration [M]．New York：VCH Publishers，1989.

[23] 钱华，吕春绪，叶志文．HSM-5 催化下 N_2O_5 对甲苯的选择性硝化研究 [J]．火炸药学报，2006，29（5）：9-11.

[24] 钱华，叶志文，吕春绪．绿色硝化剂 N_2O_5 对甲苯的硝化研究 [J]．化学世界，2006，47 (12)：717-719.

[25] Qian H，Lv C-X，Ye-W. HSM-5 Assisted Nitration of Toluene with Dinitrogen Pentoxide [A]. Preceedings of thr 2nd International Seminar on Industrial Explosive [C]. Beijing：The Publishing House of Ordnance Industry，2006，50-53.

[26] 葛学忠，李高明，洪峰，等．绿色硝化技术合成 HMX 的小试工艺研究 [J]．火炸药学报，2002 (1)：45-47.

[27] 何志勇，罗军，吕春绪，等．N_2O_5 硝解 DPT 制备 HMX [J]．火炸药学报，2010，33 (2)：1-4.

[28] 齐秀芳，程广斌，段雪蕾，吕春绪．Brønsted 酸性功能离子液体存在下甲苯的硝化反应 [J]．火炸药学报，2007，30 (5)：12-15.

[29] 岳彩波，魏运洋，吕敏杰．Bronsted 酸性离子液体中芳烃硝化反应的研究 [J]．含能材料，2007，15 (2)：118-127.

[30] 方东，施群荣，巩凯．离子液体催化甲苯绿色硝化反应研究 [J]．含能材料，2007，15 (2)：122-124.

[31] Fang D，Luo J，Zhou X L，et al. Mannich reaction in water using acidic ionic liquid as recoverable and reusable catalyst [J]. Catal. Lett.，2007，116 (1-2)：76-80.

[32] Fang D，Luo J，hou X-L，et al. One-pot green procedure for Biginelli reaction catalyed by novel task-specific room-temperature ionic liquids [J]. J. Mol. Catal. A：Chem.，2007，274：208-211.

[33] Laali K K，Gettwert V J. Electrophilic nitration of aromatics in ionic liquid solvents [J]. J. Org. Chem.，2001，66 (1)：35-40.

[34] Qiao K，Hagiwara H，Yokoyama C. Acidic ionic liquid modified silica gel as novel solid catalysts for esterification and nitration reactions [J]. J. Mol. Catal. A：Chem.，2006，246：65-69.

[35] Lancaster N L，Llopis-Mestre V. Aromatic nitrations in ionic liquids：the importance of cation choice [J]. Chem. Commun.，2003，22：2812-2813.

[36] Smith K，Liu S，El-Hiti G A. Regioselective mononitration of simp learomatic compounds under mild conditions in ionic liquids [J]. Ind. Eng. Chem. Res.，2005，44：8611-8615.

[37] 刘丽荣，职慧珍，罗军，吕春绪．新型酸性离子液体催化下甲苯的选择性硝化 [J]．含能材料，2009，17 (6)：717-721.

[38] 职慧珍，罗军，马伟，吕春绪．PEG 型酸性温控离子液体中芳香酸和醇的酯化反应 [J]．高等学校化学学报，2008，29 (4)：772-774.

[39] Qi X F，Cheng G B，Lu C X，et al. Nitration of toluene and chlorobenene with HNO_3/Ac_2O catalyed by caprolactam-based Brønsted acidic ionic liquids [J]. Central European Journal of Energetic Materials，2007，4 (3)：105-113.

[40] Zhi H Z，Lü C X，Hang Q，Luo J. A new PEG-1000-based dicationic ionic liquid exhibiting temperature-dependent phase behavior with toluene and its application in one-pot synthesis of benopyrans [J]. Chem Commun，2009，20：2878-2880.

[41] Cheng G，Duan X，Qi X，Lu C. Nitration of Aromatics Compounds with NO_2/Air Catalyed by Sulfonic acid-fundtionalied Ionic Liquids [J]. Catalysis Communications，2008，10 (2)：201-204.

[42] 程广斌，钱德胜，齐秀芳，吕春绪．己内酰胺对甲基苯磺酸离子液体中甲苯的选择性硝化反应 [J]．应用化学，2007，24 (11)：1255-1258.

[43] 齐秀芳，程广斌，吕春绪．酸性离子液体存在下甲苯的硝酸硝化（Ⅱ）[J]．含能材料，2008，16 (4)：398-400.

[44] Qi X，Cheng G，Lu C，Qian D. Nitration of toluene and chlorobenene with HNO_3/Ac_2O Catalyed by caprolactam-based Brønsted acidic ionic liquids [J]. Central European Journal of Energetic Materials，2007，4 (3)：105-110.

[45] Qi X，Cheng G，Lu C，Qian D. Nitration of simple aromatics with NO_2 under air atmosphere in the presence of novel Brønsted acidic ionic liquids [J]. Synth. Commun，2008，38 (4)：537-541.

[46] Zhi H，Luo J，Lü C X. The application of acidic ionic liquids and dinitrogen pentoxide in nitrolysis of DPT [J]. Chin Chem Lett，2009，20：379-382.

[47] 何志勇，罗军，吕春绪．N_2O_5 硝解 DPT 制备 HMX [J]．火炸药学报，2010，33 (2)：36-38.

[48] Cheng Guangbin，Li Xia，Qi Xiufang，Lu Chunxu. Synthesis of RDX by nitrolysis of hexamethylenetetramine in the presence of Brønsted acidic ionic liquids [M]. Theory and practice of energetic materials (VOL. VIII)，Science

Press，2009，8：48-51.
[49] 程广斌，李霞，齐秀芳，吕春绪．离子液体在乌洛托品硝解反应中的应用［J］．2008年火炸药学术研讨会论文集，2008，7-10.
[50] 石煜．黑索今的合成工艺研究［D］．南京：南京理工大学，2010.
[51] ZHI Hui-zhen，LUO Jun，FENG Guang-an，et al. An efficient method to synthesize HMX by nitrolysis of DPT with N_2O_5 and a novel ionic liquid［J］．Chinese Chemical Letters，2009，20（4）：379-382.
[52] 钱华，吕春绪，叶志文．绿色硝解合成六硝基六氮杂异伍兹烷［J］．火炸药学报，2006，29（3）：52-53.
[53] 钱华，叶志文，吕春绪．N_2O_5/HNO_3 硝化硝解TAIW制备CL-20［J］．应用化学，2008，25（3）：424-426.
[54] 钱华，吕春绪，叶志文．绿色硝化剂五氧化二氮及其在硝化中的应用［J］．精细化工，2006，23（6）：620-624.
[55] Estes Z L. Summary of chlorine free synthesis of TATB［R］．［S1］：ERA，1977.
[56] Estes Z L，Locke J G. Synthesis of chlorine-free TATB［R］．［S. 1.］：ERA，1979.
[57] Atkins R L，Nielsen A T，Norris W P. New method for preparing pentanitroaniline and triaminotrini-tro benzenes from trinitrotoluene. US，4248798［P］．1980.
[58] 魏运洋．DATB、TATB和其中间体的新合成方法［J］．兵工学报，1992（2）：79-81.
[59] 魏运洋．1,3,5-三氨基-2,4,6-三硝基苯的合成新法［J］．应用化学，1990，7（1）：70-71.
[60] Bellamy A J，Ward S J. A new synthetic route to 1,3,5-triamino-2,4,6-trinitrobenzene（TATB）［J］．Prop，Expl，Pyro，2002，27：49-58.
[61] 钱华，叶志文，吕春绪．HSM-5催化下 N_2O_5 对氯苯的选择性硝化研究［J］．含能材料，2007，15（1）：56-59.
[62] 马晓明，李斌栋，吕春绪，等．无氯TATB的合成及其热分解动力学［J］．火炸药学报，2009，32（6）：24-27.
[63] 何志勇，罗军，吕春绪．绿色硝化剂 N_2O_5 的制备方法及其应用进展［J］．火炸药学报，2010，33（1）：1-4.
[64] Harrar J E，Pearson R K. Electrosynthesis of N_2O_5 by controlled-potential oxidation of N_2O_4 in anhydrous HNO_3［J］. Journal of Electrochemical Society，1983，130（1）：108-112.
[65] WANG Qing-fa，MI Zhen-tao，WANG Ya-quan，et al，Epoxidation of allyl choride with molecular oxygen and 2-ethyl-anthrahydroquinone Catalyed by TS-1［J］J Mol Catal A：Chem，2005，229：71-75.
[66] WANG Qing-fa，SU Min，ZHANG Xing-wen，et al. Electrochemical synthesis of N_2O_5 by oxidation of N_2O_4 in nitric acid with PTFE membrane［J］．Electrochimica Acta，2007，52：3667-3672.
[67] WANG Qing-fa，HANG Xing-wen，LI Wang，et al. Kinetic of epoxidation of hydroxyl terminated polybutadiene with hydrogen peroxide under phase transfer catalysis［J］．Ind Eng Chem Res，2007，5：381.
[68] 王庆法，石飞，米镇涛，张香文，王莅. 硝酸酯的绿色合成［J］. 含能材料，2007，15（4）：416-420.
[69] 石飞，王庆法，米镇涛. 1，2-丙二醇二硝酸酯的绿色合成研究［J］. 火炸药学报，2007，132（2）：75-77.
[70] 苏敏，王庆法，张香文，王莅，米镇涛．新型绿色硝化剂 N_2O_5 的电化学合成研究进展［J］. 含能材料，2006，14（1）：66-67.
[71] 王庆法，米镇涛，张香文，王莅. 硝酸酯类含能黏合剂绿色合成研究进展［J］. 化学推进剂与高分子材料，2008，6（2）：11-15.

Study Progress on Green Nitration

Lv Chunxu

（School of Chemical Engineering，Nanjing University of Science and Technology，Nanjing 210094，China）

Abstract：The current situation of green nitrification was reviewed. Pointing out that it is of vital importance to apply some advanced technologies such as fluorous biphasic，microwave and ultrasound technology and N_2O_5-ionic liquid system in the nitration reactions，particularly in the nitrification of amines. The system is of great significance in the promotion of green nitrification with 71 references.

Key words：organic chemistry；green nitration；N_2O_5；Ionic Liquid；explosives and propellants

（注：此文原载于 火炸药学报，2011，34（1）：1-8）

Zeolite-assisted Nitration of Neat Toluene and Chlorobenzene with a Nitrogen Dioxide/molecular Oxygen System. Remarkable Enhancement of Para-selectivity

Xinhua Peng[2], Hitomi Suzukia[1], Lv Chunxu[2]

(1. School of Science, Kwansei Gakuin University, Uegahara, Nishinomiya 662-8501, Japan)

(2. Nanjing University of Science and Technology, Xiao Ling Wei, Nanjing 210094, China)

SCI: 000169311800020

Abstract: In the presence of molecular oxygen and HZSM-5, neat toluene reacted with liquid nitrogen dioxide in a regioselective manner at room temperature to yield mononitrotoluenes as the main product, where the para isomer predominated up to 90% and the ortho-para isomer ratio improved to 0.08. Chlorobenzene behaved similarly, giving a para-predominant nitration product.

Keywords: nitration; regioselection; nitrogen oxides; arenes; nitro compounds; zeolites.

Classical methodology for the preparative nitration of toluene involves mixtures of nitric and sulfuric acids. The reaction is not selective and gives rise to a mixture of ca. 60% o-nitrotoluene, ca. 37% p-nitrotoluene and less than 3% m-nitrotoluene, along with some oxidation and/or polynitration products[1]. Since p-nitrotoluene has more market demand than the o-isomer, large amounts of the unwanted products in general pose a problem in industrial manufacture. Moreover, the disposal of waste products and spent acids is environmentally unfriendly and can be costly.

The Kyodai nitration using a NO_2/O_3 system has demonstrated excellent conversion of a variety of aromatic compounds into the corresponding nitro derivatives[2]. The reaction bears the general appearance of being an electrophilic process and is characterized by some unique feature as operating under neutral conditions. Further efforts to find an alternative means toactivate nitrogen dioxide has led to the Fe (Ⅲ) -cata-lyzed nitration of aromatic compounds with a NO_2/O_2 system[3]. Rather unfortunately, the product composition differs little from those of the Kyodai nitration and conventional aromatic nitration.

Over the past few decades there has been a considerable interest in improving the para-selectivity of toluene nitration and many attempts have been made using the catalysts supported on porous inorganic solids. Laszlo has developed Claycop (Cu $(NO_3)_2$/montmorillonite-K10) that selectively nitrates toluene in acetic anhydride/CCl_4 with an isomer distribution of ortho 23, meta 1 and para 76%[4]. Smith has shown that mordenite and zeolite β, can play an important role in the selective mononitration of toluene, where the para-selectivity of 67% by using benzoyl nitrate[5] and the ortho-para isomer ratio of 0. 23 by using acetyl nitrate[6] have been achieved. Toluene also has been nitrated with isopropyl nitrate in the presence of HZSM-5, producing a product of isomeric composition ortho 5, meta 0 and para 95%[7]. Zeolite catalysts are attractive for being inexpensive and reusable, and direct use of NO_2 as the source of nitro group has considerable potential in industrial application. Thus, in search for the para-selective methodology of aromatic nitration under neutral conditions, we have investigated the reaction of neat toluene and chlorobenzene with liquid NO_2 in the presence of a variety of zeolites by using each substrate as its own solvent. We now report the activation of nitrogen dioxide on the zeolite surface that results in the para-selective nitration of these aromatic substrates.

CH_3 —NO_2, O_2→ CH_3 —NO_2 + CH_2NO_2 + CHO + CH_3 —$(NO_2)_2$

1 2 3 4 5 6

ortho meta para

Scheme1

Weakly activated toluene was quite reluctant to react with NO_2 alone, but passage of O_2 into the reaction mixture resulted in a certain degree of conversion into nitrated compounds. Unfortunately, the reaction was quite slow and unselective, giving a mixture of nitro-toluenes (ortho-1, meta-2 and para-3), phenyl-nitromethane 4, benzaldehyde 5 and dinitrotoluenes 6 (Scheme 1) . Trace amounts of a nitrophenol was also detected. In the presence of oxygen and HZSM-5, however, the reaction was promoted and the para-selec-tivity was found to increase considerably. Thus, in the hope of improving the para-selectivity further, the NO_2-O_2-zeolite system has been investigated systematically and the representative results are summarized in Table 1. Wider pore zeolite β and various types of modified zeolites were also examined and some selected results were included for comparison.

All zeolites were found to promote the reaction and the isomer proportion always moved to the direction to favor the para-isomer. HZSM-5 exhibited better para-selectivity than zeolite, but their difference was slight as to the total product yield. Zeolites modified with metal ion or heteropoly acid were not so effective to accelerate the reaction and brought about complicated variation in the product distribution. In contrast, zeolites treated with methanesulfonic acid considerably facilitated nuclear nitration and the isomer proportion approached those of conventional nitration. Increasing the amount of NO_2 led to higher conversion, but was indifferent to the ortho-para isomer ratio, while increasing the amount of zeolites worked favorably for the nuclear nitration as well as para-substitution. The rise of reaction temperature from-10 to 10 to 50°C improved both product yield and extent of para-substitution considerably, the ortho-para isomer ratio turning for the better from 0. 33 to 0. 15 to 0. 07[8]. The positive effect of elevated temperature would probably be due to the enhanced transport and interaction of the reactants inside the zeolite channels.

Our results contrast to those reported by Smith, where HZSM-5 was less effective than zeoliteβ, and the ortho and para isomers were obtained in comparable amount[6,9]. This difference may well be attributed to the absence of intervention by chlorinated solvent molecules, as well as the higher Si/Al ratio of our zeolite samples (Si/Al 1000) . Mixing of toluene with a large excess of NO_2 over HZSM-5 at room temperature led to complete consumption of the substrate, leading to the distribution of products 1-6 of 41 : 6 : 47 : 3 : 1 : 2 and an ortho-para ratio of 0. 87.

Table 1 Regioselectivity of nitration of toluene with NO_2-O_2 over various zeolite catalysts

Catalyst[a]	Yield[b] (mmol)	Product distribution(mol%)[c] 1	2	3	4	5	6	ortho-para	R/Sd[d]
–[e]	0.47(1)	7	7	5	34	35	12	1.48	0.45
–	1.15(5)	34	7	23	6	12	18	1.47	4.6
HZSM-5[f]	2.12(11)	17	4	57	10	7	5	0.30	4.9
HZSM-5[f,g]	4.62(24)	6	2	74	10	7	1	0.08	4.9
HZSM-5/$MeSO_3H$	7.30(46)	51	3	44	Trace	Trace	2	1.15	>99
BiHZSM-5[h]	2.45(14)	24	6	58	4	3	5	0.41	13
CuHZSM-5[h]	1.00(6)	26	4	69	Trace	Trace	1	0.38	>99
FeHZSM-5[h]	1.47(8)	15	5	66	4	4	6	0.22	12
ZnHZSM-5[h]	2.08(11)	30	7	47	7	5	4	0.64	7.3
P_2O_5 HZSM-5[i]	1.62(8)	21	6	51	9	8	5	0.41	4.9
$Mo_2O_3P_2O_5$ HZSM-5[i]	2.63(14)	17	7	60	5	5	6	0.29	9.0
Hβ-25[j]	4.16(22)	30	8	46	7	4	5	0.64	8.1
Hβ-150[j]	2.13(13)	36	5	52	2	3	2	0.70	19
Fe Hβ150[h]	1.10(6)	37	6	49	2	4	2	0.75	16
NaMFI-90[j]	2.63(14)	25	6	51	6	4	8	0.49	9.0

a All reactions were carried out by stirring a mixture of toluene (5.0ml), liquid NO_2 (0.5ml), catalyst (2.0g), 3 AD molecular sieves (0.5g), and cyclododecane (internal standard) at room temperature for 22h under oxygen, unless otherwise noted. Toluene was dried over 4 AD molecular sieves.

b Sum of products 1-6, estimated by GC. Numeral in parenthesis refers to total yield of nitrotoluenes based on NO_2.

c Calculated from GC data.

d Ratio of ring and side-chain reaction products.

e Under exclusion of oxygen.

f HZSM-5 is a commercial product (Acros) of Si/Al ratio 1000. All catalysts except HZSM-5/$MeSO_3H$ were calcined in air at 500°C for 8h prior to use.

g Double amount of catalyst.

h Prepared as follows; a suspension of zeolite (5g) in 0.2M solution of a given salt in water (100ml) was stirred for 24h under reflux. The solid was washed, dried at 110°C for 3h, and calcined in air at 500°C for 8h.

i Prepared by stirring HZSM-5 with aqueous phosphomolybdic acid ($H_3PMo_{12}O_{40} \cdot nH_2O$) or 85% phosphoric acid for 48-68h at room temperature. Without washing, the solid was dried at 110°C for 5h and calcined in air at 500°C for 8 h.

j A gift from the Catalytic Society of Japan.

To our knowledge, no substantiated mechanisms have been proposed to date for the nitration of arenes on the surface of solid catalyst. High ortho-para preference along with insignificant formation of m-isomer reflects the general characteristic of electrophilic substitution involving the nitronium ion. However, concurrent formation of a slight amount of side-chain nitration and oxidation products suggests a minor involvement of a radical process.

In the presence of a radical scavenger such as m-dinitrobenzene, no obvious change was observed as to the relative ratio of ring and side-chain substitution products as well as the ortho-para isomer ratio. The Brønsted active sites resulting from the hydroxylic oxygen atoms bridging between the silicon and aluminum atoms in the zeolite framework are located over 90% of total amounts in the interior of channels[7,10]. They would play a role similar to mineral acids inside the channel system of HZSM-5, leading to the para-selective nitration.

The side-chain reaction products 4 and 5 were probably formed through the capture of the benzyl radical, partly derived from toluene via the direct abstraction of hydrogen by NO_2 and partly via the electron transfer between zeolite followed by proton release from alkyl side-chain, by nitrogen dioxide and oxygen, respectively. Nitrogen dioxide has been known to trap the benzyl radical more efficiently than molecular oxygen[11], thus favoring phenylnitromethane 4 over benzaldehyde 5. Small amounts of dinitrotoluenes 6 were simulta-neously formed in spite of the presence of a large excess of toluene. Since none of the isomeric mononitrotoluenes reacted further to form dinitro derivatives under the conditions employed, these dinitro compounds are most likely to form directly from toluene via the addition-elimination sequence, as suggested previously[12].

Under similar conditions, chlorobenzene also reacted in a para-selective manner. Thus, on stirring a mixture of chlorobenzene (5.0ml), liquid NO_2 (0.5ml), HZSM-5 (2.0g) and 3 AD molecular sieves (0.5g) at room temperature for 22h under oxygen, chloronitroben-zenes (3.74 mmol) were obtained in the isomer ratio of ortho 8, meta 2 and para 90. The ortho-para ratio 0.09 was better than those (0.25-0.40) previously reported by Smith[9]. In the absence of HZSM-5, the reaction was quite sluggish and unselective, affording the nitrated product in the ratio ortho : meta : para=32 : 15 : 53.

In conclusion, the present work has revealed that the ortho-rich composition ordinarily expected from conventional nitration of toluene can be reversed to the para-dominant one by using a NO_2-O_2 system in the presence of HZSM-5, the ortho-para ratio being improved to 0.08 and the para-selectivity up to 90%. The solvent-free nitration of toluene with NO_2 is most likely to occur in the interior of zeolite channels via the electrophilic process involving the nitronium ion, partly accompanied by an ordinary radical process tht leads to side-chain reaction products.

Acknowledgements

X. P. thanks the Japan Society for the Promotion of Science for the Fellowship (no. P99282).

References

[1] (a)Olah, G. A.; Malhotra, R.; Narang, S. C. Nitration: Methods and Mechanisms; VCH: New York, 1989; (b) Schofield, K. Aromatic Nitration; Cambridge University Press: Cambridge, 1980.

[2] (a)Nonoyama, N.; Mori, T.; Suzuki, H. Zh. Org. Khim. 1998, 34, 1591-1601; (b) Suzuki, T.; Noyori, R. Chem-tracts 1997, 10, 813-815; (c) Mori, T.; Suzuki, H. Synlett 1995, 383-392.

[3] Suzuki, H.; Yonezawa, S.; Nonoyama, N.; Mori, T. J. Chem. Soc., Perkin Trans. 1 1996, 2385-2389.

[4] (a)Laszlo, P.; Vandormael, J. Chem. Lett. 1988, 1843-1846; (b) Cornelis, A.; Delaude, L.; Gerstmans, A.; Las-zlo, P. Tetrahedron Lett. 1988, 29, 5657-5660.

[5] Smith, K.; Fry, K. Tetrahedron Lett. 1989, 30, 5333-5336.

[6] (a) Smith, K.; Musson, A.; DeBoos, G. A. Chem. Com-mun. 1996, 469-470; (b) Smith, K.; Musson, A.; DeBoos, G. A. J. Org. Chem. 1998, 63, 8448-8454.

[7] Kwok, T. J.; Jayasuriya, K.; Damavarapu, R.; Brodman, B. W. J. Org. Chem. 1994, 59, 4939-4942.

[8] The reaction was carried out in a closed vessel by stirring a mixture of toluene (1. 0ml), NO_2 (0. 1ml), HZSM-5 (1. 0g), and 3 AD molecular sieves (0. 1g) for 22h under oxygen.

[9] Smith, K.; Almeer, S.; Black, S. J. Chem. Commun. 2000, 1571-1572.

[10] Tanabe, K.; Misono, M.; Ono, Y.; Hattori, H. New Solid Acids and Bases: Their Catalytic Properties; Elsevier: Amsterdam, 1989; p. 143.
[11] Goumri, A.; Elmaimouni, L.; Sawerysyn, J. P.; Devolder, P. J. Phys. Chem. 1992, 96, 5395-5400.
[12] Suzuki, H.; Mori, T. J. Chem. Soc., Perkin Trans. 2 1995, 41-44.

（注：此文原载于 Tetrahedrom Letters，2001，42（26）：4357-4359）

Nitration of Aromatic Compounds with NO_2/Air Catalyzed by Sulfonic Acid-Functionalized Ionic Liquids

Cheng Guangbin, Duan Xuelei, Qi Xiufang, Lv Chunxu
Nanjing University of Science and Technology, Nanjing 210094, China
SCI: 000260988200018; EI: 20084311655396

Abstract: Nitration of aromatic compounds with NO_2/air catalyzed by sulfonic acid-functionalized ionic liquid under solvent free conditions is reported. A variety of ionic liquids [BSPy][HSO_4], [BSPy][TfO], [BSPy][*p*TSA] ([BSPy] =*N*- (4-hydroxsulfonylbutyl) pyridinium) were prepared. Their acidities were further investigated by Hammett method and the acidity-catalytic relationship was discussed. It was found that satisfactory results were obtained with [BSPy][HSO_4]. Ionic liquids could be conveniently separated with the products and reused for four times with excellent yield of mono-nitration products and para-selectivity.

Key words: Nitration, NO_2/air, Ionic liquid

1 Introduction

Nitration of aromatic compounds is one of the most fundamental therefore important reactions in organic synthesis, which provides key organic intermediates or energetic materials. Traditionally, nitration of aromatic compounds is performed with mixture of nitric and sulfuric acids[1,2]. However, this method often provides poor selectivity for desired products and it is not environmentally benign either. Developing an efficient and economical method to dispose or regenerate used acids remain a formidable challenge. On the other hand, various clean nitration approaches have been explored which involve using nitrogen oxides such as NO_2[3,4], N_2O_4[5], N_2O_4[6] under "neutral" condition, or using recyclable catalysts such as lanthanied triflates[7,8], perfluorinated resin immobilized sulfonic acid[9], claycop or zeolites[10~13]. Although much success has been achieved, some problems still remain, for example, the use of lanthanide triflate or polymeric sulfonic acid resins as catalysts did not improve the selectivity. Furthermore, volatile chlorinated solvents were still required in some processes.

Over the past few years, ionic liquid received great attention because of their high polarity, negligible vapor pressure, and good solubility in a broad range of organic and inorganic compounds, which endorse them as useful "clean and green" solvents or catalysts for a number of reactions[14~18]. It was worthwhile to investigate the possibility applying ionic liquid in nitration. To our best knowledge, although there were several reports on the nitration of aromatic compounds using HNO_3 (67%)[19,20] or HNO_3/Ac_2O[21] in the media of ionic liquid, Study on nitration of aromatic compounds with NO_2 using ionic liquid as catalysts have not been reported yet.

R + NO_2/O_2 —ionic liquid→ R, NO_2

R=CH_3, H, Cl

Scheme 1 Nitration of simple aromatic compounds

Nitration is an electrophilic reaction. The acidity of catalyst is an important parameter related to its catalytic activity. In this paper we report the results of nitration of toluene, chlorobenzene and benzene with NO_2 under air atmosphere (Scheme 1) using novel sulfonic acid-functionalized pyridinium cation-based ionic liquid (Fig. 1) as catalysts and solvent. The acidities of the ionic liquids were determined using the Hammett method with UV-visible spectroscopy, and the acidity-catalytic activity relationship was discussed.

2 Experimental

2.1 General information

SO_3H

[BSPy][H_2SO_4] X- =HSO_4^-

[BSPy][OTf] X- =$CF_3SO_3^-$

[BSPy][pTSA] X- =H_3C—C6H4—SO_3^-

Fig. 1 Structures of three sulfonic acid-functionalized ionic liquids

All reagents were commercially available. Toluene, chlorobenzene, benzene and dichloromethane were dried over 4Å molecular sieve.

Product mixtures were analyzed on Agilent Technologies 6820 Gas Chromatography with OV-101 capillary column (30m × 0.32mm) and FID detector. The 1H NMR Spectra were recorded on a Bruker Advanced Digital 300 MHz spectrometer in D_2O and cailberated with tetramethylsilane (TMS) as internal reference. The UV-visible spectra were recorded on Beijing Ruili UV-1100 spectrophotometer.

2.2 Preparation of sulfonic acid-functionalized ionic liquids

Three sulfonic acid-functionalized ionic liquids were synthesized following the literature reported method[22]. They are all colorless liquid and entirely miscible with water.

The data of ^{1}NMR and MS spectra of the three ionic liquids were shown as follows:

[BSPy] [HSO_4]:^{1}HNMR (300MHz, D_2O): δ1.41 (m, 2H), 1.79 (m, 2H), 2.58 (t, 2H, J=7.4Hz), 4.28 (t, 2H, J=7.3Hz), 7.69 (t, 2H, J=7.0Hz), 8.17 (t, 1H, J=8.0Hz), 8.48 (d, 2H, J=6.7Hz).

[BSPy] [OTf]:^{1}HNMR (300MHz, D_2O): δ1.42 (t, 2H, J=10.4Hz), 1.78 (t, 2H, J=6.4Hz), 2.55 (m, 2H), 4.24 (m, 2H), 7.69 (t, 2H, J=5.5Hz), 8.16 (q, 1H, J=7.0Hz), 8.45 (t, 2H, J=5.3Hz).

[BSPy] [pTSA]:^{1}HNMR (300MHz, D_2O): δ1.87 (m, 2H), 2.01 (s, 3H), 2.28 (m, 2H), 3.19 (t, 2H, J=6.2Hz), 4.58 (t, 2H, J=5.3Hz), 6.92 (s, 2H), 7.47 (s, 2H), 7.70 (s, 2H), 8.14 (s, 1H), 8.53 (d, 2H, J=4.9Hz).

2.3 General nitration procedure

To a cooled (−15℃, ice-salt bath), vigorously stirred mixture of the ionic liquid (15mol% of substrate) and the substrate (10.0mmol) under air atmosphere (provided by air balloon) were added liquid nitrogen dioxide (0.5mL, ca. 14mmol). After a certain time as indicated the reaction was quenched with deionized water (ca. 5mL), extracted with hexane (5mL×4). The organic phase was isolated and organic solutions were combined and washed with saturated solution of sodium bicarbonate (5mL), water (5mL×3) and brine (5mL×2), dried over Na_2SO_4 and analyzed by GC (different internal standard, was used. (for toluene, and chlorobenzene, nitrobenzene; and for benzene, hexadecane 0.2mL). The ionic liquid was recovered from the aqueous solution by removing the water at 65℃ under vacuum for 5h and was reused without significantly lost its reactivity.

2.4 UV-visible acidity determination

The ionic liquids are all colorless liquid with good fluidity at room temperature, so the experiments were carried out at room temperature. Dichloromethane solutions of the ionic liquids were prepared from dried CH_2Cl_2 and ionic liquids (dried under vacuum at 70℃ for 2h).

3 Results and discussion

3.1 Acidities of the sulfonic acid-functionalized ionic liquids

The acidity of the ionic liquids is an important parameter relevant to the catalytic activity of acid-catalyzed reaction. In order to understand the catalytic activities of the sulfonic acid-functionalized ionic liquids, it is necessary to study their acidities. A common and effective way to evaluate the acidity of Brønsted acid is Hammett method, which was first proposed by Gilbert et al. in 2003 on the basis of the Hammett acidity function[23]. Wherein a basic indicator was used to trap the acidic proton. The Hammett function (H_0) can be expressed as:

$$H_0 = pK(I)_{aq} + \log([I]/[IH^+])$$

Where pK (I) aq is the pKa value of the indicator, [I] and [IH^+] are respectively the molar concentrations of the unprotonated and protonated forms of the indicator, which can usually be determined by UV-visible spectroscopy.

The acidities of the three ionic liquids were examined using 4-nitroanline (Hammett content is 0.99, concentration is 10mmol/L) as indicator in dichloromethane and the results are shown in Figure 2. Dichloromethane was chosen as test solvent, because it is an aprotic polar solvent, stable under acidic conditions and has considerable solubility for all tested ionic liquids. The maximal absorbance of the unprotonated form of 4-nitroanline was observed at 349nm in CH_2Cl_2 (Figure 2. spectra a). As the acidic ionic liquid was added, the absorbance of the indicator decreased. We could determine the [I]/[IH^+] ratio by measuring the absorbance when each ionic liquids was added (spectra b-d), and then the Hammett function (H_0) is calculated (see Table 1). The experiments showed that the acidity order was: [BSPy][OTf] > [BSPy][HSO_4] > [BSPy][pTSA]. It is clear that acidities of the ionic liquid depend on the anions. H. B. Xing[24] determined the

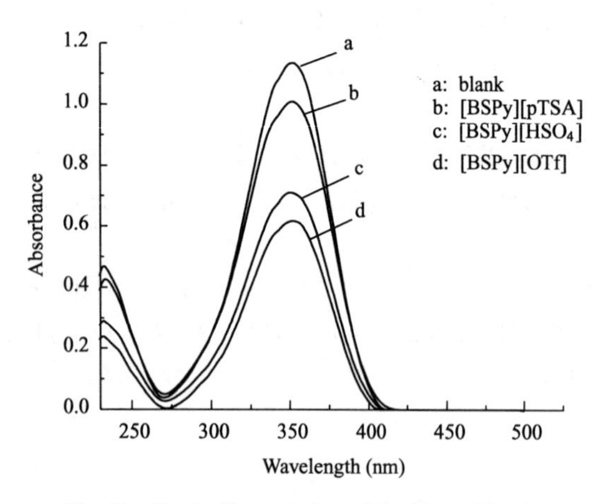

Fig. 2 Absorption spectra of 4-nitroaniline in various ionic liquids solution in dichloromethane

minimum energy geometries of these ionic liquids, the results manifested that anions had strong interaction with the sulfonic acid proton. It is considered that in addition to alkyl sulfonic acid group, the anion is likely to serve as available acid site. So the acidities and catalytic activities of the ionic liquids depend on the kinds of anion as well.

Table 1 H_0 values of ionic liquids in dichloromethane at room temperature[a]

Ionic liquids	Absorbance(AU)[b]	[I](%)	$[IH^+]$(%)	H_0
Blank	1.134	100	0	—
[BSPy][HSO_4]	0.709	62.5	37.5	1.21
[BSPy][OTf]	0.615	54.2	45.8	1.06
[BSPy][pTSA]	1.007	88.8	11.2	3.98

a Concentration: 10mmol/L; indicator: 4-nitranline.

b The maximal absorbance of the unprotonated form of 4-nitroaniline was observed at 349nm in CH_2Cl_2.

3.2 Nitration of different conditions using [BSPy] [HSO_4] as catalyst

Toluene was chosen as a test substrate and [BSPy] [HSO_4] was used as catalyst for the investigations on nitration reaction conditions with NO_2/air, the results are summarized in table 2.

The reaction proceeded well and mono-nitrotoluene was gained in moderate yield with good ortho/para ratio (Table 2, entry 3) . When the amount of [BSPy] [HSO_4] increase (Table 2, entries 1-4), the yield was also increased (Table 2, entry 3), and so did ortho/para ratio. When it reached up to 20%, the conversion and yield decreased slightly. It may be due to the solvating effect. It seems that concentration of [BSPy] [HSO_4] has little effect on the para-selectivity. The equivalents of NO_2 used in the reaction had a significant effect on conversion and yield, but little effect on the distribution of product (Table 2, entries 5-7) . When 0.5mL (ca. 14mmol) NO_2 was used, which was a little excessive, gained the best yield. We also investigated the effect of reaction temperature. At the temperature below 0℃ the reaction preceded very slow (Table 2, entry 9) . It may need much longer time to gain reasonable conversion. Therefore, we carried out the reaction at higher temperature for 20h and good yield and selectivity was obtained (Table 2, entries 3, 8) . An even better ortho/para ratio (Table 2, entry 8, ortho/para=1.13) was obtained at 45℃, however both the conversion and yield decreased. It can be seen that a period of time at low temperature and then at higher temperature are both necessary to the reaction.

From the result shown above, the best reaction condition is: toluene 10mmol, NO_2 0.5ml, ionic liquid 15mol%, −15℃/30min, then 0℃/5h, and rt/20h.

Table 2 nitration of toluene in [BSPy] [HSO_4] under different conditions[a]

Entry	[BSPy][HSO_4][d] (mol%)	NO_2 (mL)	conversion[c] (%)	Yield[e] (%)	Product distribution(%)			Ortho/ Para
					Ortho	Meta	Para	
1	5	0.5	49.9	33.0	55.4	3.9	40.7	1.36
2	10	0.5	49.0	33.3	53.7	3.7	42.6	1.26
3	15	0.5	79.4	77.0	54.2	3.3	42.5	1.27
4	20	0.5	61.7	52.8	54.1	3.5	42.4	1.27
5	15	0.2	43.9	16.7	54.2	4.3	41.5	1.31
6	15	0.8	41.9	35.9	54.3	3.7	42.0	1.29
7	15	1	31.8	28.7	54.5	3.5	42.0	1.30
8[b]	15	0.5	46.8	35.7	51.0	3.9	45.1	1.13
9[c]	15	0.5	15.9	4.6	56.0	0	44.0	1.27

Substrate 10mmol; reaction time and temperature: a－15℃/30min, then 0℃/5h, and rt/20h(when the air balloon was removed). b－15℃/30min, then 0℃/5h, and 45℃/20h (when the air balloon was removed). c－15℃/30min, then 0℃/5h. dmole ratio to toluene. e Calculated by quantitative GC.

3.3 Nitration of simple aromatic compounds catalyzed by different ionic liquid

In view of the success of new reaction, the reaction scope was explored with various ionic liquids and other aromatic compounds. For comparison, all reactions were conducted under the identical conditions indicated above. The results are summarized in table 3.

Table 3 nitration of simple aromatic compounds catalyzed by different ionic liquids

Entry	IL	R	Yield[c](%)	Product distribution(%)			Ortho/Para
				Ortho	Meta	Para	
1[a]	[BSPy][HSO_4]	CH_3	77.0	54.2	3.3	42.5	1.27
2[a]	[BSPy][OTf]	CH_3	63.6	53.0	3.3	43.7	1.21
3[a]	[BSPy][pTSA]	CH_3	30.8	53.2	6.5	40.3	1.32
31[a]	[hexPy][PTSA]	CH_3	13.4	54.5	8.0	37.5	1.45
4[b]	[BSPy][HSO_4]	Cl	34.7	22.5	2.2	75.3	0.30
5[b]	[BSPy][OTf]	Cl	32.6	19.8	0.5	79.7	0.25
6[b]	[BSPy][pTSA]	Cl	16.4	24.3	1.8	73.8	0.33
7[b]	[BSPy][HSO_4]	H	38.9			—	
8[b]	[BSPy][OTf]	H	28.6			—	
9[b]	[BSPy][pTSA]	H	15.4			—	
10[a]	—	CH_3	11.4	57.9	5.7	36.4	1.59
11[b]	—	Cl	4.5	39.8	0	60.1	0.66
12[b]	—	H	8.6			—	

Substrate 10mmol, NO_2 0.5mL; reaction time and temperature: a－15℃/30min, then 0℃/5h, and rt/20h (when the air balloon was removed). b－15℃/30min, then 0℃/5h, and rt/40h (when the air balloon was removed). c Calculated by quantitative GC.

All of the ionic liquids were found to be able to promote the NO_2/air nitration of aromatic compounds. Clearly, the use of the three sulfuric acid-functionalized ionic liquids was advantageous from the view of the yield and para-selectivity in comparison with the reaction without the acidic ionic liquid (Table 3, entries 10-12) and with ionic liquid [hexPy] [pTSA] (Table 3, entry 4).

Among the three aromatic compounds, nitration of toluene gave the best result. Nitration of chlorobenzene gained lower yield (and also benzene), it may be explained by the activity of aromatic compounds. But they all gave improved para-selectivity. As for the sulfonic acid-functionalized ionic liquid, anion has great effect on the yield or selectivity of the reaction. Reactions conducted in [BSPy] [HSO_4] and [BSPy] [OTf] were much better in yield than in [BSPy] [pTSA]. This can be explained by their acidities order: [BSPy] [OTf] > [BSPy] [HSO_4] > [BSPy] [pTSA]. However, [BSPy] [HSO_4] has the best catalytic activity, due to that [BSPy] [HSO_4] has two protons, which provide more acid source for nitration reaction. [BSPy] [OTf] showed best para-selectivity, probably due to the higher polarity of [BSPy] [OTf] which give it good affinity to aromatic compounds. In view of the foregoing, [BSPy] [HSO_4] and [BSPy] [OTf] should be suitable catalysts for the nitration.

3.4 Reusability of [BSPy] [HSO_4]

The reusability of [BSPy] [HSO_4] in nitration of toluene was examined and the results were listed in table 4. [BSPy] [HSO_4] was used for four cycles without significant change in yield and selectivity. This indicated that [BSPy] [HSO_4] has excellent reusability.

Table 4 Reusability of [BSPy] [HSO_4] for nitration of toluene[a]

Entry	times	Yieldb(%)	Product distribution(%)			Ortho/Para[b]
			Ortho	Meta	Para	
1	1	77.0	54.2	3.3	42.5	1.27
2	2	76.3	53.3	3.9	42.8	1.25
3	3	74.9	54.0	4.4	41.6	1.30
4	4	70.1	53.3	4.8	41.9	1.27

a Substrate 10mmol, −15℃/30min, then 0℃/5h, and rt/20h (when the air balloon was removed) .

b Calculated by quantitative GC.

4 Conclusions

Nitration of simple aromatic compounds in acidic ionic liquid with NO_2/air has showed higher para-selectivity and moderate yield. Furthermore, the use of NO_2/air is a clean nitration, and air is very cheap and facile. The acidic ionic liquid could be easily separated from the products and easily recycled with excellent yield and selectivity. This methodology offers significant improvements for nitration of aromatic compounds with regard to yield of products, simplicity in operation, and green aspects by avoiding toxic catalysts and solvents.

Reference

[1] K. Schofield, Aromatic Nitration, Cambridge University Press, Cambridge, 1980.
[2] G. A. Olah, R. Malhotea, S. C. Narang, Nitration: Methods and Mechanisms, VCH, New York, 1989.
[3] H. Sato, K. Hirose, Appl. Catal. 174 (1998) 77.
[4] H. Suzuki, S. Yonezawa, N. Nonoyama, T. Mori, J. Chem. Soc. : Perkin Trans 1 (1996) 2385.
[5] N. Iranpoor, H. Firouzabadi, M. A. Zolfigol, Syn. Commun. 28 (1998) 2773.
[6] R. R. Bak, A. J. Smallridge, Tetradron Lett. 42 (2001) 6767.
[7] A. G. M. Barrett, D. C. Braddock, R. Ducray, R. M. Mckinnell, F. J. Waller, Syn. Lett. (2000) 57.
[8] F. J. Waller, A. G. M. Barrett, D. C. Braddock, D. Ramprasad, Chem. Commun. (1997) 613.
[9] G. A. Olah, R. Malhotra, S. C. Narang, J. Org. Chem. 43 (1978) 4628.
[10] L. Delaude, P. Laszlo, K. Smith, Acc. Chem. Res. 26 (1993) 607.
[11] B. Gigante, A. O. Prazeres, M. J. Marcelo-Curto, J. Org. Chem. 60 (1995) 3445.
[12] S. P. Dagade, S. B. Waghmode, V. S. Kadam, M. K. Dongare, Appl. Catal. A. 226 (2002) 49.
[13] X. Peng, H. Suzuki, C. X. Lu, Tetrahedron. Lett. 42 (2001), 4357.
[14] P. Wasserscheid, W. Keim, Angew. Chem. Int. Ed. 39 (2000) 3772.
[15] R. Sheldon, Chem. Commun. (2001) 2399.
[16] A. C. Cole, J. L. Jensen, I. Ntai, K. L. T. Tran, K. J. Weaver, J. Am. Chem. Soc. 124 (2002) 5962.
[17] D. M. Li, F. Shi, J. J. Peng, S. Guo, Y. Q. Deng, J. Org. Chem. 69 (2004) 3582.
[18] E. J. Angueira, M. G. White, J. Mol. Catal. A. 227 (2005) 51.
[19] M. J. Earle, S. P. Katdare, WO. Pat. WP0230865, 2002.
[20] K. Qiao, C. Yokoyama, Chem. Lett. 33 (2004) 808.
[21] K. Smith, S. F Liu, G. A. El-Hiti, Ind. Eng. Chem. Res. 44 (2005) 8611.
[22] J. Z. Gui, X. H. Cong, D. Liu, X. T. Zhang, Z. Hu, Z. L. Sun, Catal. Comm. 5 (2004) 473.
[23] C. Thomazeau, H. Bourbigou, S. Luts, B. Gillber, J. Am. Chem. Soc. 125 (2003) 5264.
[24] H. B. Xing, T. Wang, Z. H. Zhou, Y. Y. Dai, J. Mol. Catal. A. 264 (2007) 53.

(注：此文原载于 Catalysis Communication, 2008, 10: 201-204)

乙酰苯胺选择性硝化及影响因素的分析

金铁柱，吕春绪，国振双，顾建良
（南京理工大学化工学院，江苏，南京.210094）

EI：96043124174

摘要：通过正文设计实验研究了乙酰苯胺的选择性硝化，并用灰色关联分析对影响选择性硝化的因素作了分析。

关键词：正交设计，乙酰苯胺，灰色关联分析，芳烃硝化

1 前言

乙酰苯胺选择性硝化生成的对硝基乙酰苯胺经还原得到对氨基乙酰苯胺。它是合成酸性红染料（BG200）的中间体[1]。酸性红 BG200 主要用于毛、丝、锦纶的染色和丝毛的直接印花及阳极铝、纸、肥皂的着色。

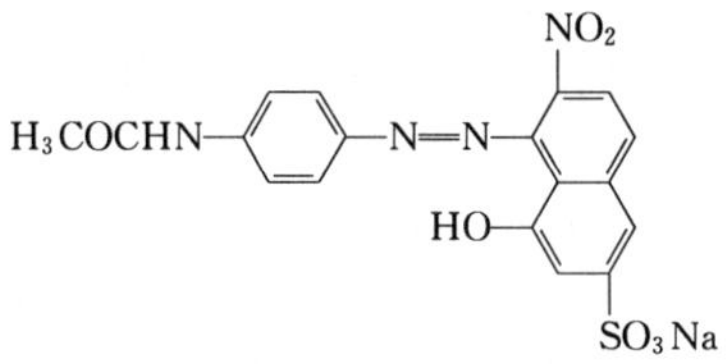

NO_2
H_3COCHN—N═N—
HO
SO_3Na

酸性红 BG200

目前工业上乙酰苯胺的硝化是用硝酸和硫酸的混酸作为硝化剂。现有工艺存在低温（<7℃）硝化、反应时间长（20～40h）、生产效率低等缺点。因此提高硝化得高率和位置选择性，找出反应主要影响因素，降低生产成本，便成丁乙酰苯胺的硝化反应中必须解决的问题。

2 实验部分

2.1 乙酰苯胺的硝化单因素影响试验

由表 1 看出，最高产率 85％所对应的最好结果是：硝酸用量 15mL，硫酸用量 45mL，反应温度 15℃，反应时间 2h。这为下一步做正交试验设计水平的选择提供了依据，也可作为因素灰色关联分析的原始数据。

2.2 正交试验及极差分析

从表 2 可以看 $B_3A_1D_3C_3$ 出为最优化试验，对硝基乙酰苯胺的得率为 85.8％，这些试验条件可作为制备产品的依据。利用极差分析可以分析出 R 值为 $9.6>2.2>1.7>1.5$ 即硫酸用量和硝酸用量为反应的主要影响因素，其次是反应时间和反应温度。

表 1 单因素影响实验

Table 1 Experiments of monofactor effect

序号	V_{NA}/(ml)X_1	V_{SA}/(ml)X_2	T/(℃)X_3	t/(h)X_4	y/(%)X_5	1/2(o/p)X_o'	mp/(℃)X_o''
1	14.4	30	25	3	60.6	0.167	205～215
2	15.0	30	25	3	69.7	0.167	200～209
3	15.7	30	25	3	68.9	0.190	201～209
4	15.0	45	25	3	73.9	0.125	209～213
5	15.0	45	15	3	77.2	0.125	207～211
6	15.0	45	5	3	76.1	0.145	209～210
7	15.0	45	15	2	85.0	0.056	209～213
8	15.0	45	15	5	81.7	0.056	206～209

注：V_{NA}，65%（质量）硝酸的体积；V_{SA}，98%（质量）硫酸的体积；T，反应温度；1/2（o/p），邻/对比为 1/2；t，反应时间；y，对硝基乙酰苯胺得率；m_p，用对硝基乙酰苯胺-乙醇饱和溶液洗过的硝化产物的熔点。

表 2 乙酰苯胺硝化反应正交试验及极差分析

Table 2 Orthogonal experiments and range anaysis of nitration reaction for acetanilide

序号	V_{NA}/ml A	V_{SA}/ml B	T/℃ C	t/h D	y/%	mp/℃
1	14.4	35	10	1.5	75.0	209～213
2	14.4	45	15	2.0	80.3	210～214
3	14.4	55	20	2.5	35.8	209～213
4	15.0	35	15	2.5	73.6	213～216
5	15.0	45	20	1.5	11.2	208～213
6	15.0	55	10	2.0	83.9	208～212
7	15.7	35	20	2.0	76.9	209～212
8	15.1	45	10	2.5	76.7	209～214
9	15.7	55	15	1.5	84.7	209～213
R_1	80.4	75.2	78.5	79.0		
R_2	78.2	78.1	79.5	80.4		
R_3	79.4	84.8	80.0	78.7		
R	2.2	9.6	1.5	1.7		

3 灰色系统分析

3.1 灰色系统分析原理[2]

（1）确定参考数列（母因素），记为 X_o（k）；

（2）确定比较数列（子因素），记为 X_i（k）；

（3）求关联系数

$$\varepsilon_i(K)=\frac{\min\limits_{i}\min\limits_{K}|X_0(K)-X_i(X)|+\theta\max\limits_{i}\max\limits_{K}|X_0(K)-X_i(X)|}{|X_0(K)-X_i(X)|+\theta\max\limits_{i}\max\limits_{K}|X_0(K)-X_i(X)|} \tag{1}$$

式中，ε_i 是比较数列 X_i 的第 k 个因素与参考数列 X_0 的第 k 个元素 X_0（k）之间的关联系数；θ 为分辨系数，$\theta\in[0,1]$，本文 $\theta=0.5$；

（4）求关联度

$$Y_i=\frac{1}{N}\sum_{K=1}^{N}\varepsilon_i(K) \tag{2}$$

式中，ε_i 为 X_i 与 X_0 的关联度，r_i 力越大则 X_i 与 X_0 越密切。

3.2 影响乙酰苯胺硝化反应因素的灰色关联分析

如前所述，硫酸用量、硝酸用量、反应时间、反应温度均为影响乙酰苯胺硝化反应的因素，这些因素可作为因素灰色关联分析的子序列，记为｛X_i｝ i=1，2，3，4；而用对硝基乙酰苯胺的产率和1/2（邻/对）比来衡量反应的效率，并作为因素关联分析的母序列，分别记为｛X_0｝｛X_i'｝。将表1中单因素影响的条件和结果作为因素灰色关联分析的原始数据。

现以 X_0 指标进行关联分析。关联计算前需首先对表1所示的数据进行初值化（即无量纲化）得到表3。

由表3计算△i（k）=｜X_0（k）-X_i（k）｜，得到表4，由表4和式（1）、式（2）得到表5。

表3 原始数据初值化结果

Table 3 Non-dlmensional results of original data

序号	X_1	X_2	X	X_4	X_0
1	1	1	1	1	1
2	1.0417	1	1	1	1.0502
3	1.0903	1	1	1	1.1370
4	1.0417	1	1	1	1.2195
5	1.0417	1.5	0.6	1	1.2558
6	1.0417	1.5	0.2	1	1.2558
7	1.0417	1.5	0.6	0.6667	1.4026
8	1.0417	1.5	0.6	1.6667	1.3482

表4 Δi（k）表

Table 4 Calculated Δi（k）

k	1	2	3	4			
1	0	0	0	0			
2	0.1085	0.1502	0.1502	0.1502			
3	0.0467	0.1370	0.1370	0.1370			
4	0.1778	0.2505	0.2195	0.2195			
5	0.2322	0.2261	0.6739	0.2739			
6	0.2142	0.2442	1.0558	0.2558			
7	0.3069	0.0974	0.8026	0.7359			
8	0.3609	0.1518	0.7482	0.3185			
min K	0	0	0	0	min i	min k	0
max k	0.3609	0.2805	1.0558	0.7359	max i	max k	1.0558

表5 关联系数和关联度

Table 5 Related coefficient and related degree

k	1	2	3	4
1	1.0	1.0	1.0	1.0
2	0.8295	0.7785	0.7785	0.7785
3	0.9187	0.7940	0.7940	0.7940
4	0.7480	0.6530	0.7063	0.7063
5	0.6945	0.7001	1.4393	0.6584
6	0.7115	0.6837	0.3333	0.6736
7	0.5940	0.8442	0.3968	0.4177
8	0.6327	0.7667	0.4137	0.6237
r_i	0.7661	0.7788	0.6077	0.7065

从表5可以看出关联度为$X_2>X_1>X_4>X_3$，即影响硝化反应的主要因素为硫酸用量、硝酸用量，其它因素依次是反应温度、反应时间。

4 结论

通过正交设计试验法优化了乙酰苯胺的硝化反应，其反应条件为硫酸用量55ml，硝酸用量14.4ml，反应温度20℃，反应时间2h。此条件与现工艺相比，提高了反应温度，缩短了保温时间，硝化得率和产物的熔点较高。通过正交设计因素极差分析与因素灰色关联分析得出了硫酸用量和硝酸用量是影响乙酰苯胺硝化的主要因素。灰色关联分析作为乙酰苯胺硝化反应因素分析，验证了极差分析的可靠性。

参考文献

[1] 禹茂章等. 世界精细化工手册［M］. 北京：化工部科技情报所，1988.
[2] 邓突龙. 灰色系统基本方法［M］. 武汉：华中理工大学出版社，1987.

Selective Nitration of Acetanlide and the Influence Factor Analysis Thereon

Jin Tiezhu，Lv Chunxu，Guo Zhenshuang，Gu Jianliang
(School of Chemical Engineering，Nanjing University of Science and Technology，Nanjing 210094，Jiangsu，China)

Abstract: The selective nitration of acetanilide was studied by orthogonally designed experiments. The influence factors on the selective nitration were evaluated by means of grey relation analysis.

Keywords: Orthogonal design，Acetanilide，Grey relation analysis，Aromatic nitration.

（注：此文原载于 含能材料，1994，2（4）：26-30）

用 Ip 作为 N_2O_4-O_3 硝化取代芳烃的定位判据

吕早生[1]，吕春绪[2]
(1. 武汉科技大学，湖北，武汉，430081)
(2. 南京理工大学，江苏，南京，210094)
EI：4238195444

摘要： 利用前沿轨道理论及 N_2O_4-O_3 硝化芳烃的机理推导出取代芳烃第一电离能与取代基定位方向之间关系，用第一电离能 I_p 的大小判断 N_2O_4-O_3 硝化各种取代芳烃过程中 NO_2 基团所取代的位置，提出了判断取代芳烃中取代基是对位定位基还是间位定位基的判据，此判据与实验结果相一致。

关键词： 硝化反应，N_2O_4，定位判据，第一电离能

1 问题的提出

N_2O_4-O_3 硝化芳烃化合物的过程为氧化自由基偶合过程。这个过程既不是亲电取代过程，又不同于亲核取代过程，所以传统的芳烃亲核或亲电过程中取代基定位规则能否适合该反应过程，至今尚未见有关报道。本文通过理论分析，提出了利用取代芳烃的第一电离能作为该反应取代基的定位判据，并以实验加以验证。N_2O_4-O_3 硝化芳烃化合物的过程见图 1。

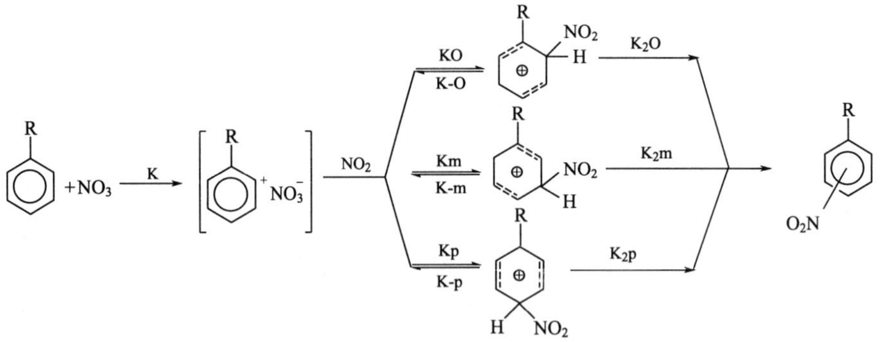

图 1 N_2O_4-O_3 硝化取代芳烃的过程[1]

2 利用前沿轨道理论讨论该反应的取代基团定位规律

可从自由基正离子前沿轨道的能量、各碳原子上未耦合自由电子密度来分析、划分取代基的性质。由图 1 可看出，决定异构体的分布为自由基偶合过程。

苯自由基正离子中未偶合电子位于在的 e_{1g} HOMO 的 2 个简并轨道 b_1 和 a_2 中，其中，b_1 为对

称轨道 S；a_2 为反对称轨道 A，见图 2。

从图 2 可见，是 S（b_1）在 1，4 碳上原子有大的电子自旋密度，而在 A（a_2）上 1，4 碳原子为一节点，总的结果是，e_{1g} 轨道上每个碳原子未偶合电子密度为 1/6。

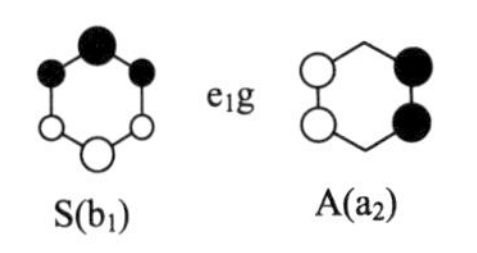

图 2　苯自由基正离子 HOMO 中两简并 e_{1g}

在苯环上引入一个取代基后，就消除了苯自由基正离子 e_{1g} 中 S（b_1），A（a_2）的简并性。当引入的取代基为吸电子基 *Y* 时，则对称轨道 S（b_1）位能下降；当引入的取代基为供电子基 *X* 时，则对称轨道 S（b_1）位能上升，见图 3。

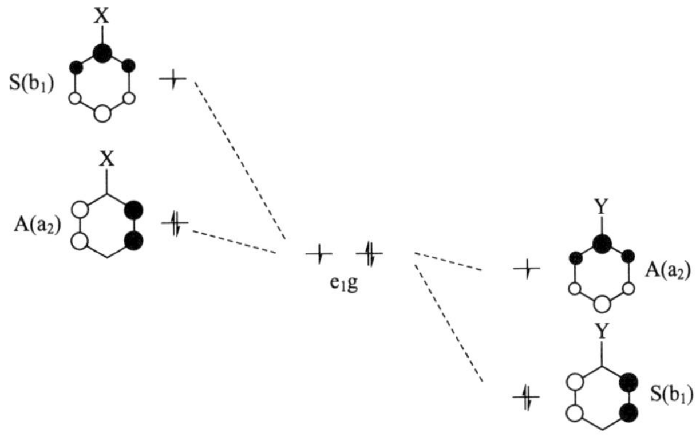

图 3　在苯中引入取代基 *X* 或 *Y* 后，自由基正离子 e_{1g} 两轨道 A（a_2）和位能 S（b_1）的变化

从图 3 可看出，当引入吸电子基 *Y* 时，取代苯的自由基正离子间位上未偶合电子密度增加，而对位上下降，其前沿轨道是 A（a_2），相对应的第一电离能 I_p 上升；当引入的取代基为供电子基 *X* 时，则取代苯的自由基正离子对位上未偶合电子密度增加，而间位上下降，其前沿轨道是 S（b_1），相对应的第一电离能 Ip 下降，这已被取代苯自由基正离子的 ESR 谱所证实[1]。当 NO_2 与带吸电子取代基的取代苯自由基正离子发生偶合反应时，则主要在间位上进行取代，故称为间位定位取代基；当 NO_2 与带供电子取代基的取代苯自由基正离子发生偶合时，则主要在对位上进行取代，故称为对位定位取代基。这样就可以按图 3 对取代基进行重新分类，其定量指标就是取代苯的第一电离能 I_p。当第一电离能 I_p 比苯的第一电离能 I_p 小时，则为对位取代基；当第一电离能 I_p 比苯的第一电离能 I_p 大时，则为间位定位基。

3　实验结果

从表 1 和表 2 可见，NO_2 进入取代基对位的取代基是 OH，OMe，*t*-Bu，*i*-Pr，Et，Me，NHAc，Ph，I，Br，Cl，F，其对应的取代芳烃的第一电离能 I_p 都小于苯的第一电离能 I_p；而间位定位基 CN，NO_2，CF_3，CO_2H，CO_2Me，CHO，$COCH_3$ 所对应的取代芳烃的第一电离能 I_p 都大于苯的第一电离能 I_p，这与实验结果相吻合。

表 1　第一电离能 I_p 小于苯的取代苯（对位定位取代基）[2]

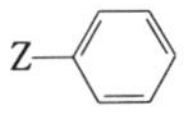

Z—C_6H_5

Z	OMe	OH	*i*-Pr	Et	Me	NHAc	Ph	I	Br	Cl	F	H
I_p/eV	8.39	8.73	8.69	8.76	8.82	8.47	8.32	8.67	9.05	9.08	9.19	9.23

表 2 第一电离能 I_p 大于苯的取代苯（间位定位取代基）[2]

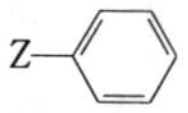

Z-C₆H₅

Z	CN	NO_2	CF_3	CO_2H	CO_2Me	CHO	$COCH_3$	H
I_p/eV	9.72	9.93	9.70	9.73	9.35	9.59	9.37	9.23

4 结果与讨论

（1）与亲电取代反应中取代基定位结果相似，是因为这两个反应过程中决定异构体分布的过程相似，N_2O_4-O_3 硝化芳烃时是在自由基偶合过程决定异构体分布的（见图 1），而在亲电取代过程中决定异构体分布是 Π 络合物向 σ 络合物转化过程（见图 4）[3]，这两个过程都与苯环的前沿轨道上电子密度分布有关。

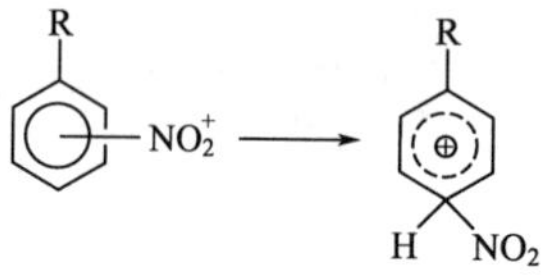

图 4 Π 络合物向 σ 络合物转化过程

（2）在亲电取代过程中，以 $\lg K^x/K_H=\sigma^+\rho$ 中的指标 σ^+ 作为划分取代基类型的理论依据，但以第一电离能作为划分取代基定位类型能更科学、合理，更能反映反应过程的实质。这种划分方法有如下两个特点：①与进攻的硝化试剂类型无关，只取决于取代芳烃；②与传统方法不同，F，Cl，Br，I 被划归到供电子取代基类，这与其硝化产物为邻-对位分布相一致。

（3）这种方法只是一定性方法，实际上产物异构体分布比率还与取代芳烃所形成 σ 络合物的稳定性有关，即热力学因素。但对有些共轭和空间因素影响不大的取代苯（如氟苯、甲苯），其硝基产物的异构体分布与其有很好的一致性。

（4）若取代芳烃的第一电离能 I_p 大于苯的第一电离能，则该芳环上的取代基为间位定位基；若取代芳烃的第一电离能 I_p 小于苯的第一电离能，则该芳环上的取代基为对位定位取代基。

参考文献

[1] S Fukuzumi，J K Kochi. Electrophilic Aromatic Substitution. Charge-Transfer Excited States and the Nature of the Activated Complex ［J］. J Am Chem Soc，1981，103：7240-7252.

[2] 吕早生. N_2O_4-O_3 硝化芳烃化合物的宏观动力学与机理的研究（博士论文）［D］. 南京：南京理工大学，2001：91-95.

[3] 吕春绪. 硝化理论［M］. 南京：江苏科技出版社，1993，59-70.

[4] Thomas H Lowry. Mechanism and Theory in Organic Chemistry ［M］. New York：Harper & Row，Publishers. 1981：291-503V.

Utilizing I_p as the Orientation Criterion of N_2O_4-O_3 Nitration Substituted Benzene

Lv Zaosheng[1]，Lv Chunxu[2]

（1. Wuhan University of Science and Technology，Wuhan 430081，China）

（2. Nanjing University of Science and Technology，Nanjing 210094，China）

Abstract：The relationship between the first ionization potential Ip of substituted benzene and the orientation rule of substituted group in N_2O_4-O_3 nitration aromatics is discussed by frontier orbital theory and reaction mechanism of N_2O_4-O_3 nitration aromatic. The NO_2 entering position in N_2O_4-O_3 nitrating substituted benzene can thus be determined by the first ionization potential Ip of the substituted benzene. The orientation rule of substituted group has been put forward, which agrees with the experiment results.

Key words: nitration reaction, nitrogen tetroxide, orientation criterion, first ionization potentia

（注：此文原载于　武汉科技大学学报，2002，25（4）：352）

硝酰阳离子和二氧化氮分子的弯曲变形研究

曹阳[1]，吕春绪[1]，吕早生[2]，蔡春[1]，魏运洋[1]，李斌栋[1]

（1. 南京理工大学化学系，南京 210094）

（2. 武汉科技大学化学与环境工程学院，武汉 430081）

SCI：000176546800011

摘要：采用密度泛函理论的 B3LYP 方法（6-311＋G*）基组：计算了 NO_2^+、NO_2 以及其它与硝化反应机理研究相关的分子、离子和激发态的结构与性质。进而研究了当键角在 90°～180°之间变化时，这些相关物质能量的变化规律，由此探讨了不同硝化机理发生的可能性，为以后进一步研究不同结构与活性的芳香化合物的硝化反应机理提供依据。

关键词：硝化反应，硝酰阳离子，弯曲变形，电子转移，密度泛函理论

硝化反应是重要的基本有机反应。多年来，人们不但积累了丰富的实践经验，还进行了深入细致的理论研究，提出了多种硝化反应的机理。但至今，对硝化反应机理的研究还存在激烈的争论。目前争论的焦点集中在硝化反应中是否存在单电子转移过程[1-5]。随着实验技术的不断完善，特别是时间分辨光谱技术的应用，电子转移机理得到更多认同[1-4]，但 Eberson[5] 指出，由于普通硝化反应与自由基偶合硝化反应的产物分布不同，因而采用混酸的硝化反应过程中不存在电子转移。此外，近年来，采用 O_3＋NO_2 体系作为硝化剂的新工艺日益受到关注。研究表明，其反应机理也涉及电子转移过程[6-7]。硝化反应机理至今未有统一结论。

随着量子化学的发展，有关硝化反应的量子化学研究也时有报导[8-11]。但以前的计算多是在较低水平下，针对较小的分子体系进行的，且主要是对经典的硝酰阳离子（NO_2^+）进攻机理的研究，有关电子转移硝化机理的量子化学研究不多见，这可能是因为电子转移机理的提出时间较晚，也可能是因为电子转移机理的量子化学研究难度较大。

不论哪种机理，进攻试剂 NO_2^+ 在反应过程中都发生了弯曲，其键角由最初的 180°变为 124.6°（硝基苯在 B3LYP/6-311＋G* 水平下的计算值）左右。可以想象，反应中的 NO_2^+ 弯曲变形过程应该包含相应的能量以及电子结构的变化，这些变化对硝化反应的机理研究起着重要的作用。我们从这一设想出发，采用近年来十分流行的密度泛函理论（DFT），首先计算各分子的几何构型，接着对 NO_2^+ 和 NO_2 分子弯曲变形过程中的能量、原子电荷、键级等进行分析，进而探讨弯曲变形对硝化机理的影响。

1 计算方法

对分子、过渡态等的平衡构型的全优化计算采用密度泛函理论 UB3LYP/6-311＋G* 的方法[12-13]（未采用更大基组是考虑到原子电荷计算的准确性以及此结果将被应用于研究芳香化合物的硝化反应），并进行振动分析（用 0.96 校正）[14]，然后将键长固定为平衡构型时的计算值，使键角在 90°～

180°之间变化，采用相同方法计算单点能、原子电荷、偶极矩等。N—O 键级则采用 AIM（atoms in molecules）[15]理论计算。全部计算工作采用 Gaussian 98 程序，在联想 PⅢ800 计算机上完成。

2 结果与讨论

2.1 平衡构型的计算

对平衡构型的计算结果列于表 1。为便于比较，表 1 还列出了 NO_2^- 离子以及硝基苯中硝基官能团的计算结果。

表 1 平衡构型的 B3LYP/6-311+G* 全优化计算结果

Table 1 Key results obtained from the optimizations at B3LYP/6-311+G* level

	E(a. u.)	Bond length (nm)	Bond angel (°)	$10^{30}\mu$/C·m	10^{19} q_N/C	10^{19} q_o/C
NO_2^+	−204.7757	0.11176	180.0	0.00	0.731	0.436
NO_2	−205.1407	0.11932	134.4	1.13	0.003	−0.002
NO_2^-	−205.2250	0.12578	116.8	1.23	−0.868	−0.367
Ex NO_2^+	−204.6622	0.12053	121.5	1.13	0.303	0.650
TS NO_2	−205.0808	0.11975	180.0	0.00	0.160	−0.080
NO_2 group	—	0.12245	124.6	—	−0.187	−0.067

表 2 平衡构型的振动频率、零点振动能及自旋密度

Table 2 The infrared frequencies, zero-point energies (ZPE) and spin densities (sd) of the title substances *

	ν_1	ν_2	ν_3	ZPE/kJ·mol^{-1}	10^{19} sd(N)/C	10^{19} sd(O)/C
NO_2^+	π_μ587(9.8)	σ_g1394(0.0)	σ_μ235(386.2)	29.47	0.000	0.000
NO_2	$A_1$734(7.4)	$A_1$1336(0.1)	$B_2$1634(459.2)	22.16	0.731	0.436
NO_2^-	$A_1$768(3.6)	$A_1$1281(13.8)	$B_2$1245(739.2)	19.70	0.000	0.000
Ex NO_2^+	$A_1$634(8.9)	$A_1$1357(0.8)	$B_2$919(0.1)	17.43	0.471	1.367
TS NO_2	π_μ443(5.2) 1240i(5.2)	σ_g1418(0.0)	σ_μ186(342.9)	20.66	0.884	0.359

* ν is the scaled frequency (in cm^{-1}) by the factor 0.96, intensity (in parentheses) is in km·mol^{-1}

NO_2^+ 为对称直线结构。比它多一个电子的 NO_2 分子为对称弯曲结构，同时总能量降低 958.2kJ·mol^{-1}。如果 NO_2 再得一个电子成为负离子，总能量还可进一步降低 221.3kJ·mol^{-1}，如考虑零点振动能，总能量降低的数值更大一些。从能量角度可以看出，NO_2^+ 是高反应活性的物质，具有强烈的夺取电子成为稳定结构的倾向。电子数的增加对分子起到稳定的作用，使得 N—O 键长增大，但增加的未成键电子对 N—O 键的排斥作用导致 O—N—O 键角减小，偶极矩增大。所有的计算结果均与前人的结论一致[16-18]，说明 DFT 方法的精确度较高。

硝基苯中的硝基官能团共带约 0.32×10^{-19}C 的负电荷，介于 NO_2 和 NO_2^- 之间，它的 O—N—O 键角也介于两者之间。这说明由直线型的 NO_2^+ 变为弯曲的硝基官能团的过程中，NO_2^+ 从芳环获得了电子。获得电子方式的不同引出了不同的硝化机理理论。如 NO_2^+ 以空轨道和芳环的 π 电子形成配位键，是经典的亲电反应理论；如果 NO_2^+ 从被硝化底物夺取一个电子并与之生成自由基对，则是单电子转移理论。

在研究过程中，我们还得到了一个弯曲的 NO_2^+ 的三重激发态（Ex NO_2^+），它两个未成对的 α 电子占据了对称性分别为 A_1 和 B_2 的两个分子轨道。氧原子上的正电荷和自旋密度都较高，表明两个未成对电子主要分布在氧原子上，使得氧原子具有较大的夺取单电子的能力。该激发态的振动频率都

是正值，说明它在能级图上是一个稳定的结构，但它的电子能比基态离子高 298.2kJ·mol^{-1}。我们还注意到，该激发态电子占据分子轨道的对称类型与 NO_2 分子完全相同，且具有相近的键长，因而也可以把它看作 NO_2 分子中次外层一个成对电子被电离的产物，两个自旋平行的未成对电子对 N—O 键的排斥作用更加显著，导致键角进一步减小。此外，该激发态键长与键角的数值与硝基官能团非常接近。所有这些说明，这个激发态在硝化反应中有可能形成并发挥作用。

我们还发现了一个直线型的 NO_2 分子的过渡态（TS NO_2），它的键长与其基态分子几乎相等，但总能量比基态分子高 157.2kJ·mol^{-1}。该过渡态有两个有趣的性质，一是该过渡态与 NO_2^+ 都是对称直线结构，过渡态的能量在能级图上是极大值，而 NO_2^+ 的能量是极小值，但极大值比极小值还要低 801.1kJ·mol^{-1}。这进一步说明 NO_2^+ 具有强烈的夺电子倾向；二是该过渡态的弯曲振动具有两个不同的振动频率，其中一个为虚频，说明未成对的电子在其周围不是平均分布的，因而向不同方向的弯曲振动频率不同。

2.2 弯曲变形过程中的能量变化

我们将键长固定为平衡构型时的数值，改变键角，在相同水平下计算了 NO_2^+ 及其三重激发态、NO_2 分子、NO_2^- 及其三重激发态（Ex NO_2^+）的总能量，比较了它们与 NO_2^+ 基态总能量的差值，并将计算结果用图形表示出来，见图 1。

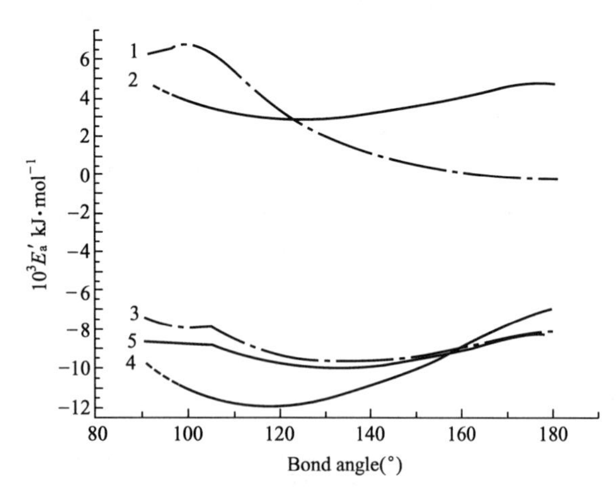

图 1 相对能量-键角关系图

Fig. 1 Relationship between relative energies (to ground state NO_2^+) and O—N—O bond angles

1) NO_2^+，2) Ex NO_2^+，3) NO_2，4) NO_2^-，5) Ex NO_2^-

由图 1 可见，180°是 NO_2^+ 的平衡键角，在能级图上属于极小值，但它并不是最稳定的。若此时获得一个电子，可使其能量降低 801.1kJ·mol^{-1}，而且此时还是能级图上的极大值，能量还可进一步降低。随着 NO_2^+ 逐渐弯曲，其能量与相同键角的 NO_2 分子的差距也增大，显示出越来越强烈的夺电子倾向，直到键角小于 NO_2 分子的平衡键角（134.4°）。

NO_2^- 可以看作是弯曲的 NO_2^+ 以空轨道接受了一对游离的电子形成的负离子，是一种特殊的配合物，它的形成可作为 NO_2^+ 发生配位反应的近似。我们注意到，当键角大于 160°时，NO_2^- 的能量高于其三重激发态以及 NO_2 分子，此时 NO_2^- 离子即使能够形成，也是具有一对自旋平行电子的激发态。所以从能量角度考虑，当键角大于 160°时，NO_2^+ 只可能夺取单个电子，而不可能以空轨道和其它分子的孤对电子或芳环的 π 电子形成配位键。此时，若芳环与其正离子自由基的能量差不大，即第一电离能（I_p）较小，则容易与 NO_2^+ 通过单电子转移机理生成硝基化合物。

当键角小于 160°时，NO_2^- 基态的能量低于其三重激发态，且能量随键角减小降低的幅度大于 NO_2 分子，这个趋势一直到其平衡键角为止。此时如弯曲的 NO_2^+ 与富电子的芳环相遇，则有可能与 π 电子形成配位键，进而发生亲电取代反应。如果芳环因被吸电子基团取代而贫电子，且第一电离能较大，这时即使 NO_2^+ 弯曲到键角小于 130°时也未必能有效成键。此时可能的过程是 NO_2^+ 将一个电子激发到高能态，形成在能级图上是稳定点的三重激发态 Ex NO_2^+，它的能量比同角度的基态 NO_2^+ 还低。由于该结构是稳定点，因而能够存在较长时间。从前面对其结构的分析可知，它能够有效地夺取单电子，从而与芳环生成自由基对。Ex NO_2^+ 也可能从吸电子的取代基夺取电子。虽然 Ex NO_2^+ 的能量高于其稳定的基态离子，但它可通过吸收光子或硝化反应释放的热形成。

从以上的分析可见，不同的芳香化合物与 NO_2^+ 反应存在不同的硝化机理。这主要取决于芳香化合物的第一电离能的大小以及芳环上富电子的情况。富电子的芳环是否一定有较小的第一电离能，目前还不能定论。在硝化反应中，作为进攻试剂的 NO_2^+ 都会发生弯曲，当其弯曲到一定程度时，就会和芳香化合物以特定的机理发生硝化反应。

2.3 NO_2^+ 弯曲变形过程中其它性质的变化

表 3 和表 4 分别列出了键角从 90°到 180°时，NO_2^+ 和 NO_2 的电子结构、偶极矩以及 N—O 键级的变化情况。

表 3 NO_2^+ 的电子结构、偶极矩以及 N—O 键级与 O—N—O 键角的关系

Table 3 Electronic structures, dipole moments, N—O bond orders of NO_2^+ at various bond angles

Bond angle(°)	$10^{19}q_N/C$	$10^{19}q_O/C$	$-E_{HOMO}/eV$	$-E_{LUMO}/eV$	$\Delta E/eV$	$10^{30}\mu/C\cdot m$	Bond order
90	0.328	0.638	19.793	16.122	3.671	0.83	1.602
100	0.272	0.665	18.835	17.154	1.181	0.33	1.640
110	0.596	0.503	18.944	16.720	2.224	3.67	2.420
120	0.561	0.521	19.613	15.916	3.697	3.00	2.452
130	0.532	0.535	20.083	15.114	4.969	2.40	2.482
140	0.508	0.546	20.416	14.286	6.130	1.90	2.508
150	0.506	0.548	20.648	13.420	7.228	1.40	2.530
160	0.551	0.526	20.800	12.541	8.259	0.93	2.546
170	0.655	0.473	20.885	11.755	9.130	0.47	2.556
180	0.731	0.436	20.911	11.388	9.523	0.00	2.559

表 4 NO_2 的电子结构、偶极矩以及 N—O 键级与 O—N—O 键角的关系

Table 4 Electronic structures, dipole moments, N—O bond orders of NO_2 at various bond angles

Bond angle(°)	$10^{19}q_N/C$	$10^{19}q_O/C$	$10^{19}sd(N)/C$	$10^{19}sd(O)/C$	$10^{30}\mu/C\cdot m$	Bond order
90	0.222	−0.011	−0.046	0.523	2.17	1.751
100	−0.045	0.022	−0.050	0.525	1.63	1.778
110	0.099	−0.050	0.336	0.332	2.50	2.154
120	0.061	−0.030	0.382	0.308	1.93	2.181
130	0.019	−0.010	0.433	0.284	1.37	2.206
140	−0.019	0.010	0.484	0.258	0.87	2.228
150	−0.038	0.019	0.532	0.234	0.37	2.246
160	−0.019	0.010	0.567	0.216	0.03	2.257
170	0.077	−0.038	0.572	0.214	0.23	2.256
180	0.167	−0.083	0.550	0.225	0.00	2.249

从表 3 可见，NO_2^+ 在弯曲变形过程中，HOMO 和 LUMO 的能级差逐渐减小，表明它的稳定性变差，反应性增强，偶极矩逐渐变大，因此极性溶剂的使用有利于 NO_2^+ 的弯曲变形。N—O 键级变小，意味着两个氧原子之间也出现了成键作用，进一步减小键角还可得到一个三角形的 NO_2^+ 异构体，O—O 键的产生缓和了由于弯曲变形引起的能量升高。由表 4 可见，弯曲变形对 NO_2 的电荷分布影响不大，氮原子上的自旋密度随键角减小而降低，N—O 键级也随键角变小，但偶极矩随键角减小而增大，说明极性溶剂也有利于形成较小键角的 NO_2。键角小于 120°时，NO_2^+ 和 NO_2 的一些性质呈另一种变化，但这已超出硝化反应研究的范围。

3 结论

本文从硝化反应中进攻试剂 NO_2^+ 发生弯曲变形的角度出发，用 DFT 方法研究了 NO_2^+、NO_2 及其它与硝化反应机理有关的物质的性质随 O—N—O 键角的变化，得到以下结论：

（1）当键角大于 160°时，NO_2^+ 只能通过单电子转移机理与 Ip 较小的芳环发生硝化反应。

（2）键角小于 160°时，NO_2^+ 也可能与富电子的芳环通过形成配合物发生经典的硝化反应。

（3）对于十分钝化的芳环，NO_2^+ 可能通过形成三重激发态来从芳环或取代基夺取单电子。

（4）极性溶剂的使用有利于弯曲变形，也就可能有利于硝化反应。

References

[1] Kochi J K. Pure & Appl. Chem.，1991，63（2）：255.
[2] Schmitt R J，Buttrill S E，Ross D S. J. Am. Chem. Soc.，1984，106：926.
[3] Haney S W A，Kochi J K. J. Am. Chem. Soc.，1987，109：5235.
[4] Kim E K，Bockman T M，Kochi J K. J. Am. Chem. Soc.，1981，103：7240.
[5] Eberson L，Hartshom M P，Radner F. Acta Chemica Scandinavia，1994，48：937.
[6] Suzuki H，Mori T. J. Chem. Soc. Perkin Trans. 2，1996：677.
[7] Suzuki H，Tatsumi A，Ishibabashi T，Mori T. J. Chem. Soc. Perkin Trans. 1，1995：339.
[8] Politzer P，Jayasuria K，Sjoberg P. Laurence. J. Am. Chem. Soc.，1985，107：1174.
[9] Szabo K J. J. Mol. Struct.（Theochem.），1988，181：1.
[10] Szabo K J，Hornfeldt A B，Gronowitz S. J. Am. Chem. Soc.，1992，114：6827.
[11] Sandall J P B. J. Chem. Soc. Perkin Trans. 2，1992：1689.
[12] Becke A D. Phys. Rev. A，1988，38：3098.
[13] Lee C，Yang W，Parr R G. Phys. Rev. B，1988，37：785.
[14] Wong M W. Chem. Phys. Lett.，1996，256：391.
[15] Cioslowski J. Chem. Phys. Lett.，1994，219：151.
[16] Lee T J. Chem. Phys. Lett.，1992，188：154.
[17] Cruickshank D W J，Eisenstein M. J. Comput. Chem.，1987，8（1）：6.
[18] Bryant G，Jiang Y，Martin M. J. Chem. Phys.，1994，101：7199.

Theoretical Studies of the Bending Transformation of Nitronium and Nitrogen Dioxide

Cao yang[1]，lv chunxu[1]，lv zaosheng[2]，cai chun[1]，wei yunyang[1]，li bindong[1]

（1. Department of Chemistry，Nanjing University of Science and Technology，Nanjing 210094）

（2. Environmental Engineering，Wuhan University of Science and Technology，Wuhan 430081）

Abstract： Density functional theory（DFT）with the B3LYP method and 6-311+G* basis set has been employed to investigate the geometries and other properties of NO_2^+、NO_2 and some other molecules，ions and their transitional or excited states that concern the mechanism of aromatic nitration. Furthermore，the relative energies and other properties of these substances have been studied with the bond angle ranging from 90° to 180° at fixed bond length at the same level. From the results obtained，the occurrence possibilities of various nitration mechanisms，including electrophilic reaction mechanism and electron transfer mechanism，have been discussed. Each possibility depends on the first ionization potential and electron richness of aromatic substrates. Polar solvents are considered to be favorable for nitration reactions. The works reported here are ready for the further theoretical and experimental studies of the mechanism of nitration.

Keywords： Nitration，Nitronium，Bending，Electron transfer，Density functional theory（DFT）

（注：此文原载于 物理化学学报，2002，18（6）：527-531）

固体铌酸催化下甲苯的硝化

刘丽荣[1,2]，吕春绪[1]，李霞[1]
（1. 南京理工大学化工学院，南京，210094）
（2. 淮海工学院，化学工程系，连云港，222005）

摘要： 在不同焙烧温度下制备了一系列固体铌酸催化剂。测定了固体铌酸催化剂的比表面积、表面酸强度及表面酸量。研究了催化剂焙烧温度、硝化反应时间、硝化反应温度、有无醋酐存在、硝酸浓度、催化剂的重复使用等因素对甲苯硝化反应影响。实验结果表明，当反应温度为40℃，反应时间为60min时，以 CCl_4 为溶剂，以质量分数为95%的硝酸为硝化剂，在醋酐存在条件下，以经300℃焙烧3h后的铌酸作为催化剂，甲苯硝化产物中异构体的邻对比达1.26，较硝硫混酸的1.67显著降低，产物得率达99.3%。且该催化剂循环使用5次。催化活性基本不变，是一种极具应用前景的绿色硝化反应催化剂。

关键词： 铌酸，催化剂，甲苯，硝化

硝基甲苯是一种重要的有机中间体，可广泛用于农药、医药、染料合成等领域。目前，一硝甲苯的工业硝化制备广泛采用硝硫混酸硝化技术，该方法工艺成熟、生产过程稳定、产物收率高，但硝化产物邻对比约为1.67，而且几乎没有区域选择性，从而造成邻位产物严重过剩，而且会产生大量的废酸和有机酸性废水，环境污染严重，无法满足可持续发展的要求。因此，寻找新型高效催化剂。建立新的硝化反应体系，以提高甲苯硝化的区域选择性，改善硝化反应环境，是研究硝化反应的重要课题[1-3]。铌酸（$Nb_2O_5 \cdot nH_2O$）是一种固体酸催化剂，将其进行低温（100℃）热处理后，其 $H_0=-5.6$，酸强度与质量分数为70%的硫酸相当。多数金属氧化物经高温（500℃）热处理后呈酸性，但在含水体系中催化剂表面酸性迅速降低或消失，而铌酸却具有在含水体系中表面仍可保持较高酸性的独特性质，因此在那些有水分子参与或释出的酸催化反应中具有很高的催化活性[4-6]。铌酸作为催化剂已广泛地用于酯化[7]、脱水[8]和缩酮[9]等反应，表现出异常的稳定性，但尚未见应用于硝化领域。本文利用固体铌酸代替硫酸作为甲苯区域选择性硝化反应的催化剂，研究了焙烧温度以及硝化反应条件对甲苯硝化反应区域选择性的影响，以期开发出高效固体酸催化剂。

1 实验部分

1.1 仪器和试剂

Agilent6820型气相色谱仪（美国），HP.INNOWAX型毛细管色谱柱（美国），0.5μm×0.2mm×30m，FID检测器，SA3100型比表面积分析仪。铌酸，纯度99.5%，醋酐及硝基苯为分析纯试剂，其它试剂均为化学纯。

1.2 表面酸度测定

表面酸强度（Ho）用 Hammet 指示剂法测定，表面酸量用丁胺滴定法测定[10]。

1.3 实验方法

在 100mL 四口烧瓶中置 15mL CCl_4、6.0mL 醋酐、5.0mL 甲苯和一定量在一定温度下焙烧 3h 的催化剂 1.0g，恒温水浴下滴加一定质量分数的硝酸 5.0mL，滴加完毕后，升温至一定温度，搅拌一定时间。反应结束后，过滤分出有机相，相继采用质量分数为 5%的 $NaHCO_3$ 水溶液和水洗涤至中性，以硝基苯作内标物进行气相色谱分析。

2 结果与讨论

2.1 催化剂的表面特性

取不同温度下焙烧 3h 后的催化剂，测定其 BET 比表面积、表面酸强度及表面酸量，所得结果列于表 1。从表 1 可知，在 100～300℃范围内，随焙烧温度的升高，铌酸比表面积略有减小，酸量则略有增加；当温度超过 300℃后，铌酸比表面积和酸量随焙烧温度的升高而迅速降低。

表 1 铌酸催化剂的表面特性

Table 1 Surface properties of niobic catalysts

Calcination Temperature /℃	Specific area /($m^2 \cdot g^{-1}$)	Acid amount /($mmol \cdot g^{-1}$)	Distribution of acidity/($mmol \cdot g^{-1}$)	
			$-8.2<H_0<-5.6$	$-5.6<H_0\leqslant 1.5$
100	262.132	0.34	0.10	0.24
200	250.533	0.36	0.12	0.24
300	234.231	0.38	0.12	0.26
400	153.326	0.20	0	0.20
500	105.216	0.16	0	0.16

2.2 催化剂的焙烧温度和反应时间对甲苯硝化反应的影响

取不同温度下焙烧的催化剂，以质量分数为 65%的硝酸为硝化剂，按本文 1.3 节实验方法，反应温度控制在 40℃，反应进行 60min，结果列于表 2。从表 2 可知，焙烧温度对催化剂的甲苯硝化反应有显著影响，升高焙烧温度，催化剂的活性增加，当焙烧温度达到 300℃时，催化剂的活性最大，硝化产物的邻对比最小，这是因为经 30℃下焙烧处理的铌酸催化剂上 Lewis 酸的强度最高的缘故[4]。焙烧温度继续升高，催化剂的活性有所下降，硝化产物的邻对比有所增加，产率也随之下降，这是由于焙烧温度继续升高，造成催化剂比表面积大幅度减小，酸量迅速降低，从而导致催化剂活性下降的缘故。

表 2 催化剂焙烧温度对甲苯硝化反应的影响

Table 2 Effects of calcination temperatures of catalyst on toluene nitration

Calcination temperature /℃	Nitrotoluene isomer/%			ortho/para	Yield/%
	ortho	meta	para		
100	54.4	3.9	41.7	1.30	29.5
200	55.6	2.8	41.6	1.33	32.5
300	54.4	3.2	42.4	1.28	48.5
400	54.9	2.7	42.4	1.30	36.5
500	56.2	2.8	41.0	1.37	24.5

取 300℃下焙烧的催化剂，以上述同样反应条件下改变反应时间，所得结果列于表 3。从表 3 可知，反应初期随着反应时间增加，硝基甲苯产物得率增加，邻对比保持不变，反应时间达 60min 时，产物邻对比最小，达 1.28，同时产物得率也达到最大值。随着反应时间继续增加，邻对比有所上升，产率明显下降，这是由于反应时间过短反应进行不完全，反应时间过长可能由于副反应的增加而使产率下降。

表 3 反应时间对甲苯硝化反应的影响

Table 3 Efects of diferent reaction time on toluene nitraiton

Reaction time/min	Nitrotoluene isomer/%			ortho/para	Yield/%
	Ortho	meta	para		
15	55.4	2.6	42.0	1.32	36.0
30	54.8	3.4	41.8	1.32	39.0
45	55.1	2.8	42.1	1.32	40.0
60	54.4	3.2	42.3	1.28	48.5
75	55.4	2.6	42.0	1.33	37.8

2.3 反应温度对甲苯硝化反应区域选择性的影响

取 300℃下焙烧的催化剂，以本文 2.2 节同样反应条件下，改变反应温度，结果列于表 4。由表 4 可见，反应温度对催化剂的甲苯硝化反应的区域选择性和催化活性有一定的影响。反应在 40℃下进行时，硝化产物的邻对比最小，产率最大。再继续升高温度，邻对比有所上升，产率有所下降，气相色谱分析结果发现，硝化产物中有二硝基甲苯存在，表明随着反应温度增加，二硝基甲苯等副产物增加。

表 4 反应温度对甲苯硝化反应的影响

Table 4 Effects of reaction temperatures on toluene nitration

Reaction temperature /℃	Nitrotolueneisomer/%			ortho/para	Yield/%
	ortho	meta	para		
20	57.5	4.0	38.5	1.49	23.7
30	55.1	2.9	42.0	1.32	32.9
40	54.4	3.2	42.3	1.28	48.5
50	55.9	3.8	40.3	1.39	44.9
60	57.3	5.8	36.9	1.56	41.5

2.4 催化剂的再生

取 300℃下焙烧的催化剂，以本文 2.2 节同样反应条件，结果列于表 5。由表 5 可见，催化剂连续使用 5 次，其催化活性变化不大，这说明铌酸催化剂有很好的重复使用性能。甲苯硝化产物的邻对比波动较小，但产率有所变化。催化活性的降低可能是由于催化剂表面吸附了有机物的缘故。

表 5 催化剂的重复使用对甲苯硝化反应的影响

Table 5 Effects of recovered catalysts on toluene nitration

Catalyst	Nitrotoluene isomer/%			ortho/para	Yield/%
	ortho	Meta	para		
1st use	54.4	3.2	42.3	1.28	48.5
2nd use	54.2	3.2	42.6	1.26	49.0
3rd use	54.1	3.7	42.2	1.28	48.1
4th use	54.6	3.2	42.2	1.30	47.3
5th use	54.9	3.1	42.0	1.32	46.7

2.5 醋酐对甲苯硝化反应区域选择性的影响

取300℃下焙烧的催化剂，以本文2.2节同样反应条件不加醋酐，结果列于表6。由表6可见，醋酐的存在对硝化反应区域选择性有较大影响，有醋酐存在条件下，硝化产物的产率比无醋酐存在时的产率有较大幅度提高，并且硝化产物的邻对比要比无醋酐存在时的邻对比低很多。在无醋酐存在条件下，硝酰阳离子是芳烃在硝酸中硝化时的活化硝化剂，同时硝酸又是强氧化剂，易使反应体系发生副反应。醋酐的作用在于此硝化体系中生成的活化硝化剂是乙酰硝酸酯、质子化的乙酰硝酸酯、五氧化二氮和硝酰阳离子，此体系能满足硝化能力的要求，又能使反应比较平缓的进行，副反应较少；此外醋酐还有脱水作用，它可使反应体系在反应后放出醋酸而不放出水，即醋酐的存在可以造成低水或非水反应环境[11]。

表6 醋酐对甲苯硝化反应的影响

Table 6 Effect of Acetic anhydride on toluene nitration

Acetic anhydride	Nitrotolueneisomer/%			ortho/para	Yield/%
	ortho	meta	para		
Acetic anhydride	54.4	3.2	42.3	1.28	48.5
—	58.5	5.4	36.1	1.61	3.76

2.6 硝酸含量对甲苯硝化反应的影响

取300℃下焙烧的催化剂，以本文2.2节同样反应条件改变硝酸含量（质量分数），结果列于表7。从表7可知，硝酸含量对甲苯硝化反应结果有很大影响，随着硝酸含量增加硝化产物的邻对比变化很小，但产物得率却大幅度提高。当硝酸的质量分数达到95%时，硝化产物的邻对比为1.26，产物得率为9.3%，接近10%。在投料量不变的情况下，增加硝酸含量，相当于增加了硝酸的比例，硝酸比例越大，与硝酸体系性质越接近，体系硝化能力越强，但由于醋酐的存在，该体系又不会像硝酸硝化剂那样具有很强的氧化性，因而，反应又相对温和，副反应相对较少[12]。

固体铌酸催化剂应用于甲苯硝化，取得了令人满意的区域选择性和产物得率。催化剂制备和分离容易，可再生循环使用，从生产源头解决了环境污染问题，是一种极具应用前景的绿色硝化反应催化剂。

表7 硝酸浓度对甲苯硝化反应的影响

Table 7 Effects of concentrations of nitric acid on toluene nitration

Concentration of nitric /%	Nitrotoluene isomer/%			ortho/para	Yield/%
	ortho	meta	para		
65	54.4	3.2	42.4	1.28	48.5
80	54.4	3.4	42.2	1.28	75.3
95	53.8	3.6	42.6	1.26	99.3

参考文献

[1] Mohanmed Haouas, Andreas Kogelbauer, Roelprins. Catallet [J], 2000, 72: 61.

[2] Pengxin-Hua, Lv Chun-Xu. Fine chem [J], 2000, 17 (1): 17.

[3] Diego Vasena, Andreas Kogelbauer, Roe Lprins. Catal Today [J], 2000, 60: 275.

[4] Tanabek, Misono M Ono Y Auths. Zhenglu-Bin, Wang gong-Wei, Zhangying-Zhen Trans. New Solid Acids Bases And Their Catalytic Properties [M] . Beijing: Chemical Industry Press, 1992: 64.

[5] Chenz, Iizukat, Tanabek. Chemlet [J], 1984, 151 (7): 1085.

[6] Tanabek. Cataltoday [J], 1990, (8): 1.

[7] Suolong-Ning, Zhoucui-Wen. Adv fine petrochem [J], 2002, 3 (3): 4.

[8] Kiyoraka Asakrua, Yasuhira Lwasawa. J Phys Chem [J], 1991, (95): 1711.

[9] Liju-Ren, Wuxiao-Hui, Tianzhi-Xin, Gongjian. Chiensejsynchem [J], 2002, 10 (1): 53.

[10] Tanabek, Misono M Ono Y Auths. Zhenglu-Bin, Wanggong-Wei, Zhangying-ZhengTrans. New Solid Acids Bases And Their Catalytic Properties [M] . Beijing: Chemical Industry Press, 1992: 4.

[11] Lvchun-Xu Auths. The Theory Of Nitration [M] . Nanjing: Jiangsu Science And Technology Press, 1993: 9.

[12] Lv Chun-Xu Auths. The Theory Of Nityrl Cation [M] . Beijing: Enginery Industry Press, 2006: 22.

Regioselective Nitration of Toluene in the Presence of Solid Niobic Acid Catalysts

Liu Lirong[1,2], Lv ChunXu[1], Li Xia[1]

(1. Department of Chemistry, Najing University of Science and Technology, Najing 210094)

(2. Department of Chemical Engineering, Huaihai Institute of Technology, Lianyungang 222005)

Abstract: A series of niobic acids at different calcination temperatures was prepared. BET, acid amount and distribution of acidity of niobic acid catalysts were investigated. The effects of all kinds of factors, just like calcination temperature of niobic acid catalysts, nitration reaction time, nitration reaction temperature, recovered catalyst, acetic anhydride and concentration of nitric acid were studied by gas chromatography (GC) . The nitration of toluene was carried out in the presence of acetic anhydride with CCl as solvent, 95% nitric acid as nitrating reagent, and niobic acid calcinated at 300℃ for 3h as catalyst for 60min at 40℃. A yield of 99.3% of mononitoluene with an ortho-para-isomer ratio of 1.26 was obtained, which is greatly below 1.67 with sulfonitric acid ascatalyst. The catalyst could be reutilized up to five times with little decrease in activity. It is a clean nitrated reaction catalyst in vast prospects of application.

Key words: niobic acid, catalyst, toluene, nitration

(注：此文原载于 应用化学，2007，24 (12)：1374-1377)

Nitration of Simple Aromatics with NO_2 under Air atmosphere in the Presence of Novel Brønsted Acidic Ionic Liquids

Qi Xiufang, Cheng Guangbin, Lv Chunxu, Qian Desheng

School of Chemical Engineering, Nanjing University of Science & Technology, Nanjing, 210094 China.

SCI: 000253042300007

Abstract: Aromatics nitration with NO_2/air catalyzed by novel Brønsted acidic ionic liquids (ILs) without any volatile chlorinated organic solvent under mild conditions. The ILs employed were caprolactam-based, [Caprolactam] X (X^- = $pTSO^-$, BSO^-, BF_4^-, NO_3^-), which are of relatively lower cost and lower toxicity than traditional imidazolium-based ILs. The nitration reactions were carried out at (−15℃) for a period of time first then at room temperature for longer time with a little excessive NO_2 (ca. 1. 4eqv.) and moderate yield was gained (for toluene). The IL could be reused for four times.

Keywords: Nitration, NO_2/air, novel Brønsted acidic ionic liquids, aromatics

Nitration of aromatics is a fundamental reaction of great industrial importance, which provides key organic intermediates or energetic materials. Unfortunately, the usual commercial process of nitration of aromatics which uses nitric and sulfuric acids is not environmentally benign with resulting in disposal problems, regeneration of used acids, and often provides poor selectivity for the desired products. Therefore, various nitration approaches have been explored to avoid the problems of the traditional mixed acid method[1]. Much success has been achieved, but some problems still exist, such as volatile chlorinated solvents required in some processes.

Because ionic liquids have strong polarity, negligible vapor pressure, wide solubility to organics and inorganics, and variable Lewis/Brøsted acidity, they have been used in a series of reactions[2] as catalysts and/or slovents (instead of volatile organic solvents) . That also drew our great attention and we desired to develop a green way for environmentally benign nitration process via using ILs as catalysts and solvents.

Because imidazolium-based ILs have relatively higher cost and toxicity[3] in general chemical applications, we made efforts to seek new Brønsted acidic ILs with lower cost and lower toxicity. Then it was found that lactam-based Brønsted acidic ILs are relatively cheaper, easily available in large amounts from industry and more environmentally friendly in comparison to imidazolium-based ILs[4] . And the additional carbonyl groups in the lactam might lead to specific functions when these lactam-based ILs are used as media or solvents. Therefore, we chose [caprolactam] BF_4 and [caprolactam] NO_3, whose Brønsted acidity are relatively stronger, and applied them to nitration of simple aromatics. Considering that addition of large aromatic groups to the molecular of ILs may increase the solubility of ILs to aromatics, we synthesized two new caprolactam-based Brønsted acidic ILs, [caprolactam] pTSO and [caprolactam] BSO with p-toluenesulfonic acid and benzenesulfonic acid as conjugated acid respectively. Then the ILs were characterized by spectrometry and their activity as catalysts for the nitration of toluene, chlorobenzene and benzene were investigated. There are several reports on aromatics nitration with HNO_3 (67%)[5] or HNO_3/Ac_2O[6] in the presence of ILs. However, there has been no report on the nitration of aromatics with NO_2 using ILs as solvents and/or catalysts so far today. And NO_2 has its potential applicable ability for nitration. In this paper we report for the first time the results of the nitration of simple aromatics with NO_2 under air atmosphere using Brønsted acidic ionic liquids as catalysts.

Experimental section

Preparation and characterization of caprolactam-based ionic liquids

The preparation of caprolactam-based ILs was illustrated in Scheme 1. The ILs were prepared according to the method in ref. 4.

^{1}HNMR spectra were recorded on a Bruker Advanced Digital 300MHz NMR spectrometer and mass spectra were recorded on a Finnigan TSQ Quantun Ultra AM spectrometer. IR spectra were recorded on a Bruker EQUINOX55 IR spectrometer, melting points were measured on a XT4 micro-melting point instrument made by Beijing Keyi Dianguang Instrument Plant.

Using pyridine as IR probe molecule, pyridine and ILs mixed in a given ratio (1 : 1 by molar) and then spread into KBr liquid films, the Brønsted acidities of these ILs were determinated by IR spectrometer.

+ HX —Water or benzene, rt→ X^-

X^- = (pTSO), (BSO), BF_4^-, NO_3^-

(1) (2) (3) (4)

Scheme 1 Preparation of caprolactam-based ILs

Nitration of aromatic compounds with NO_2 under air atmosphere in the presence of ionic liquids: General Procedure

To a cooled (0 or −15℃) vigorously stirred mixture of the IL (quantities indicated in the tables) and the substrate (10.0mmol) under air atmosphere, 0.5mL liquid nitrogen dioxide was added quickly all at once. After a given time the reaction was quenched by addition of deionized water (ca. 5mL), and then the internal standard (nitrobenzene for toluene, anisole and chlorobenzene, hexadecane for benzene, 0.2mL) was added, followed by addition of hexane (5mL×4). Two phases formed once the stirring was stopped. The bottom one was an aqueous phase containing the IL, and the upper one was an organic phase. The upper phase was removed by simple decantation from the aqueous phase. The organic layers were combined and washed with saturated solution of sodium bicarbonate (5mL), water (5mL×3) and brine (5mL×2) in turn, dried with Na_2SO_4, and then analyzed by GC. (When in preparative scale, the dried organic layers were distilled under reduced pressure to remove solvent and the unreacted substrates.) The recovered IL was dried at 65℃ under reduced pressure for 5h to remove water accumulated during the workup, and employed for further use.

Results and discussion

Preparation and characterization of caprolactam-based ionic liquids

All of the caprolactam-based ILs were prepared with high yields of 85-92%. They are moisture stable. Among of them, [Caprolactam] pTSO is a white solid with a melting point of 69℃; [Caprolactam] BSO is a yellow liquid of high viscosity; [Caprolactam] BF_4 is a colorless liquid with good fluidity; and [Caprolactam] NO_3, a white solid with a melting point of 30℃. All of them are easily soluble in water and have good moisture absorption (Scheme 2).

Data

The data of ^{1}H NMR and MS spectra of the four ILs were shown as following. The data were similar with that reported in ref. 4. In the ^{1}HNMR ($CDCl_3$) spectra, there are two active hydrogen in the range of δ6.80-16.24ppm respectively, indicating that they all formed N-protonated caprolactam cation.

R + NO_2/O_2 $\xrightarrow{IL}$ ortho + meta + para

$R=OCH_3, CH_3, Cl, H$

Scheme 2 Nitration of simple aromatics with NO_2/air catalyzed by ILs

[Caprolactam] pTSO: ^{1}HNMR ($CDCl_3$): δ1.73-1.83 (m, 6H), 2.38 (s, 3H), 2.68 (t, J=4.3, 2H), 3.46 (s, 2H), 7.22 (d, J=7.5, 2H), 7.76 (d, J=7.7, 2H), 9.92 (s, 1H), 10.91 (s, 1H). ^{1}H NMR (D_2O): 1.61-1.77 (m, 6H), 2.41 (s, 3H), 2.49 (t, J=3.6, 2H), 3.26 (t, J=4.6, 2H), 7.38 (d, J=7.9, 2H), 7.71 (d, J=7.9, 2H). MS (ESI): m/z 114.12 (Cation), calculated for cation ($C_6H_{12}NO$) 114.09.

[Caprolactam] BSO: ^{1}HNMR ($CDCl_3$): δ1.74-1.85 (m, 6H), 2.65 (d, J=9.1, 2H), 3.43 (s, 2H), 6.80 (s, 1H), 7.45 (s, 3H), 7.90 (d, J=5.9, 2H), 10.03 (s, 1H); ^{1}H NMR: (D_2O): δ1.47-1.63 (m, 6H), 2.33 (t, J=5.3, 2H), 3.11 (t, J=4.5, 2H), 7.44 (t, J=7.7, 3H), 7.68 (d, J=7.6, 2H). MS (ESI): m/z 114.13 (Cation), calculated for cation ($C_6H_{12}NO$) 114.09.

[Caprolactam] BF_4: ^{1}HNMR ($CDCl_3$): δ1.74-1.88 (m, 6H), 2.71-2.76 (q, J=5.6, 2H), 3.52-3.57 (t, J=7.1, 2H), 3.46 (s, 2H), 8.49 (s, 1H), 9.79 (s, 1H). ^{1}HNMR (D_2O): 1.37-1.58 (m, 6H), 2.29-2.33 (t, J=5.7, 2H), 3.05-3.09 (t, J=5.0, 2H). MS (ESI):

m/z 114.09 (Cation), calculated for cation ($C_6H_{12}NO$) 114.09.

[Caprolactam] NO_3: ^{1}HNMR ($CDCl_3$): δ1.64-1.84 (m, 6H), 2.62-2.65 (t, J=5.6, 2H), 3.40-3.43 (t, J=4.5, 2H), 8.54 (s, 1H), 16.24 (s, 1H); ^{1}HNMR: (D_2O): δ1.35-1.54 (m, 6H), 2.25-2.29 (t, J=5.7, 2H), 3.01-3.04 (t, J=5.0, 2H). MS (ESI): m/z 114.10 (Cation), calculated for cation ($C_6H_{12}NO$) 114.09.

Using pyridine as IR probe molecule, the IR spectra of the caprolactam-based ILs were obtained. According to refs. 7 and 8, the presence of a band near $1450cm^{-1}$ is indicative of pyridine coordinated to Lewis acid sites; whilst a band near $1540cm^{-1}$ is an indication of the formation of pyridinium ions resulting from the presence of Brønsted acidic sites. From each probe IR spectra of these ILs, a band is observed near $1540cm^{-1}$, confirming that Brønsted acidic sites are all present in the ILs synthesized. The wavenumber of the band corresponding to coordination at Brønsted acidic sites increases from $1541.59cm^{-1}$ for [Caprolactam] BSO to $1543.69cm^{-1}$ for [Caprolactam] pTSO, $1544.61cm^{-1}$ for [Caprolactam] NO_3 and $1549.39cm^{-1}$ for [Caprolactam] BF_4, indicating that the Brønsted acidity increases in the order: [Caprolactam] BSO< [Caprolactam] pTSO< [Caprolactam] NO_3< [Caprolactam] BF_4.

NO_2 nitration of simple aromatics under air atmosphere in the presence of Brønsted acidic ILs

Some results of NO_2 nitration of simple aromatics under air atmosphere in the presence of Brønsted acidic ILs were listed in Table 1.

The data in Table 1 indicated that the caprolactam-based ILs prepared all had catalytic acitivity for NO_2/air nitration of aromatics; and among of them, [Caprolactam] pTSO was the best one because of better affinity to aromatics than the others. Comparing the result of Entry 1 with the result of Entry 2, the latter showed a little lower yield but a slight higher para-selectivity than the former. This could be explained by the polarization ability of IL2 stronger than IL1's which was resulted by BSO's stronger polarity than pTSO's. Under the condition[b] indicated in the table, nitration of chlorobenzene gained a smaller yield (and also benzene), but gave improved para-selectivity than the case without IL. From analyzing the results of NO_2/air nitration of toluene catalyzed by different ILs (Entry 1-4 in Table 1) and the Brønsted acidity of the ILs (which increases in the order [Caprolactam] BSO< [Caprolactam] pTSO< [Caprolactam] NO_3< [Caprolactam] BF_4), it could be drawn that the acidity and affinity to aromatics of ILs have important roles on their catalytic activity.

Table 1 Results of NO_2-air nitration of simple aromatics catalyzed by caprolactam-based ILs

Entry	Ionic liquid	R	Yield (%)[c]	Product distribution(%)			Ortho/para
				ortho	meta	para	
1[a]	IL1	CH_3	61.4	57.1	3.9	39.0	1.46
2[a]	IL2	CH_3	44.7	49.9	6.3	43.8	1.14
3[a]	IL3	CH_3	46.5	56.8	4.7	38.4	1.48
4[a]	IL4	CH_3	31.8	56.2	7.9	35.8	1.57
5[a]	IL1	OCH_3	91.3	40.4	2.3	57.3	0.70
6[b]	IL1	Cl	48.7	19.9	3.7	76.4	0.26
7[b]	IL2	Cl	37.4	19.8	0.3	79.8	0.25
8[b]	IL3	Cl	38.2	24.3	1.9	73.7	0.33
9[b]	IL4	Cl	25.3	22.5	9.0	68.4	0.33
10[b]	IL1	H	55.4				
11[b]	IL2	H	41.4				
12[b]	IL3	H	43.8				

续表

Entry	Ionic liquid	R	Yield (%)[c]	Product distribution(%) ortho	meta	para	Ortho/para
13[b]	IL4	H	29.2				
14[b]	none	H	9.2				
15[b]	none	Cl	4.9	39.8	0	60.1	0.66
16[a]	none	CH_3	13.4	57.9	5.7	36.4	1.59

a Substrate 10mmol, IL 10mol%, NO_2 0.5mL (ca. 14mmol), −15℃/30min, then 0℃/6h, and rt/24h (when the air balloon was removed).

b Substrate 10mmol, IL 10mol% of substrate, NO_2 0.5mL (ca. 14mmol), −15℃/30min, then 0℃/6h, and rt/60h (when the air balloon was removed).

c Calculated by quantitative GC.

Recycling of IL1 for NO_2/air nitration of toluene had been studied and the results were listed in Table 2 Though the yield decreased and ortho/para ratio lowered after each recycling because of the loss of IL in the process of disposing, the IL still had catalytic activity after four runs.

Table 2 Recycling of IL1 for NO_2-air nitration of toluene[a]

Run	Yield(%)[b]	Product distribution(%) ortho	meta	para	Ortho/para
1	61.4	57.1	3.9	39.0	1.46
2	60.7	56.3	4.6	39.1	1.44
3	57.3	54.2	6.8	39.0	1.39
4	57.8	54.3	7.7	37.9	1.43
5	54.8	52.6	5.7	41.7	1.26

a Substrate 10mmol, IL 10mol%, NO_2 0.5mL (ca. 14mmol), −15℃/30min, then 0℃/6h, and rt/24h (when the air balloon was removed);

b Calculated by quantitative GC.

To check the utility of this process for mass production, nitro compounds were synthesized with good yields by using this method in preparative scale. And the details were shown in Table 3.

Table 3 Synthesis of nitro compounds in preparative scale using the method upon

Entry	R	Amount of products(g)	Yield(%)[c]	Product distribution(%)[d] ortho	meta	para	Ortho/para
1[a]	OCH_3	26.3	85.9	39.6	2.9	57.5	0.69
2[a]	CH_3	17.0	62.0	54.2	3.4	42.4	1.28
3[b]	H	13.1	53.4		No isomers		
4[b]	Cl	14.6	44.6	22.3	3.5	74.2	0.30

a Substrate 0.2mol, IL1 10mol%, NO_2 10mL, −15℃/30min, then 0℃/6h, and rt/24h (when the air balloon was removed).

b Substrate 0.2mol, IL1 10mol% of substrate, NO_2 10mL, −15℃/30min, then 0℃/6h, and rt/60h (when the air balloon was removed).

c Isolated yield.

d Calculated by quantitative GC.

The mechanism of the present process at this stage is mysterious. The nitration of aromatic compounds with N_2O_5 is very reasonable. N_2O_4 is partially converted to N_2O_5 with O_2. Meanwhile, knowing that dinitrogen tetroxide is almost completely ionized in anhydrous nitric acid,[9] we are encouraged to imagine that NO_2-air nitration of simple aromatic compounds catalyzed by Brønsted acidic ILs may also proceed via the pathway showed in Figure 1. As shown in the Figure from formula (1) to formula (6), nitration products are produced and the side-product is nitric acid. When reaction temperature rises, catalyzed by Brønsted acidic ILs, NO_2^+ is generated by HNO_3 and reacts with the aromatics. Thus, the usage of nitrogen dioxide is increased. The most ideal result is that water is the only by-product.

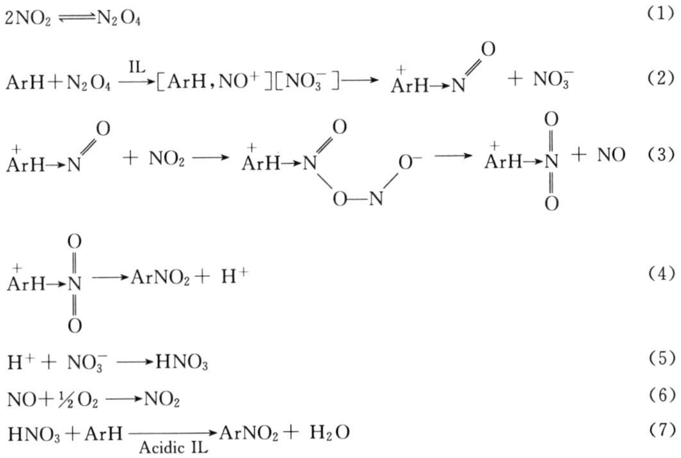

$$2NO_2 \rightleftharpoons N_2O_4 \quad (1)$$

$$ArH + N_2O_4 \xrightarrow{IL} [ArH, NO^+][NO_3^-] \longrightarrow \overset{+}{ArH} \rightarrow N{=}O + NO_3^- \quad (2)$$

$$\overset{+}{ArH} \rightarrow N{=}O + NO_2 \longrightarrow \overset{+}{ArH} \rightarrow N(=O)\text{–}O\text{–}N{=}O \longrightarrow \overset{+}{ArH} \rightarrow N(=O)_2 + NO \quad (3)$$

$$\overset{+}{ArH} \rightarrow N(=O)_2 \longrightarrow ArNO_2 + H^+ \quad (4)$$

$$H^+ + NO_3^- \longrightarrow HNO_3 \quad (5)$$

$$NO + \tfrac{1}{2}O_2 \longrightarrow NO_2 \quad (6)$$

$$HNO_3 + ArH \xrightarrow[\text{Acidic IL}]{} ArNO_2 + H_2O \quad (7)$$

Figure 1 Possible pathway for NO_2-air nitration of aromatics catalyzed by Brønsted acidic ILs

References

[1] see for example, (a) Suzuki, H. ; Tatsumi, A. ; Suzuki, H. M. ; Maeda, K. A novel non-acidic method for the preparation of 2, 2, 2-trifluoro-1- (3-nitrophenyl) ethanone and 1-nitro-3-trifluoromethylben-zene, versatile starting materials for trifluoromethyl-containing aromatic compounds. Synthesis 1995, 1353-1354; (b) Suzuki, H. ; Mori, T. ; Ozone-mediated nitration of chloro-and bromo-benzenes and some methyl derivatives with nitrogen dioxide. High ortho-directing trends of the chlorine and bromine substituents. J. Chem. Soc. Perkin trans. 2 1994, 479-484; (c) Suzuki, H. ; Mori, T. ; Maeda, K. Ozone-mediated reaction of polychlorobenzenes and some related halogeno compounds with nitrogen dioxides: A novel non-acid methodology for the selective mononitration of moderately deactivated aromatic systems. Synthesis 1994, 841-845; (d) Suzuki, H. ; Tomaru, J. ; Murashima, T. Iron (Ⅲ) -catalysed nitration of non-activated and moderately activated arenas with nitrogen dioxide-molecular oxygen under neutral conditions. J. Chem. Soc. , Perkin Trans. 1 1996, 2385-2389; (e) Barrett, A. G. M. ; Braddock, D. C. ; Ducray, R. ; Mckinnell, R. M. ; Waller, F. J. Lanthanide triflate and triflide catalyzed atom economic nitration of fluoro arenes. Synlett. 2000, 57-59; (f) Waller, F. J. ; Barrett, A. G. M. ; Braddock, D. C. ; Ramprasad, D. Lanthanide (Ⅲ) triflates as recyclable catalysts for atom economic aromatic nitration. Chem. Commun. 1997, 613-615; (g) Delaude, L. ; Laszlo, P. ; Smith, k. Heightened selectivity in aromatic nitrations and chlorinations by the use of sokid supports and catalysts. Acc. Chem. Res. 1993, 26, 607-613; (h) Gigante, B. ; Prazeres, A. O. ; Marcelo-Curto, M. J. Mild and selective nitration by claycop. J. Org. Chem. 1995, 60, 3445-3447; (i) Kwok, T. J. ; Jayasuriya, K. ; Damavarapu, R. ; Brodman, B. W. Application of H-ZSM-5 zeolite for regioselective mononitration of toluene. J. Org. Chem. 1994, 59, 4939-4942; (j) Vessena, D. ; Kogelbauer, A. ; Prins, R. Potential routes for the nitration of toluene and nitrotoluene with solid acids. Catal. Today 2000, 60, 275-287; (k) Dagade, S. P. ; Waghmode, S. B. ; Kadam, V. S; Dongare, M. K. Vapor phase nitration of toluene using dilute nitric acid and molecular modeling studies over beta zeolite. Appl. Catal. , A 2002, 226, 49-61; (l) X. Peng, H. Suzuki, Lu, C. Zeolite-assisted nitration of neat toluene and chlorobenzene with a nitrogen dioxide/molecular oxygen system. Remarkable enhancement of para-selectivity. Tetrahedron Letters 2001, 42, 4357-4359.

[2] see for example, (a) Sheldon, R. Catalytic reactions in ionic liquids. Chem. Commun. , 2001, 2399-2407; (b) Dupont, J. ; Souza, R. F. ; Suarez, P. A. Z. Ionic liquid (Molten salt) phase organometallic catalysis. Chem. Rev. 2002, 102, 3667-3692; (c) Cole, A. C. ; Jensen, J. L. ; Ntai, I. ; Tran, K. L. T. ; Weaver, K. J. Novel Brøsted acidic ionic liquids and their use as dual solvent-catalysts. J. Am. Chem. Soc. 2002, 124, 5962-5963; (d) Gui, J. Z. ; Ban, J. Y. ; Cong, X. H. ; Zhang, X. T. ; Sun, Z. L. Selective alkylation of phenol with tert-butyl

alcohol catalyzed by Brøsted acidic imidazolium salts. J. Mol. Catal. A 2005，225，27-31；(e) Joseph，T.；Sahoo，S.；Halligudi，S. B. Brøsted acidic ionic liquids：A green，efficient and reusable catalyst system and reaction medium for Fischer esterification. J. Mol Catal. A 2005，234，107-110；(f) Angueira，E. J.；White，M. G. Arene carbonylation in acidic，chloro aluminate ionic liquids. J. Mol. Catal. A 2005，227，51-58；(g) Li，D. M.；Shi，F.；Peng，J. J.；Guo，S.；Deng，Y. Q. Application of functional ionic liquids possessing two adjacent acid sites for acetalization of aldehydes. J. Org. Chem. 2004，69，3582-3585.

[3] Swatloski，R.；Holbrey，J.；Memon，S.；Caldwell，G.；Caldwell，K.；Rogers，R. Using Caenorhabditis elegans to probe toxicity of 1-alkyl-3-methylimidazolium chloride based ionic liquids. Chem. Commun. 2004，668-669.

[4] Du，Z. Y.；Li，Z. P.；Guo，S.；Zhong，J.；Zhu，L. Y.；Deng，Y. Q. Investigation of physicochemical properties of lactam-based Brøsted acidic ionic liquids. J. Phys. Chem.，B 2005，109，19542-19546.

[5] Qiao，K.；Yokoyama，C. Nitration of aromatic compounds with nitric acid catalyzed by ionic liquids. Chem. Lett.，2004，33，808-809.

[6] Smith，K.；Liu，S. F；El-Hiti，G. A. Regioselective mononitration of simple aromatic compounds under mild conditions in ionic liquids. Ind. Eng. Chem. Res. 2005，44，8611-8615.

[7] Yang，Y. L.；Kou，Y. Determination of the Lewis acidity of ionic liquids by means of an IR spectroscopic probe. Chem. Commun. 2004，226-227.

[8] Parry，E. P. An infrared of pyridine adsorbed on acidic solids，characterization of surface acidity. J. Catal. 1963，2，371-379.

[9] Schofield，K. Aromatic nitration. Cambridge University Press，1980；p. 77.

(注：此文原载于 Synthetic Communications，2008，38：537-545)

Regioselective Mononitration of Aromatic Compounds Using Keggin Heteropolyacid Anion Based Brønsted Acidic Ionic Salts

Yang Hongwei[1], Qi Xiufang[1], Wen Liang[2], Lv Chunxu[1], Cheng Guangbin[1]
(1. Nanjing University of Science & Technology, Nanjing, 210094, China.)
(2. Southwest University of Science & Technology, Mianyang 621010, China)
SCI: 000295237000058

Abstract: Novel Keggin heteropolyacid anion based Brønsted acidic ionic salts $[SO_3H(CH_2)_4Mim]_nH_{3-n}PMo_{12}O_{40}$ (n=1,2,3) were synthesized and used as catalysts in the nitration of aromatic compound in HNO_3 (67%). The catalytic activity was in the order of $[(CH_2)_4SO_3HMim]_3PMo_{12}O_{40} > [(CH_2)_4SO_3HMim]_2HPMo_{12}O_{40} > [(CH_2)_4SO_3HMim]H_2PMo_{12}O_{40}$. Effect of catalyst loading, quantity of nitro acid and temperature on nitration of toluene was discussed, and 87.7% yield and good para selectivity with 1.19 ratio of ortho to para can be obtained with optimized catalyst $[(CH_2)_4SO_3HMim]_3PMo_{12}O_{40}$. The catalyst can be easily separated from the products and reused for three times without significant loss of activity.

Keywords: Nitration-aromatic compound-Keggin heteropolyacid anion based Brønsted acidic ionic salt-catalysis

Introduction

Electronic nitration of aromatic compound is one of the most important unit reactions in chemical industry, which attribute to nitro-product role of key organic intermediates or energetic materials. The past few years have witnessed a growing interest in ionic liquids (ILs) due to their high polarity, negligible vapor pressure, and good solubility in a broad range of organic and inorganic compounds, which endorse them as useful "clean and green" solvents or catalysts for a number of reactions[1-4]. From environment standpoint, ionic liquid was received more and more attention in nitration of aromatics to avoid the problems of the traditional mixed-acid method[5-11].

Imidazolium based ionic liquid have been widely used as solvent and catalyst for nitration of aromatic compounds[12-15]. However, the overall results of nitration with diluted nitric acid in the presence of ionic liquids were not satisfactory with slow reaction rate, low yield and much side reaction. Some works in our group have been done to investigate the use of different ILs for nitration of aromatic compound, and we found that the anion has significant effect on the ionic liquid catalyzed nitration reaction[16-18].

While numerous reports described the ILs was used as catalyst for nitration system such as [bimim] BF_4 and [bdmin] BF_4 and [emim] [NO_3] . But Brønsted acidic based ILs is reported very limited as catalyst in nitration of aromatic compound. [15,19]. Keggin heteropolyacids have attracted much attention in catalysis area because of their special catalytic activities[20-22] such as the similar structure as molecular shieves and strong acidities[23]. In this paper, we report our continuous efforts at exploring the new imidazolium based modified Brønsted acidic ionic liquid alternative catalyzed nitration reaction. We have synthesized novel Keggin heteropolyacid anion based Brønsted acidic ionic salts(KHBISs)[$SO_3H(CH_2)_4Mim]_nH_{3-n}PMo_{12}O_{40}$ (n=1,2,3) which was using Keggin heteropolyacid $H_3PMo_{12}O_{40}$ as anion source and zwitterion[Mim(CH_2)$_4SO_3$]as cation source. Moreover, their catalytic activities were investigated in nitration of aromatic compounds.

Experimental

2.1 General information

All reagents were commercially available. IR spectra were recorded on a Bruker EQUINOX55 IR spectrometer. Product mixtures were analyzed on Agilent Technologies 6820 Gas Chromatography with an OV-101 capillary column (0.32mm×30m) and a FID detector.

2.2 Preparation of keggin heteropolyoxotungstate anion based brϕnsted acidic ionic salt

The whole procedure for the synthesis of the KHBISs was shown in Fig. 1.

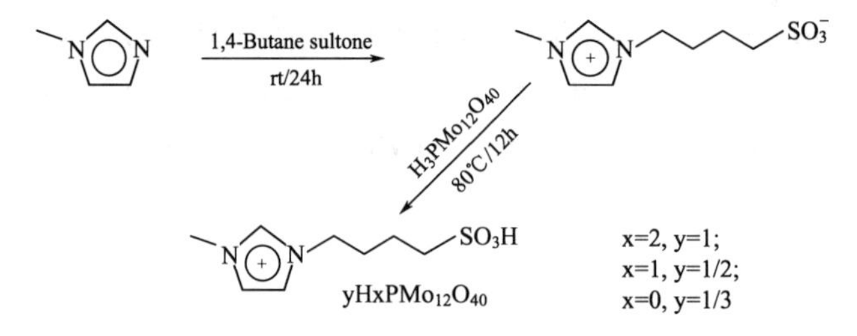

Fig. 1 Synthesis of Keggin heteropolyacid anion based Brønsted acidic ionic salts

The zwitterion[Mim(CH_2)$_4SO_3$][3] and Keggin heteropolyacid $H_3PMo_{12}O_{40}$[24] were prepared according to the literature method, and the KHBISs were synthesized as following.

Zwitterion [Mim (CH_2)$_4SO_3$] and Keggin heteropolyacid $H_3PMo_{12}O_{40}$ were dissolved in a minimum amount of deionized water in mol ratio of 1∶1, 2∶1, 3∶1 of zwitterion to heteropolyacid at room temperature. The mixture was refluxed at 80℃ for 6h. The solid was obtained from aqueous solution by removing water and dried under vacuum at 75℃ for 6h. Three of ionic salts [(CH_2)$_4SO_3$HMim]$H_2PMo_{12}O_{40}$, [(CH_2)$_4SO_3$HMim]$_2HPMo_{12}O_{40}$ and[(CH_2)$_4SO_3$HMim]$_3PMo_{12}O_{40}$ were in green with color deepened in order as solid, respectively, and soluble in water and methanol, partially soluble in acetone but not soluble in toluene.

The spectral data for [(CH_2)$_4SO_3$HMim] $H_2PMo_{12}O_{40}$: FT-IR (KBr): 3428.5, 1623.3, 1173.5, 1065.5, 965.6, 879.6, 791.6cm^{-1}. ^{1}HNMR (500 Hz): δ 1.66 (m, 2H), 1.95 (m, 2H), 2.86 (t, 2H, J=7.7Hz), 3.83 (s, 3H), 4.19 (t, 2H, J=7.1Hz), 7.37 (s, 1H), 7.43 (s, 1H), 8.65 (s, 1H) ppm. ^{13}C NMR (500 Hz): δ 20.3, 27.5, 35.0, 48.4, 49.3,

121.6, 123.1, 134.7 ppm. The spectral data for [$(CH_2)_4SO_3HMim]_2HPMo_{12}O_{40}$: FT-IR (KBr): 3428.5, 1633.3, 1163.5, 1065.5, 955.6, 889.6, 801.6cm^{-1}. ^{1}HNMR (500Hz): δ 1.72 (m, 2H), 2.05 (m, 2H), 2.91 (t, 2H, J=7.6 Hz), 3.98 (s, 3H), 4.28 (t, 2H, J=7.2 Hz), 7.45 (s, 1H), 7.51 (s, 1H), 8.65 (s, 1H) ppm. ^{13}CNMR (500 Hz): δ 20.4, 27.6, 35.1, 48.4, 49.4, 121.6, 123.1, 134.8 ppm. The spectral data for $[(CH_2)_4SO_3HMim]_3PMo_{12}O_{40}$: FT-IR (KBr): 3438.5, 1633.3, 1173.5, 1065.5, 965.6, 889.6, 801.6cm^{-1}. ^{1}HNMR (500Hz): δ 1.74 (m, 2H), 2.06 (m, 2H), 2.92 (t, 2H, J=7.6Hz), 3.97 (s, 3H), 4.28 (t, 2H, J=7.0 Hz), 7.46 (s, 1H), 7.52 (s, 1H), 8.67 (s, 1H) ppm. ^{13}C NMR (500Hz) δ 20.5, 27.6, 35.0, 48.2, 49.4, 121.6, 123.1, 134.8ppm.

2.3 General nitration procedure

To vigorously stirred mixture of given amount of ionic salt and substrate was added 67% nitric acid at room temperature. The reaction was warmed up to higher temperature which is indicated in the table. After 10h, the reaction was quenched with deionized water, extracted with hexane (5mL×3). The organic phase was isolated and organic solutions were combined and washed with saturated sodium bicarbonate (5mL), water (5mL×3) and brine (5mL×2), dried over Na_2SO_4 and analyzed by GC (different internal standard was used for toluene, and chlorobenzene, nitrobenzene; and for benzene, hexadecane 0.2mL[8]) . The ionic salt as catalyst was recovered from aqueous solution by removing water at 70℃ under vacuum for 5h for reusing.

Results and discussion

3.1 Effect of the composition of the catalysts on the nitration

Three kinds of KHBISs [$(CH_2)_4SO_3HMim$] $H_2PMo_{12}O_{40}$, [$(CH_2)_4SO_3HMim]_2HPMo_{12}O_{40}$ and [$(CH_2)_4SO_3HMim]_3PMo_{12}O_{40}$ were obtained from the reaction of zwitterion to heteropolyacid with mol ratio of 1∶1, 2∶1, 3∶1.

In our preliminary investigation, we carried out electrophilic nitration of toluene with 1 mol% loading of three different ionic as catalyst, 3 equiv. of 67% nitric acid at 50℃ for 10h. As comparing, the nitration reactions were also carried out with the Zwitterion [Mim $(CH_2)_4SO_3$] and Keggin heteropolyacid $H_3PMo_{12}O_{40}$ as catalyst respectively in the catalytic amount of offering the same quantity of acidic proton. The results were shown in Table 1. Without catalyst, the reaction can give only 51.3% conversion, 41.7% yield of mono-nitrobenzene and *ortho* to *para* ratio of 1.49 (entry 6, Table 1) . Compared with the system without catalyst, yield of mono-nitrobenzene was increasing to 44.2% with 54.6% conversion when Keggin heteropolyacid $H_3PMo_{12}O_{40}$ using as catalyst (entry 5, Table 1) . While conversion can be reach up to 74.8%, but only 49.6% of yield of mono-nitro product was obtained with accompanying formation of more byproduct when Zwitterion [Mim $(CH_2)_4SO_3$] was used as catalyst (entry 4, Table 1) . The zwitterion showed better catalytic activity than $H_3PMo_{12}O_{40}$, which may be attributed to its phase-transferring role. Three KHBISs all showed better catalytic activity than the corresponding the system with Zwitterion [Mim $(CH_2)_4SO_3$] or Keggin heteropolyacid $H_3PMo_{12}O_{40}$ as catalyst alone, which indicated that the catalytic activities of the salts were the combination of the activities of the cations and anions. And enhancing of proportion of the cation in the composition of the salt would be good to improve the catalytic activity in yield and *para* selectivity. Under offering same quantity of acidic protons by the catalysts, the catalytic activities of the five compounds were in the order of $[(CH_2)_4SO_3HMim]_3PMo_{12}O_{40}>[(CH_2)_4SO_3HMim]_2HPMo_{12}O_{40}>[(CH_2)_4SO_3HMim]H_2PMo_{12}O_{40}>Mim(CH_2)_4SO_3>H_3PMo_{12}O_{40}$. The yields of nitro products were 58.8%, 56.1%, 50.4%, 49.6% and 44.2% respectively. So $[(CH_2)_4SO_3HMim]_3PMo_{12}O_{40}$ is optimized catalyst we investigated in nitration of toluene.

Table 1 Effect of the compositionu of the catalysts on the nitration[a]

Entry	Catalysts	Conversion (%)[c]	Yield (%)[c]	Product distribution(%)			Ortho/para
				ortho	meta	para	
1	$[(CH_2)_4SO_3HMim]H_2PMo_{12}O_{40}$	58.4	50.4	55.2	4.8	40.0	1.38
2	$[(CH_2)_4SO_3HMim]_2HPMo_{12}O_{40}$	74.3	56.1	56.9	4.8	38.3	1.39
3	$[(CH_2)_4SO_3HMim]_3PMo_{12}O_{40}$	74.9	58.8	53.4	4.7	41.9	1.27
4[b]	$Mim(CH_2)_4SO_3$	74.8	49.6	55.6	4.9	39.5	1.41
5	$H_3PMo_{12}O_{40}$	54.6	44.2	57.6	4.7	37.8	1.52
6	0	51.3	41.7	57.1	4.5	38.4	1.49

a Mol ratio of catalyst to toluene was 1.0%; mol ratio of nitric acid (67%) to toluene was 3∶1, toluene: 5.0 mmol; 50℃/10h. b Mol ratio of catalyst to toluene was 3%. c Calculated by quantitative GC.

3.2 Effect of the quantity of nitric acid (67%) on the nitration

Effect of the amount of nitric acid (67%) on the nitration was investigated with 1.0mol% of $[(CH_2)_4SO_3HMim]_3PMo_{12}O_{40}$ at 50℃ for 10h. The results were listed in Table 2. It could be seen that the yield of nitro product increased with the increasing of amount of nitric acid. When the mol ratio of nitric acid (67%) to toluene was 3∶1, the reaction gave 74.9% conversion and 58.8% yield of mono-nitrotoluene (entry 3, Table 2). When the mol ratio of nitric acid (67%) to toluene was increased to 4∶1, both conversion and yield improved to 82.0% and 68.6%, but para-regioselectivity was decreased from 1.27 to 1.34 (entry 4, Table 2). In addition, it also could be seen from Table 3 that side reaction happened accompanying with the nitration reaction. And the increasing of the amount of nitric acid could improve the yield of nitro product and abate the proportion of side reaction.

Table 2 Effect of the quantity of nitric acid (67%) on the nitration[a]

Entry	Mol ratio of nitric acid to toluene	Conversion (%)[b]	Yield (%)[b]	Product distribution(%)			Ortho/para
				ortho	meta	para	
1	1∶1	56.7	30.9	56.5	4.5	39.0	1.29
2	2∶1	66.2	43.2	55.1	4.6	40.3	1.28
3	3∶1	74.9	58.8	53.4	4.7	41.9	1.27
4	4∶1	82.0	68.6	54.2	4.4	40.4	1.34

a Mol ratio of the catalyst to toluene was 1.0%, toluene: 5.0mol, 50℃/10h. b Calculated by quantitative GC.

3.3 Effect of the amount of catalyst on the nitration

The effect of the amount of catalyst on the nitration of toluene was investigated with $[(CH_2)_4SO_3HMim]_3PMo_{12}O_{40}$ in the range of 1.0～10.0mol% and 3 equiv. of 67% nitric acid at 50℃ for 10 h. The results were shown in Table 3.

The nitration reaction of toluene can provide only 41.7% of yield, 51.3% conversions and 1.49 of *ortho to para* ratio without any ionic salt using (entry 1, Table 3). While the ionic salt (1 mol%) indicated good catalytic activity, which afford 58.8% of yield, 74.9% conversions and 1.27 of *para* regioselectivity (entry 2, Table 3). By further increasing the catalyst loading from 1.0～10.0mol%, yield of mono-nitro product increased and reach up to 92.5% with 10 mol% catalyst loading. Both conversion and *para*-selectivity also was improved correspondingly to 100.0% and 1.16, respectively (entry 6, Table 3). It shows that the conversion of the toluene, the yield of product mononitrotoluene and the *para*-regioselectivity increased with the increase of the amount of catalyst used in the range of 1.0~10.0mol%. The yield increasing with the amount of catalyst can attribute to more proton acid provided by more Brønsted acidic salts and stronger the role of phase-transferring of the catalyst. While *Para*-selectivity of the nitration reaction was improved with the amount of catalyst increasing, which may be ascribed to the increasing of electron spin density on 4-C of benzene ring under the ionic environment offered by the catalyst.

Table 3 Effect of the amount of catalyst on the nitration[a]

Entry	Amount of catalyst (mol%)	Conversion (%)[b]	Yield (%)[b]	Product distribution(%)			Ortho/para
				ortho	meta	para	
1	0	51.3	41.7	57.1	4.5	38.4	1.49
2	1.0	74.9	58.8	53.4	4.7	41.9	1.27
3	2.5	88.8	78.4	52.8	4.1	43.1	1.23
4	5.0	94.4	84.6	52.5	4.0	43.5	1.21
5	7.5	99.6	89.2	52.1	4.0	43.8	1.19
6	10.0	100.0	92.5	51.6	3.9	44.5	1.16

a Catalyst loading, Mol ratio of HNO_3 (67%) to toluene: 3∶1; b Calculated by quantitative GC.

3.4 Nitration of different aromatic compounds catalyzed by ionic salt

The effect of reaction temperature on the nitration was also investigated with 5.0mol% $[(CH_2)_4SO_3HMim]_3PMo_{12}O_{40}$. From the results listed in Table 4, it could be seen that both the conversion and yield of toluene increased following the temperature rising. When the reaction was carried out at 70℃ for 10h, conversion and yield can be reached to 100.0% and 87.7%, respectively. There was about 12.3% of toluene consumed by side reaction. The regioselectivity was improved with increasing temperature and to 1.19 ratio of *ortho/para* nitro isomers were obtained at further elevated temperature to 70℃ (entry 3, Table 4).

On the basis of condition we investigated, we investigated nitration of benzene, chlorobenzene and nitrobenzene with 5 mol% optimized catalyst to examine the scope of this catalytic system. Only 13.8% and 5.2% of yield can be obtained respectively for nitration of benzene and chlorobenzene without any catalyst (entries 5, 7, Table 4). In comparison, yield was significantly increased to 75.4% and 51.5% respectively for nitration of benzene and chlorobenzene with catalyst using (entries 4, 6, Table 4). Under identical conditions, nitration of toluene can give best result in terms of conversion and yield. Activity of electronic nitration of aromatic compound with electron-donating groups was higher than that with electron-withdraw groups in *para*-position in this catalytic system. The yields of nitration of chlorobenzene and benzene are lower compared to toluene, which show the fact is less electron rich in aromatic ring of benzene and chlorobenzene result in deactivated by electron induction (cf. entries 4, 6 with entry 3, Table 4). Nitration of chlorobenzene show *para* position is more active than *ortho* position, which can afford significant higher *para*-selectivity with 0.25 ratio of *ortho* to *para*. Our catalyst system has limitation in nitration of nitrobenzene, the reaction was sluggish and no product could be detected by GC analysis (entry 8, Table 4). It can attribute to decreasing of the density of electron cloud of aromatic when aromatic ring substituted by electron-withdraw groups, which is harder for NO_2^+ to attack the aromatic ring.

R + HNO_3 —Brφsted Acidic Ionic Salts→ R(NO_2) + H_2O

Table 4 nitration of different aromatic compounds catalyst by ionic salt

Entry	R	Amount of catalyst(mol%)	Temperature (℃)	Conversion (%)ᵇ	Yield (%)ᵇ	Product distribution(%)			Ortho/para
						ortho	meta	para	
1	CH_3	5	30	80.3	68.2	52.8	4.1	43.1	1.23
2	CH_3	5	50	94.4	84.6	52.5	4.0	43.5	1.21
3	CH_3	5	70	100.0	87.7	52.1	4.2	43.7	1.19
4	Cl	5	70	57.6	51.5	19.9	0.3	79.8	0.25
5	Cl	none	70	5.2	5.2	32.4	trace	67.6	0.48
6	H	5	70	82.1	75.4				
7	H	none	70	16.4	13.8				
8	NO_2	5	70	<1					

a Mol ratio of nitric acid (67%) to toluene was 3∶1, aromatic compound: 5.0mmol, nitric acid (67%): 15.0mmol; reaction time was 10h. b Calculated by quantitative GC.

3.5 Effect of the catalyst reused on the nitration

The results of reusing of the catalyst [$(CH_2)_4SO_3HMim]_3PMo_{12}O_{40}$ were listed in Table 5. In the first run, the reaction was carried out with the amount of catalyst of 5.0mol%, 3.0 equiv. nitric acid (67%) at 70℃ for 16h. The yield for the first run was 87.7%, and 12.3% of toluene was consumed by side reaction (entry 1, Table 5). The recovered catalyst without purification was used for the second and third runs. On the second run, the yield of nitro product was decreased only by 2.4%. After three cycles, yield and *para*-selectivity was decreased respectively to 79.5 and 1.23 (entry 3, Table 5). This indicated that the catalyst has good reusability. Compared with the second run, the yield of the third run decreased by 5.8%, which may be ascribe to the byproducts absorbed on the catalyst. The byproducts hampered the effective contacting of the catalyst with reactants also result in decreasing the catalytic activity of the catalyst.

Table 5 Effect of the catalyst reused on the nitration

Entry	Time	Conversion (%)ᵇ	Yield (%)ᵇ	Product distribution(%)ᵇ			Ortho/Para
				ortho	meta	para	
1	1a	100.0	87.7	52.1	4.2	43.7	1.19
2	2	93.5	85.3	52.6	4.3	43.1	1.22
3	3	90.2	79.5	52.1	4.5	43.4	1.23

a Mol ratio of nitric acid (67%) to toluene was 3∶1, toluene: 5.0mmol, nitric acid (67%): 15.0mmol; mol ratio of catalyst to toluene was 5.0%; 70℃/10h. b Calculated by quantitative GC.

Conclusions

KHBISs$[SO_3H(CH_2)_4Mim]_nH_{3-n}PMo_{12}O_{40}$ (n=1,2,3) showed good catalytic activity in the nitration of toluene with 67% HNO_3 in high *para* selectivity and yield. The catalytic activity of the salts was in the order of$[(CH_2)_4SO_3HMim]_3PMo_{12}O_{40}$ > $[(CH_2)_4SO_3HMim]_2HPMo_{12}O_{40}$ > $[(CH_2)_4SO_3HMim]H_2PMo_{12}O_{40}$. And three KHBISs all show better catalytic activity than Zwitterion $[Mim(CH_2)_4SO_3]$ and Keggin heteropolyacid $H_3PMo_{12}O_{40}$. In an attempt to improve the yield and *para*-selectivity of mono-nitro product we screened a host of different reaction conditions using different catalyst loading, the quantity of nitric acid, and temperature. Nitration of benzene, chlorobenzene and nitrobenzene also was investigated with optimized catalyst of$[(CH_2)_4SO_3HMim]_3PMo_{12}O_{40}$ to examined scope of this catalyst system. Compared to 1.67 ratio of *ortho*/*para* of nitration of toluene in HNO_3/H_2SO_4[25], 87.7% yield and good *para* selectivity with 1.19 ratio of *ortho*/*para* is obtained in our catalytic system. Reactivity of electronic nitration of aromatic compound with electron-donating groups was higher than that with electron-withdraw groups in *para*-position in this catalytic system. The ionic salt can be easily separated from the products and easily recycled with the nitration yield lost about 8.2% after three cycles. It shows good reusability. Our study will provide the basis for studies on the application of ionic salt in nitration of aromatic acid.

Reference

[1] Wasserscheid, P.; Keim, W. Ionic liquids—new "solutions" for transition metal catalysis. Angew. Chem. Int. Ed. 2000, 39, 3772-3789.

[2] Sheldon, R. Catalytic reactions in ionic liquids. Chem. Commun. 2001, 2399-2407.

[3] Cole, A. C.; Jensen, J. L.; Ntai, I.; Tran, K. L. T.; Weaver, K. J.; Forbes, D. C. Davis, J. H. Jr. Novel Brønsted acidic ionic liquids and their use as dual solvent-catalysts. J. Am. Chem. Soc., 2002, 124, 5962-5963.

[4] Li, D.; Shi, F.; Peng, J.; Guo, S.; Deng, Y. Application of functional ionic liquids possessing two adjacent acid sites for acetalization of aldehydes. J. Org. Chem. 2004, 69, 3582-3585.

[5] Laali, K. K.; Gettwert, V. J. Electrophilic nitration of aromatics in ionic liquid solvents. J. Org. Chem., 2001, 66: 35-40.

[6] Earl, M. J.; Katdare, S. P.; Seddon, K. R. Paradigm confirmed: the first use of ionic liquids to dramatically influence the outcome of chemical reactions. Org. Lett., 2004, 6, 707-710.

[7] Martyn, E. J.; Katdare, S. P. Aromatic nitration reaction. WO 02/30865, 2002.

[8] Smith, K.; Liu, S.; EL-Hiti, G.. A. Regioselective mononitration of simple aromatic compounds under mild conditions in ionic liquids. Ind. Eng. Chem. Res., 2005, 44, 8611-8615.

[9] Lancaster, N. L.; Verónica, L. M. Aromatic nitrations in ionic liquids: the importance of cation choice. Chem. Commun., 2003, 2812-2813.

[10] Qiao, K.; Yokoyama, C. Nitration of aromatic compounds with nitric acid catalyzed by ionic liquids. Chem. Lett., 2004, 33, 808-809.

[11] Rajagopal, R.; Srinivasan, K. V. Ultrasound promoted para-selective nitration of phenols in ionic liquid. Ultrasonics Sonochem., 2003, 10, 41-43.

[12] Freemantle, M. Designer solvents-ionic liquids may boost clean technology development Chem. Eng. News 1998, 76, 32-37.

[13] Freemantle, M. Ionic Liquids Show Promise for Clean Separation Technology. Chem. Eng. News, 1998, 34, 12-13.

[14] Freemantle, M. Buffered ionic liquid boosts catalyst activity. Chem. Eng. News, 1999, 77, 11-12.

[15] Qiao, K., Hagiwara, H., Yokoyama, C. Acidic ionic liquid modified silica gel as novel solid catalysts for esterification and nitration reactions. J. Mol. Catal. A: Chem., 2006, 246: 65-69.

[16] Cheng, G.; Duan, X.; Qi, X.; Lu, C. Nitration of aromatic compounds with NO_2/air catalyzed by sulfonic acid-functionalized ionic liquids. Catal. Commun. 2008, 10, 201-204.

[17] Qi, X.; Cheng, G.;. Lu, C.; Qian, D. Nitration of simple aromatics with NO_2 under air atmosphere in the presence of novel Brønsted acidic ionic liquids. Synth. Commun., 2008, 38, 537-545.

[18] Qi, X.; Cheng, G.;. Lu, C.; Qian, D. Nitration of toluene and chlorobenzene with HNO_3/AC_2O catalyzed by caprolactam-based Brønsted acidic ionic liquids. Cent. Eur. J. Energ. Mater, 2007, 4, 105-113.

[19] Fang, D.; Shi, Q.-R.; Cheng, J.; Gong, K.; Liu, Z.-L. Regioselective mononitration of aromatic compounds using Brønsted acidic ionic liquids as recoverable catalysts, Appl. Catal. A: General, 2008, 345, 158-163.

[20] Nisikin, M.; Nakamura, R.; Niiyama, H. An unusual suppression of trimethlyamine formation in the high temperature alkylation of ammonia with methanol over $[NH_4]_3PW_{12}O_{40}$. Chem. Lett., 1993: 209-212.

[21] Nishimura, T.; Okuhara, T.; Misono, M. High catalytic activities of pseudoliquid phase of dodecatungstophosphoric acid for reaction of polar molecules. Chem. Lett., 1991: 1695-1698.

[22] Sato, H.; Nagai, K.; Yoshioka, H.; et al. Vapor-phase nitration of benzene over silid acid catalysts: nitration with nitric acid; mixed metal oxide treated with sulfuric acid and heteropolyacid partially neutralized. Appl. Catal. A: General, 1998, 175: 209-213.

[23] Wang, E. B.; Hu, C. L.; Xu, L. Introduction to Polyacid Chemistry [M]. Beijing: Chemical Industry Press. 1998.

[24] Cheng, G..; Lu, C.; Peng, X. Regioselective nitration of toluene catalyzed by $H_3PMo_{12}O_{40}$. Chin. J. Energ. Mater, 2004, 12, 110-112.

[25] Schofield, K. Aromatic Nitration. Cambridge University Press, Cambridge, 1980.

（注：此文原载于 Ind. Eng. Chem. Res，2011，50（19）：11440-11444）

第二部分

膨化硝酸铵与膨化硝铵炸药自敏化理论

Research on Expanding Mechanism of Ammonium Nitrate

Lv Chunxu, Liu Zuliang, Ye Zhiwen
Nanjing University of Science & Technology 210094

Abstract: The composition and characteristics of a new type of powder industrial explosive (rock expanded ammonium nitrate (AN) explosive) are briefly described in this paper. Expanding mechanism of AN is proposed. Influence factors and results of expansion of AN with or without surfactants are discussed.

Keywords: Ammonium Nitrate, Ammonium Nitrate explosive, expanding mechanism, surfactant

1 Introduction

Rock expanded AN explosive is a new type of power industrial explosive. It does not contain TNT. Its explosion properties are similar those of No. 2 rock Ammonite and its physical, safety and store properties are better than those of No. 2 rock Ammonite. Its quality has reached to advanced level of international standard of the same kind of product. The technology of expanded AN possesses novel and creative characteristics. New ways of research and application of powder industrial explosive have been opened up. We are the first to invent this new technology at home and abroad.

Because no TNT is contained, cost of new product is decreased and harm to human body and pollution of enviroment are avoided. Remarkable economic and social benefit have been obtained. To a factory with annual output of 8000 tons, the cost of crude material is 3 million yuan less than that of a factory originally producing rock AN-TNT-oil explosive (contained TNT 7%) and 4.68 million yuan less than that of a factory originally producting rock Ammonite explosive(contained TNT 11%). Obviously, range of decreased cost is quite cousiderable[1-3].

The technology of vertical, horizontal and continuous expanding AN is suitable, the equipment is reliable, the design is stable. Practical technic conditions and examine method, which can derectly guide production of expanded AN, have been established[4].

Rock expanded AN explosive and other two kinds of industrial explosives, which are present main industrial explosives in China, are tested through random sampling, Tab. 1, Tab. 2 and Tab. 3 show their comparative results of compositions and property parameters[2].

Because of the above-mentioned remarkable advantages, rock expanded AN explosives have been deeply welcomed by many factories and users. Up till now, more than 30 factories have adopted this

Table 1 Composition of three kinds of powder explosives

Name of Explosives	AN /%	TNT /%	Wood powder /%	Oil /%	Additives
Expanded AN explosive	92.0±2.0	0	4.0±0.5	4.0±0.5	0.120±0.005
No. 2 rock Ammonite	85.0±1.5	11.0±1.0	4.0±0.5		
No. 2 rock Ammonite-oil	87.5±1.5	7.0±0.7	4.0±0.5	1.5±0.3	0.100±0.005

Table 2 Explosion properties of three kinds of powder explosives

		Expanded AN explosive	No. 2 rock Ammonite	No. 2 rock Ammonite-oil
Density	$g \cdot cm^{-3}$	0.85~0.95	0.95~1.10	0.95~1.10
Transimmsion	cm	4	5	4
Brisance	mm	12	12	12
Strength	ml	320	320	320
Detonation velocity	$m \cdot s^{-1}$	3200	3200	3200
Moisture	%	0.30	0.30	0.30
Storage lifetime	month	6	6	6

Table 3 The crude material cost comparision of three kinds of powder explosives (on the basis of prices in 1995)

Crude material	Unit price /yuan	Expanded AN explosive		No. 2 rock Ammonite		No. 2 rock Ammonite-oil	
		Amount /$kg \cdot t^{-1}$	Cost /yuan	Amount /$kg \cdot t^{-1}$	Cost /yuan	Amount /$kg \cdot t^{-1}$	Cost /yuan
AN	2000	920	1840	850	1700	875	1750
TNT	8000			110	880	70	560
Wood powder	300	40	12	40	12	40	12
Oil	3000	40	120			15	45
Additive	15000					1	15
Expanded agent	28800	1.2	35				
Total($yuan \cdot t^{-1}$)		2007		2892		2382	

new technique to produce the industrial explosive, which is one of the major development projects of "95 programme in china", one of the spread items of national significant scientific and technic achievements state science and technology commission. Concrete and further spread works have been started.

The key technique of expanded AN explosive is expansion of AN, the substance of expansion is surface activity technique, in this paper, expanding mechanism of AN is discussed as follows:

2 Expanding mechanism of AN

Through studies of influence of technological conditions and foamy property of surfactants on results of expansion of AN, at present we consider that mechanism of expansion of AN is as follows[5]: under certain temperature and vacuum conditions, water in saturated solution of AN is vaporized, to form water vapor bubble. Under action of surfactants, a large number of relatively stable foams is instantaneously formed. When temperature and vacuum degree reached a specified value, foams are rapidly busted, steam quickly escaped, forming a great quantity of holes and cavities in the crystal of AN. Surface of particles of AN rapidly crystallized is not level and smooth. Linking form between particles of AN is extremely irregular, so light, porous and bulky AN is formed.

3 Study on the expanding mechanism of AN without surfactants

The factors influencing expansion of AN include variety and amount of surfactants and vacuum evaporation crystallization conditions of saturated solution of AN (concentration of solution, temperature and vacuum degree of crystallization) . The successful expansion of AN by vacuum evaporation crystallization without the presence of any surfactant is a basis for us to study expanding mechanism of AN. To study on expansion of AN under action of surfactants is important content of expanding mechanism research. Concentration of solution, temperature, vacuum degree and their orthogonal test have significantly influenced on results of expansion of AN. Influence of initial concentration of solution is shown in Tab. 4[5].

Table 4 Influence of initial concentration on expanding results of AN

Concentration of solution/%	Dissolution time/min	Crystallization Time/min	Volume expansion multiple	Situation of expansion
<85	<15	>20	2	AN hard and dense basically not expanded
85～88	15	10～14	3. 2	a little expanded, hard
88～95	15～27	3～5	5. 0～5. 5	expanded, a little loosen a lot of holes
>95	>28	1～2	4. 3	AN hard, partially expanded

If concentration of solution is low, rapid crystallization can not take place smoothly, formation of irregular crystal of AN is not favorable. Moreover if water wapour is not removed, crystal of well expanded AN will redissolve and "Liquid bridge" between partides of AN will form. After water is continuously vapourated, "salt bridge" will form, which will results in forming of hard dense crystal of AN. If concentration of solution of AN 13 too large, content of water will be low and amount of water vapor will be less, formation of air chink in the crystal of expanded AN will be difficult, so the results of expansion is relatively worse. A lot of experiments indicate that under given conditions of initial concentration, temperature and vacuum degree, volume expansion multiple of expanded AN can reach more than 5. Through observation with Scanning Electron Microanalyzer (SEM) microphotography of micro-structure of common AN and expanded AN without surfactant is shown in Fig. 1[5-8]:

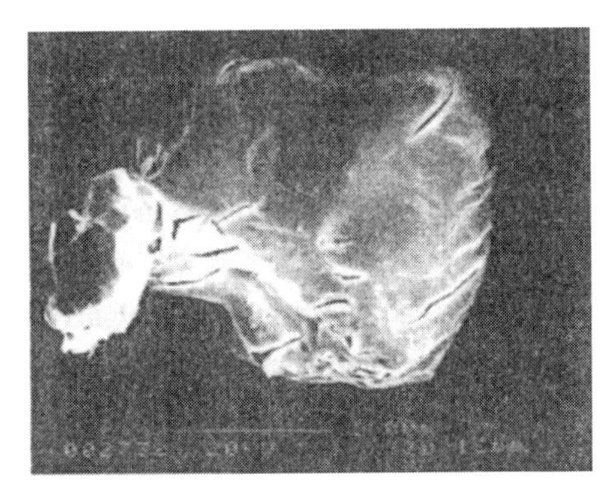

(a) Common AN

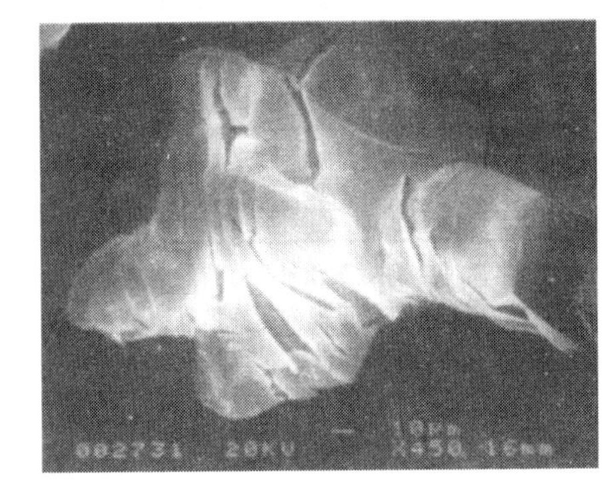

(b) Expanded AN without surfactant

Fig. 1 To compare of micro-structure of common AN and expanded AN without surfactant

According to the photography of SEM we consider that rift of expanded AN without surfactant is more than that of common AN, and expanded AN possesses large "rough degree" and more edge and corner in surface. This can illustrate that expanded AN without surfactant possesses preliminary modified structure of expansion. From this we can say that without surfactant and only adopted vacuum evaporation, expansion of AN have merely been appeared limited effects.

4 Study on the expanding mechanism of AN with surfactant

To make use of foamy properties of surfactant and specified technological conditions of expansion, result of expansion is enormously better than that of without surfactant.

Affect of different surfactants on results of preparing expanded AN is showed in Tab. 5[5]:

Table 5 Affect of different surfactants on results of preparing expanded AN

surfactant	Volume expansion multiple	Accumulative density	Situation of expansion
Nothing	5.5	0.62	Expansion, rift
S1	10.7	0.49	Strong expansion
T1	9.1	0.52	Strong expansion
O1	8.6	0.55	Strong expansion
O1/S11 : 1	11.8	0.42	Extremely strong expansion large and more holes in middle part
O1/T11 : 1	9.5	0.51	Strong expansion
S1/T11 : 1	12.2	0.42	Extremely strong expansion large and more holes in middle part
O1/H11 : 1	8.5	0.54	Strong expansion

SEM microphotography of micro-structure of expanded AN with surfactant is shown in Fig. 2[5-8]:

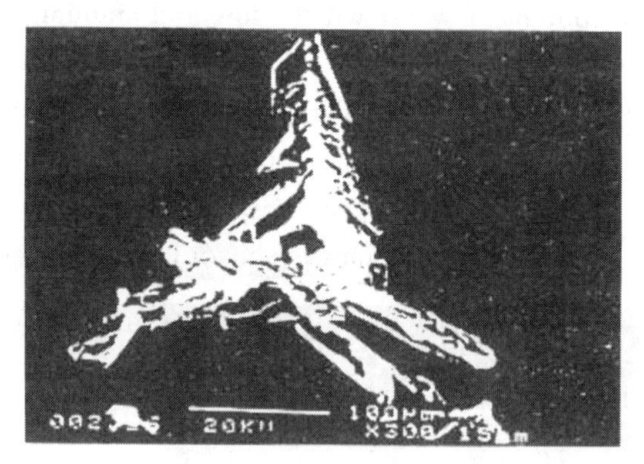

Fig. 2 Micro-structure of expanded AN with surfactants

Expanding AN with surfactants is extremely irregular crystal of AN with more edge, corner, large holes and rifts. This indicates that expansion is successful.

The results of research show that: vacuum evaporation may itself affect on expansion of AN at given degree, and foamy property of surfactant will even more strengthen expanding results of AN. The results of experiments show that: results of expansion of AN accord basically with foamy power of surfactant.

The results of research also show that saturated solution of AN with composite surfactants possesses better foamy property, among them surface activity of cationic-anionic surfactant mixture is largest and formability best. The results of experiments also show that composite surfactants in saturated solution of AN possess better mutually synergism. The parameters of interaction of cationic-anionic surfactant in saturated solution of AN is the largest, interactivity strongest. The surfactants possess comprehensive functions of foaming, coating and crystal shape modification during expanding process of AN.

5 Conclusion

(1) Rock expanded AN explosive without TNT possesses excellent explosive and physical property. Quality of product has reached or surpassed the standard of No. 2 rock Ammonite. The new technology possesses remarkable economic and social benefit, which begins to spread and is welcomed by numerous factories and users.

(2) The key technique of rock expanded AN explosive is expansion of AN. The mechanism of expansion of AN is a physical-chemical process. Under given conditions of temperature and vacuum de-

gree in the presence of certain surfactants, saturated solution of AN is foamed, then molecule of water is fast escaped out from bubble and expanded AN id rapidly crystallized.

(3) Expanding process is first depend on technological conditions, such as initial concentration, temperature, vacuum degree, and especially their orthogonal test. Surfactant is determing factor which can ensure successful expansion. The effect of composite surfactants on expanding process is better than that of single surfactant.

(4) The surfactant possesses functions of foaming, coating and crystal shape modification during expanding process of AN. And it makes expanded AN possess better properties of anti-hygroscopic and anti-caking.

Reference

[1] Liu Zuliang, Hui Junming, Lv Chunxu, Wang Yilin. Research on Light, porous and bulky AN and HF series power AN explosives. Explosive Materials, 1991 (5): 5.

[2] Liu Zuliang, Lu Chunxu, Lv Ming, Chen Tianyun. Research and application of expanded AN Proceedings of ISPE. Beijing: 1995, 220.

[3] Hui Junming, Liu Zuliang, Lv Chunxu, Wang Yilin. Powder AN explosive and its preparetion CN 91107051.6, 1991.

[4] Yang Tuong. Investigation of technique of rock expanded AN explosive. Explosive Materials. 1996, (3): 30.

[5] Ye Zhiwen. Expanded Mechanism of AN [Thesis] . Nanjing: Nanjing University of science and technology, 1995.

[6] Lv Chunxu. Research on surface chemistry of expanded AN. Journal of Nanjing Universtry, 1995, (31): 286.

[7] Ding Yun. Improving properties of AN and its explosives using surfactants [Thesis] Nanjing: East China Institute of Technology, 1990.

[8] Chen Tianyun. Research on surface characteristics and application of AN [Thesis], Nanjing: East China Institute of Technology, 1992.

(注：此文原载于 Proceedings of the 23th International Pyrotechnics Seminar, 1997: 483-490)

Research on the Cap Sensitivity of Expanded Ammonium Nitrate

Lv Chunxu, Lu Ming, Liu Zuliang, Ye Zeiwen
(Nanjing University of Science and Technology, Nanjing, 210094, China)
ISTP: 000085212600027

Abstract: In this paper, a new type of industrial powder expanded Ammonium Nitrate (AN) explosive is briefly introduced, its cap sensitivity was studied in detail. The research results showed that under some technical conditions expanded AN will lose its cap sensitivity, so this sensitivity can be controlled. The manufacturing safety can be guaranteed.

Key words: Cap sensitivity, Expanded Ammonium Nitrate, Poweder industrial explosive

1 Introduction

Expanded AN is the most main raw materials of expanded AN explosive.

Expanded AN explosive, which was invented at home and abroad by us, is a new type of powder industrial explosive. It doesn't contain TNT. Its explosion properties are similar to those of No. 2 rock Ammonite and other properties, such as physical, safety, and store properties and so on, are better than those of No. 2 rock Ammonite. Its quality has reached to the advanced level of international standard for the same kind of product.

Because no TNT is contained, the cost of the new product is decreased and the harm to human body and enviromental pollution are avoided. Remarkable economic and social benefits have been obtained. Expanded AN explosive and other two kinds of industrial explosive which are main industrial explosives in China are tested through random sampling. Table 1 shows their comparative results of compositions, performance and the raw material cost[1-4].

Because of the remarkable advantages mentional above, expanded AN explosives have been deeply welcomed by many factories and users. Up to now more than 40 factories have adopted this new technique to produce the powder industsrial explosive. Further spreading of the technique of the technique of this explosive have been started.

The key technique of expanded AN explosives is the expansion of AN. The expanding technology of AN possesses novel and creative characteristics. The expanded AN possesses light, porous and bulky structure.

We are very much interested in the properties of expanded AN, expecially its cap sensitivity which is the embodiment of safety property. In the paper, cap sensitivity and its influeuce are discussed.

Table 1 Comparative results of three kinds of powder explosive (on the basis of prices 1997)

Explosive	Composition				Characteristics				Raw material cost
	AN %	TNT %	Wood powder %	Oil %	Density ρ $g \cdot cm^{-1}$	Detonation velocity V_D $m \cdot s^{-1}$	Strength V_S ml	Brisance H_{Lc} mm	$Yuan \cdot t^{-1}$
Expanded AN explosive	92.0±2.0	0	4.0±0.5	4.0±0.5	0.95~1.00	3200~3500	360~380	14.5~16.0	2007
No. 2 rock Ammonite	85.0±1.5	11.0±1.0	4.0±0.5	0	0.95~1.10	3000~3200	≥320	≥12	2892
No. 2 rock Ammonite-oil	87.5±1.5	7.0±0.7	4.0±0.5	1.5±0.3	0.95~1.05	3000~3300	≥340	≥12	2382

2 Apparatus and materials

Cap sensitivity expresses the sensitiveness of substance to the shock from a standard detonator or blasting cap. So sensitivity of expanded AN to detonator is one of important marks of safety. Especially in the different densities and temperratures, sensitivity of expanded AN to detonator is of great significance for its manufacturing safety.

According to United Nation recommendation on Test and Criteria of Transport of Dangerous Goods, section 36 Test 5(a)[5], the Cap Sensitivity Test of expanded AN was carried out. The substance under test (expanded AN and common AN) was filled into a tube which was made of cardboard with minimum diameter 80mm, minimum length 160mm and a wall thickness of maximum 1.5mm. Initiation was done by a standard detonator inserted coaxially into the top of the expanded AN to a depth equal to its length. Below the tube was the witness plate. The results were assessed by whether the expanded AN was completely detonated. Results of test are shown in following sections[6].

3 Experimental results and discussions

3.1 Results of cap sensitivity test of expanded AN for pure surfactants

Table 2 Sensitivity of expanded AN to detonator (Amount of surfactant 0.1%)

No.	Density $/g \cdot cm^{-3}$	Diameter of particle of expanded AN/μm	Temperature /℃	Result	No.	Density $/g \cdot cm^{-3}$	Diameter of particle of expanded AN/μm	Temperature /℃	Result
1	0.34	400	29	—	9	0.61	300	20	—
2	0.52	400	29	—	10	0.62	300	20	—
3	0.68	400	29	—	11	0.70	300	20	—
4	0.43	400	28	—	12	0.67	300	20	—
5	0.34	400	28	—	13	0.69	200	21	—
6	0.30	400	27	—	14	0.70	200	21	—
7	0.47	400	20	—	15	0.76	200	21	—
8	0.50	300	20	—	16	0.84	200	21	—

Table 3 Sensitivity of expanded AN to detonator (Amount of surfactant 0.15%)

No.	Density/g・cm^{-3}	Diameter of particle of expanded AN/μm	Temperature/℃	Result
1	0.25	400	70	—
2	0.26	400	70	—
3	0.27	400	70	—
4	0.30	400	70	—
5	0.50	300	20	—
6	0.62	300	20	—
7	0.66	300	70	—
8	0.67	300	70	—
9	0.54	300	20	—
10	0.57	300	20	—
11	0.69	300	20	—
12	0.70	300	70	—
13	0.61	300	70	—
14	0.54	300	70	—
15	0.63	200	20	—
16	0.67	200	20	—
17	0.78	200	20	—
18	0.81	200	20	—
19	0.70	200	70	—
20	0.76	200	70	—
21	0.85	200	70	—

The result "sensitivity to detonator" is signed by a "+" and "non-senitivity to detonator" is signed by a "—".

Table 4 Sensitivity of expanded AN to detonator (Amount of surfactant 0.2%)

No.	Density/g・cm^{-3}	Diameter of particle of expanded AN/μm	Temperature/℃	Result
1	0.21	400	45	—
2	0.36	400	45	+
3	0.40	400	45	+
4	0.40	400	45	+
5	0.45	400	45	+
6	0.52	400	45	+
7	0.22	400	35	—
8	0.23	400	35	—
9	0.25	400	35	—
10	0.36	400	35	—
11	0.40	400	35	+
12	0.42	400	35	+
13	0.53	400	35	+
14	0.63	400	35	—
15	0.63	400	35	—
16	0.63	400	35	—
17	0.49	200	25	—
18	0.50	200	25	+
19	0.56	200	25	+
20	0.58	200	25	+
21	0.60	200	25	+
22	0.65	200	25	—
23	0.70	200	25	—

Table 5 Sensitivity of expanded AN to detonator (Amount of complex surfactant 0.2%)

No.	Density /g · cm^{-3}	Temperature /℃	Result	No.	Density /g · cm^{-3}	Temperature /℃	Result
1	0.62	50	—	5	0.63	79	—
2	0.62	55	—	6	0.65	85	+
3	0.63	60	—	7	0.63	90	+
4	0.65	75	—	8	0.65	100	+

4 Conclusions

Through a series of cap sensitivity tests, conclusions are reached as follows:

(1) The amount of surfactant will intensely influence on cap sensitivity of expanded AN. When the amount of surfactant is less than 0.12%, cap sensitivity of expanded AN will approach to 0.

(2) When the density of expanded AN is lower than 0.35g · cm^{-3} or larger than 0.6 g · cm^{-3}, the sensitivity to detonator of the expanded AN will be lost.

(3) Temperature will affect cap sensitivity of expanded AN. When its temperature decreases, its sensitivity to detonator will be decreased.

(4) When the amount of surfactant is less than 0.12%, density is lower than 0.35g · cm^{-3} or larger than 0.6 g · cm^{-3}, temperature is lower than 60℃, it will lose sensitivity to detonator. So we can decrease its sensitivity to detonator, and guarantee its safety in manufacturing process by controlling the technical conditions.

Reference

[1] Liu Zuliang, Hui Junming, Lu Chunxu, Wang Yilin: Research on light, porous and bulky AN and HF series powder AN explosives. Explosive Materials, 1991 (5): 5.

[2] Hui Junming, Liu Zuliang, Lu Chunxu, Wang Yilin: Powder AN explosives and its prepartion CN 91107051.6, 1991.

[3] Yang Tuong: Investigation of technique of rock expanded AN explosive. Explosive Materials, 1996. (3): 30.

[4] Lu Chunxu, Liu Zuliang, Hui Junming, Wang Yilin: Research on rock expanded AN explosive. Explosive Materials, 1997 (1): 5.

[5] United Nations, Recommendations on the transport of dangerous goods, Test and Criteria (Second edition) New York, 1990.

[6] Hu Bingcheng, Liu Zuliang: Study on the safety property of expanded AN and explosives containing it. Explosive Materials, 1997 (6): 13.

(注：此文原载于 Proceedings of the 24th International Pyrotechnics Seminar. California, 1998: 357-364)

Status and Development of Industrial Explosives in China

Lv Chunxu
Nanjing University of Science & Technology 210094
EI: 98124506644

1 Introduction

Prof Lu Chunxu at the Yong AN Chemical Plant, which pro-

Variety, output amounts, new techniques and theory of industrial explosives in China have developed rapidly in recent years. A complete scientific and technical research system has been established. About ten kinds of industrial explosives are now manufacture in our country with annual output amounts to 1.3 million tons.

The China institute of industrial explosive materials (China IIEM) established in april. 1985, is the sole special institute of the explosive materials occupation in china. It also serves as the window and base of NORINCO (north industries corporation) and CEMTA (China explosive materials trade association) in the south of China, it plays an important role in the development of explosive materials in the field of scientific research, on the basis of our research data, the status and development of industrial explosive in china follows.

2 Status of industrial explosive in China

The first factory manufacturing industrial explosives was set up in china in 1949. The annual output amount of industrial explosives has increased yearly, from 70000 tons in 1955 to 1.3 million tons in 1996. Fig. 1 shows the change of output of industrial explosives in China over the years.

2.1 Industrial explosives powder[1-7]

Industrial explosives powder was one of the earliest industrial explosives widely used in China, ammonite is the main industrial explosive in our country, it can be classified into rock explosives, explosives for open-pit operation, and permissible explosives, the main varieties of ammonite include, among these No. 2 rock ammonite was the widest manufactured and used trial explosive in china.

Ammonite containing TNT has obvious disadvantage since TNT requires explosive raw materials and has serious effects on the health of the human body-with the aid of surfactant techniques, non-TNT or Oligo-TNT Ammonium Nitrate (AN) explosives was invented by china IIEM and ChangSha RIMM (ChangSha) research institute of mining and Metallurgy in 1989.

Expand AN explosive, which was developed at home and abroad by China IIEM is a new type of industrial explosive powder. It does not contain TNT. its explosive properties are similar to those of No. 2 rock. ammonite and its physical properties cause it to be safer and better to store than those of No. 2 rock Ammonite. Its quality has reached a higher level of international standards for the same kind of product. Because no TNT is contained, the cost of the new product is decreased and the harm to human and industrial environmental pollution are avoided, remarkable economic and social benefits have been obtained. In addition to expanded AN explosives two other kinds of industrial explosives in china are tested randomly Table 1 shows their comparative compositions, performance and the raw material cost. Because of the remarkable advantages mentioned above expanded AN explosives has been deeply welcomed by many factories and users. Up to now, more than 40 factories have adopted this new technique to produce the industrial explosive powders. The techniques associated with this explosive have been spreading.

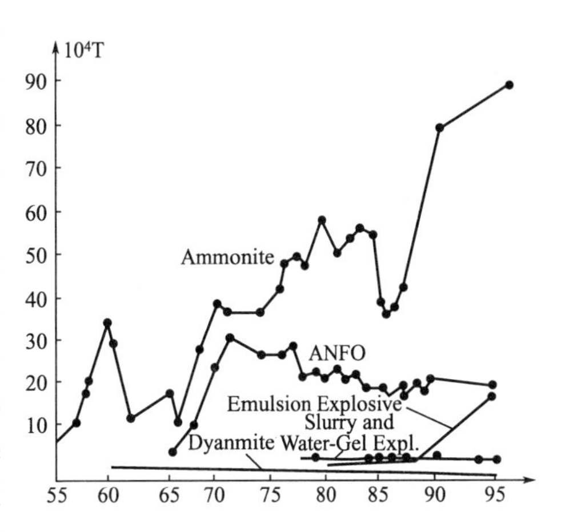

Fig. 1 Output of industrial explosive over the years

The key technique of expanded AN explosive is the expansion of AN. The expanding technology of AN possesses a light potous and bulky structure. By means of scanning electro-Microanalyzer (SEM), the mico photographs of micro-structure of common AN and expanded AN can be researched and compared.

Table 1 Comparative results of three kinds of powder explosives (on the basis of prices in 1995)

Explosives	Composition AN /%	TNT /%	Wood powder /%	Oil /%	Characteristics Density ρ /g·cm⁻¹	Detonation velocity VD /ml	Strength Vs /ml	Brissnce HLe /mm	Raw material cost /Yuan·t⁻¹
Expanded AN explosive	92.0±2.0	0	4.0±0.5	4.0±0.5	0.95~1.00	3200~3500	360~380	14.5~16.0	2007
NO. 2 rock Ammonite	85.0±1.5	11.0±1.0	4.0±0.5	0	0.95±1.10	3000~3200	≥320	≥12	2592
NO. 2 rock Ammonite-oil	87.5±1.5	7.0±0.7	4.0±0.5	1.5±0.3	0.95±1.05	3000~3300	≥340	≥12	2382

ANFO (ammonium Nitrate Fuel Oil) is one of the earliest industrial explosives used in china. Because of conditions of usage and habit of user, annual output amount was only 190. 000 tons in 1996. the main variety of ANFO is porous particle ANFO explosives.

2.2 Dynamite[8,9]

Dynamite has been manufactured since 1955 in china. with the output of 1000 tons in that year. In 1960, output increased to 8000 tons, because its problems of exudation, freezing aging and security, the range of usage has greatly decreased, it has been replaced by other industrial explosive, so at present its yearly output is only about ten thousand tons.

Dynamite include No. 1, No. 2 common dynamite and No. 1, No. 2 antifreezing dynamite. Powder dynamite includes No. 1, No. 2 powder dynamite.

2.3 Slurry explosives

Slurry explosives were the first generation of water-containing explosives in our country. It began to be researched in the fifties and to be applied in metal mines since the seventies. Yearly output was 2000 tons in 1976. and 20000 tons in 1982. But at present, its yearly output has enormously decreased to under ten thousand tons. Typical formulas are slurry No. 5, No. 10, Gel No. 1 etc.

2.4 Water gel explosives

Water Gel Explosives are the second generation water-containing explosives. Research on water gels began in 1975 and have been manufactured successfully since 1979. The production line of Dupont Co. with a yearly output of 6000 tons. Was introduced in 1983 from the United States Up until now, the yearly output has been more than ten thousand tons.

At present, the water gel explosives, manufactured by our country include SHJ (rock, permissible, for open-pit operation) . CS30, M201 and American formula Tovex series etc.

2.5 Emulsion explosives

Emulsions are a new type of water-containing explosive. It began to be developed in the seventies. Different types of emulsion explosives have primarily been researched by Beijing Researched Institute of Mining and Metallurgy (Beijing RIMM) and Fuxin No. 12 Factory since the eighties. From then on, emulsion explosives were quickly developed. Emulsifier theory and emulsion mechanism have been widely and profoundly investigated. Independent in china and integrated systems of manufacturing and research of water-resisting industrial explosives have been established in china. The yearly output is about 185000 tons. According to its usage it can be classified into types for rock, open-pit operation, high strength, permissible etc. Table 2 and Table 3 show formulas and their properties.

Emulsion powder explosives. Which was researched by our China IIEM, is a new type of emulsion explosive. Its characteristics are the same to those of emulsion explosives, but the appearance of the final product is similar to powder. So sometimes it is considered to be industrial explosive powder. The detonation velocity of this product is more than 3400m/s'. The diameter of the charge is 32±1 mm and the bulk is 600 g bag4 or 24kg bag4.

2.6 Special industrial explosives

Special industrial explosives made in china include liquid explosives, heat-resistant explosives and low detonation velocity explosives. Their composition and technology have been widely researched some of them possess obvious advantages and are adopted by a large number of factories and users.

Table 2 Formula of main emulsion explosives. Number represent percentages

44	EI102	EI105	RJ-12	RJ-52	Permissible Explosive Fuxin No. 12 Factory	WR(M) Wuhan IIS	MD(1) Maan shan RIM
	Beijing RIMM		Changsha IMR				
AN	56.5~73.5	42.7~58.5	58.0~61.3	64~67.3	55~65	50~70	60~70
NN	10~12	8~12	10.0	12.0	10~16	10~15	10~20

续表

44	EI102	EI105	RJ-12	RJ-52	Permissible Explosive Fuxin No. 12 Factory	WR(M) Wuhan IIS	MD(1) Maan shan RIM
	Beijing RIMM		Changsha IMR				
MMAN		15～25	12～25				
Urea	1.5～2.5	1.0～2.5	3	3		2～4	
Water	8～12	8～10	10～13	14	8～13	10～13	9～14
Emulsifier	0.5～1.5	0.5～1.5	1.0	1.0～1.5	8～1.2	0.8～2.0	1～2.5
Stablizer	0.1～0.3	0.1～0.3	0.5～1.0	0.3～0.5	0.8～1.2	0.8～2.0	1～2.5
Oil	1～2	0.5～1.0	1.0	1.25	3～5	1～2	0.5～2
Wax	2～3	2.5～3.5	3.0	3.75		2～3	2～4
At powder	1～3	0.5～1.0			1～5		0～1
Flame-deepressant					5～10		

Table 3 properties of main emulsion explosives

Properties	EI Beijing RIMM	RJ Chansha IMR	Permissible Fuxin No. 12 Factory	WR Wuhan IIS	MD Maan Shan RIM
Density	1.05～1.35	1.0～1.30	1.0～1.25	1.0～1.20	1.0～1.25
Detonation Velocity	4000～4700	4500～5400	3891	4700～5800	4500～5100
Gap Distance	11～15	7～8	6～8	5～10	7～8
Critical Diameter	16～20	13	20～25	12～16	20
Brisance	15～17	16～128	12～17	18～20	16～12
Storage Life/month	5	3	3～4	3	3

2.7 Development of industrial explosives in China

Future research on industrial Explosives will focus on and include following:

1) Research on new types and series of industrial explosives to satisfy the requirements of domestic and abroad markets.

2) Research on automation and consecutiveness in the manufacture of industrial explosives, to raise the technical level of our country up to advanced international standards.

3) Research on the basic theory of industrial explosives, to enhance the whole level of the explosives industry.

4) Research on novel techniques of analysis and testing, to improve the quality of industrial explosives and to promote scientific research to a new level.

Professor Chunxu Lu is the Director of the China Institute of Industrial Explosive Materials and Vice president of Nanjing University of science and Technology in the people's Republic of China.

References

[1] Liu Zuliang, Hui Junming, Lü Chunxu, Wan Yilin. Research on Light Porous and Bulky AN and HF Serials Powder AN Exoplosive. Exlosive Materials, 1991, 20 (5): 5-8.

[2] Lü Chunxu. Research on Surface Chemistry of Expand AN. Journal of Nanjing University, 1995 (31): 286.

[3] Liu Zuliang, Lü Chunxu, Lu ming, Chen Tianyun. Research and Applocation of Expanded AN. Proceedings of ISPE. Beijing: 1995, 220.

[4] Hui Junming, Liu Zuliang, Lü Chunxu, Lü Chunxu. Powder AN Explosive and its Preparation. Cn91107051. 6, 1991.

[5] Ye zhiwen. Mechanism of Expansion of AN. [thesis], Nanjing University of Science and Technology, 1995.

[6] Din Yun. Modification of AN with Surfactant. [thesis], Nanjing University of Science and Technology, 1900.

[7] Chen Tianyun. Surface characteristics of AN. [thesis], East China Institute of Technology, 1995.
[8] Lv Chunxu, Liu Zuliang and Ni Ouqi. Industrial Explosives. Beijing: 1994.
[9] Jiao Shuyan, Forecast of Industrial Explosives in China in 2000. Explosive Materials. 1986 (3): 30.
[10] Wei Yunyang. History and Forecast of Industrial Explosives. Explosive Materials. 1990 (2): 34.
[11] Zhang Guoshun. Development and Forecast of Industrial Explosives. Proceedings of Symposium on China industrial Explosive Materials Association. Nanjing: 1988.
[12] Fang Xiouxia. Production Situation and Analysis of industrial Explosive Materials in 1994. Explosive Materials. 1995 (3): 36.

(注：此文原载于 J. of Explosive Engineering，1998，15 (6)：36-38)

Research and Development of Powder Industrial Explosives in China

Lv Chunxu

China Institute of Industrial Explosive Materials, Nanjing, 210094 (P. R. China)

EI: 99064703937

Forschung und entwicklung von pulverförmigen industrie sprengstoffen

In dieser Arbeit wird der Stand pulverformiger Industriesprengstoffe in China beschrienben. Basierend auf wissenschaftlichen Forschungsarbeiten des China-Instituts Industrial Explosive Materials (China IIEM) werden Ammoniumnitrat-Sprengstoffe, besonders Geste inssprengstoffe mit voluminösem (expanded) Ammoniumnitrat, ohne und mit geringen Mengen an TNT ausführlich diskutiert und unter verschiedenen Gesichtspunkten die weitere Entwicklung der pulverförmigen Industriesperngstoffe in China dargestellt.

Recherche et développement d'explosifs industriels pulvérulents

La présent étude décrit l'état de l'art des explosifs industriels pulvérulents en Chine. Sur la base de travaux de recherches scirntifiques de l'institut de Chine Industrial Explosive Materials (China IIEM), on a discuté en détail des explosifs nitrate d ammonium, en particulier des explosifs rocheux *à* nitrate d'ammonium volumineux (expanded), sans ou avec faible quantitié de TNT et présenté sous différents aspects le developpment dés explosifs industriels pulvérulents en Chine.

Summary

In this paper, status of powder industrial explosives in China is described. On the basis of scientific research of China Institute Explosive Materials (China IIEM) non-TNT or oligo-TNT Ammonium Nitrate explosive, especially rock expanded Ammonium Nitrate explosive, is emphatically discussde and several points of view about further development of powder industrial explosives in our country are stated.

1 Introduction

Powder industrial explosive is present main industrial explosive in China. About several kinds of industrial explosives are now manufactured in our country and annual output amounts to 1. 1 million tons. A complete scientific and technical research system has been established. Variety, technology, theory and application of powder industrial explosives in China have quickly been developed recently. New products, new techniques, new technology and new equipment are unceasingly emerging. But, with the deepening of economic reform and open-door policy, systematization of structure variety. Uniformity of product properties, especially level of production technology and controlment of production process show the state of backward. We must make still great efforts to raise the whole level of our country powder industrial explosives.

2 Research of powder industrial explosives in China

The manufacture of powder industrial explosives has begun since 1949 in China. The output of powder industrial explosives has been increased by a factor 50 over the 40 years, from 20000 tons in 1953 to 1.1 million in 1996.

The main powder industrial explosives in our country are as follows[1-5].

2.1 Ammonite: ammonium nitrate (AN) explosives with TNT

Ammonite has been the main industrial explosive in our country. It possesses good properties in explosion, initiation and stability. It can be classified into rock explosives, explosives for open-pit operation, permissible explosives. Main varieties of ammonite include No. 1, No. 2 and No. 3 rock ammonite, No. 4 water-resisting rock ammonite. No. 1, No. 2 and No. 3 permitted explosives, safe sheathed explosives and ion exchange type safety explosives. Among them, No. 2 rock ammonite is the widest applicated kind of industrial explosives in China.

2.2 Oligo-TNT or non-TNT ammonium nitrate explosive

Ammonium nitrate explosives with TNT possess remarkable disadvantages so that TNT seriously affects the health of human body. With the aid of disperse and surfactants techniques, oligo-TNT or non-TNT AN explosives have been invented by Changsha Research Institute of Mining and Metallurgy (Changsha RIMM) and China IIEM in 1989[6-8]. Tables 1 and 2 show composition and main properties of oligo-TNT AN explosives.

Table 1 Composition of Main Oligo-TNT Explosives in China

Composition (%)	No. 2 Rock Ammonite	No. 2 Rock New Ammonite	RF Explosive Chang Sha RIMM	HF Explosive China IIEM
AN	85.0	88.0	88.0	90.5
TNT	11.0	7.0	5.5	2.5
Wood powder	4.0	3.0	4.4	3.5
Complex oil		2.0	2.1	3.5

Non-TNT AN explosive, invented by China IIEM, is known as expended ammonium nitrate explosive. It is a new type of non-TNT explosive.

Expanded ammonium nitrate is prepared through vacuum crystallization technology with the aid of surfactants from common AN. It is light, porous and bulky. It possesses very large specific surface area; specific surface area of common AN and expanded AN being 471.95$cm^2 \cdot cm^{-3}$ (or 258.76$cm^2 \cdot g^{-1}$), 1454.57$cm^2 \cdot cm^{-3}$ (or 3328.54$cm^2 \cdot g^{-1}$), respectively. Therefore the specific surface area is increased by a factor of 4 after vacuum crystallization. There are many micro-holes in the particles of expanded AN. So its apparent density is very low. When it is mixed with wood powder and fuel oil. contact areas of components will increase obviously.

The composition and main properties of non-TNT expanded AN explosive are shown in Tables 3 and 4.

Table 2 Properties of Main Oligo-TNT Explosives in China

Properties	No. 2 Rock Ammonite	No. 2 Rock New Ammonite	RF Explosive Chang Sha RIMM	HF Explosive China IIEM
$\rho(g \cdot cm^{-3})$	0.95~1.10	0.95~1.05	0.95~1.10	0.95~1.10
$H_{Lc}(mm)$	12	12	14	13
$V_D(m \cdot s^{-1})$	3200	3000~3200	3200	3300
$S_D(cm)$	5	3	6	5
$V_S(ml)$	320	320	350	320

Table 3 Composition and Properties of Expanded AN Explosive

Composition			Main properties				
AN	Wood powder	Complex oil	$\rho(g \cdot cm^3)$	$H_{Lc}(mm)$	$V_D(m \cdot s^{-1})$	$S_D(cm)$	$V_S(ml)$
92.0	4.0	4.0	0.85~0.95	12	3200	4	320

Table 4 The crude Material Cost Comparison of Three Kinds of Powder Explosives (on the basis of prices in 1995)

Crude Material	Unit Price (Yuan)	Expanded AN explosive		No. 2 rock Ammonite		No. 2 rock New Ammonite	
		Amount $(kg \cdot t^{-1})$	Cost (Yuan)	Amount $(kg \cdot t^{-1})$	Cost (Yuan)	Amount $(kg \cdot t^{-1})$	Cost (Yuan)
AN	2000	920	1840	850	1700	875	1950
TNT	8000			110	880	70	560
Wood Powder	300	40	12	40	12	40	12
Oil	3000	40	120			15	45
additive	15000					1	15
Expanded agent	28800	1.2	3.5				
Total$(Yuan \cdot t^{-1})$		2007		2592		2382	

Such expanded AN explosive, which can get rid of TNT, is still of suitable sensitivity to initiation, violent explosion properties, well thermal stability. Technology of expanded AN possesses creativeness and brings up new way for research and application of powder industrial explosives. The technology has passed technical evaluation by Ministry and has been adopted by many factories.

Micro-structure of crystal of expanded AN, surface chemistry of expansion of AN and expansion mechanism have been deeply studied by IIEM[9-12].

Changsha RIMM has researched oligo-TNT and non-TNT ammonium nitrate explosives (RF explosive) through technology of dispersion and got successful results, the technology has been certificated by Ministry and adopted by factories, too.

2.3 Ammonium nitrate fuel oil mixture (ANFO)

ANFO is one of the industrial explosives used in China early, with the output of 20000 tons in 1965300000 tons in 1972200000 tons in 1990 and 190000 tons in 1996.

Main varieties of ANFO are No1, No2 and No 3 ANFO and porous particle ANFO explosive.

Other varieties of ANFO are AN-rosin-wax explosive and AN-asphalt-wax explosives which are manufactured and used locally.

2.4 Powder emulsion explosive

Powder emulsion explosive is a new type of emulsion explosive in China. It is a breakthrough of traditional ideas of emulsion explosive. Appearance of final product is not emulsion and similar to powder explosive. But the characteristics of close and full contact between oxidizer and combustible material in system of emulsion explosives is not changed, and its physical and explosive properties are obviously improved. The concrete properties parameters are shown in Table 5.

Table 5 Main Properties of Powder emulsion explosive

Density ($g \cdot cm^{-3}$)	Moisture (%)	Transmission (cm)	Detonation velocity ($m \cdot s^{-1}$)	Brisance (mm)	Strength (ml)	Store lifetime (month)
1.85～1.05	≤5%	>5	>3400	>13	>320	>6

Formula, technology and properties of powder emulsion explosive were thoroughly researched by China IIEM and Beijing RIMM. Valuable achievements are obtained. They regard formulas containing 3～5% water in powder emulsion explosive to be the best, select polymolecular emulsifier to enhance ability of emulsion in high temperature, and adopt technology of "spraying" and "air current" for forming to powder.

3 Development of powder industrial explosives in China

(1) Non-TNT powder expanded AN explosive is one of the major development projects of "95 programme in China" and one of the spread items of national significant scientific and technic achievements by State Science and Technology Commission. Non-TNT powder industrial explosive is focal of development in China.

(2) New type and series of powder industrial explosive will be researched, to satisfy requirement of domestic and abroad market.

(3) Basic theory of powder industrial explosive will be researched, to enhance the whole level of powder industrial explosive.

(4) Automation and consecutiveness of manufacture of powder industrial explosives will be researched, to facilitate technical level of our country up into international advance standards.

References

[1] Lv Chunxu, Liu Zuliang and Ni Ouqi. Industrial Explosives. Beijing: 1994.

[2] Jiao Shuyan, Forecast of Industrial Explosives in China in 2000. Explosive Materials. 1986 (3): 30.

[3] Wei Yunyang. History and Forecast of Industrial Explosives. Explosive Materials. 1990 (2): 34.

[4] Zhang Guoshun. Development and Forecast of Industrial Explosives. Proceedings of Symposium on China industrial Explosive Materials Association. Nanjing: 1988.

[5] Fang Xiouxia. Production Situation and Analysis of industrial Explosive Materials in 1994. Explosive Materials. 1995 (3): 36.

[6] Hui Junming, Liu Zuliang and Lv Chunxu. Powder Ammonium Nitrate Explosive and its Manufacture. Chinese Patent. 911070516, 1991.

[7] Hong Youqiu. Summary of Development of Modern Industrial Explosives. Explosion and Shock. 1982 (4): 75.

[8] Hong Youqiu and Huang Youxin. Manufacture and Technology of Novel Type of Powder Ammonium Nitrate Explosive. Chinese Patent. 85100399, 1986.

[9] Lv Chunxu. Research on Surface Chemistry Characteristics of Crystallization of AN. Explosive Materials. 1995: 6.

[10] Lv Chunxu. Research on Surface Chemistry Characteristics of Expanded AN. Journal of Nanjing University. 1995 (31): 286.

[11] Liu Zuliang and Lv Chunxu. Research and Application of Expanded Ammonium Nitrate. The Third ISPE. China Ordnance Society. Beijing. 1995: 220.

[12] Lv Chunxu. Research on Expanded AN Explosive. Explosive Materials. 1996 (1): 5.

(注：此文原载于 Propellants, Explosives, Pyrotechnics, 1999, 24 (1): 27-29)

硝酸铵膨化技术及应用

吕春绪

（中国兵器工业民用爆破器材研究所，南京 210094）

摘要： 硝酸铵膨化技术是一创新技术，创新设计的指导思想是硝酸铵自敏化，硝酸铵自敏化的提出是对国内外传统方法的突破。实施自敏化的技术途径是硝酸铵的膨化，其实质是表面活性技术在粉状炸药中的应用，是一个强制析晶的物理化学过程。文章重点讨论了硝酸铵膨化机理及膨化硝酸铵的技术特征，显示其独特的优点。硝酸铵膨化技术主要应用是岩石膨化硝铵炸药，给出了岩石膨化硝铵炸药的爆炸与物理特征数据，并与其它工业炸药做了比较。同时也推广应用在煤矿许用型炸药中。

关键词： 膨化硝酸铵，岩石，炸药，微结构，机理

1　硝酸铵膨化创新设计与技术途径

1.1　研制背景

尽管含水乳化炸药、重 ANFO、ANFO 在世界范围内获得了重大进展，市场占有份额越来越大，但是以 TNT 或 NG 敏化的粉状硝铵炸药仍占有一定的比例，这类炸药存在以下缺点：（1）含有较多的 TNT，毒性较大，易造成对职工身体的毒害和对环境的污染；（2）产品成本较高；（3）生产安全性较差；（4）产品的物理性能不良，吸湿结块严重。

一百多年来，国内外的许多研究者一直致力于此问题的研究，先后提出添加掺粉剂和晶型改变剂、包覆表面活性剂和憎水剂、改进装药和包装等方法来改善硝酸铵和炸药的物理性能，采用黑索今、硝酸脲和金属粉等物质试图取代梯恩梯。尽管这些途径有的是有效果的，但却是片面的和不经济的，没有从根本上解决问题。也就是说，这些问题长期以来急需解决，而又未能解决。

1.2　基本设计思想

经过对国内外粉状工业炸药技术分析和多种方案探索以后，我们认为：只有硝酸铵自身敏化才能使粉状硝铵炸药发生质的突破，才是实现粉状硝铵炸药无梯化的根本途径。其理论依据是：硝铵炸药的爆轰是一种非理想爆轰。在外界能量作用下，首先在炸药某些不均匀的局部形成能量集中的“热点”，“热点”内炸药快速分解放出热量，再引起周围炸药反应，直至形成整个体系的爆炸。“热点”可以由体系内的“气泡”绝热压缩，或颗粒棱角之间的剧烈摩擦和碰撞形成。这就启示我们，设法在硝酸铵颗粒中引入“微气泡”，同时使颗粒“畸形”化、“粗糙”化，当受到外界强烈激发作用时，这些不均匀的局部就可能形成高温高压的“热点”，进而发展成为爆炸，达到自敏化目的。

1.3　自敏化理论的提出

炸药爆炸理论中，“热点”起爆机理是一致公认和极其重要的。它不但较好地解释了炸药在极少

的外界能量作用下能够爆炸的原因，而且指导了调整炸药安全性的途径、指导了液体炸药、乳化炸药等粘塑性、粘流性爆炸体系的感度设计。

采用特殊技术途径将“微气泡”引入硝酸铵晶粒中，当起爆能量作用时，这些“微气泡”瞬间受到绝热压缩形成高温高压的“热点”；同时“畸形”化颗粒的棱角之间剧烈“摩擦”和“碰撞”，也导致“热点”形成。正是这种晶粒内部自敏化的作用取代了敏化剂 TNT 的敏化作用，使硝酸铵及其炸药的起爆感度发生了质的变化。

1.4　自敏化实施技术途径是硝酸铵的膨化

要使硝酸钹粒子内含有“微气泡”和形成“歧形”颗粒只能使硝酸铵在高度“沸腾”状态下快速析晶。在深入研究表面活性技术基础上，经反复试验合成出一种新型复合表面活性剂，能有效降低硝酸铵溶液及固体的表面张力，在减压高温工艺条件下，可在硝酸铵晶粒内部形成大量的“孔、洞、隙”，在外部形成极不规则的“活性表面”——这就是膨化硝酸铵。实践表明，硝酸铰自敏化的设计思想和控制析晶的技术途径是正确、有效的。

无梯膨化硝铵炸药的关键技术是硝酸铵的膨化，膨化实质是表面活性技术，是一个包含了在复合表面活性剂作用下强制发泡析晶的物理化学过程。

硝酸铵改性国内外一直是走包覆、外加组分、分散乳化等途径，笔者冲出这个范围走了一条崭新的路，从而使硝酸铵自敏化获得成功。

2　硝酸铵膨化机理

通过对工艺条件及表面活性剂起泡性对硝酸铵膨化效果影响的研究，目前认为硝酸铵膨化机理可能是[1-4]：在一定的温度及真空度作用下，硝酸铵饱和溶液中的水分汽化，生成气泡，在表面活性剂的作用下形成大量相对稳定的泡沫。但当温度、真空度达到特定值时，泡沫破裂，水蒸气快速蒸发，在硝酸铵晶体内形成大量“孔、洞、隙”，快速晶析出的硝酸铵粒子表面也极不规整，造成硝酸铵粒子之间也以极不规则的方式连接，形成一种轻质、多孔、疏松的膨化硝酸铵。

研究表明，减压蒸发本身对硝酸铵膨化有一定的作用，而表面活性剂的起泡性更增强了硝酸铵膨化效果。实验表明，硝酸铵膨化效果同表面活性剂起泡性的好坏基本对应。

研究还表明，复配表面活性剂对硝酸铵饱和溶液具有较好的起泡性能，其中，以复配两性表面活性剂体系的表面活性最高，起泡性最好。实验结果还表明，复配表面活性剂在硝酸铵饱和溶液中具有较好的相互增效作用，其中复配两性表面活性剂在硝酸铵饱和溶液中相互作用参数最大，作用最强。表面活性剂在硝酸铵膨化改性处理过程中，主要起了发泡剂、包覆剂和晶形改变剂的作用。

3　膨化硝酸铵的技术特征

通过表面活性剂处理而获得的膨化硝酸铵具有蜂窝状的多孔结构，用电镜显微观察的膨化硝铵的微结构如图 1 所示[4,5]。

与普通粉状、粒状及多孔粒状硝酸铵相比较，膨化硝酸铵具有以下特征：

(1) 具有相当大的比表面。用阻力降法测定膨化硝酸铵比表面为 3328.54 $cm^2 \cdot g^{-1}$ 与其相比较，相同测试条件下普通硝酸铵是 758.76 $cm^2 \cdot g^{-1}$，比表面增大近 4.4 倍。

(2) 在颗粒内部有相当多的微气孔。从图中可以非常清楚地看出，膨化硝酸钹内部孔隙多、裂纹多，在表面多毛刺、多棱角。这是膨化硝酸铵自身气泡敏化，可以不加 TNT 的物质基础。

(3) 吸湿结块性有所降低[6]。20℃，相对湿度 93%时测得的吸湿增量数据如表 1。

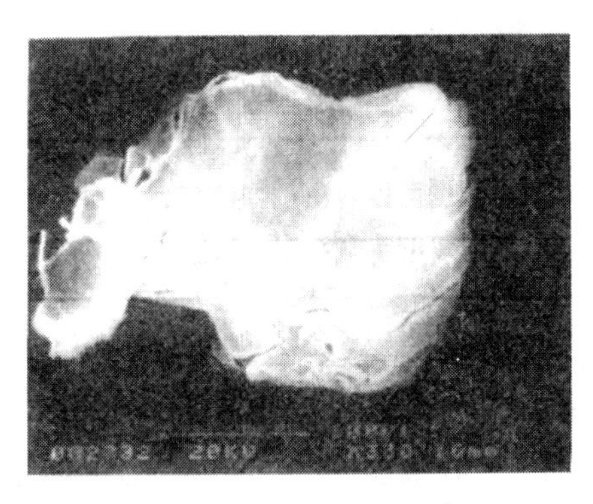
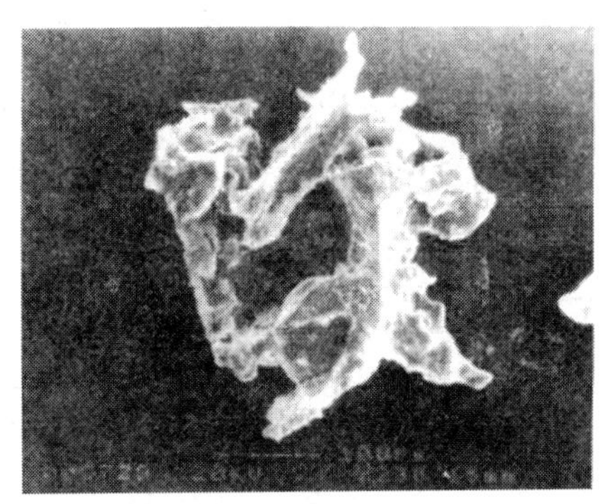

图 1　硝酸铵的微结构图

普通硝酸铵（放大 900 倍）；膨化硝酸铵（放大 850 倍）

Fig. 1 Micro-structure of Ammonium Nitrate（AN）

Common AN（900 times enlarged）; Expanded AN（850 times enlarged）

表 1　硝酸铵的吸湿增量（%）

Table 1　Hygrocopic increment of Ammonium Nitrate（%）

吸湿时间/h	2	4	6	8	10
纯硝酸铵	0.5832	1.1084	1.5792	2.0660	2.5370
膨化硝酸铵	0.4258	0.7697	1.1212	1.4829	1.8746
2 号岩石梯炸药	0.5569	1.1100	1.5456	2.0001	2.4811
岩石膨化硝胺炸药	0.3958	0.6432	0.9397	1.3025	1.5904

从表 1 数据可以看出，膨化硝酸铵平均吸湿速度是普通硝酸铵的 1/2 左右。膨化硝酸铵与普通硝酸铵在高温 80℃放置 3h，再冷却到室温，反复实验 5 次后测定其所能承受的最大压力，具体数据如表 2 所示。

表 2　循环实验后药柱所能承受的压力

Table 2　The pressure undergone by charge through cycling test

装药	所承受的最大压力/Mpa				所承受的平均压力/Mpa
普通硝酸铵	8.06	8.11	7.99	8.02	8.06
膨化硝酸铵	0.81	0.81	0.80	0.83	0.813

从表 2 数据可以看出，膨化硝酸铵承受的最大压力是普通硝酸铵的 1/8～1/9，所以我们认为膨化硝酸铵有非常好的抗结块能力。

(4) 具有质脆性，极易加工成微细粉末。刚制得的膨化硝酸铵，随着温度的降低而变得又脆又易破碎，这样直接进轮碾机只要轻轻碾压 20min，90%的膨化硝酸铵粒子将小于 250μm，较好地满足硝酸铵细度要求。

4　膨化硝酸铵技术的应用

4.1　岩石膨化硝铵炸药

0.15%表面活性剂 LC-2 被溶解在普通硝酸铵中以形成溶液，该溶液蒸发、析晶的同时，就是一个膨化过程，而得到膨化硝酸铵。在 60℃下，92%石膨化硝酸铵、4%木粉和 4%复合油相被加入到轮碾机中，混合 20℃，便得到岩石膨化硝铵炸药[7]。

它与 2 号岩石铵梯炸药以及 2 号岩石铵梯油炸药的配方及性能参数数据如表 3、表 4 所示。

表 3 三种粉状工业炸药的配方

Table 3 Composition of three kinds of powder explosive

炸药名称	ω(硝酸铵)/%	ω(梯恩梯)/%	ω(木粉)/%	ω(油相)/%	ω(外加添加剂)/%
2 号岩石铵梯炸药	85±1.5	11.0±1.0	4.0±0.5	—	—
2 号岩石铵梯油炸药	87.5±1.5	7.0±0.7	4.0±0.5	1.5±0.3	0.100±0.005
岩石膨化硝胺炸药	92.0±2.0	—	4.0±0.5	4.0±0.5	0.120±0.005

表 4 三种粉状工业炸药的爆炸性能

Table 4 Explosion prooerties of three kinds of powder explosives

项目	2 号岩石铵梯炸药	2 号岩石铵梯油炸药	岩石膨化硝铵炸药
药卷密度/$(g\cdot cm^{-2})$	0.95～1.10	0.95～1.05	0.85～1.00
殉爆距离/cm	≥4	≥4	≥5
猛度/mm	≥12	≥12	14.5～16.0
作功能力/ml	≥320	≥320	330～380
爆速/$(m\cdot s^{-1})$	≥3200	≥3300	3200～3500
ω(水分)/%	≤0.30	≤0.30	≤0.30
炸药有效期/a	0.5	0.5	0.5

从表 4 明显看出，岩石膨化硝铵炸药的爆炸特性数据达到和超过 2 号岩石铵梯炸药和 2 号岩石铵梯油炸药性能数据指标。

岩石膨化硝铵炸药另一个重要特色是它具有非常显著的经济效益。因为去除了梯恩梯，原材料成降低 300～500 元/t，产品的利润大幅度增加。此外，由于不使用梯恩梯，还省去了梯恩梯球磨工序的设备厂房投资、废水处理站的投资，生产人员及动力消耗等。

岩石膨化硝铵炸药，去除了对人体有害成分梯恩梯，减少了对环境的污染，具有显著的社会效益。

4.2 煤矿许用膨化硝铵炸药

煤矿许用膨化硝铵炸药是膨化硝铵炸药系列化产品，配方为膨化硝酸铵 79%、油 3%、食盐 15%、木粉 3%。

目前该项技术已有徐州矿务局化工厂等煤矿炸药厂进行了试生产，该项技术受到煤炭工业局的重视，已被列为“煤炭行业‘九五’重点推广计划项目”。

4.3 膨化硝铵震源药柱

研究表明，可以用膨化硝酸铵代替震源药柱中的普通硝酸铵。在不改变原有爆炸性能及其它性能的前提下，膨化硝铵震源药柱的成本大幅度降低，提高生产安全性，减轻环境污染，具有显著的经济效益和社会效益。

膨化硝铵震源药柱已在湖北襄沙化工厂建成 5000t 生产线，并已生产近 4000t 产品，经过工业试验和用户使用证明，性能优良，稳定可靠。

5 结论

(1) 根据硝酸铵自敏化总体设计思想，采用表面活性技术，在复合表面活性剂作用下制得膨化硝酸铵，它具有多微孔、大比表面和不易吸湿、结块等特性；膨化技术具有新颖性、创造性，为粉状工业炸药研究和应用开辟了新途径，属国内外首创。

由膨化硝铵制得岩石膨化硝铵炸药，尽管不含梯恩梯，却具有良好的爆炸与物理特性，达到国际先进水平。

（2）岩石膨化硝铵炸药具有原材料来源丰富，成本低廉且生产过程简单等特点，特别是由于成本降低产生较大的经济效益及由于去除毒性梯恩梯带来显著的社会效益。

（3）岩石膨化硝铵炸药，已被60多个工厂所采用。全国粉状铵梯炸药年生产量约850kt，有213个生产企业，膨化硝铵炸药技术将具有非常广阔的推广应用前景。

参考文献

[1] 吕春绪．膨化硝酸铵表面化学研究［J］．南京大学学报，1995（31）：286.
[2] 丁芸．表面活性剂对硝酸铵改性的影响［D］．南京：华东工学院，1990.
[3] 惠君明，刘祖亮，吕春绪．粉状硝胺炸药及其制法［P］．DN91107051. 1991-06.
[4] 叶志文．硝酸铵膨化机理研究［D］．南京：南京理工大学，1995.
[5] Liu Zuliang，Lü Chunxu，Lu ming，Chen Tianyun. Research and application of expanded AN［A］. Proceedings of ISPE［C］. Beijing，1995：220.
[6] 陈天云．硝酸铵表面特性研究［D］．南京：华东工学院，1992.
[7] 刘祖亮，惠君明，吕春绪，王林．轻质硝酸铵及HF型粉状硝酸铵炸药［J］．爆破器材，1991（5）：5.
[8] 吕春绪，刘祖亮，倪欧琪．工业炸药［M］．北京：兵器工业出版社，1994.

Expanded Technology of Ammonium Nitrate and its Application

Lv Chunxu
(Institute of Industrial Explosive Materials，China Ordnance Industry，Nanjing，210094，China)

Abstract: The expansion of Ammonium Nitrate (AN) is novel technology. The guideline of this innovation is the self-sensitization of AN. It is a breakthrough form classic method. The approach to self-sensitization is the expansion of AN. Its essence is a surface active technology applied to powder industrial explosive，and it is a physical chemistry process under coercive crystallization.

In this paper，The mechanism and technical characteristics of expansion are discussed and its unique advantage is shown. The expanding technology of AN is mainly applied to rock expanded AN explosive and its comparison with other industry explosives is given. The expanding technology is also used in manufacturing of permitted explosive for coal mine.

Key words: expanded ammonium nitrate，rock，explosive，micro structure，mechanism

（注：此文原载于　中国工程科学，2000，2（2）：60-63）

膨化硝酸铵自敏化理论研究

吕春绪
（国防科技工业民用爆破器材研究所，南京 210094）

摘要：膨化硝酸铵自敏化理论是膨化技术发明的核心，它是基于热点理论把微气泡植入硝酸铵炸药中。微气泡临界热点温度热力学计算为自敏化理论提供理论依据。与普通硝酸铵及珍珠岩相比较，微气泡在膨化硝酸铵内适当的分布为自敏化成功提供了有力的保证。

关键词：膨化硝酸铵，自敏化，微气泡

1 膨化硝酸铵自敏化理论设计

利用乳胶炸药敏化气泡的热点作用，提高其起爆感度及冲击波感度的手段是成功而有效的。这些对开拓粉状炸药敏化方法起到重大启发及促进作用。

能否把气泡引进粉状炸药中，显然要把气泡植入木粉或油相中是难以实现的，但可以在硝酸铵晶体内产生气泡并连同晶体歧形化使其成为热点的载体，这种构思就是硝酸铵自敏化设计的雏形。

硝铵炸药的爆轰时一种非理想爆轰。在外界能量作用下，首先在炸药某些不均匀的局部形成能量集中的热点，热点内炸药快速分解释放热量，再引起周围炸药反应，直至形成整个体系的爆炸。热点可以由体系内的微气泡绝热压缩，或颗粒棱角之间的剧烈摩擦和碰撞形成。这就启示我们，设法向硝酸铵颗粒中植入微气泡，同时使颗粒歧形化、粗糙化，当受到外界强烈激发作用时，这些不均匀的局部就可能形成高温高压的热点，进而发展成为爆炸。

经过对国内外粉状工业炸药技术分析和多种方案探索以后，大量实践表明：使硝酸铵自身敏化是实现粉状硝酸炸药五梯化的途径之一。

正是这种晶粒内部自敏化作用取代了敏化剂 TNT 的敏化作用，使膨化硝酸铵及其炸药的起爆感度及冲击波感度发生了较大的突变[1-5]。

2 膨化硝酸铵自敏化临界热点温度的热力学计算

膨化硝酸铵自敏化主要是依靠其内含的大量微气泡，这些微气泡按热力学计算可能达到的临界温度是非常重要的，也是人们十分关心的问题[6]。

由热力学公式

$$dS = C_P \frac{dT}{T} - (C_P - C_0)\frac{dP}{P}$$

绝热过程中 dS=0，

则

$$C_P \frac{dT}{T} = (C_P - C_0)\frac{dP}{P}$$

$$\frac{C_P}{C_0}=r$$

则

$$\frac{dT}{T}=\frac{r-1}{r}\times\frac{dP}{P}$$

积分得：

$$\frac{T}{T_0}=(\frac{P}{P_0})^{\frac{r-1}{r}} \tag{1}$$

若膨化硝酸铵微气泡中的空气近似地看似是氮气，则 $r=1.4$，设 $T_0=25℃=298.2K$，起爆冲击波与微气泡的初始压力之比 p/p_0 为 1000，则该微气泡作为热点其临界温度为：

$$T=T_0(\frac{P}{P_0})^{\frac{r-1}{r}}=2145.5K+1872.3℃$$

研究表明，温度介于 185～270℃时，硝酸铵热分解的主要形式是：

$$NH_4NO_3 = N_2O+2H_2O \quad \Delta H=+36.80kJ \tag{2}$$

$$2NH_4NO_3 = 2N_2+O_2+4H_2O \quad \Delta H=+238.30kJ \tag{3}$$

400℃以上时，则依下列反应式以爆炸形式进行热分解反应：

$$4NH_4NO_3 = 3N_2+2N_2O+8H_2O \quad \Delta H=+408kJ \tag{4}$$

$$8NH_4NO_3 = 2NO_2+4NO+5N_2+16H_2O \quad \Delta H=+490.0kJ \tag{5}$$

研究还表明，硝酸铵的高温热分解是动力学二级反应和自催化反应，反应中放出大量热量。

这样，由于膨化工艺在硝酸铵内部造成的大量的、一定分布的微气泡，在起爆冲击作用的绝热压缩中将全部作为热点，温升至 1872℃，足以引起热点附近的硝酸铵被加热、分解、引爆以致爆轰。显然，该热点的临界温度计算及分析是硝酸铵自敏化理论设计的依据与基础。

3 微气泡在膨化硝酸铵中的分布状态及规律

膨化工艺使膨化硝酸铵晶体内造成大量微孔，晶体表面呈现不规则的棱角、裂纹及毛刺，这些在起爆冲击波作用下都是形成热点的源泉。人们除特别关注大量微孔的表面结构特征外[1,7]，还十分重视它的分散状态及规律[8,9]。

用美国 Coulter 公司 OMNISORP（TM）表面孔容测定仪详细而系统地研究了膨化硝酸铵中微孔的分布状态及规律。不同吸附压力下测定的微孔 Kelvin 半径、吸附层厚度、孔径、解吸气体的体积、解吸体积增量、孔表面积及总表面积百分数等列于表 1。

表 1 膨化硝酸铵微孔分布状态及规律

Table 1 Distribution state and regularity of micro-pore of expanded ammonium nitrate (AN)

吸附压比	Kelvin 半径	吸附层厚度	孔半径	解吸气体积	解吸体积增量	孔表面积	总表面积百分数
p/p_0	$R_K/10^{-10}m$	$d_0/10^{-10}m$	$R_P/10^{-10}m$	$V_d/ml\cdot g^{-1}$	$V_{\Delta d}/ml\cdot g^{-1}$	$S_{dp}/m^2\cdot g^{-1}$	$S_{\Delta dp}/\%$
0.9376	147.97	15.024	162.99	0.2402	0.00045	0.0544	2.117
0.9193	105.89	13.834	119.72	0.2033	0.00083	0.1482	7.889
0.8899	81.684	12.853	94.537	0.2119	0.00125	0.2545	17.80
0.8664	66.473	12.054	78.527	0.1941	0.00165	0.3910	33.02
0.8418	55.354	11.340	66.693	0.1665	0.00199	0.5840	55.77
0.8184	47.581	10.752	58.333	0.1520	0.00230	0.8819	90.11
0.7940	41.325	10.210	51.536	0.1484	0.00262	1.342	100.0
0.7697	36.427	9.7340	46.161	0.1480	0.00294	2.024	100.0
0.7454	32.449	9.3049	41.754	0.1338	0.00323	2.981	100.0
0.7214	29.185	8.9194	38.105	0.1255	0.00350	4.342	100.0
0.6975	26.465	8.5712	35.036	0.1093	0.00374	6.197	100.0
0.6729	24.064	8.2401	32.304	0.1053	0.00396	8.693	100.0

续表

吸附压比	Kelvin 半径	吸附层厚度	孔半径	解吸气体积	解吸体积增量	孔表面积	总表面积百分数
0.6493	22.071	7.9461	30.017	0.0961	0.00417	11.90	100.0
0.6246	20.257	7.6609	27.918	0.0982	0.00438	15.84	100.0
0.6011	18.730	7.4046	26.136	0.1022	0.00461	20.47	100.0
0.5768	17.322	7.1586	24.430	0.0940	0.00482	25.28	100.0
0.5518	16.034	6.9195	22.954	0.0868	0.00500	29.71	100.0
0.5277	14.914	6.7006	21.614	0.0795	0.00517	33.06	100.0
0.5032	13.880	6.4890	20.369	0.0590	0.00527	34.01	100.0
0.4794	12.966	6.2928	19.259	0.0551	0.00536	31.90	100.0
0.4546	12.093	6.0970	18.190	0.0610	0.00547	25.28	100.0
0.4311	11.328	5.9179	17.246	0.0386	0.00550	14.74	100.0
0.4065	10.590	5.7378	16.327	0.4015	0.00555	0.2149	100.0
0.3811	9.8821	5.7378	16.327	0.4015	0.00560	0.2149	100.0
0.3570	9.2548	5.7378	16.327	0.4015	0.00564	0.2149	100.0

还测定了膨化硝酸铵孔径与孔容的关系，具体数据如表 2。

表 2 膨化硝酸铵孔径与孔容的关系

Table 2 The relation between pore radium and pore volume of expanded AN

孔径/10^{-10}m	孔容/ml·g^{-1}	占总容积的百分比/%	孔径/10^{-10}m	孔容/ml·g^{-1}	占总容积的百分比/%
>200	0.00000	0.00	40～30	0.00094	16.6
200～100	0.00087	15.4	30～20	0.00109	19.3
100～50	0.00178	31.5	20～10	0.00037	6.6
50～40	0.00061	10.8			

孔径 R 大于 10×10^{-10}m 的孔的总容积为 0.00568ml·g^{-1}。

根据 IUPAC 孔径分级的标准认为：

$R\geqslant500\times10^{-10}$m 的孔为粗孔；

$R=20\times10^{-10}\sim500\times10^{-10}$m 的孔位介孔；

$R=7\times10^{-10}\sim20\times10^{-10}$m 的孔位超细孔；

$R\leqslant7\times10^{-10}$m 的孔位极细孔。

可以看到，膨化硝酸铵中 95%的微孔在介孔范围内。因此，更准确地说应是介孔，但我们习惯还称为微孔。

从膨化硝酸铵自敏化理论而言，细孔（超细及极细孔）是无效的。因为组成膨化硝酸铵炸药时，膨化硝酸铵还必须与木粉及油相混合，孔径小于 20×10^{-10}m 的细孔，即使形成也会被细木粉，特别是油相灌满而堵塞、使孔不复存在。粗孔也是要避免的，显然，太多的粗孔将会影响成孔率，成孔率太低会直接影响到热点数、对自敏化也是不利的。因此，介孔是最理想的孔径范围。它既不会被细木粉及油相堵塞，而却使细木粉及油相均匀直接分布在含氧化剂孔的表面上，形成氧化剂与可燃物大面积紧密相贴的最佳混合状态，同时又能保持足够孔数，以保证足够的热点数，使之处于最佳起爆感度及爆轰的状态，这就是自敏化理论获得成功并被实际应用的物质基础。

测试了膨化硝酸铵静态吸附及解析等温线，如图 1 所示：它不存在平台，表示它不在细孔内，而刚好落于介孔区域中。

图 2 是膨化硝酸铵静态吸附及解析等温线与普通硝酸铵的对比情况：

由图 2 明显看出，膨化硝酸铵在同样比压下，有更大的吸附体积，比普通硝酸铵有更多的有效介孔，从而成为绝热压缩条件下能够形成热点的基础和保证。

继而研究了孔容与孔径的关系，如图 3 所示。

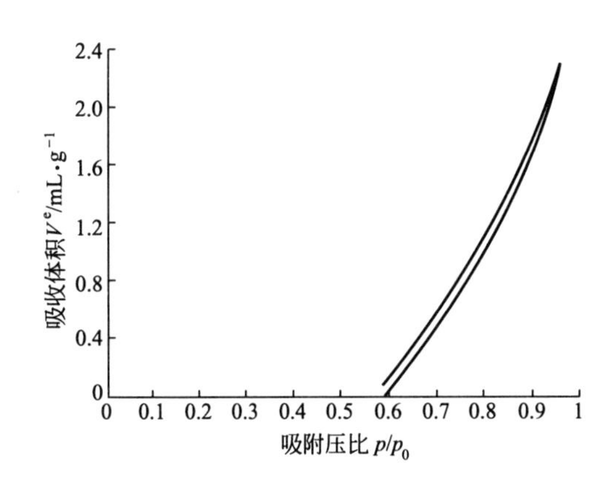

图1　膨化硝酸铵静态吸附及解析等温线

Fig. 1　The static adsorption and desorption isotherms of expanded AN

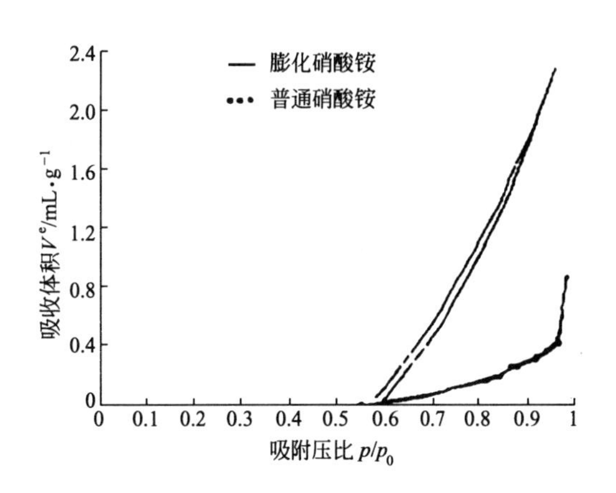

图2　膨化硝酸铵静态吸附及解析等温线与普通硝酸铵比较图

Fig. 2　The comparison between static adsorption and desorption isotherms of expanded AN and common AN

从图3可看出，膨化硝酸铵的孔大部分落在10×10^{-10}～200×10^{-10}m孔径的范围内。图4给出了孔容对孔径微分值与孔径的关系曲线，更直观地看出孔径的分布状态。

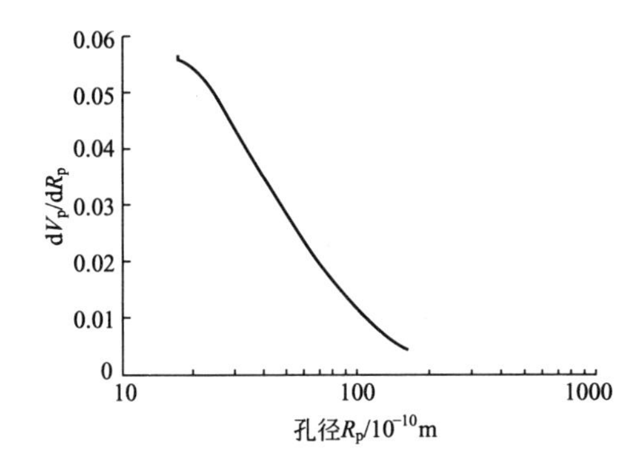

图3　膨化硝酸铵累计孔容与孔径关系

Fig. 3　Accumulated pore volume vs pore radiumof expanded AN

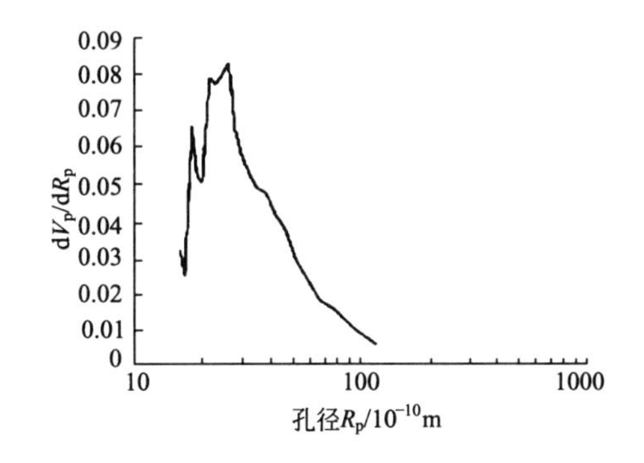

图4　膨化硝酸铵孔径分布状态图

Fig. 4　Pore radium distribution of expanded AN

为了深入理解膨化硝酸铵内孔径的分布状态及其规律，进一步认识膨化硝酸铵与普通硝酸铵的区别，还研究了普通硝酸铵孔分布状态及规律，并对它们的孔容及孔径分布进行了比较，见表3、图5。

从图5可看出：膨化硝酸铵比普通硝酸铵具有大得多的孔容，且95%以上的有效孔径落在介孔范围内，对有效起爆及完全爆轰时十分有利的。

为引入敏化气泡。膨胀珍珠岩也是一个很好的载体，人们十分关心它们的孔径及微孔总体积，经测试，结果如下：

膨化硝酸铵颗粒内的微孔孔径分布在10×10^{-10}～200×10^{-10}m的占99.5%以上，而孔径在10×10^{-10}m以上微孔的总体积为0.00568ml·g^{-1}，而膨胀珍珠微

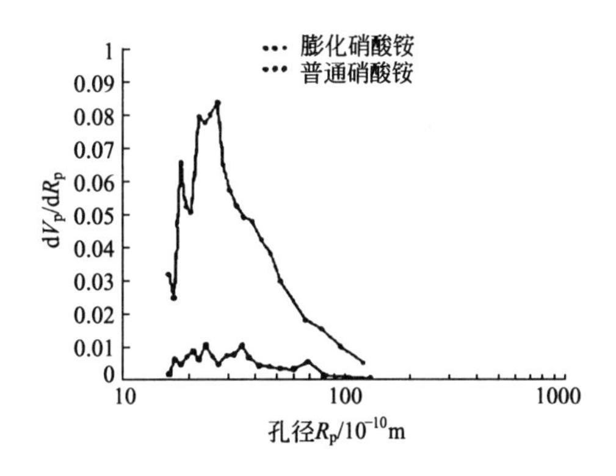

图5　膨化硝酸铵与普通硝酸铵微孔分布状态比较图

Fig. 5　Comparison between expanded AN and common AN for micro pore size distribution state

孔在 10×10^{-10}m 以上的微孔约占总孔数的 90%左右，其总体积约为 0.00331ml·g^{-1}。显然，膨化硝酸铵的微孔数及有效微孔体积高于膨胀珍珠岩。

表 3 普通硝酸铵微孔分布状态及规律表

Table 3 Distyibution and regularity of micro-pore of common AN

吸附压比	Kelvin 半径	吸附层厚度	孔半径	解吸气体积	解吸体积增量	孔表面积	总表面积百分数
p/p_0	$R_K/10^{-10}$m	$d_0/10^{-10}$m	$R_P/10^{-10}$m	V_d/ml·g^{-1}	$V_{\Delta d}$/ml·g^{-1}	$S_{dp}/m^2\cdot g^{-1}$	$S_{\Delta dp}$/%
0.9455	170.09	15.485	185.58	0.07528	0.00013	0.01422	3.854
0.9205	115.04	14.138	129.17	0.04857	0.00022	0.03305	12.81
0.8953	86.227	13.060	99.288	0.03739	0.00029	0.04712	25.58
0.8704	68.692	12.182	80.874	0.03090	0.00035	0.06967	44.45
0.8453	56.714	11.434	68.148	0.05768	0.00047	0.1422	83.00
0.8208	48.272	10.808	59.080	0.02634	0.00053	0.1785	100.0
0.7945	41.434	20.221	51.655	0.02265	0.00057	0.2664	100.0
0.7689	36.277	9.7186	45.996	0.01995	0.00061	0.3952	100.0
0.7443	32.281	9.2858	41.567	0.01679	0.00064	0.5781	100.0
0.7178	28.747	8.8651	37.613	0.02058	0.00069	0.8487	100.0
0.6922	25.915	8.4974	34.412	0.02397	0.00074	1.2183	100.0
0.6677	23.596	8.1728	31.769	0.01642	0.00078	1.6782	100.0
0.6411	21.441	7.8490	29.290	0.01564	0.00081	2.2627	100.0
0.6151	19.614	7.5550	27.169	0.01052	0.00083	2.9365	100.0
0.5894	18.034	7.2857	25.320	0.01197	0.00085	3.6883	100.0
0.5642	16.654	7.0361	23.690	0.01332	0.00088	4.4253	100.0
0.5381	15.384	6.7938	22.178	0.00959	0.00089	4.9724	100.0
0.5120	14.241	6.5640	20.805	0.01067	0.00091	5.1923	100.0
0.4870	13.247	6.3540	19.601	0.00868	0.00093	4.9277	100.0
0.4610	12.310	6.1466	18.457	0.00726	0.00094	3.9588	100.0
0.4345	11.437	5.9439	17.380	0.00779	0.00095	2.2324	100.0
0.4096	10.679	5.7602	16.440	0.00530	0.00095	0.0274	100.0
0.3827	9.9242	5.7602	16.440	0.00530	0.00095	0.0274	100.0
0.3568	9.2489	5.7602	16.440	0.00530	0.00095	0.0274	100.0

参考文献

[1] 丁芸．表面活性剂对硝酸铵改性的影响［D］．南京：华东工学院，1990.
[2] 吕春绪．表面活性理论与技术［M］．南京：江苏科学技术出版社，1991.
[3] 慧君明，刘祖亮，吕春绪．粉状硝铵炸药及其制法［P］．CN91107051，1991.
[4] 刘祖亮．惠君明，吕春绪等．轻质硝酸铵及 HF 型硝铵炸药［J］．爆破器材，1991（5）：5.
[5] 陈天云．硝酸铵表面特性研究［D］．南京：华东工学院，1992.
[6] 陆明．膨化硝铵炸药研究［D］．南京：南京理工大学，1999.
[7] 叶志文．硝酸铵膨化机理研究［D］．南京：南京理工大学，1995.
[8] 吕春绪．膨化硝酸铵表面化学研究［J］．南京大学学报，1995（31）：286.
[9] 陈天云，吕春绪，叶志文．改性硝酸铵性能研究［J］．含能材料，1996（4）：169.

Research on Self-sensitization Theory of Expanded Ammonium Nitrate

Lv Chunxu

（Insititute of Industrial Explosive Materials，Nanjing 210094，China）

Abstract：Self-sensitization theory of expanded ammonium nitrate（AN） is the key point of the present innovation. On the basis of heat points initiating mechanism of explosion，micro air bubble is embedded in explosive of expanded AN. The theoretical basis of self-sensitization is provided through calculation of critical heat point temperature of micro air bubble. Comparing expanded AN with common AN and peralstone, the micro air bubble distribution and its regularity in expanded AN are given which provide powerful basis assurance for its success of self-sensitization.

Key words: expanded AN, self-sensitization theory, micro air bubble

（注：此文原载于 中国工程科学，2000，2（11）：73-77.）

Self-sensitivity Theory Design of Expanded Ammonium Nitrate Explosive

Lv Chunxu
(Institute of Industrial Explosive Materials, China Ordnance Industry (Nanjing, 210094))

Abstract: Expanded Ammonium Nitrate (AN) explosive, which is a new kind of powder industry explosive based on innovative technology of expansion, has been awarded the National Prize and listed into the plan of Spreading Great Scientific Achievement of China Ninth five year Plan. It has been finished the design and application in more than 60 factories. In the paper, its characteristics and advantages are briefly given, comparison with the other industrial explosive is also presented and reasons for the excellent property are analyzed. In this paper, the details of its self-sensitivity theory are discussed: basic designing idea, proposal of sensitization, approach of self-sensitization, detonation characteristics of expanded AN explosive and so on are minutely stated.

Keywords: expanded AN, explosive, self-sensitization

1 Introduction

Expansion of AN is a innovative technology. Expanded AN explosive based on expansion technology is a new kind of industrial explosive.

Some explosive experts think that: technology of expanded AN is original and innovative. It opens a new approach to explosive study and reaches the advanced level of the world. Expansion technology and expanded AN was awarded Special Scientific Achievement Prized of China Ordnance Industry Company (ministry rank) in 1997 and National Scientific Achievement Prize Grade Two in 1998.

Expanded AN explosive has been arranged into List of Spreading Great Scientific Achievement of China Ninth five year Plan. National explosive line is going to replace some other powder explosive with expanded AN as a new generation product.

The achievement of expanded AN explosive is a revolution of industrial explosive. Owing to its advanced technology, it has remarkable social and economic benefits and welcome to many customers. In China, there are more than 60 factories manufacturing expanded AN explosive. American Austin Explosive Company, Japanese MITSUBISHI Chemical Company and Indian SUA Explosive Company are all interested in this technology. Some technical cooperation is under negotiation. Because of its application, the national output value has increased for 660 million Chinese Yuan and profits before tax has increased for 115 million Chinese Yuan until 1998. The output of powder AN-TNT with expanded AN. So its further market is promising.

2 The innovative idea of expanded AN self sensitivity theory and its technical approaches

2.1 Background of research

Industrial powder explosives is mainly sensitized by TNT or NG.. These explosive have serious defect: (1) Toxicity of TNT impairs the workers's health and pollutes environment. (2) Cost is relatively high. (3) Safety of manufacture is not very good. (4) Physical property, for example, hydroscopicity and caking, is not ideal.

For more than 100 years, many researchers focus on these problems. They propose adding crystal pattern modifiers, surfactant, hydrophobic agent into explosive or improving charge or package methods to enhance explosive physical property. They also attempt to replace TNT with RDX or metal powder. Although some of these methods are effective, they are partial or not economical so have not resolved the problem radically.

2.2 Basic designing idea

After technical analysis of many powdered explosive and exploring lots of approaches, we think that only through self-sensitization of AN can powder AN explosive get its breakthrough for eradicating TNT.

The idea of self-sensitization of AN is a innovation, never considered by explosive researchers.

The idea of self-sensitization is based on theory as follows: the detonation of AN is non-ideal detonation. Under action of external energy, "heat points" concentrated with energy are formed in some heterogeneous section initially; the explosive in "heat points" then decomposes rapidly, giving out lots of heat which initiates explosive of adjacent explosive, then explosion of all explosive system. "heat points" can be formed from pressing adiabatically air bubble or form friction and collision between particles. All this inform us that: introducing micro air bubble into AN particles, roughening particles surface will sensitized AN under external initiating energy.

2.3 Proposal of sensitization

In explosive theory, "heat points" initiating mechanism is generally acknowledged and very important. It not only well explains the reason of initiation of explosive under very little external energy but also directs us to adjust the safety of explosive and to design sensitization of liquid or emulsion explosive system.

Micro air bubble which is introduced into AN crystal will form "heat points" of high pressure high temperature rapidly under external energy; at the same time. Friction and collision among roughened particles also lead to forming "heat points" . Just this sensitization function from inner crystal takes place the sensitization of TNT, making initiation of AN explosive enhanced radically.

The theoretical valerie and level of self-sensitization applied in powder industrial explosive is self-evident and generally acknowledged.

Self-sensitization theory is key point of our innovation.

2.4 Approach of self sensitization expandation

In order to introduce micro air bubble into AN crystal and roughen the crystal surface, only can we make AN crystallize rapidly under high boiling sate. The best approach to it is controlled crystallization under the help pf surfactant.

After research on surface activity theory and lots of exploration, a new kind of composite surfactant is obtained, which can decrease surface tension of AN solution and AN crystal. Under subpressure and high temperature, AN will crystallize out with lots of "hole", "cleavage", "channel" in it, this is just expanded AN. It has been proved that: idea of self-sensitization of AN and technical approach of controlled crystallization is correct and effective, a technical innovation of industrial explosive.

The key technology of non-TNT expanded AN explosive is expansion of AN. The essence of expansion is surface activity techniques, a physical chemical process of compelled foaming crystallization under composite surfactant action.

Many approaches focus on coating, adding modifiers, dispersing and so on to modify the AN. We break through this frame of idea, get successful self-sensitization of AN.

3 Mechanism of expansion of self-sensitivity

After research on effect of process condition and surfactant on expansion, we think mechanism of expansion probably is that[1-4]: under certain temperature and subpressure, water in satured AN solution vaporizes and form stable foams with the help of surfactant. When temperature and subpressure reach a certain value, foams break up and water steam flees out of AN solution rapidly, leaving lots of "hole", "cleavage", "channel" . AN crystal which crystallizes out rapidly is rough on its surface, and connected with each other in irregular way, forming a kind of light, porous, bulky expanded AN.

It is showed in research that: vaporization under subpressure helps to expand AN by itself, while the foaming property of surfactant further enhances the expansion effect. Experiments show that: expansion effect is correspondent to foaming property of surfactant.

Research also shows that AN solution with composite surfactant possesses relatively better foaming performance, among which amphiprotic composite surfactant has the best surface activity and foaming performance. It also showed that composite and surfactant can be mutually beneficial in foaming and the amphiprotic composite and mutual benefit. In modifying AN, surfactant mainly plays the role of foaming agent, coating and crystal pattern modifier.

4 Technical characteristics of self sensitivity expanded AN

Expanded AN treated by surfactant has porous honeycomb like structure. Microstructure of expanded AN photoed by SEM is showed in Fig. 1[4,5].

Compared with common powder AN and grainy porous AN, expanded AN has these characteristics:

(1) Large specific area. Determined by the same method, specific area of expanded AN is

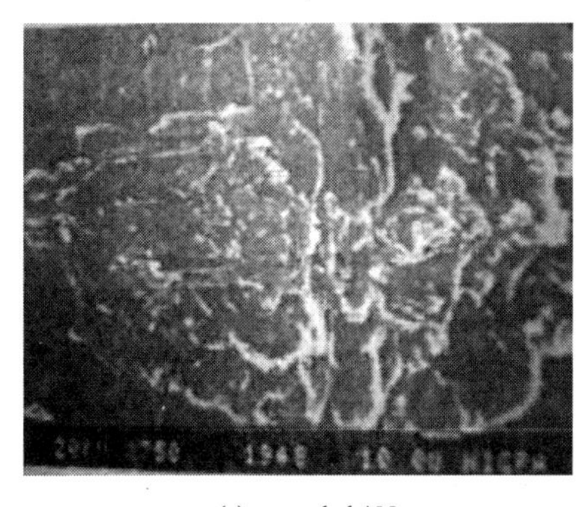

(a) expanded AN

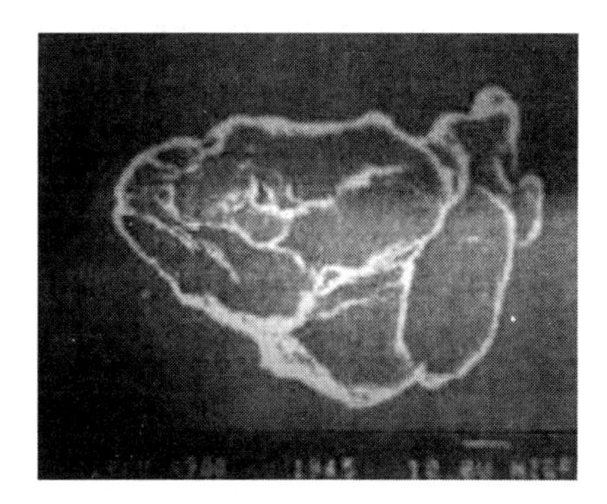

(b) Commmon AN

Fig. 1 Picture of expanded An with SEM

$3328.54cm^2 \cdot g^{-1}$, 4.4 times larger than that of common AN, $758.76cm^2 \cdot g^{-1}$.

(2) Lots of micro air bubble in AN crystal. It is showed in Fig1 that there is much of cleavage and crack in AN crystal and spins or edges on its surface. This is physical fundament of self-sensitization of AN.

(3) Relatively lower hygroscopicity and caking.

Hygroscopic increment of AN under 20℃, humidity 93% is showed in Tab. 1.

Data in Tab 1 show that average hygroscopic velocity of expanded AN is about half less than that of common AN. Putting expanded AN and common AN under 80℃ for 3 hours, then cooling it; after 5 this heating and cooling cycling, determining its undergone expanded AN is 1/8~1/9 less than that of common AN. So expanded AN is thought to process very good anti caking property.

(4) Easy process because of brittleness. Just manufactured expanded AN is very brittle and easy to be crushed. So only for 20 min, size of 90% AN powder discharged from edge mill is smaller than 250×10^{-6}m, reaching the certain required granularity of AN.

Tab. 1 Hygroscopic increment of Ammonium Nitrate

Time/h	2	4	6	8	10
Pure AN	0.5832	1.1084	1.5792	2.0660	2.5370
Expanded AN	0.4258	0.7697	1.1212	1.4829	1.8746
No. 2 Rock AN TNT explosive	0.5569	1.1100	1.5456	2.0001	2.4811
Rock expanded AN explosive	0.3958	0.6432	0.9397	1.3025	1.5904

Tab. 2 The pressure undergone by chrough cycling test

	Undergone pressure/N				Average/N
Common AN	805.9	811.1	799.3	802.3	804.7
Expanded AN	81.3	80.9	80.4	82.5	81.3

Tab. 3 Compositon of three kinds of powder explosive

Explosive	AN×100	TNT×100	Wood powder×100	Fuel×100	Additive×100
No. 2 rock AN-TNT explosive	85.0±1.5	11.0±1.0	4.0±0.5	—	—
No. 2 rock AN-TNT fuel explosive	87.5±1.5	7.0±0.7	40.±0.5	1.5±0.3	0.100±0.005
Rock expanded AN Explosive	92.0±2.0	—	4.0±0.5	4.0±0.5	0.120±0.005

Tab. 4 The comparison between rock EAN explosive and other powder industrial explosive

Explosive	Composition AN %	TNT %	Wood powder %	Oil %	Additive %	Characteristics Density ρ g·cm^{-3}	Detonation Velocity VDm·s^{-1}	Strength VL ml	Brisance HLE mm	Moisture Content %	Storage Lifetime month
EAN explosive	92.0±2.0	0	4.0±0.5	4.0±0.5	0.12±0.005	0.95~1.00	3200~3500	360~380	14.5~16.0	0.3	6
No. 2 rock Ammonite	85.0±1.5	11.0±1.0	4.0±0.5	0	0	0.95~1.10	3000~3200	≥320	≥12	0.3	6
No. 2 rock Ammonite-Oil	87.5±1.2	7.0±0.7	4.0±0.5	1.5±0.3	0.10±0.005	0.95~1.05	3000~3300	≥340	≥12	0.3	6

5 Application of expanded AN self sensitivity theory

5.1 Rock expanded AN explosive

AN solution with 0.15% surfactant LC 2 is vaporized and then crystallize. This is just the expansion of AN. Under 60℃, 92% expanded AN, 4% wood powder, 4% composite fuel are put into edge mill[7].

Rock expanded AN explosive, No. 2 rock AN TNT explosive and No. 2 rock AN TNT fuel explosive are main powder explosive in China. Their composition and explosion performance are showed in Tab. 3 and Tab. 4.

Tab 4 indicates that performance of rock expanded AN explosive reaches and even exceeds that of two other explosive.

Tab. 5 The comparison of crude material cost for three powder explosive
(On the bases of price in 1995) US Dollar

Crude material	Unit price	Rock EAN Explosive Amount /kg·t^{-1}	Cost	No. 2 rock Ammonite Amount /kg·t^{-1}	Cost	No. 2 rock Ammonite-Oil Amount /kg·t^{-1}	Cost
AN	235.3	920	216.5	850	200	875	205.9
TNT	941.2			110	103.5	70	65.9
Wood powder	35.3	40	1.4	40	1.4	40	1.4
Oil	352.9	40	14.1			15	5.3
Additive	1764.7					1	1.8
Expanded agent	3388.2	1.2	4.1				
Total		236.1		304.9		280.3	

According to detonation theory, mechanism of industrial explosive detonation includes surface section mechanism and composite reaction mechanism. So under external energy action, the easy and quantity of heat point formation is a main factor determining the detonation. There is much of "micro air hole" in expanded AN, under external energy action, micro air hole undergoes adiabatic compression and becomes initiating center "heat point" . So expanded AN has good detonation characteristics undoubtedly.

Another advantage of expanded AN is remarkable economic profit. Because TNT is not required, raw material cost decreases and profit enhances consequently.

Besides decreased cost, labor-consuming and investment of workshop concerning TNT processing and wasted water treatment are also saved. Occupational disease and pollution resulting from TNT is eradicated, bringing out lots of social profits.

5.2 Permitted expanded AN explosive

Permitted expanded AN explosive is the first serials product of expanded AN. Composition is as follows: expanded AN 79%, fuel 3%, salt 15%, wood powder 3%.

The key point of permitted explosive is safety property. Under firedamp condition, the explosive will not deflagrate. We focus on its flame depressing mechanism and make breakthrough on it.

This technology finalized the design in Dec. 1997 directed by Industrial Explosive Materials Bureau of China Ordnance Industry, and passed the appraisal of manufacturing in Sept. 1998.

AT present, this technology has been transferred to Xuzhou Mine Chemical Factory. China Coal Ministrylists it into "Project of Spreading Achievements of Ninth five year Plan of Coal Industry" and will spread it to 50 factories of Coal Ministry and 40 local explosive factories.

5.3 Seismic explosive

Expanded AN can replace common AN in seismic explosive. The cost of seismic explosive made from AN explosive decreases remarkably but with the relatively same performance. With increasing manufacture safety and less pollution, it has good social and economic profits.

A manufacturing line whit 5000 tons capacity of expanded AN seismic-explosive has been built in Hubei Rang sha Chemical Factory with annual output 4000 tons. The product has good stable performance and welcome to many customers.

China has 8 seismic explosive factories. Annual output amounts to 30000 tons. So this technology has good promising market.

6 Conclusion

(1) Based on general idea of self-sensitization of AN and surface activity theory, expanded AN with porous structure is developed. Expanded AN has larger specific area, low hydroscopicity and caking property. Expansion technology is original, innovative and gives a new way for explosive research.

Rock expanded AN explosion has good explosive and physical property although it has no TNT in it.

This achievement is awarded National Scientific Achievement Prize Grade Two and Special Scientific Achievement Prize of China Ordnance Industry.

(2) Self-sensitivity expanded AN has low cost, safe manufacturing, easy available row materials characteristics. So it is welcome to many factories. Those factories with about 10000 tons output have get net profit 3.5~5.0 million Chinese Yuan.

Until 1998, the national output value has increased for 660 million Chinese Yuan and profit before fax has increased 115 million Chinese Yuan.

(3) Rock expanded AN explosive has been finalized its design and applied in more than 60 factories. It also has been list into Plan of Spreading Achievement by Industrial Explosive Materials Bureau of China Ordnance Industry and Ministry of Science and Technology.

(4) National consume of powder AN TNT amounts to 850000 tons and there are 213 explosive factory. So expanded AN will have its promising market and contribute to replacing old explosive,

making Chinese industrial explosive reach advanced level of the world.

(5) Research on expanded AN explosive serials is now on the way, and applied successfully in permitted velocitty or high strength expanded AN explosive is also under research.

Reference

[1] Lü Chunxu. Research on Surface Chemistry of Expanded AN. Journal of Nanjing University, 1995 (31): 286.

[2] Liu Zuliang, Lü Chunxu, Lu ming, Chen Tianyun. Research and Applocation of Expanded AN. Proceedings of ISPE. Beijing: 1995, 220.

[3] Hui Junming, Liu Zuliang, Lü Chunxu. Powder AN Explosive and its Preparation. CN91107051. 6, 1991.

[4] Ye zhiwen. Mechanism of Expansion of AN. [thesis], Nanjing University of Science and Technology, 1995.

[5] Din Yun. Modification of AN with Surfactant. [thesis], Nanjing University of Science and Technology, 1990.

[6] Chen Tianyun. Surface characteristics of AN. [thesis], East China Institute of Technology, 1995.

[7] Liu Zuliang, Hui Junming, Lü Chunxu, Wan Yilin. Research on Light Porous and Bulky AN and HF Serials Powder AN Explosive. Exlosive Materials, 1991, 20 (5): 5-8.

[8] Lü Chunxu, Liu Zuliang, Ni Quqi. Industrial Explosive. Beijing: Ordnance Industry Press, 1994.

(注：此文原载于 Proceedings of International Conference on Engineering and Technological Sciences, 2000: 874-878)

膨化硝酸铵自敏化理论的微气泡研究

吕春绪，陆明，陈天云，叶志文
（南京理工大学，江苏南京，210094）
EI：02026825571

摘要： 膨化硝酸铵自敏化理论设计的依据与基础是热点学说，是把微气泡作为热点植入膨化硝酸铵中，实践表明是成功的。本文重点讨论了微气泡在膨化硝酸铵中的形成、保护、表面结构特征及其在膨化硝酸铵中的分布状态及规律。
关键词： 膨化硝酸铵，自敏化理论，微气泡，热点，孔径，孔容

1 膨化硝酸铵自敏化理论设计依据与基础

硝铵炸药的爆轰是一种非理想爆轰，其机理是在起爆冲击波作用下炸药受到强烈压缩，使炸药层温升不均匀，达到临界反应温度时，其快速化学反应将首先从这些被称为起爆中心——热点的地方开始，进而被加速并转变为爆轰在整个炸药中传播。通常该热点被认为是由于炸药颗粒表面歧形化或炸药层中所含微气泡造成的，尤其是后者引起人们的关注并设法将其引进硝铵炸药体系中。

显然将微气泡引进木粉或油相中是难以实现的，但可通过一种特殊工艺在硝酸铵（AN）晶体中产生微气泡并作为起爆热点敏化硝铵炸药（不需要加入单质炸药作为敏化剂）。这就是硝酸铵自敏化理论的设计思想。

在复合表面活性剂（膨化剂）作用下，以最佳温度、真空度使硝酸铵饱和溶液强制析晶。从而获得含有大量微气泡、表面粗糙、多裂纹且轻质、多孔和疏松的膨化硝酸铵，这就是硝酸铵膨化工艺。膨化硝酸铵中的微气泡，在起爆冲击作用的绝热压缩中将全部作为热点，足以引起其附近的硝酸铵被加热、分解、引爆以致爆轰。热点理论是硝酸铵自敏化理论设计的依据与基础[1-5]。

2 膨化硝酸铵自敏化微气泡的形成及保护

微气泡是在硝酸铵膨化这一物理化学过程中形成的。由于膨化剂作用在一个极短的时间内旧相消失、新相产生，巨大的相变热足以使水分蒸发与汽化，与此同时在膨化硝酸铵内产生大量微气泡，其形成过程如下[5,6]：

在高真空状态下（－0.085～－0.095MPa），硝酸铵由不饱和状态到饱和状态，再到过饱和状态。进而迅速膨胀析晶；当大部分析晶出来之后，体系进入干燥状态。在上述过程中，大孔径的微气泡、空隙是在体系膨胀过程中形成的，而内部的小微气泡，则是在干燥过程中形成的。

整个膨化过程中水的逸出状态及速度对析晶阶段微气泡形成起关键作用，水的逸出又直接受溶液温度的影响，溶液温度及水分的变化情况如图 2.1 和图 2.2 所示（实验条件真空度：0.092 MPa；AN 100g，H_2O 10g，膨化剂 0.15g）。

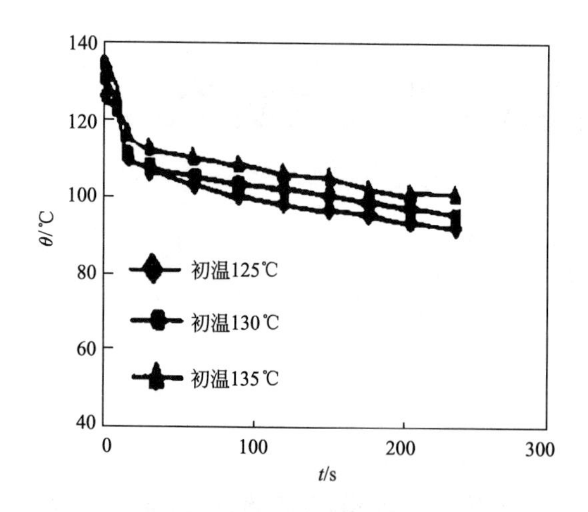

图 2.1 不同初温硝酸铵饱和溶液的温度变化情况

Fig. 2.1 Variation of temperature of saturated solution of AN under various initial temperatures

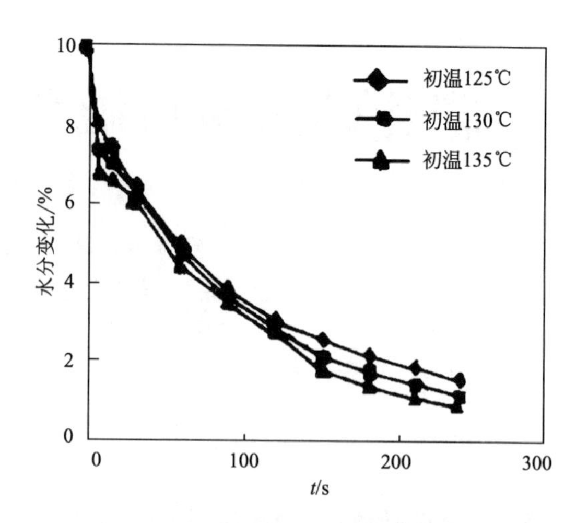

图 2.2 不同初温硝酸铵饱和溶液的水分变化情况

Fig. 2.2 Variation of moisture content of saturated solution of AN under various initial temperatures

从图 2.1 和图 2.2 明显看出，溶液起始温度较高，水分的蒸发逸出速度较快，析出的时间提前，析出的膨化硝酸铵的温度也较高。从另一方面讲，溶液起始温度较高，在真空作用下，溶液的沸腾越激烈，导致膨胀体积增加，微气泡形成数量增多，硝酸铵膨化敏化的效果也就好。

水分由硝酸铵颗粒内部向表面迁移进而失去，使其微粒发生体积收缩，导致微粒中心空穴（小微气泡）的形成。处于硝酸铵结晶内的水分子，在真空状态下，也蒸发或由里向外迁移，终形成许多微细孔道。体积收缩使硝酸铵颗粒的表面和颗粒内部在拉伸应力上产生差异，导致颗粒中裂缝的形成。在高真空和较高温度下，水分子由硝酸铵颗粒内部向外部迁移或蒸发的速度比一般硝酸铵气流干燥大大加快，使膨化硝酸铵的孔隙率远远高于多孔硝酸铵。

综上所述，在析晶及真空干燥中，所产生的中心空穴、微细孔道和裂缝，最后就是膨化硝酸铵中的微气泡，它们是嘭化硝酸铵具有较高起爆感度即自身敏化的前提和条件。

膨化硝酸铵水分过大，则微气泡被水分子阻塞；堆积密度大，则膨化硝酸铵中的微气泡被“压死”而数目减少，造成起爆时的“热点”少，则起爆感度低。当膨化硝酸铵中水分含量大于 0.5%或膨化硝酸铵的堆积密度大于 0.60g/cm^3 时，其起爆感度大大降低，几乎趋于零。

一般从膨化结晶机中出来的膨化硝酸铵，其水分含量小于 0.1%，而堆积密度小于 0.5g/cm^3。为保护其中的微气泡，应尽量避免其吸湿和长时间的挤压、贮存、一般当班生产的膨化硝酸铵，应当班混制成炸药，这样困炸药中含有 3%～4%的油相材料，使整个炸药的吸湿结块性有显著改进。

在硝酸铵膨化过程中，其溶液的初始温度不宜过高，过高会使膨化硝酸铵的出料温度提高，硝酸铵的热塑性增加，使其发粘、不脆而易被挤压，造成堆积密度提高，微气泡减少。

3 膨化硝酸铵自敏化微气泡表面结构特征

在硝酸铵膨化过程中，膨化剂起到了重要改性作用，使膨化晶体生成及形成具有显著特征的微气泡表面结构，不同膨化剂作用得到的膨化硝酸铵中微气泡外形特征如图 3.1（（a）AOCTA/AW-80 膨化剂（十八胺盐）（失水山梨醇酯）；（b）AOCTA/鉎 ST 膨化剂（硬脂酸））所示[7-9]。扫描电镜（SEM）可以把微气泡表面形貌直观可靠拍摄下来，为研究其表面特征提供有力的保证。

从图中显见膨化硝酸铵颗粒中的微气泡大小不一，且边缘极不规则、多棱角、多孔隙。表面形状

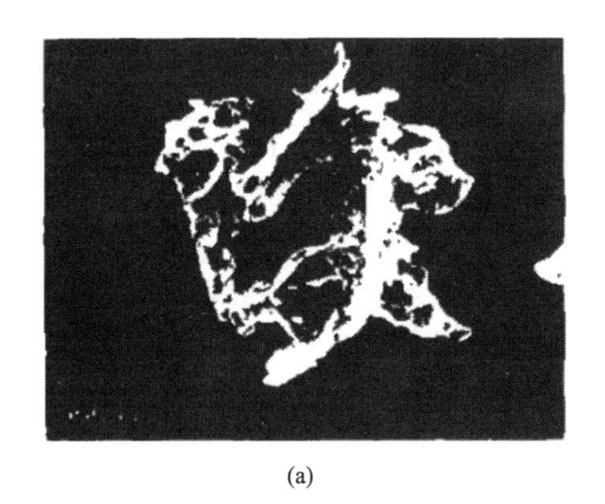

(a)

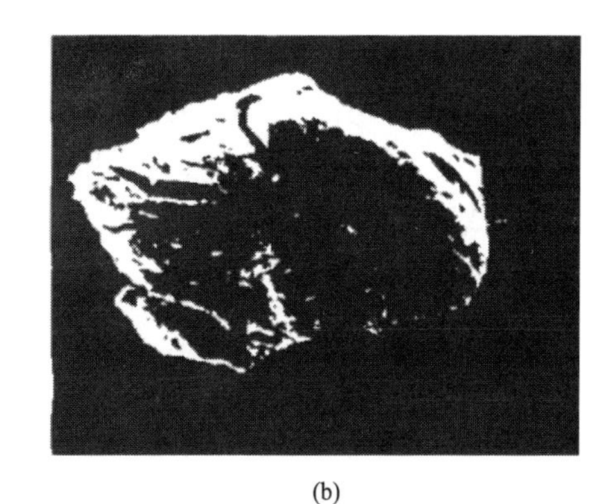

(b)

图 3.1 不同膨化剂得到的膨化硝酸铵中微气泡表面结构图

Fig. 3.1 Surface structure of micro air bubble in expanded AN manufactured with different expanding agents

歧化、毛刺多，导致硝酸铵粒子之间连接不规则，必然形成一种膨松化体系。

硝酸铵在膨化剂作用下膨化，产生大量微气泡，从而使膨化硝酸铵的比表面积剧增，它与普通硝酸铵相比较其比表面积大小如表 3.1 所示[1,7,10]。

表 3.1 不同工艺条件硝酸铵的比表面积（膨化剂用量 0.2%）

Tab. 3.1 Specific surface of AN through different conditions of processing

工艺条件	普通硝酸铵	工艺 A 膨化的硝酸铵		工艺 B 膨化的硝酸铵	
		(A_1)	(B_1)	(A_1)	(B_1)
比表面积/$cm^2 \cdot cm^{-3}$	471.95	1076.21	1101.26	1474.80	1454.57
比表面积/$cm^2 \cdot g^{-1}$	758.76	2265.71	2167.83	3810.85	3328.54

注：1. A_1 及 B_1 为不同表面活性剂组成的膨化剂；

2. 工艺 A 是指降温析晶工艺；

3. 工艺 B 是指真空蒸发析晶工艺。

4 膨化硝酸铵自敏化微气泡容积与半径分布特征

由膨化工艺使膨化硝酸铵晶体内形成大量微气泡，且晶体表面呈极不规则的棱角、裂纹及毛刺。这些都是在起爆冲击波作用下形成热点的源泉。人们除特别关注大量微气泡的表面结构特征外[1,7,10]，还十分重视它的分布特征[10,11]。

最近，在美国 Coulter 公司生产的 OMNISORP（TM）表面孔容测定仪上研究了膨化硝酸铵中微气泡的分布状态及规律。在不同吸附压力下测定了微气泡的 Kelvin 半径，吸附层厚度、孔径、解吸气体的体积、解吸体积增量、孔表面积及总表面积百分数等。

还测定了膨化硝酸铵微气泡的孔径与孔容的关系，具体数据如表 4.1 所示。

表 4.1 膨化硝酸铵微气泡的孔径与孔容的关系

Tab. 4.1 Relationship between pore radius and pore volume for micro air bubbles in expanded AN

$R/10^{-10}m$	$V/mL \cdot g^{-1}$	占总容积的百分比%	$R/10^{-10}m$	$V/mL \cdot g^{-1}$	占总容积的百分比%
500～400	0.00000	0.00	50～40	0.00061	10.80
400～300	0.00000	0.00	40～30	0.00094	16.65
300～200	0.00000	0.00	30～20	0.00109	19.30
200～100	0.00087	15.41	20～10	0.00037	6.55
100～50	0.00178	31.52			

由表 4.1 中数据表面：孔径大于 $10 \times 10^{-10}m$ 的孔径的总容积为 $0.00568mL \cdot g^{-1}$。

根据 IUPAC 孔径分级的标准认为：孔径 $\geqslant 500 \times 10^{-10}m$ 为粗孔；孔径 $20 \times 10^{-10}m \sim 500 \times$

10^{-10} m为介孔；孔径7×10^{-10} m～20×10^{-10} m为超细孔；孔径≤7×10^{-10} m为极细孔。

就膨化硝酸铵自敏化理论而言，其中的细孔（超细及极细孔）是无效的。因为组成膨化硝酸铵炸药时，它还必须与木粉及油相混合，小20×10^{-10} m的细孔，即使形成也会被细木粉，或油相灌满而堵塞，使孔不复存在。太多的粗孔将会影响成孔率，成孔率太低直接影响到热点数，对自敏化理论不利。因此，介孔是最理想的孔径范围。它能使细木粉及油相均匀直接分布在孔的表面上。形成零氧平衡接近分子状态的氧化剂与可燃物大比表面积紧密相贴的最佳状态混合，使膨化硝铵炸药处于最佳起爆感度及最完全爆轰的状态，这就是自敏化理论获得成功并被实际应用的物质基础。

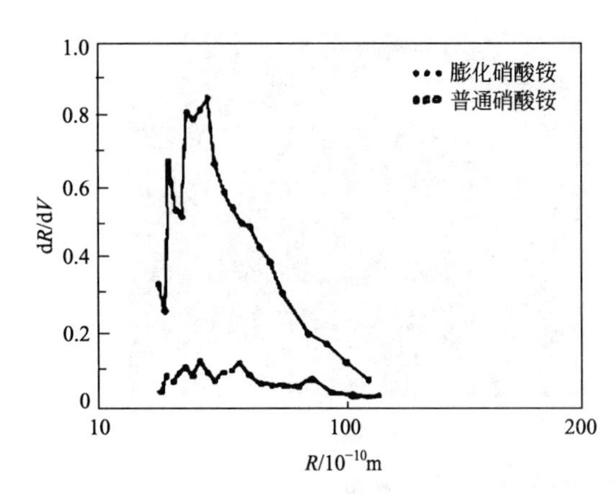

图4.1 膨化硝酸铵与普通硝酸铵微孔分布状态比较图

Fig. 4. 1 Plot of comparison between expanded AN and common AN for their state of distribution in micro pore sizes

为了进一步认识膨化硝酸铵与普通硝酸铵区别，还研究了它们的孔容大小及孔径分布状态图，并作了比较，如图4.1所示。由图可见，在自敏化理论总体设计思想指导下，由特定的膨化工艺制造的膨化硝酸铵比普通硝酸铵具有大得多的孔容。有效孔径95%以上坐落在介孔范围内，对有效起爆及完全爆轰十分有利。

5 结论

（1）膨化硝酸铵自敏化理论设计的依据与基础是热点学说。膨化工艺成功地把大量微气泡作为起爆热点植入膨化硝酸铵中，组成零氧平衡且不含有TNT组分的膨化硝铵炸药，其爆炸性超过2号岩石铵梯炸药。

（2）微气泡在硝酸铵膨化这一物理化学过程中形成，膨化过程中水的逸出状态及速度对析晶阶段微气泡的形成起关键作用；在干燥过程中生成的中心空穴、微细孔道和裂缝等也形成小微气泡。这些微气泡是膨化硝酸铵具有较高起爆感度——自敏化的前提及保障。

（3）从扫描电镜照片（SEM）清晰可知，膨化硝酸铵颗粒边缘极不规整，多棱角、多毛刺。

（4）用OMNISORP表面孔容测定仪，研究了膨化硝酸铵中微气泡的分布状态及规律。膨化硝酸铵95%微气泡坐落在介孔范围内。介孔对膨化硝酸铵自敏化是最有效的孔径，这也是自敏化理论获得成功并被实际使用的物质基础。

参考文献

[1] 丁芸．表面活性剂对硝酸铵改性的影响：[硕士学位论文]．南京：华东工学院，1990.

[2] 吕春绪．表面活性理论与技术．南京：江苏科技出版社，1991.

[3] 惠君明，刘祖亮，吕春绪，王依林．轻质硝酸铵及HF型硝铵炸药．爆破器材．1991，(5)：5.

[4] 陈天云．硝酸铵表面特性研究：[硕士学位论文]．南京：南京理工大学，1992.

[5] 陆明．膨化硝铵炸药研究：[博士学位论文]．南京：南京理工大学，1999.

[6] Lu Ming，Lu Chunxu，Liu Zuliang. Study on the formation of microholes in expanded ammonium nitrate. In：Proc of the 26th International Pyrotechnics Seminar. China Nanjing：1999，357.

[7] 叶志文．硝酸铵膨化机理研究：[硕士学位论文]．南京：南京理工大学，1995.

[8] Ye Zhiwen，Zhou Xinli. Research on thermodynamical properties of the micellization of mixed surfactant in ammonium nitrate solution. In：Proc of the 26th Internatioanl Pyrotechnics Seminar. China Nanjing：1999，573.

[9] 吕春绪，刘祖亮，叶志文．硝酸铵膨化机理研究．火炸药学报，1998，(2)：16.

[10] 吕春绪．膨化硝酸铵界面化学研究．南京大学学报，1995，(31)：286.
[11] 陈天云，吕春绪，叶志文．改性硝酸铵性能研究．含能材料，1996，(4)：169.

A Study on Micro Air-Bubbles in the Self-sensitization Theory of Expanded Ammonium Nitrate

Lv Chunxu, Lu Ming, Chen Tianyun, Ye Zhiwen
(Nanjing University of Science and Technology, Nanjing, 210094)

Abstract: The foundation of the self-sensitization theory of expanded ammonium nitrate (AN) lies in the hot-spot hypothesis of such explosives, where the hot-spots in the initiation of expanded ammonium nitrate concur with micro sized air bubbles. In this paper are discussed the formation and protection of micro air bubbles in expanded AN, characteristics of the surface structure of micro air bubbles, their state of distribution and regularity in expanded AN.

Key words: expanded AN, self-sensitization theory, micro air bubbles, hot-spot, pore radius, pore volume

(注：此文原载于 兵工学报，2001，22 (4)：485-488)

硝酸铵膨化理论研究

吕春绪

（国防科技工业民用爆破器材研究所，南京.210094）

摘要： 描述了硝酸铵膨化过程，计算与测定了其物理化学参数，研究了膨化过程的影响因素，并给出了膨化过程可能的机理。

关键词： 膨化硝酸铵，膨化理论，物理化学参数，影响因素

1 引言

硝酸铵膨化的实质是表面化学与结晶化学技术在粉状工业炸药上的实际应用，它是一个在复合表面活性剂作用下强制发泡析晶，以达到自敏化的物理化学过程[1,2]。硝酸铵膨化理论是我们非常关心的主要问题之一，对它的深入研究有利于进一步认识膨化技术，并使其在生产实践中得到进一步的完善。

2 硝酸铵膨化过程的描述

在 1000ml 磨口烧瓶中加入 200ml 的 H_2O，0.3g 专用膨化表面活性剂，一边用加热套加热，一边摇动烧瓶，同时慢慢加入经过粉碎的普通工业硝酸铵 200g，待全部溶解后，将溶液升温至 130℃，此时，打开真空泵电源开关，开始抽真空，经 30s 后磨口烧瓶内的真空度达到 0.095MPa。在抽真空的同时，用摄像机拍摄整个膨化过程。将所拍摄的录像带，经编辑后用监视器播放。

从播放画面可以观察到硝酸铵膨化的一个近似微观过程。当真空容器内的压力低于硝酸铵溶液的饱和蒸气压时，溶液沸腾，由于表面活性剂在硝酸铵溶液中起降低表面张力和发泡剂的作用，更增加了沸腾的剧烈状态。沸腾有利于水分的逸出，随着水分的不断逸出，体系硝酸铵的浓度增大，溶液的表面张力增加，因此体系的沸腾状态减弱。随着水分的逸出和由此带来的温度下降，促使硝酸铵溶液由不饱和状态转化到饱和状态，进而达到过饱和状态。当达到过饱和状态后，硝酸铵开始迅速结晶，伴随结晶过程中硝酸铵结晶热的释放，促使水分迅速蒸发，导致整个体系体积迅速膨胀至原来的 6～10 倍，从而使体系中形成许多间隙和空洞。随着绝大部分硝酸铵的析出，体系进入硝酸铵的干燥阶段。

3 硝酸铵膨化过程中物理化学参数研究[3,5]

3.1 硝酸铵溶液比热容 c 的计算

在硝酸铵的膨化过程中，考虑到真空和较高的温度环境（100℃），用实验的方法和溶液析晶时能量衡算的原理来求取 c 值，以减小计算误差。

所进行的实验是实验室或工业生产中通常采用的工艺条件。假设硝酸铵析晶前后其 c 值相同，由体系水的蒸发潜热应等于体系的温度下降所释放的热量，可求得 c 值，即：

$$m_v Q_{H_2O}=c(398-T)(100+m_{H_2O})$$

$$c=\frac{m_v Q_{H_2O}}{(398-T)(100+m_{H_2O})} \tag{1}$$

式中 m_v——水的蒸发量，g；

Q_{H_2O}——单位质量水的蒸发潜热，$2260kJ \cdot kg^{-1}$；

T——析晶点的温度，K；

m_{H_2O}——硝酸铵溶液中水的初始质量，g。

则

$$c=\frac{2.0\times 2260}{(398-378.8)(100+10)}=2.14kJ \cdot kg^{-1} \cdot K^{-1}$$

3.2 硝酸铵膨化过程中有关结晶参数的测定和计算

硝酸铵膨化结晶过程中，质量与能量相互关联，应将物料和热量一并计算。从溶质物料平衡上看，应有溶质的总量等于析出的溶质量与残留母液中溶质量之和。则在时间间隔 $\Delta t=t_1-t_2$ 时间内有下列关系存在：

$$m_e w_1=\Delta m_Y+(m_e-\Delta m_v-\Delta m_Y)w_2 \tag{2}$$

式中 w_1——t_1 时刻溶液中硝酸铵的质量分数，%；

w_2——t_2 时刻硝酸铵的质量分数，%；

m_e——t_1 时刻硝酸铵溶液的质量，g；

Δm_v——Δt 时间内水的蒸发量，g；

Δm_Y——Δt 时间内硝酸铵的结晶量，g。

从能量守恒上看，溶剂水的蒸发潜热应等于溶液的显热降、硝酸铵的结晶热和已结晶硝酸铵的显热降之和。同样在 $\Delta t=t_1-t_2$ 时间间隔内，有

$$\Delta m_v Q_{H_2O}=c m_e \Delta T+Q_{AN}\Delta m_Y+c_{AN}m_Y\Delta T \tag{3}$$

式中 c_{AN}——硝酸铵晶体的比热容，近似为 $1.76kJ \cdot kg^{-1} \cdot K^{-1}$；

ΔT——Δt 时间间隔内的温度降，K；

Q_{AN}——硝酸铵结晶热，近似等于硝酸铵的熔融热和在 398K 左右时的晶变热之和，即：

$$Q_{AN}=12.24+16.7=28.96kcal \cdot kg^{-1}=121.0kJ \cdot kg^{-1}$$

m_Y——为已结晶的硝酸铵的总量，g。

将有关数值代入式（3），解出

$$\Delta m_Y=\frac{2260\Delta m_v-(2.14m_e\Delta T+1.76m_Y\Delta T)}{121.0} \tag{4}$$

由式（2）可解出 w_2

$$\omega_2=\frac{m_e\omega_2-\Delta m_Y}{m_e-\Delta m_v-\Delta m_Y} \tag{5}$$

则绝对过饱和度：

$$\Delta\omega_s=\omega_2-\omega_s \tag{6}$$

相对过饱和度：

$$\sigma=\frac{\Delta\omega_s}{\omega_s}=\frac{\omega_2-\omega_s}{\omega_s} \tag{7}$$

式中，w_s 为 t_2 时刻对应温度 T_2 时硝酸铵溶液的饱和浓度，可查表而知。

联立式（2）～式（5），只要测量各时刻的温度和水的蒸发量，就可解得结晶过程中一系列参数。有关测量和计算结果见表 1 和表 2。

表 1 初温 125℃ 时硝酸铵膨化过程结晶参数

Table 1 The parameters of crystallization during process of AN expansion at initial temperature of 125℃

时间 /s	温度 T /K	失水量 m_v/g	结晶量 m_Y/g	溶液质量 m_e/g	硝酸铵质量分数 w/%	硝酸铵饱和质量分数 w_s/%	绝对过饱和度 Δw_s/%
8.4	378.8	2.0	0	108	92.59	92.00	0.59
15	377.0	2.6	9.42	97.98	92.45	91.70	0.75
30	375.0	3.6	25.03	81.37	92.13	91.40	0.73
60	372.4	5.0	45.65	59.35	91.58	90.67	0.91
90	369.8	6.1	61.66	42.24	90.77	90.07	0.70
120	367.5	6.8	70.90	30.30	90.01	89.58	0.43
150	365.8	7.4	79.36	23.24	89.81	89.55	0.26
180	364.2	7.8	84.31	17.89	88.91	88.81	0.10
210	363.0	8.1	88.05	13.85	87.95	88.43	−0.48

注：表中温度 T、失水量 m_v 为实测值，硝酸铵饱和质量分数 w_s 为文献查取值，实验用硝酸铵 100g、水 10g、膨化剂 0.15%，真空度 0.092MPa。

表 2 初温 135℃ 时硝酸铵膨化过程结晶参数

Table 2 The parameters of crystallization during process of AN expansion at initial temperature of 135℃

时间 /s	温度/T /K	失水量 m_v/g	结晶量 m_Y/g	溶液质量 m_e/g	硝酸铵质量分数 w/%	硝酸铵饱和质量分数 w_s/%	绝对过饱和度 Δw_s/%
7.5	383.2	3.2	0	106.8	93.63	93.05	0.58
15	382.4	3.7	7.78	98.52	93.60	92.89	0.71
30	381.0	4.7	23.78	81.51	93.50	92.60	0.90
60	378.4	6.4	50.79	52.81	93.18	92.27	0.91
90	377.2	7.1	71.38	41.28	92.84	91.80	1.04
120	376.2	7.7	79.44	30.92	92.56	91.58	0.98
150	375.4	8.2	79.44	22.36	91.94	91.40	0.54
180	374.6	8.6	89.40	12.00	91.33	91.21	0.12
210	374.0	9.1	93.71	7.19	90.05	90.73	−0.23

注：表中温度 T、失水量 m_v 为实测值，硝酸铵饱和质量分数 w_s 为文献值，实验用硝酸铵 100g、水 10g、膨化剂 0.15%，真空度 0.092MPa。

比较表 1 和表 2 可知，硝酸铵溶液的初始温度较高，则在相同真空条件下水的蒸发速度快，相应的体系的绝对过饱和度大，体系中硝酸铵的结晶速度就快。当绝大部分硝酸铵结晶后，这时体系可供结晶的晶核特别多，溶液过饱和度下降，甚至出现负值，说明此时硝酸铵的干燥成为主体，体系由结晶阶段进入干燥阶段。当然干燥时也有少量结晶产生。

4 硝酸铵膨化机理及影响因素[2, 6-9]

4.1 空白膨化实验研究

空白膨化是指不加任何表面活性剂条件下强制析晶过程。我们研究了其溶液浓度、温度和真空度等对膨化的影响，其实验结果如表 3～表 5 所示：

由表 3 的实验结果可知，只有当溶液浓度增大到一定程度时，在一定的温度及真空度下，硝酸铵溶液中的水分立即汽化成水蒸气，水蒸气挥发后形成气缝，同时快速结晶出来的硝酸铵晶体互相不规则连接，使硝酸铵结晶体产生了部分“气穴”及“孔隙”，结晶硝酸铵表面略显“粗糙”，初步形成多孔膨化硝酸铵。此即为所提出的无表面活性剂存在时的空白硝酸铵膨化机理。

表 3 溶液初始浓度对硝酸铵空白膨化效果的影响

Table 3 The iufluence of solut ion initial concent rat ion on effect of AN expansion without surfactant

硝酸铵质量分数/%	溶解时间/min	析晶时间/min	膨化后 AN 体积膨胀倍数	膨化情况
<5	<15	>20	2	AN 硬而密实，基本无膨化
85~88	15	10~14	3.2	稍膨化，仍硬
88~92	15~16	3~5	5.5	膨化，稍疏松
92~95	24~27	3	5.0	膨化，但不疏松
>95	>28	1~2	4.3	AN 硬，部分膨化

注：实验是在溶液温度为 130℃、真空度为 0.092Mpa 情况下，减压析晶。

同样，表 4，表 5 的实验结果亦可基于此膨化机理得到解释。

表 4 结果表明，一定浓度及真空度下，溶液温度的提高，有利于硝酸铵饱和溶液水分的快速蒸发，缩短析晶时间，避免硝酸铵晶体之间"盐桥"的出现，使硝酸铵膨化效果较好。而当溶液温度提高到一定程度后，硝酸铵结晶的同时，要放出结晶热，这部分结晶热同时参与硝酸铵晶体内部水分的汽化蒸发，所以继续提高溶液温度对膨化效果影响不大。

表 4 溶液温度对硝酸铵空白膨化的影响

Table 4 The influence of solution temperature on effect of AN expansion without surfactant

溶液温度 T/K	析晶时间 /min	膨化后 AN 体积膨胀倍数	膨化情况	溶液温度 T/K	析晶时间 /min	膨化后 AN 体积膨胀倍数	膨化情况
378	16	2.5	AN 硬而实，基本无膨化	403	3	5.6	膨化，稍疏松
383	12	3.0	稍膨化，较硬	408	2~3	5.5	膨化，稍疏松
388	5~7	3.7	稍膨化，较硬	413	2	5.7	膨化，稍疏松
398	3	5.3	膨化，但不疏松				

注：实验是在初始溶液硝酸铵质量分数为 90%，真空度为 0.092Mpa 情况下，减压蒸发结晶。

表 5 结果表明，只有在较高真空度下，才能使硝酸铵结晶体形成相对较好的多孔膨松状。因为高真空度下有利于水分的快速蒸发及抽走。否则，低真空度下导致析晶时间较长，硝酸铵颗粒之间出现"盐桥"，致使硝酸铵膨化效果差。

表 5 真空度对硝酸铵空白膨化的影响

Table 5 The influence of vacuum on effect of AN expansion without surfactant

真空度 /MPa	析晶时间 /min	AN 膨化后体积膨胀倍数	膨化情况	真空度 /MPa	析晶时间 /min	AN 膨化后体积膨胀倍数	膨化情况
0.080	15~17	2.6	AN 硬实，基本无膨化	0.092	3~5	5.3	膨化，但不疏松
0.085	7~10	23.5	较硬，稍膨化	0.094	3	5.3	膨化，但不疏松
0.088	5~6	4.6	较硬，部分膨化	0.096	1~3	5.5	膨化，稍疏松
0.090	3~5	5.1	膨化，但不疏松				

注：实验是在溶液初始硝酸铵分数为 90%，溶液温度为 130℃下的减压蒸发结晶。

实验研究表明，在溶液初始硝酸铵质量分数为 90%。溶液温度为 130℃，结晶真空度为 0.092MPa 时，硝酸铵膨化后体积膨胀倍数可达 5 倍之多。扫描电镜（SEM）观察此条件下空白膨化硝酸铵（无表面活性剂作用）颗粒形貌，表面仅有裂纹及棱角，而无大量孔穴（见图 1）。

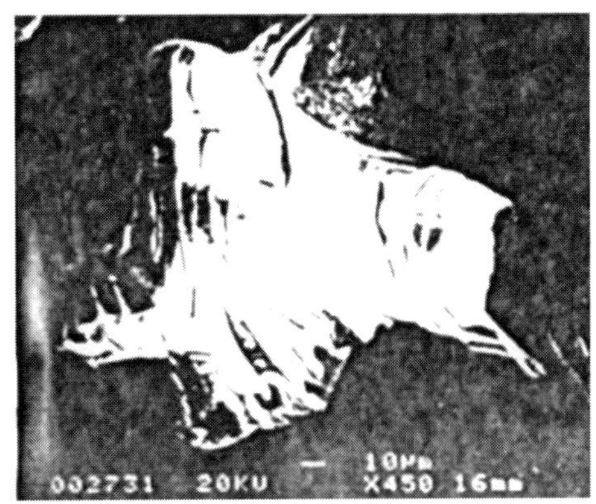

图 1 空白膨化硝酸铵

Fig. 1 Micro-structure of expanded AN without surfactant

因此，基于所提出的无表面活性剂作用的空白硝酸铵膨化机理及上述分析，一般将硝酸铵膨化处理工艺即减压蒸发结晶过程的工艺条件选择为：溶液中硝酸铵质量分数为 86%~94%、溶液温度为 110~140℃、结晶真空度≥0.085MPa。

尽管空白膨化硝酸铵质地不疏松、较硬，表

面多裂缝、少孔，但它已是我们膨化硝酸铵的雏型，上述空白膨化工艺条件都为后来形成的最佳膨化工艺条件奠定了基础。

4.2 膨化剂用量对膨化的影响

膨化剂在硝酸铵结晶过程中能有效地降低溶液的表面能，使溶液体系的介稳区变窄，并“干扰”晶体的生长，使结晶过程中产生大量的晶形“歧化”的细晶。尤其在水分快速蒸发时，会在其晶体上留下空隙并使表面更加粗糙，有较大的比表面，同时将微气泡引入到硝酸铵晶体内，造成起爆热点。由此可见膨化剂对硝酸铵的膨化起着至关重要的作用。膨化剂用量对膨化效果的影响见表 6。

表 6 膨化剂用量对膨化效果的影响

Table 6 The influence of expanding agent dosage on effect of AN expansion

膨化剂质量分数/%	<0.10	0.10	0.12	0.15	>0.15
膨化倍数	<6	7～8	8～9	9～10	9～10
比表面积/$cm^2 \cdot g^{-1}$	<1590	1710～2180	2140～2770	2550～3120	2550～3120

在表 6 中，膨化剂用量指加入的膨化剂作硝酸铵的质量百分比；膨化倍数是指硝酸铵膨化后的体积与膨化前溶液体积之比值；比表面积是采用透气法测定出来的，普通硝酸铵的比表面积为 758 $cm^2 \cdot g^{-1}$（文献值）。膨化过程中真空度为 0.092MPa，膨化硝酸铵中水的质量分数均低于 0.15%。

4.3 溶液浓度、温度和真空度对膨化的影响

硝酸铵的膨化是硝酸铵饱和溶液在膨化剂作用下的真空结晶过程。根据溶解度——超溶解度曲线，晶体的析出与溶液浓度、温度有一定的关系，结晶只能发生在过饱和状态下。在一定真空条件下，当外界压力低于硝酸铵溶液的平衡饱和蒸发压时，水即从溶液中快速蒸发出来，并吸收汽化热，使溶液温度下降和浓度提高。当温度降至低于饱和溶液相应的温度时，溶液就进入过饱和状态，并逐步析出结晶，同时释放出结晶热，又促使溶液蒸发。这样在不断蒸发水的同时，不断产生结晶。大量实验表明，硝酸铵质量分数为 90%的溶液在温度为 125～135℃时不需要另外加热，即可减压蒸发制得水的质量分数小于 0.15%的膨化硝酸铵。溶液浓度偏低则膨化硝酸铵的水分含量超标，浓度偏高则要求相应的溶解度升高，会造成硝酸铵膨化后发粘，硬度变大，不利于后续加工。

硝酸铵在膨化的同时也被干燥。真空条件一方面使硝酸铵溶液的沸点降低以利于析晶，另一方面将其中的水分带走，因此真空是硝酸铵膨化必须具备的条件。从理论上讲，真空度越高，能促使硝酸铵溶液的沸点降低并加速水分的蒸发，从而缩短膨化过程所需时间，膨化效果越好。但高真空条件往往对设备的要求较高，因此在膨化硝酸铵的生产中需要给出合适的真空度下限（见表 7)。

表 7 真空度对膨化效果的影响

T able 7 The influence of vacuum on effect of AN expansion

真空度/MPa	≤0.070	0.075	0.080	0.085	0.090	≥0.092
膨化倍数	1	2～3	4～5	7～8	9～10	9～10
水质量分数/%	≥0.80	0.50	0.20	0.12	0.07	0.06

注：实验膨化剂质量分数均为 0.15%

综上所述，作者认为硝酸铵膨化的最佳工艺条件应为：

饱和溶液的硝酸铵质量分数在 86%～94%之间；温度控制在 110～140℃；膨化剂质量分数 0.12%～0.15%，兼顾生产安全性与膨化效果，在实际生产中膨化剂质量分数控制在 0.12%左右；真空度 0.085～0.095MPa 就可以获得最佳膨化效果，并使膨化硝酸铵中水的质量分数小于 0.15%以满足炸药生产的要求。

4.4 硝酸铵膨化过程可能的机理

通过对工艺条件及表面活性剂起泡性对硝酸铵膨化效果影响的研究，目前认为硝酸铵膨化机理可能是：在一定的温度及真空度作用下，硝酸铵饱和溶液中的水分汽化，生成汽泡，在表面活性剂的作用下形成大量的相对稳定的泡沫。但当温度、真空度达到特定值情况下，泡沫破裂，水蒸汽快速蒸发，由于水分子快速逸出，在硝酸铵晶体内形成大量“孔、洞、隙”，快速析晶出的硝酸铵粒子表面也极不规整，造成硝酸铵粒子之间以极不规则的方式连接，形成一种轻质、多孔、疏松的膨化硝酸铵。

5 结语

膨化理论研究，特别是它的物理化学参数计算与测定、表面活性剂的作用、膨化影响因素及膨化机理研究等为膨化工艺途径提出及确定提供重要理论基础，为膨化技术的有效实施提供可靠保证。

硝酸铵膨化技术已经比较成熟，是一种完全能够被工厂接受的，受操作工人欢迎的“绿色”生产技术，在进一步完善工艺过程的连续化及遥控化、连续混药工艺及自动化包装技术以后，必将使我国粉状硝铵炸药生产技术走在世界前列。

参考文献

[1] 吕春绪．表面活性剂理论与技术［M］．南京：江苏科学技术出版社，1991.

[2] 吕春绪．膨化硝酸铵表面化学研究［J］．南京大学学报，1995（31）：286.

[3] 陆明．膨化硝铵炸药研究［D］．南京：南京理工大学，1999.

[4] 陆明，吕春绪，刘祖亮．膨化硝酸铵的膨化过程研究［J］．火炸药学报．1999（3）：15.

[5] 陆明，吕春绪，刘祖亮．硝酸铵膨化过程中的结晶参数研究［J］．含能材料，1999（2）：79.

[6] 吕春绪，刘祖亮，惠君明．岩石膨化硝铵炸药研究［J］．爆破器材，1997（1）：5.

[7] Lv Chunxu，Liu Zuliang. Research on Expanding Mechanism of Ammonium Nitrate［A］. Proceedings of the 23th International Pyrotechnics Seminar（IPS）［C］. Tsukuba，Japan. 1997：483.

[8] Lv Chunxu. Research on Expansion Technology of Ammonium Nitrate［A］. Proceedings of the 26th IPS［C］. Nanjing China. 1999：344.

[9] 胡炳成，刘祖亮，吕春绪．影响膨化硝酸铵性能的因素［J］．爆破器材，2000（2）：16.

Research on Expansion Theory of Ammonium Nitrate

Lv Chunxu

（Institute of Industrial Explosive Materials，Nanjing 210094，China）

Abstract：In this paper，expansion process of AN is described，its physical chemical parameters are calculated and determined，the influencing factors of AN expansion process are studied and a possible mechanism of expansion process is also given.

Key words：expanded ammonium nitrate；expansion theory；physical chemical parameter；influencing factor

（注：此文原载于 中国工程科学，2001，3（9）：58-63）

膨化硝酸铵晶体特性研究

吕春绪

（南京理工大学，江苏南京，210094）

EI：2002427143949

摘要：膨化硝酸铵是一种改性硝酸铵晶体，具有显著自敏化特征，成功应用于各类膨化硝铵炸药及震源药柱中。在膨化剂（表面活性剂）的作用下改善了膨化硝酸铵的物理性能和晶体结构及其特征。本文通过硝酸铵固体表面接触角的测量，算出了膨化硝酸晶体表面能；利用差热分析（DSC）技术给出了膨化剂对硝酸铵晶变的影响；采用X射线衍射分析了膨化硝酸铵晶体的晶格特性。

关键词：膨化硝酸铵，晶体特征，表面能，晶变，晶格

膨化硝酸铵是一种经过膨化工艺获得的具有自敏化特征、轻质、多孔、疏松的改性硝酸铵晶体。膨化工艺的实质是表面活性技术，是膨化剂（表面活性剂）作用下强制析晶的物理化学过程。

用膨化剂对硝酸铵进行膨化处理时，膨化剂会均匀地分散在硝酸铵颗粒表面，其分子中的极性基和硝酸铵分子相互吸引，而疏水性的非极性基则指向硝酸铵分子外部。这样形成了一层刁水薄膜，防止了外界水分的侵入。根据表面化学理论，硝酸铵颗粒的表面能降低，同时膨化剂亦能改善硝酸铵的晶变。因此，膨化剂不仅可以降低硝酸铵的吸湿性，而且具有防结块作用。另外，在膨化剂作用下，硝酸铵晶格特征同样也是非常重要的。

1 膨化硝酸铵晶体表面接触角以及表面自由能测算

根据表面化学理论，对于低能表面物质，可以用单液法测接触角，从而间接求得表面能及其分量[1]。但对于硝酸铵粒子这种高能表面物质，由于液体会迅速在其表面上铺展，用单液法无法测得接触角数据，必须用双液系接触角测定法进行测定。用各类膨化剂处理得到的膨化硝酸铵接触角数据如表 1.1 所示[5]。

表 1.1 各类膨化剂处理的膨化硝酸铵的接触角

Tab. 1. 1 Contact angles of expanded ammonium nitrate (AN) following treatment with various expanding agents

膨化剂	含量%	乙二醇/(°)	丙三醇/(°)
TW-80	0.10	13.0	38.0
	0.15	21.0	43.5
	0.20	24.5	46.0
SDS	0.10	14.0	39.5
	0.15	22.5	45.5
	0.20	27.0	48.5

续表

膨化剂	含量%	乙二醇/(°)	丙三醇/(°)
SOCTA	0.10	15.5	40.5
	0.15	25.0	49.0
	0.20	29.5	52.0
TW-80/SDS	0.10	14.5	39.5
	0.15	22.5	46.0
	0.20	27.5	49.5
TW-80/SOCTA 1∶1	0.10	16.0	41.0
	0.15	24.5	49.0
	0.20	30.0	53.5
SOCTA/SDS 1∶1	0.10	19.5	44.0
	0.15	30.0	52.5
	0.20	34.0	55.0
SOCTA/HST 3∶1	0.10	20.5	45.0
	0.15	29.0	52.5
	0.20	33.0	55.0

注：1. 普通硝酸铵在测度液乙二醇中接触角为10°，丙二醇中为37°。

2. 所有测试均选用环己烷为参比液。

利用接触角的实验数据及固体表面自由能及表面张力公式，并依据一定计算程序可算出各类膨化剂作用获得的不同膨化硝酸铵晶体的表面张力，其结果如表1.2所示[5]。

表1.2 各类膨化剂处理的膨化硝酸铵的表面张力

Tab. 1.2 Surface tension of expanded ANs following treatment with various expanding agents

膨化剂	含量%	γ_s^d/mJ·m^{-2}	γ_s^p/mJ·m^{-2}	γ_s/mJ·m^{-2}
TW-80	0.10	81.26	16.17	97.43
	0.15	69.60	15.89	85.49
	0.20	65.83	15.61	81.44
	0.10	76.46	16.27	92.73
SDS	0.15	63.88	15.93	79.81
	0.20	59.52	15.51	75.03
	0.10	73.72	16.27	89.99
SOCTA	0.15	53.63	16.02	69.65
	0.20	56.60	15.57	72.17
TW-80/SDS 1∶1	0.10	77.06	16.22	93.28
	0.15	61.81	16.01	77.82
	0.20	56.60	15.57	72.17
TW-80/SOCTA 1∶1	0.10	72.70	16.25	88.95
	0.15	52.63	16.14	68.77
	0.20	44.76	15.71	60.47
SOCTA/SDS 1∶1	0.10	64.84	16.22	81.06
	0.15	49.04	15.53	64.57
	0.20	44.92	15.07	59.99
SOCTA/HST 1∶1	0.10	62.32	16.22	88.54
	0.15	46.82	15.78	62.60
	0.20	44.91	15.24	60.15

从表1.2可以看出下面几点[2-5]：

（1）用膨化剂处理得到的膨化硝酸铵的表面能都有不同程度的降低。

（2）膨化剂含量是从0.10%升到0.15%时，硝酸铵固体表面能降低得较多（10～20mJ·m^{-2}），而当膨化剂含量从0.15%上升到0.20%时，一面能降低得较少（1～mJ·m^{-2}）。

（3）含有阳离子SOCTA的表面活性剂体系，都具有降低硝酸铵表面能的良好作用，SOCTA/

SDS、SOCTA/HST、SOCTA/TW-80、SOCTA 处理的硝酸铵表面能均低于用 SDS/TW-80、SDS、TW-80 处理的硝酸铵表面能。说明 SOCTA 为较好的硝酸铵表面包覆剂。

（4）复配体系处理的硝酸铵表面能均低于相应的单一体系处理的硝酸铵表面能，其中以 SOCTA/SDS 为最佳。SOCTA/HST 同 SOCTA/SDS 效果相当。

2 膨化硝酸铵晶体晶变的研究

硝酸铵具有多晶性，常压下，它具有五种热力学稳定的晶体结构，而每种结构仅在一定的温度范围内存在，当温度发生变化时，各晶型之间可以相互转变[1]。硝酸铵的晶型变化，不仅结构发生改变，同时还伴有一系列物理性能的变化，如密度、比容、膨胀系数等都会变化，特别是在 32℃ 当硝酸铵晶型Ⅳ转变为晶型Ⅲ时，其体积变化最明显，为 3.6%。由于此时正是在硝酸铵正常储存、使用的温度范围内，而且其晶型改变所引起的硝酸铵体积变化最明显，因而硝酸铵极易结块。

各种膨化剂作用下的膨化硝酸铵的 DSC 谱图和晶变点数据的测试结果见图 2.1［（a）TW-80/SDS 膨化处理 AN；（b）SOCTA/SDS 膨化处理 AN］及表 2.1[5]。

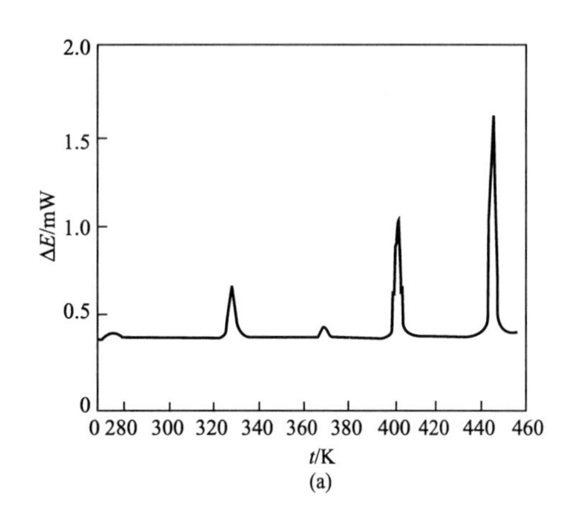

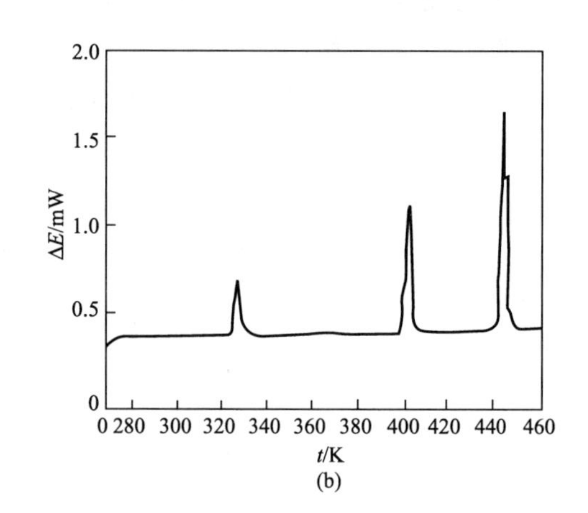

图 2.1 各类膨化硝酸铵的 DSC 谱图

Fig. 2.1 DSC diagram of various expanded AN

表 2.1 不同膨化剂对硝酸铵晶变的影响

Tab. 2.1 Influence of different expanding agents on the crystalline transfer of ANs

膨化剂	晶变点/k		晶变热/J·g^{-1}	
	Ⅳ⇌Ⅲ	Ⅲ⇌Ⅱ	Ⅳ⇌Ⅲ	Ⅲ⇌Ⅱ
SDS	327.30	368.37	21.07	9.91
TW-80	329.66	365.86	24.12	9.45
SOCTA	329.57	364.22	24.41	1.30
SDS/TW-80(1∶1)	329.59	264.69	23.99	6.44
SOCTA/TW-80(1∶1)	328.90	366.24	24.20	1.88
SOCTA/SDS(1∶1)	328.96	363.03	23.66	0.33
SOCTA/HST(3∶1)	329.29	367.10	23.70	1.42

注：膨化剂用量为 0.15%

从 DSC 图谱及表 2.1 可以看出：

（1）膨化剂膨化处理的硝酸铵，Ⅲ⇌Ⅱ相变吸收峰明显变小，晶变热明显降低（普通纯硝铵Ⅲ⇌Ⅱ的晶变热为 16.72J·g^{-1}），且晶变点后移。

（2）用含 SOCTA 的膨化剂体系处理的硝酸铵于Ⅲ⇌Ⅱ的相变时的晶变热极小（0～1.88J·g^{-1}），说明硝酸铵在 328K 附近的晶变实际上主要是Ⅳ⇌Ⅱ转变，只有很少一部分Ⅳ⇌Ⅱ转变，说明 SOCTA 是改善硝酸铵晶变点的良好添加剂。

(3) 用 SOCTA/SDS 处理的硝酸铵 328K 的吸收峰极其微弱，Ⅲ⇌Ⅱ相变热仅为 0.33J·g^{-1}。说明硝酸铵晶体中基本没有Ⅲ⇌Ⅱ相变。因而可以判定硝酸铵晶体中基本不存在Ⅲ相相变。

通过分析可以得出：硝酸铵的晶体结构组成是由硝酸铵分子的 NH_4^+ 和 NO_3^- 两种离子所决定的，而硝酸铵的热力学多晶现象就是由这两种基本质点的相互作用而造成。用膨化剂处理的硝酸铵，由于膨化剂分子中的极性基与硝酸铵分子中的离子之间可以通过静电作用一定程度上“扰乱”了原来硝酸铵分子中 NO_3^- 和 NH_4^+ 之间的作用，达到了随温度变化，硝酸铵不易产生因有晶变的目的。由于 SOCTA 中的 RNH_3^+ 同 NH_4^+ 具有相似的结构，它的“扰乱度”更大。而 SOCTA/SDS 正负离子表面活性剂同硝酸铵具有更强的静电吸引作用，其结果是 SOCTA/SDS 改善硝酸铵晶变的效果最好。

3 膨化硝酸铵晶体改善吸湿性和结块性研究

对膨化硝酸铵晶体改善其吸湿及结块进行了大量实验研究[6-12]，并掌握了一些变化规律。经膨化处理的膨化硝酸铵晶体表面包覆了一层疏水薄膜，导致硝酸铵表面能的降低，改善了膨化硝酸铵的吸湿性。选择了普通硝酸铵及各种膨化剂处理后的膨化硝酸铵体系来进行吸湿性研究。所测定的各类硝酸铵吸湿增重结果如表 3.1 所示。

表 3.1 各类膨化硝酸铵晶体的吸湿性（相对百分量%）

Tab. 3.1 Hygroscopicity of various AN crystals after expansion (relative percent %)

硝酸铵体系		样品质量/g	吸湿时间/h							
			1	2	3	4	5	6	7	9
膨化硝酸铵	普通硝酸铵	2.0032	0.30	0.52	0.71	0.86	0.98	1.07	1.15	1.21
	SDS 膨化剂	1.9765	0.22	0.41	0.58	0.75	0.88	1.00	1.09	1.18
	TW-80 膨化剂	1.9992	0.18	0.34	0.52	0.65	0.78	0.89	1.00	1.14
	SOCTA 膨化剂	2.0130	0.13	0.27	0.41	0.53	0.65	0.75	0.83	0.95
	SOCTA/SDS 膨化剂	1.9477	0.11	0.21	0.30	0.41	0.52	0.61	0.69	0.76

注：膨化剂用量为 0.15%，温度 18℃，相对湿度 76%，各类膨化硝酸铵 60 目以下。

从表 3.1 中数据充分显示：膨化剂改善硝酸铵吸湿的效果同降低硝酸铵固体表面能的效果有很好的一致性；各类膨化剂改善硝酸铵结块性效果同改善硝酸铵晶变及吸湿性的效果也具有良好的一致性。

4 膨化硝酸铵晶体晶格的研究

膨化剂可以影响膨化硝酸铵晶体的成长速度及成长方向，改变晶体晶格构成，进而影响晶体很多其它特性。用扫描电镜（SEM）及 X 射线衍射可以观察分析膨化硝酸铵晶体的这些性质[5]。对普通硝酸铵、空白膨化硝酸铵、SOCTA/SDS 膨化的硝酸铵及 SDS/TW-80 膨化的硝酸铵作其 X 射线衍射图，并相互比较，如图 4.1 [（a）普通 AN；（b）空白膨化 AN；（c）SOCTA/SDS 膨化 AN；（d）SDS/TW-80 膨化 AN] 所示。

可见，四种硝酸铵的 X 射线衍射谱具有相同的衍射峰区及 I/I_0 值。因此，可认为它们的晶格结构相同。说明硝酸铵膨化过程，不改变其晶体结构的几何图形和结点空间排布及其规律性。SEM 观察结果也证明这一现象。

5 结论

(1) 用各种膨化剂处理制得的膨化硝酸铵的表面能都有不同程度的降低，其中阳离子型膨化剂

SOCTA 效果较好，且复配体系有更好的效果。

（2）膨化剂在其表面形成疏水性薄膜，同时是降低硝酸铵晶变热、改善晶变点的良好添加剂，SOCTA 表现尤佳，从而有效地改善了硝酸铵晶体的吸湿性和结块性。

（3）膨化剂仅仅改变硝酸铵的晶形而不改变其晶格。

参考文献

[1] 吕春绪．表面活性理论与技术．南京：江苏科学技术出版社，1991. 24-25.

[2] 丁芸．表面活性剂改善硝酸铵及其炸药性能的研究：[硕士学位论文]．南京：华东工学院，1990.

[3] 吕春绪，刘祖亮，倪欧琪．工业炸药．北京：兵器工业出版社，1994. 155-162.

[4] 吕春绪．硝酸铵晶析的表面化学特性研究．中国民爆器材学会第四届年会论文集．北京：兵器工业出版社，1995. 6.

[5] 叶志文．硝酸铵膨化机理研究：[硕士学位论文]．南京：南京理工大学，1995.

[6] 陈天云．硝酸铵表面特性及其应用研究：[硕士学位论文]．南京：南京理工大学，1993.

[7] 胡炳成．表面活性剂在硝酸铵改性中的应用研究：[硕士学位论文]．南京：南京理工大学，1994.

[8] 吕春绪．膨化硝酸铵界面化学研究．南京：南京大学学报，1995，(31)：286.

[9] 陈天云，吕春绪．表面活性剂和添加剂降低硝酸铵吸湿性研究．火炸药学报．1998，(4)：19.

[10] 胡炳成，刘祖亮，陆明，陈天云．表面活性剂改善硝酸铵结块性研究．中国民爆器材学会第四届年会论文集．北京：兵器工业出版社，1995. 74.

[11] Chen Tianyun，Liu Zuliang，Lv Chunxu，Hui Junming，Lu Ming. Research on Expanded Ammonium Nitrate Explosive. Proceedings of the Third Beijing International Symposium on Pyrotechnics and Explosives. Beijing China. 1995. 99.

[12] 刘祖亮，叶志文，胡炳成，王依林．多功能复配型硝酸铵膨化剂的研究．中国民用爆破器材学会第五届年会论文集．苏州．2000.

A Study on Crystalline Characteristics of Expanded Ammonium Nitrate

Lv Chunxu

(Nanjing University of Science and Technology，Nanjing，210094)

Abstract: Expanded ammonium nitrate (AN) is a kind of new improved AN，possessing outstanding self-sensitized characteristics，and is successfully applied to various kinds of expanded AN explosives and seismic explosive columns. Under the action of expanding agents (surfactants)，its physical properties，crystalline structures and characteristics are remarkably improved. In this paper，through determination of solid surface contact angles，the surface energy of expanded AN crystal lattice characteristics of expanding agents on crystalline change is obtained through DSC，and the crystal lattice characteristics of expanded AN are analyzed through SEM.

Key words: expanded AN，crystal characteristic，surface energy，crystalline change，crystal lattice

（注：此文原载于 兵工学报，2002，23（3)：316-319）

Expansion Technology of Ammonium Nitrate and its Application

Lv Chunxu[❶]
Nanjing University of Science & Technology (Nanjing 210094)
Proceedings of 1st International Symposium on Energetic Materials and their Applications (1th ISEM), Tokyo, 2002: PL1

Abstract: The expansion of ammonium nitrate (AN) is a novel technology. The guideline of this innnovation is the self-sensitization of AN. It is a breakthrough to classic methods. The approach to self-sensitization is through the expansion of AN. Its essence is a surfactant technology applied to powdery industrial explosive, which is a physical chemical process under coercive crystallization. In this paper, the mechanism and technical characteristics of expansion are discussed, as well as its unique advantage. The expansion technology of AN is mainly applied to rock expansive and its comparison with other indystrial explosives is also given. The expansion technology is also used in manufacturing of permitted explosive.

Key words: expanded ammonium nitrate; rock explosive; micro structure; mechanism

1 The technology of expanded AN

1.1 The essence of technology of expanded AN

Expansion of AN is a new technology invented by our research group.

Expanded AN explosive based on expansion technology is a new kind of industrial explosive.

The essence of expansion is surface activity, a physical chemical process of compelled foaming crystallization under the action of composite surfactant.

1.2 The self-sensitization theory

We have broken through the frame of traditional methods, successfully creating a theory of self-sensitization of AN. The self-sensitization of AN is to introduce micro air bubble into AN crystal to roughen crystal surface.

---

❶ 应日本首届含能材料及其应用国际研讨会（1st International Symposium on Energetic Materials and Their Applications (ISEM) 15-17 May 2002 Tokyo，Japan）邀请，吕春绪教授作为顾问委员会委员（International Advisory Committee），并特邀发言（Invited Speaker）作大会报告（Plenary Lecture）。

1.3 The basis of self-sensitization theory of AN

Detonation mechanism of industrial explosive——the heat point theory is the foundation of self-sensitization theory of AN.

The production of porous AN is on the basis of increasing specific surface area.

The sensitizing of emulsion explosive can be obtained by means of introducing micro-bubble into the body of emulsion explosive.

In order to make micro-bubble into AN particles, heat point must be formed under the strong action from outside. This is self-sensitization theory of AN.

1.4 Self-sensitization of AN

After undergone a physical chemical process, forcibly foaming and crystallizing in vacuum under the action of composite surfactant, micro-bubbles would appear. The modified AN then would have a great number of pores, cavities and cracks. So the whole surface would look extremely irregular and the modified AN would be light, porous and loose. We call this kind of modified AN expanded AN[1-5].

1.5 The expanding mechanism of AN

(1) Describing of expanding mechanism of AN

Mechanism of expansion of AN is as follows: water in saturated solution of AN is vaporized, to form water vapor bubbles under certain temperature and vacuum conditions. Then through action of surfactants, a large number of relatively stable foams are instantaneously formed. When temperature and vacuum degree reach the specified value, foams are rapidly bursted, steam quickly escaped, forming a great quantity of holes and cavities in the crystal of AN. Surface of particles of AN rapidly crystallized is not horizontal and smooth. Linking form between particles of AN is extremely irregular, therefore light, porous and bulky AN is formed. That is expanded AN.

(2) The process of AN expansion

H_2O, special surfactants and industrial AN are added to a 1000ml glass flask, the industrial AN is dissolved at certain temperature in a vacuum, the change of process of AN solution is pictured with camera. Through technical treatment, a group of pictures about the expansion process of AN is obtained as follows[6-8].

When the pressure in the glass flask is less than that of the saturated vapor pressure of AN solution, it begins to boil. With the boiling of AN solution, water molecules in AN solution is volatilized out from the glass flask accordingly. The vaporization of water absorbs heat, hence the temperature of system drops. The vaporization of water and the drop of temperature make AN solution change from a unsaturated state to a saturated state and further to a super saturated state. Expanded AN begins crystallization. At the same time, the volume of system expands quickly to 6～10 times more than that of common AN.

1.6 The structure characteristics of expanded AN

(1) The micro-structure of expanded AN

The shape of expanded AN is observed through Scanning Electronic Microscope. Compared with common AN, surface of expanded AN is quite irregular, with many moles and cracks on its

AN solution before vacuum treatment

The boiling state of AN solution under vacuum for 10s

The boiling state of AN solution under vacuum for 20s

The boiling state of AN solution under vacuum for 30s

The state of system when expanded AN being crystallization

The state of system 10s after EAN being crystallization

The state of system 20s after EAN being crystallization

The state of system 30s after EAN being crystallization

Fig. 1 The process of AN explosion

body. Crystal shape disproportionates and contains a large number of micro-gas pocket.

According to "heat point" theory of explosive, under action of explosion shock wave form outside, micro-gas pockets in expanded AN particles are adiabaticly compressed, then form "heat point" of high temperature and high pressure. The "heat point" increase sensitivity of expanded AN.

(2) The distribution state and regularity of micro-bubble in expanded AN

The relation between pure radius and volume of expanded AN and distribution of expanded AN pore radius are tested with pore radius and volume instrument.

In addition, pore volume and radius distribution of expanded AN and common AN is researched deeply.

From test data, we clearly see that pore volume of expanded AN is bigger than that of common AN, we also know that 95% of valid pore radius is in range of mesopore. This is very useful and important and benefits for valid initiation and complete blast of expanded AN.

(3) Particle size distribution

Particle size distribution of modified AN and common AN are tested respectively by a laser particle size instrument, tested results are shown in the following figure.

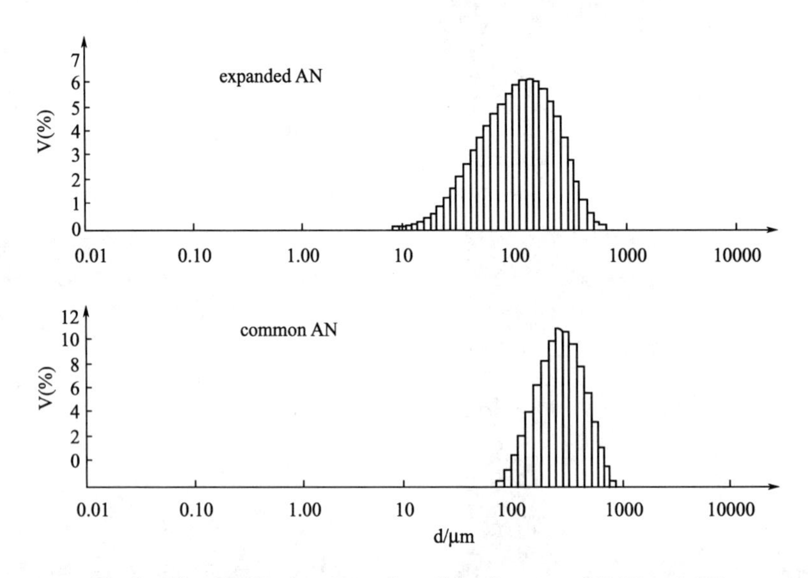

Fig. 2 The distribution state of particle diameter of EAN and AN

From above figure, we can know that particle size of expanded AN is mainly distributed in scope of 10μm-100μm, and that of common AN is mainly distributed in scope of 100μm-1000μm, average particle size of expanded AN is a smaller than that of common AN. Compared to common AN, expanded AN has wide distribution of particle size, more micro-powder, better particle size gradation, higher uniform degree of mixture with combustible agents. According to mixing react mechanism of industrial explosive, these characteristics are good for improving explosion properties of explosive.

(4) Larger specific surface area

Specific area of expanded AN is 4.38 times larger than that of common AN. It's shown in Fig. 3.

(5) Modification of surface performance of expanded AN

The rate of moisture adsorption decreases and the final product is not liable to cake[9-10].

① Research on hygroscopicty of expanded AN

Dry expanded AN and common AN, then weigh the hygroscopic increments of both. The hygro-

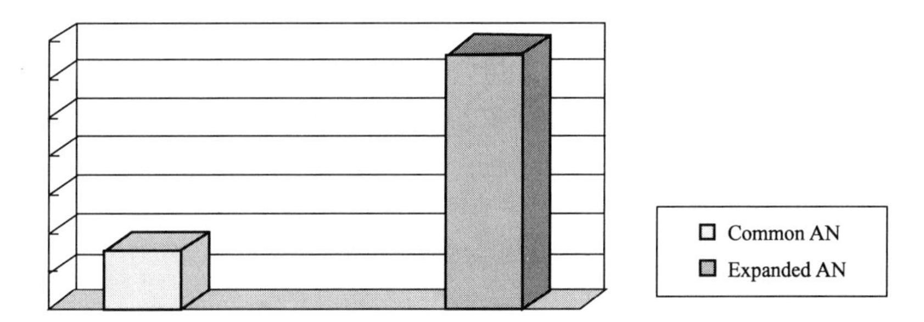

Fig. 3 Specific area of expanded AN and common AN

scopic rates are calculated and the results are shown in Table 1.

Table 1 Hygroscopic rates of expanded AN and common AN

Samples	Average Mass/g	Relative humidity (%)	Average(time/h) 2	4	6	8	20	30
Common AN	2.4671	70	0.61	1.19	1.960	2.48	6.37	8.93
	2.5012	90	2.30	4.23	6.11	8.35	14.13	17.10
Expanded AN	2.4173	70	0.24	0.45	0.72	0.93	2.22	2.94
	2.3995	90	0.47	1.60	2.47	3.35	4.80	5.27

Table 1 indicates that hygroscopic rate of expanded AN is 60%, which is less than that of common AN.

② Research on anti-caking property of AN

a. Crystal pattern transition of expanded AN studied by DSC

Crystal pattern transition and extent of AN can be determined according to heat change of AN in heating process by DSC analysis, the results are shown in Table 2.

The experimental results indicate crystal pattern transition number of expanded AN declines, crystal pattern transition position mores back compared with common AN.

Table 2 Data in DSC picture of AN

Peak number	Common AN Position of peak/℃ Start	End	High	Transition heat/J·g^{-1}	Expanded AN Position of peak/℃ Start	End	High	Transition heat/J·g^{-1}
1	46.5	62.8	54.4	21.2	50.9	64.8	56.5	23.7
2	89.4	101.4	95.5	9.9	126.6	135.8	129.8	55.1
3	125.2	133.8	129.6	55.7	166.9	178.1	172.8	76.3

b. Anti-caking property measurement of expanded AN

Anti-caking property of expanded AN and common AN are studied respectively. The results are shown in the following Table 3.

The experimental results indicate anti-caking property of expanded AN is significantly superior to that of common AN.

Table 3 Data of resistance to compression of caking AN

Sample	1	2	3	4	5	Average
Common AN	479.2	480.2	477.5	475.9	477.4	478.0
Expanded AN	86.9	87.7	83.4	81.2	84.1	84.7

Compared with common AN, crystal shape of expanded AN treated with composite surfactants disproportionates, crystal contains a large number of micro-gas pocket, has a larger specific surface area. With less hygroscopicity, better anti-caking property, expanded AN is more suitable for serving as a main oxidizer of commercial explosives.

(6) The process of machining to fine powder is extremely easy.

1.7 The safety property of expanded AN

The core of our invention is self-sensitization theory. The key of self-sensitization is safety property of expanded AN. No safety property, No self-sensitization.

We are very much interested in the properties of expanded AN, especially the cap sensitivity which is the embodiment of safety property. So cap sensitivity and its influence are detailed and deeply researched.

1.8 Conclusions about expanding AN

(1) The key technique of rock expanded AN explosive is expansion of AN. The mechanism of expansion of AN is a physical-chemical process.

(2) Expanding process firstly depend on technological conditions. Surfactant is determining factor which can ensure a successful expansion. The effect of composite surfactants on expanding process is better than that of single surfactant.

(3) The surfactant possesses functions of foaming, coating and crystal shape modification during expanding process of AN. And it makes expanded AN possess better properties of anti-hygroscopic and anti-caking.

2 Rock expanded AN explosive

There are three kinds of technology on the production process of rock expanded AN explosive.

(1) The production process by vertical expanding machine

(2) So far, this process has been attorned to more than 10 plants in China. The production process by horizontal expanding machine

Up to now, this process has been attorned to more than 50 plants in China. It is shown in Fig. 5.

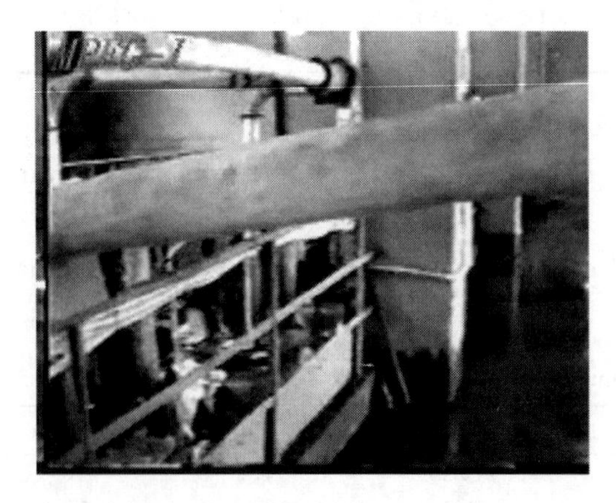

Fig. 4 Vertical expanding machine

Fig. 5 Horizontal expanding machine

(3) The production process by continuous expanding machine

This technology benefits from autocontrol and has been attorned to 10 plants in China till now.

Fig. 6 Continuous expanding machine

3 Series products of expanded AN explosives[9-12]

3.1 A new type of ANFO explosive with expanded AN

ANFO (ammonium nitrate and fuel oil) explosive has been widely used in mining, blasting engineering and construction.

Advantages of ANFO are as follows:

——It is low cost.

——The method of making ANFO explosive is simple.

Disadvantage of ANFO are as follows:

——The sensitivity to initiation is low.

——Detonating properties of ANFO explosive are lower than that of AN-TNT explosive.

——Its shelf life is shorter than that of AN-TNT explosive.

The new type of ANFO explosive is characterized by good performance, water resistance, long shelf life and low cost. The technique of this kind of ANFO explosive is simple, high efficient and continuous. Consequently, this kind of new ANFO explosive will receive welcome among the makers and users of explosive in China.

3.2 Study on the safe coal-mining expanded AN explosive

It is used in the engineering detonation under the condition of gas and powder. The main categories are 2# permitted expanded AN explosive, 2# water-resistant permitted expanded AN explosive, 3# permitted expanded AN explosive, 3# water-resistant permitted expanded AN explosive.

Powder coal-mining AN explosive and coal-mining emulsion explosive are mainly used in coal-mining exploitation in China. Coal-mining emulsion explosive has many disadvantages, such as low power and inconvenience in manual charge. It is of vital importance to study on high safe coal-mining AN explosive, namely safe coal-mining expanded AN explosive.

Serial safe permissible AN explosive was obtained from self-sensitized expanded AN, wood powder and safe fuel oil.

Safe permissible AN explosive possesses good explosion performance, safety property and water-resistance.

Process of manufacturing safe permissible AN explosive is simple and convenient.

3.3 Seismic expanded AN explosive

When used in geologic, it proves to be able to produce seismic wave.

The comparison of detonation velocity seismic of three kinds is shown in the following Table 4.

Table 4 The comparison of Detonation Velocity

	High	Middle	Low
Detonation velocity v_D/m · s^{-1}	≥5000	≥4300	≥3500
Densityρ/g · cm^{-3}	1.00～1.20	1.00～1.10	0.85～0.9
Capability of initiation to detonation	100%	100%	100%

3.4 Research on expanded AN explosive with low velocity

It is used in controlled detonation, explosion complex, low detonation velocity seismic charge and so on.

Expanded AN type of powder low detonation velocity explosive, whose cost is low, composed mainly of expanded AN through certain technology, has detonation velocity of about 2152m · s^{-1}and propagating explosion distance of above 50mm (its charge diameter is 32mm) . It solves the problem of property of propagating explosion of low detonation velocity explosive, and proves to have good effect when tested in oil wells.

The explosion composition of low detonation velocity explosive is expanded AN explosive, featuring the characteristics of reliable performance and low cost.

Expanded AN type of low detonation velocity explosive possesses good properties.

This kind of low detonation velocity explosive has been tried out by the government department concerned in the oil wells of Xinjiang Talimu Oil Field, and it is proved to have good effect with the desired results.

3.5 Orchard-oriented explosive

It is a new agent with expanded AN for both scarification and addition of nutrient to fruit trees.

3.6 Water-resistant rock expanded AN explosive

It is used in the condition existing a small quantity of water.

3.7 High strength expanded AN explosive

It is used in the very hard rock.

3.8 The second generation of expanded AN explosive

It is a new high strength powdery AN explosive. Its production process absorbes the advanced technology of current expanded AN explosive.

4 The application of expanded AN explosive[11-12]

4.1 The actuality of the application

Since 2001, Expanded AN explosive has been attorned to almost 70 plants and 46 production lines were established in more than twenty provinces of China.

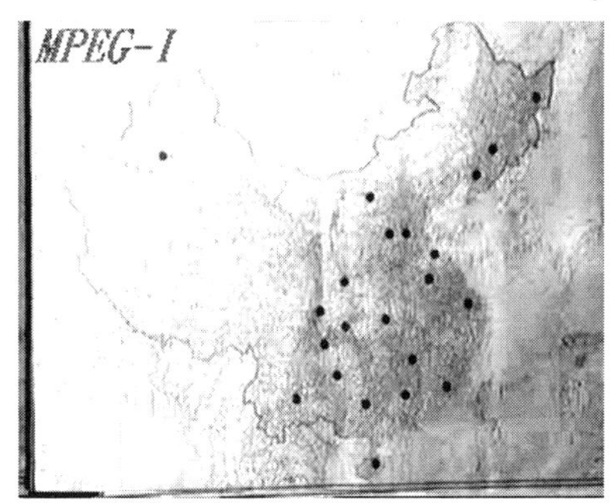

Fig. 7 The actuality of the application of expanded AN technology

Fig. 8 The symposium on the production technology of expanded AN in Hengshan of Hunan province in 2001

For the sake of promoting the development of the production technology of expanded AN, three symposiums have been held on the production technology of expanded AN.

5 Technological appraisement of expanded AN[11-12]

5.1 Main point of this invent

To introduce "micro-bubble" in powdery AN particles to sensitize it.

To invent special composite surfactant.

To find a sort of composite oil materials.

To find the prescription and production conditions of the series products of expanded AN.

To invent vertical, horizontal & continuous expansion equipments for the production of expanded AN.

5.2 Main technological appraisement

Expanded AN explosive has obtained successively many awards in China as follows:

National 2nd award for development of science and technology;

Gold award for invention of ten great inventions in China;

Emphasized spreading item of "the 9th five-year-plan" in China;

The National Torch Plan item;

The top award for development of science & technology of the Ordnance Industry Ministry;

Fig. 9 The national 2nd award for development of science and technology

Two times for the third award for development of science and technology of Jiangsu province;

One time for the third award for development of science and technology of Hubei province;

One time for the second award for development of science and technology of Hebei province;

One time for ten great invention items of Jiangsu province.

Expanded AN and explosives deriving from it contribute a great deal to overtake the world in China and they are to be perfected and promoted.

Reference

[1] Ding Yun. Improving properties of AN and its explosives using surfactant (degree thesis). Nanjing: East China Institute of Technology, 1990.

[2] Hui Junming, Liu Zuliang, Lv Chunxu. Powder AN and its preparation, China Patent, CN 91107075.6, 1991.

[3] Liu Zuliang, Hui Junming, Lv Chunxu. Light, porous AN and its manufacture, China Patent, CN 921078730, 1992.

[4] Liu Zuliang, Hui Junming, Lv Chunxu. Research on light, porous and bulky AN and HF series powder AN explosive. Explosive Materials, 1991 (2): 5-10.

[5] Lv Chunxu, Liu Zuliang, Hui Junming. Rock EAN explosive. Explosive Materials, 1997 (1): 5-10.

[6] Ye Zhiwen. Expanded mechanism of ammonium nitrate (degree thesis). Nanjing: Nanjing University of Science and Technology, 1995.

[7] Lu Ming. Study on expanded ammonium nitrate explosive (degree thesis). Nanjing: Nanjing University of Science and Technology, 1999.

[8] Lv Chunxu. Research on surface chemistry of EAN. Journal of Nanjing University, 1995 (31): 286-290.

[9] Chen Tianyun. Research on surface characteristics and application of ammonium nitrate (degree thesis). Nanjing: East China Institute of Technlolgy, 1992.

[10] Chen Tianyun. Study on the ammonium nitrate self-sensitization and its application (degree thesis). Nanjing: Nanjing University of Science and Technology, 2000.

[11] Lv Chunxu et al. Expanded Ammonium Nitrate Explosive. Beijing: The Publishing House of Ordnance Industry, 2001.

[12] Lv Chunxu et al. The theory of industrial explosive. Beijing: The Publishing House of Ordnance Industry, 2002.

ISEM2002

火薬学会

The 1st International Symposium on

Japan Explosives society

Energetic Materials and their Applications

15-17 May, 2002 Tokyo, Japan

第一届含能材料及其应用国际会议（The 1st International Symposium on Energetic Materials and their Applications）2002年5月15～17日在日本东京举行。该会议由日本火药学会主办，会议主席为火药学会会长、日本东京大学田村昌三教授 Prof. Masamitsu Tamura（Tokoyo University, President of the Japan Explosives Society）。会议执行委员会主席为日本横滨国立大学小川辉繁教授 Prof. Tesushige Ogawa（Yokohama National University）。

来自中国、印度、印度尼西亚、韩国、荷兰、巴基斯坦、波兰、俄罗斯、泰国、美国及日本的近120名代表参加会议，4篇大会特邀报告及68篇论文在会上发表。

会议邀请四位国际著名的火炸药专家作了重要报告，会议涉及含能材料，烟火制造术，爆轰燃烧，爆破，冲击压缩，危害评估，安全与风险，规章、标准及其它。现介绍如下（按时间顺序）：

（1）Expansion technology of Ammonium Nitrate and it's application

Prof. Lv Chunxu（Nanjing University of Science and technology，P. R. China）

硝酸铵的膨化是一种新技术。自敏化理论是该技术发明点的核心。它是对传统方法的突破，它实现是通过 AN 的膨化。它的本质是表面活性剂技术在粉状工业炸药中的应用。它是在强制析晶下的物理化学过程。在这篇论文中讨论了膨化机理、技术特征及优势。首先应用在岩石膨化硝铵炸药中，也应用在煤矿许用炸药中。与其它工业炸药比较，它具有突出的优点。

（2）Transport regulations and safety management of energetic materials

Dr. Charles H. Ke（Chief，Science Group；Office of Hazardous Materials Technology；Research and Special Programs Administration；U. S. Department of Transportation）

含能材料包括一大类材料。在规章上和工业标准上，关于什么事含能材料没有明确的定义。笼统地说，为了方便运输，含能材料被分为以下几个不同，比如：炸药、易燃液体、氧化物、有机过氧化合物和自身活泼的物质等。运输规定包含四种基本的特征：①根据工业部门评估材料的危险性质来分类；②通过标签和标识来识别含能材料的危险性；③用适当的容器和设备来容纳这些材料；④在适当的操作条件下处理该材料。这些特征也用在其它含能材料的安全管理中。就此而言，运输规章可以为安全处理含能材料方面提供基本的指导。

（3）Development of an automated design program for tunel blasting

Chung-In Lee，Youg-Hun Jong and Tae-Hyang Kim（School of Civil，Urban and Geosystem Engineering，School National University，Korea）

在这项研究中，隧道爆破模式的电脑程序设计已经有所发展。这个程序包括两个部分：其一为隧道爆破模式的设计，其二为估计最高粒子速度，挖掘损坏区域及裂缝分布的爆破模型。

通过对 23 次隧道爆破测试数据的分析，得出的结论是他们之间的相互关系非常的好，能很好的应用于掏槽的设计。为了检测改进方案和实际实施的有效性，在韩国两座不同的隧道建筑中实施了爆破测试。结果令人非常满意，平均成功率达 90%，稍许的损坏没有造成额外的损失。

（4）Phenomentogical charactersitics of highly energetic non-ideal explosive composition

Prof. Allen J. Tulis（Applied Research Associates Inc and International Pyrotechnics Seminars，USA）

分子炸药如 TNT 和 RDX 的爆炸性能与 Chapman-Jouguet（CJ）理论公式计算的结果高度吻合，因此我们认为它们属于理想炸药。进而人们致力于合成性能更优良、爆炸威力更强的分子炸药，例如，最近人们合成出八硝基立方烷。然而，当考虑到爆炸过程中释放能量的多少时，高能量混合物（通常定义为烟火或推进剂）在爆炸过程中的释放能量要比理想炸药高很多。尽管我们可以用高能量混合物获得良好的 CJ 爆炸，但通常它们的爆炸模式是非理想的。

观察证实发生了非理想反应，并且观测到上述的反应现象特征，并且还观测到证实存在多种振动波前锋、细胞结构和旋陷爆震的证据，但是这种证据并不是所有的实验中都能观测到的，即使实验情况类似也不一定。

本文还着重介绍分散相（燃料/空气）和压缩相增强反应（在可反应介质中），以及温压药剂（燃料/氧化剂/空气）型的炸药。

ISEM2002 荣幸地得到来自旭化成（料）海有限公司，日本化药有限公司，日本油脂株式会社，日本 carlit 有限公司，日本科基有限公司，关西化药有限公司，大赛路化学有限公司，石川岛航空有限公司，科学研究美国空军办事处及航空航天研究及发展亚洲办事处，以及其他公司资助。

（李新蕊，钱华，金序兰）

A Study on Self-Sensitization Theory of Expanded Ammonium Nitrate Explosive

Lv Chunxu, Liu Zuliang, Lu Ming & Chen Tianyun
(China Institute of Industrial Explosive Materials, Nanjing 210094, Jiangsu, China)
EI: 04208166552

Abstract: Expanded Ammonium Nitrate (EAN) explosive is a new type of powder industrial explosive. In is our invention self-sensitization theory of expanded AN is the key technique of our invention. On the basis of mechanism of explosion initiated by hot spot, micro-pore is introduced in expanded AN explosive. The theoretical foundation of self-sensitization is provided by the calculation of critical hot spot temperature of micro-pore. The comparison of expanded AN with common AN and pearlite is also studied in this paper. The micro-pore distribution and its regularity of expanded AN during expansion are given, which provided powerful assurance of its success of self-sensitization.

Keywords: expanded AN, self-sensitization theory, micro-pore

1 Introduction

EAN explosive, a new type of powder industrial explosive, does not contain TNT but its explosion properties are similar to that of No. 2 rock Ammonite. The physical properties, and safety and storage natures of the rock EAN explosive are better than that of No. 2 rock Ammonite. Its quality has reached to advanced level of international standard of the same kind of product. The technology of EAN possesses novel and creative characteristics.

We are the first at home and abroad who invented this new technology and developed the EAN explosive[1~6].

Because no TNT is needed, the cost of the new product would be decreased and the harm to human body and the pollution to environment would be avoided. Therefore remarkable economic and social benefits have been obtained. To a factory with annual output of 8000 tons, the cost of raw material about 353 thousands US Dollar, if adopts the EAN technique, will be less than that of the factory originally producing rock AN-TNT-oil explosive (contained TNT 7%) and about 550 thousands US Dollar less than that of a factory originally producing rock Ammonite explosive (contained TNT 11%). Obviously, the decreased amplitude to the cost is quite considerable.

Because of the above-mentioned advantages, the EAN explosive has been favorably accepted by many factories and users. Up till now, more than 60 factories in China have adopted this new technique to manufacture their own products, which is one of the spread items of national significant scientific and technique achievements of 95 programme in China, by State Science and Technology Commission in 1996. Concrete and further spread works have been carried on.

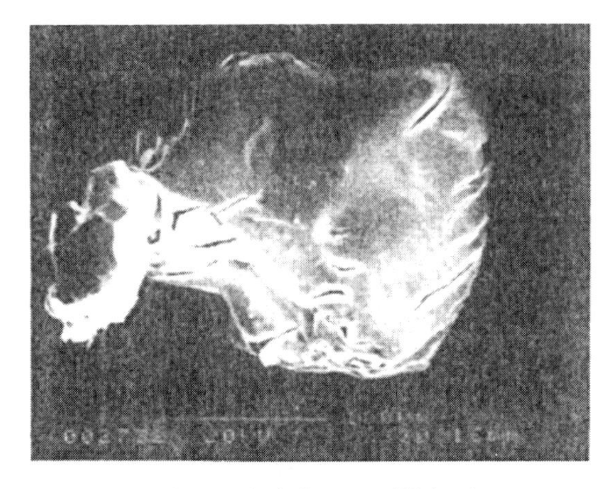

Common AN(enlargement 900 times)

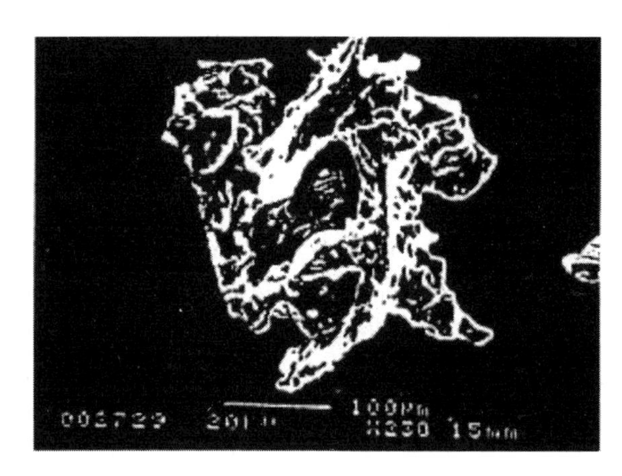

EAN (enlargement 750 times)

Fig. 1 Micro-structure of EAN

EAN manufactured by surfactants treatment possesses porous structure of sponge-like. Fig. 1 shows micro structure of EAN under electronicro-graphy:[5~7]

To compare with common powder, granular and porous granular AN, EAN possesses macro-characteristics as follows:

(1) Large specific surface area: its specific surface area measured by method of resistance-down is 3328.54$cm^2 \cdot g^{-1}$. As a comparison, the corresponding value of common powder AN is 758.76$cm^2 \cdot g^{-1}$. Relative ration is about 4.4 : 1.

(2) Many micro-pores in the particle: structure of holes of EAN can be obviously observed from Fig 1. By means of a microscope it may be clearly observed that EAN is full of micro-pore.

(3) Irregular shape: shape of particle of EAN is "branched" with many edges, corners and burrs, namely, extremely irregular in shape.

(4) Lower hygroscopicity and nearly non-caking.

Some condusions is as follows:[8~11]

① By using surface active technology, under action of properly selected surfactants, EAN, which possesses porosity, large specific surface area and non caking, has been obtained. EAN explosive is manufactured from it.

② EAN explosive, which free from TNT, possesses excellent explosion and physical properties. Quality of products has reached or surpassed to that of No. 2 rock Ammonite.

③ EAN explosive made from EAN, sawdust and composite fuel oil, possesses characteristics of abundant raw material, low cost and simple production process. Especially in view of economic benefit produced by low cost and social benefit brought by elimination of toxic TNT this product favorably accepted by factories and final users.

④ Technology of EAN explosive has been certificated by Ministry and adopted by many factories. Laboratory research on permissible EAN explosive and water-resistant EAN explosive and so on have been accomplished and the study of pilot manufacturing will be in progress.

2 Design of self - sensitization theory of expanded ammonium nitrate

The application of micro-pore, acting as hot spot, can efficiently promote the detonation performance and shock sensitivity of emulsion explosive, which contributes to find new ways to sensitize powder explosive.

It is extremely important to introduce micro-pore in powder explosive. But obviously, introducing micro-pore in sawdust and oil is very difficult and whose amount can be neglected. Then the only practical way is to introduce it in the crystal of ammonium nitrate. This is the basis to design self-sensitization of expanded ammonium nitrate.

The detonation initiated by exterior energy of AN-based explosive is a kind of non-ideal detonation. Under the action of exterior energy, some rough parts in explosive would form hot spot with a great deal of energy, which can lead to rapid decomposition of explosive in these hot spot and then proceed to initiate reaction of other explosive till the whole system to explode. Hot spot can be formed by adiabatic compression on micro-pore or violent friction or collision between edges of explosive particles. What showed above inspire us that it is reasonable to introduce micro-pore in AN and roughen its crystal. This kind of design makes it possible to form hot spot with high temperature and pressure to initiate explosive under the strong action of exterior factors. This is so-called self-sensitization design.

Through analyzing production technologies and methods of powder industrial explosive, a great number of results can be obtained, which indicate that self-sensitization of AN is one of ways to realize producing powder explosive free from TNT.

This kind of interior self-sensitization of AN can take place sensitizer TNT that sensitize explosive, which give rise to tremendous change of detonation performance and shock sensitivity of expanded AN and explosive derived from it.

3 Thermodynamic calculation of critical temperature of self - sensitization hot spot of expanded AN

As what showed above, self-sensitization of expanded ammonium nitrate depends mainly on a great many micro-pores in it. So it is very important to calculate critical temperature of these micro-pores by thermodynamic method and this is one important problem people concerned about[12].

Thermodynamic equation is showed as following;

$$ds=c_p\frac{dT}{T}-(c_p-c_v)\frac{dp}{p}$$

In adiabatic process, $ds=0$

Then
$$c_p\frac{dT}{T}-(c_p-c_v)\frac{dp}{p}=0$$

Define
$$\frac{c_p}{c_v}=r$$

Thus
$$\frac{dT}{T}=\frac{r-1}{r}\frac{dp}{p}$$

Integrate above equation and get
$$\frac{T}{T_0}=\left(\frac{P}{P_0}\right)^{\frac{r-1}{r}}$$

Proposing that the gas in micro-pore is N_2, then r=1.4, and proposing again that $T_0=25℃=298.2K$ and pressure ratio P/P_0 of detonation shock wave to initial pressure of micro-pore is 1000, then the critical temperature of micro-pore is

$$\frac{T}{T_0}=(\frac{P}{P_0})^{\frac{r-1}{r}}=2145.5K=1872.3℃. \quad (1)$$

The research results show that if the temperature is 185～270 °C, the main thermal decomposition of AN is:

$$NH_4NO_3 = N_2O + 2H_2O \quad \Delta H=+36.80kJ \quad (2)$$

$$2NH_4NO_3 = 2N_2 + O_2 + 4H_2O \quad \Delta H=+238.30kJ \quad (3)$$

When the temperature is higher than 400 °C, the main thermal decomposition of AN is:

$$4NH_4NO_3 = 3N_2 + 2N_2O + 8H_2O \quad \Delta H=+408.0kJ \quad (4)$$

$$8NH_4NO_3 = 2N_2O + 4NO + 5N_2 + 16H_2O \quad \Delta H=+490.0kJ \quad (5)$$

The research results also demonstrate that thermal decomposition of AN under high temperature is a second order and auto-catalyzed reaction and a great deal of reaction heat produced.

Therefore, innumerable micro-pore distributed equably in expanded AN would convert to hot spot with a temperature of 1872° C under the action of adiabatic compression from detonation shock. And the temperature is so high as to make AN around them heated, decomposition, initiated and even detonated.

Obviously, calculation and analysis of critical temperature of hot spot is the foundation and basis to design self-sensitization theory of AN.

4 Distribution and regularity of micro-pore of expanded AN

Expanding process makes a great of pore, crack, burrs and irregular edges and corners on crystal surface. And these factors would lead to the formation of hot spot under the action of detonation shock. Besides concerning surface characteristics of micro-pore, its distribution and regularity also have been studied[13,14].

We have used OMNISORP (TM) specific surface area and total volume apparatus made by Amer ican.

Coulter Corp. to study distribution and regularity of micro-pore of expanded AN under different adsorption pressure and obtained its Kelvin radius, thickness of adsorption layer, pore radius, volume increment of desorption, surface area of pore and percentage of total surface area. The results listed in Table 1.

Table 1 Distribution and regularity of micro-pore in expanded ammonium nitrate

Pressure ratio of adsorption P/P_0	Kelvin radius $R_k/10^{-10}$m	Thickness of adsorption layer d $/10^{-10}$m	Pore radius $R_p/10^{-10}$m	Volume of desorption V_d/ml·g^{-1}	Volume increment of desorption V_{ad}/ml·g^{-1}	Surface area of pore S_{dp}/m·g^{-1}	Percentage of total surface area S_{zdp}%
0.9376	147.97	15.024	162.99	0.2402	0.00045	0.05436	2.1171
0.91913	105.89	13.834	119.72	0.2033	0.00083	0.1482	7.8888
0.8899	81.684	12.853	94.537	0.2119	0.00125	0.2545	17.798
0.8664	66.473	12.054	78.527	0.1941	0.00165	0.391	33.023
0.8418	55.354	11.34	66.693	0.1665	0.00199	0.584	55.765
0.8184	47,581	10.752	58.333	0.152	0.00230	0.8819	90.111
0.794	41.325	10.210	51.536	0.1484	0.00262	1.3421	100.0

续表

Pressure ratio of adsorption P/P_0	Kelvin radius $R_k/10^{-10}$m	Thickness of adsorption layer d $/10^{-10}$m	Pore radius $R_p/10^{-10}$m	Volume of desorption V_d/ml·g^{-1}	Volume increment of desorption V_{ad}/ml·g^{-1}	Surface area of pore S_{dp}/m·g^{-1}	Percentage of total surface area $S_{\Sigma dp}$%
0.7697	36.427	9.734	41.161	0.148	0.00294	2.0238	100.0
0.7454	32.449	9.3049	41.754	0.1338	0.00323	2.9808	100.0
0.7214	29.185	8.9194	38.105	0.1255	0.00350	4.3421	100.0
0.6975	26.465	8.5712	35.036	0.1093	0.00374	6.1974	100.0
0.6729	24.064	8.2401	32.304	0.1053	0.00396	8.693	100.0
0.6493	22.071	7.9461	30.017	0.0961	0.00417	11.900	100.0
0.6246	20.257	7.6609	27.918	0.0982	0.00438	15.837	100.0
0.6011	18.73	7.4046	26.136	0.1022	0.00461	20.446	100.0
0.5768	17.322	7.1586	24.430	0.0940	0.00482	25.278	100.0
0.5518	16.034	6.9195	22.954	0.0868	0.00500	29.711	100.0
0.05277	14.914	6.7006	21.614	0.0795	0.00517	33.057	100.0
0.5032	13.880	6.489	20.369	0.0590	0.00527	34.006	100.0
0.4794	12.966	6.2928	19.259	0.0551	0.00536	31.897	100.0
0.4546	12.093	6.097	18.19	0.0610	0.00547	25.278	100.0
0.4311	11.328	5.9179	17.246	0.0386	0.00550	14.743	100.0
0.4065	10.59	5.7378	16.327	0.0415	0.00550	0.2149	100.0
0.3811	9.8821	5.7378	16.327	0.0415	0.00560	0.2149	100.0
0.357	9.2548	5.7378	16.327	0.0415	0.00564	0.2149	100.0

The relation between pore radius and volume of expanded AN is also determined and the results are listed in Table 2.

Table 2 The relation between pore radius and pore volume of expanded AN

pore radius /×10^{-10}m	Pore volume /ml·g^{-1}	Percentage over total volume/%	pore radius /×10^{-10}m	Pore volume /ml·g^{-1}	Percentage over total volume/%
>200	0.00000	0.00	40-30	0.00094	16.65
200-100	0.00087	15.41	30-20	0.00109	19.30
100-50	0.00178	31.52	20-10	0.00037	6.55
50-40	0.00061	10.80			

The total volume of pore with a radius R bigger than 10×10^{-10}m is 0.00568ml·g^{-1}.

According to the standard for pore radius distribution from IUPAC, the pores are classified respectively as following:

$R\geqslant500\times10^{-10}$: macro pore;

$R=20\times10^{-10}\text{m}\sim500\times10^{-10}\text{m}$: mesopore;

$R=7\times10^{-10}\text{m}\sim20\times10^{-10}\text{m}$: micro-pore;

$R\leqslant7\times10^{-10}\text{m}$: imperceptible-pore

From this standard, 95% of pore in expanded AN belongs to mesopores. So it is defined as mesopore exactly, whereas we call it micro-pore customarily.

According to self-sensitization theory of expanded AN, fine pore (micro-pore and imperceptible micro-pore) is invalid. Because it is inevitable to mix expanded AN, sawdust and oil together when manufacturing expanded ammonium nitrate explosive so that any pore with a radius smaller than 20×10^{-10}m is sure to be filled by sawdust and oil. On the other hand, macro pore should also be avoided because nimiety macro pores would decrease pore ratio so as to influence the amount of hot spot, which is disadvantage to self-sensitization. In a word, mesopore is the best satisfactory for self-sensitization. Because it not only can not be filled by sawdust and oil but also can make sawdust and oil disperse evenly on the surface of pore, so as to oxide and combustible substance can be blended tightly. In addition, it can maintain adequate pore, namely adequate hot spot, so as to make sure it can reach the best detonation sensitivity and explosion performance, which is the substantial basis of self-sensitization to succeed and be applied practically. The isotherm of static adsorption of expanded AN is determined and showed in Fig. 2.

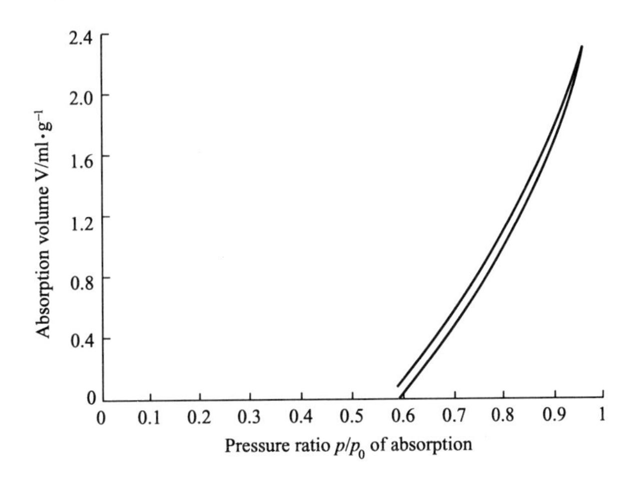

Fig. 2 The static adsorption and desorption isotherm for expanded AN

It is clear from Fig. 1 that there is not flat roof, which demonstrate that the pores are not among the range of fine pores but among that of mesopores. Fig. 3 shows the comparative conditions of static adsorption and desorption isotherm of expanded AN and common AN.

It is shown in Fig. 3 that expanded AN has a bigger adsorption volume and more effective mesopores than common AN under same pressure ratio, which guarantees to form enough hotspot under adiabatic compression

The following study is concerning the relationship between pore volume and radius. Cumulative pore volume of different radius is showed in Fig. 4.

According to Fig. 4, pore radius of expanded AN mainly distributes in 10×10^{-10} m $\sim 200\times10^{-10}$ m.

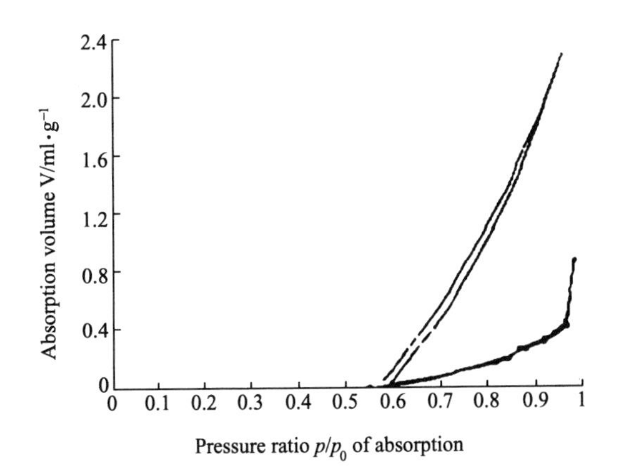

Fig. 3 The plot of comparison between expanded AN and common AN for static adsorption and desorption isotherm

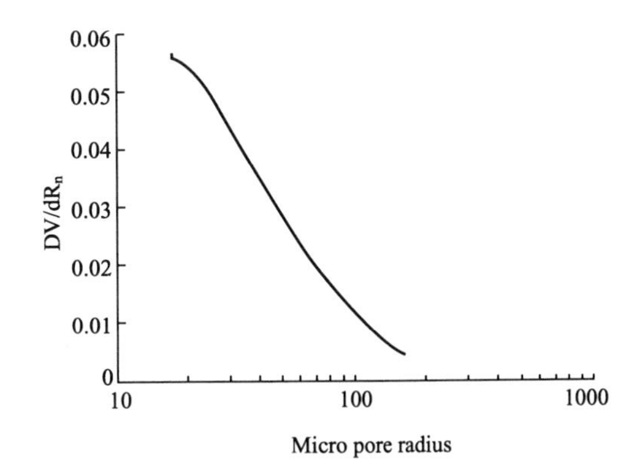

Fig. 4 The plot of pore volume vs pore radius for expanded AN

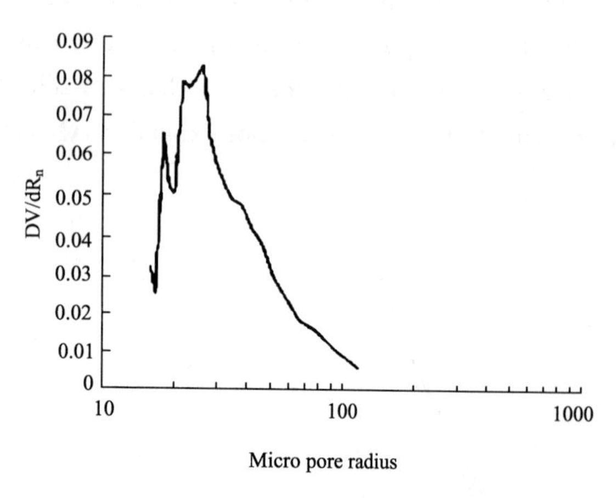

Fig. 5 The plot of pore radius distribution in expanded AN

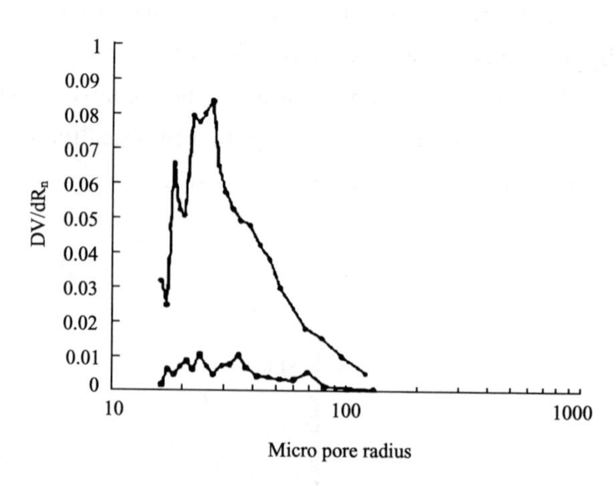

Fig. 6 The plot of comparison between expanded AN and common AN for size distribution of micro-pore

It can be observed more obviously the distribution of pore radius from relationship plot between pore volume and radius as showed in Fig 5.

In order to comprehend the distribution and regularity of pore radius of expanded AN deeply, we also have studied that of common AN for comparison. The date is showed in Table 3.

Table 3 Distribution and regularity of micro-pore in common AN

Pressure ratio of adsorption P/P_0	Kelvin radius $R_k/10^{-10}$m	Thickness of adsorption layer d $/10^{-10}$m	Pore radius $R_p/10^{-10}$m	Volume of desorption V_d/ml·g^{-1}	Volume increment of desorption V_{ad}/ml·g^{-1}	Surface area of pore S_{dp}/m·g^{-1}	Percentage of total surface area S_{zdp}%
0.9455	170.09	15.485	185.58	0.07258	0.00013	0.04122	3.8537
0.9205	115.04	14.138	129.17	0.04857	0.00022	0.03305	12.809
0.8953	86.227	13.060	99.288	0.03739	0.00029	0.04712	25.576
0.8704	68.692	12.182	80.784	0.0309	0.00035	0.06967	44.452
0.8453	56.714	11.434	68.148	0.05768	0.00047	0.1422	82.995
0.8028	48.272	10.808	59.080	0.02634	0.00053	0.01785	100.0
0.7945	41.434	10.221	51.655	0.02265	0.00057	0.2664	100.0
0.7689	36.277	9.7186	45.996	0.01995	0.00061	0.3952	100.0
0.7443	32.281	9.2858	41.567	0.01679	0.00064	0.5781	100.0
0.7178	28.747	8.8651	37.613	0.02058	0.00069	0.8487	100.0
0.6922	25.915	8.4794	34.412	0.02397	0.00074	1.2183	100.0
0.6677	23.596	8.1728	31.769	0.01642	0.00081	2.2627	100.0
0.6411	21.441	7.849	29.290	0.01564	0.00081	2.2627	100.0
0.6151	19.614	7.555	27.169	0.01052	0.00083	2.9365	100.0
0.5894	18.034	7.2857	25.320	0.01197	0.00085	3.6883	100.0
0.5642	16.654	7.0361	23.690	0.01332	0.00088	4.4253	100.0
0.5381	15.384	6.7938	22.178	0.00959	0.00089	4.9724	100.0
0.512	14.241	6.564	20.805	0.01067	0.00091	5.1923	100.0
0.487	13.247	6.354	19.601	0.00868	0.00093	4.9277	100.0
0.461	12.310	6.1466	18.457	0.00726	0.00094	3.9588	100.0
0.4345	11.437	5.9439	17.380	0.00779	0.00095	2.2324	100.0
0.4096	10.679	5.7620	16.440	0.00530	0.00095	0.0274	100.0
0.3827	9.9242	5.7602	16.440	0.00530	0.00095	0.0274	100.0
0.3568	9.2489	5.7602	16.440	0.00530	0.00095	0.0274	100.0

In the interest of comprehending the difference between expanded AN and common AN, it is necessary to compare pore volume with pore radius distribution of pore and results showed in Fig. 6.

Some results are obtained from Fig 5 that expanded AN has a much bigger pore volume than common AN and 95% of its valid pore belongs to micro-pore, which is advantageous to effective and complete detonation.

Expanded pearlite is also a satisfactory carrier for introducing pore in ammonium nitrate explosive for sensitization. So it is interesting to give attention to its pore radius and total volume. Comparing it with expanded AN and obtain some results as following:

99.5% of the pore in expanded AN distribution has a radius bigger than 10×10^{-10} m and total volume of pore with a radius bigger than 10×10^{-10} m is 0.00568 ml · g^{-1}, while micro-pore with a radius bigger than 10×10^{-10} m in expanded pearlite only has percentage of about 90% and its total volume is 0.00331 ml · g^{-1}. So it is obvious that the quantity and valid volume of pore in expanded AN are bigger than that of in expanded pearlite.

References

[1] Ding Yun. Improving Properties of AN and it's Explosives Using Surfactants. [Thesis]. Nanjing: East China Institute of Technology, 1990.

[2] Lv Chunxu. Theory and technology of surfactant Nanjing China: Jiangsu Science press, 1991.

[3] Hui Junming, Liu Zuliang, Lv Chunxu, Wang Yilin. Powder AN explosive and its preparation, Chinese Patent, CN 91107051. 6. 1991.

[4] Liu Zuliang, Hui Junming, Lv Chunxu, Wang Yilin. Light, porous AN and its manufacture, Chinese Patent, CN 921078730, 1992.

[5] Liu Zuliang, Hui Junming, Lv Chunxu, Wang Yilin. Research on light, porou and bulky AN and HF series powder AN explosive. Explosive Materials, 1991 (5): 5.

[6] Lv Chunxu, Liu Zuliang, Hui Junming, Wang Yilin. Rock EAN explosive. Explosive Malerials, 1997 (1): 5.

[7] Ye Zewen. Expanded Mechanism of Ammonium Nitrate. [Thesis]. Nanjing: Nanjing University of Science and Technology, 1995.

[8] Lv Chunxu. Research on surface chemistry of EAN. Journal of Nanjing University 1995, (31): 286.

[9] Chen Tianyun. Research on Surface Characteristics and Application of Ammonium Nitrate. [Thesis], Nanjing East China Institute of Technology, 1992.

[10] Yang Tong. Investigation of technique of rock EAN explosive. Explosive Materials, 1996, (3): 30.

[11] Lv Chunxu, Liu Zuliang, Ni Ouqi. Industrial Explosives. Beijing: Ordnance Industry Press, 1994.

[12] Lu Ming. Study of expanded ammonium nitrate explosive [Thesis]. Nanjing China: Nanjing university of science & technology, 1999.

[13] Ye Zhiwen. Study of expanding mechanism of ammonium nitrate [Thesis]. Nanjing China: Nanjing University of science & technology, 1995.

[14] Chen Tianyun, Lv Chunxu, Ye Zhiwen. Study of performance of modified ammonium nitrate. Energetic Materials, 1996, (4).

（注：此文原载于 Proceedings of 2003 International Autumn Seminar on Propellants, Explosives and Pyrotechnic, 2003: 724-731）

A Study and Development of Expanded Ammonium Nitrate Explosive

Lv Chunxu, Liu Zuliang, Chen Tianyun, Lu Ming
Nanjing University of Science and Technology (Nanjing, 210094)

Abstract: Expanded ammonium nitrate (AN) explosive, which is a new kind of powdered industry explosive based on innovative technology of expansion, has been awarded national prize and listed into plan of Spreading Great Scientific Achievement of China Ninth-fire-year Plan. It has now manufactured in more than 70 factories in China. Owing to its application, the national output value has increased for 2254 million Chinese Yuan and profit including tax has increased for 405million Chinese Yuan until 2002.

The guideline of innovation is self-sensitization of AN. It is a breakthrough for classic method. The approach to self-sensitization is expansion of AN, and essence of expansion is surface activity technique, a physicochemical process under action of composite surfactant.

In this paper, the self-sensitization theory of expanded ammonium nitrate the mechanism and technical characteristics of expansion is discussed, its unique advantage is showed. Some data about rock expanded AN explosive and its comparison with other industry explosive are given. The expansion technology is mainly applied in rock explosive-permitted explosive seismic explosive and expanded AN explosives with low detonation velocity.

Key words: expanded ammonium nitrate, self-sensitization, micro structure, mechanism, characteristics, technology

1 Introduction

Expansion of AN is a innovative technology. Expanded AN based on expansion technology is a new kind of industrial explosive.

Some explosive experts think that: technology of expanded AN is original and innovative. It opens a new approach to explosive research and reaches the advanced level of the world.

Expansion technology and expanded AN was awarded Special Scientific Achievement Prize of China Ordance Industry Company (ministry rank) in 1997 and National Scientific Achievement Prize Grade Two in 1998.

Expanded AN explosive has been arranged into List of Spreading Great Scientific Achievement of China Ninth-five-year Plan. National explosive line is going to replace some other powder explosive with expanded AN as a new generation product.

The achievement of expanded AN explosive is a revolution of industial explosive. Owing to its advanced technology, it has remarkable social and economic benefits and welcome to many customers. In China, there are more than 70 factories manufacturing expanded AN explosive. American Austin Explosive Company, Japanese MITSUBISHI Chemical Company and Indian SUA Explosive Company are all interested in this technology. Some technical cooperation is under negotiation. Because of its application, the national output value has increased for 2254 million Chinese Yuan and profits before tax has increased for 405 million Chinese Yuan until 2002. The output of powder AN-TNT explosive is 850000 tons in China, and there are 213 explosive factories, which will repalace their AN-TNT with expanded AN. So its further market is promising.

2 The innovative idea and design of self - sensitiz-ation theory of expanded ammonium nitrate[1-4]

2.1 Background of research

Industrial powder explosives is mainly sensitized by TNT or NG. These explosives have serious defect: (1) toxity of TNT impairs the workers' health and pullutes environment. (2) Cost is relatively high. (3) Safety of manufacture is not very good. (4) Phisical property, for example, hygroscopicity and caking, is not ideal.

For more than 100 years, many researchers focus on these problems. They propose adding crystal pattern modifiers-surfactant-hydrophobic agent into explosive or improving charge or package methods to improve explosive physical property. They also attempt to replace TNT with RDX or metal powder. Although some of these methods are effective, they are partial or not economical so have not resolve the problem radically.

2.2 Basic designing idea

After technical analysis of many powdered explosive and exploring lots of approaches. We think that only through self-sensitization of AN can make powder AN explosive get its breakthrough for eradicating TNT.

The idea of self-sensitization of AN is a innovation, never considered by explosive researchers.

The idea of self-sensitization is based on theory as follows: the detonation of AN is non-ideal detonation. Under action of external energy, "hot spot" concentrated with energy are formed in some heterogeneous section initially; the explosive in "hot spot" then decomposes rapidly, giving out lots of heat which initiates explosion of adjacent explosive, then explosion of all explosive system. "hot spot" can be formed from pressing adiabatically micro-pore or form friction and collision between particles. All this inform us that: introducing micro pore into AN particles, roughening particles surface will sensitize AN under external initiating energy.

2.3 Proposal of sensitization

In explosion theory, "hot spot" -initiating mechanism is generally acknowledged and very important. It not only well explains the reason of initiation of explosive under very little external energy but also directs us to adjust the safety of explosive and to design sensitization of liquid or emulsion explosive system.

Micro pore which is introduced into AN crystal will form "hot spot" of high pressure high temperature rapidly under external energy; at the same time, friction and collision among roughened particles also lead to forming "hot spot" . Just this sensitization function from inner crystal takes place the sensitization of TNT, making initiation of AN explosive enhanced radically. The theoretical value and level of self-sensitization applied in powder industrial explosive is self-evident and generally acknowledged.

Self-sensitization theory is key point of our innovation.

2.4 Design of self-sensitization theory of expanded ammonium nitrate

The application of micro-pore, acting as hot spot, can efficiently promote the detonation performance and shock sensitivity of emulsion explosive, which contributes to find new ways to sensitize powder explosive.

It is extremely important to introduce micro-pore in powder explosive. But obviously, introducing micro-pore in sawdust and oil is very difficult and whose amount can be neglected. Then the only practical way is to introduce it in the scrystal of ammonium nitrate. This is the basis to design self-sensitization of expanded ammonium nitrate.

The detonation initiated by exterior energy of AN-based explosive is a kind of non-ideal detonation. Under the action of exterior energy, some rough parts in explosive would form hot spot with a great deal of energy, which can lead to rapid decomposition of explosive in these hot spot can be formed by adiabatic compression on micro-pore or violent friction or collision between edges of explosive particles. What showed above inspire us that it is reasonable to introduce micro-pore in AN and roughen its crystal. This kind of design makes it possible to form hot spot with high temperature and pressure to initiate explosive under the strong action of exterior factors. This is so-called self-sensitization design.

Through analyzing production technologies and methods of powder industrial explosive, a great number of results can be obtained, which indicate that self-sensitization of AN is one of ways to realize producing powder explosive free from TNT.

This kind of interior self-sensitization of AN can take place sensitizer TNT that sensitize explosive, which give rise to tremendous change of detonation performance and shock sensitivity of expanded AN and explosive derived from it.

3 Mechanism of expansion of AN[5,6]

After research on offect of process condition and surfactant on expansion, we think mechanism of expansion probably is that: under given temperature and subpressure, water in saturated AN solution vaporizes and form stable foams with the help of surfactant. When temperature and subpressure reach a certain value, foams break up and water steam flees out of AN solution rapidly, leaving lots of "hole-cleavage-channel" . AN crystal which crystallizes out rapidly is rough on its surface, and connected with each other in irregular way, forming a kind of light-porous-bulky expanded AN.

It is showed in research that: vaporization under subpressure helps to expand AN by itself, while the foaming property of surfactant further enhances the expansion effect. Experiments show that: expansion effect is correspondent to foaming property of surfactant.

Research also shows that AN solution with composite surfactant possesses relatively better foaming performance, among which amphiprotic composite surfactant has the best surface activity and foaming performance. It also shows that composite surfactant can be mutually beneficial in foaming and the amphiprotic composite surfactant has the greatest interactive parameter and mutual benefit. In modifying AN, surfactant mainly plays the role of foaming agent-coating and crystal pattern modif ier.

4 Technical characteristics of expanded AN[5,6]

Expanded AN treated by surfactant has porous honeycomb-like structure. Micro-structure of expanded AN photoed by SEM is showed in Fig. 1.

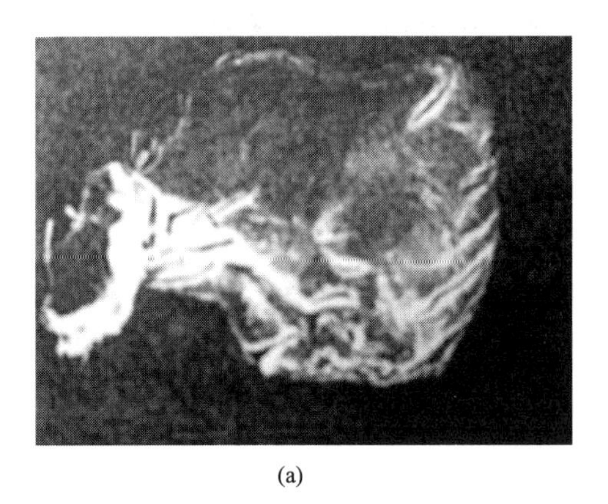

(a)

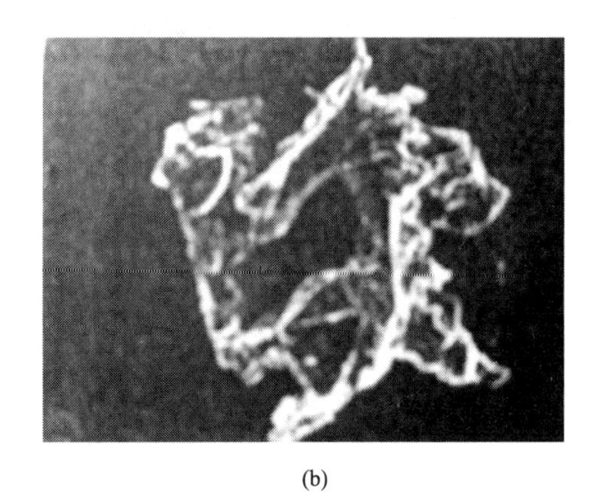

(b)

Fig. 1 Micro-structure of Expanded AN

Where (a) common AN (enlarged 900 times); (b) Expanded AN (enlarged 850 times) Compared with common powder AN and granular porous AN, expanded AN has these characteristics:

(1) Large specific surface area. Determined by the same method, specific surface area of expanded AN is 3328.54cm^2 · g^{-1}, 4.4 times larger than that of common AN, 758.76 cm^2 · g^{-1}.

(2) Lots of micro pore in AN crystal. It is showed in Fig. 1 that there is much of cleavage and crack in AN crystal and spins or edges on its surface. This is physical fundament of self-sensitization of AN.

(3) Relatively lower hygroscopicity and caking.

Hygroscopicity increment of AN under 20-, humidity 93% is showed in Table 1.

Table 1 Hygroscopicity increment of Ammonium Nitrate

Time/h	2	4	6	8	10
Pure AN	0.5832	1.1084	1.5792	2.0660	2.5370
Expanded AN	0.4258	10.7697	1.1212	1.8746	1.8746
No. 2 Rock AN-TNT explosive	0.5569	1.1100	1.5456	2.0001	2.4811
Rock expanded AN explosive	0.3958	0.6432	0.9397	1.3025	1.5904

Data in Table 1 shoes that average hygroscopic velocity of expanded AN is about half less than that of common AN. Putting expanded AN and common AN under 80-for 3 hours, then cooling it; after 5 this heating and cooling cycling, determinating its undergone pressure.

Table 2 The pressure undergone by charge through cycling test

	Undergone pressure/N				Average/N
Common AN	805.9	811.1	799.3	802.3	804.7
Expanded AN	81.3	8.9	80.4	82.5	81.3

Table 2 indicates that undergone pressure of expanded AN is 1/8～1/9 less than that of common AN. So expanded AN is thought to process very good anti-caking property.

(4) Easy process because of brittleness. Just manufactured expanded AN is very brittle and easy to be crushed. So only for 20 minutes, size of 90% AN power discharged from edge mill is smaller than 250×10^{-6}m, reaching the certain required granularity of AN.

5 Application of expanded AN technology[7～10]

5.1 Rock expanded AN explosive[7]

Rock expanded AN explosive, a new type of power industrial explosive, does not contain TNT but its explosion properties are similar to those of No. 2 rock Ammonite. The physical properties, and safety and storage natures of the rock expanded AN explosive are better than those of No. 2 rock Ammonite. Its quality has reached to advanced level of international standard of the same kind of product. The technology of expanded AN explosive novel and creative characteristics.

We are the first at home and abroad who invented this new technology and developed the expanded AN explosive.

The key technique for preparing rock expanded AN explosive is preparation of expanded AN.

No. 2 rock Ammonite, No. 2 rock Ammonite-oil and rock expanded AN explosive, which are present main industrial explosive in China, are tested through random sampling, Table 1 show their results.

Table 3 The comparison between rock expanded AN explosive and other power Industrial explosives

Explosive	Composition					Characteristics						
	AN /%	TNT /%	Wood Power /%	Oil /%	Additive /%	Density /(g/cm³)	Detonation Velocity VD /(m/s)	Strength VL /ml	Brisance HLe* /mm	Moisture Content /%	Storage Lifetime month
Expanded AN explosive	92.0±2.0	0	4.0±0.5	4.0±0.5	0.12±0.005	0.95～1.00	3200～3500	360～380	14.5～16.0	0.3	6
No. 2 rock Ammonite	85.0±1.5	11.0±1.0	4.0±0.5	0	0	0.95～1.10	3000～3200	≥3200	≥12	0.3	6
No. 2 rock Ammonite-oil	87.5±1.5	7.0±0.7	4.0±0.5	1.5±0.3	0.10±0.005	0.95～1.05	3000～3300	≥340	≥12	0.3	6

HLe*: height of lead column compressed

It is indicated from Tab. 3 that explosive property of rock expanded AN explosive has reached or surpassed that of No. 2 rock Ammonite and No. 2 rock Ammonite-oil.

Another important advantage of rock expanded AN explosive is its remarkable economic benefit. Because expensive sensitizer TNT can be got rid of, cost of crude materials is enormously decreased. Beside, production process of explosive is simple, and so productive cost will not be increased. Thus total the cost of rock expanded AN explosive will be decreased by 35～38US Dollar/t as compared to that of original Ammonite. Table 4 shows their current specific costs.

The achievement of expanded AN explosive is a revolution of industrial explosive. Owing to its advanced technology, it has remarkable social and economic benefits and welcome to many customers. In China, there are more than 70 factories manufacturing expanded AN explosive.

Table 4 The comparison of crude material cost for three powder explosives (on the base of prices in 1995)

Crude material	Unit price /US Dollar	Rock EAN Explosive		No. 2 rock Ammonite		No. 2 rock Ammonite-oil	
		Amount /(kg/t)	Cost/US Dollar	Amount /(kg/t)	Cost/US Dolla	Amount /(kg/t)	Cost/US Dollar
AN	235. 3	920	216. 5	850	200	875	205. 9
TNT	941. 2			110	103. 5	70	65. 9
Wood power	35. 3	40	1. 4	40	1. 4	40	1. 4
Oil	352. 9	40	14. 1			15	5. 3
Additive	1764. 7					1	1. 8
Expanded agent	3388. 2	1. 2	4. 1				
Total(US Dollar)			236. 1		304. 9		280. 3

5. 2 Permitted expanded AN explosive[8]

Permitted explosive AN explosive is the first serials product AN.

The key point of permitted explosive is safty property. Under firedamp condition, the explosive will not deflagrate. We focus on its flame-depressing mechanism and make breakthrough on it.

The composition and properties are as follows: expanded AN 79%, fud 15%, sawdust 3%; Detonation velocity (VD) ≥2800m/s, Strength (VL) ≥240ml, Brisance (HLe) ≥11mm, Gap distance≥3cm, Density (ρ) is 0. 95～1. 05g/cm^3, Moisture content≤0. 3%, Storage lifetime≥6 months.

At present, this technology has been transferred to Xuzhou Mine Chemical Factory. China Coal Ministry lists it into "Project of Spreading Achievements of Ninth-Five-year Plan of Coal Industry" and will spread it to factories of Coal Ministry and local explosive factories.

5. 3 Seismic explosive[9]

Explosive AN can replace common AN in seismic explosive. The cost of seismic explosive made from expanded AN explosive decreases remarkablely but with the relatively same performance. With increasing manufacture safety and less pollution, it has good social and economic profits.

A manufacturing line with 5000 tons capacity of expanded AN seimic-explosive has been built in Hubei Rang-sha Chemical Factory with annual output 4000 tons. The product has good stable performance and welcome to many customers.

China has 8 seismic explosive factories. Annual output amounts to 30000 tons.

So this technology has good promising market

The composition and characteristics are as follows:

Tab. 5 The comparision of three kinds of Seismics

Seismics	Composition				Characteristics		
	Expanded AN/%	TNT /%	Sawdust /%	Oil /%	Detonation Velocity VD/m · s^{-1}	Density g · /cm^{-3}	Capability of Initiation to detonation
Expanded AN High Detonation Velocity	73. 0±2. 0	24. 0±1. 0	1. 5±0. 5	1. 5±0. 5	≥5000	1. 00～1. 20	100
Expanded AN Middle Detonation Velocity Seismic	86. 0±2. 0	11. 0±1. 0	1. 5±0. 5	1. 5±0. 5	≥4300	1. 00～1. 10	100
Expanded AN Low Detonation Velocity Seismic	80. 5±2. 0	5. 0±1. 0	13. 0	1. 5±0. 5	≥3500	0. 85～0. 9	100

5.4 Low detonation velocity of expanded AN explosive[10]

Low detonation velocity explosive is widespreadly used in explosive welding (compounding), smooth explosion, and recently it has been used more and more in the field of petroleum research.

Concrete compositions are as follows: Expanded AN 72～83%, sawdust 2.5～3.5% composite oil phase 2.5～3.5%, pearlite 5～9, salt 5～8, stabilizer 2～4%.

The characteristics of this explosive are as follows: detonation velocity 2152 $m \cdot s^{-1}$, sensitivity to initiatial detonation 4cm, charge density 0.62 $g \cdot cm^{-3}$.

The fruit tree blasting agent is also kinds of low detonation velocity explosive, contained "nutrient" in it which can promote fruit tree grow up.

The technology of fruit tree blasting agent is passed the appraisal of design by Industrial Explosive Materials Bureau of China Ordance Industry in Spet. 2000.

The compositions and characteristics are as follows: expanded AN 80.0～85.0, composite oil phase 2.0～4.0, "nutrient" 12.0～14.0; Moisture content≤0.30%, Detonation Velocity (VD) ≥ 2600$m \cdot s^{-1}$, Brisance (HLe) ≥10mm, strength (VL) ≥280ml, Diameter of loosen the soil by blasting≥1.1m, Nitrogen Content≥15.0%, Other "nutrient" content≥10.0%, Sensitivity to impact≥0%, sensitivity to friction≤0%, toxious gas content after blasting≤100$L \cdot kg^{-1}$.

Because of it's low cost, high safety and less pollution, the technology will extensively spread and be warmly welcome by many factories and users.

Reference

[1] Ding Yun. Improving Properties of AN and it's Explosives Using Surfactants. [Thesis]. Nanjing: East China Institute of Technology, 1990.

[2] Lü Chunxu. Theory and technology of surfactant. Nanjing China: Jiangsu science Press. 1991.

[3] Liu Zuliang, Hui Junming, Lü Chunxu, Wang Yilin. Light, Porous AN and its manufacture. Chinese Patent, CN 921078730, 1992.

[4] Yang Tong. Investigation of technique of rock EAN explosive. Explosive Materials, 1996, (3): 30.

[5] Ye Zhiwen. Expanded Mechanism of Ammonium Nitrate. [Thesis]. Nanjing: Nanjing University of Science and Technology, 1995.

[6] Lu Ming. Research on Expanded Mechanism of Ammonium Nitrate. Bing Gong Xue Bao, 2002 (1): 23.

[7] Lü Chunxu, Liu Zuliang, Hui Junming, Wang Yilin. Rock EAN explosive. Explosive Materials, 1997 (1): 5.

[8] Chen Tianyun. Research on surface characteristics and application of Ammonium Nitrate [Thesis]. Nanjing: East China Institute of Technology, 1992.

[9] Lu Ming, Lü Chunxu. Research on Seimics of non-TNT expanded AN. Journal of Nanjing University of Science and Technology, 1999 (3): 261.

[10] Xu Binchen, Lü Chunxu. Research on Compositions of Low detonation velocity expanded AN explosive. Journal of Propellants and explosives. 2000 (3): 1.

(注：此文原载于 Proceedings of 35th International Annual Conference of Institute of Chemical Technology, Germany, 2004, 130: 1-10)

Research on Safety Properties of Expanded Ammonium Nitrate

Lv Chunxu, Liu Zuliang, Hu Bingcheng
(Nanjing University of Science & Technology, Nanjing 210094, Jiangsu, China)
EI: 05058814382; ISTP: 000226863500283

Abstract: New industrial powder expanded Ammonium Nitrate (AN) explosive is briefly introduced, the safety properties of expanded AN is minutely studied. The results show that the safety properties of expanded AN are similar to that of common AN. Because of its other excellent advances, it is widely used to manufacture powder industrial explosives in China.
Keywords: Safety property; expanded Ammonium Nitrate; powder industrial explosive

1 Introduction

Expanded AN explosive is a new type of power industrial explosive. It is free from TNT. Its explosive properties are similar to those of No. 2 rock Ammonite and storage properties are better than those of No. 2 rock Ammonite. Its quality has reached to the advanced level of international standard for the same kind of product. The technology of expanded AN possesses novel and creative characteristics. A new way of research application of industrial power explosive has been opened up. This new technology is originated from our research group at home and abroad. Because no TNT is contained, the cost of new product is decreased and harm to human body and environmental pollution are avoided. Remarkable economic and social benefits have been obtained.

Expanded AN explosive and other two kinds of industrial explosive which are main industrial explosive in China are tested through random sampling. Table 1 shows their comparative results of compositions, performance and the raw material cost.

Table 1 Comparative results of three kinds of powder explosives

Explosives	Composition				Characteristics				Raw material cost
	AN /%	TNT /%	Sawdust /%	Oil /%	Density $\rho/(g\cdot cm^{-3})$	Detonation Velocity V_D $/(m\cdot s^{-1})$	Strength V_S/mL	Brisance H_{Le}/mm	RMB /(Yuan · t^{-1})
Expanded AN explosive	92.0±2.0	0	4.0±0.5	4.0±0.5	0.85~0.95	3200	320	12.0	2007
No. 2 rock Ammonite	85.0±1.5	11.0±1.0	4.0±0.5	0	0.95~1.10	3200	320	12.0	2592
No. 2 rock Ammonite-Oil	88.0±1.5	7.0±0.7	3.0±0.5	2.0±0.3	0.95~1.05	3000~3200	320	12.0	2382

Because of the remarkable advantages mentioned above, expanded AN explosive have been deeply welcomed by many factories and final users. Up till now more than 40 factories have adopted this new technique to product the industrial explosive. Further spreading of this technique have been started.

The key technique of expanded AN explosive is expansion of AN. Expanded AN possesses light, porous and bulky structure. Though observation with Scanning Electric Microscope (SEM) microphotographs of the micro-structure of common AN and expanded AN are shown in Fig. 1.

We are very much interested in the properties of expanded AN, especially its safety property. In this paper, safety properties and its influence are discussed as follows:

2 Experiments and Discussions

2.1 Mechanical Sensitivity

During production, transportation, storage and application, expanded AN explosive undergoes mechanical effect of impact or friction. So research on safety of expanded AN during above process is of great significance.

2.1.1 Sensitivity to impact

According to W J 1053-81 the test to determine the sensitivity of solid explosive to impact is carried on weight of drop hammer (10000±10) g, falling height (250±1) mm, Amount of explosive (50±2) mg.

Two group of parallel tests are done for every sample. The explosion percentage of sample is determined. Sensitivity of expanded AN is compared with that of common AN. The results are shown in Table 2.

Table 2 Sensitivity to impact for expanded AN

Sample	Temperature /℃	Weight of drop hammer/kg	Falling height /mm	Amount of explosive/mg	Percentage of explosive/%
Common AN	25	10	250	50	10
Expanded AN	25	10	250	50	0
Common AN	100	10	250	50	0
Expanded AN	100	10	250	50	0
Common AN	100	10	500	50	2
Expanded AN	100	10	500	50	4

2.1.2 Sensitively to friction

According to WJ 1054-81 the test to determine the sensitivity of solid explosive to friction is carried on length of the pendulum 760mm, angel of the pendulum 96°, apparent pressure 4.92MPa, Amount of explosive (30±1) mg. Two groups of parallel testes are done for every sample. The explosive percentage of sample determined. Sensitivity of expanded AN is compared with that of common AN. The results are shown in Table 3.

2.2 Sensitivity to Electric Spark

During the process of production, the shape of expanded AN and its explosive is powdery. This is similar to common AN and its explosive. Especially in the technique of air flow there are large amount of dusts. A great quantity of dusts are concentrated to form clouds with static. If the static should discharge, electric spark would be produced to cause danger.

Table 3 Sensitivity to friction for expanded AN

Sample	Temperature /℃	Weight of drop hammer/kg	Angle of pendulum/℃	Amount of explosive/mg	Percentage of explosive/%
Common AN	25	1.5	96	30	0
Expanded AN	25	1.5	96	30	0
Common AN	100	1.5	96	30	0
Expanded AN	100	1.5	96	30	0

The energy of dusts may be determined by the determined by the method of spark discharge. So we designed and manufactured a set of facility which would determine the sensitivity of explosive to electric spark. Under flow state, sensitivity of explosive to electric spark is determined the by standard condition, the results are shown in Table 4.

Table 4 Safety of expanded AN and its mixture to statics

Sample	Voltage of Discharge/V	800	120	1600	2000	2400	2800	3200	3600	4000
	Energy of electric spark/J	0.7	1.4	2.3	3.6	5.0	6.9	9.0	11.6	14.7
Common AN		×	×	×	×	×	×	×	×	×
Expanded AN		×	×	×	×	×	×	×	×	×
Oil		×	√	√	√	√	√	√	√	√
Sawdust		×	×	×	×	×	×	√	√	√
Expanded AN+Sawdust		×	×	×	×	×	×	×	×	×
Expanded AN+Oil		×	×	×	×	×	×	×	×	×
Expanded AN explosive		×	×	×	×	×	×	×	×	×

× means that material doesn't ignite; √ means that deflagration takes place

2.3 Sensitivity to Detonator

Sensitivity to detonator expresses the sensitiveness of substance to the shock from a standard detonator or blasting cap. Sensitivity of expanded AN explosive to detonator is one of important mark of its of great significance for its safe production.

According to United Nations recommendation on test and Criteria of Transport of Dangerous Goods, Section 36 Test 5 (a) Cap Sensitivity Test, test is carried on. The substance under test is filled into the tube which consists of a cardboard tube of minimum diameter 80mm, minimum length 160mm with a wall thickness of maximum 1.5mm. Initiation is done by a standard detonator inserted coaxially into the top of the explosive to a depth equal to its length. Below the tube is the witness plate. The results are assessed by whether the explosive is completely detonated. Results of test are shown in Table 5.

Table 5 Sensitivity of expanded AN to detonator

Diameter of particle of expanded AN/μm	Density /(g·cm^{-3})	Temperature /℃	Result	Diameter of particle of expanded AN/μm	Density /(g·cm^{-3})	Temperature /℃	Result
400	0.25	40-45	—	400	0.42	29	+
400	0.25	40-45	—	400	0.42	29	+
400	0.30	40-45	—	400	0.53	40-45	+
400	0.34	33.5	—	400	0.63	29	—
400	0.37	33.5	+	400	0.69	29	—
400	0.40	33.5	+	400	0.70	29	—

The result "sensitivity to detonator" is signed by a "+"

3 Conclusions

Though a series of sensitivity test, conclusions are reached as follows:

(1) Explosive AN similar to common AN is in sensitive to mechanical function. From results of Tab. 2, only under stern condition of temperature 100℃ and falling height 500mm, sensitivity of expanded AN to impact is less higher than that of common AN.

(2) Form datum of Table. 4 under function of strong static (Energy of electric spark 0.7～14.7J) expanded AN and its explosive compares with common AN and its explosive, its sensitivity to electric spark is the same as common AN.

(3) When density of expanded AN is larger than 0.3g • cm^{-3} it may possess sensitivity to detonator. But when density of expanded AN is too larger (more than 0.6g • cm^{-3}) it will lose sensitivity to detonator. So we can artificially control its density, decrease its sensitivity to detonator, guarantee its safety in production process.

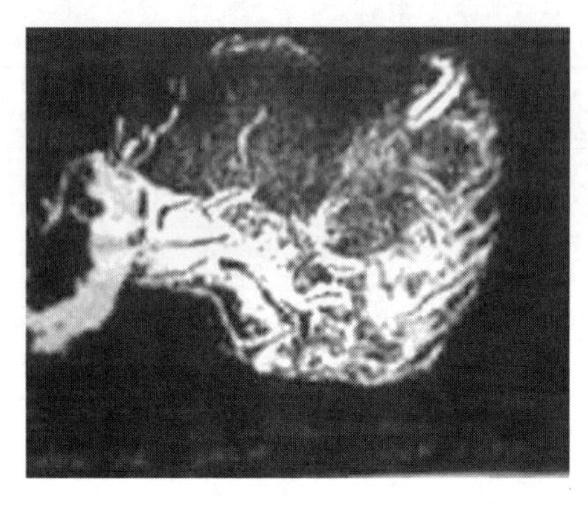

(a) Common AN

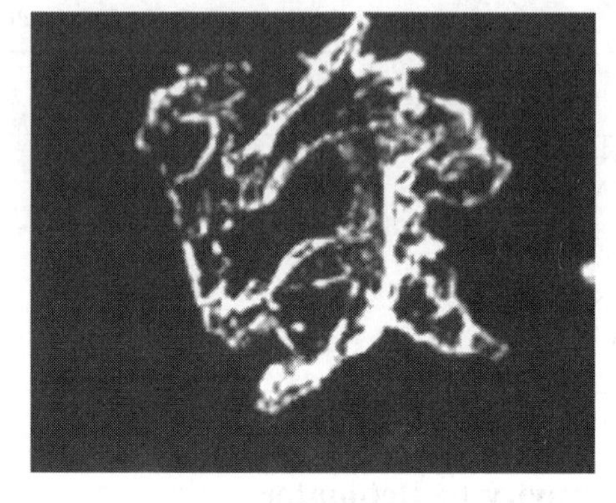

(b) Expanded AN

Fig. 1 Micro-structure of common AN and expanded AN

Sensitivity of expanded AN and its explosive to detonator and to shock wave is little bit higher than that of common AN and its explosive, so this kind of explosive is much more easily initiated, reliability of function is obviously increased; meanwhile sensitivity of expanded AN to impact, friction and electric spark is similar to that of common AN, so it is very safe in the manufacture and usage. Therefore expanded AN explosive is kind of new industrial explosive with excellent safety.

Reference

[1] Liu Zuliang, Hui Junming, Lu Chunxu, Wang Yilin. Research on light, porous and bulky AN and HF series power AN explosive. Explosive Materials, 1991, (5): 5.

[2] Hui Junming, Liu Zuliang, Lu Chunxu, Wang Yilin. Power AN explosive and its preparation. CN 91107051.6, 1991.

[3] Yang Tong. Investigation of technique of rock expanded AN explosive. Explosive Materials, 1996, (3): 30.

[4] Lu Chunxu, Liu Zuliang, Hui Junming, Wang Yilin. Research on rock expanded AN explosive. Explosive Materials, 1997, (1): 5.

[5] Lu Chunxu, Liu Zuliang, Ni Quqi. Industrial Explosive. Beijing: Ordnance Industry press, 1994.

[6] Hu Bingcheng, Liu Zuliang: Study on the safety property of expanded AN and explosive containing it. Explosive Materials, 1997, (6): 13.

(注：此文原载于 Proceedings of the 2004 International Symposium on Safety Science and Technology, 2004: 1604-1607)

Theoretical Research of Expanded AN Explosive

Lv Chunxu[❶]
Nanjing University of Science & Technology (Nanjing 210094)
Proceedings of 2nd International Symposium on Energetic Materials and Their Applications. 2005. PL2 (L-31)
Explosion 2005, 15 (3): 107-119
EI: 06039643004

Abstract: Expanded AN explosive is a novel powder industrial explosive with high performance, low energy consumption, low costs, better safety and with our own patent. The paper explained the status and development and theoretical research of expanded AN explosive, and mainly introduced the self-sensitization theory, status and development of the theoretical research of the expanded AN explosive.

Key Words: Expanded AN explosive, theoretical research, self-sensitization theory, status, development

1 The status and development of expanded AN explosive

1.1 Technology transfer of expanded AN explosive

Expanded AN explosive is a novel powder industrial explosive with high performance, low energy consumption, low cost, better safety and with our own patent. It is our Chinese pride. The theory and technology of expanded AN is a patent technology, which is invented by Chinese scientists and possess our own knowledge property right. It has obtained many awards including the top award for development of science & technology of the Ordance Industry Ministry, National 2nd award for development of science and technology. The scientific and technological monogragh of "Expanded AN Explosives" obtained scientific working committee of national defense and production facilities of expanded AN explosive obtained awards for development of science and technology including Jiangsu province; Hubei province and so on. In 1999, the technology of expanded explosive AN acquired the Gold award for invention of ten great inventions in China. This technology is planed on the emphasized spreading item of "the 9th and 10th five-year-plan" in China and the national torch plan item. Self-sensitization theory is the core of the invention and the expanded technology is the key.

---

❶ 应日本第二届含能材料及其应用国际研讨会(The 2nd International Symposium on Energetic Materials and Their Applications (ISEM) 26-27 May 2005 Tokyo, Japan) 邀请,吕春绪教授被作为顾问委员会委员 (International Advisory Committee),并特邀发言 (Invited Speaker) 作大会报告 (Plenary Lecture)。

Expanded AN technology has been attorned to more than 80 plants in China. Up to 2003, 50 production lines has been established. The products of expanded AN explosive have been used all around the country. The total production value is 3 billion yuan (RMB) and the profit is 500 million yuan (RMB) . It has got outstanding economic and social benefits.

Recently, four typical new production lines of expanded AN have been established. They are Hubei Kailong, Zhejiang Yongjin, Hunan Nanchen and Qinghai Haixi dongnuo. The four lines all show high performance, low energy consumption, low cost and better safety of expanded AN explosive.

The continuous and automation production line of expanded AN explosive in Hubei Kailong, chemical group limited-liability company passed the check and accept for the change of production line of the labor union organization in Hubei province. The line uses automatic operation, control, regulation, surveillance and record to the dissolving and crystallizing stations which have busy operatic and higher technology accuracy by computer automatic control system. It has increased the product quality. The key points of the line were all set up detection, limited-alarm and safeguard, and chain safeguard equipments, so it improves the reliability and safety of the whole production line. The line invested 9.21 million yuan (RMB) altogether. It started to establish in Sep. 2003 and its building projects erection and debugging of the facilities and people straining finished in Apr, 2004. In May 10, 2004, it is formally produced. Up to July 10, 2004, it produced thirty-six groups, 800 tons products. Their gap distance is 6 to 7 cm. Detnation velocity is 3000-3500m/s. Brisance is 13 to 15mm. They all reach the national industrial standards.

On the basis of introduced patent technology of expanded AN explosive of Nanjing University of Science & technology, by means of math brought forth new ideas, reformed tradition technology, introduced liquid pipeline, transported powder continuity and developed new type special purpose equipment, connected production technology of expanded AN explosive disconnection Yongjin company of Zhejiang carry out expandation and mix combination and by introduced advanced JX300DCS system applied domination technology of short for forward position field and carried out mechanism and meter and so on. They made combination of expandation and mix combination technology, reseached domination system of expanded AN explosive in the course of continuous production and fist built computer domination continuous production of expqandation and mix combination in our own country, thereby continuous, stable, safe and high-yield operation in the course of production explosive, which whole increased technology equipment and production technology standard and as well enhanced contribution of native industrial explosive standard.

Expanded explosive production-line of Hunan Nanling civil blasting appliance limited company Qidong subsidiary company is first expanded explosive production-line in Hunan and first introduced the cooperative technology of Nanjing University of Science & Technology and Guangdong 309 chemical plant in our own country, this key built vertical disconnection expanding-expandation and mix combination production-line of spiral continuous mix filling, which lie in Qindong county, Hunan province, it was built at the end of 2003, and computed debug at the beginning of 2004. New production-line is trying out, it has technology compaction, technology easy and smooth, high serviceability, function stable etc. During trying out production have gained approve of a lot of new and old user on putting into market and production emerge situation of demand exceeds supply supply. New company is improving and perfecting technology our company and Nanjing University of Science & technology, expanded explosive production-line of Nanling civil blasting company is bound to built expandation and mixture combination explosive production-line of native first-class and flower garden. They welcome lead at all levels and the same trade direction jobs, and together promote expanded explosive technology advance.

Qinghai Haixi dongnuo chemical limited company introduced improving vertical expanding technology, expanding process domination system, automatic continuous roller system and continuous mix charge domination system. The company production technology of combination roller-compress and servotab charge mixture mete. In it horizontal continuous roller-mix system have many functions of mixture, shatter transport, heat transport, mass transfer and technical characteristic of servotab copulation, sensitivity regulation, precision fixed quantity. the production of product-line and powder AN explosive are match in high product density, even blending, steady quality, blasting function and use effect. Determination consider that the technology belong to initial throughout the world, it has significance and prospect of wide spread and apply. At the same time, a roller-mix machine productivity equal six or eight edge runner, It has characteristics of low consumption, a litter personnel, high safety and nothing pollution.

1.2 Theory of expanded AN explosive

Our main theory points include:

(1) The self-sensitization of micro-pores introduced into the power AN explosive particles;

(2) The structure characteristic and distribution of micro-pore of expanded AN;

(3) The mechanism and technology of expansion of AN;

(4) Publish two monograghs: "Expanded AN Explosives" and "Theory of Industrial Explosive". They explained industrial explosive systematically, especially the self-sensitization theory of expanded AN explosive and theory and application for expansion.

The theory of expanded AN explosive has one light point and two strong supports. One light point: expanded AN explosive is invented by Chinese scientists and we own the patent; it is obtained "gold award for patent of ten great inventions in China" joint given by the United Nations Patent Organization and the Patent Office of China in 1985. Two strong supports: self-sensitization theory and implementing expansion theory and technology.

2 Self - sensitization theory of expanded AN explosive

Try to draw into trace bubble in AN pellets besides, pellets besides, pellets should dismutation reaction and course on the time of suffering strong outside stimulation, these non-even parts can form high temperature and high pressure hot spot, and then develop blasting, reach to the aim of self-sensitization, which is the AN design of self-sensitization.

Self-sensitization theory of expanded AN is the core of our main invention. After Self-sensitization and expansion, AN is honeycomb platellite crystal, which has loose structure, rough surface, a great number of fissures. This crystal structure contains appropriate amount of micro-pores and has a large surface, meanwhile there is a little surfactant in its exterior and interior, with these the surface is activated. Synthesis of these characteristic surely makes AN fully sensitized. AN is self-sensitized by the integrate applications of all kinds of sensitive techniques.

Detonation mechanism of industrial explosive-the hotspot theory is the foundation of self-sensitization theory of AN.

Self-sensitization of expanded AN mainly depends on a large number of bubbles in itself. The crystal temperature of these micro-pores calculated according to thermodynamics is very important, and it is also the problem which is concerned about. In expanded technology, a large number of micro-pores, which are distributed on interior surface of AN, completely becomes hotspot in the insulating condensation. When temperature reaches 1870℃, AN around the hotspot is heated, decomposed, ignited and then detonated. The calculation and analyzation of the critical temperature of hotspot is the basis and foundation of self-sensitization theory of AN.

The superficial character of micro-pores of expanded AN is as follows Fig. 1.

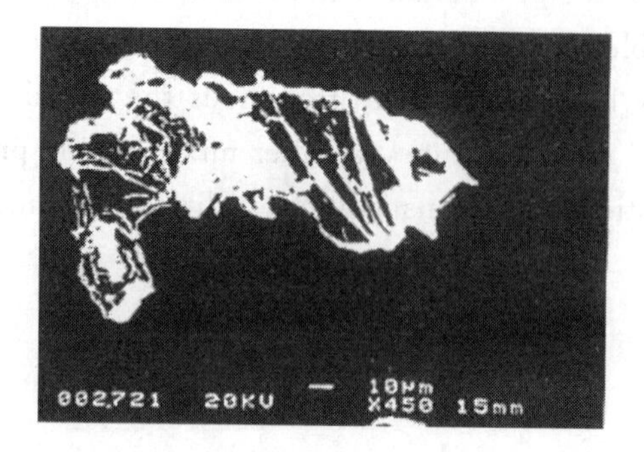

SDS Expanded AN with many arris and renventages

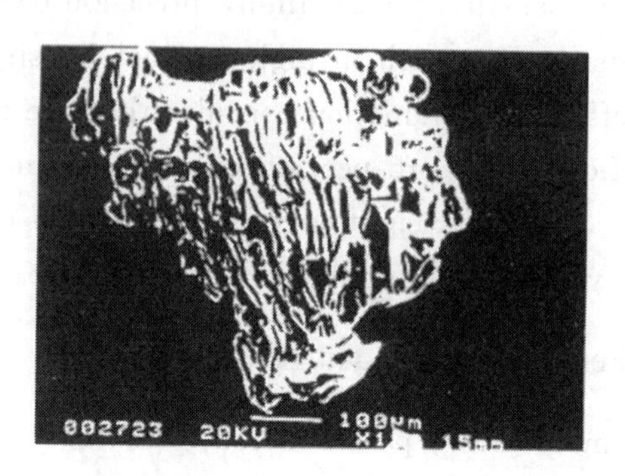

TW-80 Expanded AN with many arris and fissures

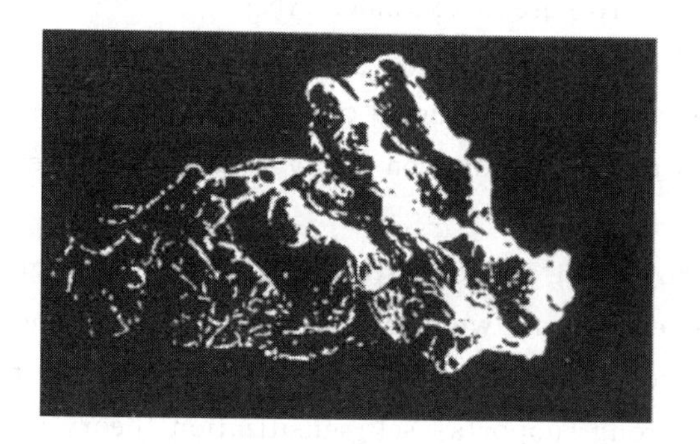

SOCTA Expanded AN with irregular surface and large ventage

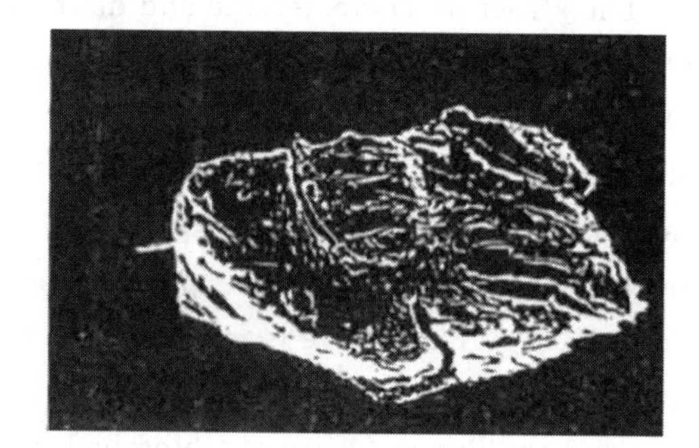

SOCTA/HST Expanded AN with rough surface and have Fissures, ventages

Fig. 1 The micro-structure of expanded AN

Compared with common AN, expanded AN is possessed larger specific area, it is shown as follows Table 1.

Table 1 Specific surface of AN in different crystallization conditions is listed in the table:

Crystallization condition	Common AN crystal	AN crystallization from A technique		AN crystallization from B technique	
		(A1)	(B1)	(A1)	(B1)
Specific surface/$cm^2 \cdot cm^{-3}$	471.95	1076.21	1101.26	1474.80	1454.57
Specific surface/$cm^2 \cdot g^{-1}$	758.76	2265.71	2167.83	3810.85	3328.54

Expanded technics lead expanded AN crystal to form a large number of micro-pores inside, and crystal surface have extremely irregular corners, fissures and sentusi. These are sources of hotspot in the effect of detonation chock wave. We not only interest in surface structure characteristic of micro-pores, but also interest in distribution and its law. we have studied distribution and its law of micro-pores in expanded AN.

We are lucky to discovery that 95% micro-pores of Expanded AN were in critical hole range. So, an accurate name of micro-pores of expanded AN must be critical hole, but we called it micro-pores in habits.

According to self-sensitization theory of expanded AN, stoma is of no use. Bulky pores also should be avoided. Critical hole is the best radius range. It can not be blocked by thin wood powder and oil, and the powder and oil can be well distributed on surface of the pore, from a best mixed state with big specific surface and high dispersion by mixing zero-oxygen-balance reaching numerator state oxidant with combustible, meanwhile it can keep enough pores, and go step further to keep enough hotspot, and make it keep in best initiating sensitivity and fully detonation.

According to self-sensitization theory of expanded AN, stoma is of no use. Bulky pores also should be avoided. Critical hole is the best radius range. It can not be blocked by thin wood powder and oil, and the powder and oil can be well distributed on surface of the pore, from a best mixed state with big specific surface and high dispersion by mixing zero-oxygen-balance reaching numerator state oxidant with combustible, meanwhile it can keep enough pores, and go step further to keep enough hotspot, and make it keep in best initiating sensitivity and fully detonation.

This is material basic of success and application of self-sensitization theory.

In the same time, we also studied the detonation character of expanded AN.

Expanded AN is a type of modified AN came from forced crystallization through a series of conditions. Its initiating sensitivity changes essentially, and has self-sensitization character.

The scanning locus pictures and velocity and pressure of detonation of expanded AN explosive is as follows:

Fig. 2 Scanning locus picture of Rock Expanded AN

Table 2 Results of detonation velocity of rock expanded AN and other industrial explosives

Name of explosive	ρ_0/g · cm^{-3}	D/mm · μs^{-1}	Name of explosive	ρ_0/g · cm^{-3}	D/mm · μs^{-1}
Rock Expanded AN explosive	0.997	3.556	Ammonite	1.067	3.196
Coal Expanded AN explosive	0.987	3.148	Emulsion explosive	1.026	4.826

Table 3 Results of detonation pressure of rock expanded AN and other industrial explosives

Name of explosive	K	A	Us/mm · μs^{-1}	P_{CJ}/GPa
Rock expanded AN explosive	-4.750×10^{-11}	-4.485×10^{9}	4.694	5.968
	-4.750×10^{-11}	-4.423×10^{9}	4.760	6.221
Coal expanded AN explosive	-4.750×10^{-11}	-4.510×10^{9}	4.668	5.583
	-4.750×10^{-11}	-4.734×10^{9}	4.447	4.791
Ammonite	-4.750×10^{-11}	-4.331×10^{9}	4.861	6.521
	-4.750×10^{-11}	-4.254×10^{9}	4.948	6.872
Emulsion explosive	-4.750×10^{-11}	-3.797×10^{9}	4.694	10.857

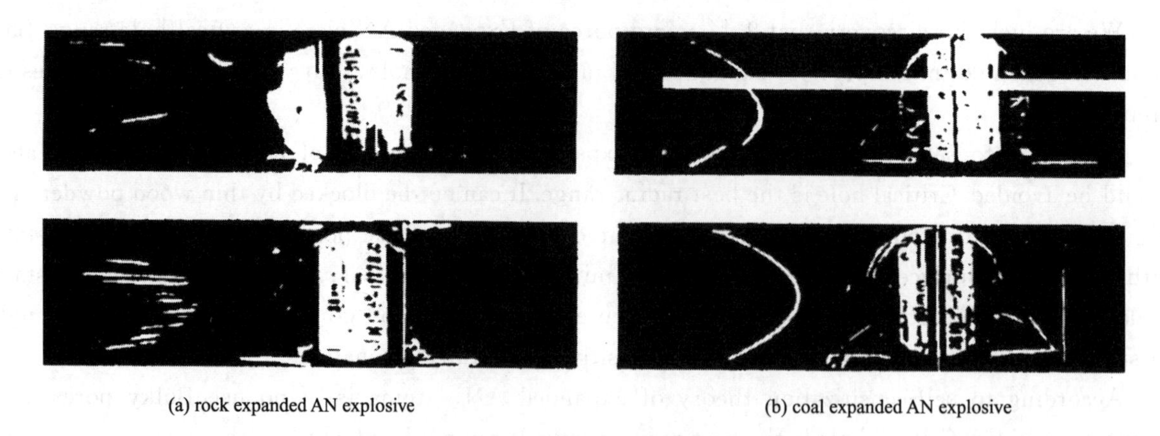
(a) rock expanded AN explosive (b) coal expanded AN explosive

Fig. 3 The wavefront tract of expanded AN detonation

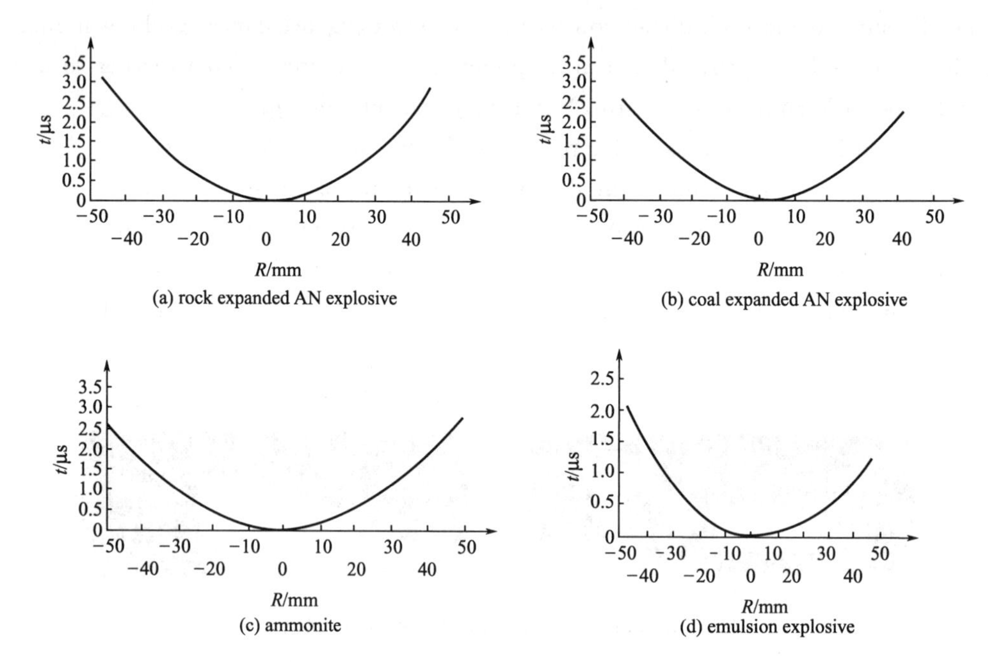

(a) rock expanded AN explosive (b) coal expanded AN explosive

(c) ammonite (d) emulsion explosive

Fig. 4 The fitting curve of output wavefornt of rock expanded AN explosive and other industrial explosives

Table 4 Detonation status of rock expanded AN and other industrial explosives

Name of explosive	Time dispersion from origin to−30mm/μs	Time dispersion from origin to+30mm/μs
Rock expanded 1-3	1.4	1.1
Ammonite1-4	0.8	1.25
Rock expanded 2-1	1.52	1.40
Ammonite2-2	1.40	1.25
Ammonite3-1	0.95	1.1
Explosvie 3-2	0.80	1.25
Emulsion explosive 4-1	0.8	0.53
Perlite 4-2	1.0	0.72

From the wave picture of detonation wave through explosive underside, we can conclude that wave of expanded AN is the same with wave of ammonite and emulsion explosive, time dispersion from origin is also fundamentally close, according to formula, technique designed by self-sensitization theory, expanded AN produced can fully detonated, and has a good detonating character.

3 Theory and technics of expansion of AN

Expansion of AN is "ABC" line. It not only has a lower temperature process, but also has a densification process. Surdissolubility curve is the fundamental of expansion.

Both lower temperature crystallization process and vacuum evaporation process can obviously improve specific surface and pulverization of AN, also they can cause some shortcoming in crystal such as sentus and corners, which can change firing and detonation character of AN explosive. Compared to particles of AN dealing with lower temperature process, AN dealing with decompression evaporation crystallization have a more rough surface, and in crystal there are great number of bubbles and ventages, in result, forms light more-pores Expanded AN. consequently, AN dealing with decompression evaporated crystallization have a better exploding character, and can be easily carried out, have a higher efficiency, but energy consumption is lower. So, technology of AN's expansion is: in the effect of bulking substance, under particular temperature, consistency and vacuum, saturation AN solution is enforced crystallized through decompression evaporation.

What should be noted is that: in system, surfactant treated as bulking substance extremely changes the process of enforced crystallization; as soon as crystallization started, crystallization energy is enough to ensure added-water be clearly evaporated, without any extra energy from exterior. Two points are secret and succeeded assurance of expanded technology.

Key technology of expanded AN is expansion. Essence of expansion is comprehensive utilization of surfactant technology, crystallization chemistry, explosive detonation mechanism, it is a physical and chemical process of enforced crystallization in the effect of complex surfactants.

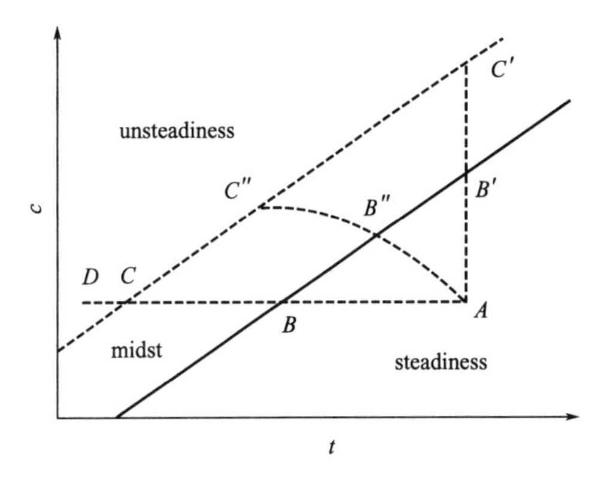

Fig. 5 Curve of dissulobility-surdissulobility

The expansion mechanism of AN maybe is that: in the effect of particular temperature and vacuum, water of AN saturation solution vaporizes, forms pores, and then forms many relatively balanced foam in the effect of surfactant. But when temperature and vacuum reaches another particular lever, foam breaks up, steam vaporizes rapidly, and then in AN crystal there are many "holes, cavities, gaps" . surface of AN separated out quickly is also extremely irregular, which makes the bonds among AN particles very irregular, it forms a type of light, more-pores, loose expanded AN. following is process of AN expansion.

Following is process of AN expansion:

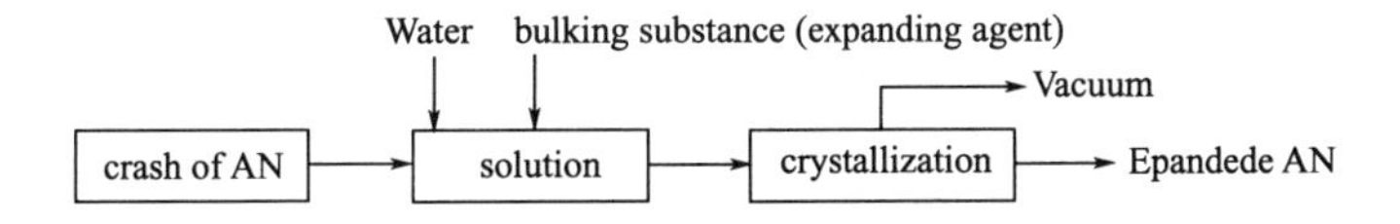

Effect of reserved complex surfactant—bulking substance in AN expansion.

(1) Bulking substance lower surface tension of AN solution in the expansion of AN.

① With a little bulking substance, surface tension of AN solution can be lowered tremendously.

② When bulking substance reaches a particular consistency, surface tension is lowered mildly.

③ Temperature has a definite influence to surface tension of AN solution, when temperature rises, surface tension drops.

④ In the same mass percentage of bulking substance, when AN mass percentage rises, surface activation is improved, and surface tension drops.

(2) Bulking substance played as foaming agent in AN expansion

① When mass percentage of bulking substance rises, foaming ability of solution is improved because of surface tension of solution dropped

② When mass percentage of AN solution rises, action among numerators strengthens, solution density increases, water in solution decreases, so foaming ability drops;

(3) Bulking substance makes crystal embryos easily be formed in AN enforced crystallization. Expansion enforced crystallization is composed by following steps:

First is system reaches saturated state; second is forming crystal embryos, third is crystal grows.

r_c is critical radius of embryos. Only if $r>r_c$, embryos can spontaneously grow up to be large crystal. Because AN solution has 0.15% bulking substance (surfactant), which can reduce specific surface area energy of solid-solution interface, r_c reduces, in another way embryos forms easily, as a result, AN enforced crystallization is accelerated.

In the whole AN expansion, as soon as crystal forms, latent-energy of crystallization can maintain the whole evaporation, it is constant temperature process in theory, meanwhile water is effusing continuously, at the same time, expanded AN is in a dry process until expansion ends, when water mass percentage reaches target. But different first temperatures have different influence to the process so to say.

Solution consistency, temperature, and vacuum have obviously influence to expansion:

AN expansion is vacuum crystallization of AN saturated solution in the effect of bulking substance. According to solubility—persolubility curve, crystal separation has some relation with solution consistency and temperature, crystallization only happens in saturated solution. In a particular vacuum condition, when outside pressure is lower than balance-saturated stream pressure of AN solution, water could evaporate from solution immediately, absorbing gasify energy, as a result, temperature of solution decreases while density improves.

When temperature decreases below temperature of saturated solution, solution is in over-saturated state, and crystallizes gradually, releasing crystallization energy to meet with energy needed by evaporation, and keep evaporation continue. This way, water evaporates and crystal forms. It's proved that: between 125℃～135℃, without extra heating, AN solution with 90% mass percentage can be decompressed evaporated to expanded AN with water lower than 0.15%. if solution density lower, then water content of expanded AN higher, but higher density need high solubility, which could lead AN become sticky and hard, not beneficial to following process.

Expansion and dry of AN happens in the same time vacuum not only decreases the boiling point of AN solution which is helpful to crystallization, but also carries water away. Vacuum is one of must condition of AN expansion. In theory, vacuum higher, speed of depression of boiling point of AN solution and evaporation of water faster. In result, time of expansion shorten, and the effect of expansion is better. But high vacuum condition needs a harsh requirement to equipment, so in production of expanded AN, a suitable low limit of vacuum is necessary. Following is requirement of vacuum:

Table 5 The effect of vacuum to expansion

Vacuum/MPa	≤0.070	0.075	0.080	0.085	0.090	≥0.092
Expansion times	1	2~3	4~5	7~8	9~10	9~10
Mass percentage of water(%)	≥0.80	0.50	0.20	0.12	0.07	0.06

To sum up, the best technique condition is:

Mass percentage of saturated AN solution is above 80%~94%, temperature is 110℃~140℃, mass percentage of bulking substance is 0.12%~0.15%, considering safety and effect of expansion, amount of bulking substance in fact is about 0.12%, vacuum is 0.085~0.095MPa, so it will gain the best effect, and water in expanded Anis lower than 0.15% which is satisfied with demand of detonation.

4 Status of expanded AN in the theory

In recent ten years, we published a lot of disquisitions about the self-sensitization theory, expanding techniques, prescription design and production series. There are 12 thesis published in the first-class magazines, four thesis cited by EI, 31 thesis in key magazines, 20 thesis in international conferences.

The first-class magazine:

• Lv Chunxu Expanded technology of AN and its application. Engineering Science. 2000, 2 (2): 60

• Lv Chunxu Research on self-sensitizaion theory of Expanded AN. Engineering Science. 2000, 2 (11): 73

• Lv Chunxu Research on expanded theory of AN. Engineering Science. 2001, 3 (9): 58

• Lv Chunxu A study on micro air-bubbles in the self-sensitization theory of expanded ammomium nitrate. Bing Gong Xue Bao. 2001, 22 (4): 485

• Lv Chunxu. A study on crystalline characteristics of expanded AN. Bing Gong Xue Bao. 2002, 23 (3): 316

• Lu Ming. Stoichiometrical design and optimization of expanded AN seismic explosive columns. Bing Gong Xue Bao. 2000, 21 (3): 217

• Lu Ming. The mathematical model for the prescription design of powdery commercial explosive. Explode and Impact. 2001, 21 (3): 210

• Lu Ming. A study on the expansion mechanism of AN. Bing Gong Xue Bao. 2002, 23 (1): 30

• Lu Ming. The atom economical analysis of industrial explosive explosion and shock waves. Explode and Impact. 2003, 23 (1): 86

• Ye Zhiwen. Self-sensitizable characteritics of modified AN. Applied Chemistry. 2002, 19 (2): 130

• Ye Zhiwen. Crystal properties of modified AN. Bing Gong Xue Bao. 2002, 23 (3): 412

• Ye Zhiwen. Study on modification of AN by Gemini surfactants. Chemistry. 2004, 67 (1): W004

Papers embodied by EI:

• Ye Zhiwen. Study of the thermodynamical properties of the micellar structure of composite surfactant in AN solution. Explosive materials. 1996, [6]: 1

• Ye Zhiwen. Research on improving crystal pattern transition of expanded AN by srufactants. Explosive materials. 1998, 27 (4): 8

• Ye Zhiwen. Caculation and analysisi of solid surface free energy of expanded AN. Explosive materials. 2000, 29 (5): 1

• Ye Zhiwen. Research on microstructure of expanded AN. Explosive materials. 2002, 31 (1): 7

We published two monographs: 《Expanded AN Explosives》 and 《Theory of Industrial Explosive》. These two monographs is supported by the Fund of National Science and Technology Press. 《Expanded AN Explosives》 won the second prize of National Department.

Five master dissertations and five doctorate dissertation about expanded AN:

• Dingyun. <master dissertation>. A study on AN and its explosion performances improved by Srufactants. 1990, 12

• Chen Tianyun. <master dissertation>. The external performance and applying investigation of AN. 1992, 12

• Hu Bincheng. <master dissertation>. Application investigation on AN modified by Surfactant. 1993, 12

• Ye Zhiwen. <master dissertation>. Research on the expansion mechanism of AN. 1995, 12

• Zhao Guoqiang. <master dissertation>. A study of continuous work lines of expanded AN. 2000

• Lu Ming. <doctor dissertation>. Study on expanded AN. 1999, 7

• Chen Tianyun. <doctor dissertation>. Self-sensitization of expanded AN and its application. 2000. 10

• Gao Dayuan. <doctor dissertation>. A study of detonation and safety of mixed explosives. 2003, 6

• Zhou Xingli. <doctor dissertation>. Research on theory of self-sensitization of expanded AN and physical properties of its explosive and modification. 2003, 6

• Tang Shuangling. <doctor dissertation>. A study of desensizition of AN. 2004, 4

We hold many science communication around Expanded AN:

• Hosted the 26th International Pyrotechnics Seminar and 1th International Seminar on Industrial Explosive Materials in Nanjing

• Invite famousProf of Sweden University of Lund. Lindman and others to give lectures.

• Hold producing technic seminar and authenticate seminar about expanded AN in Xiameng and other places.

• Take part in international seminars four times in the USA, Japan, Australia and Germany.

• Vist the USA, Sweden, Norway and India to review the status and development of industrial explosives.

• Train two foreign students in energetic materials.

5 Development of expanded AN explosive theory research

(1) The mathematical model for the prescription design of powdery commercial explosive

(2) Research of structure and character of micro-pores in self-sensitization theory

(3) Research of relation of medium-hole's distribution and detonation character in self-sensitization theory

(4) Research of randomicity of AN hotspot

(5) Research of crystal state and transition point of expanded AN

(6) Research of particles of expanded AN

(7) Research of chosen of best technical condition of expansion

(8) Research of expansion mechanism and influence

Expanded AN explosive theory is the base of expanded AN explosive. With its development, technology of expanded AN will be improved greatly. We continue to do our best to research and development on Expanded AN explosives and it's theory, and let's do things better!

Reference

[1] Dingyun. Influence of surfactant to AN's modification [master dissertation] . Nanjing: Engineering college in east china, 1990.

[2] Lvchunxu. Theory and technology of surfactant. Nanjing: Jiangsu science and technology publishing company, 1991.

[3] Hui junming, Liu zhuliang, Lv chunxu. Powdery AN explosive and its production. CN91107051. 1991.

[4] Liu zhuliang, hui junming, Lv chunxu. Light AN and HF AN explosive. Explosive materials, 1991 (5): 5.

[5] Chen tianyun. Research of surface character of AN [master dissertation] . Nanjing: Engineering college in east china, 1992.

[6] Luming. Research of expanded AN explosive [doctor dissertation] . Nanjing: NUST, 1999.

[7] Yc zhiwen. Research of expansion mechanism of AN [master dissertation] . Nanjing: NUST, 1995.

[8] Lv chunxu. Surface chemical researchof expanded AN. Transaction of Nanjing university, 1995 (31): 286.

[9] Chen tianyun, Lv chunxu, Ye zhiwen. Character research of modified AN. Energetic materials, 1996 (4): 169.

[10] Lv chunxu, Liu zhuliang, Hui junming, Wang yilin. research of rock expanded AN explosive. Explosive materials, 1997 (1): 5.

[11] Lv chunxu, Liu zhuliang, Hui junming. Form and development of self-sensitization theory of AN. Transaction of powder and explosive, 2000 (4): 1.

[12] Lv chunxu, Liu zhuliang. Basis and experimental research of self-sensitization of expanded AN. Explosive materials, 2000 (4): 1.

[13] Lv chunxu. Research of self-sensitization of expanded AN. Engineering science, 2000 (11): 73.

[14] A Study on micro air-bubbles in the Self-sensitization theory of expanded ammomium nitrate. Bing Gong Xue Bao. 2001 Vol. 22 No. 4: 485.

[15] Lv chunxu. Research of Expanded AN crystal. Bing Gong Xue Bao, 2002 (3): 316.

[16] Ye zhiwen, liu zuliang, Lv chunxu. Study of an AN crystal. 2002 (3): 412.

[17] Chen tianyun. Study of the application of sensitization of AN [doctor dissertation] . Nanjng: NUST, 2000.

[18] Zhou xingli. Research on theory of self-sensitization of expanded AN and physical properties of its explosive and modification. [doctor dissertation] . Nanjng: NUST, 2003. 6.

[19] Gao Dayuan. A study of detonation and safety of mixed explosives. [doctor dissertation] . Nanjng: NUST, 2003. 6.

[20] Gao Dayuan, Lv chunxu, Dong haishan. Study on the underside output waveo of industrial explosives. Explosive materials, 2002 (6): 1.

[21] Gao Dayuan, Lv chunxu, Dong haishan. Study on the detonation of industrial explosives. Transaction of explosive, 2003 (1): 8.

第二届含能材料及其应用国际会议（The 2nd international Symposium on Energetic Materials and their Applications）2005年5月26-27日在日本东京举行。该会议由日本火药学会主办，会议主席为火药学会会长，日本横滨国立大学小川辉繁教授 Prof. Terushige Ogawa（Yokohama National University，President of the Japan Explosives Society），会议执行主席为东京大学越光男教授 Prof. Mitsuo Koshi（University of Tokyo），日本产业综合研究所 Dr. Shuzo Fujiwara（Director，Research Center for Explosion and Safety，AIST）。

日本火药学会成立于1939年，长期以来为日本经济的发展与进步作出了重要的贡献。该学会的宗旨是含能材料科学的发展，快速燃烧和爆炸现象的技术应用；减缓含能材料及其应用对环境的不利影响。

在会议发表的论文包括含能材料，烟火，爆轰及爆炸，冲击，压力震荡，灾害，安全，安全措施和灾害管理，政策与法规，及其他专题。来自中国，中国台湾，韩国，印度，捷克，俄罗斯，泰国，比利时，以色列，伊朗，法国，美国，英国，日本等的研究者发表78篇论文（包括口头及贴图发表）。其后经审查合格论文将被收录在《火药学会志》杂志上。

会议邀请五位国际上享有盛名的火药专家作了重要的演讲，分别介绍如下（按时间顺序）：

（1）Managing hazardous materials safety and security risks

Dr. Charles H. Ke（U. S. Department of Transportation，U. S. A.）

回顾了历史上化学工业中发生的重大事故，如50年前 Texas City 造成576人死亡的硝铵肥料的爆炸事件；1984年印度造成2000人死亡，2万人中毒的联合碳化公司（美）甲基异氰酸酯的泄露事件；1992年墨西哥184人死亡，600人受伤的LPG气体爆炸事件……直到美国发生的911事件。由此，危险物安全管理应该得到各国及国际间高度的重视。作者详细介绍了美国安全运输部着力于制定危险物管理法规，并将其与国际危险物管理法规靠拢。

（2）Theoretical research of expanded AN explosive

Prof. Lu Chun Xu（Nanjing University of Science and Technology，China）

作者在第一届含能材料及其应用国际会议（ISEM2002，日本东京）介绍了膨化硝铵的膨化技术及其应用，得到与会者的广泛兴趣。本次会议上进一步介绍了作为这种性能高，耗能低，造价低，安全性能好的粉状工业炸药的近期推广（尤其是在中国西北部地区）的状况，并且介绍了膨化硝铵技术的理论发展水平，包括自敏化理论的现状及发展前景。

（3）Current trend of R&D in the field of high energy materials（HEMs）-an overview

Dr. Singh Haridwar（High Energy Material Research Laboratory，India）

本文综述了各种高含能材料在印度民用，宇航及军事上的应用。在印度有65个炸药厂生产百万吨级民用炸药。

（4）Chemical kinetic modeling of open burning and open detonation of munitions

Dr. Charles K. Westbrook and Dr. William J. Pitz（Lawrence Livermore National Laboratory，U. S. A.）

燃烧模型，包括化学反应动力学机理，广泛应用于烃类及其他燃料的反应过程分析中。然而类似模型用于分析或者预测含能材料或者火炸药的燃烧方面的工作还很少。这归因于这类反应的极高速率，反应中间体的难观测性，以及分子结构的复杂性。作者详细地介绍了其研究工作是基于含能材料如RDX，TNT等的原子/分子结构来描述其反应动力学，并由此模拟其燃烧现象的燃烧模型。揭示了烟生成对于燃烧影响的原理。建立的模型可以解释RDX及HMX燃烧中烟生成很少或不生成，而TNT燃烧中有大量烟生成的原因。由此通过计算模拟，然后调整燃烧模型，可以大大降低TNT基炸药燃烧的烟生成量。模型还能扩展到分析用于军需品的大量有机磷（如沙林）的燃烧现象。

（5）The future trend of risk management in chemical industries-for the drastic reduction of chemical accidents

Pro. En Sup Yoon（Seoul National University，Korea）

介绍了韩国化学工业危险物管理的现状及今后的发展趋势。随着工业的发展，涉及安全，健康，环境（SHE）的全社会灾害管理系统的开发也变得越来越重要。作者介绍了汉城大学利用 OECD 模型建立了韩国化学石油工业安全审查系统及其在对政府，社会及工业之间的协调中所起的作用。

ISEM2005 由 Asian Office of Aerospace Research and Development（AOARD），U. S. Air Force Office of Scientific Research（AFOSR）及日本化学公司如日本油脂，旭日化成等协助承办。

（钱华，金序兰）

A Computer Model for Formulation of ANFO Explosives

Lu Ming, Lv Chunxu

(Nanjing University of Science and Technology, Nanjing, P. R. China 210094)

SCI: 00025861000004, EI: 20074610922487

Abstract: A novel computer model has been designed to aid in the formulation of powdered Ammonium Nitrate/Fuel Oil (ANFO) explosives. This mathematical model calculates heats of explosion, oxygen balances and raw material costs as a function of explosive ingredients. The best theoretical formulation consists of 91.5%～92.0% ammonium nitrate (AN), 4.5%～5.0% wood powder (WP), 3.0%～3.5% composite fuel oil (CFO). This formulation results in an oxygen balance of zero or close to zero.

Keywords: ANFO explosive, ammonium nitrate, computer aid, formulation design

1 Introduction

One of the primary industrial explosives employed in China consists of ANFO. In order to decrease raw material costs, Fuel Oil (FO) is occasionally replaced with WP by explosive manufacturers. The net result, however, is that non-optimum explosive formulations are manufactured which results in: decreased fuel oxidation; poor blast performance; and an increase in noxious fumes, all of which are detrimental to the overall blasting objectives.

This paper reports on the following: 1) a novel computer aided method for the formulation of powdered ANFO explosives; 2) the establishment of a mathematical model for the formulation of ANFO explosives; 3) the computer aided formulation design and optimization of ANFO explosive and, 4) numerous reasonable formulations with different raw material costs.

2 A mathematical model for the formulation design of ANFO explosives containing C, H, O, and N

Explosive properties as a function of cost are usually the primary consideration during the formulation of industrial explosives. However, the ratios of Heats of Explosion as a function of cost should be the determining factor, not just cost alone. Explosive properties are a function of the heat of explosion (Qv) of an explosive, so the heat of explosion is selected as the primary figure of merit for our mathematical model. However, since cost is also an important factor in the manufacturing of explosives, the ratio of the amount of energy per unit cost (how much bang for the buck) should be the decisive factor as long as raw materials are the only cost consideration. However, while manpower should also be included in the equation, our model does not take this into manpower factors into consideration as these values vary with geographic location. Powdered ANFO explosives consist of Carbon (C), Hydrogen (H), Oxygen (O) and Nitrogen (N) . The general empirical formula for ANFO is "$C_aH_bO_cN_d$" . Assuming the specific empirical formula is, "$C_{ai}H_{bi}O_{ci}N_{di}$", then the molar mass of component "i" is given as "m_i" and the molar number of component "i" in 1kg explosive is $10x_i/m_i$, where "x_i" is the percent mass content of component "i".

The equation for the reaction of an explosive containing C, H, O, N is given in equation 1.

$$C_aH_bO_cN_d=\Sigma 10x_i/m_iC_{ai}H_{bi}O_{ci}N_{di}\longrightarrow b/2H_2O+(c-b/2-a)CO_2+(2a-c+b/2)CO+d/2\ N_2 \quad (1)$$

Where $a=\Sigma 10a_ix_i/m_i$, $b=\Sigma 10b_ix_i/m_i$, $c=\Sigma 10c_ix_i/m_i$, and $d=\Sigma 10d_ix_i/m_i$.

The mathematical formula (where the primary figure of merit is Heat of Explosion) is shown in equation 2.

$$\max Q_v=\Delta H_{f(H_2O)}*g_1+\Delta H_{f(N_2)}*g_2+\Delta H_{f(CO_2)}*g_3+\Delta H_{f(CO)}*g_4-\Sigma 10x_i\Delta H_{f(i)}/m_i \quad (2)$$

$$\text{Where } g_1=b*2^{-1},g_2=d*2^{-1},g_3=(c-b*2^{-1}-a),g_4=2a-c+b*2^{-1}$$

The mathematical expressions for the other figures of merit are:

Oxygen Balance (OB) $=\Sigma a_ix_i=100\eta$

Variable Sum (VS) $=\Sigma x_i=100$

Cost Sum (CS) $=\Sigma p_ix_i=p*10^{-1}$

Lower limits and upper limits$=s_i<x_i<t_i$, where I=1, 2, …, n.

In this mathematical model the following definitions are employed:

ΔH_f Heat of formation for the ingredients, the entire explosive or explosive products ($kJ\cdot mol^{-1}$);

g the molar number of the explosion products;

x_i the content of component "i" of the explosive;

m_i the molar mass of component "i";

a_i the oxygen balance of component "i" (%);

η the oxygen balance of the explosive (%);

p_i raw material cost of component "i", RMBy t^{-1};

p raw material cost of explosive, RMBy t^{-1};

s_i lower limit of component "i"; (%);

t_i upper limit of component "i"; (%).

3 Mathematical expression for ANFO explosive formulations

3.1 Commercial ANFO explosives ingredients and their chemical parameters

The oxidizer for ANFO explosives usually consists of: common Ammonium Nitrate; improved Ammonium Nitrate; or porous Ammonium Nitrate. The fuel is typically WP or CFO (which generally consists of a mixture of fuel oil and wax) . The formulas, molar mass, heats of formation, oxygen balances and raw material costs of these components are listed in Table 1[1-5].

Table 1 Physical chemistry parameters of ANFO explosive raw materials

No.	Empirical formula	Ingredient	MW (g · mole^{-1})	Hf (kJ · mole^{-1})	OB (%)	Cost (RMBy · t^{-1})
1	$N_2O_3H_4$	AN	80	353.46	+20	1400
2	$C_{15}H_{22}O_{10}$	WP	362	1649.43	−137	400
3	$C_{16}H_{32}$	FO	224	660.44	−342	2000
4	$C_{18}H_{36}$	Wax	254	558.03	−346	4000

3.2 Mathematical expression for formulation of ANFO explosives

If the amount of AN, WP, FO and wax is respectively x_1, x_2, x_3, and x_4, then:

$g_1 = 0.250 * x_1 + 0.304 * x_2 + 0.714 * x_3 + 0.748 * x_4$

$g_2 = 0.125 * x_1$

$g_3 = 0.125 * x_1 - 0.442 * x_2 - 1.492 * x_3 - 1.457 * x_4$

$g_4 = -0.125 * x_1 + 0.856 * x_2 + 2.143 * x_3 + 2.165 * x_4$

and, formula (3) gives the general formula for Heat of Formation for ANFO explosives.

$$\Sigma 10 x_i \Delta H_{f(i)} / m_i = 44.183 * x_1 + 45.562 * x_2 + 29.486 * x_3 + 21.970 * x_4 \quad (3)$$

Then the primary figure of merit (Maximum Heat of Explosion) is given in formula (4).

$$\text{Max } Qv = 51.084 * x_1 - 50.653 * x_2 - 180.317 * x_3 - 173.169 * x_4 \quad (4)$$

Formula (4) assumes the following conditions:

$20 * x_1 - 137 * x_2 - 342 * x_3 - 346 * x_4 = 100\eta$

$x_1 + x_2 + x_3 + x_4 = 100$

$1.4 * x_1 + 0.4 * x_2 + 2.0 * x_3 + 4.0 * x_4 \leqslant P * 10^{-1}$

$89.0 \leqslant x_1 \leqslant 95.0$

$2.5 \leqslant x_2 \leqslant 7.0$

$1.0 \leqslant x_3 \leqslant 3.0$

$0 \leqslant x_4 \leqslant 1.0$

For formula (4), it is assumed that in general $\eta \leqslant 0$ and the values η and P may be calculated.

4 Computer solutions

To arrive at a solution, a computer is employed. The mathematical formulas previously referenced are input into a MS Excel program. The mathematical solution is different with different oxygen balance values (η) and raw material costs. The solutions are detailed in Tables 2 and 3.

Table 2 Calculated properties of ANFO explosives at $\eta = 0$

Formula No.	%AN	%WP	% FO+ Wax	Qv (kJ · kg^{-1})	OB (%)	Cost (RMBy t^{-1})	Hf/Cost^{-1} kJt kg^{-1}RMBy
1	92.80	3.00	4.20	3845.2	0	1435.2	2.68
2	92.00	4.00	4.00	3787.8	0	1416	2.68
3	91.74	4.85	3.41	3836.9	0	1400	2.74
4	91.65	5.00	3.35	3832.7	0	1390.1	2.76
5	91.18	5.82	3.00	3821.9	0	1359.8	2.81
6	90.50	7.00	2.50	3820.3	0	1345	2.84

From Table 2, we observe that Formulation No. 6 has the higher ratio of heat of explosion to raw material cost than that of other formulations. However, the content of WP in Formulation No. 6 is higher than in the other formulations, resulting in a lower charge density compared to the other formulations, thus the energy per unit volume explosive is lower. Therefore, Formulation No. 4 has the overall best theoretical performance when charge density is considered. As AN content increases, the raw material cost increases, the heat of explosion increases, and overall explosive properties are improved based on formula (4) . Since the value $x_1 > 0$, as AN increases, the theoretical heat of explosion increases. If the values of x_2, x_3, and x_4 increase (because of the multiples of x_2, x_3 and x_4 are < 0), then the heat of explosion for ANFO explosives decreases.

Table 3 Properties of ANFO explosives formulations at $\eta < 0$

No.	%AN	%WP	% FO+ Wax	Hf (kJ · kg^{-1})	OB (%)	Cost (RMBy t^{-1})	□Hf/Cost^{-1} kJt kgRMBy
1	91.03	6.00	2.97	3816.2	−0.2	1372.9	2.78
2	90.98	6.00	3.02	3803.7	−0.4	1373.8	2.77
3	90.92	6.00	3.08	3791.1	−0.6	1374.6	2.76
4	90.87	6.00	3.13	3778.6	−0.8	1375.5	2.75

From Table 3, we see that, (for conditions when oxygen balance is less than zero ($\eta < 0$)), as oxygen balance becomes more negative, then: the AN weight percent decreases; the weight percent of fuel oil and wax increases; and the ANFO heat of explosion decreases. On the other hand (for oxygen balance values greater than zero ($\eta < 0$)), the volume of toxic gases (i. e., CO) produced after the explosion increases. Therefore, it is important that oxygen balances be zero (or close to zero) in order that formulations have high heats of explosion and low toxic gas volumes.

References

[1] Lu Ming and Lu Chunxu. J. of Chin. Explosive Mat. 28 (4), 1-5 (1999).
[2] Yuan, Zuhiu. J. of Chin. Explosive Mat., 9 (2), 1-6 (1980).
[3] Lu Chunxu; Liu Zuliang and Li Ouqi. "Commercial Explosives", Beijing China, Ordnance Industry Press, p136 (1994).
[4] Lu Ming. "Formulation Design of Commercial Explosives", Beijing China, Ordnance Industry Press, p199 (2002).
[5] Lu Ming and Lu Chunxu. J. of Chin. Explosion and Shock Waves. 23 (1), 86-90 (2003).

(注：此文原载于 Sci. Tech. Energetic Materials, 2007, 117-119)

Research and Application of Expanded Ammonium Nitrate

Liu Zuliang, Lv Chunxu, Lu Ming, Chen Tianyun
(Nanjing University of Science and Technology, Nanjing 210094, China)
EI: 96043144889

Abstract: The theory of preparation and its characteristics of expanded AN is discussed. The best formula of zero oxygen balance is given. On the basis of remarkable advantanges, technology of expanded AN is widely used by many factories.

1 Introduction

Ammonite (Ammonium nitrate (AN) explosive sensitized with TNT) has been one of the chiefest industrial explosives, with the output proportion of 80% in China. It possesses two remarkable disadvantages: (1) TNT content is high, toxicity is large, cost of production is expensive and environmental pollution is serious. (2) physical property is unstable, hygroscopicity and caking capacity are high. so storage life is short. In order to overcome above-mentioned fault, light porous expanded AN has been researched by us. Non-TNT rock expanded AN explosive is obtained from expanded AN. The explosive possesses more violent explosion strength and stable physical properties, and is widely welcomed by a great quantity of users and factories.

2 Explanded AN and it's characteristics

2.1 Principle of preparation

Under the action of surfactant in the range of certain concentration. Molecule of AN in the solution is regularly arranged as "state of layer upon layer" . Under comprehensive action of surfactant, temperature and vaccum. Solution of AN is presented as "state of boiling" . Molecule of water is quickly escaped out, bulky state and honey-comb-like of AN is crystalifed from the solution, to give expanded AN. Fig. 1 shows characteristics of micro structure of expanded AN through electromicrography[1]:

2.2 Characteristics of expanded AN[2]

Expanded AN obtained through surfactants treatment possesses porous structure of sponge-like. To compare with common powder, granular and porous granular AN, its characteristics are as

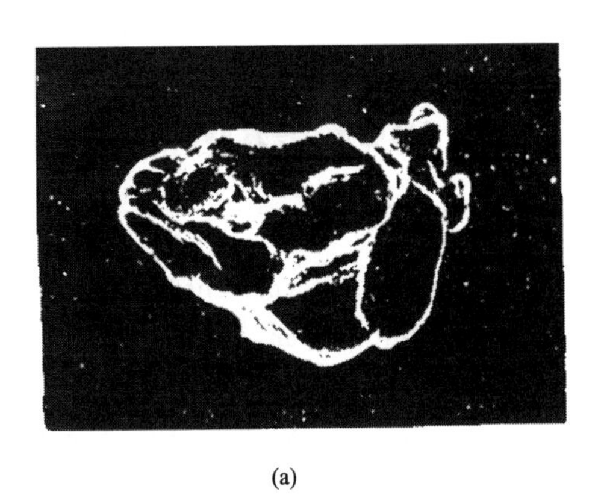

(a)

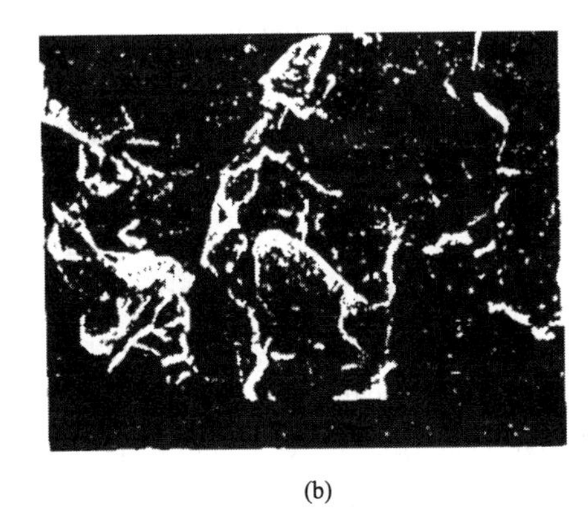

(b)

Fig. 1 Structure of expanded AN

where (a) Common AN (enlarge 900 times); (b) Expanded AN (enlarge 750 times)

follows:

(1) Larger specific surface area: It's specific surface area measured by method of resistance-down is 3328.54cm^2 • g^{-1}. To compare with it, specific surface area of common powder AN (425μ~250μ) is 758.76cm^2 • g^{-1}. Relative ratio is about 4.4 : 1.

(2) Many holes in the particle: structure of holes of expanded AN can be obviously observed from Fig. 1. Under a microscope it may be clearly observed that of expanded AN is full of a lot of small opening.

(3) Irregular shape, shape of particle of expanded AN is "brauched" with many edges, corners and burrs. Extremely irregular form exists.

(4) Nearly non-caking. Expanded AN moisture 0.3% is pressed to shape under pressure of 1.47×10^5 Pa. The cylinder charge of the explosive is placed in oven and is heat at 80℃ for 3h. After cooing it to room temprerature, compressive strength (caking strength) is measured by machine of material test. Tab. 1 shows experimental datum of caking strength of expanded AN through 5 times high-low temperature cycling test. The results indicate clearly that caking capacity of expanded AN is lower than powder common AN (600-425μ).

Table 1 Caking capacity of expanded AN through 5 times high-low temperature cycling test

Sample	Pure AN		Mixture of AN : Wood powder : Oil 92 : 4 : 4	
	Powder AN	Expanded AN	Powder AN	Expanded AN
Content of water %	0.3	0.3	0.3	0.3
Caking capacity N	474.0	93.6	470.1	77.9

(5) Easy crush: Through experiment of laboratory and practice of industrial production it is found that light porous expanded AN is extraodinary fragile and easier crush. It may be slightly pulverized by hand. So crude and fine pulverizing aren't needed in production. If it is directly added into crushing will, and rolled to 10~20mm, size of 90% particles of expanded AN will be smaller than 250μ. The effect of crush is better in ball will pulverizer than in crussing mills.

(6) Small hygroscopic rate: Despite expanded AN possesses quite largy specific surface area, its hygroscopic rate is only 70%~90% of that of common powder AN owing to hydrophobic action of surfactants. If a small amount of oil phase materials added, its hygroscopic rate can be decreased by 50%.

(7) Good explosion properties.

3 Application of expanded AN - rock expanded AN explosive

Classical explosion theory of explosive suggests that action of external energy explosion takes place from explosive, because external energy is not evenly distributed over a whole of explosive, but is concentrate on the local areas, with the formation of "heating point" of high temperature and pressure. Explosive near the "heating point" is first decomposed and certain quantity of heat is developed, which is sufficient to make second layer of explosive decomposition. In this way uninterrupted circalation occurred, quantity of heat unceasing accumulated, decompolition is constantly accelerated and explosion finally takes place. The theory analysis and research results point out that: "heating point" is formed mainly through three ways-adiabatic compression of air bubble. heating by friction and heating by viscous flow.

According to the explosion mechanism and characteristics of microhole, "brauching" and high specific surface area of expanded AN, we think that expanded AN should be self-sensibilized. If we use it as oxydizer and suitable combustible material, a new AN explosive with excellent explosion properties should be manufactured. Experimental results prove above analysis and infer to be true. Through a lot of experimental research and industrial trial for long time, rock expanded AN explosive without any sensitizer has been successfully manufactured from expanded AN, wood powder and composite full oil. Concrete formulation is as follows:

Expanded AN 92%±2.0%

Wood powder 4%±1.0%

Composite full oil 4%±1.0%

4 Propeties and characteristics of rock expanded an explosive

4.1 Physical property[3]

To compare with common AN, original hygroscopicity and caking capacity of expanded AN are decreased very much, while explosive is manufactured from expanded AN, more oil phase material and wood powder are added, so that physical properties is further improved. Results of test illustrate that : caking strength of rock expanded AN explosive is 6~8 times as low as that of common powder AN-wood powder-oil explosive, and is 15 times as low as that of common ammonite. Hygroscopic rate is decreased by 50%. As "oil and water aren't compatible", general problem of ANFO explosive is that the phenomenon of exudation is serious, and storage life is short. Because specific surface area of expanded AN is large, and surface of expanded AN is covered with amphoteric surfactants, so its oil-absorption rate is high (it may be reached to 30%) and stability is well. Some of results given by test of thermo-exudation is as follows, when expanded AN and oil phase mixture is heated for 2h at 60℃, not any signs of exudation appear, but under the same conditions, area of exudation of porous particle ANFO explosive is 1.5~2.0 times larger than area of cross section of charge. Experimental results of products of storage for long time indicate that its stability is remarkable better than Ammonite.

Table 2 Sensitivity and detonation velocity of rock expanded AN explosive

Item	Target	Condition of test	Common AN-wood-oil Explosive	Expanded AN-wood-oil Explosive
Sensitivity to shock wave	Detonation distance cm	Main charge $\varphi 42\times 22$(50g) Tetryl(ρ=1.57)	16.9	21.0
	Superpressure$\triangle P$/Pa	Tested explosive charge$\varphi 32\times 150$g(ρ=0.95)	6.3×10^4	5.0×10^4
Sensitivity to impact	Special falling height/%	Falling hammer 10kg amount of explosive 50mg	>170cm	126.0cm
Explosive characteristic	Detonation velocity $m\cdot s^{-1}$	$\varphi 32\times 150$g ρ=0.98~1.0	2300~2600	3190~3460

4.2 Explosion property

Notable explosion property of rock expanded AN explosive is that explosion strength is large, sensitivity to initiation is high, sensitivity to mechanical action is low, amount of toxic gases is less. Tab. 2, Tab. 3 and Tab. 4 show test and comparting datum of partical explosion properties.

Table 3 The explosion properties comparison between rock expanded AN explosive and other industrial explosives

Explosives	Density ρ $g\cdot cm^{-3}$	Moisture %	Detonation velocity V_D $m\cdot s^{-1}$	Strength V_S ml	Brisance H_{le} mm	Transimission distance S_D cm	Sensitivity to impact %	Sensitivity to friction %
Rock explosive	0.95~1.00	≤0.2	3200~3500	360~380	14.5~16.0	4~6	0	0
No. 2 Rock Ammonite	0.95~1.10	≤0.3	3000~3200	≥320	≥12	≥5	0	0
* No. 2 Rock new Ammonite	0.95~1.05	≤0.3	3000~3300	≥340	≥12	3~6	0	0

*: AN-TNT-oil explosive contained 7% TNT

Table 4 Test datum of National Industrial Explosive Test Centre

No.	Item	Unit	Non-TNT expanded AN explosive	Oligo-TNT expanded* AN explosive
1	moisture	%	0.1	0.1
2	density	$g\cdot cm^{-3}$	0.91	1.00
3	detonation relocity	$m\cdot s^{-1}$	3330±80	3560±92
4	strength	ml	349	406
5	transimission distance	cm	4	4
6	amount of toxic gases	$L\cdot kg^{-1}$	68.6	69.5
7	brisance	mm	15.3	16.2
8	sensitivity to shock	%	0	0
9	sensitivity to friction	%	0	0

* composition of Oligo-TNT expanded AN explosive is expanded AN 91.5%, TNT2.5%, wood powder3% and composite oil 3%.

4.3 Other characteristics

Another improtant advantage of rock expanded AN explosive is remarkable economic result. Because expensive sensitivizer TNT can be got rid of, cost of crude meterials is enormously decreased. Production process of explosive is simple and productive cost need not increase. So total cost of expanded AN explosive has decreased by 300~500 Yuan $\cdot t^{-1}$ as compared with that of original Ammonite and ANFO, Tab. 5 shows their concrete datum.

In addition, sensitivity of expanded AN explosive to mechanical action and heat and so on are all low quietly, and production process can get rid of TNT. So the production may be secure, non-toxic and non-dust.

Table 5 The crude material cost comparison of several indeutrial explosive

Exploaives	Formula				Cost of crude[①] merterial yuan • t^{-1}	Save cost[②] yuan • t^{-1}
	AN	Wood powder	Full oil	TNT		
Rock expanded AN explosive	92.0	4.0	4.0	0	1038[③]	482
No. 2 Rock explosive	85.0	4.0	0	11.0	1520	—
No. 2 Rock new explosive	88.0	3.0	2.0	7.0	1387	133

① Calculated on the basis of AN 900 yuan • t^{-1}、TNT6500 yuan • t^{-1}、and Wood powder 1000 yuan • t^{-1}

② 2 Rock Ammonite if considered as standard comparison

③ In cluding cost of additive about 50 yuan • t^{-1}

5 Conclusions

(1) To adopt surface active technology, under action of properly selected surfactants, expanded AN is obtained which possesses characteristics of many small holes, large specific surface area and non-caking. Expanded AN explosive is manufactured from it. Sensitivity to initiation. detonation velocity and physical and chemical property of the expanded AN explosive are available enhanced.

(2) Rock expanded AN explosive without TNT possesses excellent explosion and physical property. Quality of product has reached to target of No. 2 Rock Ammonite.

(3) Rock expanded AN explosive made from expanded AN. wood powder and composite full oil possesses characteristics of crude material being rich, cost low and production process simple. Especially, economic benefit produced by cost decrease and social benefit brought by elimination of toxic TNT are welcomed by factories and users.

(4) Technology of rock expanded AN explosive has been certificated by ministry and adopted by many factories. Loboratory research of permissible expanded AN explosive and. anti-water expanded AN explosive and so on have been accomplished, and pilot plant studies will be in progress.

References

[1] Yun ding. Improving Properties of AN and its Explosives Using Surfactants. [Thesis] . Nanjing: Nanjing University of Science and Technology, 1990.

[2] Zuliang Liu, Junming Hui, Chunxu Lu. Research on Light Porous and Bulky AN and HF Series Powder AN Explosives. Explosive Materials. 1991. (5): 5-8.

[3] Bingchen Hu. Application Studies of Surfactants in Improving Properties of AN. [Thesis] . Nanjing: Nanjing University of Science and Technology, 1993.

(注：此文原载于 Proceedings of the 3th Beijing International Symposium on Pyrotechnics and Explosives, 1995: 220-224)

Self-sensitizable Characteristics of Modified Ammonium Nitrate

Ye Zhiwen, Qian Hua, Lv Chunxu and Liu Zuliang
(Nanjing University of Science and Technology, School of Chemical Engineering, Nanjing, 210094, P. R. China)

Abstract: Ammonium nitrate has been modified by evaporative recrystallization from its mixture solution with hexadecyl trimethyl ammonium bromide and sodium dodecyl sulfate. The physical property of the modified compound has been investigated by SEM, DSC, specific surface area and particle size measurements. The results show that in comparation with common ammonium nitrate, the modified ammonium nitrate has an irregular crystal shape, greater specific surface area, better particle size distribution, lower heat of crystal pattern transition and higher transition temperature. These indicate the modified ammonium nitrate has a mesoporous structure with good self-sensitizable, anti-hygroscopic and anticaking performances as an oxidizer of industrial explosive.

Keywords: ammonium nitrate, modification, explosive properties

Introduction

Ammonium nitrate (AN) is extensively used as non smog oxidizer of industrial explosives and propellant because of low cost, wide resources of AN, and little smog produced by burning AN. However, lower sensitivity, serious caking and hygroscopicity of AN influence available properties of propellants and industrial explosives which are made from AN[1].

In order to overcome above shortages, many research staffs have tried their best to obtain some valid methods in which utilizing surface-active technology is effective in practice[2], namely, AN can be recrystallized from saturated solution of AN under some certain process by action of surfactants. The recrystallized AN is light, porous, so physical properties of the modified AN is different from that of common AN[3].

Experimental

Preparation of modified AN

The surfactants of 0.5% mass concentration are added to the 88～99% AN solution with temperature 125-130℃. The system is dried under vacuum of −0.085～−0.095MPa for 15 minutes. The modified AN, which has light porous structure, is obtained (water concentration < 0.03%).

Results and discussion

Shape analysis

The shapes of AN are observed through SEM. Compared with common AN, surface of the modified AN is quite irregular, its body has many holes and cracks, crystal shape disproportionates and contains a large number of micro-gas pocket. The photos of SEM are shown in Figure 1.

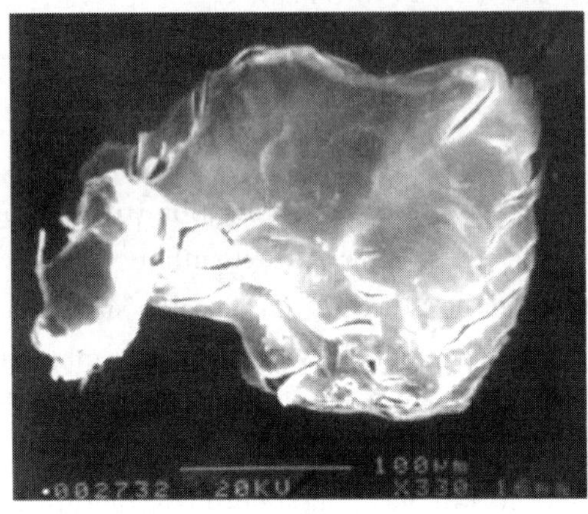

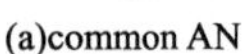
(a)common AN

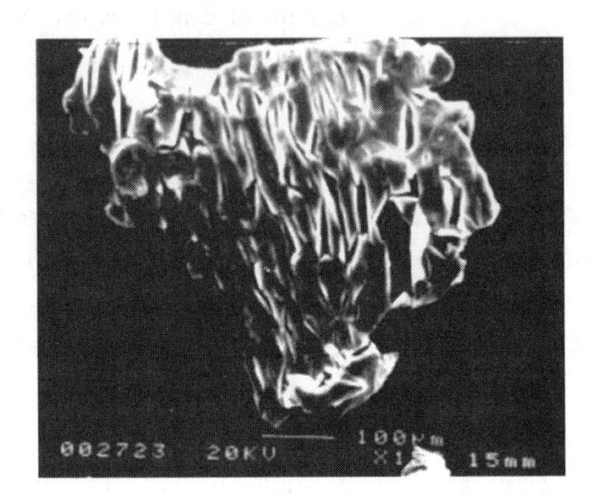

(b)modified AN

Figure 1 SEM photos of AN.

Size ratio of crystal faces which absorb surfactants results in crystal shapes disproportionation. The irregular shape characteristics of the modified AN treated by surfactants is obviously seen from Figure 1.

According to "heat point" theory[4], under action of explosion shock wave from outside, micro-gas pockets in the modified AN particles are compressed in adiabatic condition, then form "heat points" of high temperature and high pressure. These "heat points" increase sensitivity of the modified AN.

The distributions of pore radius and pore volume

Determination of the relation between pore radius and pore volume of the modified ammonium nitrate and the distribution of pore radius with NOVA1000 specific surface instrument. The specific surface data are represented in Table 1.

Table 1 Pore radius and pore volume of the modified ammonium nitrate

Pore radius	Pore volume	Pore volume	Pore radius	Pore volume	Pore volume
20～10	0.87	15.41	4～3	0.94	16.65
10～5	1.78	31.52	3～2	1.09	19.30
5～4	6.10	10.80	2～1	0.37	6.55

Note: In the modified ammonium nitrate the pore having radius greater than 20 nm was not observed.

The total volume of pore, whose pore radius R_p is more than 1.0nm, is 5.68mL·g^{-1}.

According to the standard of IUPAC pore radius grade: the pore whose $R_p \geqslant 50$nm is wide pore, the pore whose R_p=2-50nm is mesopore, the pore whose R_p=0.7-2nm is supper laciness, the pore whose $R_p \leqslant 0.7$nm is extreme laciness[5]. It is shown from Table 1 that the 95% micropore of the modified ammonium nitrate is mesopore. The exact version should be mesopore, but it is used to being called micropore.

On the basis of self-sensitizable effect of the modified ammonium nitrate, laciness (super and extreme laciness) is invalid. In order to produce the modified ammonium nitrate explosive, the modified ammonium nitrate must be mixed with wood powders and fuel oil, so even though the micropore of RPless than 2.0nm is formed, it can be blocked with the fine wood powders, especially with the fuel oil and it cant exist. So wide pore also should be avoided. Obviously, too many wide pores will influence the porosity. If the porosity is too low, it will affect the heat point numbers directly and wont be favorable for the self-sensitizable effect. Therefore, mesopore is the better pore radius range. It wont be blocked by fine wood powders and oil, Moreover, it can make the fine wood powders and oil well-distributed on the pore surface, and then the oxidizing agent with big specific surface and large mesopore numbers are mixed with combustible substances in the state of zero-oxygen balance to form industrial explosive. The explosive with the enough pores and heat points has the best initiation sensitivity and the most complete state of detonation.

The distributed results of pore volume and pore radius of the modified ammonium nitrate and common ammonium nitrate are shown in Figure 2.

Its clear to know from Figure 2 that the pore volume of the modified Ammonium nitrate is larger than that of common ammonium nitrate. Moreover, its shown from Table 1 that more than 95% valid pore radius of modified ammonium nitrate is in mesopore pore range, which is extremely beneficial to initiative effectiveness of the modified ammonium nitrate.

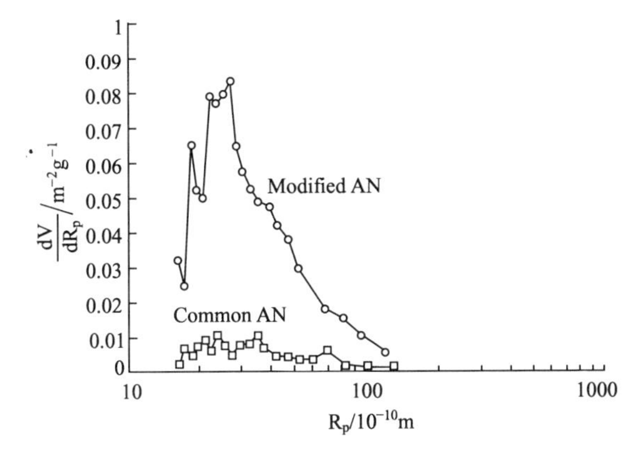

Figure 2 The plot of comparison between the modified AN and common AN for size distribution of micro-pore.

Distribution of particle size

Distributions of particle size for two kinds of AN are tested separately. The results are shown in Figure 3.

It is shown from Figure 3 that particle size of the modified AN mainly distributes in range from 10 to 1000μm. Average particle size of the modified AN treated with surfactants (87.6μm) is lower than that of common AN. Compared with common AN, the modified AN has wider particle size distribution, more micro-powders and better grade arrangement. If the modified AN is mixed with combustible agents, well-distributed degree of mixed system can beneficially increase explosion veloci-

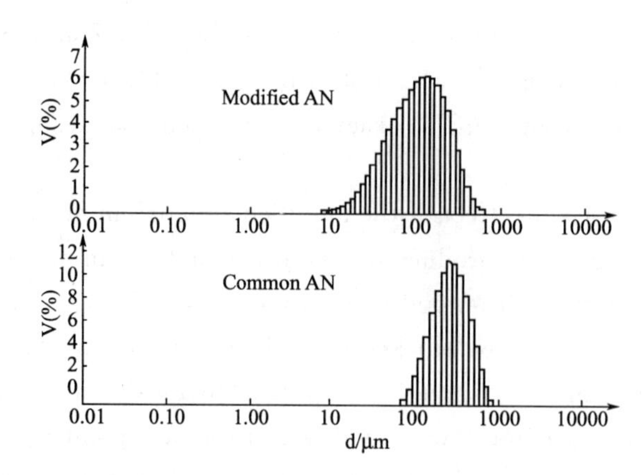

Figure 3 Particle size distribution curves of AN.

ty. Properties of mixed explosive according to mixing reaction mechanism of industrial explosive will be increased.

Specific surface area measurement

Specific surface area of the modified AN (3328.54$cm^2 \cdot g^{-1}$) is 4.4 times bigger than that of common AN (756.48$cm^2 \cdot g^{-1}$). So it benefits to increase velocity of explosion reaction and complete extent of explosion reaction.

Crystal pattern transition of modified AN studied by DSC

Under 1 atm, AN has five thermodynamically stable crystal structures. Each structure exists at temperature range[6], if temperature changes, it is convertible between different crystal patterns. The transition is one of main caking reasons. Crystal pattern transition is accompanied by heat change. So crystal pattern transition and extent of AN can be determined according to heat change of AN in temperature by DSC analysis. The results are shown in Table 2.

Table 2 Data in DSC picture of AN

Peak	Common AN Position of peak[℃]			Transition	Modified AN Position of peak[℃]			Transition
	Start	End	High		Start	End	High	
1	46.5	62.8	54.4	21.2	50.9	64.8	56.5	23.7
2	89.4	101.4	95.5	9.9	126.6	135.8	129.8	55.1
3	125.2	133.8	129.6	55.7	166.9	178.1	172.8	76.3
4	163.4	175.2	171.3	76.4				

The experimental results indicate that crystal pattern transition number of the modified AN declines, crystal pattern transition position moves back compared with common AN. It is because of combination between polar groups of surfactants and ion of AN through electrostatic action. It harasses action between ammonium ions and nitrate radical ions in AN molecules to some extent, so intrinsic crystal pattern transitions of AN don't easily happen with temperature change. Some transitions are held back, and lead that number of transition declines and transition position moves back.

Research on hygroscopicity of modified AN.

Dried modified AN and common AN with particle size ranging from 0.045～0.200mm are weighed, and separately placed into condition under which relative humidity is 70% and 90% respectively. Over a period of time, their hygroscopic increments are also weighed, their hygroscopic rates are calculated and the results are also shown in Table 3.

Table 3 Hygroscopic rates of common AN and modified AN

Samples	Average Mass[g]	Relative	Average time[h]					
			2	4	6	8	20	30
Common AN	2.4671	70	0.61	1.19	1.960	2.48	6.37	8.93
	2.5012	90	2.30	4.23	6.11	8.35	14.13	17.10
Modified AN	2.4173	70	0.24	0.45	0.72	0.93	2.22	2.94
	2.3995	90	0.47	1.60	2.47	3.35	4.80	5.27

Table 3 indicates that hygroscopic rates of the modified AN are greatly lower than that of common AN. It is reasoned that the modified AN is treated with selected surfactants, polar groups of surfactants combine with ion in AN surface, and unpolar groups of surfactants form a close hydrophobic film, the thin film prevents validly from contacting between AN molecules and water molecules outside. So hygroscopicity of the modified AN is low.

Anti-caking property measurement of the modified AN

Anti-caking property of the modified AN and common AN respectively pressed and formed in the same model under 1.47MPa pressure are tested, their break forces are regarded as indices of anti-caking property. If there is a big break force, AN is easily to cake. Otherwise, AN is not easily to cake. The results are shown in Table 4.

Table 4 Data of resistance to compression of caking AN (N)

Average	1	2	3	4	5	Average
Common AN	479.2	480.2	477.5	475.9	477.4	478.0
Expanded AN	86.9	87.7	83.4	81.2	84.1	84.7

The experimental results indicate that anti-caking property of the modified AN is significantly superior to that of common AN. From photos of SEM (Figure 1), it can be seen that common AN crystal is more tight, crystal particles closely knit through "salt bridge", intrinsic crystal pattern transition and hygroscopicity keep it firm, so common AN easily cakes, the modified AN particles connect each other by means of ting branch shape structure. This structure is loose. Moreover, hygroscopicity and crystal pattern transition of the modified AN are improved by action of surfactants, therefore, the modified AN has better anti-caking property.

Explosion properties

Common AN and modified AN used as oxidizer are mixed with wood powders and fuel oil used as reduction agent in mill machine to prepare industrial explosive according to mass ratio 92 : 4 : 4. The explosion properties are listed in Table 5.

Table 5 Explosion properties

Oxidizer in formula	Charge density[$g \cdot cm^{-3}$]	Explosion velocity[$m \cdot s^{-1}$]	Gas distance/[cm]	Brisance[mm]	Storage[month]
Common AN	0.90～0.95	2600±50	2	9±1	6
Modified AN	0.90～0.95	3200±50	4	13±1	10

From Table 5, it can be seen that explosion properties of industrial explosive produced by modified AN is better than that of industrial explosive produced by common AN.

Summary

Compared with the common AN, modified AN has more irregular crystal shape, larger numbers of micro gas-pocket, wider particle size distribution and better particle size grade arrangement, smaller particle size of average value, bigger specific surface area, more excellent anti-hygroscopicity and anti-caking property. The industrial explosive produced by the modified AN has more excellent explosion and physical properties.

Acknowledgements

The authors would like to thank the Nanjing University of Science and Technology for the financial and technical support.

References

[1] Lu Chunxu *et al.*, Industrial Explosive Theory, *Weapon Industry Press*, Beijing 2003.

[2] Li Lianmong, Application of Complex Anti-Caking Agent in Ammonium Nitrate Explosive, *Explosive Materials*, 1998, 2, 18-21.

[3] Ye Zhiwen, Calculation and Analysis of Solid Surface Free Energy of Expanded Ammonium Nitrate, *ibid.*, 2000, 5, 1-4.

[4] Hui Junming, Chen Tianyun, *Explosive Explosion Theory*, Jiangsu Science and Technology Press, Nanjing 1995.

[5] Fu Xiancai, Shen Wenxiu, Yao Tianyang, *Physical Chemistry*, Last, Higher Education Press, Beijing 1996.

[6] ALefuski B. M, *Ammonium Nitrate Technology*, Chemical Industry Press, Beijing 1983.

（注：此文原载于 Central European Journal of Energetic Materials，2005，2（4）：55-62）

改性硝酸铵性能研究

陈天云，吕春绪，叶志文，王依林

（南京理工大学化工学院，南京．210094）

EI：97033563554

摘要：应用表面活性剂技术对工业硝酸铵进行处理制得了改性硝酸铵，研究了改性硝酸铵的物理性能和爆炸性能。研究结果表明，改性硝酸铵比工业硝酸铵具有更优良的性能，更适合于作为微烟、少烟工业炸药的氧化剂。

关键词：硝酸铵，吸湿，结块，吸油，爆炸

引言

硝酸铵（AN）是一种成本低廉、来源广泛、低爆轰性、燃烧时无烟或少烟的物质，它可用作微烟或少烟工业炸药的氧化剂，但由于硝酸铵容易吸湿结块，影响以其作为主要原料制成的固体推进剂和工业炸药的使用性能及应用，严重时还会出现燃烧不完全或拒爆。为了克服硝酸铵的上述缺点，许多科技工作者做了大量的研究工作，采取了一些行之有效的措施[1-3]，但至今仍未能彻底解决这一问题。

应用表面活性理论和技术，将复合型表面活性剂加入硝酸铵溶液中在一定条件下，使硝酸铵重新结晶。由于表面活性剂的作用，硝酸铵的结构发生了变化[4]。这种改性硝酸铵的吸湿性和结块性明显降低，用它代替工业硝酸铵所制得的粉状工业炸药具有优良的防潮性能和爆炸性能[5]。

1　实验

1.1　改性硝酸铵的制备

在温度为125℃、质量浓度为90%的工业硝酸铵溶液中加入4%的复合型表面活性剂，在绝热蒸发的条件下使硝酸铵重新结晶，结晶后的硝酸铵具有多微孔蓬松的片状结构，脆性大，易粉碎，这种硝酸铵即为改性硝酸铵。

1.2　改性硝酸铵吸湿性研究

将改性硝酸铵和工业硝酸铵粉碎过40目筛，在烘箱中干燥至恒重，称量一定质量的干燥过的改性硝酸铵和工业硝酸铵，在一定温度下，分别放入相对湿度为70%和90%的环境中，间隔一定时间，称量它们的吸湿增量，计算出各自的吸湿百分率。

实验结果表明，改性硝酸铵的吸湿率比工业硝酸铵的稀释率降低60%，如表1所示。

表 1 在 25℃ 下两种硝酸铵的吸湿率

Table 1 Absorptivity of different AN at 25℃

样品名称	平均质量/g	相对湿度/%	吸湿时间/h 1	2	4	6	8	12	20	30
工业 AN	2.4671	70	0.31	0.60	1.19	1.90	2.48	4.14	6.37	8.93
	2.5012	90	1.20	2.31	4.23	6.11	8.35	10.72	14.13	17.01
改性 AN	2.4173	70	0.12	0.24	0.45	0.72	0.93	1.47	2.22	2.94
	2.3995	90	0.45	0.47	1.60	2.47	3.35	3.86	4.80	5.27

1.3 改性硝酸铵抗结块性研究

1.3.1 DSC 差热分析测定改性硝酸铵的晶变

在常压下，硝酸铵具有五种热力学稳定的晶体结构[6]，每一种结构仅在一定的温度范围内存在，若温度发生变化，则各晶型之间可以相互转变。当晶型改变时，硝酸铵的密度、比容等一系列温度性质也随之变化，因此，可以利用 DSC 差热分析仪来测定硝酸铵随温度变化的情况，根据无吸热峰或放热峰以及峰的大小就可以判断硝酸铵是否发生晶变以及晶变的程度。实验结果表明，随着温度的变化，改性硝酸铵出现晶变的数目减少、晶变点后移，实验结果如表 2 所示。

表 2 硝酸铵 DSC 图谱的数据

Table 2 DSC spectrum of AN

峰号	工业 AN 峰的位置/℃ 起点	终点	最大点	晶变热 (J/g)	改性 AN 峰的位置/℃ 起点	终点	最大点	晶变热 (J/g)
1	46.5	62.8	54.4	21.1	50.9	64.8	56.5	23.7
2	89.4	101.4	95.5	9.9	126.6	135.8	129.8	55.1
3	125.2	133.8	129.6	55.7	166.9	178.1	172.8	76.3
4	163.4	175.2	171.3	76.4				

注：样品的水分含量为 0.03%，参比物为 Al_2O_3。

1.3.2 改性硝酸铵抗结块性试验

在 1.47MPa 压力下，将工业硝酸铵和改性硝酸铵在同一模具中压制成型，放入烘箱中干燥数小时后再放入干燥器中冷却至室温，然后取出进行抗压试验，其破坏力可作为抗结块性指标。若破坏力大，则说明易结块；反之，则不易结块。实验结果表明，改性硝酸铵抗结块性明显优于工业硝酸铵，如表 3 所示。

表 3 结块硝酸铵的抗压数据

Table 3 Compression resistance of different AN

样品名称	破坏力/(N) 1	2	3	4	5	平均
工业 AN	479.2	480.2	477.5	475.9	477.4	478.0
改性 AN	86.9	87.7	83.4	81.2	84.1	84.7

1.4 改性硝酸铵吸油性和渗油性研究

硝酸铵可作为铵油炸药、2 号岩石铵梯油炸药以及岩石膨化硝铵炸药的主要原料、由于这些炸药中均含有一定量的油相，为了保证其具有稳定的爆炸性能，要求这些炸药中的硝酸铵具有良好的吸油性和较低的渗油性。

1.4.1 硝酸铵吸油性实验

分别将改性和工业硝酸铵粉碎烘干，过 40 目筛，然后分别称取 100g，在搅拌条件下加入纯柴油，柴油全部浸湿并覆盖硝酸铵，10min 后，真空抽滤 10min，然后称其质量并计算其吸油百分含量。实验结果表明，改性硝酸铵具有很高的吸油性，如表 4 所示。

表 4 在 20℃ 下硝酸铵的吸油率

Table 4 Oil absorbability of AN at 20℃

样品名称	吸油率/(%)					
	1	2	3	4	5	平均
工业 AN	8.8	9.1	8.4	8.7	9.5	8.9
改性 AN	54.7	56.3	55.5	56.5	56.0	55.8

1.4.2 硝酸铵渗油性试验

分别称取已干燥且过 40 目筛的改性和工业硝酸铵各 100g，在充分搅拌的条件下加入 8g 纯柴油，用直径和高均为 40mm 的一端敞口的容器装满上述硝酸铵，并使其密度达到 0.95-1.00g/cm^3，然后将容器的敞口端放在滤纸上，在一定温度下，每隔一定时间测量滤纸上油迹的面积。油迹的面积越大，则渗油性越大。实验结果表明，改性硝酸铵的渗油性很小，几乎不渗油，如表 5 所示。

表 5 浸油硝酸铵的浸油面积

Table 5 Oil-exuding area of oiled AN ($\times 10^3$ mm^2)

样品名称	温度(℃)	放置时间/(h)				
		1	6	7	24	45
工业 AN	20	49	79	108	188	299
	35	53	90	122	265	440
改性 AN	20	0	0	0	0	0
	35	0	0	0	0	0

1.5 硝酸铵雷管感度试验

感度雷管衡量可爆炸物对冲击波敏感性的重要指标。对于工业炸药来说，希望硝酸铵具有适当的雷管感度，这样，在制备工业炸药时就不需加入其它的敏化剂来敏化硝酸铵。

按照雷管感度的试验方法[7]，对工业硝酸铵和改性硝酸铵的雷管感度进行了测试。试样密度为 0.7-0.9 g/cm^3。试验结果表明，改性硝酸铵雷管感度高，易于起爆，如表 6 所示。

表 6 硝酸铵雷管感度

Table 6 Initiation sensitivity of AN

样品名称	温度(℃)	实验结果		
		1	2	3
工业 AN	25	—	—	—
	80	—	—	—
改性 AN	25	+	+	+
	80	+	+	+

注：“+”表示对雷管敏感，“—”表示对雷管不敏感。

2 结果与讨论

(1) 采用特殊的处理方法将阴离子表面活性剂和阳离子表面活性剂进行复合而制得的复合表面活性剂，克服了单一型离子表面活性剂的缺点，具有很强的表面活性。用复合型表面活性剂处理硝酸铵，表面活性剂分子中的极性基团与硝酸铵分子中的离子相结合，而非极性基团便在硝酸铵分子周围形成了一层致密的憎水薄膜，这层薄膜有效地阻止了外界的水分子与硝酸铵分子的接触，因此，用复合型表面活性剂处理后的改性硝酸铵具有很低的吸湿性。

(2) 用复合型表面活性剂处理后的改性硝酸铵，由于表面活性剂分子中的极性基团通过静电作用而与硝酸铵分子中的离子紧密结合起来，从而在一定程度上“搅乱”了原先硝酸铵分子中的 NH_4^+ 与 NO_3^- 离子之间的作用，使得硝酸铵分子随温度的变化不易产生原先固有的晶变，从而阻止了硝酸铵某些晶相转变，使得改性硝酸铵晶变的数量减少，晶变点后移。

（3）对工业硝酸铵和改性硝酸铵样品进行放大观察后发现，在未用表面活性剂处理的工业硝酸铵中，其颗粒之间是通过“柱状”形式连接的，当外界条件发生变化时，硝酸铵晶型的改变使其体积发生变化，晶体的内凝聚力作用使“柱状”连接更加牢固，强度也增加，这种硝酸铵在外界作用下不易粉碎，因而易结块。而用表面活性剂处理后的改性硝酸铵。其颗粒之间是通过细小的‘枝状阳’形式连接的，这种，“枝状”结构很琉松、不牢固，在外界作用下很容易被粉碎、强度低，因而不易结块，因此，用复合型表面活性剂处理的硝酸铵抗结块性好。

（4）改性硝酸铵具有多徽微孔膨松结构（如图 1 所示），其比表面积较大（实验侧得每克改性硝酸铵的比表面积是工业硝酸铵的 4～5 倍）。此外，表面活性剂分子中的非极性基团在硝酸铵顺位表面形成了一层憎水薄膜。当加入油相时，油相分子既可以在硝酸铵颗粒表面通过润湿而吸附，又可以和颗粒表面的憎水薄膜通过范德华力作用被吸，而后者的吸附是较牢固的，也是主要的。当外界条件发生变化时．被吸附的油相也不易渗透下来，因此，改性后硝酸铵比工业硝酸铵具有较大的吸油性和较低的渗油性。

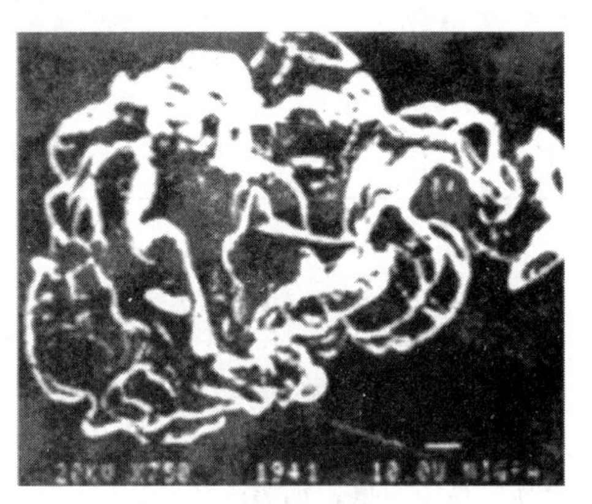

(a) 放大倍数为750
(a) 750 times enlarged

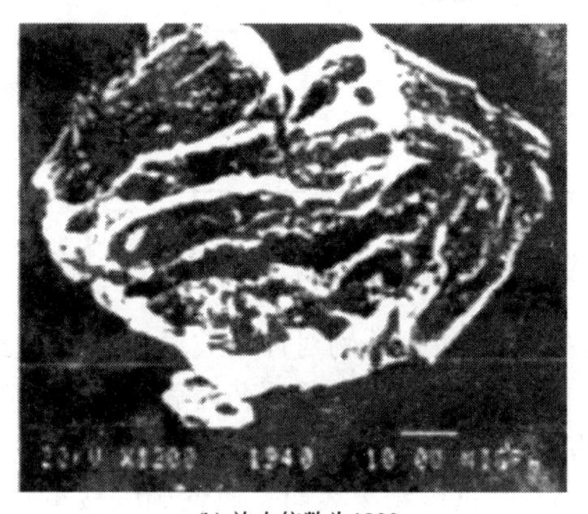

(b) 放大倍数为1200
(b) 1200 times enlarged

图 1　改性硝酸铵颗粒电子显微照相

Fig. 1　Photomicrograph of modified ammonium nitrate

（5）改性后的硝酸铵，由于颗粒内部含有大量的微气孔，在外界能量作用下很容易形成微气泡。当改性硝酸铵受到爆炸冲击作用，颗粒内的微气泡就会被绝热压缩而形成高温高压的“热点”。使其体系敏感度提高，即改性硝酸铁由于受到自身颗粒内部气泡的敏化而具有一定的雷管感度

3 结论

用复合型表面活性剂处理的硝酸铵与未经处理的硝酸铵相比具有较小的吸湿性、较好的抗结块性、较大的吸油性和较低的渗油性，具有较高的雷管感度可以用作工业炸药中的氧化剂。

参考文献

[1] Tsuchiya Y N. 硝酸铵晶体性能的改进研究．工业火药，1960（21）：18.
[2] 包昌火．国外硝铵炸药防潮和防结块的概况．爆破材料．1965（2）：39-43.
[3] 混合炸药编写组．猛炸药的化学与工艺学（下册）．北京：国防工业出版社．1983.
[4] 陈天云．硝酸铵表面特性及其应用研究［D］．南京：华东工学院．1992.

[5] 刘祖亮等．轻质多孔膨松硝酸铵及其粉末硝铵炸药．爆破器材，1991，20（5）：24-26.
[6] 阿列夫斯基 B M. 硝酸铵工艺学，王令仪等译．北京：化学工业出版社．1983.
[7] 联合国编．危险货物的运输．华东工学院民爆器材研究所译．北京：国防工业出版社．1988.

Properties of Modified Ammonium Nitrate

Chen Tianyun, Lv Chunxu, Ye Zhiwen, Wang Yilin
(School of Chemical Engineering, NUST, Nanjing 210094)

Abstract: The commercial ammonium nitrate (AN) was modified by using surfactants and the physical and explosive properties of the modified product were experimentally investigated. The results show that the properties of modified AN are better than those of the commercial one and can be used as a good oxygenant in solid propellant and industrial explosives.

Keywords: absorptivity, ammonium nitrate, caking, explosive, oil-absorbing.

（注：此文原载于 含能材料，1996，2（4）：169-174）

FTIR光谱遥测红外药剂的燃烧温度

周新利，李燕，刘祖亮，朱长江，王俊德，吕春绪
（南京理工大学化工学院，江苏南京 210094）
SCI：000179099400018；EI：03027311705

摘要： 本文利用遥感FTIR光谱，对红外药剂的燃烧特性进行了研究。在分辨率为 $4cm^{-1}$ 时，收集 $4700 \sim 740cm^{-1}$ 波段的光谱。从HF等燃烧产物发射的分子振转基带精细结构的谱线强度分布，可以对燃烧温度进行遥感测定，并给出了燃烧温度随时间的变化关系，实验结果表明燃烧表面附近温度梯度很大，存在着急剧的变化温度场，同时也说明，在不干扰火焰温度场的情况下，利用遥感FTIR光谱对剧烈的、非稳态快速燃烧的火焰温度进行连续实时的遥感测量，是一种快速、准确、灵敏度高的测温疗法，显示了它在燃烧温度测量、产物浓度测试以及燃烧机理研究等方面的应用前景。

关键词： 遥感测量，FTIR，温度测量，烟火药

引言

含有氟、氯等元素的有机化合物燃烧时将热分解，产生大量的含有碳、氧、氟、氯的气体化合物，这样我们就利用遥感FTIR光谱来遥测HCl、HF、CO_2 和CO这些气体分子的热激发产生的分子基带振动光谱[1-3]。利用这些分子的转振带光谱，就可以对剧烈的、非稳态快速燃烧的火焰温度进行直接的测量，而不干扰温度场。

燃烧温度是研究燃烧过程的一个重要参数，通过燃烧温度可以研究燃烧反应的特征、燃烧机理、能量分布及气体产物的组成等。本文利用遥感FTIR光谱对含有聚四氟乙烯的红外药剂的燃烧温度进行了测定。通过对燃烧产物HF光谱特性的分析，可以计算出燃烧的火焰温度，并给出了燃烧温度随时间的变化。实验结果说明，遥感FTIR光谱法在测量燃烧温度等方面具有很大的优势。

1 基本原理

利用某些气体转振谱带的R-分支和/或P-分支光谱辐射强度，可以求得燃烧的火焰温度，根据文献［4］有如下方程式：

$$Y = AX + b \tag{1}$$

式中

$$Y = \ln\left(J \frac{\nu_0^4}{I_{relative}}\right) \tag{2}$$

$$X = J(J' + 1) \tag{3}$$

R-分支取 J=J′，P-分支取 J=J′，P-分支取 J=J′+1。

$$A=B_{\nu}\frac{hc}{kT} \tag{4}$$

从方程（4）可得到火焰温度

$$T=B_{\nu}\frac{hc}{kA}$$

其中 J 是谱线的下态角量子数，ν_0 是分子振转基带精细结构的谱线中心频率（cm^{-1}），h 为 Planck 常数，k 为 Boltzmann 常数，c 为光速，B_{ν} 为分子的转动常数，A 为 X 与 Y 直线方程的斜率。

从所测量的光谱，可以得到分子振转基带精细结构中的每条光谱线的相对强度 $I_{relative}$，他们对应的角量子数 J 和每条谱线的频率 ν_0，斜率 A 很容易通过最小二乘法计算得到。

2 实验

2.1 装置

用带有 Dall-Kirkham 反射式望远镜的 Bruker55 遥感 FTIR 光谱仪，收集来自红外药剂燃烧源的红外发射信号，此仪器采用 KBr 分束器和液氮冷却的 MCT 检测器，光谱分辨率为 $4cm^{-1}$。

2.2 燃烧样品

用于测试的红外药剂主要由 30%的镁粉，20%的聚四氟乙烯和 35%的高氯酸铵等组成。样品重 30g，压装成直径 20mm，高 15mm 圆柱形装药。样品离 FTIR 光谱仪约 30m。

3 结果与讨论

3.1 时间分辨光谱图

根据不同时刻收集的辐射光谱。作出燃烧产物 HF 在 4037～$4400cm^{-1}$ 波段的时间分辨光谱的 3D 图，如图 1 所示。

3.2 燃烧温度的计算

利用燃烧产物 HF 在 4037～$4400cm^{-1}$ 波段 R-分支的精细结构来计算火焰的燃烧温度。以燃烧时间在 8s 时，收集到的 HF 在 4037～$4400cm^{-1}$ 波段 R-分支的精细结构光谱图为例，求出 HF 转振带 R-分支的精细结构中每条谱线的频率和相对光谱强度，以 $Y=\ln\left(J\frac{\nu_0^4}{I_{relative}}\right)$ 对 J′J′(+1)作图，并用最小二乘法拟合得一直线，如图 2 所示，所得直线方程为 $\ln\left(J\frac{\nu_0^4}{I_{relative}}\right)=0.0152J'J'(+1)+31.039$，斜率 $A=0.0152$，截距 $b=31.039$，相关系数 $r=0.9698$，由此计算的温度 $T-1982.03K$，由文献［5，6］可知，相关系数 $r=0.9698$ 时，该实验的可信度将大于 99.5%。

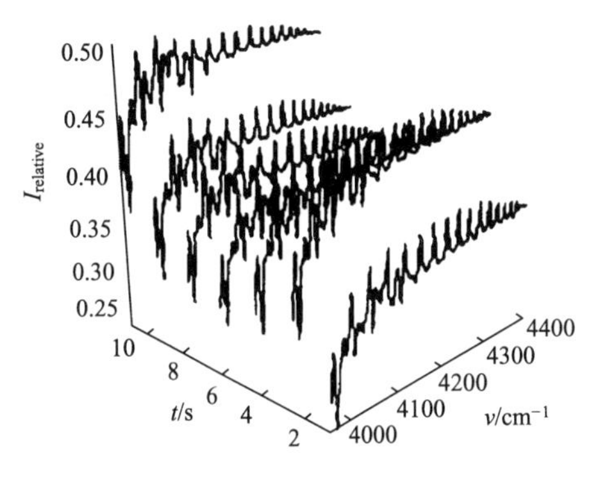

图 1 HF 分辨光谱 3D 图

Fig. 1 3D plot of HF emission spectra

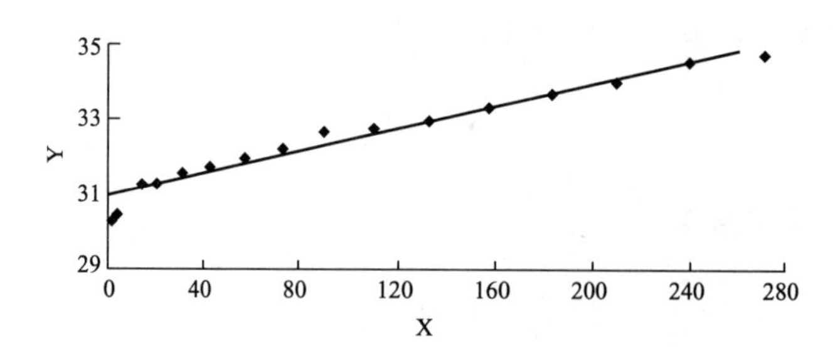

图 2 HF 在 R-分支的精细结构上 $Y=\ln(J\frac{v_0^4}{I_{relative}})$ 与 J′J′(+1)的关系图

Fig. 2 Plot of $Y=\ln(J\frac{v_0^4}{I_{relative}})$ VS J′J′(+1) of fine structure of R-branch of HF

3.3 火焰温度与时间的关系

用同样的方法，由每条 HF 谱线的频率和相对谱线强度，计算出不同时刻的燃烧温度，如图 3 所示。由图可知，燃烧温度随着燃烧时间的延长先升高后降低，这是因为随着燃烧过程的进行及火焰的扩展，红外药剂表面燃烧产生出更多的炽热燃气，并且伴随着热量的积累，导致温度的升高；随着燃烧时间的延长及红外药剂的减少，气体产物的扩散和热量的辐射导致温度的降低。反应燃烧表面附近温度梯度很大，存在着急剧的变化温度场。

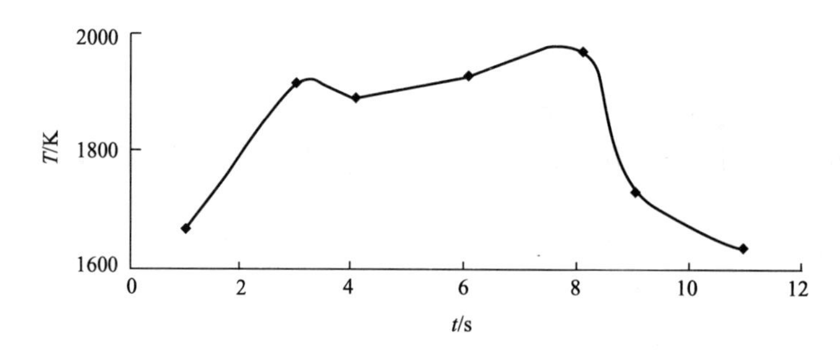

图 3 关于燃烧时间特性曲线

Fig. 3 Combustion temperature-time distrubtion curve

4 结论

根据上述实验方法和结果可以得出：利用遥感 FTIR 光谱可以很好地对红外药剂的燃烧温度进行遥测，该方法的特点：在不干扰火焰温度场的情况下，对火焰温度进行连续实时的遥感测量，是一种快速、准确、灵敏度高的测温方法：显示了它在燃烧温度测量、产物浓度测试以及燃烧机理研究等方面的应用前景。

参考文献

[1] DC Tilotta，K Busch and M A Buch. Appl，Spectrosc.，1989，43：704.

[2] 王俊德，陈作如，罗蕴花，黄梅，康建霞和俞柏恒．某些有机化合物燃烧火焰的 FTIR 发射光谱研究，汪尔康主编．分析化学进展，南京：南京大学出版社，1994，449.

[3] LI Yan，WANG Jun-de. Instrum. Sci and Tech.，2002，in press.

[4] Tianshu Wang，Changjiang Zhu，Junde Wang，Fuming Xu，Zuoru Chen and Yunhua Luo. Anal，Chim，Acta，1995，336：294.

[5] Xuetie Zhou，Yan Li，Junde Wang and Zhonghua Huang. IEEE Trans，on Plasma Science，2001，29：306.

[6] J Mandel, The Statistical Analysis of Experimental Data, New York: Wiley, 1964.

Combustion Temperature Measurement of Pyrotechnic Composition Using Remote Sensing Fourier Transform Infrared Spectrometry

Zhou Xinli, Li Yan, Liu Zuliang, Zhu Changjiang, Wang Junde and Lv Chunxu

(School of Chemical Engineering, Nanjing University of Science and Technology, Nanjing 210094, Jiangsu, China)

Abstract: In this paper, combustion characterization of pyrotechnic composition is investigated using a remote sensing Fourier transform infrared spectrometry. The emission spectra have been recorded between 4700 and 740cm^{-1} with a spectral resolution of 4cm^{-1}. The combustion temperature can be determined remotely from spectral line intensity distribution of the fine structure of the emission fundamental band of gaseous products such as HF. The relationship between combustion temperature and combustion time has been given. Results show that there is a violent mutative temperature field with bigger temperature gradient near combustion surface. It reveals that the method of temperature measurement using remote sensing FTIR for flame temperature of unstable, violent and short time combustion on real time is a rapid, accurate and sensitive technique with interference the flame temperature field, Potential prospects of temperature measurement gas product concentration measurement and combustion mechanism are also revealed.

Key words: Remote sensing measurement, FTIR, Temperature measurement, Pyrotechnic composition.

（注：此文原载于 光谱学与光谱分析，2002，22 (5)：764-766)

改性硝酸铵爆炸安全性研究Ⅱ. 无机化学肥料对硝酸铵爆炸安全性的影响

唐双凌，吕春绪，周新利，王依林，刘祖亮
（南京理工大学化工学院，南京，210094）

摘要： 将磷酸氢二铵、磷酸二氢铵、硫酸铵、硫酸钾和氯化铵等无机化学肥料改性的硝酸铵按照工业炸药配方配制成铵油炸药，对硝酸铵的爆炸安全性进行了评价。采用恒温热分解方法和差示扫描量热法，研究了改性硝酸铵及铵油炸药在密封的毛细管和高压密封样品池中的热分解行为。结果表明，当硝酸铵中含质量分数（下同）为20%以上磷酸氢二铵、或25%以上的硫酸铵、或25%以上的硫酸钾、或40%以上的磷酸二氢铵时，配制的铵油炸药不能被8号雷管起爆。热分解结果与爆炸结果有如下关系：所有不能起爆的改性硝酸铵在370℃下加热1min的热分解率都低于50%。由于氯离子催化硝酸铵的分解，含质量分数为5%氯化铵就能使硝酸铵在340℃下加热1min接近完全分解，故氯化铵的添加量高达30%仍能被8号雷管起爆。

关键词： 硝酸铵，无机化学肥料，热分解，爆炸安全性

前文[1]已述及为了防止恐怖分子利用硝酸铵从事破坏活动，现有的农用硝酸铵必须没有爆炸性，不能制成炸药。解决途径之一是与各种无机化肥混合，以复合肥的形式继续在农业上使用。但存在以下问题：硝酸铵在复合肥中的比例为多少就不会有爆炸性？何种化肥降低硝酸铵爆炸特性的效果最佳？是否有的肥料在一定添加量范围内，不但不能降低硝酸铵的爆炸特性，还会促使其爆炸？这些问题的解答有助于硝酸铵生产线的改造。本文按照爆轰感度较高的铵油工业炸药配方，将磷酸氢二铵、磷酸二氢铵、硫酸铵、硫酸钾和氯化铵等改性的硝酸铵制成工业炸药，用8号电雷管起爆，根据起爆情况考察含硝酸铵复合肥的爆轰安全性。自行设计了一种恒温热分析方法，可测定硝酸铵在密封状态下的高温短时热分解行为，以模拟爆轰过程的快速化学反应，结合差示扫描量热法（DSC），试图建立改性硝酸铵的热稳定性与铵油炸药爆轰感度之间的关系。

1 实验部分

1.1 无机化肥改性硝酸铵和铵油炸药的恒温热分解实验

改性硝酸铵的制备：一定量的无机化肥溶于水中，加入硝酸铵（工业级）并加热溶解，搅拌均匀，冷却后粉碎，过0.42mm筛孔，在烘箱中于80℃干燥6h后备用。

铵油炸药的制备：94份硝酸铵与6份0#柴油碾磨混合。以改性硝酸铵配制时换算成纯硝酸铵加入。

恒温热分解实验：准确称量3mg样品封装在毛细管内，自制恒温加热设备，温度误差±5℃，分别在260、340和370℃，加热1～30min后，将毛细管破碎、蒸馏水浸洗，洗液转移到50mL容量瓶

中定容，测定硝酸根的浓度，即可算出硝酸铵的分解率。

硝酸根的测定方法：根据 NO_3^- 在 210～240nm 紫外区的强吸收，采取双波长分光光度法测定，以消除无机化肥和分解产物的干扰，测试波长选择 225 和 235nm。标准工作曲线的回归方程为：$c(NO_3^-)=38.374\times(A_{225nm}-A_{235nm})-1.0128$；相关系数 $R=0.9997$。

1.2 改性铵油炸药的 DSC 分析及改性硝酸铵的爆轰实验

日本岛津 DSC-50 型差热分析仪，高压密封坩埚样品池，升温速率 20℃/min，样品量约为 1.0mg。改性硝酸铵的爆轰实验参见前文[1]。

2 结果与讨论

2.1 改性硝酸铵和铵油炸药的恒温热分解实验结果

实验结果参见表 1 和表 2。

表 1 在不同加热条件下改性硝酸铵的热分解率

Table 1 Decomposition of modified ammonium nitrate at different heating conditions

Sample	Decomposition/% (260℃,30min)	Decomposition/% (340℃,1min)	Decomposition/% (370℃,1min)
Neat AN	56.9	76.4	96.8
AN+5% * $NH_4H_2PO_4$	59.7	76.0	96.7
AN+15% * $NH_4H_2PO_4$	35.6	57.3	83.4
AN+40% * $NH_4H_2PO_4$	23.7	37.3	45.5
AN+5% * $(NH_4)_2HPO_4$	45.3	75.5	94.1
AN+15% * $(NH_4)_2HPO_4$	28.0	60.9	73.8
AN+40% * $(NH_4)_2HPO_4$	18.9	31.2	40.5
AN+5% * $(NH_4)_2SO_4$	47.6	77.9	93.2
AN+15% * $(NH_4)_2SO_4$	34.8	68.5	87.1
AN+25% * $(NH_4)_2SO_4$	33.4	33.7	40.1
AN+40% * $(NH_4)_2SO_4$	23.4	28.6	28.4
AN+25% * K_2SO_4	26.6	37.4	42.6
AN+40% * K_2SO_4	21.4	25.7	31.2
AN+5% * NH_4Cl	98.9	98.6	99.5

* The valus before compound added denotes its mass fraction to 100g AN.

表 2 在不同加热条件下改性铵油炸药的热分解率

Table 2 Decomposition of modified ANFO explosives at different heating conditions

Sample	Decomposition/% (260℃,30min)	Decomposition/% (340℃,1min)	Decomposition/% (370℃,1min)
Neat AN+6%oil	43.9	74.3	92.1
AN+5%$NH_4H_2PO_4$+6%oil	46.8	83.6	95.6
AN+15%$NH_4H_2PO_4$+6%oil	31.5	49.3	83.1
AN+40%$NH_4H_2PO_4$+6%oil	25.6	35.4	46.8
AN+5%$(NH_4)_2HPO_4$+6%oil	41.2	78.3	89.8
AN+25%$(NH_4)_2HPO_4$+6%oil	30.1	38.5	47.2
AN+40%$(NH_4)_2HPO_4$+6%oil	22.3	28.7	34.7
AN+5%$(NH_4)_2SO_4$+6%oil	39.7	81.6	93.2
AN+15%$(NH_4)_2SO_4$+6%oil	29.3	41.7	57.0
AN+25%$(NH_4)_2SO_4$+6%oil	25.3	32.0	37.0
AN+40%$(NH_4)_2SO_4$+6%oil	22.6	32.1	28.4
AN+25%K_2SO_4+6%oil	26.2	32.7	44.5
AN+40%K_2SO_4+6%oil	27.6	29.5	25.1
AN+5%NH4Cl+6%oil	95.8	97.5	97.3

从表1中结果可看出，纯硝酸铵在340和370℃的分解远快于在260℃的分解硝酸铵在290℃以下的分解是离子反应机理[2]，决定速率的步骤为NO_2^+的生成反应，随后NO_2^+参与的分解反应较慢：

$$NH_4NO_3 \rightleftharpoons NH_3+HNO_3$$

$$HNO_3+HA \xrightarrow{-A^-} H_2ONO_2^+ \xrightarrow{slow} NO_2^+ +H_2O(HA=NH_4^+, H_3O^+, HNO_3)$$

$$NO_2^+ +NH_3 \longrightarrow [NH_3NO_2^+] \longrightarrow N_2O+H_3O^+$$

在290℃以上的分解是自由基反应机理，决定速率的步骤为HNO_3的均裂反应，在自由基参与的后续反应较快：

$$NH_4NO_3 \longrightarrow NH_3+HONH_2;\ HONO_2 \xrightarrow{slow} HO^\cdot +NO_2^\cdot$$

$$HO^\cdot +NH_3 \longrightarrow H_2O+NH_2^\cdot$$

$$NH_3^\cdot +NO_2^\cdot \longrightarrow NH_2NO_2 \xrightarrow{fast} N_2O+H_2O$$

爆炸是非常快速的反应（1×10^{-6}s），硝酸铵高温短时（370℃×1min）的热分解比低温长时（260℃×30min）的热分解应更接近爆炸反应过程。

磷酸氢二铵、磷酸二氢铵、硫酸铵和硫酸钾等无机盐对硝酸铵的热分解随添加量的增加有比较明显的延缓作用。由于不涉及复杂的氧化还原反应，磷酸氢二铵、磷酸二氢铵、硫酸铵等铵盐的热分解是比较缓慢的，硫酸钾的热稳定性更高，它们可认为是惰性物质。这些无机盐与硝酸铵形成复盐，将NH_4^+和NO_3^-隔开，不利于电子从－3价的铵态氮向＋5价的硝态氮转移，故硝酸铵的热分解被延缓。

同为磷酸盐，磷酸氢二铵比磷酸二氢铵更能提高硝酸铵的热稳定性，原因是磷酸二氢铵具有酸性，H^+可以促进硝酸铵的分解[3]；硫酸铵和硫酸钾对硝酸铵热稳定性的影响无明显差异；氯化铵可显著加速硝酸铵的热分解，氯离子对硝酸铵的分解具有催化作用[4-6]：

$$NO_2^+ +Cl^- \longrightarrow NO_2+Cl^\cdot;\ NO^+ +Cl^- \longrightarrow NO+Cl^\cdot$$

$$NH_4^+ +Cl^\cdot \longrightarrow NH_3^+ +HCl;\ NH_3+Cl^\cdot \longrightarrow NH_2^\cdot +HCl$$

$$NO_2+NH_3^+ \longrightarrow [O_2NNH_3^+] \xrightarrow{fast} N_2O+H_3O^+$$

$$NH_3^\cdot +NO \longrightarrow [H_2NNO] \xrightarrow{fast} N_2+H_2O$$

由硝酸铵分解产生的少量硝酰阳离子（NO_2^+）和亚硝酰阳离子（NO^+）将氯离子氧化成活泼的氯原子，氯原子通过抽氢反应，使NH_4^+和NH_3生成NH_3^+和NH_2^*，后二者再分别与NO_2和NO形成不稳定的中间态快速分解为N_2O和N_2，从而加速了硝酸铵的分解。

比较表1和表2可看出，含质量分数为6%柴油的硝酸铵在260℃下的热分解有明显降低，340和370℃的分解基本无变化；含质量分数为6%柴油的改性硝酸铵，在添加剂含量较低情况下，260℃下的热分解也有明显降低，添加量大的样品的热分解没有明显不同。硝酸铵的分解是一个自催化反应，初始分解产物可加速硝酸铵分解，柴油通过捕捉初始分解产物，可降低硝酸铵在260℃的分解速度。改性硝酸铵在添加量大的情况下分解率已有大幅度降低，再添加质量分数为6%的柴油，分解率降低不明显。

2.2 改性铵油炸药的DSC测试结果

实验结果见图1～图6。由图中可见，DSC曲线上的放热峰均为硝酸铵分解为氮氧化合物以及与木粉和柴油发生复杂的氧化还原反应的结果，这是爆炸反应的根本原因之一。

DSC分析与恒温热分解的结果具有可比性，即：由于$NH_4H_2PO_4$具有酸性，放热峰温度较$(NH_4)_2HPO_4$为低；NH_4Cl对硝酸铵的分解具有很强的催化能力，放热峰温度最低为275.55℃。同为硫酸盐，由于K^+不参与氧化还原反应，质量分数为25%K_2SO_4较质量分数为25%$(NH_4)_2SO_4$改性的铵油炸药放热量为低。

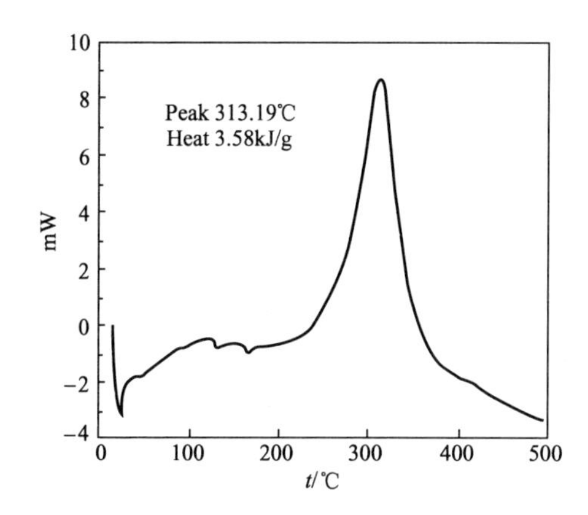

图 1 ANFO 的 DSC 分析

Fig. 1 DSC curve of ANFO

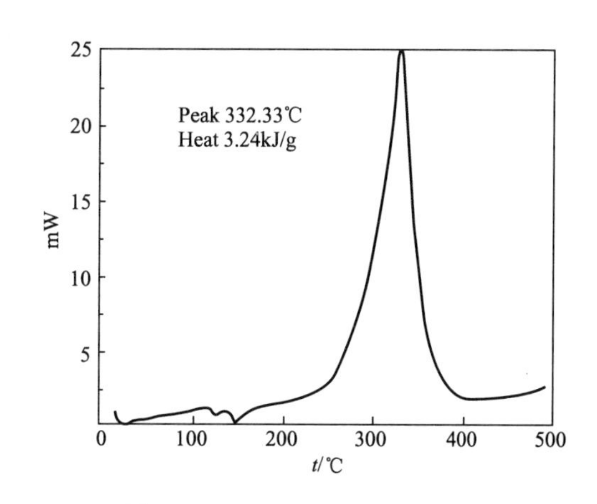

图 2 15% $(NH_4)_2HPO_4$ 改性 ANFO 的 DSC 分析

Fig. 2 DSC curve of 15% $(NH_4)_2HPO_4$ modified ANFO

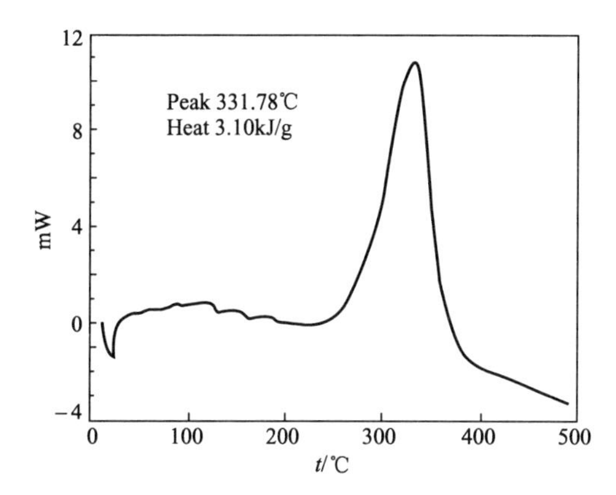

图 3 15% $NH_4H_2PO_4$ 改性 ANFO 的 DSC 分析

Fig. 3 DSC curve of 15% $NH_4H_2PO_4$ modified ANFO

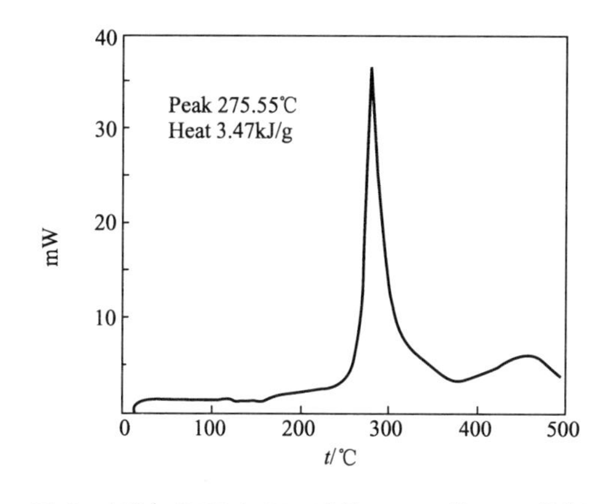

图 4 25% $(NH_4)_2SO_4$ 改性 ANFO 的 DSC 分析

Fig. 4 DSC curve of 25% $(NH_4)_2SO_4$ modified ANFO

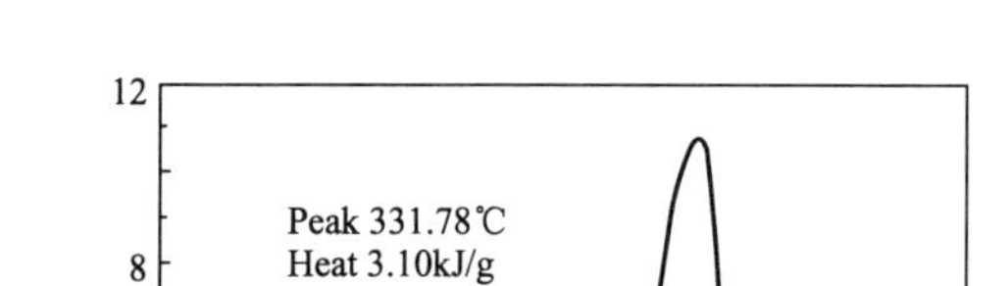

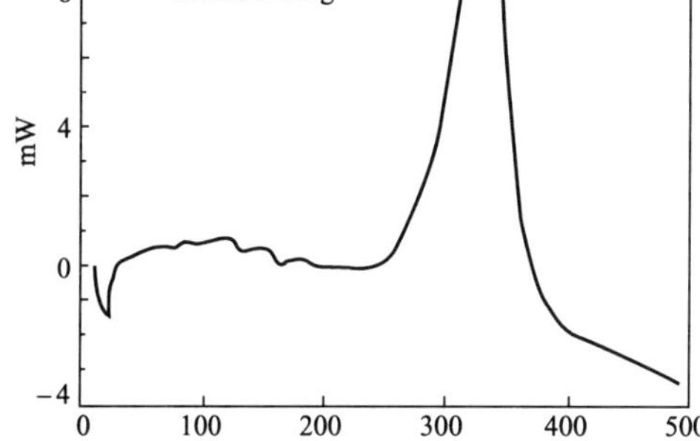

图 5 25% K_2SO_4 改性 ANFO 的 DSC 分析

Fig. 5 DSC curve of 25% K_2SO_4 modified ANFO

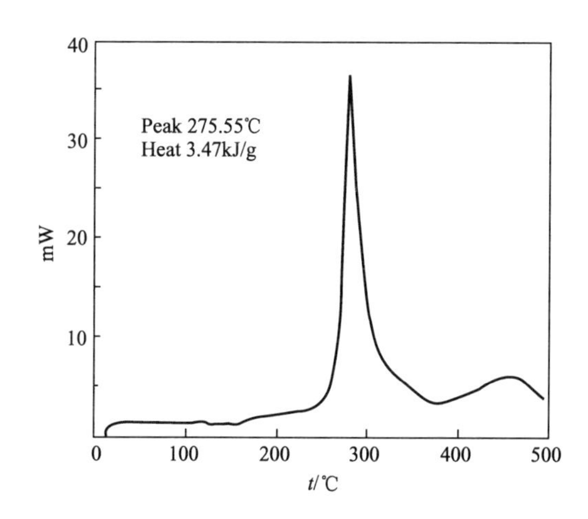

图 6 5%NH_4Cl 改性 ANFO 的 DSC 分析

Fig. 6 DSC curve of 5% NH_4Cl modified ANFO

2.3　无机化学肥料改性后的铵油炸药爆轰实验结果

实验结果参见表 3。从表 3 可以看出，炸药装药直径越大，越容易起爆。同为磷酸盐，$NH_4H_2PO_4$ 的阻爆效果较（$(NH_4)_2HPO_4$）为差，65mm 大直径药卷阻爆的添加质量分数分别为 40%和 20%。硫酸铵和硫酸钾的阻爆效果相差无几，65mm 大直径药卷阻爆的添加质量分数均为 25%；氯化铵在添加质量分数 30%以内对铵油炸药没有任何阻爆作用。爆轰实验结果与硝酸铵高温短时间的热分解结果具有可比性，即：在 370℃加热 1min，纯硝酸铵和铵油炸药的热分解率分别为 96.8%和 92.1%，不能被 8 号雷管起爆的改性硝酸铵和铵油炸药的热分解率则低于 50%。

表 3　无机化肥改性的铵油炸药起爆情况

Table 3　Detonation results of inorganic chemical fertilizer modified ammonium nitrate fuel oil explosive

Additive	ω(Additive)/%	Diameter of explosive charger		
		32 mm	45mm	65mm
$NH_4H_2PO_4$	5	√[a]	√	√
	10	×	√	√
	15	×	√	√
	30	×	×	√
	40	×	×	×
$(NH_4)_2HPO_4$	5	×	×	√
	10	×	×	√
	20	×	×	×
	25	×	×	×
	40	×	×	×
$(NH_4)_2SO_4$	5	×	√	√
	10	×	√	√
	20	×	×	√
	25	×	×	×
	40	×	×	×
K_2SO_4	5	×	√	√
	15	×	×	√
	25	×	×	×
NH_4Cl	10	√	√	√
	20	√	√	√
	30	√	√	√

磷酸氢二铵、磷酸二氢铵、硫酸铵和硫酸钾等无机盐在一定添加量以上都可使铵油炸药失去雷管感度。由于炸药的起爆是一个非常快速的链式反应，上述无机盐与硝酸铵形成复盐，降低了还原性的铵离子和氧化性的硝酸根离子直接发生氧化还原反应的有效碰撞几率，可提高硝酸铵的热稳定性，阻止或延缓链式反应的快速进行，从而达到阻爆目的。氯化铵由于对硝酸铵的分解具有非常强的催化作用，添加质量分数高达 30%仍不能消除铵油炸药的雷管感度。

参考文献

[1] TANG Shuang-Ling（唐双凌），LIU Zu-Liang（刘祖亮），Lú Chun-Xu（吕春绪），*et al*. *Chin J Appl Chem*（应用化学）[J]，in press.

[2] Brower K R，Oxley J C，Tewari M P. *J Phys Chem* [J]，1989，93：4029.

[3] Lowry T H，Richardson K S，Auths. Mechanism and Theory in Organic Chemistry [M] . New York ：Harper and Row，1987：241.

[4] Keenan A G，Notz K，Franco N B. *J Am Chem Soc* [J]，1969，91：3 168.

[5] Kelmers A D，Maya L，Browning D N，*et al*. *J Inorg Nucl Chem* [J]，1979，41：1 583.

[6] Macneil J H，Zhang H T，Berseth P，*et al*. *J Am Chem Soc* [J]，1997，119：9 738.

Studies on the Detonation Safety of Modified Ammonium Nitrate Ⅱ. The Influence of Inorganic Chemical Fertilizer

Tang Shuangling, Lv Chunxu, Zhou Xinl, Wang Yilin, Liu Zuliang

(School of Chemical Enginerring, Nanjing University of Science & Technology, Nanjing, 210094)

Abstrct: Ammonium nitrate (AN) is modified by chemical fertilizer such as diammonium phosphate, monoammonium phosphate, ammonium sulfate, potassium sulfate and ammonium chloride, and then mixed with fuel oil and sawdust to make ANFO explosives. The explosives were initiated by No. 8 detonator. The thermal decomposition of the modified AN and ANFO explosives have been studied by isothermal and differential scanning calorimetry (DSC) in sealed capillary and anti-pressure sealed cell, respectively. It is found that the detonation of ANFO explosives can not be initiated with more than 40% monoammonium phosphate, 20% diammonium phosphate, 25% ammonium sulfate or 25% potassium sulfate in the modified AN. Detonation results are consistent with isothermal data, namely all the undetonated explosives possess the feature of decomposition percent less than 50% on heating at 370℃ for 1 minute. Chloride ion catalyzes the decomposition of AN, so the modified AN could be detonated with 30% ammonium chloride and decomposition percent is nearly 100% on heating at 370℃ for 1 minute.

Keywords: ammonium nitrate, chemical fertilizer, thermal decomposition, detonation safety

（注：此文原载于　应用化学，2004，21（4）：400-404）

聚合物对硝酸铵相转变的影响

叶方青[1]，曾贵玉[1,2]，吕春绪[1]，黄辉[2]
（1. 南京理工大学化工学院，江苏南京，210094）
（2. 中国工程物理研究院化工材料研究所，四川绵阳，621900）
EI：20081311171192

摘要： 硝酸铵（AN）在常压下有五种晶型，温度变化时晶型要发生转变，导致使用性能降低。为抑制或防止 AN 发生某种相转变，在 AN 中添加 0.99%的六种不同类型聚合物，用差式扫描量热法（DSC）研究聚合物对 AN 相转变的影响。结果表明，聚对苯乙烯磺酸钠-丙烯酸聚合物对 AN 的相转变影响较小，聚对苯乙烯磺酸钠-丙烯酸十二酯聚合物可有效防止 AN 发生Ⅳ-Ⅲ相转变，聚对苯乙烯磺酸钠、丙烯酸十二酯、聚对苯乙烯磺酸钠-N,N-二甲基二烯丙基氯化铵和聚对苯乙烯磺酸钠-丙烯酸十二酯-N,N-二甲基二烯丙基氯化铵可有效防止 AN 发生Ⅲ-Ⅱ相转变。聚合物对 AN 相转变的影响在于聚合物可加强 AN 的氢键网络并阻碍 AN 分子中 NO_3^- 的转动。

关键词： 物理化学，硝酸铵（AN），相转变，聚合物，DSC

1 引言

硝酸铵（AN）是一种来源广泛、成本低廉的物质，在化肥、炸药等领域有着广泛使用，是目前大多数工业炸药的主要组分[1]。另外由于其燃烧产物无烟且对环境友好，因此在无氯推进剂中它是氧化剂高氯酸铵（AP）的良好替代物之一[2-3]。但 AN 存在晶型转变的缺陷：常压下 AN 在－17～170℃的温度范围内有五种热力学稳定的晶型，每种晶型仅在一定温度范围内存在。温度变化，晶型也将发生转变，从而使 AN 的晶体结构和晶格体积发生变化，导致 AN 药柱的体积不断变化且性能下降，这使 AN 在工业炸药及推进剂领域的应用受到了很大限制[4]。因此必须采取有效措施尽可能抑制或防止 AN 在使用温度范围内发生相转变。

已有文献[5-8]报道了添加剂对 AN 相稳定的影响，但采用的添加剂用量较大（一般需 2%以上方能起到较好的相稳定效果），研究的添加剂种类也不够多。针对 AN 相稳定问题，前文[9]曾报道了 20 余种不同类型无机添加剂对 AN 相转变温度的影响，本研究探讨了六种聚合物对 AN 相转变的影响。

2 实验

2.1 原材料与仪器

硝酸铵：分析纯，天津市东升化学试剂厂；几种聚合物均自制；DZTW 型电热套，北京市永光明医疗仪器厂；NTEZSCH STA449C 型微分扫描式量热仪。

2.2 样品制备

将硝酸铵与聚合物按 100∶1（质量比）的比例称料，加入适量蒸馏水，混合后加热，使组分溶解并混合均匀，然后冷却结晶。最后在120℃、−0.085MPa真空度下干燥4h，粉碎、过筛即得到相稳定改性的硝酸铵（PSAN）样品，聚合物含量为0.99%。

2.3 相转变温度测试

按GJB772A—97标准方法502.1差热分析和差式扫描量热法（DSC）测试PSAN样品的焓变温度及焓变值，得到PSAN的相转变温度。DSC升温速率为10.0℃·min^{-1}，常压氮气气氛下进行测试，氮气流量0.02L·min^{-1}。

3 结果与讨论

实验考察了聚对苯乙烯磺酸钠、丙烯酸十二酯及聚对苯乙烯磺酸钠分别与丙烯酸、丙烯酸十二酯、N,N-二甲基二烯丙基氯化铵、丙烯酸十二酯-N,N-二甲基二烯丙基氯化铵形成的聚合物对AN相转变的影响，DSC测试结果见表1和图1。

表1 聚合物对AN相转变的影响

Table 1 The effect of polymers on the phase transition of AN

samples	additives	Peak1		Peak2		Peak3		Peak4	
		T/℃	ΔH/J·g^{-1}	T/℃	ΔH/J·g^{-1}	T/℃	ΔH/J·g^{-1}	T/℃	ΔH/J·g^{-1}
PSAN-0	/	52.46	13	90.75	11	127.80	53	169.03	72
PSAN-1	sodium *p*-ethylene benzene sulfonate-acrylic acid polymer	52.98	21	90.69	4	126.89	56	168.49	68
PSAN-2	Sodium *p*-ethylene benzene sulfonate-dodecyl acrylate polymer			89.35	14	127.24	49	167.50	64
PSAN-3	sodium *p*-ethylene benzene sulfonate polymer	52.92	22			127.14	51	167.22	57
PSAN-4	dodecyl acrylate	52.49	25			127.00	59	169.38	81
PSAN-5	sodium *p*-ethylene benzene sulfonate-N,N-dimethyldiallyl ammonia chloride polymer	53.04	24			127.11	54	167.70	60
PSAN-6	sodium *p*-ethylene benzene sulfonate-dodecyl acrylate-N,N-dimethyldiallyl ammonia chloride polymer	53.40	24			127.19	55	168.46	65

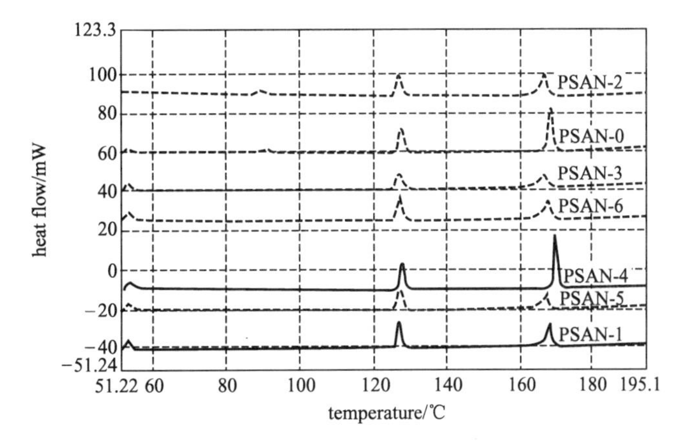

图1 聚合物对AN相转变温度的影响

Fig. 1 Effect of polymers on the phase transtion temperature of AN

从表1可见，在本实验条件下，AN样品在原32℃左右的相转变温度提高到了53℃左右，原84℃左右的相转变温度提高到了90℃左右，原125℃左右的相转变温度提高到127℃左右。在峰的归属上，53℃左右的吸热峰是AN晶体从Ⅳ型（即α斜方晶型）向Ⅲ型（即β斜方晶型）转变的晶相转变吸热峰，90℃左右的吸热峰是AN晶体从Ⅲ型向Ⅱ型（即四方晶体）转变的晶相转变吸热，127℃左右的吸热峰是AN晶体从Ⅱ型向Ⅰ型（即立方晶体）转变的晶相转变吸热峰，168℃附近的吸热峰是AN晶体熔化及分解吸热峰。

从图1可知，经六种自制聚合物改性得到的PSAN样品在AN发生Ⅱ-Ⅰ晶型转变过程中，晶型转变峰的位置、形状基本相似，说明聚合物对AN的Ⅱ-Ⅰ晶型转变过程没有影响或影响很小；从表1得到的AN晶相转变焓变值可知，所有AN样品在168℃左右熔化及分解过程所产生的焓变值均比AN发生相转变时的焓变值大，说明AN熔化及分解过程所需要的热量比发生晶型转变所需要的热量多，而发生Ⅱ-Ⅰ相转变所需焓变值较发生Ⅲ-Ⅱ和Ⅳ-Ⅲ相转变所需焓变值大很多，也说明要减轻或防止AN发生Ⅱ-Ⅰ相转变是比较困难的。

从DSC数据也可看出，不同聚合物对Ⅳ型、Ⅲ型及Ⅱ型间的相转变影响是不同的。在六种聚合物中，聚对苯乙烯磺酸钠-丙烯酸聚合物对抑制AN发生Ⅲ-Ⅱ相转变有一定作用，其相转变焓变值明显减小，但相转变峰仍然存在，同时它对其它晶型间的相转变影响很小，因此该聚合物对AN的相转变影响较小；聚对苯乙烯磺酸钠-丙烯酸十二酯聚合物对抑制AN发生Ⅳ-Ⅲ相转变有显著影响，已基本消除了Ⅳ-Ⅲ相转变，表现在53℃左右未出现相转变峰，同时它对其它晶型间的相转变影响小，因此该聚合物是常温下有效的AN相稳定剂；聚对苯乙烯磺酸钠、丙烯酸十二酯、聚对苯乙烯磺酸钠-N,N-二甲基二烯丙基氯化铵聚合物和聚对苯乙烯磺酸钠-丙烯酸十二酯-N,N-二甲基二烯丙基氯化铵聚合物均可有效防止AN发生Ⅲ-Ⅱ相转变，而对其它晶型间的相转变影响较小。

聚合物对AN相转变的影响在于其对AN晶体空间结构产生作用的程度。AN的晶体结构由NH_4^+和NO_3^-两种离子在空间的分布所决定，AN的多晶型现象就是这两种质点相互作用的结果，AN晶型转变过程主要包括NO_3^-的转动、氢键的破坏和重组。温度升高，分子运动加快，加速了NO_3^-的转动并削弱了AN体系的氢键网络，有可能发生晶型转变。因此，为了抑制或防止AN发生某种相转变，就要设法加强AN的氢键网络并阻碍NO_3^-发生转动，使AN晶体结构无法转变为某种空间分布，从而达到稳定某种晶型的目的。聚合物对AN相转变的影响就在于聚合物分子中的极性基团可以通过氢键等作用力与AN分子表面的原子或原子团紧密结合在一起，其非极性部分在AN分子表面形成一层薄而致密、有弹性和韧性的包覆层，这不仅加强了AN体系的氢键网络，且一定程度上扰乱了AN分子中NH_4^+和NO_3^-之间的相互作用，阻碍了NO_3^-的转动，因此当温度发生变化时，AN晶型不易发生转变，从而有利于减轻或阻止AN的某些相转变，达到防止AN发生某种相转变、稳定某种晶型的目的。聚合物对AN相转变的影响很大程度上取决于聚合物的结构和分子量，聚合物结构不同或分子量不同，则所稳定的AN晶型可能就不同，因此不同聚合物抑制或防止AN发生相转变的种类不同。目前这方面的研究还比较少，没有一个定性的结论，需要进一步研究。

4　结论

所研究的六种聚合物对AN相转变均有一定影响，但不同聚合物对AN相转变的影响不同。聚对苯乙烯磺酸钠-丙烯酸聚合物仅可一定程度上抑制AN发生Ⅲ-Ⅱ相转变；聚对苯乙烯磺酸钠-丙烯酸十二酯聚合物基本上消除了Ⅳ-Ⅲ相转变，是常温下良好的相稳定剂；聚对苯乙烯磺酸钠、丙烯酸十二酯、聚对苯乙烯磺酸钠-N,N-二甲基二烯丙基氯化铵聚合物、聚对苯乙烯磺酸钠-丙烯酸十二酯-N,N-二甲基二烯丙基氯化铵聚合物可有效防止AN发生Ⅲ-Ⅱ相转变。聚合物对AN晶型转变的影响与聚合物可加强AN体系氢键网络和阻碍NO_3^-的转动有关。

致谢：本研究得到中国工程物理研究院化工材料研究所聂福德、周建华、沈永兴、夏敬琼等许多同事及南京理工大学户安军博士的帮助，在此表示感谢！

参考文献

[1] 吕春绪．工业炸药理论 [M]．北京：兵器工业出版社，2003.
Lü Chun-xu. Industrial Explosive Theory [M]. Beijing: Ordnance Industry Press, 2003.
[2] A O Remya Sudhakar, Suresh Mathew. Thermal behaviour of CuO doped phase-stabilised ammonium nitrate [J]. Thermochimica Acta, 2006, 451: 5-9.
[3] 张海燕，陈红．低特征信号推进剂的氧化剂：相稳定硝酸铵 [J]．飞航导弹，1996，2：38-41.
[4] Oommen C, Jain S R. Ammonium nitrate: A promising rocket propellant oxidizer [J]. Journal of Hazardous Materials, 1999: 253-281.
[5] 张杰，杨荣杰，刘云飞等．聚乙烯醇缩丁醛包覆硝酸铵的性能研究 [J]．火炸药学报，2001，1：41-43.
[6] 蔡敏敏，陈天云，黄建祯等．无机盐添加剂对硝酸铵相转变及结块性的影响 [J]．南京理工大学学报，2000，24 (1)：76-79.
[7] Jimmie C Oxley, James L Smith, Evan Rogers, et al. Ammonium nitrate: Thermal stability and explosivity modifiers [J]. Thermochimica Acta, 2002, 384: 23-45.
[8] 李艺，惠君明．几种添加剂对硝酸铵热稳定性的影响 [J]．火炸药学报，2005，28 (1)：76-80.
[9] 曾贵玉，周建华，吕春绪，等．无机物对硝酸铵相转变的影响 [J]．含能材料，2007，15 (4)：400-403.

Effects of Polymers on Phase Transition of Ammonium Nitrate

Ye Fangqing[1], Zeng Guiyu[1,2], Lv Chunxu[1], Huang Hui[2]
(1. School of Chemical Engineering, Nanjing University of Science & Technology, Nanjing 210094, China)
(2. INsitute of Chemical Materials, CAEP, Mianyang 621900, China)

Abstract: In order to restain or prevent AN phase transition, six kinds of polymers were added to AN samples and the effects of polymers on AN phase transition were studied by differential scanning calorimetry (DSC) method. The results show that sodium p-ethylene benzenesulfonate-acrylic acid polymer has little effect on AN phase transition, while sodium p-ethylene benzenesulfonate-dodecylacrylate poltmer prevents the transition of Ⅳ-Ⅲ effectively. And sodium p-ethylene benzenesulfanate polymer, dodecyl acrylate, sodium p-ethylenebenzenesulfonate-N,N-dimethyldiallyl ammonia chloride polymer and sodium p-ethylenebenzenesulfonate-dodecyl acrylate-N,N-dimethyldiallyl ammonia chloride polymer prevent the transition of Ⅲ-Ⅱ effectively. The effect of polymer on AN phase transition results from that polymer strwngthens the hydrogen-bond network and baffle the turn of NO_3^-.

Key words: physical chemistry, ammonium nitrate (AN), phase transition, polymer, DSC

（注：此文原载于　含能材料，2008，16 (1)：77-79）

Self-sensitization Structure of Expanded Ammonium Nitrate and its Effect on ANFO Detonation Properties

Zeng Guiyu, Lv Chunxu, Hung Hui

(1. Institute of Chemical Materials, China Academy Engineering Physics, Mianyang 621900 Sichuan, China)

(2. Nanjing University of Science and Technology, Nanjing 210094 Jiangsu, China)

Abstract: The expanded ammonium nitrate (EAN) samples with different states were prepared by using a vacuum crystallizing technology. The structure characters, such as porosity, pore structure, specific surface area, particle surface shape and surface defects, and detonator initiation sensitivity and explosion power, of common ammonium nitrate (AN) and EAN were tested using density measuring, N_2 adsorbing, scanning electron microscope (SEM) and plate trace test methods. The tested results show that the particle surface of common AN is smoother, denser, lower porosity and specific surface area than those tested of EAN. The particle surface of EAN is irregular, which has edges, protuberance and severely distorted crystal form, and its specific surface area and porosity are larger than those of un-expanded AN. EAN has typical self-sensitization structure characters. The detonator initiation sensitivity and explosion power of ammonium nitrate-fuel oil (ANFO) made of different states of EAN are related to the self-sensitization structures of EAN, and expanded ANFO sample has higher detonator initiation sensitivity and explosion power compared with un-expanded ANFO sample. The characterization techniques can be used to reveal the self-sensitization structure of EAN.

Key Words: applied chemistry; self-sensitization structure; detonation property; expanded ammonium nitrate

Introduction

Ammonium nitrate (AN) is a low cost and widely available substance which is smokeless or low smoke at burning and has been widely used to prepare industrial ammonium nitrate explosives and propellant. Since the initiation sensitivity of AN based industrial explosive is rather low, the application of AN in the area of industrial explosives and propellant has been affected. It is known from the hotspot theory of explosive initiation tht, when pore, defect, hard impurity and abnormity shape exist in the explosives, they can form the initiation hotspots easily under external stimuli and their properties, especially initiation property, may change significantly. Therefore, when pores and defects are introduced into AN crystal or the particle surface of AN is distorted, the microstructure of AN can be changed to improve the detonator initiation sensitivity of AN based explosives and increase the explosion power of corresponding explosives. The design, development and successful application of expanded ammonium nitrate (EAN) are just the practical case of hotspot theory application and development[1-3]. In other words, the increase in initiation sensitivity of EAN based explosives is dependent on its microstructure improvement, which is self-sensitization, its connotation is that AN particle has a large number of pores and/or a severely distorted surface. The preparation technology and structure characters of EAN have been studied deeply. However, the self-sensitization structure of EAN with different states hasn't been contrastively investigated, and the effect of EAN self-sensitization structure on the initiation property and explosion power of corresponding ANFO (ammonium nitrate-fuel oil) also hasn't been studied. In this paper, the self-sensitization structure and detonation properties of EAN are investigated by the manifold techniques and compared with common AN to understand the self-sensitization theory further. Therefore, this study has important theoretic and practical significance for comprehending the shock initiation mechanism and detonation properties of ammonium nitrate-based explosives.

1 Experiment

1.1 Main materials

Ammonium nitrate used for commercial products; self-made expanded agent; distilled water.

1.2 Preparation of Samples

Ammonium nitrate, water and expanded agent were added into a reaction still in the proportion of 100: 10: 0.15 and heated to 130-132℃, and the still was connected with a vacuum system of 0.09MPa. The ammonium nitrate solution was expanded rapidly under high vacuum condition. When the crystallizing time reached 15min, the completely expanded sample EAN-1 was obtained. Other different state EAN samples could be got by changing the expanded conditions, which are EAN-2 sample at 130-132℃ for 2min of crystallizing time; EAN-3 sample at 116-120℃ for 15min of crystallizing time. The un-expanded industrial sample is AN-0.

Industrial AN and EAN samples were mixed with fuel oil in the proportion as 94.5: 5.5 respectively at 80℃, the ANFO samples with different self-sensitization structures were obtained after dried completely, which are EANFO-1, EANFO-2, EANFO-3 and ANFO-0.

All the samples were filtered with 40 mesh sieve, and the materials passing through sieve pore were taken as experimental samples.

1.3 Testing method

The bulk density of sample was measured according to GJB772A-97 Item 402.1, the specific surface areas and pore structures of different state samples were tested with NOVA2000 specific surface area instrument, and their particle size and surface shape were observed by using KYKY2800 scanning electron microscope (SEM).

2 Results

2.1 Porosity

The porosity reflects the structure information of pores amount and particle surface roughness etc, is the integrative reflection of EAN self-sensitization structure, and is tightly linked with sample density. Thus the density data can be used to compare the sample's porosity. The bulk density data of different state AN samples are presented in Table 1.

Table 1 Bulk densities of the different state AN samples

sample	AN-0	EAN-1	EAN-2	EAN-3
$\rho_P/(g \cdot cm^{-3})$	1.02	0.88	0.90	0.96

From Table 1, it is obvious that the bulk densities of ammonium nitrate is correlated closely with the sample states. The un-expanded AN sample has larger bulk densitity, lower pore and porosity than those of EAN. After AN is expanded, its bulk densities are reduced, the pore volume is increased, and the porosity is improved. The results indicate that the pore amount of EAN samples is increased during the process of expanding. In other words, the self-sensitizalion structure of EAN sample is strengthened during the process of expanding, this is helpful to increasing the initiation sensitivity of corresponding explosive.

It also can be seen from Table 1 that the pore amount introduced during the process of expanding is related with the expanded extent. The fully expanded sample EAN-1 has a low bulk density. This indicates that EAN-1 has more pores, larger pore volume and higher porosity; the incompletely expanded samples EAN-2 and EAN-3 have less pores than that of EAN-1, their pore volumes and porosities are between those of EAN-1 and AN-0 respectively. The order of bulk densities of different state ANs is as follows: EAN-1<EAN-2<EAN-3<AN-0, that is, the order of porosity is EAN-1>EAN-2>EAN-3>AN-0. It can be known from the bulk density data of EAN-1 and EAN-3 that pores, which is introduced into the bulk at high temperature expanding, is more than that at low temperature expanding under the conditions of the same vacuum and reaction time. This can be explained that water in the sample is quicker vaporized and boiled off at higher temperature to lead it to be crystallized and formed a new phase with higher expanding extent, so larger pores can be produced and remained in EAN sample. In addition, it can be observed from the bulk density data of EAN-1 and EAN-2 that the prolonging of crystallization time can introduce more pores into AN sample under the conditions of the same vacuum and crystallization temperature. The reason is that the remnant water can be more completely removed from the sample due to the longer crystallization time, consequently, more pores are left in sample.

2.2 Pore structures

An important character of EAN self-sensitization is that the abundant pores exist in AN crystal[4]. The pore structures, such as specific surface area, pore volume and pore distribution etc, reflect the pore size, porosity, and the distortion degree of particle surface. The study of pore structures of different state AN samples is an indispensable premise for realizing and understanding EAN self-sensitization theory.

The typical adsorption isotherms and de-adsorption isotherms of un-expanded AN and EAN are shown in Fig. 1 and 2.

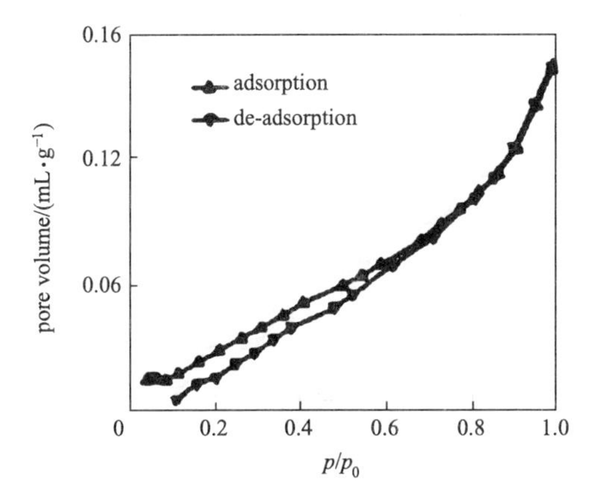

Fig. 1 Adsorption and de-adsorption isotherms of the AN-0

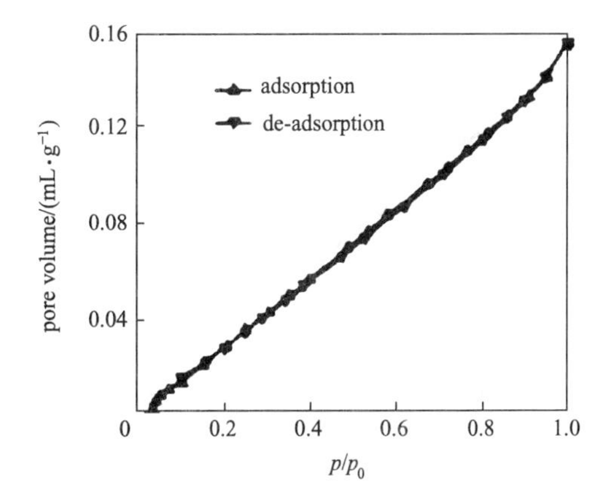

Fig. 2 Adsorption and de-adsorption isotherms of the EAN-1

Figure 1 and Fig. 2 the curves of un-expanded AN which protrude upwards show that the adsorption and EAN are similar, slightly and exhibit monomolecule layer adsorption; With increase of pressure, the multilayer adsorption arises gradually, which rise gently and have no obvious saltation, and no visible circumscription between monolayer adsorption and multilayer adsorption in the mid-pressure region; when p/p_0 reaches to 1, which sharply climb and don't show a saturation adsorption. This reveals that AN sample maybe have lots of big pores. The adsorption curve of AN typically represents the adsorption curve of class II materials.

The specific surface areas, pore volumes and fractal dimensions (FM) of different state ANs can be calculated from adsorption curve and de-adsorption curve. The results are shown in Table 2.

Table 2 Specific surface alias, pore volumes and fractal dimensions of the different state AN samples

Sample	AN-0	EAN-1	EAN-2	EAN-3
specific surface area/($m^2 \cdot g^{-1}$)	0.1702	2.5570	0.9200	0.6259
pore volume/(10^4 mL $\cdot$ g^{-1})	2.5	23.9	15.5	6.2
fractal dimension	2.2834	2.4312	2.5258	2.6771

The data in Table 2 show that the states of AN have evident effects on specific surface area and pore volume. Un-expanded AN sample has less pores and small specific surface area. After expanded, the specific surface area and pore volume of AN are simultaneously increased so that EAN samples have much evident self-sensitization structure characters and the detonator initiation sensitivity of corresponding explosive can be enhanced. It can be also seen from Table 2 that the pore volume order obtained with N_2 adsorption method is consistent with the porosity order obtained with the density measuring method, i. e. EAN-1 > EAN-2 > EAN-3 > AN-0, and the results of two methods confirm each other.

For the characterization of particle shape, the fractal geometry has proved itself to be an effective and feasible method to describe the erose particle shape, and the particle surface roughness can be quantitatively described by using fractal dimension[5]. The FD data in Table 2 show that the particle surface roughnesses (distortion degree) of different state AN samples are rather different and can be relatively compared using FD value, i. e. EAN-3>EAN-2>EAN-1>AN-0. It is obvious that the particle roughness order is not consistent with that of pore volume, both the parameters reflect the self-sensitization structure of AN from various aspects.

2.3 Crystal surface character

SEM technique was used to study the crystal shapes and particle surface characters of different state ANs. Fig. 3 and 4 are the SEM images of un-expanded sample AN-0 and expanded sample EAN-3, respectively.

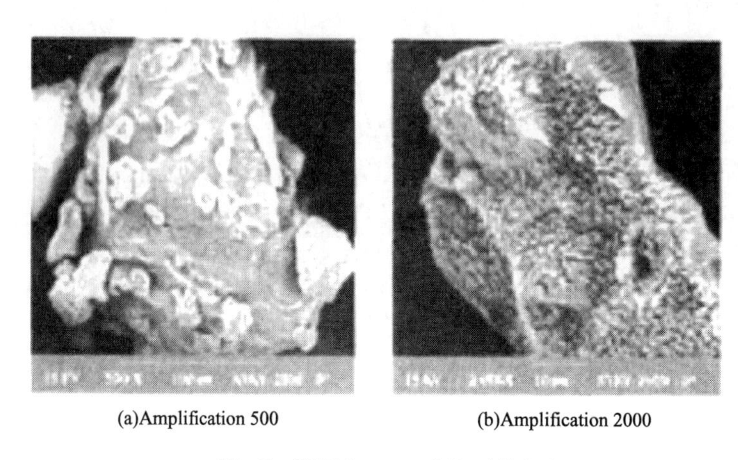

(a)Amplification 500 (b)Amplification 2000

Fig. 3 SEM images of the AN-0

The images in Fig. 3 and 4 show that the un-expanded AN has a little edge angles and protuberances on the particle surface, its particle surface shape is more regular and smoother than that of EAN, and the distortion degree is low; the particle surface topography of EAN is irregular, and has many edge angles, protuberances, dents and cracks on it, and furthermore, many small particles may be accumulated on and stuck to the surfaces of large particles. The structure characters result in EAN particles being not stacked tightly and the abundant cavities existing in particles, forming a swelling system. According to the hotspot theory of explosive initiation, these cavities can easily become the initiation hotspots under exterior action and increase the initiation sensitivity of corresponding explosive. The SEM images visually reflect the self-sensitization structure characters of EAN.

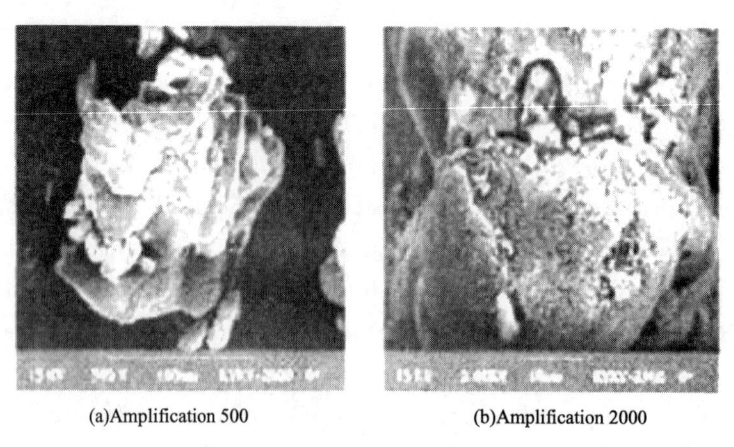

(a)Amplification 500 (b)Amplification 2000

Fig. 4 SEM images of the EAN-3

2.4 Effect of self-sensitization structure on initiation sensitivity and explosion power

The shock initiation sensitivities of ANFOs with different structures were tested by using 8# high-pressure electric detonator. The detonation traces of four samples are showed in Fig. 5-Fig. 8.

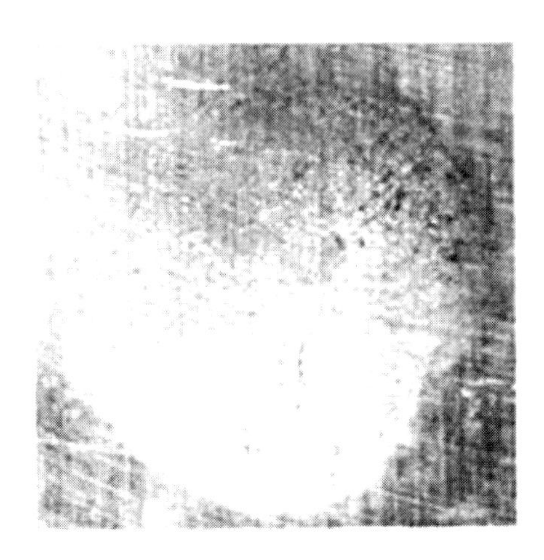

Fig. 5 The trace of the ANFO-0 left on a witness plate

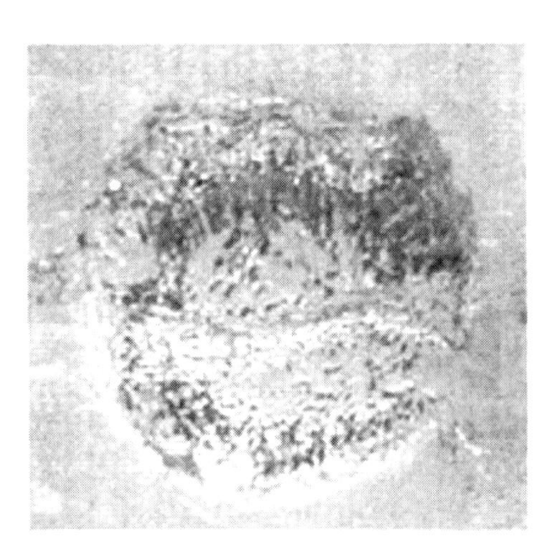

Fig. 6 The trace of the ANFO-1 left on a witness plate

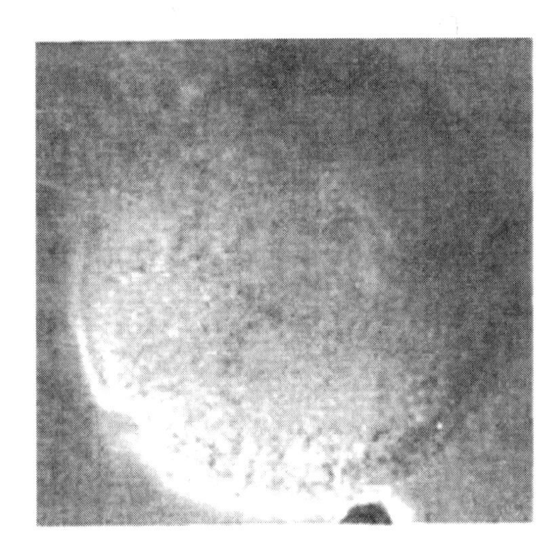

Fig. 7 The trace of the ANFO-2 left on a witness plate

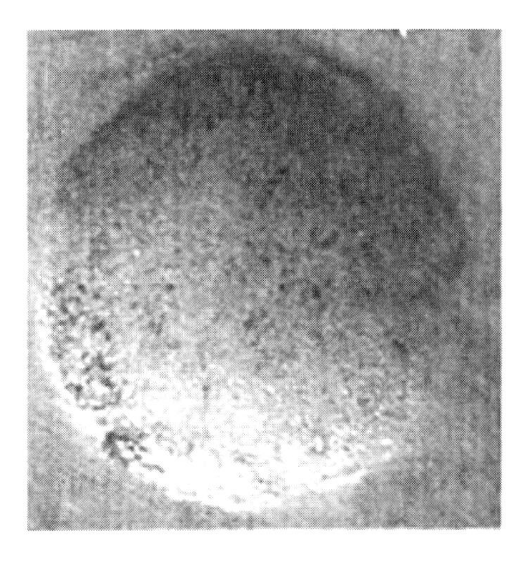

Fig. 8 The trace of the ANFO-3 left on a witness plate

It can be seen from the images that the obvious combusting vestige of un-expanded ANFO sample was left on the witness plate after the initiation action of the 8# high-pressure electric detonator, the indent was very shallow. This shows that the un-expanded AN is rather insensitive to the action of 8# high-pressure electric detonator, and ANFO explodes incompletely under the action of detonator. All the three expanded ANFO samples exploded under the shock wave action of 8# high-pressure electric detonator, and left an evident pit and had no combusting vestige on the witness plate. The results show that the self-sensitization structure of EAN observably improves the detonator initiation sensitivity of expanded ANFO, which is the macroscopic exhibition of EAN self-sensitization structure in detonator initiation sensitivity and the experimental confirmation of EAN self-sensitization theory.

The depths of dint left by ANFO samples on the witness plate after detonation are shown in Table 3. The dint depth reflects the relative explosion power and can be used to compare the detonation capacities of ANFO samples. It can be seen from Table 3 that the detonation capacities of ANFO samples made of different self-sensitization structures ANs and fuel oil are much different. The dint depth left by un-expanded ANFO is shallow, while the dint depth left by expanded ANFO increases markedly. This shows that the self-sensitization structure of EAN significantly improves the explosion power of corresponding ANFO. In spite of this, the different self-sensitization structures have different effects on the explosion power of ANFO, and the explosion power is the resuit of multi-factors. The explosion power order of four ANFO samples can be achieved from dint depth data and expressed as ANFO-3>ANFO-2>ANFO-1>ANFO-0. This order is in accordance with the order of AN particle surface roughness, but is not in agreement with the orders of pore volume and specific surface area. This is a significant finding, which shows that the effect of AN particle surface roughness on explosion power exceeds those of porosity and specific surface area, and the particle surface roughness is a prominent factor which affects the explosion power of ANFO.

Table 3 Dint depths of the ANFO samples after detonating

sample	ANFO-0	ANFO-1	ANFO-2	ANFO-3
dint depth/mm	0.474	0.557	0.904	1.634

For ANFO is a heterogeneous explosive, the shock initiation and detonation process contain two different reaction phases: hotspot forming and ignition phase, and hotspot growing and developing phase. The hotspot forming and ignition phase is the key to understand ANFO shock initiation mechanism and behavior. The main influencing factor is porosity, the more the pores and bubbles are, the more the hotspots form, and the easier explosives ignite and explode. Thus the expanded ANFO samples with more pores exhibit higher initiation sensitivity. The hotspot growing and developing phase is the key to comprehend the detonation capacity of ANFO. In this phase, the dominating chemical reaction is the combustion of particles, and the primary influencing factor is the particle surface roughness and specific surface area. Compared with the regular structure, the rough structure, such as concave pits, edge angle and protuberance, not only possesses higher potential energy which is easier to be combusted, but also holds larger specific surface area which is helpful to increasing combustion velocity. Therefore, the rougher the ANFO particle structure is, the easier the matter combusts, the quicker the combustion velocity rises and the combustion turns to detonation, resulting in higher detonation power. Hereby, the detonation capacity of ANFO is closely related with the particle roughness (FM) of explosive.

3 Conclusions

1) N_2 adsorption test results show that the pore volume of AN sample is largely increased in the expanding process, the FM data reveal that EAN sampie has rougher particle surface. Both larger pore volume and rougher particle surface strengthen the self-sensitization structure characters of EAN.

2) The particle surface of industrial AN sample is relative regular, and has less edge angles and

cracks. The particle surface of EAN is rather irregular, and has many edge angles and protuberances, server distortion crystal form and typical self-sensitization structure characters.

3) The self-sensitization structure of EAN increases the detonator initiation sensitivity and explosion power of corresponding ANFO, and the effect of EAN particle surface roughness on the explosion power exceeds that of porosity and specific surface area.

Acknowledgements

The authors greatly appreciate Prof. Liu Zuliang and Prof. Lu Ming of Nanjing University Science and Technology, Dr. Gao Dayuan, Han Yong and Li Wei of Institute of Chemical Materials, China Academy Engineering Physics for their suggestions and advice.

References

[1] Lv Chunxu, Expanded ammonium nitrate explosives [M] . Beijing: The Publishing House of Ordance Industry, 2001 : 150-158.

[2] Lv Chunxu, Liu Zuliang, Hui Junming, The advancement and development of self-sensitization theory for expanded ammonium nitrate [J] . Chinese Journal of Explosives&Prepellants, 2000, (4): 1-4.

[3] Chen Tianyun. The self-sensitization and appilications of Ammonium Nitrate [D] . Nanjing: Nanjing University of Science and Technology, 2000.

[4] Ye Zhiwen,Lv Chunxu, Liu Zuliang, et al. Self-sensitizable characteristics of modified ammonium nitrate [J]. Chinese Journal of Appied Chemistry, 2002, 19 (2): 130-134.

[5] Chen Haiyang, Wang Ren, Li Jianguo, et al. Grain shape analysis of calcareous soil [J] . Rock and Soil Mechanics, 2005, 26 (9): 1389-1392.

(注：此文原载于 Journal of China Ordnance, 2009, 5 (3): 209-214)

硝酸铵自敏化结构与爆轰性能

梅震华[1,2]，曾贵玉[1,3]，钱华[1]，吕春绪[1]

（1. 南京理工大学化工学院，江苏南京，210094）

（2. 南京市公安局，江苏南京，210005）

（3. 中国工程物理研究院化工材料研究所，四川绵阳，621900）

摘要： 采用压汞法、扫描电镜法测试四种硝铵炸药用硝酸铵（AN）的孔隙结构、比表面积和粒子表面形貌，进而研究工业硝铵炸药用硝酸铵的自敏化结构特征。结果表明：南理工膨化硝酸铵（EAN）样品存在大量孔隙、且有效热点范围内的孔隙数量较多，颗粒表面存在大量棱角、突起、晶形严重歧化，自敏化特征明显；长沙EAN样品中也存在较多的孔隙，且孔径较小、孔隙分布范围窄；深圳EAN样品和普通AN样品相似，孔隙率和表面积较小，有效热点范围内的孔隙数量较少。

关键词： 孔隙结构，晶形歧化，自敏化特征，工业炸药，硝酸铵

硝酸铵（AN）成本低廉、来源广泛，使得以硝酸铵为主要成分的硝铵炸药成为应用最广泛的工业炸药之一。但以AN为基的炸药起爆感度低，影响了其在工业炸药领域的应用[1]。由炸药起爆热点理论可知，当炸药中存在孔隙、气泡、缺陷、不规则形貌等结构特征时，炸药在外界刺激下更容易形成起爆热点，炸药性能尤其是起爆感度性能可能发生显著改变[2,3]。因此，设法在AN晶体中引入孔隙、气泡、缺陷或使颗粒表面歧化，使AN微观结构发生改变，将能显著改善AN炸药的性能。膨化硝酸铵（EAN）的敏化就是热点理论应用和发展的具体案例[4-6]，是由南京理工大学研发的具有我国自主知识产权的新型敏化技术，荣获国家发明专利金奖。这种通过改变自身微观结构来达到提高起爆感度目的的技术途径称为自敏化，自敏化特征的内涵就是颗粒表面歧化且颗粒中含有大量尺寸适度的孔隙。因此，本文采用孔隙测试技术和颗粒形貌检测技术研究不同来源工业AN的微观结构特征，以进一步理解自敏化理论，同时对深入认识工业AN炸药微观结构、掌握AN炸药起爆性能也有着重要的现实意义。

1 实验部分

1.1 试剂和仪器

AN，工业品，由各生产厂家提供。

Autopore IV 950压汞仪（美国Micromeritics公司）；LEO440扫描电镜（英国Zeiss公司）。

1.2 测试方法

采用Autopore IV 950压汞仪测试不同AN样品的比表面积及孔隙结构；采用LEO440扫描电镜（SEM）观察AN样品表面形貌和缺陷。

2 结果与讨论

2.1 孔隙结构

AN自敏化的重要特征之一是AN晶体内部存在大量孔隙[7]，因此研究不同来源工业AN炸药用AN样品的孔隙结构，是认识和理解工业AN炸药敏化的重要前提。

采用压汞仪测试几种AN的孔隙体积及孔隙分布，结果见表1。

表1 不同来源硝酸铵的孔隙结构

Table 1 The pore structure of different origin ammonium nitrate

样品	普通AN	南理工EAN	长沙AN	深圳AN
总孔面积/$m^2 \cdot g^{-1}$	0.082	0.421	2.680	0.119
孔体积/$ml \cdot g^{-1}$	0.9060	2.2467	1.4123	0.9276
中值孔直径/nm	1146.0	350.7	10.0	2214.6
假密度/$g \cdot cm^{-3}$	0.6609	0.3496	0.4995	0.6634
真密度/$g \cdot cm^{-3}$	1.6470	1.6288	1.6956	1.7247
孔隙率/%	59.8751	78.5383	70.5433	61.5355

从表1各AN样品的孔隙数据可以看出，不同来源AN样品的孔隙结构相差较大：(1) 按孔体积大小排序，有：南理工EAN样品>长沙EAN样品>深圳EAN样品>普通AN样品，南理工EAN样品孔体积最大，其次是长沙EAN样品，深圳样品只比普通AN样品稍多一点。(2) 按孔面积大小排序，有：长沙EAN样品>南理工EAN样品>深圳EAN样品>普通AN样品，其中长沙EAN样品的孔面积远大于其它样品，说明该样品中含有较多孔隙或粒子粒径较小，普通AN样品孔面积是所有样品中最小的，而深圳EAN样品的孔面积仅略高于普通AN样品。(3) 按孔隙率大小排序，有：南理工EAN样品>长沙EAN样品>普通AN样品>深圳EAN样品，南理工EAN样品的孔隙率是所有AN样品中最大的，同样，深圳EAN样品和普通AN样品的假密度和孔隙率相近，尤其是深圳EAN样品的真密度已与AN晶体密度（$1.72g \cdot cm^{-3}$）非常相近。由此可以判断：深圳EAN样品孔隙结构与普通工业AN样品相似。(4) 从平均孔径（面积）看，深圳EAN样品>普通AN样品>南理工EAN样品>长沙EAN样品。

综合上述数据，可以看出：南理工EAN样品和长沙EAN样品均含有大量孔隙，但南理工EAN样品孔隙直径较长沙EAN样品大很多，长沙样品孔面积高和孔隙率较大而孔径较小、分布较窄；普通AN样品和深圳EAN样品中的孔隙相对较少、孔径较大，因而总孔面积低、假密度较高。

各硝酸铵样品的孔隙分布见图1。

从图1几个EAN样品孔隙体积的微分分布图可知：普通AN样品孔径较大，所有孔都分布在600nm以上，其中大量孔分布在600nm～1000nm之间，另有较多孔分布在10m以上；深圳AN样品孔隙分布与普通AN相似，只不过在2μm～6μm期间还分布了大量孔隙，10μm以上的孔隙也较多；长沙EAN样品的孔隙分布非常独特：孔隙范围很窄，几乎所有孔都局限在10nm附近，其它直径的孔隙很少；南理工EAN样品的孔隙分布较宽，但分布区间明显不同于其它几个AN样品：大部分孔隙分布在200nm～500nm之间。

按炸药起爆热点理论，并不是所有热点均能发生爆炸，除温度、作用时间和放出的热量外，还要求尺寸要适度，即热点半径为100nm～10 m的孔才能逐渐发展为爆炸[1]。因此，如果仅从孔径角度考虑，南理工EAN样品在100nm～10μm分布有大量孔隙，则相应EAN炸药起爆性能将可能优于其它EAN炸药；长沙EAN样品在100nm～10μm区间的孔隙极少，其EAN炸药起爆可能比较困难；深圳EAN炸药起爆性能可能稍好于工业AN炸药。

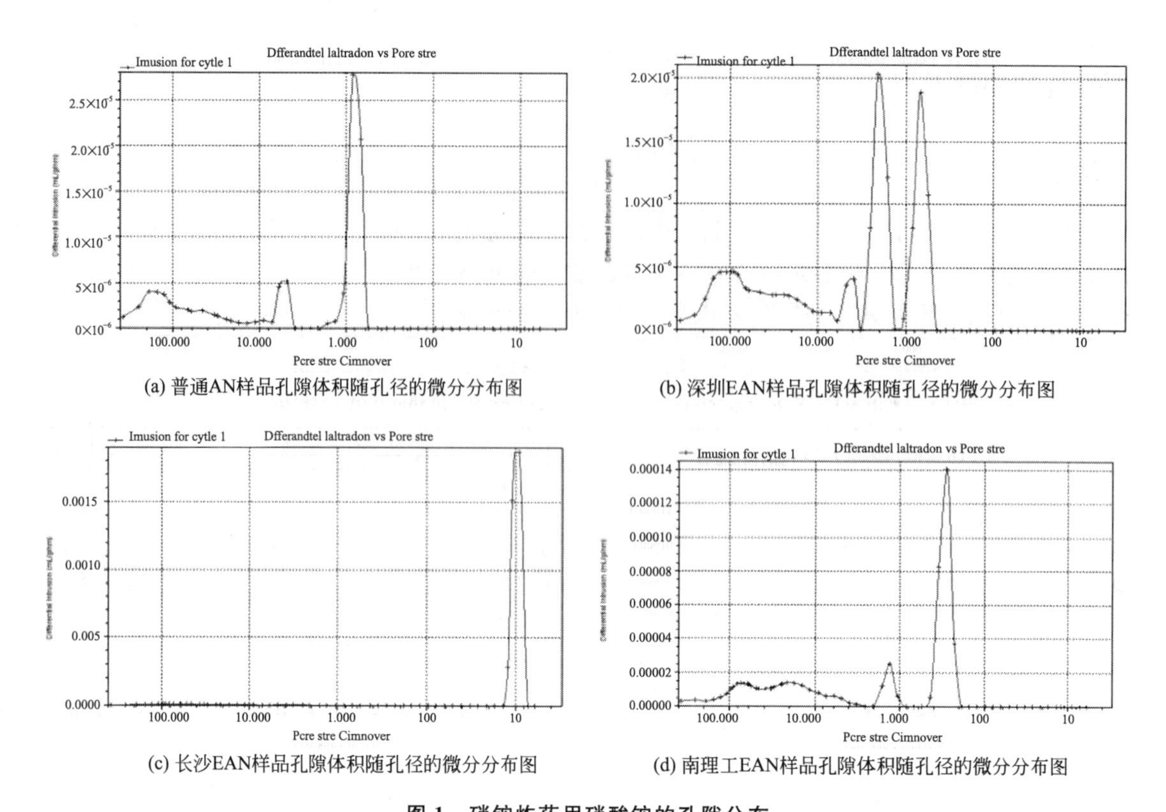

(a) 普通AN样品孔隙体积随孔径的微分分布图　　(b) 深圳EAN样品孔隙体积随孔径的微分分布图

(c) 长沙EAN样品孔隙体积随孔径的微分分布图　　(d) 南理工EAN样品孔隙体积随孔径的微分分布图

图 1　硝铵炸药用硝酸铵的孔隙分布

Fig. 1　The pore distribution of ammonium nitrate used for ammonia dynamite

2.2　表面特征

采用 SEM 方法研究各硝酸铵样品的晶体形貌和表面特征，500 倍数下各硝酸铵样品的表面特征见图 2。

从图 2 低倍数下的 SEM 图片可知，几种来源 EAN 样品表面均呈凹凸不平的不规则形状。相对而言，工业普通 AN 颗粒较大，颗粒表面有少许裂纹，棱角、突起相对较少，表面形貌较规则、歧化程度较低一些；长沙和深圳 EAN 样品颗粒不规则程度较接近，长沙 EAN 样品颗粒歧化相对要严重一些，深圳 EAN 样品颗粒相对要大一些；南理工 EAN 样品颗粒歧化严重，颗粒表面存在大量突起、棱角、凹痕，形貌极不规则，粗糙度大，这些结构特征导致颗粒之间无法紧密堆积，颗粒间必然存在大量空隙，成为一种膨松体系。按颗粒表面歧化程度排序，有：南理工 EAN＞长沙 EAN＞深圳

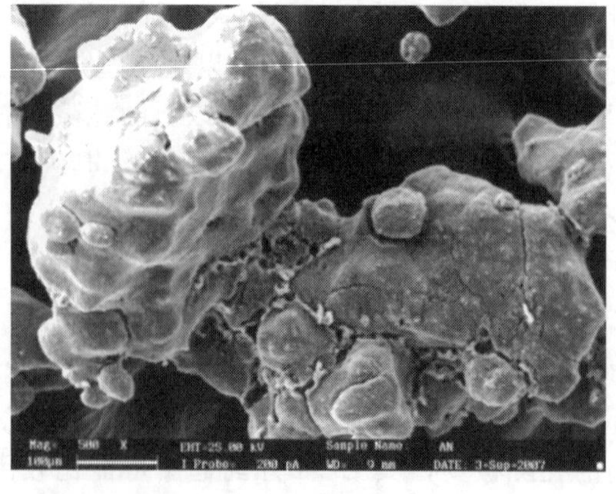

(a) 工业普通AN

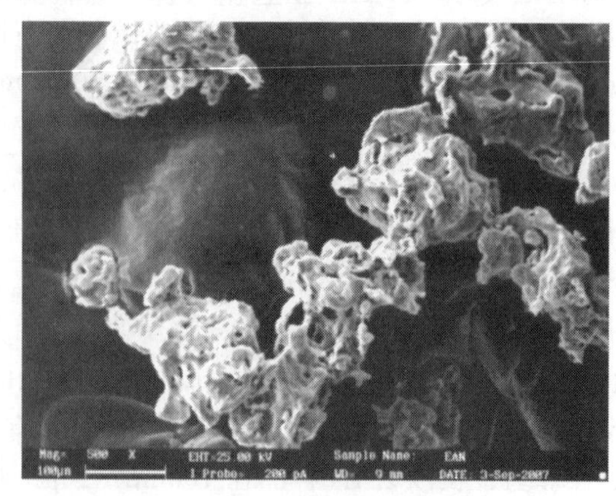

(b) 南理工EAN

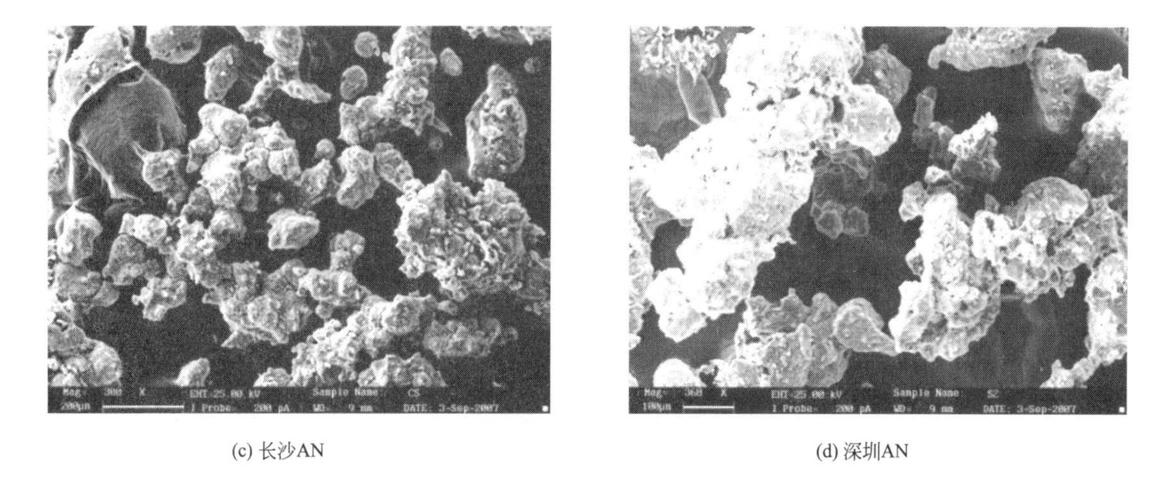

(c) 长沙AN (d) 深圳AN

图 2 不同来源硝酸铵样品在 500 倍下的 SEM 照片

Fig. 2 SEM image of different origin ammonium nitrate samples in 500times

EAN＞普通 AN。根据炸药起爆“热点”理论，这些表面歧化的颗粒及颗粒间的空隙在外界作用下很容易形成起爆热点，有利于提高硝酸铵炸药的冲击起爆感度，因此 SEM 图片直观反映了工业 AN 炸药的自敏化特征。5000 倍下的 SEM 照片见图 3。

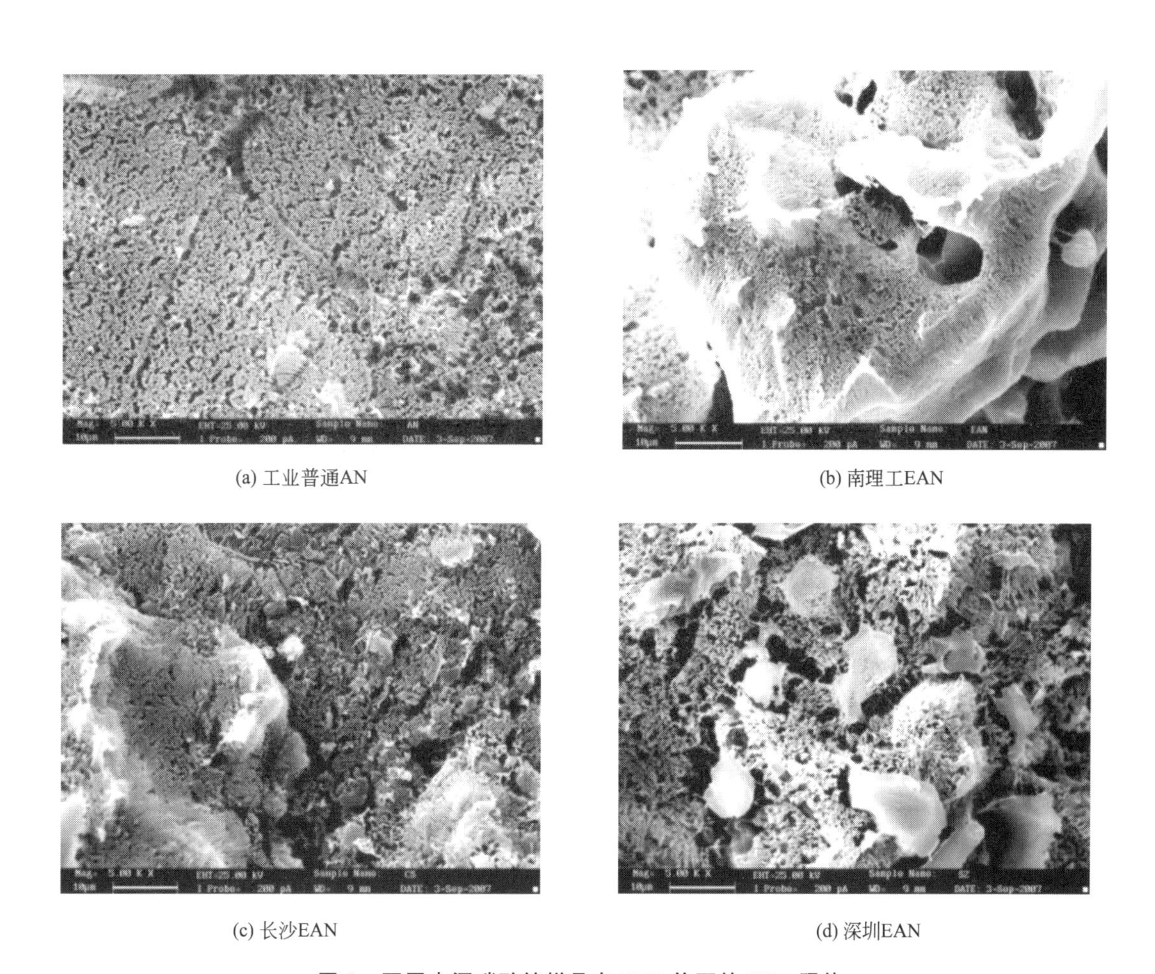

(a) 工业普通AN (b) 南理工EAN

(c) 长沙EAN (d) 深圳EAN

图 3 不同来源硝酸铵样品在 5000 倍下的 SEM 照片

Fig. 3 SEM image of different origin ammonium nitrate samples in 5000 times

从图 3 各样品在 5000 倍下的电镜照片可知，各 EAN 样品均存在大小不一的毛孔。相对而言，普通 AN 颗粒较密实，孔隙较少；南理工 EAN 样品颗粒歧化极为严重，呈风化石状，颗粒中存在相当多的毛孔；长沙和深圳 EAN 样品的毛孔数量相对要比南理工 EAN 样品少一些，且长沙 EAN 样品的毛孔较小。这与压汞测试数据相一致。

结论

孔隙结构测试和 SEM 均表明：南理工 EAN 样品中含有大量孔隙，且孔径在有效热点范围内的孔隙较多，自敏化特征显著；普通 AN 和深圳 EAN 样品含的孔隙相对较少、分布范围相似；长沙 EAN 样品孔隙分布很窄、孔径小、有效热点范围内的孔隙较少。AN 孔隙结构的差异反映出 AN 自敏化特征的不同，这将使相应 AN 炸药的起爆性能存在差异。各样品相对应的起爆性能的研究正在进行中。

参考文献

[1] 吕春绪．工业炸药理论 [M]．北京：兵器工业出版社，2003.
[2] Horvath G.，Kawazoe K. Method for the Calculation of Effective Pore Size Distribution in Molecular Sieve Carbon [J]. J. Chem. Eng. (Japan). 1983，16 (5)：470-475.
[3] Frank A，Chau H，Lee R. Reaction Zones in Ultrafine TATB [J]. Propellants，Explosives，Pyrotechnics，2003，28 (5)：259-264.
[4] 吕春绪. 北京：兵器工业出版社. 膨化硝铵炸药 [M]. 2001.
[5] 陈天云. 硝酸铵的自敏化及应用研究 [D]. 南京：南京理工大学，2000.
[6] 叶志文，吕春绪，刘祖亮，等. 改性硝酸铵自敏化结构特征应用化学 [J]，2002，19 (2)：130-134.
[7] Ommen C，Jain S. Ammonium nitrate：a promising rocket propellant oxidizer [J]. Journal of Hazard Materials，1999，141 (1)：253-281.

The Self-sensitizing Characters of Ammonium Nitrate

Mei Zhenhua[1,2]，Zeng Guiyu[1,3]，Qian Hua[1]，Lv Chunxu[1]
(1. College of chemical engineering，Nanjing University of Science & Technology，Nanjing 210094;)
(2. Nanjing public security bureau，Nanjing 210005)
(3. Institute of chemical materials，CAEP，Sichuan Mianyang 621900)

Abstract: In order to study the self-sensitizing structure characters of ammonium nitrate (AN) used for industrial ammonia dynamite，the pore structure characters (such as porosity，pore diameter，specific surface area，pore distribution) and particles shape were investigated by mercury injection method and scanning electric microscopy (SEM) method. The results show that there are lots of pores and the pore quantity whose size is within the effective hotspots range is much high in Nanjing University of science and technology's expanded ammonium nitrate (EAN) sample，its self-sensitizing character is obvious. In Changsha's EAN sample，there are also many pores，but the pore size is much smaller and the pore distribution is narrow. The Shenzhen's EAN sample is similar to the common AN，i. e. the porosity and face area are much less and the pore quantity whose size is within the effective hotspots range is also less.

Keywords: pore structure; crystal branch change; self-sensitizing character; industrial explosive; ammonium nitrate

（注：此文原载于 含能材料，2011，19 (1)：33-36）

第三部分

炸药合成及应用

SJY 炸药及其应用

吕春绪，胡刚，吴腾芳

（1984 年 5 月 15 日收到）

摘要： SJY 炸药是一种由硝酸肼/水合肼/添加剂所组成的新型液体混合炸药。它的比重 1.401/30℃，氧平衡-5.8%，撞击感度 20%，爆速 8475 米·秒$^{-1}$/30℃，特别是具有较高的猛度，铅柱压缩值大于 32 毫米（半量）。它的综合性能优于奥斯屈莱特 G（Astrolite G）。SJY 炸药主体成分硝酸肼的制备工艺简单。在线型聚能切割及煤田勘探方面进行了应用实验，效果良好。

液体炸药一般具有良好的能量特性及流动特性，适合于某些特殊应用的需要。如液体炸药的爆热高，硝酸肼-肼系统的爆热为 1600 千卡·公斤$^{-1}$，而奥克托金的爆热为 1356 千卡·公斤$^{-1}$，体积能量高，硝酸肼-肼系统为 2.54 千卡·厘米$^{-3}$，而梯恩梯为 1.45 千卡·厘米$^{-3}$，奥克托金为 2.50 千卡·厘米$^{-3}$；它可流动，易于装填用于各神复杂弹型及环境[1]。

液体炸药在军事上有很大的价值，可装填航弹、爆破筒、各种雷等兵器，也可用于扫雷，清理战场及开辟通道等。第二次世界大战期间，液体炸药就在航弹上使用过，其性能优于梯恩梯。在民用上无论从技术上还是从经济效益上均有可取之处，它广泛应用于坚硬岩石爆破、聚能切割钢板及海下水管、海洋工程爆破、沉船打捞、拆船工程以及物田勘探[2]。

液体炸药国内外研究得比较多，有硝酸、氮的氧化物、硝基甲烷及双氧水液体炸药[3]，含高氯酸脲的液体炸药[4,5]，氨基酸（醛）类的液体炸药[6,7]，含有硝酸肼的液体炸药等[8]。这些液体炸药有其优点，但均在不同程度上存在这样或那样的缺点。我们开展了 SJY 炸药即含有硝酸肼的水合肼型高爆速液体炸药的研究工作，获得较好配方，并提出一套适合我国国情的工艺。

一、 硝酸肼及其制备工艺

硝酸肼是 SJY 炸药的主体成分，它本身也是一种高能猛炸药。国外多是在甲醇惰性介质存在下，将无水肼或水合肼用 65%稀硝酸低温盐化而成[9,10]。无水肼价格昂贵，甲醇毒性较大。

我们的制备工艺不用无水肼，而用工业水合肼的水溶液以及炸药厂的废硝酸，过程简单，综合利用，有较大的经济效益。制备出来的硝酸肼。其结构、性能与资料报道的数据及理论计算的数据基本符合。在 UM-3 型万能元素测定仪上测定了中型生产的硝酸肼的元素组成。硝酸肼的结构式 $N_2H_4 \cdot HNO_3$，分子式 $H_5N_3O_3$ 测定结果如下表：

不难看出，实测值与理论值及文献数据基一致。

在 SP2000 型红外分光光度计上用 KBr 压片法进行测定，得到的红外光谱图与美国 NASA 报告上硝酸肼的红外标准谱图[9]基本符合：如图 1、图 2 所示。

熔点、撞击感度及摩擦感度、猛度、瘸动实验、真空安定度及热差分析等与资料数据也基本符合。

表 1 硝酸肼元素组成测定数据表

组成成分%	N	H	O
理论值	44.21	5.30	50.49
文献值[11]	44.20		
实测值	44.30	5.28	50.42

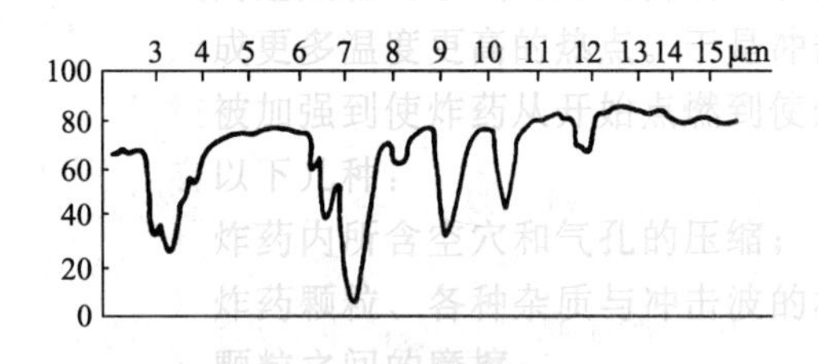

图 1 硝酸肼的测定红外图谱

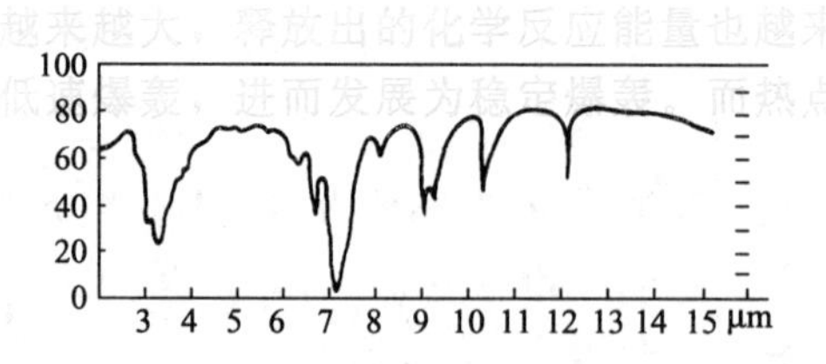

图 2 硝酸肼的标准红外图谱

二、 SJY 炸药的性能

SJY 炸药是硝酸肼、水合肼及添加剂的液体混合物。基于氧平衡、爆轰参数和某些物理性能参数（如溶解度、冰点及密度等）的计算及测定，进行配方设计。各种配方的爆轰参数是用斥力位能公式[11]进行计算的，优选之后再测定其冰点，密度等物理参数加以修正，最后获得最佳配方。

1. 理化性能

SJY-Ⅰ炸药是一个系列。SJY-Ⅰ型炸药为无色透明液体，比重 1.401/30℃，氧平衡 −5.08%，PH 值为中性。

2. 爆轰性能

SJY-Ⅰ型炸药用 E323 数字测时仪测定，测试时间精度 0.02μs，其爆速为 8475 米 · 秒$^{-1}$/25℃。撞击感度：标准落锤试验 20%。火焰感度：明火及导火索均不能使它点燃；1500W 电炉丝一段．侵入该液体炸药中。通电 2 分钟后才开始燃烧，但没有发生爆轰：枪击感度：100 克液体炸药装在 60 毫升塑料瓶中，25 米处用半自动步枪射击 6 发，不燃不爆。爆热：用恒温式量热计测定为 1323.7 千卡/公斤，猛度：用铅柱压缩法半量试验大于 32 毫米。铅柱压缩结果示于图 3。

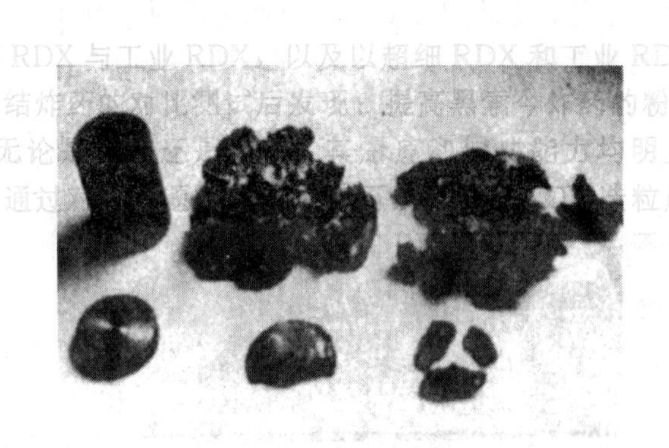

图 3 铅柱压缩试件照片

3. 坠落实验

在 76 榴弹空弹体中进行坠落实验：弹重 5.193 公斤，长 300 毫米，直径 76 毫米：药量为 5 克：加速度传感器装在头部引信处，于 3 米及 10 米高处落于铸铁板上。结果，炸药是安全的。原估计该炸药至少能承受 30000g 的过载，实测结果达 50000g 以上．实验给出了冲击加速度 a（t）波形。

4. 冲击波压力测定实验

为了研究该液体炸药距爆心不同距离上的超压，取 16 公斤 SJY-Ⅰ型炸药进行静炸实验。并与 TNT 相对比。采用 YD-15 应变仪，BPL-2 型传感器及 SC-16 示波器和 YD-6 应变仪，磁带机同时记录。不同点上的超压记录值，经 TRS-80 计算机用最小二乘法拟合进行数据处理，获得超压方程：

$$\Delta P=21.19-206.16\frac{W^{1/3}}{R}+732.928\left(\frac{W^{1/3}}{R}\right)^2-1034\left(\frac{W^{1/3}}{R}\right)^3+556.15\left(\frac{W^{1/3}}{R}\right)^4$$

式中：ΔP-超压（公斤）：W-液体炸药的装药重（公斤）：R-距爆心的距离（米）在距爆心 3 米处，该液体炸药超压为 22.86 公斤。为梯恩梯的 1.583 倍。离爆心越近，液体炸药超压与梯恩梯超压的比值越大。

SJY-Ⅰ炸药与其它几种主要炸药性能比较如下表所示：

表 2　SJY-Ⅰ炸药与奥斯屈莱特 G 综合性能比较

炸药	爆速/(米/秒)	冰点/℃	特点
SJY 炸药	8200～8475	－20～－40	威力大、感度小、安全性好、腐蚀小、毒性小、工艺简单、成本较低
奥斯屈莱特 G	8100～8600	－20～＋40	威力大、肼毒、安全性差、工艺复杂、成本较高

表 3　SJY-Ⅰ型炸药与几种弹药的性能比较

性　能	SJY-Ⅰ型炸药	梯恩梯	B 炸药	A3 炸药	奥克托儿
组分	纯	纯	TNT/黑索金 40/80	钝化黑索金（9%蜡）	HMX/TNT 75/25
密度（克/厘米$^{-3}$）	－1.401/30℃	1.59	1.68	1.59	1.81
爆速（米/秒）	847.5	6825	7900	8100	8480
猛度 铅柱压缩（毫米）	大于 32(半量)	12.5(全量)			
爆热（千卡/公斤）	1327.3	1000	1180	1100	1400
比容（升/公斤）	1088(计算)	685	820	900	850

三、 SJY-Ⅰ炸药的初步应用

SJY-Ⅰ炸药在军事上可以较好地开挖平底坑、坦克掩体以及单人掩体，装填反坦克地雷及穿孔弹。在民用上可以装填线型聚能初割器切割钢板，作地震震源用于物田勘探以及用于控制爆破等。现在，就聚能切割钢板及煤田勘探两方面的应用实验情况讨论如下：

1. 线型聚能切割器切割钢板

线型切割钢板在拆船工程及沉船打捞工程上有重要实用意义。切割器结构及尺寸如图 4 所示：

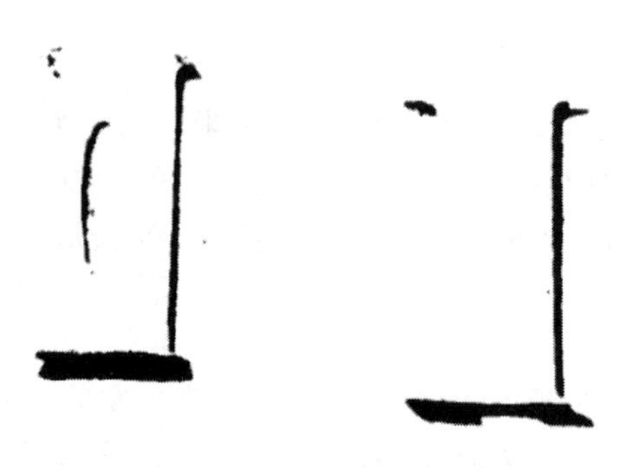

图 4 线型聚能切割器

图 5 钢板切割状况图

定量的 SJY-Ⅰ型炸药装在铜聚能罩（半圆形、φ30×100 毫米）的盛器内，炸高 30 毫米，可安全切割 25 毫米厚的钢板，结果切口深度大于 25 毫米，切口长度 115 毫米，切口上宽 25 毫米。

2. 煤田勘探

炸药在地下爆炸时，周围介质产生永久形变及弹性形变即地震波，地震波由爆炸中心以全方位向外传播，当遇到某些不均匀界面时，产生地震折射、反射波。在地面上用专门仪器装置拾取并记录地震折射、反射波信息，经过计算站计算机处理，确定地下煤田分布。

试验系使用 DZ-701 型模拟磁带地震仪，自然频率为 27Hz 检波器，仪器采用公控压制系统，门槛值为 300 毫伏，起始增益－20～－25 分贝，最终增益-7 分贝，延迟时间 0.18～0.23 秒，不同药量对比采用相同的仪器条件，回放滤波通带 40～120Hz，滤波 30%。使用单个检波的单边接收，检波距 5 米和 10 米，炮检距 60 米和 100 米。用 100 克、500 克、1 公斤及 2 公斤 SJY-Ⅰ型炸药作Ⅰ源，与硝铵炸药及钝化黑索金对比，试验结果如图 6 所示：

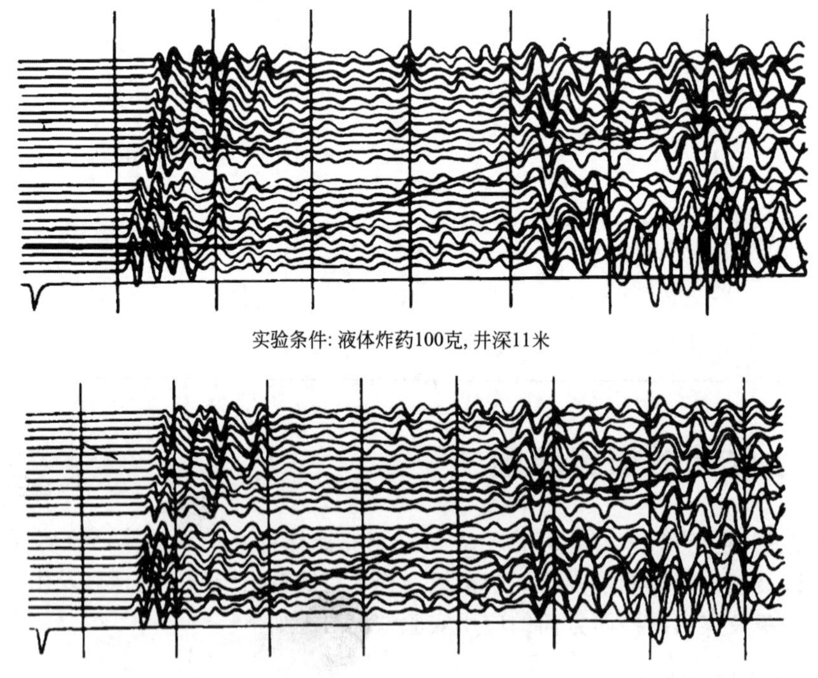

实验条件: 液体炸药100克, 井深11米

实验条件: 钝化黑索金100克, 井深11米

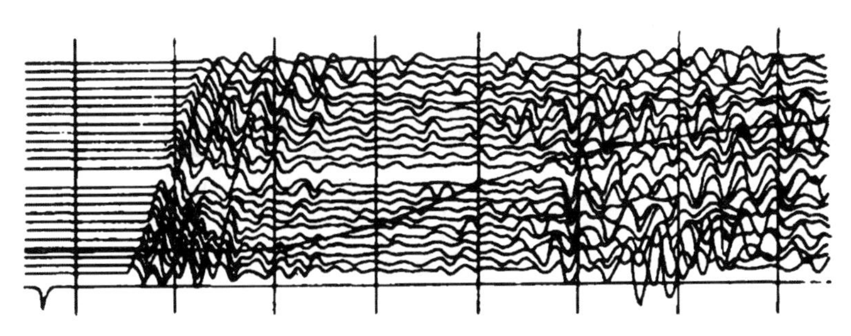

实验条件: 2号岩石硝铵炸药100克, 井深11米

图 6　三种炸药地震记录曲线

在江苏煤田地质物探测量队二队共作了 28 组对比试验，获得了 77 张记录。试验过程及结果表明：液体炸药的激发波较强，讯噪比高，装药方便，不受作业场地限制，适于野外流动作业使用；它不易点燃，无腐蚀，无污染，安全可靠。因此，该液体炸药是一种优点突出的很有使用前途的震源材料：

本文得到孙名振副教授和张熙和副教授的指导与帮助，特此致谢！

参考文献

[1] 包昌火．火炸药发展的现状，前景和建议．私人通信．(1982.7).
[2] 藤平修三．日本公开特许．74134811 (1974).
[3] 兵器工业部 204 所．火炸药手册（增订本）．第一分册 (1981).
[4] 藤平修三．昭 50-25720 (1975).
[5] 藤平修三．昭 49-134812 (1978).
[6] 藤平修三．昭 49-13313 (1974).
[7] 胜田武．昭 55-85498 (1980).
[8] Mical. E. M.， USP 3419443.
[9] James. H. K.， N 70-20440 (1970).
[10] 椎野和夫．工业火药 40 (2) (1979) . 74.
[11] 吴雄．混合炸药爆轰参数计算．私人通信．(1975).

SJY Explosive and its Applications

Lv Chunxu，Hu Gang，Wu Tengfang

Abstract: SJY explosive is a new liquid composite explosive, composed of hydrazine nitrate, hydrazine hydrate and additives. It' s specific gravity is 1.401/30℃, oxygen balance is-5.08%, sensitivity to impact — 20%, detonation velocity 8475m · s^{-1}/30℃. SJY has a high brisance value, it' s lead cylinder compression value is larger than 32mm (semi-quantity). The overall property of SJY is better than Astrolite G. The manufacture technique of hydrazine nitrate (the main component of SJY) is simple. Application test was carried out linear shaped charge cutting off and coalfield prospecting. Good results were obtained.

（注：此文原载于　爆炸与冲击，1985，(3)：54-59）

液体炸药线型切割器设计与应用研究

吕春绪

摘要：本文阐述了线型聚能切割的基本原理。在实验基础上，设计了具有最佳结构参数的液体炸药线型切割器，并确定了某些最佳使用条件，为液体炸药线型切割器的实际应用，提供了依据。

液体炸药线型切割器主要用来切割钢板或者其它坚硬材料。其基本原理是利用有罩聚能效应。

有罩聚能装药中的药型罩（铜）的作用是将炸药的爆轰能量转换成罩的动能，从而提高聚能作用，即装药爆轰后，药型罩内壁附近的金属在堆成平面上形成向着装药底部以高速运动的片状射流（称之为聚能刀）来切割钢板。

由起爆点传播出来的爆轰波到达药型罩时，压力达十万个大气压以上，在此高压作用下，金属罩被压垮，并不断地“挤”出一股连续的高速金属射流，而且迅速向轴线方向运动，射流速度一般可以达到3000米/秒，计算表明：该装填SJY炸药的有罩聚能装药的能量密度为无罩聚能装药的1.72倍，这说明了使用铜罩的优越性，它具有无罩聚能效应所无法比拟的切割能力。

药型罩的结构尺寸具有重要的意义，我们的切割器的结构设计主要是通过实验及数学处理获得（铜）罩的最佳参数的。

一、 液体炸药线型切割器最佳结构参数设计

液体炸药线型切割器是一种采用楔形的金属罩、平面对称型的聚能装药。其主要结构的参数如下图所示。

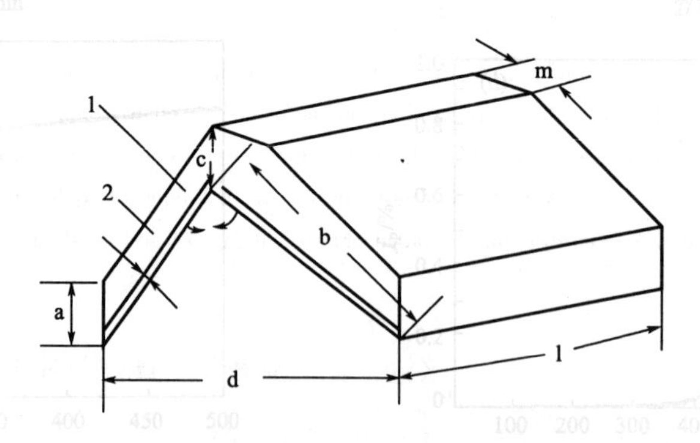

图1 液体炸药线型切割器结构示意图

1. 装药 2. 金属罩

a. 药厚 b. 罩宽 c. 药高 d. 罩口宽

l. 罩长 m. 装药上口宽

根据过去的实验结果，我们的液体炸药线型切割器选用了紫铜作罩材，以 A_3 钢为主要靶板。实验研究了罩厚、罩顶角、罩宽、药高、药厚等单因素对切割深度的影响，获得了大量切割数据，数据采用标准误差处理后，取得了各项因素在最大穿深时的最佳值；对于一些比较规则的曲线，通过回归方程得到最佳点。例如：我们在固定其它结构参数的前提下，实验研究了罩厚 δ 为 0.8、1.2、1.5、2.0 及 2.5 毫米等对穿深（A）的影响，优选后获得三条规整的曲线，经过回归处理，得到三条曲线的近似方程为：

$$A=-14.38\delta^2+47.5\delta-19.2,\quad \frac{dA}{d\delta}=-28.76\delta+47.5$$

$$A=-14.38\delta^2+38.75\delta-0.71,\quad \frac{dA}{d\delta}=-28.58\delta+38.75$$

$$A=-7.14\delta^2+19.29\delta+12.14,\quad \frac{dA}{d\delta}=-14.28\delta+19.29$$

解上述微分式，分别求得 δ 为 1.65、1.36 及 1.35 时的极值，也就是说对 SJY 液体炸药装药来说，在其它最佳条件固定前提下，罩厚与罩口宽之比（δ/d）的最佳值为 0.034～0.043。即此厚度的金属罩切割钢板具有穿深的最佳值。

通过正交实验，可以进一步验证单项因素最佳点在其它因素影响条件下的可靠性，同时还可以确定诸因素中，哪个因素是主要因素。如：我们进行了罩厚、罩宽及炸高（H）三因素三水平的正交实验，使用 SJY 液体炸药（比重：1.360/15℃，撞击感度 16%，爆速 8140 米/秒），A_3 钢为靶板，正交实验及数据处理结果如下：（紫铜罩，a=90°，a=6 毫米，l=100 毫米，靶板厚 30 毫米，每组 4 发）

表 1　正交实验实施表（单位：毫米）

组号	罩宽 (b)	罩口宽 (d)	炸高 (*H*)	罩厚 (δ)	药高 (c)	单位长药重克 (β)	穿深 *A*(平均值)	穿深 *A*/d	穿深 *A*/β
1	17	24.0	14	1.2	15.5	0.630	21.7±1.4	0.90	34.4
2	17	24.0	18	1.5	15.0	0.630	23.6±1.1	0.98	37.5
3	17	24.0	22	1.8	14.0	0.630	21.5±0	0.90	34.1
4	21	29.7	18	1.8	14.5	0.705	22.0±0.6	0.74	31.2
5	21	29.7	22	1.2	14.0	0.705	22.8±1.5	0.77	32.3
6	21	29.7	14	1.5	15.0	0.705	24.8±0.5	0.84	35.2
7	25	35.4	22	1.5	15.0	0.846	30.0±0	0.85	35.5
8	25	35.4	14	1.8	15.5	0.846	20.5±1.8	0.58	24.5
9	25	35.4	18	1.2	16.0	0.846	28.8±1.0	0.81	34.3

表 2　正交实验数据处理表（单位：毫米）

	A			*A*/d			*A*/β		
	δ	b	*H*	δ	b	*H*	δ	b	*H*
Ⅰ	73.3	66.8	67.0	2.48	2.78	2.32	101.0	106.0	94.1
Ⅱ	78.4	69.6	74.4	2.67	2.35	2.53	108.2	98.7	103.0
Ⅲ	64.0	79.3	74.3	2.22	2.24	2.52	89.8	94.3	101.9
k_1	24.4	22.3	22.3	0.83	0.93	0.77	33.7	35.3	31.4
k_2	26.1	23.2	24.8	0.89	0.78	0.84	36.0	32.9	43.3
k_3	21.3	26.4	24.8	0.74	0.75	0.84	29.9	31.4	34.0
R							6.1	3.9	2.9

表中 k 为正交实验各因素的效果值，对于三个指标 *A*、*A*/d 及 *A*/β 均可作图，但 *A*/β 值更为合理。

从表 1、表 2 明显看出，罩厚的极差最大达 6.1，炸高的极差最小为 2.9，即这组正交实验中罩厚的影响是主要因素，相对比较而言，炸高的影响是次要因素。根据极差确定出相互影响的三个因素

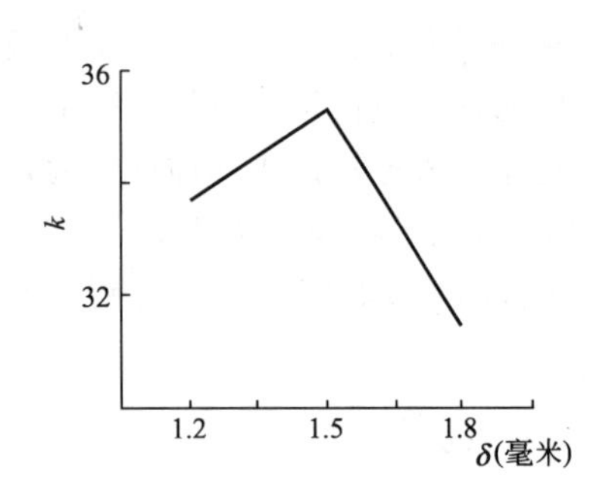

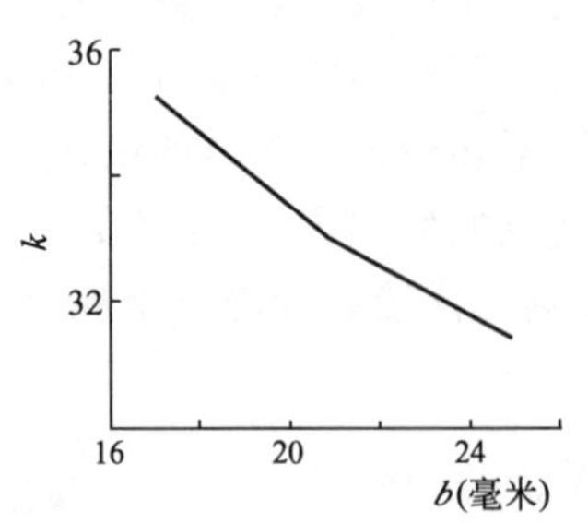

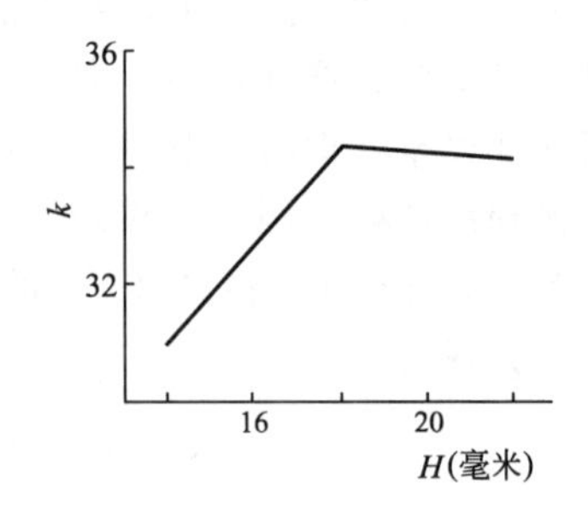

图 2 按 A/δ 的效果和因素关系图

各自的最佳值。

同时也作了罩顶角（a）、罩宽（b）及顶高（c）的相互影响的正交实验，根据数据处理的结果，确定出该三因素中罩顶角是对穿深影响的主要因素，而顶高是次要因素。

对于液体炸药线型切割器影响穿深的结构因素有：罩厚（δ），罩顶角（a），罩宽（b），炸高（H），药高（c）及药厚（a）等因素，这六个因素称为独立物理量，只有一个长度量纲。由Ⅱ定理可以组成的无量纲参数，为 a、δ/b、c/b、a/b、H/b。

我们对 a＝90°、H/b＝1.0、c/b＝1.0 的设计在 45＃钢板上进行了几何放大实验，求出穿深（A）与罩宽（b）的关系，然后对其数据利用最小二乘法处理，得到直线方程：

$$A=m_1 b+n_1, m_1 \text{ 与 } n_1 \text{ 为常数}$$

作图表明 A-b 线性关系较好。

我们也获得了所设计的液体炸药切割器的比药量（A）与穿深（β）之间的关系式：

$$\beta=m_2 A^2+n_2 A+Q \quad (16 \leqslant A \leqslant 60)$$

上式中：m_2、n_2 及 Q 为常数；β（比药量）；单位长度需药量（公斤/米）；A（穿深）；钢板被切深度（毫米）。

该关系式在工程上有较大价值，它可以直接估算出欲切割任何厚度的钢板所需要的单位长度上的药量。

基于上述实验研究结果，我们设计的液体炸药线型切割结果示意图 3、图 4。

图 3 液体炸药线型切割器切割钢板实验图

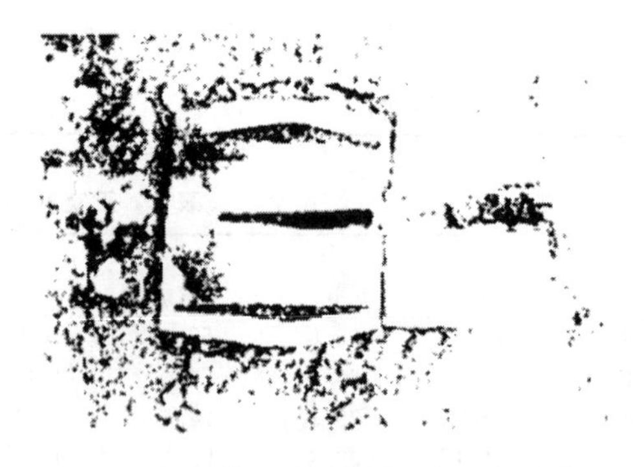

图 4 液体炸药线型切割器切割钢板实验图

二、 液体炸药线型切割器应用条件研究

具有最佳设计参数的液体炸药线型切割器的效能能发挥得是否完全与使用条件密切相关。因此，只有在最佳使用条件下，才能获得液体炸药线型切割器的最佳效果。它的使用条件包括炸高、靶材及其安放方式、起爆点位置以及传爆药柱的选择等等。

1. 最佳炸高的确定

在其它固定条件不变的情况下，使用 SJY 炸药，25 毫米 A3 钢板研究了炸高对穿深的影响，得到的部分实验数据如下表所示。

表 3　炸高对穿深的影响（5 毫米≤炸高≤45 毫米）

序号	炸高（毫米）	穿深(毫米) A	A/d	A/β	备注	序号	炸高（毫米）	穿深(毫米) A	A/d	A/β	备注
1	5	16.5	0.47	16.7		5	25	25.0	0.71	25.3	穿透
2	10	17.0	0.49	17.2		6	30	18.0	0.51	18.2	穿透
3	15	25.0	0.71	25.3		7	35	20.0	0.57	20.3	
4	20	25.0	0.71	25.3	穿透	8	45	18.0	0.51	18.2	

使用 30 毫米 A_3 钢板，在炸高为 13 到 25 毫米范围内作出的实验结果示于表 4。

表 4　炸高对穿深的影响（13 毫米≤炸高≤25 毫米）

序号	炸高(毫米)	穿深(毫米) A	A/β	备注
1	13	21.4±1.4	25.5	
2	15	26.6±2.1	30.9	钢板背后鼓包
3	18	27.5±0.2	31.7	钢板背后鼓包,裂纹
4	20	24.8±1.8	26.7	
5	22	22.8±1.6	26.4	
6	25	23.4±1.8	26.8	

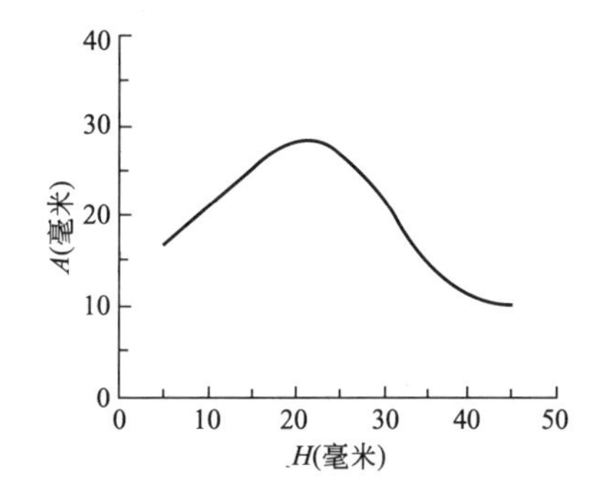

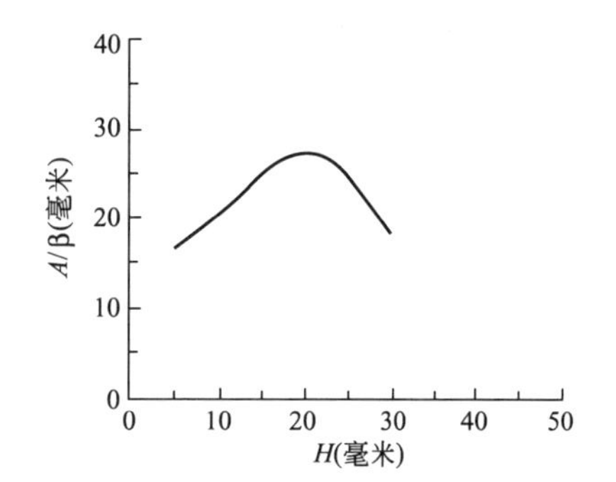

图 5　炸高与穿深的关系图

从上述实验数据可以看出，炸高在 15～18 毫米范围内穿深效果较好，能形成比较稳定的射流。炸高大于 20 毫米，虽然炸高增加，射流伸长，对穿深有一定好处，但是，伸长的射流也会产生径向分散和摆动，能量损失严重，从而造成射流断裂或飞散而影响穿深；炸高小于 15 毫米，有时因金属流与杵体没有完全脱离，故此切缝中均有杵堵现象，还有时炸高太低，金属罩虽被压垮，但还没来得及挤出金属射流就接触到钢板，对钢板仅产生崩落现象，而不能有效穿深钢板。下图是炸高不合理时切割状况。炸高不合理，射流飞散，根本没有穿深钢板。

图 6　不黑炸高穿深结果图（射流飞散）

炸高参数：1#：5 毫米；2#：0 毫米；3#：30 毫米；4#：35 毫米

考虑到其它因素的综合作用，又把炸高与其它因素如罩厚及罩宽等进行正交实验，有效炸高仍然在 15～18 毫米时具有最佳结果。

2. 靶材及其安放方式的实验研究

实际使用中所遇到的被切钢板（靶材）的型号可能不一样，被切钢板所处的位置也不尽相同，为了满足切割各种钢板的需要，研究它们之间的差别有实际意义。不同钢板及不同安放条件的切割实验结果如下表所示。

表 5 不同钢板（靶材）对穿深的影响

序号	钢板型号	破坏强度极限 σ_b(千克/毫米)	屈服强度极限 $\sigma_{0.2}$(千克/毫米)	厚 度 (毫米)	单位长药重 β(克/毫米)	穿 深 A(毫米)
1	A_3 钢	40.3	23.8	25	0.705	24.0±1.0
2	45# 钢	63.9	36.6	25	0.705	22.0±0.0
3	603 装甲钢	115～125	90～110	25	0.705	18.1±1.0

表 6 不同钢板安放方式对穿深的影响

序号	钢板安放方式	钢板厚度(毫米)	单位长药重 β(克/毫米)	穿深 A(毫米)
1	架 空(离地 20 毫米)	30	0.888	27.0±0
2	地 上(可压缩性土壤)	30	0.888	25.1±0.4
3	垫钢板(厚度 25 毫米)	30	0.888	21.5±2.0

在 603 装甲钢上切割结果如下图（四条长形切口为切割器所切）

图 7 在合金缸（603 装甲钢）上切割结果图（外长条切口与本文无关）

被切钢板的密度及强度对切深影响较大。按照定常不可压缩流体理论，切深（A）有如下关系式：

$$A=l\sqrt{\frac{\rho i}{\rho t}}$$

即切深与射流有效长度（l）成正比，与药型罩材料密度（ρ_i）的平方根成正比，而与被切钢板的密度（ρ_t）的平方根成反比。但是，实际上不考虑钢板的强度是不行的，因为被切钢板的密度相差无几，但强度却相差很大，特别在射流微元速度小于 5000 米/秒时，强度影响更加显著，强度愈高，与流体相差愈大，此时穿深速度下降就愈大。从表 4 数据明显看出，随被切钢板强度增加，穿深下降。这一点告诉我们使用切割器时，必须根据被切材料性质，按几何相似率放大或缩小切割器的几何尺寸，以适应实际需要。

钢板安放方式的实质也是被切钢板的强度问题。垫上一块钢板相当于给被切钢板增加了抗力，特别在射流作用的后期，抗力影响更加突出，抗力小的射流易穿透钢板，而抗力大的射流不易再穿深。这一点启示我们对待实际工程中有扶强结构或复合钢板时，切割器设计要比正常情况（同样厚度）要有较大的强化措施。

3. 起爆点位置及传爆药的正确选择

实验发现，起爆点位置及传爆药柱种类等对穿深有影响，但影响不大，超过极限起爆药量以后，装药在临界直径以上，基本上均能稳定爆轰。起爆点位置及传爆药柱状况实验数据如下（使用 SJY 液体炸药，靶材为一定厚度的 A_3 钢紫铜罩，标准切割器）。

在实际工程上均是使用长药条，其起爆点及传爆药柱状况的影响更小，只要保证完全起爆，稳定爆轰，切割便可很顺利地进行。大量长药条的稳定实验也完全证明了这一点。长药条切割结果如下图所示：

表 7 起爆点位置对穿深的影响（钝化黑索金传爆药柱，5克，靶板25毫米）

序号	起爆点	比药量β克/毫米	穿深(毫米) A	A/d	A/β	备注
1	药柱直接接触铜罩	0.785	25.0±0.0	1.0	31.9	全　穿
2	药柱离开铜罩10毫米	0.771	24.7±0.4	0.99	32.0	全穿或靶板背面鼓包
3	药柱刚接触液体炸药药面	0.761	23.2±0.8	0.93	30.5	

表 8 传爆药柱状况对穿深的影响（药柱离罩10毫米，靶板24毫米）

序号	传爆药柱状况	比药量β克/毫米	穿深(毫米) A	A/d	A/β	备注
1	5克黑素金小药柱平躺起爆	0.740	23.4±1.8	0.98	31.6	靶板背面裂开、鼓包
2	5克黑素金小药柱垂直起爆	0.750	21.8±1.4	0.91	29.1	
3	5克黑素金小药柱斜放起爆	0.754	20.3±1.0	0.85	26.9	
4	5克黑素金大药柱斜放起爆	0.754	21.2±0.9	0.88	28.1	
5	5克太安小药柱平躺起爆	0.744	23.7±0.2	0.99	31.9	
6	5克 C_4 塑性炸药平躺起爆	0.757	23.6±0.9	0.98	31.2	靶板背面裂开、近穿
7	5克 C_4 塑性炸药垂直起爆	0.757	22.5±0.7	0.94	29.7	靶板背面裂开、鼓包

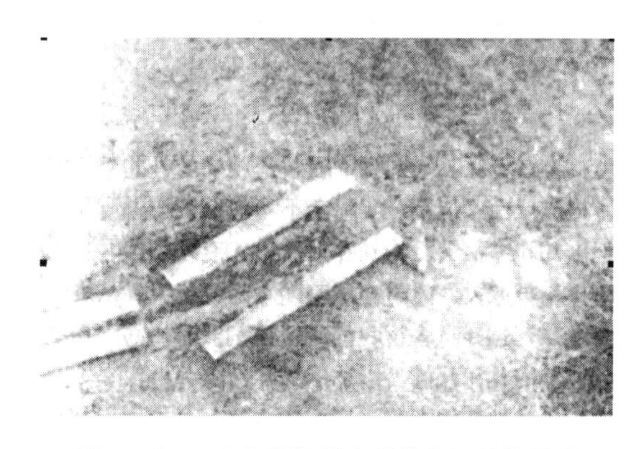

图 8 在14毫米薄钢板上长药条切割结果图

三、 几点结论

(1) 液体炸药线型切割器是一种强有力的切割手段，在某些方面，它与电割、气割相比有其长处。适用于拆船及特种海洋工程等。

(2) 液体炸药线型切割器的装药是液体炸药，它与固体炸药相比，在某些方面具有特色，即密度均匀，爆轰性能稳定；具有流动性可充填任意形状的容器；装药方便，可在安装工人撤离现场后再灌药，减少带药操作时间，增加安全性。

(3) 液体炸药线型切割器结构尺寸服从几何相似率。根据被切钢板的性质选择最优尺寸设计，再在最佳使用条件下（如有效炸高等）应用，定会收到最优设计效果。

(4) 液体炸药线型切割器应用条件研究表明：影响穿深重要因素中被切钢板的强度是不可忽视的因素，强度越高，穿深越低。起爆点及传爆药柱的影响较小，只要起爆点沉在液体炸药中，传爆药柱平躺起爆，定会收到良好切割效果。

本文得到孙名振副教授的指导与帮助，特此致谢。

参考文献

[1] 汤明钧．爆破器材．1928，(1)：12.
[2]《爆炸及其作用》编写组．爆炸及其作用（下册）．北京：国防工业出版社，1979，141.
[3] 袁伯珍（译）．聚能现象的理论与实验研究．北京：国防工业出版社，1957.

[4]《炸药理论》编写组. 炸药理论. 北京：国防工业出版社，1982，278.

Research of the Design and Application of Liquid Explosive Linear Shaper Charge Cutter

Lv Chunxu

Abstract: In this paper, the fundamental principle of linear shaper charge cutter is described. Based on a lot of experimental results, liquid explosive linear shaped charge cutter with the best structure paramenters is designed. Some of the best conditions a-vailable is determined as the basis of practical applications.

（注：此文原载于 爆炸与冲击，1987，7（1）：57）

The Development and Present Situation of Heat-Resistant Explosives

Lv Chunxu, Wu Jianzhou

Abstract: In this paper, the development and present situation of heat-resistant explosive compounds are reviewed, and the methods and techniques of synthesis for 9 typical heat-resistant explosives are discussed. By making an analysis of molecular structure of heat-resistant explosives, the extension of a conjugate molecular system, the introduction of amino group in a molecule and salt formation are considered to be the effective methods in increasing heat resistance of an explosive. At last, recent studies of some heat-resistant explosives, synthesized by the author, are presented.

1 Brief introduction

Heat-resistant explosives cover the synthesized heat-resistant explosives and the compound heat-resistant explosives. In this paper, the synthesized heat-resistant explosives are exclusively presented.

Generally, the synthesized explosives are composed in the combination of different rate of carbon, hydrogen, oxygen and nitrogen or in different modes of nitro combinations ($C\text{-}NO_2$, $N\text{-}NO_2$, $O\text{-}NO_2$) or in different combinations of functional groups, and they have the different.

Among the synthesized explosives, the one with good thermo stability is named the heat-resistant explosive. Namely, the heat-resistant explosive features the high melting point and low vapor pressure, which ensures it to explode after long duration of heating and cooling with a proper shocking momentum and high energy emission. The heat-resistant explosive is invented out due to the development of the astronautic aviation technology and the acceleration of the missile velocity. The high temperature and low pressure required in the astronautic aviation bring out the emergence of the heat-resistant explosive. The high-temperature underground explosion and drill require a stuff of the ammunition technique products such as the heat-resistant detonator to carry the heat-resistant explosive and the explosion fuse, under this condition, the common explosive can not meet the requirements. Hence, all countries start to attach importance to the research in the manufacturing technique of new heat-resistant explosives and the application of the heat-resistant explosives. Such a research has been listed in the key study objects of all countries.

2 The structural characteristics of heat-resistant explosive

Recently, most heat-resistant explosives are characterized with the trinitrotoluene (TNT) in their molecular structures. Among the molecular structures there exist several combinations: the trinitrotoluene is connected with the amino group; the trinitrotoluene is connected with each other (poly nitro poly benzene compound); the trinitrotoluene is connected with the nitrogen-link compounds or the florin functional group. These biphenyl compounds or poly phenyl compounds or poly nitro di-benzene ethylene can form the enormous conjugation group. Through the analysis of the molecular structure and the quantum chemistry calculation, it can be seen that these compounds not only have the high conjugation energy but also have a high melting point. The aromatic amino compound is in the family of compound explosives and with good thermo stability, and now it has developed to be the key material of heat-resistant explosives.

From the perspective of functional group of the compounds, the introduction of the amino into the explosive molecular structure can form the hydrogen bound between the hydrogen in the amino and the oxygen in the nitro. The hydrogen bound in the molecular structure can strengthen the density (ρ_0) of the explosive. The hydrogen bound between the molecules can shorten the distance between the molecules and accordingly increase the lattice energy of the molecule. For instance, the stack coefficient of TATB is 0.763 and it is the maximum stack coefficient of the organic chemical compounds discovered in recent time. The melting point of TATB is 360℃. The introduction of the amino into the explosive molecular structure is an efficient method to increase the heat-resistant performance of the explosive.

The salt deposition, especially the potassium, can increase the melting point and the thermal stability without lowering the energy of the explosives.

3 The present situation of heat - resistant explosive

DATB is the initial heat-resistant explosive, which was invented by an American Picatinny Arsenal in the late of 1950's. TATB is an insensitive explosive with high density and high energy emission, which was synthesized by Jackson in 1887. In 1961, L. A. Kaplan acquired the using patent of TATB. In the 1960's, China started to study the synthesis and manufacturing technique of TATB. 3,3'-diamino 2,2', 4,4', 6,6'-six nitro phenyl is one kind of heat-resistant explosive invented by US Navy and its characteristics and manufacturing technique were not completely open to public until 1968. Followed by the emergence of 3,3'-diamino-2,2', 4,4', 6,6'-six nitro di-phenyl, a large number of poly nitro poly phenyl compounds such as ONT, NONA, Dodeca are developed[6-8].

HNS is a heat-resistant explosive applied in recent time, developed by K. G. Shipp in Naval Weapons Institute of United States in 1964. Until 1966, Shipp borrowed the TNT oxidation method by using sub-potassium natrium to make the HNS-Ⅰ, which later was known as the famous Shipp method. Through adding HNS-Ⅰ into the mixed solvent of dii toluene and acetonitrile to reflux, HNS-Ⅱ is made. Because of HNS-Ⅱ with the melting point 316-317℃ and the stack density 0.5～1.0g/cm^3[11], and further for its good compatibility with poly tetrachlorin ethylene at the temperature between −195℃ and 325℃, HNS-Ⅱ is widely used in the heat-resistant explosive components and other ammunition components for the Appollo spacecraft experiment and the moon surface experiment. Since 1970, China has been engaged in the study of new manufacturing technique of HNS-Ⅰ and HNS-Ⅱ and made great achievements in the production yield and the crystal form of HNS-Ⅰ and HNS-Ⅱ. Later the Chinese HNS samples were also sold to other countries.

Through the conjugation reaction and the condensation reaction of Ullmann, it can make

TPT[12], TPM[13], two (TNP)-thiophene, 2, 5-bi (tri nitrobenzene) thi azole[15], 2, 5-bi (tri nitrobenzene) toluene[16]. China has also invented the synthesis of TMP, and has acquired the cheap manufacturing technique with a certain degree of production technical capacity.

Recently, Los Alamos National Institute in the United States published the explosive synthesis situation in the latest 20 years, among which the PXY (2,6-bi (c-nitro aniline)-3,5-di nitro pyridine) has to be started the production in the United States and as far as to 1972, its production yield amounted to 3000 pounds, termed as the new heat-resistant explosive in place of HNS. Other important compounds include PATO (3-tri nitro aniline 1, 2, 4-triazole), also called the alternative TATB explosive. And other heat-resistant explosives for special purposes have also been introduced.

4 Research on heat-resistant explosives

4.1 Research on the study of the new manufacturing technique of HNS synthesis

HNS is synthesized by the method of Shipp. That is, the TNT is added to the solvent of THF and ethanol and then through the reaction with 5% sodium hypochlorite, and the crude HNS products are formed. The crude products can be refluxed for several times with acetone to get the HNS-Ⅰ[18] and refined products with the melting point of 315℃.

China has proposed two manufacturing techniques of HNS synthesis by a lot of experiments. These manufacturing techniques have been authenticated by experts to have the advantages of scientific design, cheap solvent, convenient manufacture, high production yield (69%), etc.

4.2 The synthesis of poly-trinitroamino aromatic compounds synthesis

In order to synthesize poly-tri nitro amino aromatic compounds, firstly the tri nitro chlorine benzene[20], tirnitro dichlorine benzene[21] and the tri nitro-m-chlorine amino benzene[22] should be synthesized out. By using the strong nitration agent, the high yield products of tri nitro chlorine benzene and tir nitro di chlorine benzene can be produced. By the reaction of the ethylene amino and the tri nitro chlorine benzene, the sub-tri nitro benzene-m-chlorine amino benzene[23] can be produced with the melting point of 231℃, which is higher by 44℃[24] than the melting point 275℃ of the previously described sub-(trinitro amino)-vinylamino (Ⅰ) produced by the reaction of ethylene amino and the tri nitro-m-chlorine benzene.

(Ⅰ)

When using the sub (TNP)-ethylene amino manufacturing the sub (TNP)-amino-potassium ethylene, the resultant's melting point is 300℃, which has increased by 69℃.

The author uses the sub tri nitro-m-di chlorine benzene and m-chlorine aniline to synthesize out the 3,3′-diamino-2,2′4,4′, 6,6′-six nitro diphenylamine K (Ⅱ). The reaction is as follows:

a. The synthesis of 3,3′-dichlorine-2, 4-di nitro di aniline (condensation)

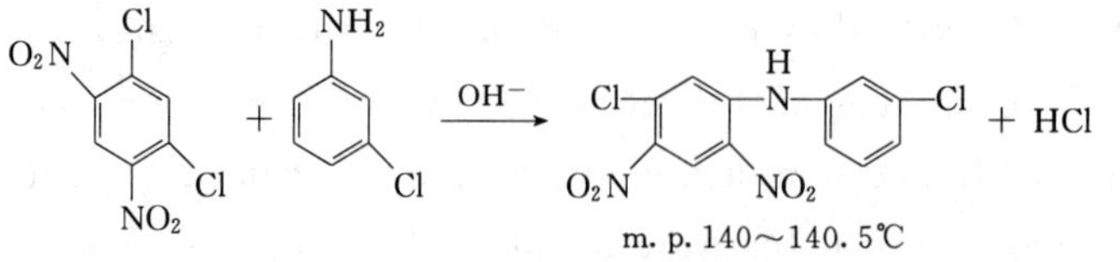

$$2,4\text{-dinitro-1,5-dichlorobenzene} + m\text{-}ClC_6H_4NH_2 \xrightarrow{OH^-} \text{product} + HCl$$

m. p. 140～140.5℃

b. The synthesis of 3,3′-diamino-2,2′, 4,4′, 6,6′-six nitro dip aniline (nitration)

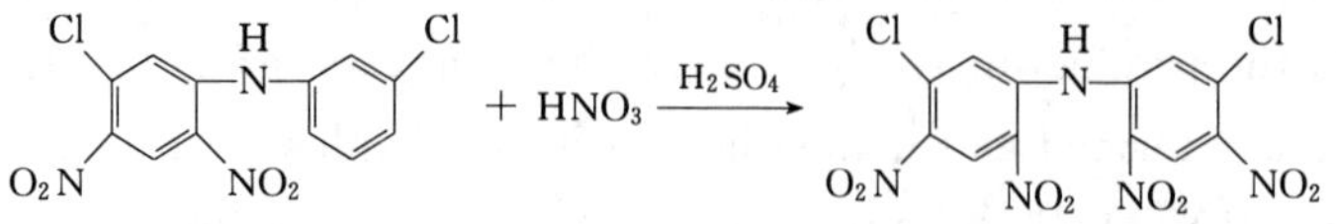

$$\text{reactant} + HNO_3 \xrightarrow{H_2SO_4} \text{product}$$

m. p. 92～94℃

c. The synthesis of 3,3′-diamino-2,2′, 4,4′, 6,6′-six nitro-dianiline (ammonization)

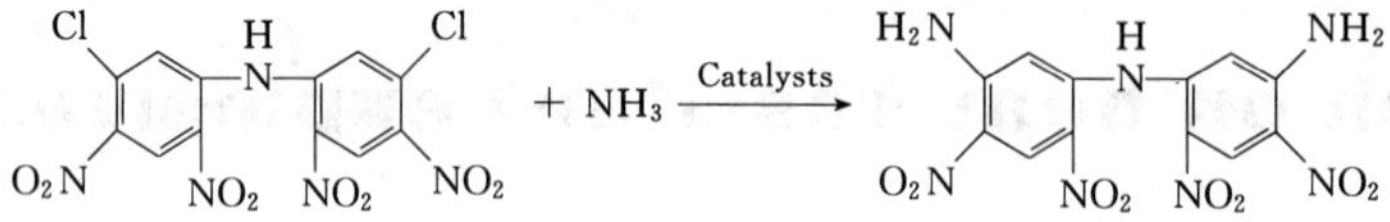

$$\text{reactant} + NH_3 \xrightarrow{\text{Catalysts}} \text{product}$$

m. p. 232～237℃

d. 3, 3″-diamino-2,2″, 4,4′, 6,6′-six nitro diphenylamine (salt deposition)

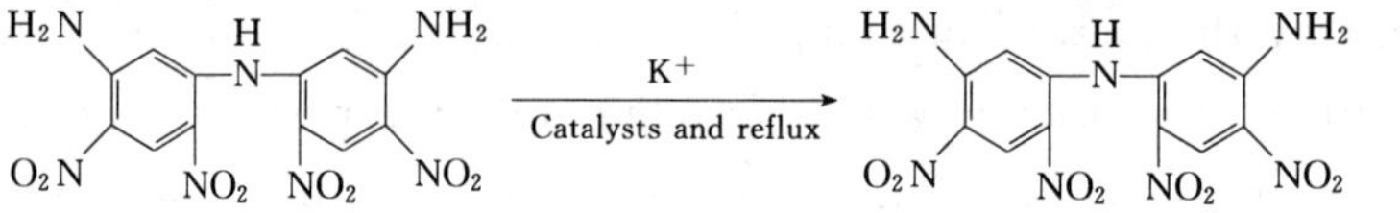

$$\text{reactant} \xrightarrow[\text{Catalysts and reflux}]{K^+} \text{product} \quad (\text{Ⅱ})$$

m. p. 333～334℃

The final resultant (Ⅱ) is a yellowish-brown crystal with a melting point of 334℃. Its stack density is 0.526g/cm^3; its impact sensitivity is 100% explosion and its spark ignition sensitivity is 100% with a high burning rate. The resultant takes effect when it is used in combination with the oxidant, which is an ideal ignition powder[26].

Through the combination of tri nitro-di chlorine benzene and m-chlorine aniline, the 1,3′-sub (3′-amino-2′,4′,6′-trinitro amine)-2,4,6-tri nitro benzene (Ⅲ)[27] can be synthesized. The reaction is as follows:

a. The synthesis of 1,3-sub (3′-chlorine aniline)-2,4,6-tri nitrobenzene (condensation)

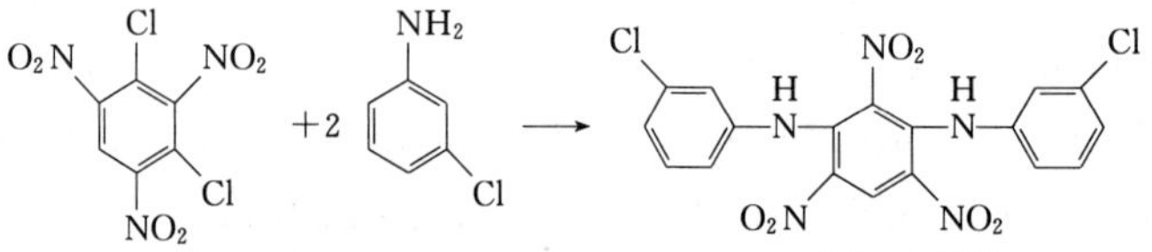

$$\text{reactant} + 2\ m\text{-}ClC_6H_4NH_2 \longrightarrow \text{product}$$

m. p. 179～180℃

b. The synthesis of 1,3-sub (3′-chlorine-2′,6′-dinitroaniline)-2,4,6-tri nitrobenzene (one time nitration)

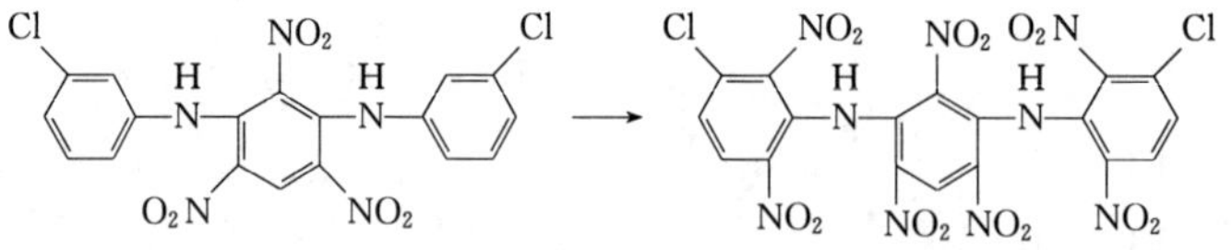

$$\text{reactant} \longrightarrow \text{product}$$

c. The synthesis of 1,3′-sub-(3′-chlorine-2′,4′,6′-trinitrophenyl amine)-2,4,6-tri nitrobenzene (super nitration)

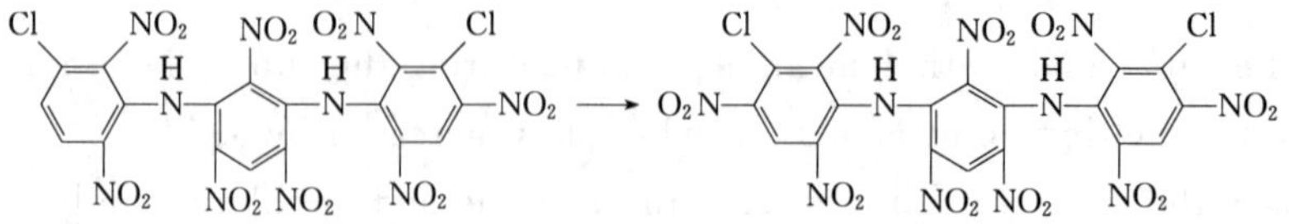

$$\text{reactant} \longrightarrow \text{product}$$

d. The synthesis of 1, 3-sub (3′-amino-2′, 5′, 6′-trinitrophenyl amine)-2, 4, 6-trinitrobenzene (ammonization)

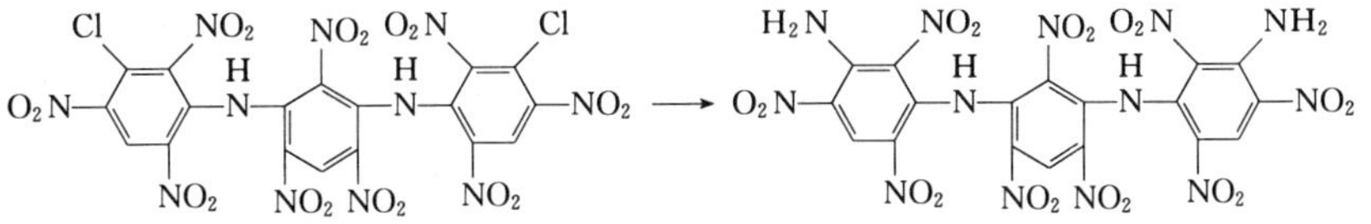

(Ⅲ)

5 The development of heat - resistant explosive

From the viewpoint of the functional group, the developing trend of the heat-resistant explosive goes to the poly amino, nitro and poly ethylene direction; from the viewpoint of molecular structure, the developing trend of the heat-resistant explosive goes from a single link to double links, multi links and large link direction.

HNS is a heat resistant explosive with good heat resistant performance. Based on the fact that the method by introducing the nitro into the explosive molecular structure can improve the heat resistance of explosives, we have designed the 2,2′, 4,4′, 6,6′-six nitro diphenylamine di amino ethylene (Ⅳ).

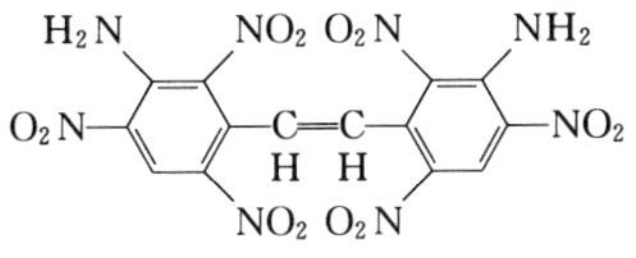

(Ⅳ)

The melting point of this compound is expected to be high. 5, 5′-diamino-2,2′, 4,4′-four nitro-dibenzene ethylene is already synthesized out and its melting point is 345-346℃, and its production yield is 92. 6%[29].

2,6-sub (tri-nitro aniline)-3,5-di nitro pyridine (PYX) is a new type of heat resistant explosive in place of HNS, and its melting point is 360℃. Its DTA analysis shows that its stability performs best within the temperature below350℃. The vacuum stability test data is 0. 1ml/g/48 hours. We designed a 2,6-sub (trinitro diamino aniline)-3,5-dinitro pyridine (Ⅴ)[22].

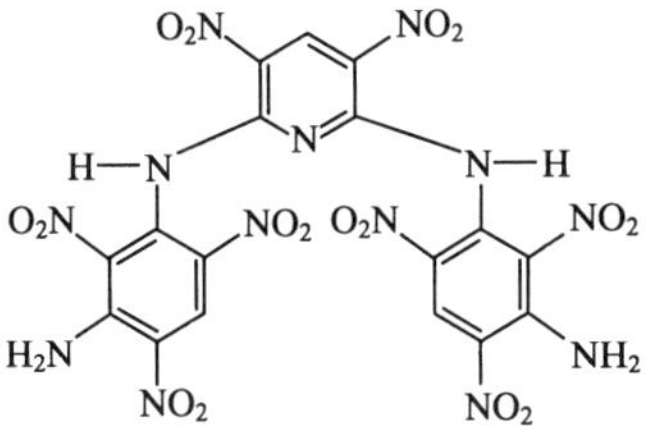

(Ⅴ)

The structural analysis gives the prediction that the compound's melting point is 400℃.

2,5,8-triphenyl-two of poly nitro azobenzene compound (Ⅵ) is an important heat-resistant explosive[30].

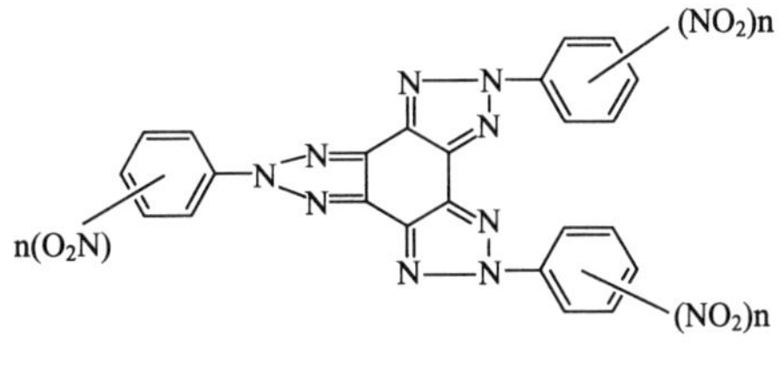

(Ⅵ)

Industrial Powder

This compound is synthesized through the oxidation by $CuSO_4$ after the coupling reaction between the 1,3,5-tri nitro benzene and the compound obtained from the re-diazotized nitro aniline. The melting point of the resultant from ρ-nitro aniline is 370℃; The melting point of the resultant from dinitro aniline is 400℃.

Macro-link poly-nitro compound is also an important heat-resistant explosive and it is a macro-link ten nitro explosive with high thermo stability.

(Ⅷ) is characterized with very good heat resistance[31].

This compound is synthesized by using m-dichlorine benzene and resorcinol as the raw materials and its melting point is above 350℃.

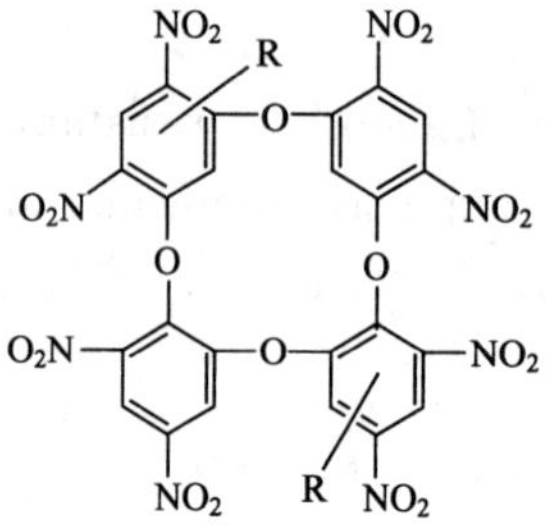

(Ⅷ)

Reference

[1] M. Warman. J. Org. Chem., 1961 (28): 2997.

[2] C. L. Jackson. J. A. C. S., 1887 (9): 354.

[3] L. A. Kaplan. U. S. P. 3002998, 1961.

[4] E. E. Kilmer. J. Spacecrait and Rockets, 1968 (10): 1216.

[5] R. E. Oesterling, U. S. P. 3404184, 1968.

[6] J. C. Dacons. U. S. P. 3592860, 1971.

[7] J. C. Dacons. U. S. P. 3755471, 1973.

[8] J. C. Dacons. U. S. P. 3450778, 1969.

[9] K. G. Shipp. J. Org. Chem., 1964 (29): 2620.

[10] K. G. Shipp. J. Org. Chem., 1966 (31): 857.

[11] J. J. Syrop, U. S. P. 3699176, 1970.

[12] J. C. Dacons. U. S. P. 3755321, 1973.

[13] M. D. Coburn. U. S. P. 3414570, 1968.

[14] M. E. Sitzmann. U. S. P. 3923830, 1976.

[15] J. C. Dacons. U. S. P. 4061658, 1977.

[16] M. E. Sitzmann. U. S. P. 3922279, 1975.

[17] M. D. Coburn. Ind. Eng. Chem. Prod. Res. Dev., 25, 68, 1986.

[18] Foreign Scientific Technology Material (Chemistry), 43, 1975.

[19] Explosive Materials, 1988 (2): 16.

[20] Lü Chunxu, Chen Tianyun. Acta Armamentarii (Powder and Explosive), 1988 (1).

[21] Lü Chunxu, Deng Aimin, Zheng Mingen. Academic Journal of East China Institute of Technology, 1989 (1): 18.

[22] Wang Sifang. Graduate Thesis of East China Institute of Technology, 1987.

[23] Lü Chunxu. Journal of Explosives and Propellants, 1989 (2): 8.

[24] Lü Chunxu, Wang Enfang, Li Guangxue. Explosive Materials, 1989 (4): 13-15.

[25] Lü Chunxu, Li Guangxue. Explosive Materials, 1988 (3): 12.

[26] Lü Zaosheng. Graduate Thesis of East China Institute of Technology, 1987.

[27] Deng Aimin. Graduate Thesis of East China Institute of Technology, 1987.
[28] Lü Chunxu, The Theory and Practice on Heat-resistant Explosive synthesis, East China Institute of Technology, 1989.
[29] Peng Xinhua. Graduate Thesis of East China Institute of Technology, 1987.
[30] T. Urbanski. Chemistry and Technology of Explosives, 1984 (4).
[31] E. E. Gilbert. U. S. P. 39941812, 1976.

（注：此文原载于 Journal of the Industrial Explosives Society Japan 1990, 51 (5): 275-278)

Research on Condensation Reaction of 3,3′-Diamino-2,2′,4,4′,6,6′-Hexnitro Diphenylamine Potassium in Biosynthesis Process

Lv Chunxu, Lv Zao Sheng, Deng Ai Min and Wu Jianzhou

Abstract: Potassium 3,3′-diamino-2,2′, 4,4′, 6,6′-hexanitro diphenylamide is a new kind of heat resistant explosive, which has not yet been reported in the literatures. Main properties of the explosive were tested. It is possessed of fine properties of ignition and burning. It's used as complex ignition agent has brought about good results. Condensation reaction of the explosive synthesis is mainly discussed in this paper. The best technological conditions of condensation reaction and the structure of the product have been thoroughly studied. It is shown that product of condensation reaction was confirmed as the design compound.

1 Summary

Heat-resistant explosive is with good thermal stability and high melting point and low vapor pressure characteristics.

Through the amino group of explosive molecules, it will be in the formation of hydrogen bonds between the hydrogen in the amino acid and the oxygen in the nitro ($-NO_2$). Molecules existing in the hydrogen bond is to improve the functional density of explosive (ρ_0). Hydrogen bonds which are in molecules can not only shorten the distance between molecules, but also increase the lattice energy between molecules. Symmetrical molecular structure of explosive is better than similar stability of explosive. These rules can not only apply to aromatic nitro compounds, but also the N-nitro compounds. Salt deposition is not in the premise of reducing explosive energy to enhance the activity of decomposition of explosive energy, and thereby it can reduce the vapor pressure and increase the melting point.

Based on the above point of view, we obtain 3,3′-diamino-2,2′, 4,4′, 6,6′-six nitrodiphenylamine K.

NO_2 K NO_2 H_2N NH_2 N O_2N NO_2 NO_2 NO_2

The molecules of this explosive is symmetric, because it contains the amidogen sediment. Therefore it will be perfect heat-resisting explosive.

According to the testing of its property, this explosive has a good ability of heat-resisting; It can be blended with oxidant and produce a new and unique amorce.

This explosive is a kind of tawny crystal with a melting point of 334℃, which is 18℃ higher than HNS.

Accumulating density: 0.526g/cm^3.

Dissolving ability: It can be easily dissolved with acetone, dimethylfomamide, dimethyl sulfoxide, and slightly dissolved with water, ethannol, carbinol and so on.

Hydrolysis stability: when add into boiling water, it will not dissolve even after 24h.

Shock sensitivity: When 10kg hanging is dropped into 0.003g testing material from the height of 25cm, the rate of detonation is 100%.

Spark sensibility: Thermal conductivity experiment, lighting rate is 100%.

Burning speed: making a pole explosive with the press of 2000kg/cm^3, length of 8mm, and diameter of 5mm's from 300mg testing material.

Blending this explosive and oxidant in the proportion of 30/70, we can produce a new explosive.

A quantity of 450mg of this explosive can be shaped as a pillar explosive with a length of 12mm and diameter of 5mm under the pressure of 2000kg/cm^3. The burning speed of this explosive is 3s/cm. This explosive is an ideal heat-resisting one.

m-dinitro-dichlorobenzenem-diamino-chlorobenzene is used as raw material for 3,3′-diamino 2,2′, 4,4′, 6,6′-six nitro diphenylamine Potassium, which is obtained trough condensation, amination, nitration, and salt deposition reaction.

2 The synthesis of 3,3′- diamino - 2,2′,4,4′,6,6′- sixnitr - odiphenylamine potassium

2.1 The synthesis of 3,3′-dichloro-2, 4-dinitro-diaminobenzene (condensation)

Cl, O_2N, NO_2, Cl + NH_2, Cl $\xrightarrow{OH^-}$ Cl–(ring with O_2N, NO_2)–N(H)–(ring)–Cl + HCl

2.2 3,3′-dichloro-2,2′, 4,4′, 6,6′-hexnitro-diaminobenzene (nitration)

Cl–(ring with O_2N, NO_2)–N(H)–(ring)–Cl + HNO_3 $\xrightarrow[0\sim5℃]{H_2SO_4}$ Cl–(ring with NO_2, O_2N, NO_2)–N(H)–(ring with NO_2, NO_2, NO_2)–Cl

m. p. 92~94℃

2.3 3,3′-dichloro-2,2′, 4,4′, 6,6′-hexnitro-diaminobenzene (amination)

Cl–(ring with NO_2, O_2N, NO_2)–N(H)–(ring with NO_2, NO_2, NO_2)–Cl + NH_2 $\xrightarrow{40,1atm}$ H_2N–(ring with NO_2, O_2N, NO_2)–N(H)–(ring with NO_2, NO_2, NO_2)–NH_2

m. p. 232~237℃

2.4 3,3′-dichloro-2,2′, 4,4′, 6,6′-hexnitro-diaminobenzene (salt deposition)

$$\text{3,3'-dichloro-2,2',4,4',6,6'-hexanitrodiphenylamine} \xrightarrow{CH_3COOK} \text{3,3'-diamino-2,2',4,4',6,6'-hexanitrodiphenylamine}$$

m. p. 333～334℃

3 The most suitable reaction conditions of condensation reaction in the process of synthesis and identification of the resultant

We have researched the synthesis of condensation, nitration, amination, and salt deposition reaction, and also made identification of the reaction temperature, reaction time and the impact on the response factors. Through orthogonal experimental design, we have proved the optimal reaction conditions for each reaction. Each product of the reaction was identified and confirmed by elemental analysis, mass spectrometry, nuclear magnetic resonance, that it is the target compounds.

This thesis is only for the condensation reaction, which discuss the most suitable reaction conditions of condensation reaction in the process of synthesis and identification of the resultant.

In the literature[5-6] we have researched the condensation reaction and kinetics of the Uhlmann. We have referred to the study and other literature[7-8], through a large number of experiments, and finally obtained the optimal reaction conditions.

3.1 The research on synthesis of 3,3′-dichloro-2,4-dinitro-diaminobenzene under optimum conditions

Add the catalyst into ethanol solution which is with m-di nitro m-dichlorobenzene and m-chloroaminobenzene. Then it will react under certain conditions, it will obtain resultant after filtered, washed, dried. The impact on the response factors are reaction temperature, reaction time, types of catalysts and method of crystallization.

3.1.1 Effects of reaction temperature

The effects of reaction temperature on output is shown in Table 1. It shows the trend of more higher temperature and higher output, The reaction catalyst is CH_3COONa (1g).

Table 1 Effects of reaction temperature

RUN	Cl, Cl, O_2N, NO_2 (g)	Cl, NH_2 (mL)	CH_3CH_2OH (mL)	Temp (℃)	Time (h)	m. p. (℃)	Yield (%)
1	5	2.5	160	75	2	140-143	57.8
2	5	2.5	160	70	2	140-143	44.9
3	5	2.5	160	60	2	141-144	40.6
4	5	2.5	160	55	2	141-144	33.2
5	5	2.5	160	40	2	141-144	12.5
6	5	2.5	160	30	2	140-142	4.9

3.1.2 Effects of reaction time

The effects of reaction time on output is shown in Table 2. It shows the trend of longer time and higher output, the resultant is in higher melting point. The reaction catalyst is $NaCO_3$ (1g), reaction temperature is 70℃.

Table 2 Effects of reaction time

RUN	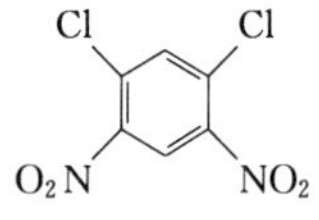 Cl, Cl, O_2N, NO_2 (g)	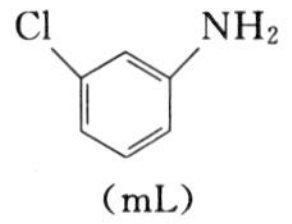 Cl, NH_2 (mL)	CH_3CH_2OH (mL)	Time (h)	m. p. (℃)	Yield (%)
1	5	2.5	160	2	140-143	57.8
2	5	2.5	160	1.5	140-143	44.9
3	5	2.5	160	1.0	141-144	40.6
4	5	2.5	160	0.5	141-144	33.2
5	5	2.5	160	0.2	141-144	12.5

3.1.3 Effects of catalysts

We consider that the catalyst influencing reaction with two reasons. One is removing the HCl produced during the reaction; another is reducing the activation energy of the reaction. Alkali which is used for the reaction catalyst in our research, only can remove the HCl produced during the reaction, it can not reduce the activation energy of the reaction. Moreover, the alkaline used as a catalyst should be in appropriate degree. When the alkalineis is too higher, it will be mixed with 2.4-Dinitro-m-Dichloroaniline. When the alkalineis too lower, it can not remove the HCl.

The using output of various catalysts is shown in table 3.

Table 3 Effects of various catalysts

RUN	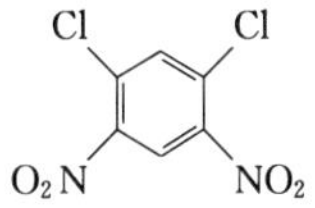 Cl, Cl, O_2N, NO_2 (g)	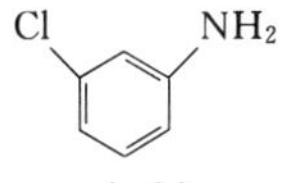 Cl, NH_2 (mL)	CH_3CH_2OH (mL)	Catalysts		Temp (℃)	Time (h)	m. p. (℃)	Yield (%)
1	5	2.5	160	CH_3COONa	1.73(g)	75	2	140-143	57.8
2	5	2.5	160	$NaHCO_3$	1.73(g)	70	2	140-143	44.9
3	5	2.5	160	Na_2CO_3	1.12(g)	60	2	141-144	40.6
4	5	2.5	160	NaOH	0.86(g)	55	2	141-144	33.2
5	5	2.5	160	CaO	0.50(g)	40	2	141-144	12.5
6	5	2.5	160	 N	1.70(mL)	30	2	140-142	4.9
7	5	2.5	160	$Mg(CO_3)_4 \cdot Mg(OH)_2 \cdot 5H_2O$	2.56(g)	30	2	140-142	4.9

The usage of catalyst is with the best quality. When NaOH is the catalyst, because of high alkaline and low production, the resultant is also in low melting point. When $Mg(CO_3)_4 \cdot Mg(OH)_2 \cdot 5H_2O$ is the catalyst, the resultant is in highest production and melting point.

Table 4 Effects of various catalysts at different temperature (2h)

RUN	Cl, Cl, O_2N, NO_2 (g)	Cl, NH_2 (mL)	Temp (℃)	Na_2CO_3 m. p. (℃)	Na_2CO_3 Yield (%)	CH_3COONa m. p. (℃)	CH_3COONa Yield (%)	$Mg(CO_3)_4 \cdot Mg(OH)_2 \cdot 5H_2O$ m. p. (℃)	$Mg(CO_3)_4 \cdot Mg(OH)_2 \cdot 5H_2O$ Yield (%)
1	5	2. 5	75	139-141	41. 9	140-143	57. 8	139-143	59. 2
2	5	2. 5	70	140-143	40. 4	140-143	44. 9	140-144	47. 7
3	5	2. 5	60	140-144	37. 5	141-144	40. 6	140-143	39. 0
4	5	2. 5	55	140-143	24. 5	141-144	33. 2	139-142	34. 6
5	5	2. 5	40	140-142	13. 0	141-144	12. 5	140-142	31. 8

3. 1. 4 Effects of catalysts at different temperature

The output and melting point of production under different temperatures is shown in table 4. The highest output will be produced, when it is with Mg $(CO_3)_4 \cdot$ Mg $(OH)_2 \cdot 5H_2O$ as catalyst and under 75℃.

3. 1. 5 Effects of crystallization method

The two crystalline melting point methods of resultant is shown in Table 5. Compared to water crystallization method, the melting point of direct crystallization method is higher.

Table 5 Effects of Crystallization method

RUN	Cl, Cl, O_2N, NO_2 (g)	Cl, NH_2 (mL)	CH_3CH_2OH (mL)	Temp (℃)	m. p. (℃)	Crystallization method
1	5	2. 5	150	75	84-104	Crystallization in cold water
2	5	2. 5	150	75	85-90	Crystallization in cold water
3	5	2. 5	150	75	85-124	Crystallization in cold water
4	5	2. 5	150	75	140-141	Crystallization in reaction solution
5	5	2. 5	150	75	140-141	Crystallization in reaction solution

3. 2 The synthesis of 3,3′-dichloro-2,4-dinitro-niaminobenzene

Adding 10g 2, 4-dinitro-m-dichlorobenzene and 2g catalyst into 250ml three-neck flask, then adding 160ml alcohol until heat and dissolved, adding 30ml ethanol solution which contains 5ml m-Chloro aminobenzene. Heating the reaction solution until the water is boiling. After 4 hour's circulation, stop heating. Placing the filter after 12h and washing 30ml ethanol about 3 times. We will obtain the resultant after drying. The compounds will be more than 75% yield and melting point is 140. 0-140. 5℃.

3. 3 The synthesis of 3,3′-dichloro-2,4-dinitro-diaminobenzene

The resultant is a refined product after condensation reaction. It can be confirmed through NMR analysis, elementary analysis and quality analysis.

3. 3. 1 NMR analysis

$\delta_1 = 9$ppm $\delta_2 = 7.1 \sim 7.5$ppm

$\delta_1 : \delta_2 = 1 : 6$ (Area ratio)

δ_1 is the chemical shift of hydrogen which is in two N-H. δ_2 is the chemical shift of hydrogen which is in benzenes ring. The results of NMR and the target is the same resultant.

3.3.2 Elementary analysis

Analysis value:	C 43.67%	H 2.10%	N 12.03%
Calculated value:	C 43.92%	H 2.13%	N 12.80%

3.3.3 Quality analysis

The results quality analysis is shown in Table 6.

Table 6 Data of mass spectrometry

m/e	M 327	M+1 328	M+2 329	M+3 330	M+4 331	M+5 332
Relative intensity of experimental values	100%	14.30%	65.60%	9.20%	11.00%	1.40%
Relative intensity of calculated values	100%	13.47%	65.50%	8.64%	10.75%	1.28%

Through making an analysis of above-mentioned NMR, elementary and quality, we make confirmation of the compounds produced in the reaction which is the target compounds.

Reference

[1] M. D. Coburn. Ind. Eng. Chem. Prod. Res. Dev., 1986, (25): 68.
[2] T. Urbanski. Chem. and Technol. Of Expl., Vol. 4 (1984) Warzawa.
[3] Lü Chun Xu. Synthesis of theory and practice of heat-resistant explosive East China Institute of Technology, 1986.
[4] Lü Chun Xu. Journal of explosives and propellants, 1989, (2): 8.
[5] T. D. Tuong, Bull. Chem. Soc. Japan. 1970, (43): 1763.
[6] T. D. Tuong, Bull. Chem. Soc. Japan. 1971, (44): 765.
[7] M. D. Coburn. U. S. P. 3414570, 1968.
[8] M. D. Coburn. U. S. P. 3678061, 1972.

(注：此文原载于 Journal of the Industrial Explosives Society, Japan 1990, 51 (5): 281-285)

Investigation on Condensation Reaction and Its Kinetics for 3,3′ Prime-Dichloro-4,6-Dinitrodiphenylamine

Lv Chunxu, Lv Zaoseng, Cai Chun
(East China Institute of Technology, Nanjing, 210094, China)
EI: 92011202622

Abstract: The synthesis of a new heat-resistant explosive is described in this paper, and the kinetics of the condensation reaction of the important intermediate 3,3′-dichloro-4, 6-dinitrodiphenylamine was studied by UV/Visible method. The results are better understood than those obtained by conductivity measurement.

General

3,3′-dichloro-4, 6-dinotrodiphenylamine is an important intermediate for potassium 3,3′-diamino-2,2′, 4,4′, 6,6′-hexanitrodiphenylamine. The latter is an new heat-resistant explosive which has not been reported. Potassium 3,3′-diamino-2,2′, 4,4′, 6,6′-hexanitrodiphenylamine, a dark coffee coloured power, melting at 334℃, is soluble in DMF, acetone, DMSO, slightly soluble in water, ethanol and methanol. Bolting in water for 24h did not cause decomposition, which showed a good hydrolysis stability of this compound. Its flame sensitivity is similar to that impact is 100%. It can be ignited by blasting fuse reliably. Its flame sensitivity is similar to that of normal igniting composition. When ignited with fire, it can deflagrate at a high speed. The deflagration state is between high explosive and primary explosive. The distinguished feature of this compound is a high deflagration velocity. When it is mixed with oxidant at the ratio of 30/70, a new igniting composition forms. When 405mg of the compound was pressed under pressure of 196Mpa to give a charge of 5mm diameter. The charge was put into a delay element, and the deflagration velocity was measured. A value of 0.33 cm • s^{-1} was found.

Probation showed that this compound can be used as a good igniting composition, especially in the field of heat-resistant explosive materials. The structure of this compound has been identified. The result of elemental analysis is as below.

	N	C	H	O
found	22.47	28.88	1.13	34.61
theoretical	24.85	28.4	1.1	37.87

The result of NMR measurement also confirms the structure showing below.

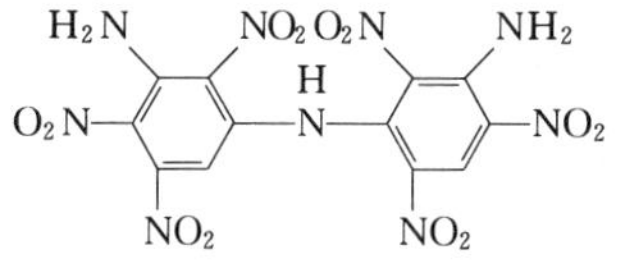

We proposed the following reaction route[1-4].

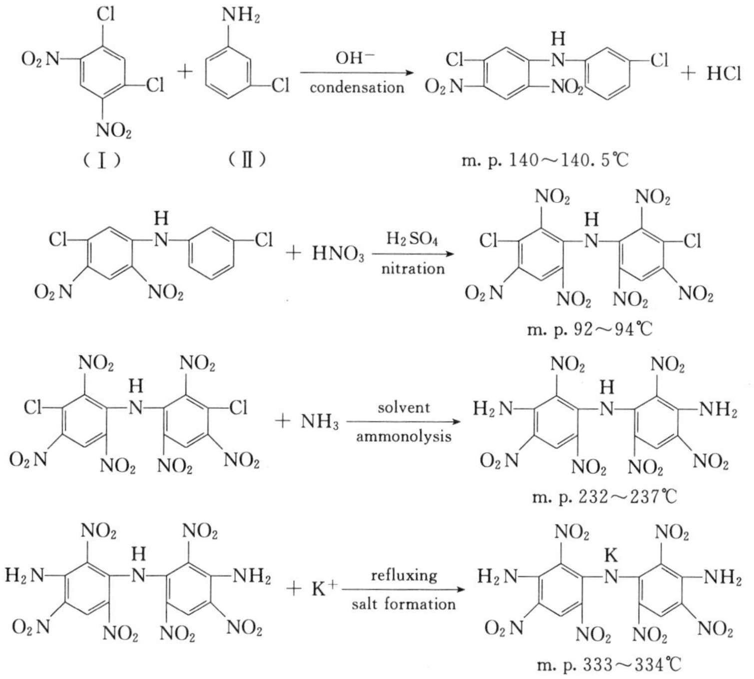

m. p. 140~140. 5℃

m. p. 92~94℃

m. p. 232~237℃

m. p. 333~334℃

Our emphasis is the kinetic study of the condensation step.

Optimization of reaction conditions of the condensation step

We found that reaction temperature, reaction time and catalysts have considerable effects on condensation reaction. The results are given in Table 1, Table 2 and Table 3.

Table 1 Effects of reaction temperature

Amount of (Ⅰ)/g	Amount of (Ⅱ)/ml	Amount of EtOH/ml	Temperature /℃	Time /h	MP of product/℃	Yield of product/%
5	2. 5	160	75	2	140~143	57. 8
5	2. 5	160	70	2	140~143	44. 9
5	2. 5	160	60	2	141~144	40. 6
5	2. 5	160	55	2	141~144	33. 2
5	2. 5	160	40	2	141~144	12. 5
5	2. 5	160	20	2	140~143	4. 0

Table 2 Effects of reaction time

Amount of (Ⅰ)/g	Amount of (Ⅱ)/ml	Amount of EtOH/ml	Time /h	MP of product/℃	Yield of product/%
5	2. 5	160	6. 3	141~144	62. 1
5	2. 5	160	5	141~143	57. 8
5	2. 5	160	4	137~140	54. 9
5	2. 5	160	2	140~144	47. 7
5	2. 5	160	1	139~142	28. 9

Table 3 Effects of catalyst

Amount of (Ⅰ)/g	Amount of (Ⅱ)/ml	Amount of EtOH/ml	Base catalyst/g		Temperature /℃	Time /h	MP of product /℃	Yield of product/%
5	2.5	160	CH_3COONa	1.73	75	4	137~140	67.8
5	2.5	160	$NaHCO_3$	1.73	75	4	138~140	63.5
5	2.5	160	Na_2CO_3	1.12	75	4	137~140	62.9
5	2.5	160	NaOH	0.86	75	4	97~98	23.8
5	2.5	160	CaO	0.5	75	4	138~140	72.3
5	2.5	160	Pyridine	1.37ml	75	4	137~139	52.4
5	2.5	160	$MgCO_3Mg(OH)_2 \cdot 3H_2O$	2.56	75	4	138~142	84.2

Thus the optimum procedure is as follows. 10g of 2,4-dinitro-m-dichlorobenzene and 2g of catalyst are added to a 250ml three-necked flask. To the obtained mixture 160ml of 95% alcohol is added. Heating the mixture dissolves the solids. Then 5ml of m-chloroaniline in alcohol is added, and the resultant mixture is refluxed for 4h. Remove the heat bath and allow the mixture to stand overnight. Filtration followed by alcohol and water washing gives orange yellow crystalline product, mp. 140~140.5℃. Yield of the product is over 75%.

Study on kinetics of the consendation step by UV/visible method

The consendation reaction between 2, 4-dinitro-m-dichlorobenzene and m-chloroaniline can be expressed as follows:

Cl, Cl, O_2N, NO_2 + 2 (Cl, NH_2) $\xrightarrow{K_1}$ (Cl, H, N, Cl, O_2N, NO_2) + (Cl, NH_3Cl)

1 The experimental expression of absorbance A[5-7]

The change in concentration of reactants or products with time can be derived from the change in absorbance with tome. Assuming that time t, the absorbance of reaction mixture at the wave length λ is A, we can write

$$A = A_1 + A_2 + A_3 + A_4 \quad (1)$$

where $A_1 = J_1 L c_1$, J is extinction coefficient.

If L=1cm, (1) becomes

$$A = J_1 c_1 + J_2 c_2 + J_3 c_3 + J_4 c_4 \quad (2)$$

Assuming that the initial concentration of Ⅰ, Ⅱ are a_0 and b_0 respectively, we can write

$$c_1 = a_0 - c_3,\ c_2 = b_0 - 2c_1,\ c_3 = c_4 \quad (3)$$

where c_1, c_2, c_3, c_4 are the concentrations of Ⅰ, Ⅱ, Ⅲ and Ⅳ at time t, J_1, J_2, J_3 and J_4 are extinction coeffcients of Ⅰ, Ⅱ, Ⅲ and Ⅳ at wave length λ.

Putting (3) into (2), we can get

$$A = (J_1 a_0 + J_2 b_0) + (J_3 + J_4 - J_1 - 2J_2) c_3 \quad (4)$$

In order to measure the change of absorbance A with wave length λ, the spectra of dilute alcohol solutions of Ⅰ, Ⅱ, Ⅲ and Ⅳ ($<10^{-5}$ mol · L^{-1}) were recorded. It is found that Ⅳ shows no absorbance at 350nm, so J_4 can be ignored. Furthermore, we obtained the following results $J_1=643$, $J_2=3$, $J_3=14678$. Obviously, the contribution of Ⅱ at 350nm can also be ignored. So we write the expression of absorbance as follows.

$$A=J_1 a_0+(J_3-J_1) c_3 \quad (5)$$

2 Determination of the rate coefficients

A suitable amount of 2, 4-dinitro-m-dichlorobenzene is dissolved in 100ml of EtOH. Raise the temperature to T℃, add Vml of m-chloroaniline, and record the time. At a certain time interval take out 1ml of reaction solution, dilute it to 50ml, and measure the absorbance A at wave length 350nm. By using equation (5), we can obtain the concentration c_3, and calculate rate coefficient. In EtOH at 70℃, the rate coefficient k was calculated to be 0.04511 • mol^{-1} • min^{-1}.

3 Factors affecting k

K is very sensitive to temperature. Table 4 shows that the higher the temperature is, the larger the k becomes.

Table 4 Effects of temperature on k

$a_0/mol \cdot l^{-1}$	$b_0/mol \cdot l^{-1}$	Temperature/℃	$k/l \cdot mol^{-1} \cdot min^{-1}$
0.01210	0.02040	75	0.08241
0.01210	0.02040	70	0.04410
0.01210	0.02040	60	0.03133
0.01210	0.02040	50	0.01970

K can also be increased rapidly by addition of polar solvent. The results are given in Table 5.

Table 5 Effects of solvent polarity on k

$a_0/mol \cdot l^{-1}$	$b_0/mol \cdot l^{-1}$	Amount of H_2O/ml	Amount of EtOH/ml	$k/l \cdot mol^{-1} \cdot min^{-1}$
0.01245	0.02086	0	100	0.045107
0.01245	0.02086	20	80	0.077396
0.01245	0.02086	30	70	0.092019
0.01245	0.02086	40	60	0.105661
0.01245	0.02086	50	50	0.111986

A plot of Y against lnk shows a straight line, the slope m of-0.25552 is the sensitivity of reaction rate vs solvent polarity (Fig. 1) . So we can use $\ln k=-0.25552\ Y+2.56503$ to get rate coefficients for other solvent (Table 6).

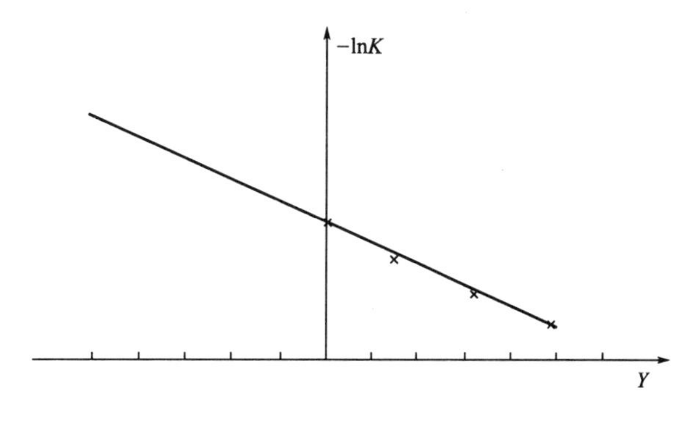

Fig. 1 Influence of solvent polarity on reaction rate coefficient

Table 6 Effects of solvent polarity on k

solvent		Y	temperature/℃	k
amount of water/%	amount of EtOH/%			/l · mol^{-1} · min^{-1}
0	100	−2.03	70	0.045107
20	80	0	70	0.077396
30	70	0.6	70	0.092019
40	60	1.12	70	0.10566
50	50	1.16	70	0.111986

Type and amount of base catalysts also have considerable effects on k. The results are tabulated in Table 7.

Table 7 Effects of catalyst on k

a_0/mol · l^{-1}	b_0/mol · l^{-1}	base catalyst		k
		compound	amount used/g	/l · mol^{-1} · min^{-1}
0.0121	0.02407	—	—	0.045107
0.0121	0.02407	NaF	0.1	0.046445
0.0121	0.02407	NaF	0.5	0.047333
0.0121	0.02407	CH_3COONa	0.2	0.046413
0.0121	0.02407	CH_3COONa	0.4	0.098588
0.0121	0.02407	CH_3COONa	0.6	0.051411
0.0121	0.02407	$NaHCO_3$	0.5	0.049398
0.0121	0.02407	Na_2CO_3	0.3	0.049686
0.0121	0.02407	Pyridine	0.2	0.060167

The date listed above coincide with experimental values obtained from conductivity measurements, and the tendencies of results by two methods are the same.

References

[1] Tuong, T. D., Bull. Chem. Soc. Japan. 43, 1763 (1970).
[2] Tuong, T. D., Bull. Chem. Soc. Japan. 44, 765 (1970).
[3] Coburn, M. D., US Pat. 3678061 (1972).
[4] Tayloy, F., US Pat 3418372 (1968).
[5] Silversteln, R. M., Spectometric Identification of Orange Compounds. John Wiler. 1974.
[6] Wang Qi, An Introduction to Chemical Kinetics, Jilin People's Press, 1982.
[7] Sheng Hongkang, Orange Acid and Base, Higher Education Press, 1978.

(注：此文原载于 Proceeding of the 17th International Pyrotechnics Seminar, 1991, (1): 205-211)

3,3′-二氨基-2,2′,4,4′,6,6′-六硝基二苯胺钾合成及其应用

吕春绪，吕早生

（华东工学院）

摘要： 3,3′-二氨基-2,2′，4,4′，6,6′-六硝基二苯胺钾是个新型耐热炸药，该化合物至今未见报导。本文研究了该炸药合成的最佳工艺条件及其中间体、产物的结构鉴定，重点讨论了缩合反应。对该炸药主要性能进行了测试，它具有良好的点火和燃烧性能，在点火药中复配试用效果良好。并表明，将会在耐热爆破器材的实际应用中提供一种新型耐热点火药剂。

关键词： 耐热炸药，合成，缩合反应，应用

1 前言

3,3′-二氨基-2,2′，4,4′，6,6′-六硝基二苯胺钾是个新型耐热炸药。

NO_2 K NO_2 H_2N N NH_2 O_2N NO_2 O_2N NO_2

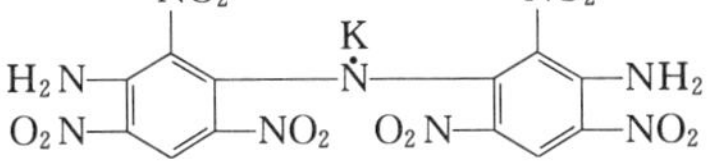

从耐热炸药分子结构特点看，分子中引入—NH_2 基是有利的，它的作用在于可与炸药中—NO_2 基里的氧形成氢键，如形成分子内氢键可使其特征密度（ρ_0）增大，如形成分子间氢键可缩短分子间距而使堆积系数（κ 值）升高；使分子的晶格能增加，提高化合物的熔点和分解点；—NH_2 基是供电子基，无论是共轭效应，还是诱导效应，都可提高苯环上的电子云密度，对提高化合物的稳定性有利。成盐会在不降低炸药其它能量示性数的前提下，使化合物熔点大幅度升高。同类炸药中，对称结构的具有较好的热安定性；结构恢复对称性的，安定性也立刻提高，在芳烃硝基化合物和杂环硝胺中都有此规律[1,2]。

3,3′-二氨基-2,2′，4,4′，6,6′-六硝基二苯胺钾的分子结构刚好就具有含有氨基、成盐和对称等特征，因此它是一个设计较为合理的耐热炸药。它是以二硝基间二氯苯和间氯苯胺为原料，经过缩合、氨基化、硝化和成盐等步骤而获得。主要性能数据测试表明：它是一种良好的新型耐热炸药，还可与其它氧化剂组分混合组成新型耐热点火药。

2 3,3′-二氨基-2,2′,4,4′,6,6′-六硝基二苯胺钾合成反应原理[3-6]

（1）3,3′-二氯-2，4-二硝基二苯胺的合成（缩合）

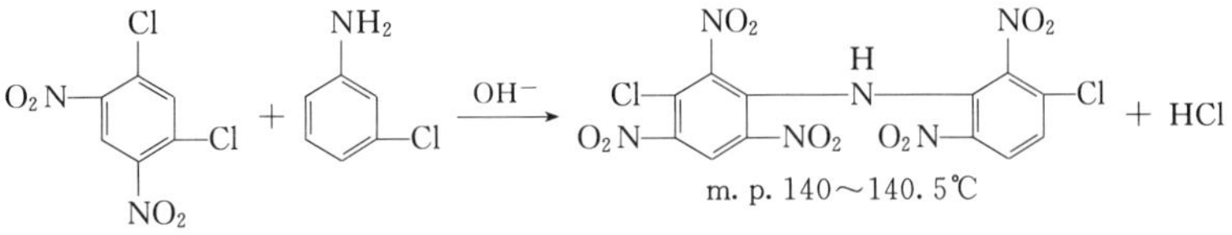

Cl NH_2 NO_2 H NO_2 O_2N + Cl $\xrightarrow{OH^-}$ Cl N Cl + HCl Cl O_2N NO_2 O_2N NO_2

m. p. 140～140. 5℃

（2）3,3′-二氯-2,2′，4,4′，6,6′-六硝基二苯胺的合成（硝化）

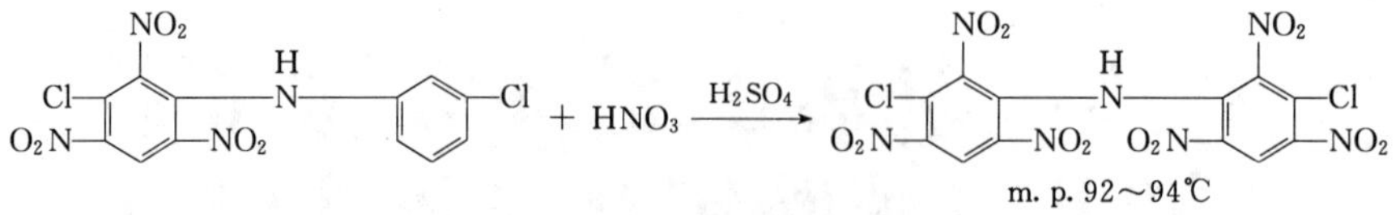

$+ HNO_3 \xrightarrow{H_2SO_4}$

m. p. 92～94℃

（3）3,3′-二氨基-2,2′，4,4′，6,6′-六硝基二苯胺的合成（氨基化）

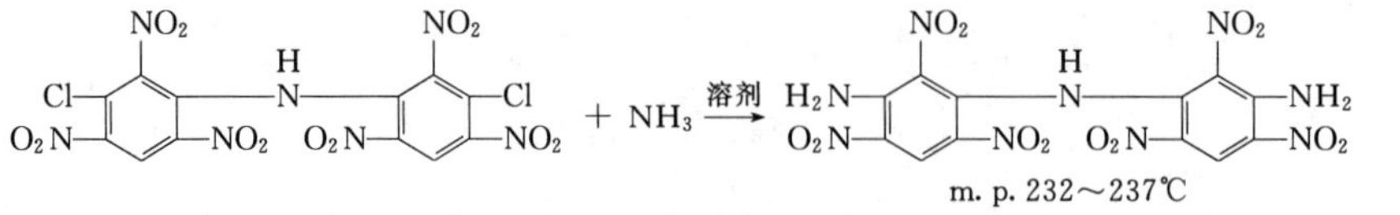

$+ NH_3 \xrightarrow{溶剂}$

m. p. 232～237℃

（4）3,3′-二氨基-2,2′，4,4′，6,6′-六硝基二苯胺钾的合成（成盐）

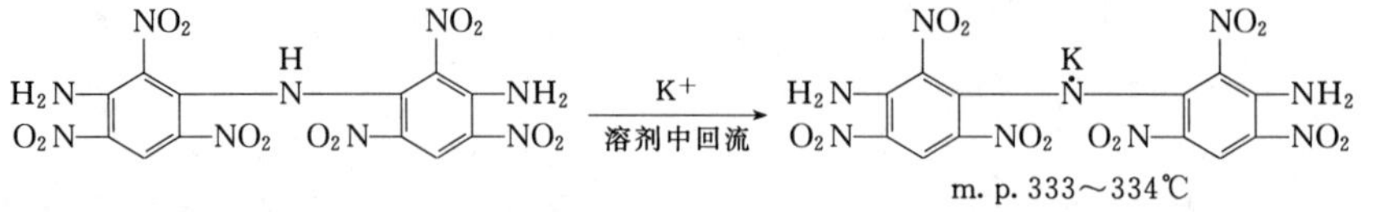

$\xrightarrow[溶剂中回流]{K^+}$

m. p. 333～334℃

3　实验研究

3.1　3,3′-二氨基-2,2′,4,4′,6,6′-六硝基二苯胺钾合成最佳工艺条件研究

我们研究了上述缩合、硝化、氨基化和成盐等反应和反应的温度、时间、料比等单因素的影响，进行了反应参数相互影响的正交实验，确定出各反应的最佳工艺条件。本文仅以缩合反应为例，阐述其最佳工艺条件研究和产物结构鉴定结果。

3.1.1　3,3′-二氯-2, 4-二硝基二苯胺缩合最佳工艺条件研究

2,4-二硝基间二氯苯与间氯苯胺在乙醇介质中加入催化剂后，一定温度下进行反应，经过滤、洗涤、烘干后得到产品。反应参数的影响如下：

（1）反应温度的影响

反应取 CH_3COONa 作为催化剂，结果见表 3.1。显见，反应得率随温度升高而增高。

（2）反应时间的影响

反应用 Na_2CO_3 作催化剂，反应温度 70℃，结果如表 3.2 所示。由表显见，一般情况下反应时间越长，收率越高，产品纯度也越高。

（3）催化剂的影响

催化剂的影响可从两方面考虑：一是可除去反应产生的 HCl，保证间氯苯胺呈游离态；另一是可降低反应控制步骤的活化能。但从动力学研究的结果表明，本反应中的碱只起游离反应物间氯苯胺的作用，不能降低反应过程的活化能，即没有催化活性。所使用的碱不能太强，也不能太弱，否则要么与 2,4-二硝基间二氯苯反应，要么起不到游离间氯苯胺的作用。实验结果列于表 3.3。由表 3.3 可见，用 NaOH 作为催化剂，由于碱性太强，产品熔点和收率都很低；而用 $MgCO_3 \cdot Mg(OH)_2 \cdot 3H_2O$ 作为催化剂，收率很高，熔点也最高，说明它的效果最好。

表 3.1　反应温度对缩合反应熔点和得率的影响

Tab. 3.1　Effect of reaction temperature on the melting point and yield of condensation

二硝间二氯苯/g	间氯苯胺/ml	乙醇/ml	反应温度/℃	反应时间/h	熔点/℃	得率/%
5	2.5	160	75	2	140～143	57.8
5	2.5	160	70	2	140～143	44.9
5	2.5	160	60	2	141～144	40.6
5	2.5	160	55	2	141～144	33.2
5	2.5	160	40	2	141～144	12.5
5	2.5	160	30	2	140～142	4.9

表 3.2 反应时间对缩合反应熔点和得率的影响

Tab. 3.2 Effect of reaction time on the melting point and yield of condensation

二硝间二氯苯/g	间氯苯胺/ml	乙醇/ml	反应时间/h	熔点/℃	得率/%
5	2.5	160	6.3	141～144	62.1
5	2.5	160	5	141～143	57.8
5	2.5	160	4	137～140	54.8
5	2.5	160	2	140～144	47.7
5	2.5	160	1	139～142	28.9

表 3.3 催化剂对缩合反应熔点和得率的影响

Tab. 3.3 Effect of catalyst on the melting point and yield of condensation

二硝间二氯苯/g	间氯苯胺/ml	乙醇/ml	催化剂碱/g		反应温度/℃	反应时间/h	熔点/℃	得率/(%)
5	2.5	160	CH_3COONa	1.73	75	4	137～140	67.8
5	2.5	160	$NaHCO_3$	1.73	75	4	138～140	73.5
5	2.5	160	Na_2CO_3	1.12	75	4	137～140	62.9
5	2.5	160	NaOH	0.86	75	4	97～98	23.8
5	2.5	160	CaO	0.5	75	4	138～140	72.3
5	2.5	160	N	1.7ml	75	4	137～139	52.4
5	2.5	160	$MgCO_3 \cdot Mg(OH)_2 \cdot 3H_2O$	2.56	75	4	138～142	84.2

(4) 不同温度下催化剂的影响

使用不同碱，在不同温度下，得率变化幅度也相当大。从表 3.4 可以看出：在所使用的反应条件内，$MgCO_3 \cdot Mg(OH)_2 \cdot 3H_2O$ 在不同温度下催化效果都最好，在反应温度 75℃时具有最高值。

表 3.4 不同温度下各催化剂对缩合反应熔点和得率的影响

Tab. 3.4 Effect of catalysts on the melting point and yield of condensssation under different temperature

二硝间二氯苯/g	间氯苯胺/ml	乙醇/ml	反应温度/℃	反应时间/h	Na_2CO_3 熔点/℃	Na_2CO_3 得率/(%)	CH_3COONa 熔点/℃	CH_3COONa 得率/(%)	$MgCO_3 \cdot Mg(OH)_2 \cdot 3H_2O$ 熔点/℃	$MgCO_3 \cdot Mg(OH)_2 \cdot 3H_2O$ 得率/(%)
5	2.5	160	75	2	139～141	41.9	140～143	57.8	139～143	59.2
5	2.5	160	70	2	140～143	40.4	140～143	44.9	140～144	47.7
5	2.5	160	60	2	140～144	37.5	141～144	40.6	140～143	39.0
5	2.5	160	55	2	140～143	24.5	141～144	33.2	139～142	34.6
5	2.5	160	40	2	140～142	13.0	141～144	12.5	140～142	31.8

(5) 析出方式的影响

实验表明，直接析出比稀释析出有较高的熔点。结果见表 3.5。

表 3.5 析出方式对缩合反应熔点的影响

Tab. 3.5 Effect of the manner of separation on the melting point of condensation

二硝间二氯苯/g	间氯苯胺/ml	乙醇/ml	反应温度/℃	熔点/℃	析出方式
5	2.2	150	75	84～104	冷水冲析
5	2.2	150	75	85～90	冷水冲析
5	2.2	150	75	82～124	冷水冲析
5	2.2	150	75	140～141	直接过滤
5	2.2	150	75	140～141	直接过滤

3.1.2 3,3′-二氯-2,4-二氯-2,4-二硝基二苯胺缩合反应最佳工艺

取 2,4-二硝基间二氯苯 10g，催化剂 2g，一起加入 250ml 三口烧瓶中。加入 95%酒精 160ml，加热，使其溶解，再加入含 5ml 间氯苯胺的酒精溶液 30ml。在水浴上加热至沸腾，回流 4h，停止加

热，自然降温，放置过夜后过滤，用30ml酒精分三次洗涤，再用30ml水分三次洗涤后抽干。干燥得产品为橙黄色晶体，熔点140～140.5℃，得率75%以上。

3.1.3 3,3′-二氯-2,4-二硝基二苯胺结构鉴定

按3.1.2缩合工艺制备的产品分别用核磁共振、元素分析和质谱等进行结构鉴定。

(1) 核磁共振分析（南京大学现代分析中心）

δ_1=9ppm单峰，δ_2=7.1～7.5ppm多重峰，峰面积δ_1∶δ_2=1∶6

可以认定δ_1为 —N(H)— 上氢的化学位移，δ_2为苯环上氢所产生的化学位移，其质子数之比正好是核磁共振谱上两峰面积之比1∶6。这与3,3′-二氯-2,4-二硝基二苯胺结构相符。

(2) 元素分析（北京中国科学院化学所）

理论值（%）：C 43.92 H 2.13 N 12.80

实测值（%）：C 43.67 H 2.10 N 12.03

(3) 质谱分析（南京大学现代分析中心）

从3,3′-二氯-2,4-二硝基二苯胺分子结构式计算，其分子离子峰应为327；质谱图上，在327处确有一强峰，其相对强度100%。其主要峰值和相对强度值如表3.6所示。

从核磁共振、元素分析和质谱结果可以确定，该缩合产物与设计化合物基本相符。

表3.6 缩合反应产物的质谱峰值和相对强度

Tab.3.6 Peak value and relative strength of mass spectra of the condensation product

	M	M+1	M+2	M+3	M+4	M+5
质 量/e	327	328	329	330	331	332
实测相对强度	100%	14.30%	65.60%	9.20%	11.00%	1.4%
计算相对强度	100%	13.47%	65.50%	8.64%	10.75%	1.28%

3.2 3,3′-二氨基-2,2′,4,4′,6,6′-六硝基二苯胺钾结构鉴定

上述缩合产物经硝化、氨基化、成盐得3,3′-二氨基-2,2′,4,4′,6,6′-六硝基二苯胺钾。

成盐就是在二氨基六硝基二苯胺中用一个钾离子取代一个氢质子而成为二氨基六硝基二苯胺钾。显然，反映氢质子变化状态的核磁共振谱，对该产品结构鉴定可提供充分依据。

该产品经南京大学现代分析中心进行核磁共振分析：

δ_1=7.9ppm为一宽单峰，δ_2=8.8ppm为一单峰，峰面积δ_1：δ_2=4：2。

δ_1为（NH_2）上氢所产生的化学位移，由于氢键原因所以该峰很宽；δ_2为苯环上两个等价氢所产生的化学位移；氢质子总数之比正好是δ_1与δ_2处的积分面积4∶2之比。

核磁共振结果表明，产品与3,3′-二氨基-2,2′,4,4′,6,6′-六硝基二苯胺钾结构完全相符。对该产品还进行了元素分析、质谱等其它结构鉴定，结论与核磁共振相吻合。

3.3 3,3′-二氨基-2,2′,4,4′,6,6′-六硝基二苯胺钾主要性能和初步应用

该产品为咖啡色晶体，熔点334℃；假密度526kg/m³；溶解性：可溶于二甲亚砜、丙酮、二甲基甲酰胺，微溶于水、乙醇、甲醇；水解安定性：在沸水中煮沸24h可破坏；撞击感度：10g落锤、25cm落高、药量0.03g爆炸百分数100%；火焰感度：导火索实验可靠着火，相当于常用点火药的火焰感度；燃烧状况：明火点燃后相当快地燃烧，燃烧状态介于猛炸药与起爆药之间；燃速：药量300mg，在196Mp压力下成型为长度8mm、直径5mm的药柱，在延期管中燃烧，测时仪测时，测得燃速为0.078cm/s。

该产品与常用氧化剂以30/70比例混合，组成新点火药，药量450mg，在196MPa压力下成型为长度12mm、直径5mm的药柱，用与上相同的方法测得燃速为0.33cm/s，且燃烧时无残渣。试用表

明，它是一个很好的点火药剂，特别在耐热爆破器材的实际应用中，将是一个理想的新型耐热点火药剂。

4 几点结论

（1）3,3′-二氨基-2,2′，4,4′，6,6′-六硝基二苯胺钾是个新型耐热炸药。

（2）所合成的中间体和最终化合物经元素分析、红外光谱、核磁共振和质谱分析等结构鉴定，证明与设计的化合物基本相符。

（3）所设计的合成反应路线，实验表明是合理的。对各单元反应经大量单因素实验和正交实验，确定出反应的最佳工艺条件。

（4）经性能测试和组成点火药剂试用表明，点火可靠，燃速高且稳定，燃烧后无残渣，因此可以作为一种新型耐热点火药剂。

参考文献

[1] Coburn M D. Ind Eng Chen prod dev. 1986，25：68.
[2] Urbanskl T. Chem. And Techno Of Expl. Warzawa. 1984，4.
[3] Tuong T D. Bull Chem. Soc. Japan，1970，43：1763.
[4] Coburn T D. Bull Chem. Soc. Japan，1971，44：765.
[5] Coburn M D. U. S. Patent，3678061. 1972.
[6] Taylor F. U. S. Patent，348372. 1968.

Synthesis of Potassium 3,3′-Diamino-2,2′,4,4′,6,6′-Hexanitrodiphenylamide and its Applications

Lv Chunxu，Lv Zaosheng
(East China Institute of Technology)

Abstract: Potassium 3,3′-dimaino-2,2′，4,4′，6,6′-hexanitrodiphenylamide is a new king of heat-resistant explosive，not yet seen reported in the literature. In this paper，the optimum technological conditions for the synthesis of this explosive are studied in detail and structure of the intermediates and of the final product identidied. The condensation reaction is discussed in particular. Chief properties of the explosive have been tested. It possesses fine characteristics on ignition and combustion; its use as a complex ignition agent has brought about satisfactory results. It has been shown that in practical application as heat-resistant detonation materials，it will prove to be a new kind of heat-resistant ignition agent.

Key words: heat-resistant explosive，synthesis，condensation，application

（注：此文原载于 兵工学报，1992，(2)：49-54）

1,3-二(3′-氯苯胺基)-2,4,6-三硝基苯的合成研究

吕春绪，邓爱民

（华东工学院）

摘要： 1,3-二（3′-氯苯胺基)-2,4,6-三硝基苯是新型耐热炸药的重要中间体，是由三硝基间二氯苯与间氯苯胺经 Ullmann 缩合反应而成。本文主要研究了该缩合反应和反应温度、反应时间、反应料比等单因素的影响，以及他们的正交实验，从而确定出其制备的最佳工艺条件。该工艺条件下获得的产品经红外、质谱和元素分析等结构鉴定表明与设计化合物基本相符。

关键词： 1,3-二（3′-氯苯胺基)-2,4,6-三硝基苯，三硝基间二氯苯，间氯苯胺，缩合

1 引言

1,3-二（3′-氯苯胺基)-2,4,6-三硝基苯是个未见报道的新化合物。它是新型耐热炸药 1,3-二（3′-氨基-2,4,6-三硝基苯胺基)-2,4,6-三硝基苯的重要中间体[1-2]，是由三硝基间二氯苯与间氯苯胺经 Ullmann 缩合反应而制得[3-5]。

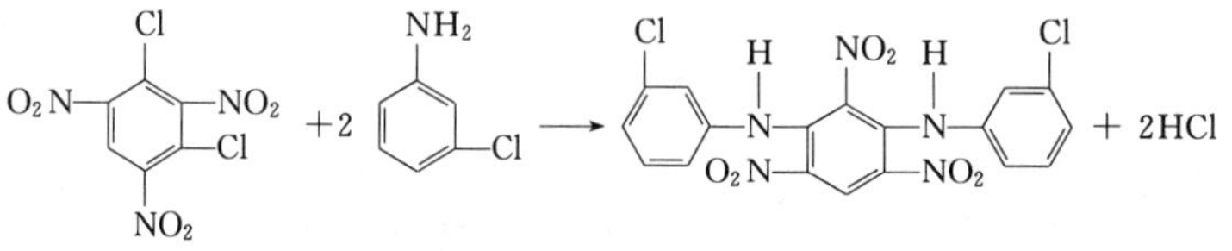

Cl　O_2N　NO_2　Cl　NO_2　+2　NH_2　Cl　⟶　Cl　H　NO_2　H　Cl　N　N　O_2N　NO_2　+ 2HCl

2 实验研究

2.1 1,3-二（3′-氯苯胺基)-2,4,6-三硝基苯合成工艺条件的单因素研究

1,3-二（3′-氯苯胺基)-2,4,6-三硝基苯是由三硝基间二氯苯与间氯苯胺在乙醇溶剂中反应而制得。本文首先研究了反应温度、时间和料比等单因素的影响。

2.1.1 反应温度的影响

取 1.41g（0.005mol）三硝基间氯二苯，在搅拌下加入 1.5ml（0.015mol）间氯苯胺和 50ml 95%乙醇组成的溶液中，再加入 1g 碳酸氢钠。待三硝基间二氯苯和碳酸氢钠全部溶解后，在不同的反应温度下反应 1.5h，冷却至室温，过滤，洗涤，烘干。不同反应温度对产品得率和熔点的影响见表 1。低温度对于反应不利，较高温度下具有较高得率。

2.1.2 反应时间的影响

前述反应条件下，当温度为 60℃，不同反应时间对产品得率和熔点的影响如表 2。随着反应时间的增长，熔点基本不变，但得率却显著增加。

表 1 反应温度的影响

序号	温度/℃	熔点/℃	得率/(%)	序号	温度/℃	熔点/℃	得率/(%)
1	80	178～180	99.0	4	50	176～179	70.7
2	70	179～181	90.5	5	40	175～179	52.2
3	60	177～179	82.8				

表 2 不同反应时间的影响

序号	反应时间/h	熔点/℃	得率/(%)	序号	反应时间/h	熔点/℃	得率/(%)
1	1	178～180	60.8	3	3	178～179	88.8
2	2	177～179	77.6				

2.1.3 料比的影响

前述反应条件下，当反应温度为60℃。反应时间为1.5h，不同料比对得率和熔点的影响如表3。显见，间氯苯胺过量有利于得率提高，当料比为1∶4时具有最高得率。

表 3 反应料比的影响

序号	摩尔比	熔点/℃	得率/(%)	序号	摩尔比	熔点/℃	得率/(%)
1	1∶1	低于175	36.4	4	1∶4	178～180	88.8
2	1∶2	175～178	82.8	5	1∶5	179～181	85.3
3	1∶3	176～179	84.5				

2.2 1,3-二（3′-氯苯胺基）-2,4,6-三硝基苯合成工艺条件的正交实验研究

在研究单因素影响的基础上，为获得各反应参数的最佳匹配，用正交实验研究了它们之间的相互影响，得到了考虑综合因素后的最佳工艺条件。正交实验数据如表4。显见，摩尔比是影响产品得率的主要因素，各反应参数的最佳匹配为C3B3A2。因此，上诉反应参数最佳值为乙醇用量50ml，反应温度80℃，反应时间2.25h，料比（摩尔比）1∶4。

我们还对溶剂类型、用量，反应体系pH值和中和剂种类等单因素和正交实验进行了研究，最后确定出该缩合反应的最佳工艺条件。

表 4 正交实验结果和数据处理

序号	A 反应时间/h	B 反应温度/℃	C 摩尔比	熔点/℃	得率/(%)
1	1.5	60	1∶2	177.5	47.8
2	1.5	70	1∶3	173	95.3
3	1.5	80	1∶4	179.5	96.7
4	2.25	60	1∶3	179	69.8
5	2.25	70	1∶4	178.5	98.3
6	2.25	80	1∶2	173	81.0
7	3.0	60	1∶4	180	99.1
8	3.0	70	1∶2	178.5	56.9
9	3.0	80	1∶3	179	77.6
K_1	239.8	216.7	185.7		
K_2	249.1	250.5	242.7		
K_3	233.6	255.3	294.0		
k_1	79.9	72.2	61.9		
k_2	83.0	83.5	80.9		
k_3	77.9	85.1	98.0		
R	5.2	12.9	36.1		

3 实验结果

3.1 1,3-二(3′-氯苯胺基)-2,4,6-三硝基苯合成的最佳工艺

将1.41g(0.005mol)三硝基甲二氯苯，在搅拌下加入2.0ml(0.02mol)间氯苯胺和50ml 95%乙醇组成的溶液中，加料温度为60℃。再加入1g碳酸氢钠，不断搅拌，使三硝基间二氯苯和碳酸氢钠逐渐溶解。将反应混合物在沸水中回流2.5h，再冷却至室温，过滤，水洗，烘干，得橘黄色粉状晶体，熔点175～179℃。用冰醋酸精制后，产品熔点为179～180℃。产品可溶于丙酮、醋酐等溶剂，微溶于乙醇，不溶于水。此实验进行近百次，得率稳定在92～95%，列入表5所示。

表5 最佳工艺条件的稳定实验数据

序号	三硝基间二氯苯/mol	间氯苯胺/mol	乙醇用量/ml	反应温度/℃	反应时间/h	熔点/℃	得率/(%)
1	0.005	0.02	50	79	2.5	178～180	93.1
2	0.005	0.02	50	80	2.5	177～179	94.0
3	0.005	0.02	50	78	2.5	177～180	93.5
4	0.005	0.02	50	78	2.5	178～180	92.7
5	0.005	0.02	50	79	2.5	179～180	93.5
6	0.005	0.02	50	78	2.5	179～180	93.1

3.2 1,3-二(3′-氯苯胺基)-2,4,6-三硝基苯产品的结构鉴定

按上述最佳工艺条件所得产品经华东工学院的红外光谱仪和南京大学现代分析中心的质谱仪、元素分析仪结构鉴定，结果如下：

红外光谱 1540，1325($Ar-NO_2$)；765，680(C-Cl)；3230(N-H)；1590，1570cm^{-1}(Ar骨架)。

质谱(ml/e) 465(M)，466(M+1)，467(M+2)。

元素分析 计算值C 46.65，H 2.37，N 15.08%；

实验值C 46.43，H 2.21，N 15.07%。

从以上红外光谱图、质谱和元素分析等解析确认该产品与设计化合物基本相符。

参考文献

[1] Cogurm M D. Ind. Eng. Che. Prod. Res. Dev.，1986，(25)：68.
[2] Urbenski T. Chem. and Techno. of Explosives，Warzawa，1984，4.
[3] Tuong T D. Bull. Chem. Soc.，Japan 1970，(43)：1763.
[4] Tuong T D. Bull. Chem. Soc.，Japan 1971，(44)：765.
[5] Fanta P E. Chem. Rev.，1964：613.

Synthesis of 1,3-Bis (3′-chloro-anlino)-2,4,6-Trnitrobenzene

Lv Chunxu，Deng Aimin
(Est China Institute of Technology)

Abstract：1,3-bis (3′-chloro-anlino)-2,4,6-trnitrobenzene is an important imtermediate of a new heat-resistant explosive. It is synthesis via Ullmann condensation 1,3-Deichloro-2,4,6-trinitrobenzene and 3-chloroaniline. In this paper，primarily the condensation reaction is discussed，influence of the reaction parameters (reaction temperature，time and mass ratio) and orthogonal design are also studied. Through these the best bechnological conditions for its preparation are depicted. Structure of the product as obtained under the best technological conditions is identified by IR, Electron impact MS and Elementary Analysis. The product is shown through these means to be the one expected.

Key words: 1, 3-bis (3′-chloro-anlino)-2, 4, 6-trnitrobenzene, 1, 3-deichloro-2, 4, 6-trinitrobenzene, 3-chloroaniline, conde-Nsation

（注：此文原载于 兵工学报．火化工分册，1992，(1)：80-83）

纳米 RDX 粉体的制备与撞击感度

陈厚和，孟庆刚，曹虎，裴艳敏，诸吕锋，吕春绪
（南京理工大学，化工学院，江苏南京，210094）

摘要：将普通工业黑索今炸药溶解于混合溶剂中，通过喷雾干燥，制取纳米黑索今炸药粉体，得率达到50％以上，粒径为40～60nm。实验结果表明，随着粒径的减小，黑索今的机械撞击感度降低，纳米黑索今撞击感度的特性落高约为普通工业黑索今的两倍。

关键词：爆炸力学，冲击感度，喷雾干燥，黑索今，纳米，炸药

引言

纳米材料的制备是纳米科学研究的基础，更是纳米材料应用的前提，因此国内外都进行了大量研究[1-3]。但有关有机化合物纳米粉体，特别是有机含能纳米材料（如纳米级火药、炸药粉体）的制备技术报道不多，尤其是每批次制备量达克级，可以适合于中试和工业化生产的技术与工艺，实测的纳米火药、炸药的有关性能没有报道，主要原因在于制取纳米级火炸药方法不多，难度也较大[4-6]。

1　纳米黑索今的制备

炸药的超细化可以借鉴已广泛应用的无机化合物、金属的超细粉体技术。但炸药是有机化合物，有其自身的特性，如相对于无机化合物、金属来说，熔点较低，容易燃烧，具有爆炸性，生产批量小、纯度要求高。因此炸药的细化技术有特殊的工艺特点和安全操作要求。黑索今（RDX）是一种性能优良的军用炸药，也是世界各国广泛使用的主要军用炸药之一，广泛应用于各种武器的战斗部装药和高性能推进剂[7]。环三甲撑基三硝胺能够在一些溶剂中溶解，然后在一定的条件下重结晶。利用这一特性，制取纳米黑索今粉体，同时在溶液中加入表面活性剂，阻止微粒的团聚和结晶时晶体的成长，是一种安全、方便有效的方法。

1.1　试剂与仪器

黑索今，工业级；混合溶剂，分析纯，自制；由聚氧乙烯多元醇非离子表面活性和高醇酯类天然表面活性剂制备的复合表面活性剂，自制。日本日立 H-800 型透射电子显微镜。

1.2　黑索今的溶解度

物质在溶剂中溶解度的大小取决于物质的内部结构和溶剂的性质。工业用的黑索今是细结晶的粉末，不吸湿，并且在实际上几乎不溶于水，很难溶于乙醇，易溶于丙酮等溶剂中。与金属不互相作用。冲淡的碱和酸与黑索今不起作用，浓硫酸会分解黑索今，浓硝酸会使它溶解，但不分解它[7]。

基于黑索今的以上特性，同时考虑到制备工艺的特点，在选择溶剂时，除考虑溶剂价格和安全性等因素外，还要求：(1) 对溶质有足够大的溶解力；(2) 稳定性好；(3) 沸点适中，在一定温度条件下蒸发，黑索今重结晶时，晶核的形成速度较快，而晶体的成长速度较小。

表 1 是黑索今在一些溶剂中的溶解度。从表中可以看出，丙酮具有较好的溶解度的能力。但由于丙酮的沸点低，挥发快，因此在实际制备过程中，为了与制备工艺配合，一般使用以丙酮为主溶剂的混合溶剂。黑索今在以丙酮为主溶剂（100g）的混合溶剂中的溶解度见图 1。

表 1　黑索今的溶解度（100g 溶剂）

Table 1　The solubility of RDX in solvent (100g solvent) g

溶剂	10℃	20℃	30℃	40℃	50℃	60℃	70℃	80℃
苯	0.02	0.05	0.06	0.09	0.15	0.20	0.30	0.41
乙醇	0.07	0.10	0.15	0.25	0.36	0.58	0.90	
甲醇	0.20	0.24	0.30	0.50	0.75	1.12		
丙酮	5.29	7.22	8.56	11.34	13.21			
甲苯	0.02	0.02	0.03	0.05	0.09	0.14	0.20	0.30

1.3　纳米黑索今的制备

30℃下，将 2.5g RDX 溶解于 300g 上述混合溶剂中，加入复合表面活性剂，形成无水微乳液。在一定的压力和温度条件下将溶液喷入接受器中。可以观察到，随着溶剂的挥发，接受器上有大量细小白色的 RDX 结晶。复合表面活性剂随着溶剂的蒸发析出，包覆在黑索今粒子的表面，不仅限制了黑索今晶体的长大，而且也防止了黑索今粉体的团聚。将上述结晶用乙醇洗涤。由于复合表面活性剂具有很好的醇溶性，经过数次洗涤，包覆在 RDX 晶体表面的复合表面活性剂，基本被洗涤掉，其红外吸收光谱与普通工业 RDX 的吸收光谱一致（见图 2）。然后，在 80～100℃干燥，得到 1.4g 外观呈白色，但微显灰暗的纳米黑索今粉体。

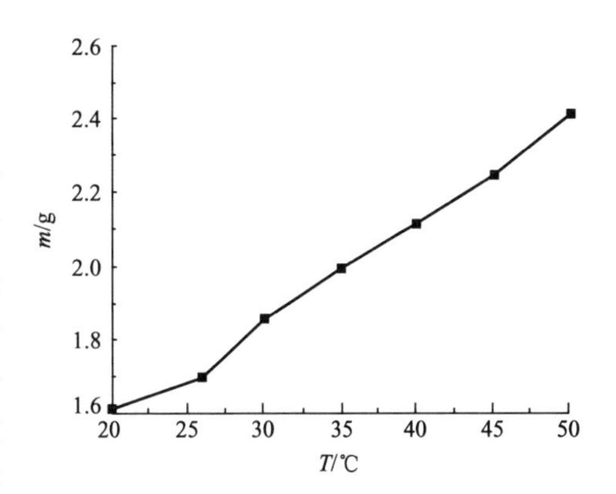

图 1　RDX 在混合溶剂中的溶解度

Fig. 1　The solubility of RDX in mixed solvent

利用日本日立 H-800 型透射电子显微镜对粉体观察拍摄。结果表明，纳米尺颗粒类球形，分散性较好，团聚现象较少，平均粒径约 60nm，结果如图 3 所示。

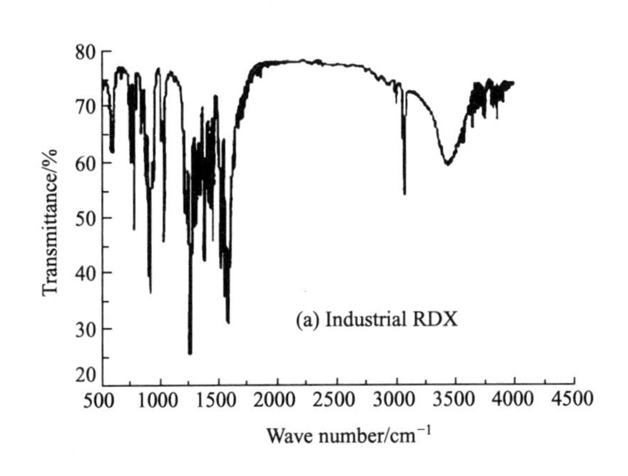

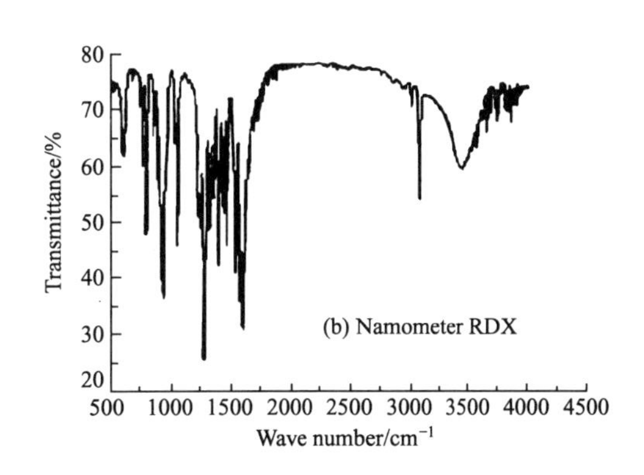

图 2　RDX 的红外光谱图

Fig. 2　The infrared spectra of RDX

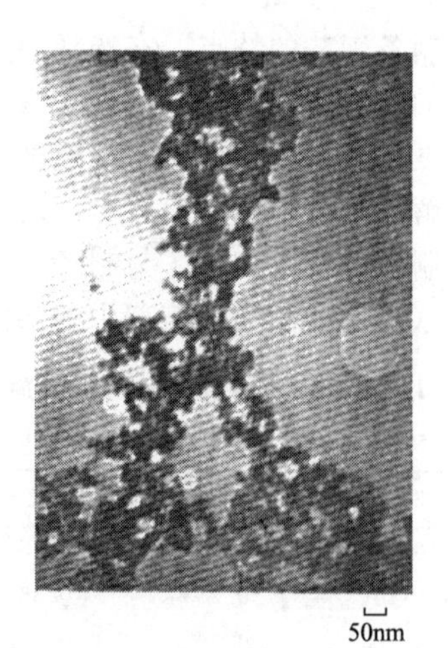

图 3 纳米 RDX 粒子投射电镜照片

Fig. 3 Photograph of transmission electron microscope of namometer particle of RDX

2 纳米黑索今的撞击感度

通过改变纳米黑索今制备过程中溶液的浓度、温度和压力等条件，可以制备不同粒径的纳米黑索今粉体，这些 RDX 颗粒都呈类球形，分散性较好，团聚现象较少。按 GJB772A-97 规定的方法，对不同粒径黑索今的机械撞击感度（落锤质量（5.000±0.005）kg）进行测试，结果为：黑索今粒径为 66000（工业）、60、120～150、40nm 时，特性落高分别为 24.2、54.1、40.6、59.8cm。从测试结果可见，黑索今粒径越小，机械撞击感度越低，纳米级黑索今的机械撞击感度，以特性落高表征，约为普通工业级黑索今（球形，平均粒径 66μm）的两倍。

3 结论

采用喷雾干燥重结晶的方法，实验室可以制备粒径为 40～60nm 的 RDX 炸药粉体，得率达到 50％以上，其红外谱图与普通工业基本一致，但特征峰更加明显。值得注意的是，长期以来，人们对微米级以上粒径的 RDX 炸药粉体的研究结论是，炸药粒径越小，撞击感度越高，而上述按 GJB772-97 规定的方法进行实际测试的结果表明，纳米 RDX 撞击感度的特性落高约高于普通工业 RDX 一倍，这对于制订纳米 RDX 炸药粉体的制备、运输和加工应用等方面的安全操作规程，具有重要的参考意义。

参考文献

[1] Shi Z，Li Y，Wang S. Self-assembled film of a new c60 derivative covalently linked to TiO_2 nanocrystaline [J]，Chemical Physics Letters，2001 (3)：264.

[2] Wu X J，Du L G，Liu H F. Synthesis and tensile property of nanocrystalline metal copper [J]，Nanostructured Materials，1999 (12)：221-224.

[3] HAO En-cai，SUN Hai-ping，ZHOU Zhen，et al. Synthesis and optical properties of CdSe and CdSe/CdS nanoparticles [J] . Chemistry of Materials，1999 (11)：3096-3102.

[4] 王卫民，赵晓利，张小宁．高速撞击流技术制备炸药超细微粉的工艺研究 [J]．火炸药学报，2001，24 (1)，52-54.

[5] 王晶禹，张景林，王保国．炸药的重结晶超细化技术研究 [J]．北京理工大学学报，2000，20 (3)：385-388.

[6] Feairheller W R, Donaldson T A, Thorpe R. Recrystallization of HNS for the preparation of detonator grades explosive material [R], DE88012862, 1998.
[7] 孙荣康，任特生，高怀玲．猛炸药的化学与工艺学 [M]．北京：国防工业出版社，1983：521-589.

Preparation and Impact Sensitivity of Nanometer Explosive Powder of RDX

Chen Houhe, Meng Qinggang, Cao Hu, Pei Yanmin, Zhu Lv feng, Lv Chunxu
(School of Chemical Engineering, Nanjing University of Science and Technology, Nanjing, 210094, Jiangsu, China)

Abstract: The nanometer explosive powder of RDX with mean diameter 40～60nm was prepared by spraying evaporation. Experimental results shown that, the mechanical impacting sensitivity of RDX play down as the particle diameter of RDX reduces. The impact sensitivity of industrial RDX is 24.2cm, but that of nanometer RDX is 54～59nm.

Key words: mechanics of explosion, impact sensitivity, sparying evaporation, RDX, nanometer, explosive

（注：此文原载于 爆炸与冲击，2004，24 (4)：382-384）

NEAK分子间炸药的热分解

赵省向[1]，张亦安[1]，胡焕性[1]，吕春绪[2]
(1. 西安近代化学研究所，陕西 西安 710065)
(2. 南京理工大学，江苏 南京 210094)

摘要：通过恒温加热失重测定和DSC分析，研究了硝基胍-乙二胺二硝酸盐-硝酸铵-硝酸钾低共溶物分子间炸药（NEAK）的热分解特性。结果表明NEAK与三元低共熔物EAK（乙二胺二硝酸盐-硝酸铵-硝酸钾）的分解性接近，加入RDX和过量NQ时，分解速率显著增加。

关键词：热稳定性，分子间炸药，低共熔物，EAK，NQ，RDX

引言

分子间炸药是相对于分子内炸药而言的一类合炸药。它是由可燃剂化合物和氧化剂化合物混合组成的混合物，该混合物的能量并不是二者能量简单的算术加和，而是比算术加和所预期的要高[1]。由于是分子之间形成的炸药体系，所以对外界刺激的敏感性较低，加之与分子内炸药相比，研制、制造成本低，周期短，因此近年来已经引起了人们广泛的关注[2]。

NEAK是在三元低共熔物EAK（EDD（乙二胺二硝酸盐）-AN（硝酸铵）-KN（硝酸钾））的基础上加入NQ（硝基胍）得到的一种分子间炸药。与EAK相比NEAK这种分子间炸药的熔点降低5℃左右，而且由于NQ的加入降低了冲击感度。但是加NQ后NEAK的热分解特性是否会受其影响，以及为了提高能量在加入高能炸药后是否影响其热安定性，是人们关心的问题。

本文通过DSQ法恒温热失重法对NEAK体系的热分解特性以及分解动力学参数进行了研究和测定。

1 实验

1.1 样品准备

将AN（化学纯）、EDD（实验室合成）、KN（化学纯）按46/46/8的比例制得EAK，EAK中加入8%的NQ制得NEAK，按NEAK：RDX（工业级，精制）为1∶1的比例制得NEAK-R；按NQ∶NEAK为1∶1的比例制得NEAK-NQ。

1.2 DSC分析

仪器采用Perkin-Elmer DSC-2C型差示扫描量热仪，样品量约1mg左右，N_2流速40mL/min，样品盘为铝盘。

1.3 恒温热失重试验

将样品在 60℃真空烘箱中真空干燥 4h，然后取样 5g 左右放入玻璃小瓶中，不加盖、在某一定温度下恒温一定时间后取出放入干燥器中，冷却后称重。

2 结果与讨论

2.1 DSC 分析结果

NEAK、EAK、NEAK-R、NEAK-NQ、NQ、RDX 的 DSC 图谱如图 2.1 所示。从图中可以看出，几乎在熔化的同时发生了分解，分解峰的峰温大约 245℃，这与文献［3］的结果是一致的。NEAK 虽然熔点比 EAK 降低 5℃左右，但是分解峰温也降低了 5℃左右，即热稳定性略有降低 NEAK 是由 92%左右的 EAK 加上 8%左右的 NQ 组成的，所以可以认为，NEAK 热稳定性比 EAK 降低是由于 NQ 作用的结果。图 2.1 中的 NEAK 曲线，其放热峰并非是一个完全的单峰，而是在峰值之外有一肩峰，表现出一旦发生分解，EAK 和 NQ 有单独分解完全的倾向，比较 NEAK、EAK 和 NQ 的 DSQ 曲线，发现 NEAK 和 NQ 的分解峰温相近，NEAK 分解峰的肩峰与 EAK 的分解峰有关。当然彼此的相互作用使得两者更容易分解。NEAK-NQ 的分解曲线更加说明了这一点。NEAK-NQ 的分解曲线出现明显的两个分解峰，峰温降至 217℃和 230℃，与 NEAK 相比分别降低 28℃和 15℃。分解峰温如此大幅度的降低说明两者相互作用是存在的，在 NEAK-NQ 的两个分解峰中，低温分解峰极有可能是由 NQ 分解引起的，另一较高温度的分解峰可能是由 EAK 引起的，因为从 EAK 和 NQ 单独的分解曲线中的分解峰值看 EAK 的热稳定性稍好于 NQ。另外，图 2.1 列出了 NEAK-R 样品的 DSC 分析结果。从此结果看，NEAK-R 的分解峰温高于 NEAK-NQ，但 NEAK-R 的起始分解温度低于 NEAK-NQ，所以从 DSQ 曲线难以看出两者热稳定性的高低。

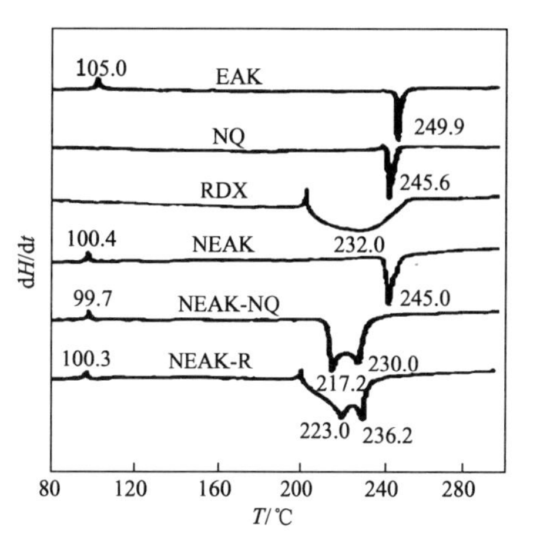

图 2.1 DSQ 热分解图谱（升温速率 5℃/min）

Fig. 2.1 DSC curves of thermal decomposition (Heating rate5℃/min)

2.2 分解动力学

图 2.2 (a) ～ (e) 分别为 107℃、124℃、140℃、150℃、160℃下测得的不同样品的分解率与时间的关系。在图 2.2 中，不同温度下的热分解率与时间的关系曲线。EAK、NEAK、和 NEAK-R 基本上是直线，而 NEAK-NQ 则并非是直线。显然在分解初期（分解率小于 7%）时，EAK、NEAK 和 NEAK-R 的热分解动力学是零级反应。NEAK-NQ 用一级反应，动力学方程积分形式 $\ln(1-a)=-kt$ 对 140℃、150℃的分解曲线进行拟合时，发现 $\ln(1-a)$ 与 t 的线性相关系数超过 0.99，所以可以认为在 150℃以下，分解率低于 7%时，NEAK-NQ 的热分解是一级反应。而在 160℃时 NEAK-NQ 的反应很难估计出反应级数。NEAK-NQ 体系，在实验过程中发现随着温度的提高，NQ 在 NEAK 中的溶解度增高，当达到 160℃时，NQ 几乎完全溶解于 NEAK 液相体系中。所以在 150℃以下的几个温度点，NEAK-NQ 体系是一个两相体系，而且 NQ 在 NEAK 中的含量随温度的不同而不同，在 160℃是一个单相体系，所以这样的体系随着温度的不同具有不同的反应历程是不奇怪的。Volk[4] 曾研究了 NQ 的分解机理，在 110～240℃，在分解反应初期，NQ 的分解反应是

一级反应，反应主要生成 NH_4、CO_2、NO_2。随着温度的升高，分解产物中，三聚氰胺及其水解产物的比例增加，而这些分解固体残渣导致分解速率降低很多，所以在分解率较大时，并非是一级反应，Lee 和 Back[5]对 NQ 的研究表明当温度达到 160℃时，NQ 的反应机理可能发生变化。以上结果显示，NEAK-NQ 的分解更接近于的分解机理，而且具有不同的分解反应机理。

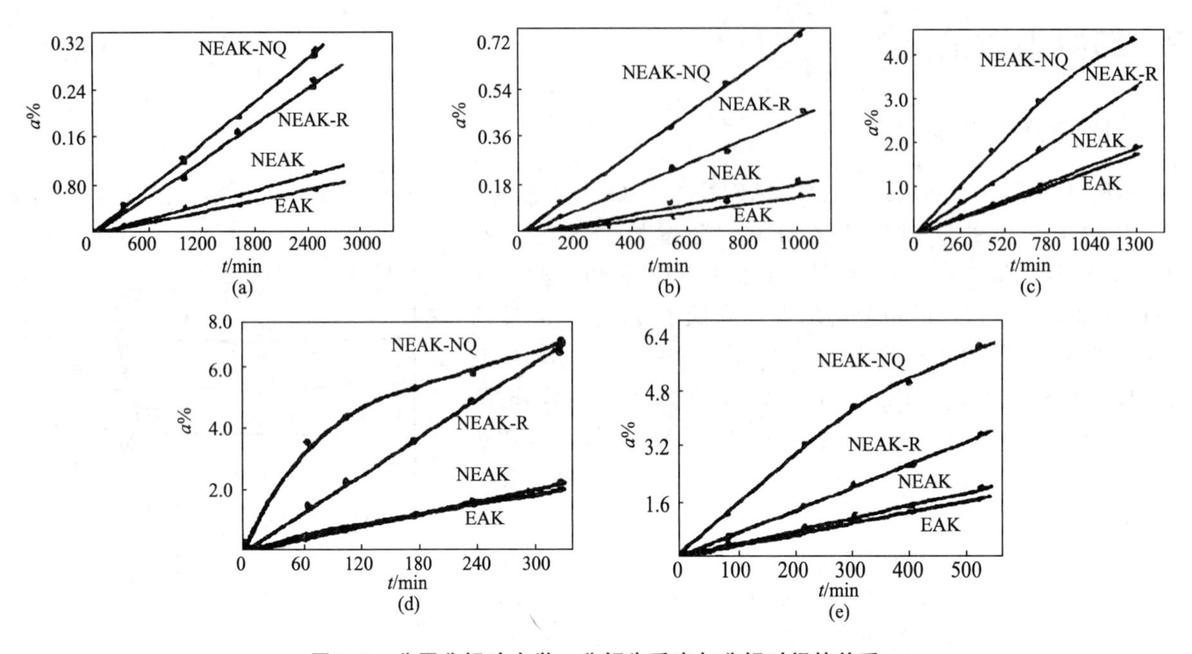

图 2.2 分图分解动力学：分解失重率与分解时间的关系

Fig. 2.2 weight-loss as a function of time at different temperates

从图 2.2 可以看出，NEAK 的热分解率高于 EAK，而在 NEAK 中加入 NQ 和 RDX 时，则使其分解率增高很多。这种情况在高温下更为明显。Debenham 和 Owen[6]对 RDX 的分解研究发现，分解产物 NH_4 对 RDX 的分解有催化作用：Volk[4]对 NQ 分解的研究也得出同样的结果，即分解产物 NH_4 对 NQ 分解有催化作用。所以 NEAK-NQ 和 NEAK-R 的分解速度比 NEAK 高出许多，正是由于 AN 和 EDD 分解产生大量的 NH_4[5]，而增加了 NQ 和 RDX 的分解速率所致。

表 2.1 分解动力学参数

Table 2.1 Decomposition kinetic parameters

样品	EAK	NEAK	NEAK-R	NEAK-NQ
活化能/kJ/mol	147.0	136.0	137.2	134.6
指前因子/s^{-1}	7.086×10^{11}	3.286×10^{10}	1.006×10^{11}	6.496×10^{10}
线性相关系数	0.9923	0.9949	0.9956	0.9996

通过图 2.2 计算的动力学参数列于表中。从动力学参数看下 NEAK 比 EAK 的活化能降低很多，说明比 EAK 有较强的反应性，这与 DSQ 的结果是一致的。NEAK-R 和 NEAK-NQ 的活化能则与 NEAK 相近，这似乎 NEAK、NEAK-R、NEAK-NQ 具有相似的反应性，其实不然。实际上，NEAK 由于含 NQ 的量很少，所以其分解的主体仍然是 EAK，即从总体反应上，两者有相似的反应历程。但 NEAK-R、NEAK-NQ 和 NEAK 具有完全不同的历程。首先 NQ 和 RDX 达到 50%的比例，所以整个体系的反应不但包括 EAK 的反应，还包括 NQ 和 RDX 单独的分解反应，以及 EAK 同 NQ 和 RDX 的相互的反应。NEAK、NEAK-R、NEAK-NQ 具有相似的活化能，并不能说明它们的反应性相似。实际，与 NQ 以及 RDX 的活化能比较，NEAK-R 的活化能远低于 RDX 的 170～220kJ/mol[7]，NEAK-NQ 的活化能远低于 NQ 的活化能 215kJ/mol[4]。

3 结论

曲线显示 NEAK 的分解峰温稍低于 EAK，通过恒温热失重法得到的分解动力学参数也显示的 NEAK 分解速率略大于 EAK 的分解速率，表明 NEAK 的分解性略大于 EAK；但两者相差不大，很接近。在该体系中加入 RDX 和过量的 NQ，显著增加了整个体系的分解性。通过恒温热失重法进行的分解动力学测定表明，NEAK、EAK、NEAK-R 在反应初期实验温度范围内是零级反应，而 NEAK-NQ 在反应初期 150℃以下是一级反应。

参考文献

[1] Zhang Yian，et al. Preparation and characterization of an an-containing IMX. In：Proceeding of China-Japan Seminar on Energetic Materials [J]. Safety and Environment，1996，75-78.

[2] Sumrall T S. Development of an economical high performance melt castable insensitive high explosive [J]. Kayaku Gakkaishi，1997，58 (2)：58-62.

[3] Lee P P. Back M H. Thermal decomposition of NQ [J]. Thermochimica Acta，1989，141：305-315.

[4] Volk F. Determination of gaseous and solid decomposition products of nltroguaniiline [J]. Prapellants Explosives Pyrotechnics，1985，10：139-146.

[5] Lee P P. Back M H. Kinetic studies of the thermal decomposition of nirroguanidine using accelerating rate calorimetry [J]. Thermochimica Acta，1980，127：89-100.

[6] Debenham D F. Owen A J. The thermal decomposition of RDX in 1,3,5-trinitrobenzene [J]. Symp Chem Problem Stab Explos，1976.4，201-220.

[7] 楚士晋. 炸药热分析 [M]. 北京：科学出版杜，1994，38-40.

Decomposition of NEAK Intermolecular Explosive

Zhao Shengxiang[1]，Zhang Yian[1]，Hu Huanxing[1]，Lv Chunxu[2]

(Xi'an Modern Chemistry Research Institute，Xi'an，710065)

(Nanjing University of Science and Technology，jinagsu Nanjing，210094)

Abstract：The decomposition of NEAK (Nitroguanidine-Ethylenediamine Dinitrate-Ammonium Nitrate-Potassium Nitrate) eutectic intermolecular explosive was investigated by non-isothermal DSC and weight loss measurement under different temperatures. It indicated that the thermal stability of NEAK approaches that of EAK (Ethylenediamine Dinitrate-Ammonium Nitrate-Potassium Nitrate)，and when NEAK was mixed with NQ or RDX，the mixtures tend to have a higher decomposition speed.

Key words：thermal stability，intermolecular explosive，eutectic，EAK，NQ，RDX

（注：此文原载于 兵工学报，2001，22 (1)：133-136）

超细 RDX 爆速和作功能力的研究与测试

刘桂涛，吕春绪，曲虹霞
南京理工大学（南京，210094）
EI：03327585042

摘要：采用纯超细 RDX 与工业 RDX，以及以超细 RDX 和工业 RDX 为主体，采用相同配方和制备工艺制取的高分子粘结炸药对比的方式，在相同的实验条件下，对爆速、作功能力进行了测试。得出超细 RDX 无论是纯品还是高分子粘结炸药，对爆速和作功能力均优于工业 RDX 的结论。

关键词：爆速，作功能力，超细

1 引言

超细材料近十几年来极大地推动了材料学科的发展。在它的推动下，炸药领域的研究人员逐渐发现：超细炸药颗粒与其同类的大颗粒相比，在一些性能上发生了变化[1]。随着超细粉体应用技术的日臻成熟，推动了超细炸药实际应用特性的研究。从检索 1988 年～2000 年间国内外相关资料看，国内外对超细黑索今（RDX）的研究尚不系统。通过对超细 RDX 机械感度、爆轰感度的研究，发现超细 RDX 无论是纯品还是高分子粘结炸药，其机械感度均明显低于工业 RDX，而爆轰感度（用极限起爆药量表征）则高于工业 RDX。从这一点可以看出，RDX 被超细后其使用过程中的安全性是有保证的，且易于起爆[2]。

在实验过程中，为了使得到的数据更具有可比性，采用纯超细 RDX 和工业 RDX 以及以超细 RDX 和工业 RDX 为主体、采用相同配方和制备工艺制取的高分子粘结炸药（以下简称造粒）对比的方式，在相同的实验条件下，对爆速、作功能力进行了测试和研究。实验中所采用的超细 RDX 的平均粒径为 5.52μm，工业 RDX 的平均粒径为 121.29μm。

2 实验部分

2.1 爆速的测定

测定爆速的方法目前常用的有道特里什法、测时仪法和高速摄影法。测时仪法是目前应用范围较广、测试精度较高（可达 10^{-7}s）、操作及原理都很简单的方法。它是以测出一定长度药柱达稳定爆轰通过两个靶线的时间，并利用时间与两靶线间距的关系达到测试爆速的目的。准确称取四种待测炸药（超细黑索今、工业黑索今、超细黑索今造粒、工业黑索今造粒），每份 7.3g，各取 20 份；选择使药柱直径达到其临界直径的合适模具，压制成形。对压制好的药柱，必须准确测出其密度，以利于下一步的挑选装配，采用排水法来测其密度（测出的密度为装药平均密度 ρ）。认真挑选四种装药平均密度相近的药柱各 10 组，编好序号进行装配。采用的测时仪型号为 MZ-C1 型；时间分辨率为 0.1μs，每次测一组药柱，得一个数据，测试结果见表 1～表 4。

表 1 纯超细 RDX 爆速测试

序号	装药平均密度 ρ/g·cm^{-3}	爆速 D/m·s^{-1}	序号	装药平均密度 ρ/g·cm^{-3}	爆速 D/m·s^{-1}
1	1.6867	8375	7	1.6877	8358
2	1.6860	8349	8	1.6870	8350
3	1.6874	8353	9	1.6865	8343
4	1.6869	8350	10	1.6873	8357
5	1.6868	8345	平均值	1.6869	8352
6	1.6872	8361			

表 2 纯工业 RDX 爆速测试

序号	装药平均密度 ρ/g·cm^{-3}	爆速 D/m·s^{-1}	序号	装药平均密度 ρ/g·cm^{-3}	爆速 D/m·s^{-1}
1	1.6870	8280	7	1.6863	8266
2	1.6871	8273	8	1.6874	8271
3	1.6866	8285	9	1.6878	8277
4	1.6873	8288	10	1.6876	8281
5	1.6875	8279	平均值	1.6871	8277
6	1.6868	8270			

表 3 超细 RDX 造粒爆速测试

序号	装药平均密度 ρ/g·cm^{-3}	爆速 D/m·s^{-1}	序号	装药平均密度 ρ/g·cm^{-3}	爆速 D/m·s^{-1}
1	1.6983	8215	7	1.7010	8135
2	1.7000	8222	8	1.6990	8212
3	1.6989	8218	9	1.6985	8229
4	1.6994	8216	10	1.6980	8208
5	1.6995	8235	平均值	1.6989	8216
6	1.6988	8206			

表 4 工业 RDX 造粒爆速测试

序号	装药平均密度 ρ/g·cm^{-3}	爆速 D/m·s^{-1}	序号	装药平均密度 ρ/g·cm^{-3}	爆速 D/m·s^{-1}
1	1.6988	8128	7	1.7010	8135
2	1.6944	8124	8	1.6990	8125
3	1.6997	8120	9	1.7001	8130
4	1.6989	8114	10	1.6987	8117
5	1.6985	8126	平均值	1.6993	8123
6	1.6991	8119			

2.2 作功能力的测定及结果

炸药爆轰时生成高温高压的爆炸产物，在对外膨胀时压缩周围的介质，使其邻近的介质发生变形、破坏和飞散而做功。因而炸药的做功能力是评价炸药性能的一个重要参数。长期以来，各国对作功能力的评定方法均有各自的具体规定，国内评定作功能力的主要方法是铅扩孔值法。制作直径 200mm、高 200mm 的圆柱形铅，中央有一直径 25mm、深 125mm 的内孔；称取炸药试样 10g，压制成带雷管孔（直径为 24mm）的密度为 1g·cm^{-3}的药柱；往雷管孔中插入 8 号雷管；其空余部分用风干的石英砂填满，装配好进行测试。测定结果见表 5。

表 5 作功能力测定

炸药种类	工业 RDX	超细 RDX	工业 RDX 造粒	超细 RDX 造粒
扩孔值/ml	408	462	354	359

3 分析讨论

关于冲击非均相炸药时形成热点，目前普遍认为是炸药受冲击或摩擦时，并不是全部遭受冲击作用的物质平均加热，而是其中很少的个别部分[3]。例如炸药内部的空穴、间隙、杂质和密度间断等处，冲击波对这些地方进行绝热压缩，将机械能转化为热能，使其温度大大高于平均温度，从而在这些区域形成热点；热点附近的炸药晶体颗粒上同时发生化学反应，反应放出的热量以热点为中心，以热爆炸或高速点燃的形式向四周传播。该能量使入射冲击波得到加强，被加强了的冲击波与密度间断作用，形成更多温度更高的热点。于是冲击波强度越来越大，释放出的化学反应能量也越来越多，直至冲击波被加强到使炸药从开始点燃到使炸药发生低速爆轰，进而发展为稳定爆轰。而热点的形成一般公认有以下几种：

（1）炸药内所含空穴和气孔的压缩；

（2）炸药颗粒、各种杂质与冲击波的相互作用；

（3）颗粒之间的摩擦；

（4）空穴或气孔的表面能转化为动能；

（5）晶体的错位和塌陷；

（6）炸药塑性流变效应。

对于粒径为微米或亚微米的超细粉体，虽然其物理化学性质与大块材料的物理化学性质相差不太大，但其比表面积增大，表面能大，表面活性高，表面与界面性质发生了很大变化[4]。通过理论近似计算表明，1kg RDX 炸药在超细到 5m 时，表面能提高近 10×4.1868J；虽然这与 RDX 炸药反应放出的能量相比是很小的数值，但从侧面也反映了炸药在超细化后总能量是提高的。

炸药在受到冲击波强烈冲击时，如果装药条件和配方完全相同，超细炸药颗粒间包络的气孔的数量较之普通炸药会增加，且各个气孔的体积趋于均匀，这就意味着各气孔的表面能相差不大，更趋于平均。在外界作用下，一旦形成爆炸热点，这些爆炸热点会迅速反应，同时将反应所释放的能量迅速传向周围的介质；由于各气孔的表面能相差不大，未形成爆炸热点的气孔只要吸收少许能量也能成为爆炸热点。因此，超细炸药与普通炸药相比，在形成爆炸热点后，能在短时间内形成数量更多的爆炸热点，故爆炸反应持续的时间会比普通炸药大大缩短。对于一种单质炸药而言，它自身的能量是确定的，但由于超细炸药爆炸反应的时间短，因此超细炸药爆炸产生的能量在爆炸过程中向外界损失的能量就少，就能形成较高的作功能力。从这个角度上说，超细炸药的爆速、作功能力较之工业同类炸药有所增强。而对于作功能力来说，由于在造粒的过程中加入了粘结剂和增塑剂，超细的优势在很大程度上被掩盖了，因此两者的作功能力相差不大。

4 结论

本课题通过对超细 RDX 与工业 RDX，以及以超细 RDX 和工业 RDX 为主体，采用相同配方和制备工艺制取的高分子粘结炸药的对比测试后发现：提高黑索今炸药的粉碎程度对其综合性能指标有明显的改善，超细 RDX 无论是纯品还是造粒，其爆速和作功能力均明显优于工业 RDX 或两者相当(造粒后的作功能力)。通过对本课题的实验研究可以看出，RDX 造粒后，在使用过程中的安全性是有保证的。从爆速和作功能力的测试数据看，超细 RDX 无论是纯品还是造粒，均优于工业 RDX。因此对超细黑索今炸药在爆轰性能及作功能力等方面的特性展开系统研究，拓展其应用范围，对于其他超细炸药的研究也具有参考价值。

参考文献

[1] 刘志建．超细材料与超细炸药技术．火炸药，1995，(2)：15-18.

[2] 刘桂涛，曲虹霞．超细 RDX 爆轰感度与机械感度的研究．南京理工大学学报，2002，(4)：410-413.
[3] 张俊秀，刘光烈，刘桂涛．爆炸及其应用技术．北京：兵器工业出版社，1998.58-65.
[4] 程传煊．表面物理化学．北京：科学技术文献出版社，1995.351-369.

Study and Test on Detonation Velocity and Power of Superfine RDX

Liu Guitao，Lv Chunxu，Qu Hongxia
Nanjing University of Science and Technology (Nanjing，210094)

Abstract: To compare the proper ties of super fine RDX with common industrial RDX，and those of granulated super fine RDX with granulated industrial RDX. The detonation velocity and power tested in the same condition. The testing results show that both pure and granulated super fine RDX are better than those of industrial RDX in the detonation velocity and power.

Keywords: detonation velocity，power，super fine

(注：此文原载于 爆破器材，2003，32 (3)：1-3)

TATB及其杂质的绝热分解研究

高大元[1,2]，徐容[1]，董海山[1]，李波涛[1]，吕春绪[2]
（1. 中国工程物理研究院化工材料研究所，四川，绵阳，621900）
（2. 南京理工大学化工学院，江苏，南京，210094）

摘要： 用加速热量仪研究了TATB及其杂质的绝热分解过程，得到了绝热分解温度和压力随时间的变化曲线以及温升速率随温度的变化曲线，根据绝热加速热量仪的温升速率方程和Arrhenius方程，计算了TATB绝热分解的动力学参数。结果表明：TATB的初始热分解温度大于290℃，反应系统最高温度大于340℃，最大压力超过1MPa，而TCTNB和TCDNB存在多次放热反应，但每次放热量都很小，只有最后一次放热速率比较大。

关键词： 爆炸力学，绝热分解，加速热量仪，TATB，杂质

1 引言

TATB是钝感炸药配方的重要组分，其质量和性能直接影响钝感炸药的安全性和可靠性[1-2]。TATB本身是非常稳定、非常安全的炸药，但在TATB产品中含有十余种有机物杂质，TCTNB（2,4,6-trichloro-1,3,5-trinitrobenzene）、TCDNB（2,4,6-trichloro-1,3-dinitrobenzene）是两种主要有机物杂质[3]，其分子结构如图1所示。从这些杂质的分子结构可以预估，其热安定性、安全性和爆轰性能都会低于TATB，杂质的存在会损害炸药部件的综合性能。为了了解它们的危害程度，必须制备出这些杂质，研究其物理化学性质、热安定性以及对各种外界刺激的感度和主要爆轰性能。在以往的含能材料热分解特性研究中，多采用差热分析或差示扫描量热分析，这两种方法中不同的程序升温速率对物质的热分解特性曲线影响很大，因而测试得到的初始分解温度值存在一定的差异。

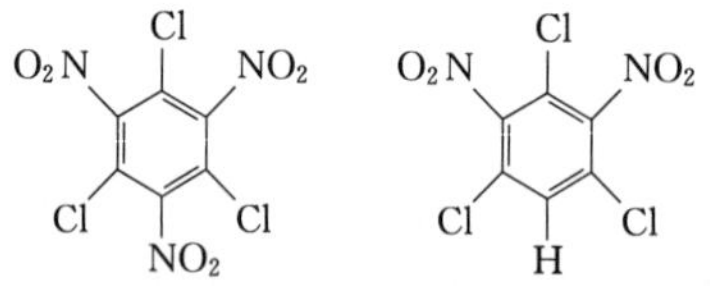

图1 TCTNB和TCDNB的分子结构

Fig. 1 Molecular structure of TCTNB and TCDNB

本文中用加速热量计（Accelerating Rate Calorimeter，ARC）[4]研究TATB及其杂质的绝热分解，以便对TATB及其杂质有一个比较全面的认识。用ARC研究物质的热分解过程，不仅可以得到绝热条件下的初始分解温度，而且还能够详细了解热分解过程中温度和压力的变化，并计算出一定温度范围内的动力学参数，这些数据为研究杂质对TATB的热安定性、安全性和爆轰性能提供理论和实验依据。

2 实验

2.1 仪器与测试原理

所用ARC为北京理工大学爆炸与安全科学国家重点实验室购买的美国Dow化学公司仪器。操作温度范围为0～500℃，压力范围为0～17MPa，测试样品量的范围为0.1～5g。

ARC的操作过程为加热-等待-搜寻，操作模式见图2。ARC首先被加热到预先设置的初始温度，等待一段时间使系统温度达到平衡后搜寻反应系统的温升速率。如果反应系统的温升速率低于预设的温升速率0.02℃/min，ARC将按照预先选择的温升幅度自动进行加热-等待-搜寻，直至探测到比预设值高的温升速率。当反应系统的温升速率超过预先设定的温升速率后，量热体系将保持绝热状态直至整个试验完成。

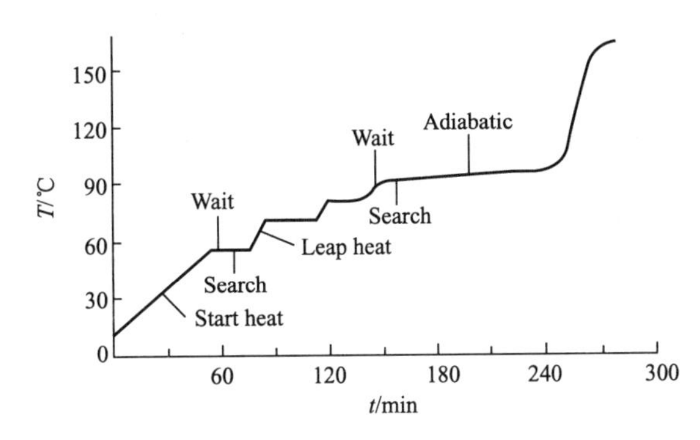

图2 加速热量仪的加热-等待-搜寻操作模式

Fig. 2 The heat-wait-search mode of ARC

2.2 试样与测试条件

TATB样品用TCTNB经氨化合成制得；TCTNB样品是从山西永济575厂购买的TCTNB在50℃、经二氯乙烷溶剂重结晶制得；TCDNB样品用三氯苯和硝酸在常温下硝化反应制得。测试样品量及测试条件见表1。表1中，M为样品质量，M_b为样品球质量，T_i为起始温度，s为温升速率灵敏度。

表1 试样量与测试条件

Table1 Mass of measured samples and measuring conditions

样品	M/g	M_b/g	T_i/℃	s/℃	样品	M/g	M_b/g	T_i/℃	s/℃
TATB-1	0.1185	6.6172	200	0.02	TCTNB	0.1740	6.6118	100	0.02
TATB-2	0.1648	6.5453	250	0.02	TCDNB	0.1532	6.6172	100	0.02

2.3 测试数据的校正

由于样品反应产生的热量不仅要用于加热自身，而且还要加热盛装样品的样品球，所以测试结果是样品与样品球所组成的整个反应系统的温度。当样品反应放出的热量全部用于加热自身时，样品的实际温升和实际温升速率都要比测量值高。

ARC实验中样品的自加热温度与测量值之间的关系为[5]

$$\Delta T_{ad}=T_f-T_0=\Phi\Delta T_{ad,s} \tag{1}$$

$$\Phi=(Mc_v+M_bC_{V,b})/(Mc_v) \tag{2}$$

在零级或准零级反应条件下（T_0附近），样品的初始温升速率

$$m_0 = \theta_{m,s} / \Phi \tag{3}$$

样品从初始分解温度 T_0 开始到达最大温升速率 m_m 所需要的时间

$$\theta_m = \theta_{m,s} / \Phi \tag{4}$$

式中：T_0 为样品的初始分解温度；T_f 为样品的最高分解温度；ΔT_{ad}为样品的绝热温升；$\Delta T_{ad,s}$为反应系统的绝热温升；φ 为热惰性因子；c_v 为样品的比热容；$c_{v,b}$为样品球的比热容；m_0 为样品初始温升速率；$m_{0,s}$为反应系统初始温升速率；θ_m 为样品最大温升速率所需时间；$\theta_{m,s}$为系统最大温升速率所需时间。

3 实验结果与讨论

3.1 TATB 的绝热分解

TATB 在起始温度分别为 200℃和 250℃的测试结果见图 3 和图 4。

由图 3 可以看出，TATB-1 在起始温度为 200℃时，初始分解温度为 294.34℃，相应的放热速率为 0.024℃/min。在 ARC 检测到放热反应后，TATB-1 反应系统的温度和压力开始缓慢上升，温升速率以较小的幅度缓慢增加，在 330.67℃出现最大温升速率 43.5℃/min，之后温升速率逐渐下降，但反应系统温度持续增加，反应系统最高温度为 340.0℃，最大反应压力为 1.076MPa。

由图 4 可以看出，TATB-2 在起始温度为 250℃时，初始分解温度为 294.87℃，相应的放热速率为 0.008℃/min。在 ARC 检测到放热反应后，TATB-2 反应系统的温度和压力开始缓慢上升，温升速率以较小的幅度缓慢增加，在 325.79℃出现最大温升速率 186.5℃/min，之后温升速率逐渐下降，但反应系统温度持续增加，反应系统最高温度为 350.74℃，最大反应压力为 1.521MP。

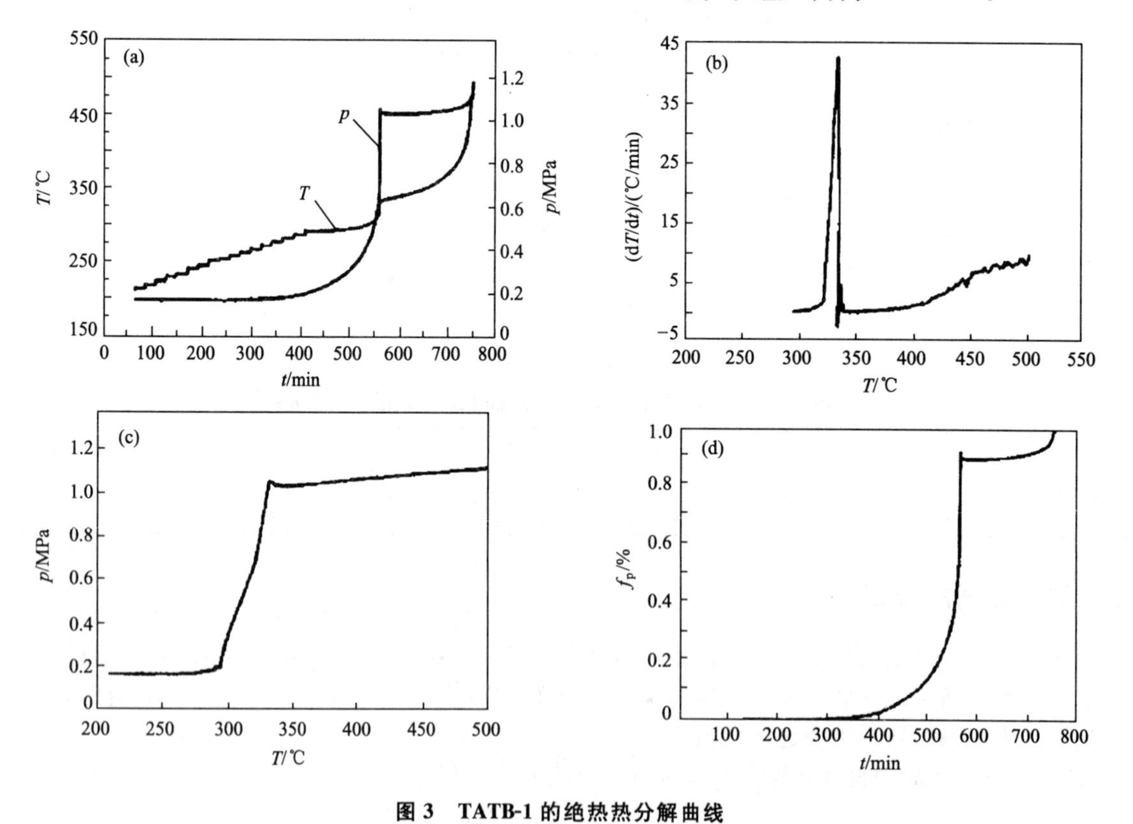

图 3 TATB-1 的绝热热分解曲线

Fig. 3 Adiabatic decomposition curves of TATB-1

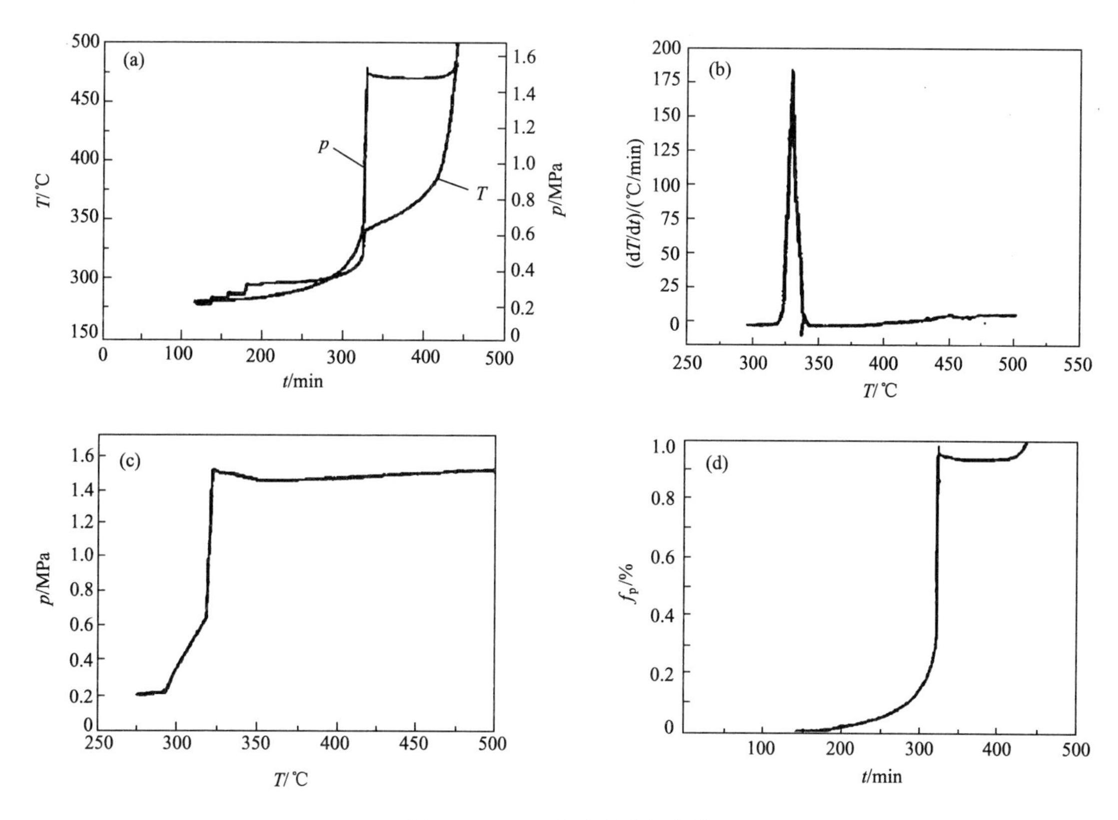

图 4　TBTA-2 的绝热分解曲线

Fig. 4　Adiabatic decomposition curves of TATB-2

表 2 列出了 TATB 测试样品以及用反应系统的热惰性参数 φ 校正的热分解特性数据。表中，测试值为系统的，而校正值则为样品的。为了进一步说明压力随时间的变化，将相对于初始分解反应发生时的压力增加与最大反应压力减去初始分解反应压力的差值之比，定义为压力转化分数 f_p，其数值范围为 0～1，作出了压力转化分数%时间曲线，见图 3（d）、图 4（d）。TATB 的压力转化分数-时间曲线关系较为简明，它在一定程度上可以反映分解反应进行的程度。

表 2　TATB 绝热分解特性参数的测试值和校正值

Table 2　Measured and modified values of the adiabatic decomposition characteristic parameters of TATB

样品	T_i/℃		φ		T_0/℃		m_0/(℃/min)		T_f/℃	
	测试值	校正值	测试值	校正值	测试值	校正值	测试值	校正值	测试值	校正值
TATB-1	200	200	—	7.064	294.34	294.34	0.024	0.170	340.0	662.54
TATB-2	250	200	—	5.313	294.87	294.87	0.008	0.0425	350.74	647.58

样品	$\Delta T_{ad,s}$/℃		M_m/(℃/min)		θ_m/min		T_m/℃		p_m/MPa	
	测试值	校正值	测试值	校正值	测试值	校正值	测试值	校正值	测试值	校正值
TATB-1	45.66	322.54	43.5	—	557.77	78.96	330.67	—	1.076	—
TATB-2	55.87	296.84	186.5	—	320.47	60.32	325.79	—	1.521	—

3.2　动力学参数的计算

根据绝热加速热量仪的温升速率方程[6]

$$m_{T,s}=\frac{dT}{dt}=\Delta T_{ad,s}k\left(\frac{T_{f,s}-T}{T_{f,s}-T_0}\right)^n \tag{5}$$

则

$$k=\frac{m_{T,s}}{\Delta T_{\mathrm{ad},s}\left(\frac{T_{f,s}-T}{\Delta T_{\mathrm{ad},s}}\right)^{n}} \tag{6}$$

由 Arrhenius 方程可得

$$\ln k=\ln A-\frac{E}{R}\left(\frac{1}{T}\right) \tag{7}$$

式中，$m_{T,s}$为反应系统的温升速率；$T_{f,s}$为反应系统的最高分解温度；T 为反应系统的温度；t 为反应时间；k 为反应速率常数；E 为反应系统表观活化能；A 为指前因子。

当反应级数 n 选取合适时，$\ln k \sim 1/T$ 为直线，由直线的斜率和截距可求活化能 E 和指前因子 A。

处理 ARC 实验数据时，根据测试得到的初始放热温度 T_0、反应系统最高温度 $T_{f,s}$以及不同时刻下反应系统的温度 T 和温升速率$m_{T,s}$数据，由方程（6）可计算出不同温度下的 k 值，作出 $\ln k \sim 1/T$ 直线。取 $n=1$ 时作线性回归，即可得斜率 E/R 和截距 $\ln A$，从而可以求出活化能 E 和指前因子 A，计算结果见表 3。

表 3　TATB 动力学参数的计算结果

Table 3　Calculated results of the kinetic parameters of TATB

样品	T_i/℃	E/(kg/mol)	A/min^{-1}
TATB-1	20	428.15	3.93×10^{35}
TATB-2	250	455.72	1.39×10^{38}

3.3　TCTNB 和 TCDNB 的绝热分解

TCTNB 和 TCDNB 在起始温度为 100℃时的测试结果见图 5、图 6。

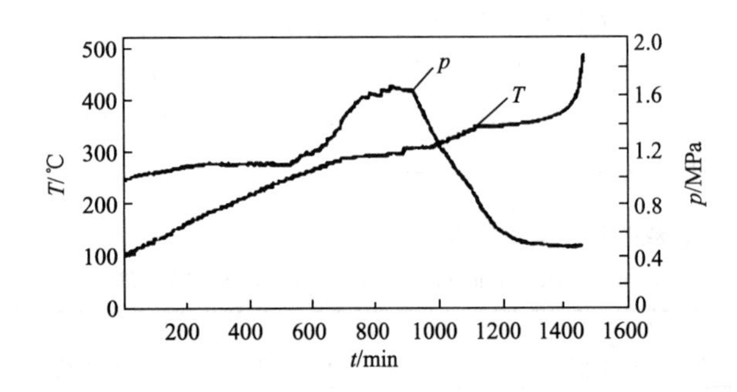

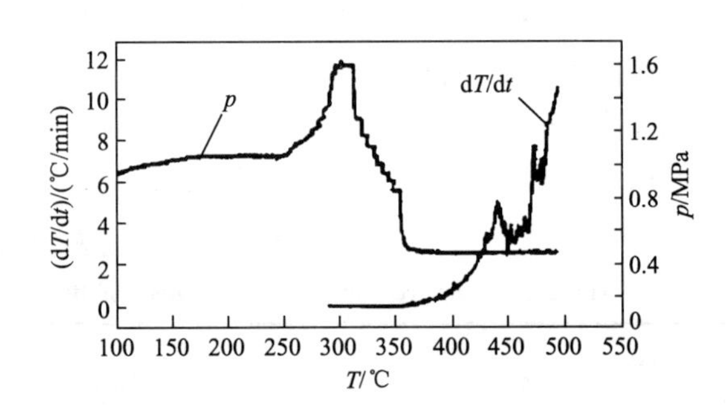

图 5　TCTNB 的绝热分解曲线

Fig. 5　Adiabatic decomposition curves of TCTNB

由图 5 可以看出，TCTNB 样品在起始温度为 100℃，初始分解温度为 290.75℃，相应的放热速率为 0.054℃/min。在 ARC 检测到放热反应后，TCTNB 样品反应系统的温度和压力开始缓慢上升，在 439.71℃出现第一次最大温升速 5.166℃，之后温升速率逐渐下降，在 471.76℃又出现第二次最大温升速率 7.916℃/min。反应系统最大反应压力为 1.614MPa。

由图 6 可以看出，TCDNB 样品在起始温度为 100℃时，初始分解温度为 285.58℃，相应的放热速率为 0.074℃/min。在 ARC 检测到放热反应后，TCDNB 样品反应系统的温度和压力开始缓慢上升，在 453.04℃出现第一次最大温升速率 5.611℃/min，之后温升速率逐渐下降，在 427.2℃出现第二次最大温升速率 8.5℃/min；在 491.06℃又出现第三次最大温升速率 9.9℃/min。反应系统最大反应压力为 0.7224MPa。

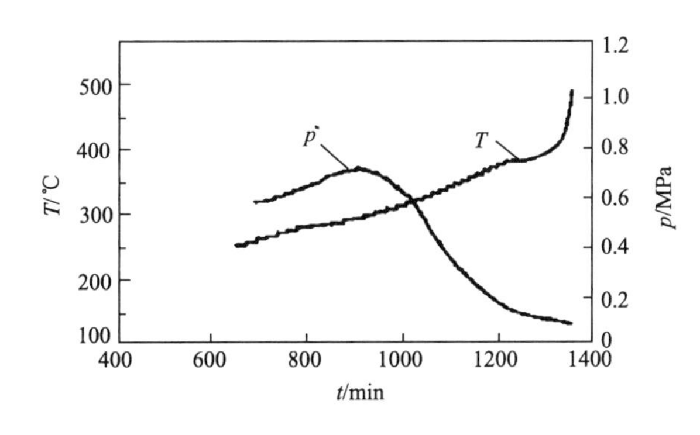

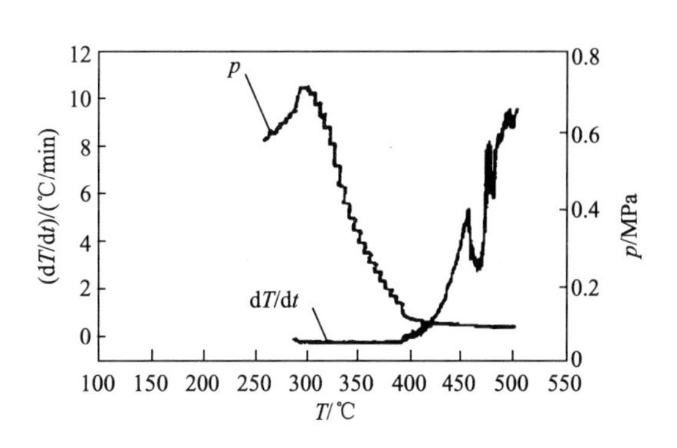

图 6 TCDNB 的绝热分解曲线

Fig. 6 Adiabatic decomposition curves of TCDNB

表 4 列出了测试 TCTNB 和 TCDNB 样品以及用反应系统的热惰性参数 φ 校正的热分解特性数据。TCTNB 和 TCDNB 样品在 ARC 实验中存在多次放热反应，但每次放热量都很小，只有最后一次放热速率比较大。TCTNB 和 TCDNB 样品的补作 ARC 实验也出现相同情况，其影响因素有待进一步研究。

表 4 TCTNB 和 TCDNB 绝热分解特性参数的测试值和校正值

Table 4 Measured and modified values of the adiabatic decomposition characteristic parameters of TCTNB and TCDNB

样品	T_i/℃		φ		T_0/℃		m_0/(℃/min)		m_{m1}/(℃/min)	
	测试值	校正值	测试值	校正值	测试值	校正值	测试值	校正值	测试值	校正值
TCTNB	100	100	—	4.386	290.75	290.75	0.054	0.237	5.166	—
TCDNB	100	100	—	3.783	285.58	285.58	0.074	0.280	5.611	—

样品	θ_1/min		T_{m1}/℃		M_{m2}/(℃/min)		θ_{m2}/min		T_{m2}/℃	
	测试值	校正值	测试值	校正值	测试值	校正值	测试值	校正值	测试值	校正值
TCTNB	1443.73	329.17	439.71	—	7.916	—	1451.99	331.05	471.76	—
TCDNB	1348.21	356.39	436.04	—	8.5	—	1353.03	357.66	472.2	—

样品	m_{m3}/(℃/min)		θ_{m3}/min		T_{m3}/℃		p_m/MPa	
	测试值	校正值	测试值	校正值	测试值	校正值	测试值	校正值
TCTNB	—	—	—	—	—	—	1.614	—
TCDNB	9.9	—	1355.37	358.28	491.06	—	0.7224	—

4 结论

(1) 在测试起始温度设置为 200℃的条件下，TATB 样品的初始热分解温度为 294.34℃。系统在 330.67℃出现最大温升速率 43.5℃/min，达到最大温升速率所需时间为 557.77min。反应系统最高温度为 340.0℃，最大反应压力为 1.076MPa。

(2) 在测试起始温度设置为 250℃的条件下，TATB 样品的初始热分解温度为 294.87℃。系统在 325.79℃出现最大温升速率 186.5℃，达到最大温升速率所需时间为 320.47min。反应系统最高温度为 350.74℃，最大反应压力为 1.521MPa。

(3) TCTNB 和 TCDNB 样品在 ARC 实验中存在多次放热反应，但每次放热量都很小，只有最后一次放热速率比较大。

在实验过程中得到北京理工大学爆炸与安全科学国家重点实验室钱新明和朱华桥同志的许多帮助，在此表示感谢！

参考文献

[1] 王耕，冯长根，郑娆．含能材料热安全性的预测方法 [J]．含能材料，2000 (03).

[2] Townsend D I; Tou J C Thermal hazard evaluation by an accelerating rate calorimeter 1980.
[3] ASTM E-27 Standard guide for assessing the thermal stability of materials by methods of accelerating rate calorimeter 1999.
[4] 李波涛，董海山，张锦云．TATB主要副产物的热性质［J］．含能材料，2000（02）.
[5] Macdougoll C S; Jacoks B J Production and identification of TATB-related species 1976.
[6] 董海山，周芬芬．高能炸药及其相关物性能，1989.

Study on Thermal Decomposition of TATB and its Impurity by Accelerating Rate Calorimeter

Gao Dayuan[1,2], Xu Rong[1], Dong Haishan[1], Li Botao[1], Lv Chunxu[2]
(1. China Academy of Engineering Physics, Mianyang 621900, Sichuan, China)
(2. Nanjing University of Science and Technology, Nanjing 210092, Jiangsu, China)

Abstract: The thermal decomposition of TATB and its impurity was investigated by accelerating rate calorimeter. The curves of thermal decomposition temperature and pressure versus time, and curves of temperature rate versus temperature for the systems are obtained. With the help of the temperature rate and Arrhenius equation, the kinetic parameters of the thermal decompositions for TATB are calculated. The results show that the onset thermal decomposition temperature of TATB is higher than 290℃; the maximal temperature of the reaction system is higher than 340℃ and the maximal pressure is higher than 1MPa. There are many times exothermal reaction in TCTNB and TCDNB, and the exothermal enthalpy is very small for each time. Only in the last time, the exothermal rate is bigger.

Key words: mechanics of explosion; adiabatic decomposition; accelerating rate calorimeter; TATB; impurity

（注：此文原载于 爆炸与冲击，2004，24（1）：69-74）

Preparation and Characterization of Reticular Nano-HMX

Zhang Yongxu, Liu Dabin* , Lv Chunxu
(Chemical Engineering School, Nanjing University of Science and Technology, Nanjing, 210094)
SCI: 000234438200009; EI: 06019628366

Abstract: Reticularly structured HMX (octahydro-1,3,5,7-tetranitro-1,3,5,7-tetrazocine) of nano-size particles was simply prepared by reprecipitation at room temperature. The sample prepared by reprecipitation was characterized by SEM, TEM, XRD, DSC, and drop weight impact. The results of SEM and TEM indicated that spherical HMX particles of about 50 nm in diameter aggregated into reticularly structured conglomerates. There are two phases (γ-and β-HMX) existing in the reticularly structured HMX as shown in the XRD pattern. It was also proved by DSC that the maximum energy release during decomposition of the reticularly structured HMX is at lower temperature. In addition, the testing result of drop weight impact showed that the reticularly structured HMX is less sensitive to impact.

Keywords: Reprecipitation, Nano-Sized Particles, HMX, Sensitivity, Reticular Structure, Recrystallization

1 Introduction

The microstructural properties of an energetic material strongly influence the combustion and explosion behavior of the formulation. These differences can be attributed in part to the strong influence of the heat and mass transport rates on the energy generation rate. The heat and mass transport rates are determined by many factors, among them, the particle size of the energetic material and the homogeneity of the formulation play the major roles. Therefore, the manipulation of these variables is commonly used to tailor the combustion and explosion properties of energetic formulations. Very fine particle size in a homogeneous formulation will help to shift the balance away from transport control toward chemical kinetic control. It is noted that this approach has motivated recent efforts to create nanoscale energetic materials.

As indicated in published results, the behavior of nanoscale energetic materials is quite different from micronsized energetic materials in many ways. The nanoscale energetic materials were found to have much higher burning rates, lower impact sensitivity, and lower temperature of the maximum energy release compared with conventional energetic materials of larger size[1-5].

To date, some techniques, such as a sol-gel method, high-speed air impaction, micro-emulsion and vacuum codeposition, have been employed to obtain nanoscale energetic materials[1,2,4,6,7]. However, only few attention was paid to nanoscale energetic materials because of the fabrication difficulty. In this paper, the simple preparation of a reticularly structured HMX (octahydro-1,3,5,7-tetranitro-1,3,5,7-tetrazocine) with nanometer size by reprecipitation at room temperature is presented.

2 Experimental

0.5mL of HMX acetone solution of different concentrations (2mmol/L, 5mmol/L, and 10mmol/L) was dropped into 50mL organic non-solvent stirring at room temperature. Then, the solution was further stirred for a few minutes. HMX micro-crystals were collected by centrifugation, and vacuum-dried at 20℃ for 12h. The particle size and crystal phase structures of HMX micro-crystals at room temperature were analyzed by TEM, SEM, and XRD.

The microstructure of the HMX particles was observed by SEM (Hitachi, X-650, Japan) and TEM (Hitachi, H-800, Japan). The crystal phase of HMX microcrystals was determined by XRD using Cu-Kα radiation (X′TRA, Thermo ARL, Switzerland). All the chemicals were analytical grade.

The DSC measurements were made with a Perkin-Elmer DSC7 differential scanning calorimeter. 0.80mg samples in a sealed Aluminum pan and a heating rate of 10K/min were used. For the impact testing, 35mg samples were subjected to an impact of a 5000kg anvil positioned at various heights. Based on the data, h_{50} was determined, which is the height required for a 50% probability of initiation of the material. A typical data set consisted of 25 replicated runs for each type of sample.

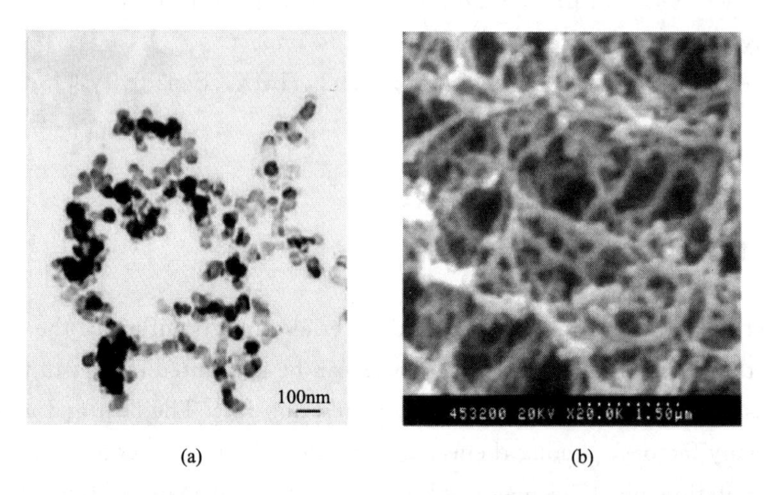

(a) (b)

Figure 1 SEM and TEM photograph of sample prepared byreprecipitation.

Preparation conditions: 0.5mL (5mmol/L) HMX acetone solution was dropped into 50 mL of stirred organic nonsolvent

3 Results and Discussion

Figure 1 (a) is a scanning electron microscopy (SEM) photograph of the sample prepared by reprecipitation, which shows that spherical particles of HMX with a diameter of about 50nm were formed in a reticularly structured aggregation. Figure 1 (b) is the transmission electron microscopy (TEM) photograph of the material. It indicates that the size of the spherical HMX particles is in the range of 40-50nm.

Figure 2 shows the XRD pattern of the reticularly structured HMX (with the 2θ-values ranging from 10℃ to 50℃ at a heating rate of 10 K/min). This result indicates that the material is of high crystallinity, although the diffraction peaks are broadened owing to the small size. The diffraction peaks cannot be indexed by a single HMX phase. The peaks at 2θ-values of 11.20°, 13.90°, and 28.04° were assigned to the γ-HMX phase which is consistent with the standard data file (JCPDS card no. 44-1621). The peaks at 2θ-values of 14.60°, 16.02°, 20.50°, 23.04°, 29.76°, and 31.90° were caused by β-HMX which is also consistent with the standard data file (JCPDS card no. 42-1768).

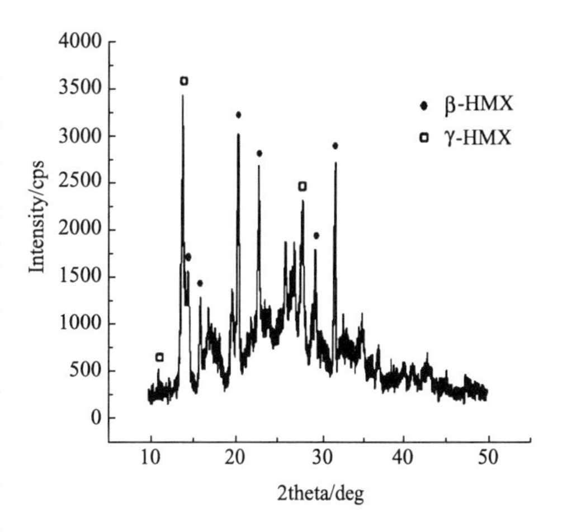

Figure 2 XRD patternsof HMX microcrystalsprepared by reprecipitation.

Preparation conditions: 0.5mL (5mmol/L) HMX acetone solution was dropped into 50mL of stirred organic nonsolvent.

A reticularly structured HMX was also prepared under other initial concentration of HMX in acetone. As it was noted, the crystallite grain size could be easily controlled by changing the initial HMX concentration in acetone, and the average crystallite grain size became smaller with decreasing HMX concentration in acetone.

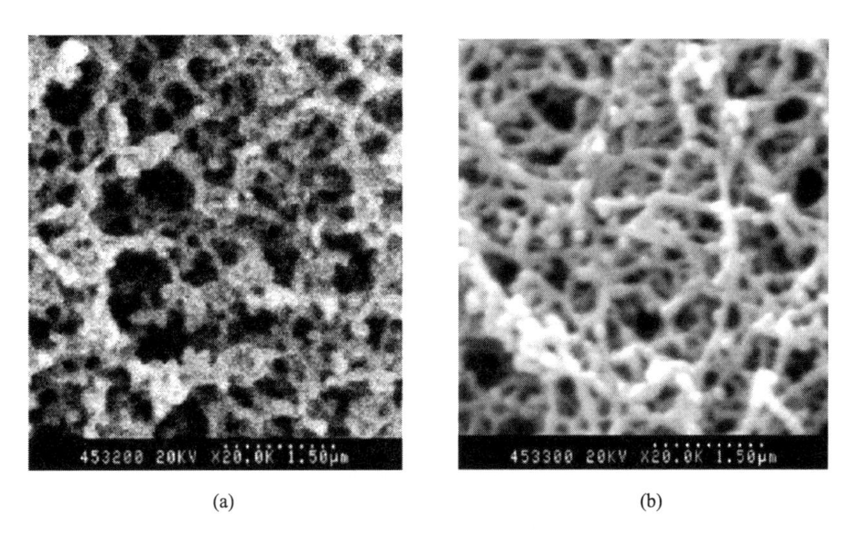

(a) (b)

Figure 3 SEM photographs of HMX microcrystals prepared by reprecipitation under different conditions.

Preparation conditions: 0.5mL HMX acetone solution was dropped into 50mL of stirred organic non-solvent. Concentration of HMX acetone solution: (a) 2mmol/L, (b) 10mmol/L. SEM (Hitachi, X-650, Japan) was used.

Figure 3 shows the SEM of reticularly structured HMX prepared by reprecipitation under different conditions. When the concentration of the HMX was 2mmol/L and 10mmol/L, the average crystallite grain sizes were 20 and 100nm, respectively.

Figure 4 shows the DSC results of reticularly structured HMX and conventional β-HMX in nitrogen at a heating rate of 10K/min. The temperature of maximum energy release during the decomposition of reticularly structured HMX is lower compared with that of the conventional β-HMX. It was seen in the DSC plot of the conventional β-HMX that the energy release of the phase transition is be-

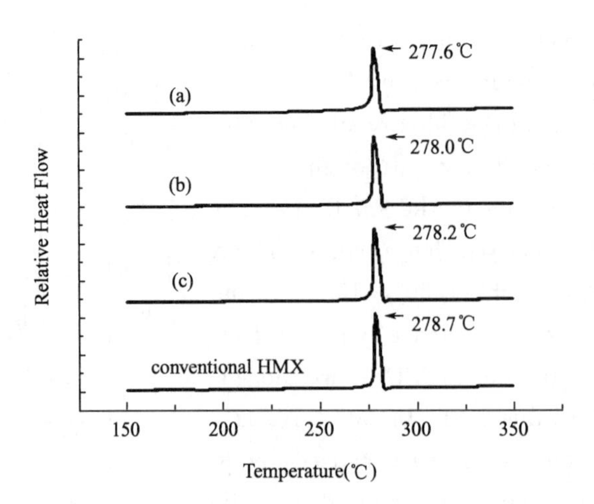

Figure 4 DSC plots of reticularly structured HMX and conventional b-HMX.

Preparation conditions: 0.5mL HMX acetone solution was dropped into 50mL of stirred organic non-solvent. Concentration of HMX acetone solution: (a) 2mmol/L, (b) 5mmol/L, (c) 10mmol/L. SEM (Hitachi, X-650, Japan) was used.

tween 165℃ and 185℃. However, the energy release of the phase transition cannot be seen in the DSC plots of reticularly structured HMX. One of the possible explanations is that a small decomposition of reticularly structured HMX occurred at these temperatures and energy release of decomposition hid the energy release of phase transition.

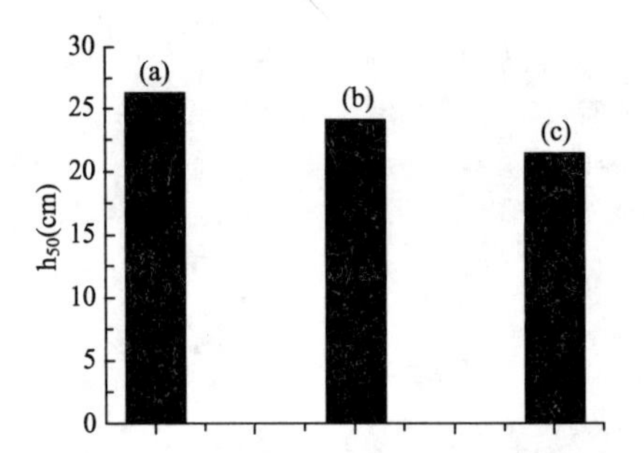

Figure 5 Impact sensitivities of reticularly structured HMX compared to conventional HMX.

Preparation conditions: 0.5mL of HMX acetone solution was dropped into 50mL of stirred organic non-solvent. Concentration of HMX acetone solution: (a) 5mmol/L, (b) 10mmol/L, (c) isconventional b-HMX

Drop weight impact testing was performed on reticularly structured and on conventional β-HMX, respectively. The h_{50} value for conventional β-HMX was 0.215m. The reticularly structured HMX had drop heights of 0.244m and 0.264m. The results indicated that the reticularly structured HMX is less sensitive to impact. Many polymorphic forms, δ, γ, α, and β, exist in HMX and the sensitivity order goes like $\delta > \gamma > \alpha > \beta$. Hence, β is the most stable phase and γ-phase is less stable. Considering the as-prepared reticularly structured HMX contain γ-HMX, the reticularly structured energetic materials of nanometer size will be very insensitive.

4 Conclusion

We have successfully prepared reticularly structured HMX by a simple reprecipitation method at room temperature. The XRD pattern showed that the reticularly structured HMX had two phases (γ-

and β-phase) . The DSC results of the reticularly structured HMX showed a lower temperature of max imum energy release during decomposition. Drop weight impact testing indicated that the reticularly structured HMX is less sensitive to impact. The average crystallite grain size in the reticularly structured HMX can be easily controlled by changing the initial concentration of HMX in acetone. The reticularly structured HMX in nanoscale will aid in achieving the effects of nanodimensionality. The reticularly structured HMX with its large specific surface is suitable as matrix of nano-composites. In these composites the fuel resides within the pores of the solid matrix while the oxidizer comprises skeletal matrix. Therefore, the balance will be away from transport control toward chemical kinetic control. Furthermore, chemical energy release rate will be very high. Over the past several years, considerable effort has been concentrated on the preparation of organic microcrystals by reprecipitation method[8-11]. Most of the reported organic microcrystals, however, are limited to non-polar compounds and some polymers, such as perylene and polydiacetylene. It is difficult to prepare microcrystals of compounds with polar groups by this method. Our result makes a great contribution to this work by using an organic non-solvent. It also demonstrated the potential of preparing organic microcrystals with polar groups by reprecipitation method.

References

[1] A. Pivkina, P. Ulyanova, Y. Frolov, S. Zavyalov, J. Schoonman, Nanomaterialsfor HeterogeneousCombustion, Propellants, Explos. , Pyrotech. 2004, 29, 39.

[2] T. M. Tillotson, L. W. Hrubesh, R. L. Simpson, R. S. Lee, R. W. Swansiger, Sol-gel Processing of Energetic Materials, J. Non-Cryst. Solids 1998, 225, 358.

[3] T. M. Tillotson, A. E. Gash, R. L. Simpson, L. W. Hrubesh, J. H. Satcher Jr. , J. F. Poco, Nanostructured Energetic Materials Using Sol-Gel Methodologies, J. Non-Cryst. Solids 2001, 285, 338.

[4] B. C. Tappan, T. B. Brill, Thermal Decomposition of Energetic Materials 85: Cryogels of Nanoscale Hydrazinium Diperchlorate in Resorcinol-Formaldehyde, Propellants, Explos. , Pyrotech. 2003, 28, 72.

[5] B. C. Tappan, T. B. Brill, Thermal Decomposition of Energetic Materials 86: Cryogel Synthesis of Nanocrystalline CL-20 Coated with Cured Nitrocellulose, Propellants, Explos. , Pyrotech. 2003, 28, 223.

[6] D. B. Liu, D. Xu, B. C. Zhao, J. H. Peng, Y. L. Gao, Q. W. Fan, Preparation of Nanometer RDX in Situ by Solvent Substitution Effect in Reverse Micelles, 26th International Pyrotechnics Seminar, Nanjing, PR China, October 1-4, 1999, p. 269.

[7] C. D. He, B. Zhang, Z. Tan, Preparation of HMX with Nanometer Particle Size and Narrow Particle Distribution, J. Energ. Mater. 2004, 12, 43, (in Chinese).

[8] H. Kasai, H. S. Nalwa, H. Oikawa, S. Okada, H. Matsuda, N. Minami, A. Kakuta, K. Ono, A. Mukoh, H. Nakanishi, A Novel Preparation Method of Organic Microcrystals, Jpn. J. Appl. Phys. Part 2 1992, 31, L1132.

[9] H. Nakanishi, Preparation and Characterization of Organic Microcrystals for Photonics, J. Photopolym. Sci. Technol. 1996, 9, 285.

[10] K. Baba, H. Kasai, S. Okada, H. Oikawa, H. Nakanishi, Novel Fabrication Process of Organic Microcrystals Using Microwave-Irradiation, Jpn. J. Appl. Phys. Part 2 2000, 39, L1256.

[11] H. Kasai, H. Oikawa, S. Okada, H. Nakanishi, Crystal Growth of Perylene Microcrystals in the Repreci pitation Method, Bull. Chem. Soc. Jpn. 1998, 71, 2597.

(注：此文原载于 Propellants, Explosives, Pyrotechnics, 2005, 30 (6): 438-441)

Ultrasonically Promoted Nitrolysis of DAPT to HMX in Ionic Liquid

Qian Hua, Ye Zhiwen, Lv Chunxu
(Chemical Engineering School, Nan Jing University of Science and Technology, Nan Jing 210094, China)
SCI: 000254443900012

Abstract: The present work aims at developing a new process to synthesize HMX from DAPT using ultrasound in ionic liquid. Reaction has been carried out in ultrasonic bath, effect of various parameters such as presence and absence of ultrasound, volume and type of solvent, temperature, concentration of nitrating agent has been investigated with an aim of obtaining the optimum conditions for the synthesis of HMX. It was observed that ultrasonically promoted nitroylsis of DAPT to HMX has exhibited significant enhancement in yield at ambient condition.
Key words: Ultrasonic, Ionic liquid, HMX, Nitrolysis

1 Introduction

HMX (1,3,5,7-tetranitro-1,3,5,7-tetraazacyclooctane) is one of the most powerful military explosives[1-3]. Just as RDX (cyclotrimelhylene trinitramine), it is prepared from inexpensive chemicals (hexamethylenetetramine and nitric acid) . However, it is about five times as costly as RDX. RDX can be made in good yields by several procedures. In contrast, only one method is available for preparing HMX that developed by Bachmann and Sheehan, and this process has a series of undesirable features[4], including high usage of acetic anhydride, slow rate of production and poor yield. Other ways such as TAT (1,3,5,7-tetraacetyl-1,3,5,7-tetraazacyclooctane) procedure and DADN (1,5-diacetyl-3,7-dinitro-1,3,5,7-tetraazacyclooctane) procedure need three or more steps to produce HMX, and they are not economical or in favor of environment.

In recent years, ultrasound has been employed in various chemical transformations with considerable enhancement in rates and yield, and in several cases facilitates organic transformations at ambient conditions which otherwise require drastic conditions of temperature and pressure, or even unachievable reactions[5-11]. The driving energy is provided by cavitations, the formation and collapse of bubbles, which liberates considerable energy in short time. The use of non-volatile solvents should provide a clue to force less volatile substrates to undergo the cavitational activation. The increasing use of non-aqueous room temperature ionic liquids (IL) for synthetic purposes corresponds to a new trend in Green Chemistry. These liquids have no vapor pressure, which should change considerably the characteristics of cavitation in this bulk[12-14]. For this reason, we envisaged to use the combination of both these concepts to induce new focuses.

One-step reaction from DAPT (3,7-diacetyl-1,3,5,7-tetraazacyclo-[3. 3. 1]-octane) to HMX can hardly be finished with very low yield (<10%) in common conditions. In this paper, this procedure can be completed with much higher yield by the ultrasonic assistance. The nitration of DAPT was carried out in a room temperature ionic liquid under sonochemical conditions using N_2O_5/HNO_3 as nitrating agent.

2 Experimental

2.1 Chemicals

Dinitrogen pentoxide is prepared by the reaction of N_2O_4 with ozone[15] and stored at −20℃ before use. IL is [BMIM] PF_6^-, which is prepared by a reported method[16], and its purity is more than 99.8% (tested by ^{1}HNMR), it is directly used in the ultrasonic nitration. Other chemicals were of research grade and were used as obtained from J&K Co. Ltd.

2.2 Equipments

The experiment was carried out in a round-bottomed flask of 25ml capacity. Ultrasonic bath was used for irradiation with ultrasound, which was produced by a ultrasonic cleaning bath (KGD-250B) at 40-80kHz. The ultrasonic cleaner had a rated input power of 120W. The tank dimensions were 350mm×1800mm×150mmwith liquid holding capacity of 8.9l. The reactions were carried out in suspended at the centre of the cleaning bath, 5cm below the surface of the liquid. The experimental setup is shown in Fig. 1.

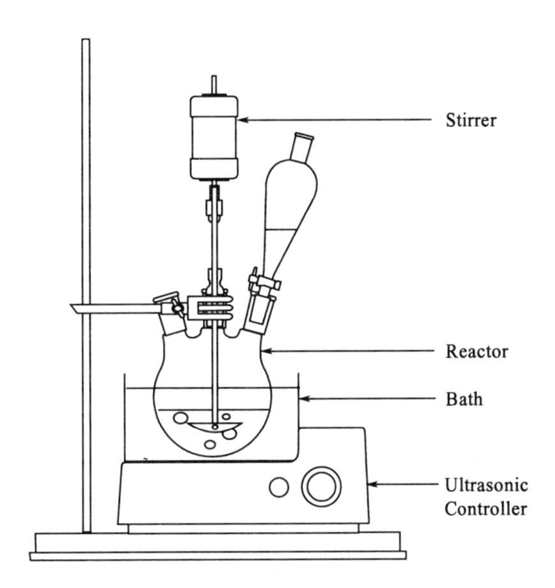

Fig. 1 Experimental setup

^{1}HNMR spectra was recorded on Bruker DRX 300MHZ.

IR spectra was recorded on MB154SFTIR using KBr pellets.

Mass spectra was recorded on Finnigan TSQ Quantumm ultra AM LC/MS spectrometer.

2.3 Typical procedure for sonochemical nitration

Quantitative IL and 1g DAPT were placed in a three-neck 25ml flask, stirred and sonicated in a thermostated ultrasonic cleaning bath at certain temperature. N_2O_5 dissolved in 98% nitric acid was cooled to −10℃ and then decanted into an addition funnel and added to the flask dropwise over a 10 minute period. The mixture was stirred for 1h and the temperature was kept the same. The ultrasonic cleaner was kept on working during the whole reaction. After reaction, the mixture was poured to 100ml ice water, cooling and crystalling. After filtrating and drying, HMX was obtained and analysed by LC/MS. IL can be recycled by simple partition.

Table 1 Nitration of DAPT under ultrasonic condition

Entry	Solvent	Frequency[a] /kHz	$n(HNO_3)$: n (N_2O_5) : n(DAPT)	Temperature (℃)[b]	Yield(%)[c]
1	none	none	24 : 3 : 1	40	9.6
2	none	40	24 : 3 : 1	40	34.5
3	5ml CH_2Cl_2	40	24 : 3 : 1	40	31.2
4	5ml IL	40	24 : 3 : 1	40	66.8
5	5ml IL	30	24 : 3 : 1	40	53.5
6	5ml IL	60	24 : 3 : 1	40	47.9
7	5ml IL	80	24 : 3 : 1	40	38.3
8	10ml IL	40	24 : 3 : 1	40	59.4
9	20ml IL	40	24 : 3 : 1	40	51.5
10	5ml IL	40	24 : 3 : 1	60	63.4
11	5ml IL	40	24 : 3 : 1	80	65.1
12	5ml IL	40	12 : 1 : 1	40	24.0
13	5ml IL	40	12 : 6 : 1	40	64.9
14	5ml IL	40	12 : 8 : 1	40	57.2
15	5ml IL	40	12 : 3 : 1	40	46.2
16	5ml IL	40	36 : 3 : 1	40	60.7
17	5ml IL	40	48 : 3 : 1	40	54.3

a The power of ultrasound transferred into the reaction depends on the frequency of ultrasound. As the dissipated power cannot be changed on the control panel, it was indirectly varied by adjusting the frequency of ultrasound.

b Temperature control is difficult, for the irradiation of ultrasound can lead to temperature rise. In order to keep the temperature invariable, the temperature on the control panel should be set a little lower than appointed temperature, the balance of the evaporation of water and the irradiation of ultrasound can keep the water in steady and needed temperature.

c Yield is calculated by DAPT.

3 Results and discussions

High price of HMX is the restriction for its application. DAPT can be gained with a yield of over 100% by acetolysis of hexamine with acetic anhydride in the presence of ammonium acetate. If HMX can directly produced from DAPT, its costs can be reduced a lot.

Initially experiments were carried out in the absence of ultrasound. One-step reaction from DAPT to HMX is almost infeasible. However, in the reactions using N_2O_5/HNO_3 as nitrating agent, the ultrasonically mediated nitrolysis has led to a high yield in an ambient condition. The effects of various operating parameters on the yield of HMX are discussed below.

$COCH_3$ / CH_3OC-N / $N-COCH_3$ / $COCH_3$ (TAT)

CH_3OC-N / $N-COCH_3$ (DAPT) ((((HMX) O_2N-N / $N-NO_2$ / NO_2 / NO_2

(DADN) $COCH_3$ / O_2N-N / $N-NO_2$ / $COCH_3$

3.1 Effect of presence of ultrasound

The pronounced effect of the use of ultrasound on this reaction can be easily demonstrated from the results shown by the entries 1-2. In the absence of ultrasound, the yield of HMX is only 9.6%. The only addition of ultrasound without IL can effectively enhance the yield up to 34.5%. The yield in the presence of ultrasound was 260% more than in the absence of ultrasound. One can also observe that ultrasound with different frequencies lead to various results (entries 4-7). Yield increased at first and then decreased with the variety of ultrasonic frequency from 30kHz to 80kHz. Only under the appropriate condition (40kHz), can the yield reach the highest (66.8%).

It is known to us all that the ultrasound assisted mass transfer is an important factor in the reaction system involving more than two phases. As a solid, DAPT is not soluble to any liquid in this system, ionic liquid and nitric acid is also not miscible, so the reaction system involves three phases. The mass transfer assisted by ultrasound is essential to promote this reaction. That's why the appearance of ultrasound can highly enhance the yield of HMX. However, this simple explanation cannot make it clear why only appropriate ultrasonic frequency leads to the highest yield. Up to now, the reason for this phenomenon is still difficult to explain. It is known that the N*-C-N* bond is firm and difficult to cleave, without ultrasound, DAPT must be transferred to DADN or TAT at first in rigorous conditions and using special reagent (the process of the cleavage of N*-C-N* bond), and then nitrated to HMX with high yield[17]. Through analyzing the product of nitration without ultrasound, many by-products with incomplete cleavage of N*-C-N* bond were found, which confirm our thought. So, the first step from DAPT to DADN or TAT is essential. That is to say the cleavage of N*-C-N* bond is crucial in the formation of HMX, and the substituting reaction over N atom is relatively easier. A plausible reason is that the ultrasound induces the formation and collapse of bubbles, and they liberate considerable energy in short time, which promotes the N*-C-N* bond cleavage, so as to enhance the yield of HMX. Of course, over high energy induced by high frequency ultrasound can also cleave other bonds, which is proved by the formation of some low melting compound, the IR spectrum of the product indicates the cleavage of the cage structure. It can also explain why yield decreased with high-frequency ultrasound.

3.2 Effect of solvent

As shown in entries 2-4, IL can effectively promoted this reaction. As we know, solvent with lower vapor pressure is in favor of the formation of bulk, so as to enhance the using efficiency of ultrasound[12-14,18]. That's why ionic liquid can highly promote the reaction. Furthermore, the nitrolysis reaction studied here is the bimolecular electrophilic substitution, in which NO_2^+ attacks the lone pair on the nitrogen atom, and the cleavage of the C-N bond and the formation of the new N-N bond occur at the same time. The dependence of the reaction yield on the concentration of N_2O_5 seems to support the idea that the elementary reaction of the nitrolysis is bimolecular. Since the electrophilic reaction involves the ionic reaction intermediates, another role of ionic liquid to enhance the reaction yield would be the stabilization of these intermediates. From entries 4 and 8-9, it can be seen that there was higher yield when lower amount of solvent was used. Use of lower amount of nitric acid also resulted in similar trends. It can be seen from entry 4 that the highest conversion obtained under these conditions was 66.8% HMX formation.

This may be because of the fact that as the volume of the reaction mixture increases, the power dissipated per unit volume i. e. power density in the reaction mixture decreases. This decreases the number of cavitational events per unit time per unit volume. From these initial studies it was decided to use minimum possible solvent (5ml IL) for further reactions.

3.3 Effect of temperature

To improve the performance of the system further, the effect of temperature on yield of HMX was studied. Experiments were carried out at 40℃, 60℃ and 80℃ as shown by entries 4 and 10-11. Here one can observe that there was not much variation in the performance of the system, yield of HMX was slightly varied from 63.4% to 66.8%. Thus all the further experiments were carried out at 40℃.

3.4 Effect of molar ratio of HNO_3, N_2O_5 and DAPT

The purpose of changing the ratio and amount of nitrating agent is to find the best synthesizing way. In the experiment, we fixed the molar number of DAPT, changed the amount of N_2O_5 and HNO_3 to find the relationship among them. As the results shown in entries 4 and 12-17, there was a significant effect on the yield when changing the molar ratio of N_2O_5. At a little higher N_2O_5 concentrations with respect to DAPT, the yield increased rapidly as the amount of N_2O_5 grew. When the amount of N_2O_5 was two times more than DAPT, more N_2O_5 even led to a lower yield. By the analysis of LC/MS, nitrating system with high N_2O_5 concentration made the ring-shaped structure cleaved. Compared with N_2O_5, the concentration of HNO_3 was less important to yield. A little increase of HNO_3 induced a higher yield. Meanwhile, just as the IL, nitric acid was also a solvent, too much nitric acid led to the decrease of the number of cavitational events per unit volume, so as to reduce the yield of HMX.

4 Conclusion

A new process to synthesize HMX from DAPT with relatively high yield has been developed using ultrasonic irradiation in IL as solvent. Various operating parameters such as temperature, volume of solvent and molar ratio of reagents have been optimized to give maximum yield of the desired product. This reaction has a wide scope in the nitrolysis of a variety of other nitrogen heterocyclics. Moreover, sonochemistry in IL as described in the present work has opened up a new domain in the area of sonochemistry and liquid interface science, and can be conceptually envisaged as a new tool in organic synthesis. Further work is in progress to extrapolate these findings to other organic transformations.

Acknowledgements

Authors would like to acknowledge the National Basic Research Program of China (613740101), Science Development Fund of NUST (2010ZYTS020) for financial support.

References

[1] V. I. Siele, US Patent 3987034 (1975).
[2] L. Silberman, S. M. Adelman, Symposium Processing Propellants, Explosives and Ingredients, Section II 1979.
[3] C. Ju, W Shaofang, Propellants, Explos., Pyrotech. 9 (1984) 58.
[4] I. J. Soloman, L. B. Silberman, US Patent 4086228 (1978).
[5] J. L. Luche, Ultrasonics 30 (3) (1992) 156.
[6] F. Ruo, Z. Yiyun, B. Ciguang, Ultrason. Sonochem. 4 (1997) 183.
[7] J. L. Luche, Synthetic Organic Sonochemistry, Plenum Press, NewYork, 1998.
[8] J. P. Mikkol, T. Salmi, Catal. Today 64 (2001) 271.
[9] G. V. Ambulgekar, B. M. Bhanage, S. D. Samant, Tetrahedron Lett. 46 (2005) 2483.
[10] A. Loupy, A. Petit, J. Hamelin, F. Texier Banlled, P. Francoise, D. Inare, Synthesis (1998) 1213.
[11] A. K. Bose, B. K. Bamik, N. Lavlinsaia, M. Jayarman, M. S. Manhas, Chem. Tec. 27 (1997) 18.
[12] A. S. Larsen, J. D. Holbrey, F. S. Tham, C. A. Reed, J. Am. Chem. Soc. 122 (2000) 7264.
[13] E. Nicholas, Leadbeater, M. T. Hanna, J. Org. Chem. 67 (2002) 3145.
[14] P. Juliusz, C. Agnieszka, P. Ryszard, Ind. Eng. Chem. Res. 40 (2001) 2379.
[15] W. Arber, A. C. Coming, J. K. Hammond, Brit U. K. Patent Appl. GB 2347670 Al, 2000.
[16] J. Fuller, R. T. Carlin, H. Long, D. Haworth, Chem. Commun 12 (3) (1994) 299.
[17] V. I. Siele, M. Warman, J. Leccacorvi, Propellants and Explosives. 6 (3) (1981) 67.
[18] S. Z. Chen, Cleaning Technology. 2 (2) (2004) 7.

(注：此文原载于 Ultrasonics Sonochemistry, 2008, 15 (4): 326-329)

无氯 TATB 的合成及其热分解动力学

马晓明，李斌栋，吕春绪，陆明
（南京理工大学化工学院，江苏 南京 210094）
EI：20101412821149

摘要：为解决传统 TATB 合成中氯杂质的影响，以间苯三酚为原料，经过五氧化二氮硝化、甲基化和氨气氨化得到 TATB，综合收率为 92%，用核磁共振光谱、红外光谱和质谱等进行了表征。采用有机溶剂法精制获得高纯度无氯 TATB，收率 75%。用 DSC 及 TG 分析了 TATB 的热性能，根据 Ozawa 公式计算出 TATB 的热分解活化能和指前因子分别为 135.28 kJ/mol 和 $2.340\times10^{9}s^{-1}$

关键词：有机化学，TATB，钝感炸药，间苯三酚，热分解动力学

引言

三氨基三硝基苯（TATB）是性能良好的高能钝感炸药，已得到广泛应用。TATB 的含氯量一直是人们关注的问题，含氯量对 TATB 的热安定性有明显影响[1]，其中杂质 NH_4Cl 的分解产物对接触材料有较强的腐蚀作用[2]，对药柱成型、药柱强度、金属弹体等也有不良作用。因此，研究无氯 TATB 合成新方法具有重要意义。

从 20 世纪 70 年代末期开始，国内外开始无氯 TATB 的合成研究。Estes[3-4] 以三硝基三丙氧基苯为中间体合成出无氯 TATB。Atkins 和 Nielson[5] 以 TNT 为原料，在二氧六环中与 H_2S 反应产生 4-氨基-2,6-二硝基甲苯，再用硝硫混酸硝化制得五硝基苯胺，然后在苯、氯化甲撑或其他合适的溶剂中与 NH3 反应得到 TATB。魏运洋[6~7] 以廉价的苯甲酸为原料经硝化、Schmidt 反应、硝化和氨化 4 步反应成功合成出 TATB。Anthony J. Bellamy 等人[8] 开发出一条合成无氯 TATB 的新路线，以间三苯酚为原料，经过硝化、烷基化、胺化合成 TATB，TATB 的综合收率为 87%，初始分解温度为 346℃，但与传统生产法相比其热稳定性和纯度不理想。本研究参考文献[8] 探索了无氯 TATB 的合成路线及热分解动力学。

1 实验

1.1 实验及仪器

试剂：间苯三酚，工业品；质量分数 98%的浓硫酸，分析纯，扬州沪宝化学试剂有限公司；质量分数 95%的硝酸，分析纯，南京化学试剂有限公司；盐酸，分析纯，扬州沪宝化学试剂有限公司；原甲酸三甲酯，分析纯，Alfa Aesar。

仪器：美国 Nicolet Instrument Corporation 公司的 Nexus870FT-IR 型红外光谱仪；美国 Finnigan 的 Tsq Quantum UltraAm 液-质联用光谱仪；德国 Bruker 公司的 Bruker DRX500MHz 核磁共振仪；瑞士 METTLER TOLEDO 公司的 TGA/SDTA 851^e 型热分析仪。

1.2 TATB 的合成路线

TATB 的合成路线如下：

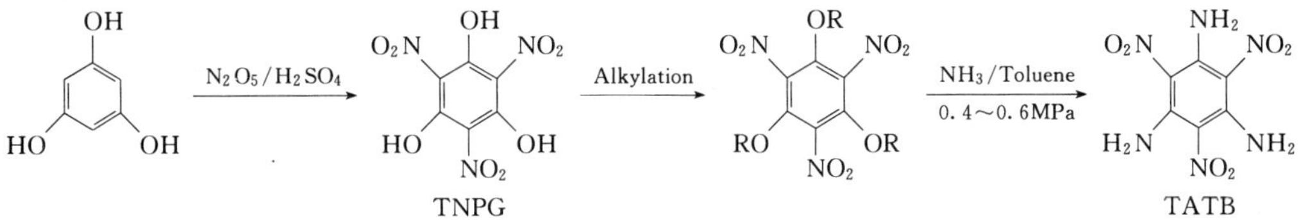

N_2O_5/H_2SO_4 → TNPG；Alkylation；NH_3/Toluene，0.4～0.6MPa → TATB

图 1 TATB 的合成路线

Fig. 1 The synthetic route of TATB

1.3 1,3,5-三羟基-2,4,6-三硝基苯（TNPG）的合成

在 1,3,5-三羟基苯（2.52g，20mmol）和质量分数 98%的硫酸（20mL）的混合体系中缓慢滴加 N_2O_5（10.8g，0.1mol）的硫酸溶液（100mL），温度低于－5℃反应 2h，将反应液倒入冰水中，待冰块融化后过滤出产生的黄色沉淀，用浓度 3mol/L 的冷盐酸洗涤，干燥得到产物 1,3,5-三羟基-2,4,6-三硝基苯 4.375g，收率 96.8%；HPLC 检测无杂质；m.p.：166.2～167℃（文献值[8]：165.3℃）。

^{1}HNMR（CH_3Cl），δ：4.5（3H，OH）；FT-IR（KBr），M（cm^{-1}）：3667，3580，3425，3126，1637，1589，1541，1453，1425，1377，1319，1222，1183，922，700。

1.4 1,3,5-甲氧基-2,4,6-三硝基苯的合成

在 30mL 原甲酸三甲酯中，加入 5.22g（20mmol）1,3,5-三羟基-2,4,6-硝基苯，溶解后在 95～100℃加热 4h。蒸馏甲醇和甲酸甲酯（b.p.：65～78℃）的共沸混合物，在 105～110℃下继续加热 1h，收集进一步蒸馏得到的馏出物，液相温度超过 100℃后停止反应。冷却后，浓缩该红色溶液得到 1,3,5-三甲氧基-2,4,6-三硝基苯 5.85g，收率 97%。用甲醇重结晶红色产物，得到黄色晶体。m.p.：73.4～74℃（文献值[8]：73.8℃）。

^{1}HNMR（CH_3Cl），δ：3.977（9H，OCH_3）；FT-IR（KBr），M（cm^{-1}）：3460，2960，2881，1593，1542，1461，1350，1197，1126，1103，941，887，844，717，663，634。

1.5 TATB 的合成

室温下在高压反应釜中，加入溶有 1,3,5-三甲氧基-2,4,6-三硝基苯（2.0g，6.6mmol）的甲苯（50mL）溶液，在 0.4～0.6MPa 压力下通入氨气，反应 4h 后将多余的氨气通入水槽中。过滤出黄色沉淀，用甲苯（25mL），丙酮（30mL×2），去离子水（40mL×2），乙醇（20mL）洗涤。干燥得到 TATB 1.62g，收率 95%。

MS（m/z）：258.1（M^+）；^{1}HNMR（DMSO-d_6）；δ：3.329（m，6H，—NH_2）；FT-IR（KBr），M（cm^{-1}）：3320，3216，1604，1571，1444，1217，1166，781，727，694。

1.6 TATB 的精制

浓硫酸法：称取 1gTATB 粗产物，溶于 20mL 质量分数为 98%的浓硫酸中，将 TATB 的硫酸溶液缓慢滴至 100g 冰块上，析出黄色固体，过滤，水洗，干燥得 TATB 0.98g，收率 98%。

有机溶剂法：称取 1gTATB 粗产物，溶于 200mL 的 DMSO（或者 250mLDMF）中，加热至沸

腾使TATB完全溶解，趁热过滤，收集滤液，冷却至室温，析出固体，过滤，水洗，干燥得TATB 0.76g，收率76%（DMF重结晶得TATB0.75g，收率75%）。

1.7 热性能测试

DSC测试：试样量2mg左右，升温速率20℃/min，氮气保护，流速20mL/min，扫描范围50～500℃。

热失重测试：试样量2mg左右，在氮气流中以10、15、20、25℃/min的升温速率升至500℃，氮气流速20mL/min。

2 结果与讨论

2.1 温度和时间对氨化反应的影响

氨化反应中，以甲苯为溶剂，在氨气压力0.4～0.6Mpa，搅拌速率800r/min的条件下，改变温度和反应时间，TATB的收率及热分解性能见表1和表2。

表1 氨化反应温度对TATB收率及热分解性质的影响

Table 1 Effect of amination reaction temperature on the yield and thermal decompositon property of TATB

序号	t/℃	η(TATB)/%	t_o/℃	t_p/℃
1	20	95.0	341.6	369.0
2	30	95.5	371.5	377.5
3	50	96.5	373.2	382.6

注：t_o为起始分解温度；t_p为峰值温度。

由表1可以看出，反应时间为4h，反应温度为20～50℃条件下，TATB的收率均不小于95%，受温度影响不大；随着温度的升高，产物的DSC分解峰值温度也升高，且只有一个放热峰，优于文献[8]的结果。当反应温度为50℃时，TATB的分解峰值温度为382.6℃，与常用TATB的分解峰值温度相当，达到了国军标的指标（≥375℃）。因此，选定50℃为反应温度。

表2 氨化反应时间对TATB收率及热分解性质的影响

Table 2 Effect of amination reaction time on the yield and thermal decompositon property of TATB

序号	t/h	η(TATB)/%	t_o/℃	t_p/℃
1	1	96	375.74	385.3
2	2	95.5	373.50	383.5
3	3	95.2	373.70	383.8
4	4	96.5	373.20	382.6

由表2可见，反应温度为50℃时，随着反应时间的增加，收率变化不大，均大于95%；分解峰值温度都高于382℃，且随着反应时间的增加，分解峰值温度有下降的趋势。因此，确定最佳反应时间为1h即可。

2.2 TATB精制方法的比较

TATB的精制方法有热精制、升华法和有机溶剂重结晶法[9]。李波涛等人[10]尝试了硫酸法精制TATB，但TATB经硫酸处理后伴随有化学变化，因此难以成功。

本研究采用分解峰值温度为377.5℃的同一批TATB粗产品，分别用浓硫酸（20mL）、DMSO（200mL）、DMF（250mL）重结晶1gTATB，收率及DSC测试结果见表3和图2。

由表3结果可知，用有机溶剂DMF和DMSO重结晶TATB的收率大约为75.0%左右，用浓硫酸重结晶可达 98.0%。但是图 2 显示，TATB 经浓硫酸精制后，热稳定性下降，放热分解。分解峰值温度（375℃）比 TATB 粗品有所降低；DMF 和 DMSO 精制后的 TATB，热稳定性提高，其放热分解峰值温度可达 394℃，比于粗品有很大的提高。因此，综合考虑制备高纯度 TATB 较为理想的方法为有机溶剂法。

表 3 TATB 的重结晶结果

Table 3 Recrystallization results of TATB

序号	溶剂	η(TATB)/%	t_p/℃
1	DMSO	76.0	393.4
2	DMF	75.0	394.0
3	浓 H_2SO_4	98.0	375.0

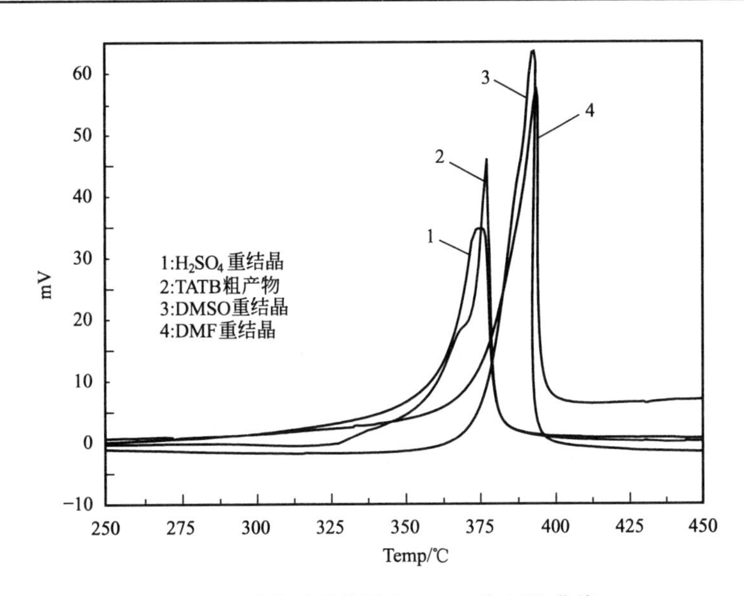

图 2 不同溶剂重结晶后 TATB 的 DSC 曲线

Fig. 2 The DSC curves of TATB purified by various solvents

2.3 TATB 的热分解动力学

根据文献[11]可知，由 Ozawa 公式推导可得：

$$\lg F(\alpha)=\lg\left(\frac{AE}{R\beta}\right)-2.315-0.4567\frac{E}{RT} \tag{1}$$

式中：α 为分解深度或转化率；A 为指前因子，s^{-1}；E 为表观活化能，J/mol；R 为理想气体常数，(J·mol^{-1}·K^{-1})；β 为升温速率，K/min；T 为温度，K。

对于任何热分解机理，lg$F$（α）和 $1/T$ 呈直线关系。由直线的斜率可以得到其活化能 E，由直线的截距可求得指前因子 A。

高大元、徐容等[12]研究了 TATB 的热分解动力学，得出 TATB 的热分解属于 $n=2/3$ 的成核和核生长机理，反应机理函数的微分形式为：

$$f(\alpha)=\frac{1}{F'(\alpha)}=\frac{3}{2}(1-\alpha)[-\ln(1-\alpha)]^{\frac{1}{3}} \tag{2}$$

由式（2）可得：$\lg F(\alpha)=\lg[-\ln(1-\alpha)]^{\frac{2}{3}}$ (3)

本研究采用经 DMSO 精制的 TATB，在升温速率分别为 10、15、20 和 25℃/min 时测试其热失重曲线（图 3），分别求取反应深度 α=0.1、0.2、0.3、0.4、0.5、0.6、0.7、0.8、0.9 时对应的反应温度 T，再求得 lg$F$（α），对 $1/T$ 进行线性回归分析，结果见表 4。

可得 TATB 的热分解活化能为 135.28kJ/mol，指前因子为 $2.340\times10^{9}s^{-1}$，热分解动力学方程为：

$$\frac{d\alpha}{dt}=kf(\alpha)=Ae^{-\frac{E}{RT}}f(\alpha)$$

$$=2.340\times10^{9}\times(1-\alpha)[-\ln(1-\alpha)]^{\frac{1}{3}}\exp\left(-\frac{1.627\times10^{4}}{T}\right)$$

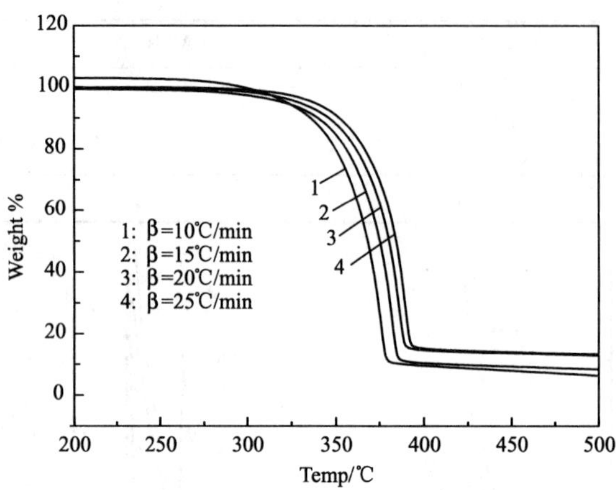

图 3 不同升温速率时 TATB 的 TG 曲线

Fig. 3 The TG curves of TATB at different heating rates

表 4 TATB 的热分解动力学参数

Table 4 Thermal decomposition kinetic data of TATB

β(℃/min)	线性方程	活化能 E(kJ/mol)	指前因子 A(s⁻¹)
10	Y=11.40625−7351.01377X	133.82	5.450×10^{8}
15	Y=10.78702−7021.44031X	127.82	2.057×10^{8}
20	Y=11.16269−7311.00332X	133.09	6.255×10^{8}
25	Y=12.21302−8041.46173X	146.39	7.982×10^{9}
平均值		135.28	2.340×10^{9}

3 结论

(1) 以间苯三酚为原料，经过硝化、甲基化、氨化反应得到 TATB，反应总收率达到 92%；以三甲氧基三硝基苯为原料进行氨化反应，反应温度 50℃，反应 1h，TATB 的收率大于 96%，热分解峰值温度 385℃，达到国军标 GJB3292298 要求。

(2) 采用 DMSO、DMF 重结晶，TATB 收率可达 75%，热分解峰值温度可达 394℃，是制备高纯度 TATB 较为理想的方法。

(3) TATB 的热分解活化能为 135.28kJ/mol，指前因子为 $2.340\times10^{9}s^{-1}$，热分解动力学方程为：

$$\frac{d\alpha}{dt}=kf(\alpha)=Ae^{-\frac{E}{RT}}f(\alpha)$$

$$=2.340\times10^{9}\times(1-\alpha)[-\ln(1-\alpha)]^{\frac{1}{3}}\exp\left(-\frac{1.627\times10^{4}}{T}\right)$$

参考文献

[1] 陆增，易景缎，邱甬生．降低 TATB 含氯量的研究 [J]．火炸药学报，1996 (4)：9-11.

[2] 浦钧鹏，唐宝霞．硫酸分解法测定TATB中无机氯含量［J］．火炸药学报，1985（1）：11-13.
[3] Estes Z L. Summary of Chlorine Free Synthesis of TATB［R］．［S. l.］：ERA，1977.
[4] Estes Z L，Locke J G. Synthesis of Chlorine-Free TATB［R］．［S. l.］：ERA，1979.
[5] Atkins R L，Nielsen A T，Norris W P. New method for preparing pentanitroaniline and triaminotrinitrobenzenes from trinitrotoluene：US，4248798［P］. 1980.
[6] 魏运洋．DATB、TATB和其中间体的新合成方法［J］．兵工学报，1992（2）：79-81.
[7] 魏运洋．1,3,5-三氨基-2,4,6-三硝基苯的合成新法［J］．应用化学，1990，7（1）：70-71.
[8] Bellamy A J，Ward S J. A new synthetic route to 1,3,5-t riamino-2,4,6-t rinitrobenzene（TATB）［J］. Propellants，Explosives，Pyrotechnics，2002，27：49-58.
[9] 李波涛．TATB的提纯与精制［J］．炸药通讯，1990（4）：57-61.
[10] 李波涛，梁叶明，达君，等．高纯度TATB的制备与表征方法研究［J］．含能材料，1993，1（2）：6-11.
[11] 高大元，董海山，李波涛，等．炸药热分解动力学研究及其应用［C］//2004年全国含能材料发展与应用学术研讨会论文集（上册）．绵阳：中国工程物理研究院化工材料研究所，2004年.
[12] 高大元，徐容，董海山，等．TATB及其杂质的热分解动力学研究［J］．爆破器材，2003，32（5）：5-9.

Synthesis and Thermal Decomposition Kinetics of TATB without Chloride

Ma Xiaoming，Li Bindong，Lv Chunxu，Lu Ming

(School of Chemical Engineering，Nanjing University of Science and Technology，Nanjing 210094，China)

Chinese Journal of Explosives and Propellants，2009，32（6），24-27

EI：20101412821149

Abstract：In order to solve the influence of chloride in the traditional synthesis of TATB，taking phoroglucinol as primary material，1,3,5-triamino-2,4,6-trinitrobeneze（TATB）was synthesized by nitration with N_2O_5，methylation and amination，and its structure was identified by NMR，IR andMS. The effect of amination reactionon the yield and thermal decomposition property of TATB was investigated，the total yield of crude TATB was 92%. High-purity TATB was obtained by organic solvent recrystallization with yield of 75%. Thermal properties of TATB was analyzed by DSC and TG，the thermal decompositon activation energy and pre-exponential factor of TATB obtained by Owaza's equation are $E_{TATB}=135.28$kJ/mol，$A_{TATB}=2.340\times10^9\,s^{-1}$.

Key words：organic chemistry；TATB；insensitive explosive；phoroglucinol；thermal decomposition kinetics

（注：此文原载于 火炸药学报，2009，32（6），24-27）

Correction and Measurement of Transmittance of Smoke in 8-14μm Waveband

Zhu Chenguang, Lv Chunxu, Wang Jun, Wei Feng
(School of Chemical Engineering, Nanjing University of Science & Technology, Nanjing 210094, China)

Abstract: To reduce the effects of smoke temperature on the transmittance measurement in 8-14μm waveband, the transmittances of infrared smoke was analyzed with and without infrared radiation source. Results show that the transmittances of 5g smoke agent is 40%-50% and 15%-20%. And the infrared smoke formed by burning leads to smoke cloud with a strong radiation. Then, a different correcting method was carried out. The transmittance is corrected by subtracting from the transmittance of infrared smoke without infrared radiation source, and transmittance of the smoke agent with infrared radiation source decrease to 25%-35%.

1 Introduction

It is well known that smoke is an effective screening agent. And the infrared smoke screen is widely used, whose transmittance is an important parameter to evaluate the obscuring performance in common infrared waveband (1-3μm, 3-5μm, 8-14μm). In modeling the effects of infrared (IR) screening smoke, one should account for the fact that the smoke itself is a strong source of IR radiation, and the degree of effect depends upon the aerosol optical thickness, emissive, and temperature. The problem has been recognized for some time by the users[1]. A smoke medium that contains isolated, heated, point-like sources called hot smoke is a more effective screening agent[2-3], so the temperature of the smoke cloud is usually higher than the ambient. However, hot smoke cloud is an impact on the transmittance measurement after smoke agent burnt. A wrong result would be aroused in 8-14μm waveband. The reason lies in the background radiation of the infrared smoke owing to temperature difference. In order to reduce the impact, instead of radiometer and spectrometer, IR-imaging equipment is used to evaluate the obscure performance of smoke[4-8], and laser equipment[9], but the sampling data is stochastic from IR-imaging equipment and the width of waveband of laser equipment is narrow. Therefore the phenomenon of infrared smoke was analyzed in the paper in 8-14μm waveband. Based on spectrometer, the relevant experiments were designed, and a correction method was presented for the smoke transmittance the 8-14μm waveband.

2 Experimental

2.1 Experiment instruments

That the spectrometer is Open path Fourier transform infrared remote sensor (OPAG33, Bruker Optics), used to measure transmittance of smoke. The scanning time of spectrometer is 32s, scans 10 times in succession. The infrared radiation source is a standard blackbody (HFY-200B blackbody) which is made in Shanghai Institute of Techno-Physics of Academy of Science of China. A standard smoke-chamber is used to discharge and flow with smoke.

2.2 Contrast testing

Test-1, 5g smoke agent was lighted up in the smoke-chamber, and the temperature of the infrared radiation source was set to 773K. Fig. 1 shows that the transmittance varies with the wavenumber. Test-2, Smoke agent was still 5g, but the infrared radiation source was shut, and the results shown in Fig. 2. Test-3, 40g smoke agent burned up in the smoke-chamber, and temperature of the infrared radiation source was set to 773K. Its spectra are shown in Fig. 3.

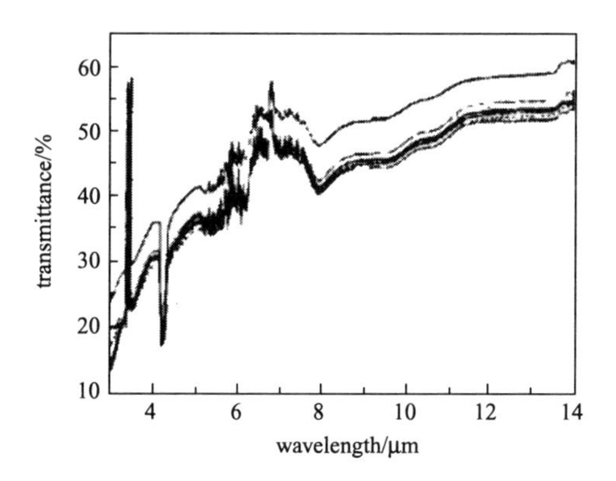

Fig. 1 The spectra of transmittance of 5g smoke agent with the infrared radiation source

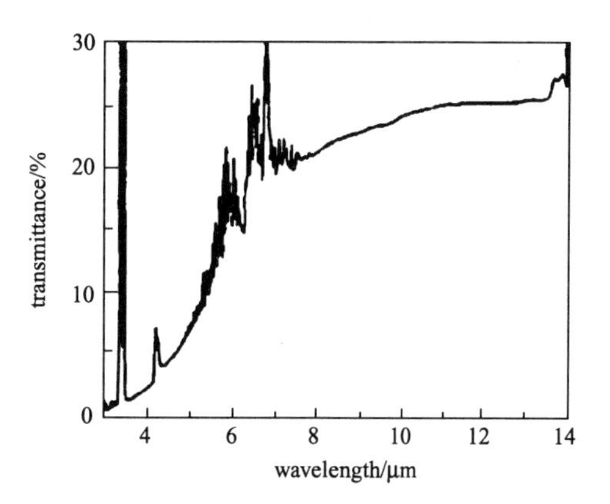

Fig. 2 The spectra of 5g smoke agent without the infrared radiation source

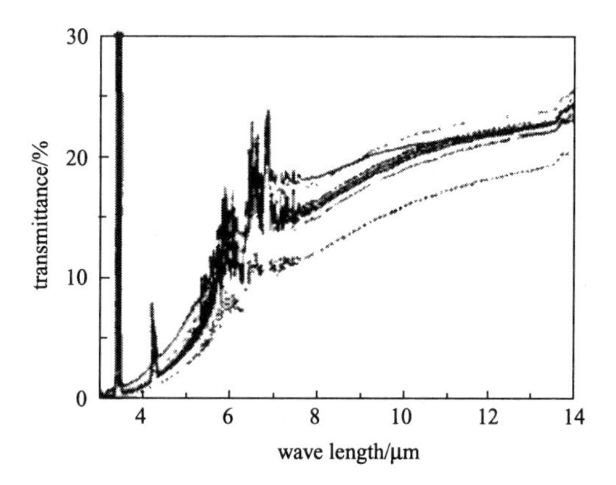

Fig. 3 The spectra of transmittance of 40g smoke agent

3 Results and discussion

Fig. 1 shows the range of the transmittance is obvious, which is from 40% to 50% in 8-14μm (1250-714cm^{-1}) waveband. In the Fig. 2, transmittance in 8-14μm should be zero because the infrared radiation source is shut, but the varying range is still shown from 15% to 20%. In testing-3, the quantity of smoke agent is increased 8 times more than that in test-l, the infrared from the radiation source should be scattered entirely by the thick smoke. , but the result same as that of test-2.

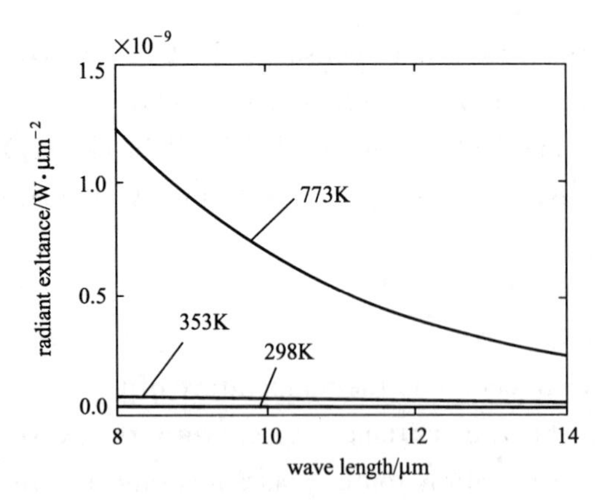

Fig. 4 The curves of radiant quantity of blackbody

The investigation indicates, the transmittance of smoke in 8-14μm waveband is affected, and the background spectrum due to higher temperature of the smoke than the ambient. Thus its results could not be used to evaluate the obscure performance of the burning-type infrared smoke before the transmittance measured is processed. In fact, the varied quantity of transmittance in Fig. 2 and Fig. 3 are the radiant from smoke. In this phenomenon, the radiant quantity may be calculated by Plank's law of radiation when the temperature of blackbody and wave length has been known, and its unit is W · μm^2. Plank's radiation formula follows as:

$$M_\lambda = c_1 \lambda^{-5} \frac{1}{e^{c_2/\lambda T} - 1} \tag{1}$$

Where, c_1, first radiant constant, 3.7418×10^{-4} W · μm^2; c_2, second radiant constant, $1.4388 \times 10^4 \mu$m · K; λ, wave length, μm; T, temperature, K.

The temperature of the burning-type smoke cloud is 343-353K and the ambient is 283-303K. The temperature of blackbody is 773K. The radiant quantity calculated based on Plank's radiation formula are shown in the Fig. 4.

The affect range in 8-14μm is estimated by the equation (2).

$$\eta = \frac{M_{\lambda \cdot \text{smoke}}}{M_{\lambda \cdot \text{b}}} \times 100\% \tag{2}$$

Where, η, The affect range of transmittance, %; $M_{\lambda \cdot \text{smoke}}$, the radiant quantity of smoke cloud, W · μm^2; $M_{\lambda \cdot \text{b}}$, the radiant quantity of the blackbody, W · μm^2.

Based on Plank's law of radiation and the temperature of smoke cloud of 353K, the range of transmittance is from 8% to 16% in 8-14μm. Therefore, a correcting method was carried out to evaluate transmittance of the infrared smoke as following.

First of all, data curves in Fig. 1 are translated into the discrete series, and put into a database as listed in Table 1. Secondly, the smoke background based on Fig. 2 is put into the databases in Table2. Finally, the mediation radiation of Table 2 are subtracted from the versus transmittance of Table 1 one by one. Its results are listed in Table 3.

According to the Table3, the spectrum curves of transmittance against waveband are plotted in Fig. 5 respectively.

Table 1 The conversion data from the original curves of transmittance in Fig. 1

wave length/μm	transmittance					
	32s	64s	96s	…	288s	320s
2.998997	0.234008	0.166108	0.168712	…	0.132606	0.12841
3.000733	0.234244	0.16634	0.168972	…	0.132868	0.128645
3.00247	0.234482	0.16659	0.169187	…	0.133117	0.128881
3.00421	0.234699	0.166812	0.169354	…	0.133319	0.129079
⋮	⋮	⋮	⋮		⋮	⋮
13.93889	0.60708	0.538828	0.556633	…	0.54668	0.530409
13.97646	0.605331	0.537563	0.555428	…	0.545517	0.529241

Table 2 The conversion from the smoke radiation curve in Fig. 2

wave length/μm	The ratio of radiation from smoke	wave length/μm	The ratio of radiation from smoke
2.998997	0.001755	⋮	⋮
3.000733	0.001861	13.93889	0.238181
3.00247	0.001937	13.97646	0.237506
3.00421	0.001985		

Table 3 The correcting data of transmittance

wave length/μm	transmittance					
	32s	64s	96s	…	288s	320s
2.998997	0.232253	0.166108	−0.06354	…	−0.19615	0.196147
3.000733	0.232383	0.16634	−0.06341	…	−0.19628	0.196279
3.00247	0.232545	0.16659	−0.06336	…	−0.19648	0.196475
3.00421	0.232714	0.166812	−0.06336	…	−0.19668	0.196679
⋮	⋮	⋮	⋮		⋮	⋮
13.93889	0.368899	0.538828	−0.187734	…	−0.35893	0.358934
13.97646	0.367852	0.537563	−0.187603	…	−0.35791	0.357914

It can be clearly seen from a comparison between Fig. 1 and Fig. 5 in 8-14μm waveband. In Fig. 1 the range of transmittance is nearly 50%, but the range in Fig. 5 by difference correcting is from 25% to 35%.

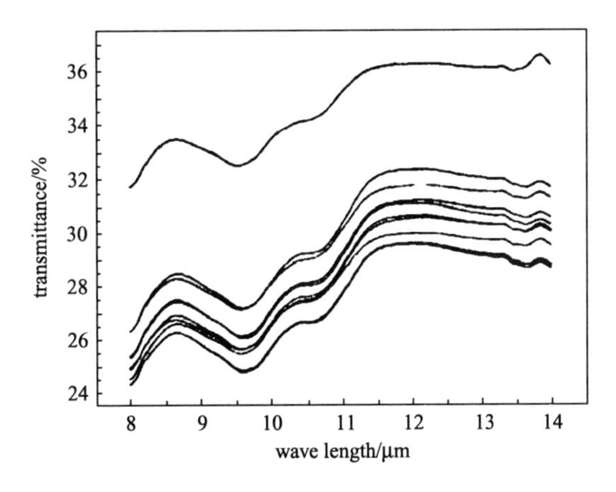

Fig. 5 The spectra of transmittance in 8-14μm wayeband after difference correcting

4 Conclusion

The higher temperature than the ambient, leads to the signal noise from IR radiation of hot smoke and the transmittance are different in 8-14μm waveband. To evaluate the infrared smoke performance, a correction method is applied to reduce the noise effect by subtracting the rate of infrared radiation of the smoke, and the range of transmittance decreases from 40%-60% to 25%-35% in 8-14μm.

The method not only is used to analyze the transmittance based on Fourier-transform-infrared spectrometer, but it is practicable to compare the data from the infrared radiometer.

References

[1] Sutherland R A, Butterfield J E. PILOT-81 VVA Experiment Analysis of the multi-path transmittance radiometer measurements [R]. ADA414693: 2003.

[2] Turner, Robert E. Investigation of emissive smoke [R]. ADA456822: 2002.

[3] Embury, Janon. Emissive versus attenuating smoke [R]. ADA400816: 2006.

[4] ZHU Chen-guang, PAN Gong-pei, WU Xiao-yun, et al. Evaluating interfered area of smoke screen using wavelet analysis [J]. Acta Armanmentarii, 2005, 26 (5): 620-623.

[5] ZHU Chen-guang, PAN Gong-pei, GUAN Hua, et al. Research of measuring method about sheltering coefficient of anti-infrared smokescreen [J]. Infrared Technology, 2004, 26 (4): 81-83.

[6] ZHOU Zun-ning, PAN Gong-pei, ZHU Chen-guang, et al. Research on interfercnee to target detection of IB imaging by smoke [J]. Acta Armanmentarii, 2006, 26 (3): 348-352.

[7] LI Ming, FAN Dong, KANG Wen, et al. A method for measurement & evaluation of infrared smoke [J]. Laser & Infrared, 2006, 36 (7): 599-603.

[8] GAO Wei. Study on evaluation of smokescreen jamming effectiveness [J]. Acta Armanmentarii, 2006, 27 (4): 681-684.

[9] LI Ming, FAN Dong-qi, YIN Chun-yong. Study on corresponding relation of laser and infrared transitivity for smoke screen [J]. J InfraredMillim Wave, 2006, 25 (2): 127-130.

(注：此文原载于 Chinese Journal of Energetic Materials, 2009, 17 (4): 482-485)

Synthesis of HMX via Nitrolysis of DPT Catalyzed by Acidic Ionic Liquids

He Zhiyong[1], Luo Jun[1], Lv Chunxu[1], Xu Rong[2], Li Jinshan[2]
(1. School of Chemical Engineering, Nanjing University of Science and Technology, Nanjing 210094, China)
(2. Institute of Chemical Materials, CAEP, Mianyang 621900, China)

Abstract: The direct nitrolysis of DPT to synthesize HMX with ionic liquids (ILs) as catalysts was investigated. The results showed that [Et_3NH] HSO_4 was the best catalyst among 18 ILs used and the yield of HMX was up to 61% against 45% without IL. The ILs could be efficiently recovered by simple distillation and extraction after reaction without any apparent loss of catalytic activity even after 10 times recycling.

Keywords: DPT, HMX, nitrolysis, ionic liquid

Introduction

1,3,5,7-Tetranitro-1,3,5,7-tetraazacyclooctane (HMX) is one of the most powerful military explosives[1-3], but its application is limited owing to high cost. The conventional methods to synthesize HMX involve the nitrolysis of Hexamine (HA)[4], 3,7-dinitro-1,3,5,7-tetraazabicyclo [3, 3, 1] nonane (DPT)[5], 1,3,5,7-tetraacetyl-1,3,5,7-tetraazacyclooctane (TAT)[6], 1, 5-diacetyl-3, 7-dinitro-1,3,5,7-tetraazacyclooctane (DADN)[7], etc. Some substrates such as TAT and DADN can afford HMX in high yields but need three or more steps, and they are not economical or environmental friendly.

DPT can be obtained with high yield from broadly available and inexpensive urea via nitrourea and/or dinitrourea[4,5,8,9], and then HMX can be produced from DPT by direct nitrolysis. This process avoids the use of large excess of acetic acid and acetic anhydride, so the total cost could be reduced remarkably. However, traditional nitrolysis of DPT by HNO_3-NH_4NO_3 only gave HMX in a low yield of 45% even when more than 30 equiv. of HNO_3 was used[10]. Some other HNO_3-$NH_4NO_3^-$ base systems were also investigated to afford HMX in moderate or low yields of 60%, 62% and 39% when MgO[11], P_2O_5[12], SO_3[13] were added, respectively. It was well known that better yields could be obtained with a mixture of acetic acid with acetic anhydride in the presence of HNO_3-NH_4NO_3[14,15]. Unfortunately, these methods inevitably showed some drawbacks such as environmental pollution and tedious workup.

Up to now, ionic liquids (ILs) have attracted much attention as environmentally benign media or catalysts thanks to many interesting properties such as wide liquid range, negligible vapor pressure, high thermal stability, and good solvating ability for a wide range of substrates and catalysts[16-18]. Qiao et al. developed a silica gel-immobilized Brønsted ionic liquid and used it as recyclable solid catalyst in the nitration of aromatic compounds, which offered practical convenience for the separation of product from catalyst[19]. Qian et al reported the ultrasonic assisted nitrolysis of DAPT in [BMIM] PF_6 to achieve HMX in a high yield of 66.8%[20]. Little attention has been paid to the nitrolysis of DPT in an acidic ionic liquid based system. Zhi et al reported a new method to synthesize HMX from DPT via nitrolysis using N_2O_5/HNO_3 catalyzed with a new dicationic acidic ionic liquid PEG_{200}-DAIL with 64% yield[21]. In this paper, we investigated the application of a series of acidic ILs in the synthesis of HMX by nitrolysis of DPT with HNO_3 (see Scheme 1).

Ionic liquids; NH_4NO_3, HNO_3

Ionic liquids:

[Hmin]X[22] X=NO_3^-, TsO^-, CF_3COO^-

[$(CH_2)_4SO_3HPy$]X[23] X=NO_3^-, TsO^-, CF_3COO^-

[$(CH_2)_4SO_3HMim$]X[24] X=NO_3^-, TsO^-, CF_3COO^-

[Caprolactam]X[25] X=NO_3^-, HSO_4^-, TsO^-

[$Et_3N(CH_2)_4SO_3H$]X[27] X=NO_3^-, HSO_4^-

[Et_3NH]X[28] X=HSO_4^-

PEG_{1000}-DAIL$(X)_2$[26] X=HSO_4^-, NO_3^-, TsO^-

Scheme 1 Nitrolysis of DPT catalyzed by ionic liquids

Materials and Methods

ILs were prepared by reported methods, DPT was prepared from urea by known method with a purity of more than 99%[4]. Melting points were determined on a Perkin-Elmer differential scanning calorimeter. The IR spectra was recorded on a Nicolete spectrometer and expressed incm^{-1} (KBr). Proton NMR was recorded on Bruker DRX300 (500MHz) spectrometer. The HPLC (Waters) was recorded on a SPD-6AV UV detector and mobile phase was methanol and water (70 : 30), velocity of flow was 1mL/min.

Typical nitrolysis procedure: 0.1g IL was dissolved in 21.9g nitric acid (98%) and cooled below 0°C, 1.9g ammonium nitrate (AN) was added, then 2g DPT was added in parts to keep the temperature below 0°C, and the result mixture was let stand at definite temperature for certain time to ensure complete reaction. The mixture was poured into an additional funnel and added dropwise into a flask containing 0.3mL of water over 20 minutes keeping the reaction temperature between 60-65℃.

The resulting mixture was stirred for another 20 minutes, then cooled to 40°C and additional water was added. After an additional stirring for 10 minutes the mixture was cooled to 20°C and let standing for 10 minutes, the solid product was collected by filtration and rinsed by water, affording pure HMX after crystallization from acetone and drying in vacuum. IR (KBr, cm^{-1}) 1560, 1462, 1395, 1203, 1144, 1086, 947, 760, 600. ^1H NMR (DMSO-d6), δ: 6.03 (s, 8H, CH_2).

Results and Discussion

Initially experiments for direct nitrolysis of DPT were carried out in NH_4NO_3/HNO_3 system in the presence of [$(CH_2)_4SO_3HPy$] NO_3. Reaction conditions such as the amount of IL and reaction time were investigated at 25°C, and the results was illustrated in Figure 1 and Figure 2. It was observed that the yield of HMX increased up to 58% when catalyzed by [$(CH_2)_4SO_3HPy$] NO_3 while only 45% was obtained without the IL (see Figure 1). However, the yield decreased with increasing IL concentration. The reduced yields may arise from an inappropriate acidity created during the nitrolysis reaction. At the same time, a shorter reaction time of 20 minutes was needed for nitrolysis in [$(CH_2)_4SO_3HPy$] NO_3 against a minimum 30 minutes for traditional nitrolysis in NH_4NO_3/HNO_3 (see Figure 2). The yield would decrease with prolonged time owing to HMX decomposition[13]. Thus, the IL [$(CH_2)_4SO_3HPy$] NO_3 proved to be an efficient catalyst which accelerate the reaction rate.

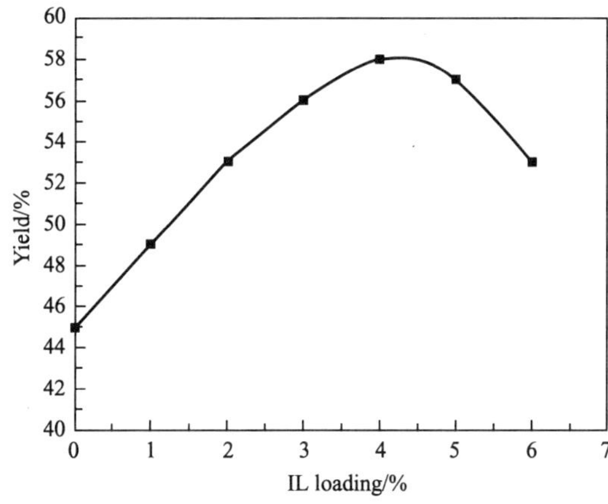

Figure 1 The effect of amount of ionic liquid on nitrolysis of DPT. Reaction conditions: 21.9g HNO_3, 2g DPT, 1.9g NH_4NO_3, 25°C, 30min

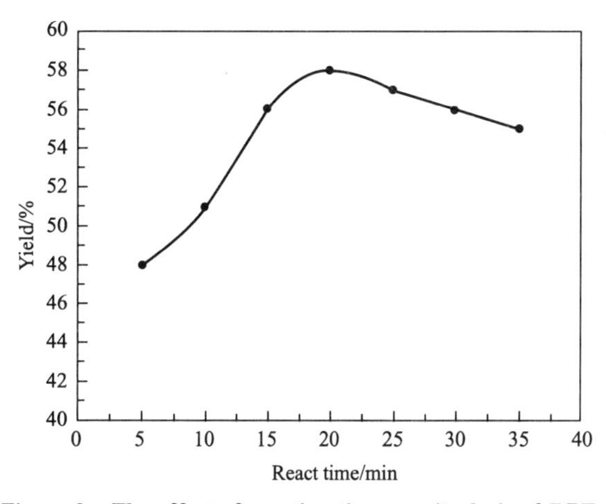

Figure 2 The effect of reaction time on nitrolysis of DPT. Reaction conditions: 21.9g HNO_3, 2g DPT, 1.9g NH_4NO_3, 25°C, 4(wt)% of [$(CH_2)_4SO_3HPy$] NO_3

A series of acidic ionic liquids were then applied as catalysts in the direct nitrolysis of DPT in NH_4NO_3/HNO_3 mixture (see Table 1). The product yields (54%~61%) were significantly better when using acidic ionic liquids as compared to control experiment (45%). The anions of ILs did not show distinct effect on the catalytic activity for nitrolysis of DPT. However, the cations had a more direct effect on the yields. The catalytic activities of [$(CH_2)_4SO_3HPy$] TsO and [Et_3NH] HSO_4 were superior than other ILs. Studies on the nitrolysis mechanism for HMX preparation has shown that esterification (O-nitration) is a competing side reaction which occurs during the formation of HMX in the nitrolysis of DPT, and that the bridge C-N bonds are the weakest ones which resulting in the formation of HMX. The choice may depend on the rapidity and completeness with which the free hydroxyl group is esterified (see Figure 3). If compound 3 is esterified quickly enough, the straight-chain product 4 would form. However, ammonium nitrate appears to favor formaldehyde removal (demethylolation) or hindered esterification, which facilitated the formation of HMX[29]. All the cations of the ILs used in this research were organic ammoniums, which might have the similar function of ammonium nitrate. This might be a reason for the nitrolysis modification affording higher yields against traditional methods.

Table 1 The effect of different ionic liquids on the nitrolysis of DPT[a]

Entrey	IL	HMX (g)	Melting Point℃	Yield (%)
1	0	1.23	274.1~274.4	45
2	[HMim]NO_3	1.47	273.6~273.9	54
3	[HMim]TsO	1.49	274.1~274.3	55
4	[HMim]CF_3COO	1.47	273.8~274.0	54
5	[$(CH_2)_4SO_3HPy$]NO_3	1.60	274.2~274.5	59
6	[$(CH_2)_4SO_3HPy$]TsO	1.63	274.4~274.6	60
7	[$(CH_2)_4SO_3HPy$]CF_3COO	1.58	274.1~274.5	58
8	[$(CH_2)_4SO_3HMim$]NO_3	1.49	274.2~274.4	55
9	[$(CH_2)_4SO_3HMim$]TsO	1.52	274.4~274.5	56
10	[$(CH_2)_4SO_3HMim$]CF_3COO	1.49	274.0~274.3	55
11	[Et_3NH]HSO_4	1.66	274.1~274.4	61
12	[$Et_3N(CH_2)_4SO_3H$]NO_3	1.49	274.1~274.4	55
13	[$Et_3N(CH_2)_4SO_3H$]HSO_4	1.49	274.1~274.4	55
14	[Caprolactam]NO_3	1.52	274.0~274.2	55
15	[Caprolactam]HSO_4	1.52	273.9~274.3	56
16	[Caprolactam]TsO	1.55	274.3~274.6	57
17	PEG_{1000}-DAIL$(NO_3)_2$	1.47	274.1~274.3	54
18	PEG_{1000}-DAIL$(HSO_4)_2$	1.47	274.0~274.3	54
19	PEG_{1000}-DAIL$(TsO)_2$	1.49	274.2~274.5	55

a 2.0g DPT, 1.9g NH_4NO_3, 21.9g HNO_3, 4 (wt)% of IL, 25℃/20 min, 65-70°C/20 min.

b Calculated by HPLC.

Figure 3 Nitration of DPT

The recoverability and recyclability of the acidic ILs was then studied. Table 2 lists the results of repeat experiments using [Et_3NH] HSO_4 as catalysts. The results indicated that the yield was reproducible.

After reaction, IL was extracted from the filtrate with 50mL methylene chloride, and it was recovered after removing methylene chloride. The recovered IL was reused in the next nitrolysis reaction after loading refresh substrates. The yield of HMX was unchanged after five recycles of acidic ionic liquid [Et_3NH] HSO_4 and then decreased slightly with continued recycling for up to ten cycles (see Figure 4).

Table 2 Repeated experiments of the nitrolysis[a]

Entry	Dosage of IL(%)	HMX(g)	Melting point ℃	Yield(%)[b]
1	4	1.67	274.4~274.6	61
2	4	1.66	274.2~274.3	61
3	4	1.67	274.3~274.5	61
4	4	1.65	274.1~274.4	61
5	4	1.66	274.3~274.5	61

a 2.0g DPT, 1.9g NH_4NO_3, 21.9g HNO_3, 4 (wt)% of [Et_3NH] HSO_4, 25°C/20 min, 65-70°C/20min.

b Calculated by quantitative HPLC.

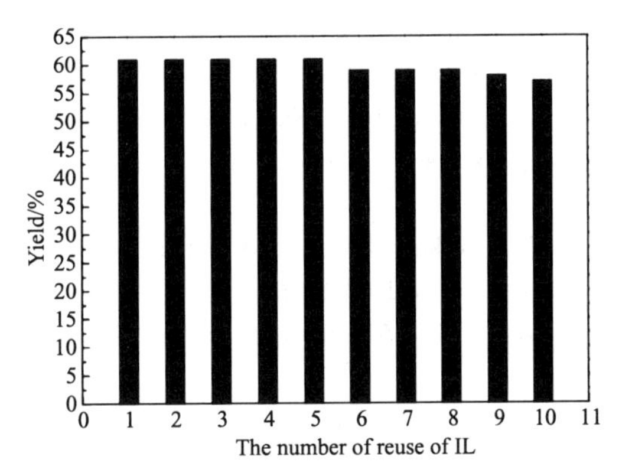

Figure 4 Repeating experiments using recovered [Et_3NH] HSO_4

Conclusions

An efficient process for the preparation of HMX from DPT in a NH_4NO_3/HNO_3 nitrolysis mixture catalyzed by a series of acidic ionic liquids was demonstrated. The ionic liquids could be reused for at least 10 cycles reproducibly affording HMX in yields up to 61%.

Acknowledgment

Financial support from Chinese National Natural Science Foundation of China-Academy of Engineering Physics (No. 10976014) is gratefully acknowledged.

References

[1] Siele V. I., Gilbert E. E., Process for Preparing 1,3,5,7-Tetranitro-1,3,5,7-tetraazacyclooctane, U. S. Patent 3939148, 1976.

[2] Chen J., Wang S. F., Use of NMR Spectrometry for Studying the Acetolysis of Hexamethylenetetramine, I. The Reaction of TAT Formation from DAPT, Propellants, Explos., Pyrotech., 1984, 9 (2), 58-63.

[3] Greenfield M., Guo Y. Q., Bernstein E. R., Ultrafast Photodissociation Dynamics of HMX/RDX from the Excited Electronic States via Femtosecond Laser Pump-Probe Techniques, Chem. Phys. Lett., 2006, 430, 277-281.

[4] Radhakrishnan S., Talawar M. B., Venugopalan S., Synthesis, Characterization and Thermolysis Studies on 3,7-Dinitro-1,3,5,7-tetraazabicyclo [3, 3, 1] nonane (DPT): A Key Precursor in the Synthesis of Most Powerful Benchmark Energetic Materials (RDX/HMX) of Today, J. Hazard. Mater., 2008, 152 (3), 1317-1324.

[5] Bachmann W. E., Horton W. J., Jenner E. L. et al., Cyclic and Linear Nitramines Formed by Nitrolysis of Hexaminel, J. Am. Chem. Soc., 1951, 6, 2769-2773.

[6] Castorina T. C., Holahan F. S., Graybush R. J. et al., Carbon-14 Tracer Studies of the Nitrolysis of Hexamethylenetetramine, J. Am. Chem. Soc., 1960, 82 (7), 1617-1623.

[7] Bachmann W. E., Sheehan J. C., A New Method of Preparing the High Explosive RDX, J. Am. Chem. Soc., 1949,

71 (5), 1842-1845.

[8] Bachmann W. E., Jenner E. L., 1-Acetoxymethyl-3, 5, 7-trinitro-1, 3, 5, 7-tetraza-cyclooctane and its Reactions. Significance in the Nitrolysis of Hexamethylene-tetramine and Related Compounds, J. Am. Chem. Soc., 1951, 73, 2773-2775.

[9] Castorina T. C., Autera J. R., Nitrogen-15 Tracer Studies of Nitrolysis of Hexamethy-lenetetramine, I&E Chemical Process development, Res. Dev., 1964, 4, 170-176.

[10] Chen L., Chen Z. M., Chen X. H., The Explore of Nitrolysis of DPT by Nitrate and Nitric Acid System, Chinese Journal of Explosives & Propellants, 1986, 3, 1-5.

[11] Li Q. L., Chen J., Wang J. L., Synthesis Craft of HMX from 1, 5-Methylene-3,7-dintrio-1,3,5,7-tetraazacyclooctane, Chinese Journal of Energetic Materials, 2007, 15, 509-510.

[12] Agrawal J. P., Hodgson R. D., Organic Chemistry of Explosives, John Wiley & Sons, Chichester 2007, 351.

[13] Cao X. M., Li F. P., High Energy Explosives of HMX and its Applications, Weapon Industry Press, Beijing 1993, 175-180.

[14] Lyushnina G. A., Bryukhanov A. Y., Turkina M. et al., Reactions of Amidosulfuric Acid Salts with Formaldehyde, Russian J. Org. Chem., 2001, 37 (7), 1030-1033.

[15] Siele V. I., Warman M., Leccacorvi J. et al., Alternative Procedures for Preparing HMX, Propellants, Explos., Pyrotech., 1981, 6 (3), 67-73.

[16] Wasserscheid P., Keim W., Ionic Liquids-new "Solutions" for Transition Metal Catalysis, Angew. Chem. Int. Ed., 2000, 39 (21), 3772-3789.

[17] DuPont J., De Souza R. F., Suarez P. A. Z., Ionic Liquid (Molten Salt) Phase Organometallic Catalysis, Chem. Rev., 2002, 102, 3667-3692.

[18] Sheldon R., Sheldon R. F., Catalytic Reactions in Ionic Liquids, Chem. Commun., 2001, 2399-2407.

[19] Qiao K., Hagiwara H., Yokoyama C. J., Acidic Ionic Liquid Modified Silica Gel as Novel Solid Catalysts for Esterification and Nitration Reactions, J. Mol. Catal. A: Chem., 2006, 246, 65-69.

[20] Qian H., Ye Z. W., Lü C. X., Ultrasonically Promoted Nitrolysis of DAPT To HMX in Ionic Liquid, Ultrason. Sonochem., 2008, 15, 326-329.

[21] Zhi H. Z., Luo J., Feng G. A. et al., An Efficient Method to Synthesize HMX by Nitrolysis of DPT with N_2O_5 and a Novel Ionic Liquid, Chin. Chem. Lett., 2009, 20, 379-382.

[22] Qi X. F., Study on Nitration (Nitrolysis) in the Presence of Brønsted Acidic Ionic Liquids, Nanjing University of Science and Technology, Nanjing 2008.

[23] Qi X. F., Cheng G. B., Duan X. L. et al., Study on Nitration of Toluene in the Presence of Brønsted Acidic Functional Ionic Liquids, Chinese Journal of Explosives & Propellants, 2007, 30, 12-14.

[24] Cole A. C., Jensen J. L., Ntai I. L. et al., Novel Brønsted Acidic Ionic Liquids and Their Use as Dual Solvent Catalysts, J. Am. Chem. Soc., 2002, 124, 5962-5963.

[25] Qi X. F., Cheng G. B., Lü C. X., et al., Regioselective Nitration of Toluene with HNO_3/Ac_2O Catalyzed by Caprolactam-Based Brønsted Acidic Ionic Liquids, Chin. J. Appl. Chem., 2008, 2, 147-151.

[26] Zhi H. Z., Luo J., Ma W. et al., Acetalization of Aromaldehyde and Ethyleneglycol in Novel PEG Thermoregulated Ionic Liquids, Chemical Journal of Chinese Universities, 2008, 29 (10), 2007-2010.

[27] Liu S., Xie C., Yu S. et al., Esterification of α-pinene and Acetic Acid Using Acidic Ionic Liquids as Catalysts, Catal. Comm., 2008, 9 (7), 1634-1638.

[28] Ganeshpure P. A., George G., Das J., Brønsted Acidic Ionic Liquids Derived from Alkylamines as Catalysts and Mediums for Fischer Esterification: Study of Structure-Activity Relationship, J. Mol. Catal. A: Chem., 2008, 279 (2), 182-186.

[29] Ren T. S., Nitramine and Nitrate Ester Explosives Chemistry and Technology, 203-223, Weapon Industry Press, Beijing, 1994.

(注：此文原载于 Central European Journal of Energetic Materials, 2011, 8 (2), 83-91)

第四部分

药物中间体及表面活性剂等的合成及工艺

乳胶炸药中乳化剂的研究

吕春绪

（302 教研室）

摘要： 本文阐述了乳胶炸药中乳化工艺及乳化剂作用的物理化学原理。基于对乳化剂的实验研究，为有效乳化和常贮安定，对乳化剂的选择提出几点意见，特别讨论了混合乳化剂的重要意义。

主题词和范畴号： 特种炸药［75F］，乳化剂［62P］

一、乳化工艺的物理化学实质

乳胶炸药通常是由氧化剂、可燃物及其它附加剂所组成，是一种膏状塑性体，它是经乳化工艺制得的。

按物理化学基本原理，认为乳化工艺是一种液溶胶形成过程。其分散相是无机氧化剂的饱和水溶液，连续相是燃料油，依靠乳化剂增加该分散体系的稳定性。它是水分散在油介质中，也称为油包水型体系，以 W/O（water in oil）表示。包复情况的显微摄影如图 1 所示。

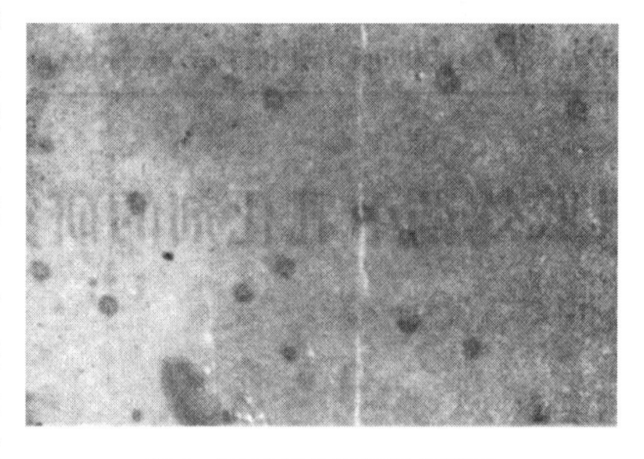

图 1　W/O 型包覆显微摄影图

凡能促进乳化和保持乳胶稳定的各种因素，均有利于乳化工艺。当然，配方设计是根本。大量实脸表明：石蜡、地蜡、复合蜡、微晶蜡、软化沥青、机油及凡士林等是较好的油相组分，其用不超过 5%（重量）。即用尽可能少量的油相，均匀地包覆水相，而成为 W/O 型乳状液。所确定的工艺条件，应是造成乳化的有力措施。

乳化是把整个水相分散在连续相—油中，形成油包水的胶粒。乳状液体系有很大的相界面，界面的大小取决于粒子的分散度。将一颗边长为 1 微米的立方体分散为边长为 1 毫微米的立方体胶粒，这时的胶粒的数目为 10^9 个，其总面积增大了 1000 倍。研究还表明：体系的分散度越大，表面积也越大。这时消耗的功也越大，因而表面能也越大。该表面能近似地可以看作增加分散相表面积时外界所给予的能量。例如，1 克水成小球时，其面积为 4.83 厘米^2，表面能为 351 尔格（约 8×10^{-6}卡），若将此小球分散成半径为 10^{-7}厘米的细小水滴，可得 2.4×10^{20}个，其总表面积为 3×10^7 厘米^2，表面能约为 50 卡。此结果表明，将 1 克水分散成半径为 10^{-7}厘米的细小水滴时，外界要供给 50 卡的能量，这 50 卡相当于使 1 克水的温度升高 50 度[1]。这个外界能量在工艺上就是由搅拌器设计及转速、乳化器的形状及大小、乳化温度等因素决定的。

乳胶是靠外界能量形成的，比表面激烈增加，是个热力学非自发过程。即在两相界面上的自由能都有自发减小的趋向。

$$\delta_{w可} = (dG)_{T,P} = \sigma \cdot dA$$

其中：

$\delta_{w可}$——以可逆方式形成新表面过程中所消耗的表面功；

$(dG)_{T,P}$——恒温、恒压下自由能增量；

σ——比界面自由能（比表面能），即$\left(\frac{\partial G}{\partial A}\right)_{T.P.m}$，温度、压力和组成一定条件下，增加一个单位面积时，体系自由能的增加，以尔格/厘米^{2}表示；

dA——界面积增量。

界面能的减少可以通过界面（A）的自动缩小，也可以通过界面张力（σ）的自动降低来达到。而加入乳化剂有利于后一种作用，使其体系稳定。已形成的胶粒聚结在一起（破乳）使比表面减少即是热力学自发过程，特别是乳状液分散相粒子较大（其直径在0.1微米以上）时，特别容易聚结，聚结的显微摄影过程如图2所示。

聚结是小珠滴的不可逆过程，大量聚结的最终结果将造成水相及油相分层（破乳），图3所示。

为了防止聚结，保持乳状液稳定以及长期贮存，要靠乳化剂（表面活性物质、电解质、固体粉末等）的作用，特别是表面活性物质，既可防止凝聚，又可降低界面张力，是稳定乳化体系必不可少的。

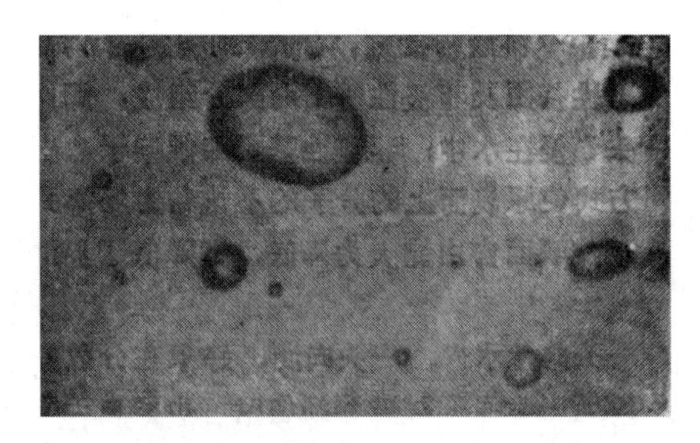
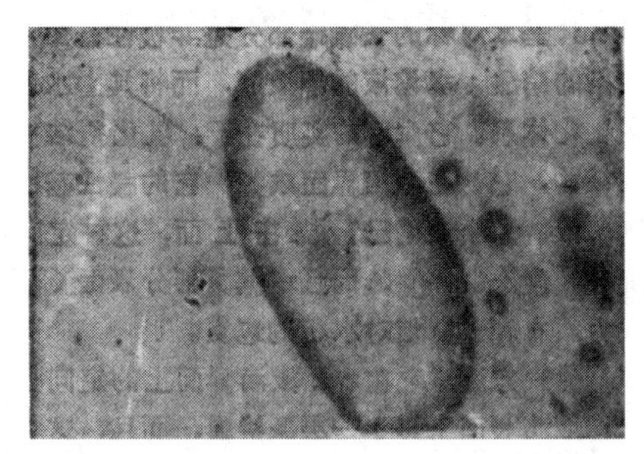

图2　乳胶微粒聚结显微摄影图

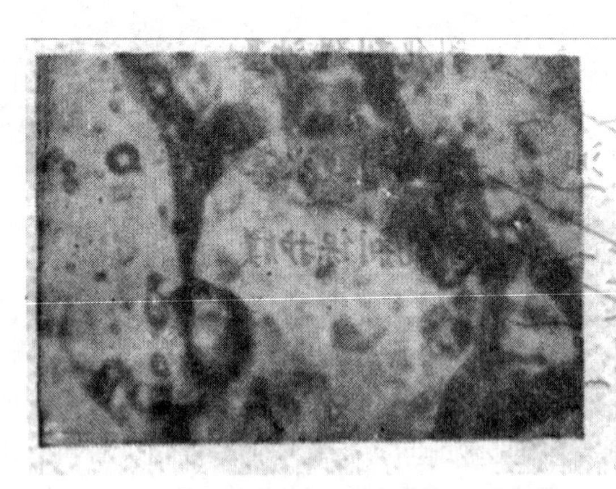
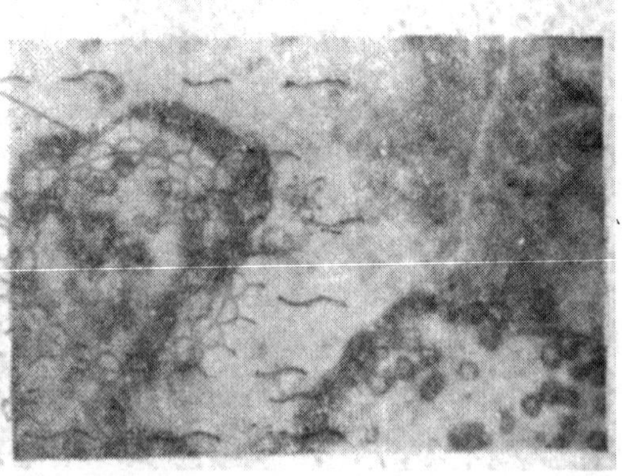

图3　乳胶破乳分层显微摄影图

二、乳化剂作用的物理化学原理

乳胶炸药中无论水相还是油相除了主氧化剂及主可燃物外，均有添加剂，如氧平衡调解剂、降低冰点剂、敏化剂、乳化促进剂及乳化剂等。要制得稳定的孔胶炸药，其中乳化剂尤为重要。

亲水基 (—OH, —O—C(=O)—) ； 憎水基 (R)

目前在乳胶炸药中得到广泛应用的乳化剂多是非离子型的含羟基的高级脂肪酸酯类，例如司盘-80 (Span-80)，即失水山梨糖醇单油酸酯，其中—OH 及—COO^- 是极性基团，因而亲水；基本对称的长碳链部分是非极性基团，是憎水的。该乳化剂可以其亲水基去包复作为氧化剂的硝酸盐水溶液的小液滴，而将其憎水基与连续相燃料油相接触，从而得到稳定的 W/O 型乳化体系。它作用的实质在于：乳化剂结构是由极性基团及非极性基团两部分组成，即两性结构，它们在两相界面浓集，它的极性基团的“头”浸在水中；非极性基团却倾向于逃出水溶液，像个“尾巴”挠在上面，这种定向排列方式使原表面上的分子的不饱和力场得到某种程度上的平衡，而且占据了部分表面（界面)，使表面自由能大为降低，即降低了界面张力，有利于油相对水相的包复。

另一方面，它在水相液滴界面上的定向吸附，一头向“水”，一头向“油”，好像在分散的水相液滴表面上裹了一层保护膜，而且，这种保护膜万一本身辰部受到损伤时，也能自动弥补“伤口”。

这层保护膜使分散液液滴相互碰撞时难于凝康，保持乳膜的稳定状态长期存在。

同一乳化剂的不同用量对乳化效果有一定影响，Trable 规则[2]认为同一乳化剂在低浓度时表面张力的降低效应与浓度成正比。但是，达到饱和吸附后，不同链长的直链有机脂肪酸的 $\sqrt{\infty}$（吸附量极值）在定温下是定值，再增加用量也不会降低表面张力。

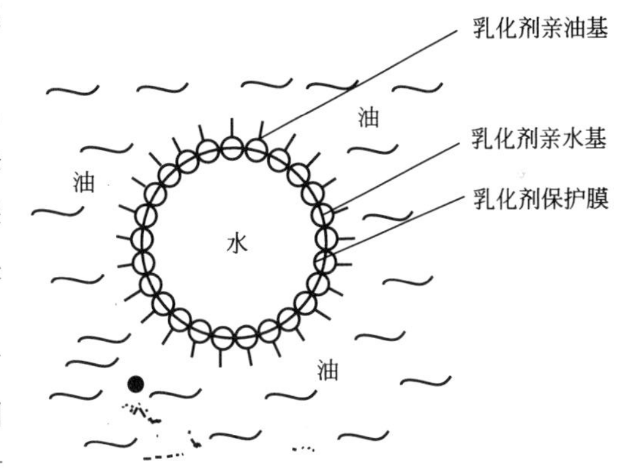

图 4　乳化剂保护膜示意图

乳化剂的乳化作用与亲水基的亲水性及憎水性有关，为了定量地表示亲水基亲水性与憎水基憎水性的相对大小，人们提出了乳化剂亲憎平衡值 HLB (Hydrophic-Lipophilic Balance) 值的概念[2,4]。HLB 值越高，表示亲水性越大；HLB 值越低，表示憎水性越大。HLB 道的大小与乳化剂分子结构有关，有人提出用如下经验公式估算 HLB 值[5,6]。

$$HLB = \sum(HLB)_{亲水基} - \sum(HLB)_{憎水基} + 7$$

式中：

$(HLB)_{亲水基}$为各种亲水基对 HLB 值的贡献

$(HLB)_{憎水基}$为各种憎水基对 HLB 值的贡献

各种基团均有其对 HLB 值的贡献值。经验计算出某些乳化剂的 HLB 值，例如，失水山梨糖醇单油酸酯的 HLB 值为 4.3，失水山梨糖醇单硬脂酸酯为 4.7，甘油单硬脂酸酯的 HLB 值为 3.8。实验发现乳化剂的 HLB 值为 3.0～6 时，最适宜作 W/O 型乳化剂。事实上，上述三种乳化剂已在乳胶炸药中广泛应用。

三、 乳化剂及其用量的实验研究

对于 W/O 型乳化剂的种类我们作了大量实验。实验进一步发现对于 W/O 型乳化，乳化剂的 HLB 位在 4.0～4.3 之间时乳化效果最佳。司盘型乳化剂乳化效果实验数据如下表所示：

表 1　乳化剂种类对乳化的影响（配方以 100 克计）

序号	NH_4NO_3	$NaNO_3$	敏化剂	水	司盘85	司盘80	改司盘80	司盘60	司盘40	石蜡	地蜡	凡士林	HLB	物态	挤压出水
1	64	10	12	8	2					2	1	1	1.8	坚硬	渗水
2	64	10	12	8		2				2	1	1	4.1	软	不易
3	64	10	12	8			2			2	1	1	4.3	较软	不易
4	64	10	12	8				2		2	1	1	4.7	坚硬	自然渗水
5	64	10	12	8					2	2	1	1	6.7	坚硬	自然渗水

研究还指出：一些常温下为固态的乳化剂，例如司盘-60，所制备的乳胶在常温下异常坚硬，自然渗水（破乳），故在乳化剂选择中，以物态的观点，应该选择常温下为液态或半液态的乳化剂，司盘-80，特别是改性司盘-80—混合乳化剂的效果特别突出。

乳化剂的用量也具有重要意义，因为它不仅关系到保护膜的致密程度，即乳胶体的稳定性，也关系到乳胶体的成本高低，实验表明：当乳化剂用最大于 0.5%时，就能够顺利地进行乳化，在 1.5%～2.0%时比较合适，若贮存时间在 120 天以上时，应以 1.8～2.0%为佳。司盘型乳化剂用量影响的实验数据如下表所示：

表 2　乳化剂用量对乳化的影响（配方以 100 克计）

	NH_4NO_3	$NaNO_3$	敏化剂	水	司盘 80	机油	石蜡	地蜡	乳化时间（秒）	挤压出水
1	63.5	10	12	8	0	1.5	3	2	不乳化	
2	63.5	10	12	8	0.5	1.0	3	2	75	易
3	63.5	10	12	8	1	0.5	3	2	60	较易
4	63.5	10	12	8	1.5	0	3	2	55	较不易
5	63.5	10	12	8	2	0	2.5	2	45	较不易

四、 乳化剂选择的几点意见

（一）复合乳化剂的使用具有较大的实际意义

大量实验表明复合乳化剂在乳化及保证长贮安定性方面有重要作用。

被乳化物要求乳化剂具有适宜的 HLB 值。最常用的油相材料：机油、凡士林、蜡等，要求适宜的 HLB 值如表 3 所示。

表 3　被乳化物要求乳化剂适宜的 HLB 值

被乳化物	要求乳化剂具有的 HLB 值		被乳化物	要求乳化剂具有的 HLB 值	
	W/O 型乳液	O/W 型乳液		W/O 型乳液	O/W 型乳液
矿物油(重质)	4	10.5	蜂蜡	5	10～16
矿物油(轻质)	4	10	石蜡	4	9
凡士林	4	10.5			

被乳化物的 HLB 值越强，与乳化剂的 HLB 值差别越大，两者的亲和力就越差，乳化效果就越不好，变化乳化剂的 HLB 值使其与被乳化物要求的 HLB 值接近，会使乳化效果有较大改善，如图 5 所示。

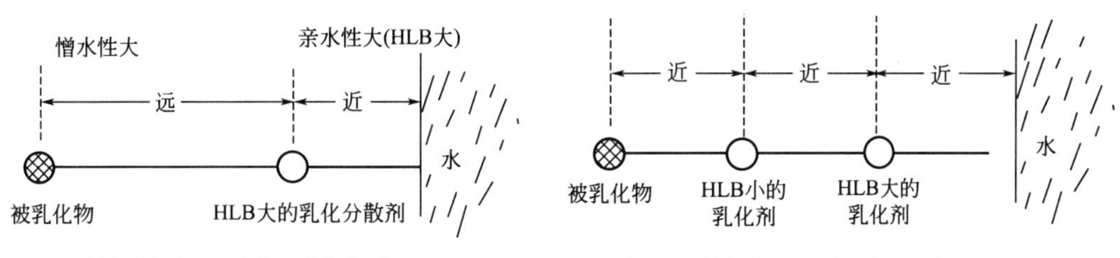

(a) 乳化剂溶于水，不能很好乳化的例子

(b) 用HLB小的乳化剂将两者连接起来稳定乳化的例子

图 5　乳化剂配合使用原理图

要满足这要求，只能采用混合（复合）乳化剂，经调节达到。实脸表明，复合乳化剂的 HLB 值可由组分中各个乳化剂的 HLB 值按重量平均算出。例如，要得到 HLB＝10 的复合乳化剂，可将司盘 60（HLB＝4.7）45％，吐温知（HLB＝14.9）55％进行复合，即可得到 HLB＝10.3 的复合乳化剂。

单一乳化剂不能满足使用要求的主要原因与分子结构有关。以司盘 80 为例，其组成成分油酸是含一个双键的不饱和酸，在同样条件下，饱和的十八酸分子所占的面积为 21～25A^2，油酸分子所占面积为 28～55A^2，显然，使用不饱和的油酸所生成的司盘在油水两相界面因定向吸附所生成的单分子膜就显得比较“膨松”。而每个油酸分子的双键处电子云分布密集产生了电荷的相斥作用。从而在双键处形成孔腔，所谓乳胶炸药破乳就是水分子从油酸分子孔腔中被挤压出所致。采用复合乳化剂，倘若在界面膜上有“复合物”生成，则此膜的强度增加，紧密堆积就最容易实现，因而液珠就不易凝结，乳状液就更稳定。显然，复合乳化剂中，辅助乳化剂与主乳化剂必须能生成“复合物”才能有较好效果。司盘 80 与木糖醇单油酸酯以一定比例混合后，收效显著。实验还发现：在司盘 80 中添加磺酸盐、固体粉末等也有一定效果。

总之，选取复合乳化剂是制取长贮安定性较好的乳胶炸药的必要条件。

（二）分子结构是选择单一乳化剂的主要依据

（1）亲水基处于憎水基末端的，具有较好的乳化能力，例如，司盘 80、司盘 60 以及十八烯醇硫酸酯钠盐等，由于亲水基位于憎水基的末端，使水相能较好地分散在连续相中。分子结构如下：

$$CH_3(CH_2)_7CH{=}CH(CH_2)_7CH_2—\underset{\underset{O}{\|}}{C}—O—CH_2CHOH\underbrace{C—CHOHCHOH—CH_2}_{O}$$

$$CH_3(CH_2)_7CH_2CH_2(CH_2)_7\underset{\underset{O}{\|}}{C}—O—CH_2CHOH\underbrace{CH—CHOHCHOH—CH_2}_{O}$$

$$CH_3(CH_2)_7CH{=}CH(CH_2)_7CH_2—OSO_3Na$$

（2）分子量较大的作为乳化分散剂效果较好，如月桂醇酯、十六醇酯及十八醇酯等。

（3）根据结构相似原理，乳化剂中的憎水基应与连续相结构相似，亲水基应与分散相结构相似，例如司盘 80 中亲水基中的—OH 与分散相 H—OH 中—OH 结构相似，憎水基中—R 基与连续相燃料油及蜡的结构相似，更有利于水相在油相中的分散及乳化。

参考文献

[1] 复旦大学化学系编：《物理化学》（下册），人民教育出版社，1978.

[2] 贝歇尔，P.：《乳状液理论与实践》，科学出版社，1978.

[3] 吉林大学编：《物理化学》（下册），人民教育出版社，1979.

[4] 华东工学院 302 教研室编：《炸药制造工艺原理》，1983.
[5] [日] 矾田孝一、藤本武彦：《表面活性剂》，轻工业出版社，1984.
[6] 赵国玺：《表面活性剂物理化学》，北京大学出版社，1984.

Special Explosives，Emulsifying Agents

Lv Chunxu

Research of the Emulsifier in the Emulsion Explosives，

J. of East China Inst. of Thechno.，Sum No. 40 (1986)，pp. 95-101

Abstract: In this paper，the physicochemical principle of emulsion technology and emulisifier role in the emulsion explosives is expounded. On the bases of experimental studies of the emulsifier，in order to emulsify effectively and store stably，some views are presented to the choice of emulsifier and，especially，important significance of mixed emulsifier is discussed.

（注：此文原载于　华东工学院学报，1986，40：95-101）

氯代苦基氯制备最佳工艺条件研究

吕春绪，邓爱民，郑明恩
（化工学院）

提　要：本文采用含有发烟硫酸的超酸硝化剂制备氯代苦基氯，这是一条新工艺路线，本文研究了反应参数（硝化温度，发烟硫酸浓度及硝酸钾用量）变化的影响，并利用正交实验研究最佳工艺条件，给出制备氯代苦基氯的最佳工艺条件，并对产品进行了结构鉴定。给出了元素分析及红外光谱数据，测试结构表明，该产物是氯代苦基氯

关键词：化合物34AA，工艺83AB，最佳化62A，氯代苦基氯

1　概述

氯代苦基氯，化学命名2,4,6-三硝基间二氯苯，为灰白色晶体，纯品熔点123℃. 易溶于乙醇等有机溶剂中，由于受三个强吸电子基团的影响，苯环上两个氯原子很活泼，可以发生Ullmann反应，生成一系列重要耐热炸药。

两个分子的氯代苦基氯Ullmann偶联反应生成二氯六硝基联苯，该化合物分子中的两个氯原子氨解后，生成著名的耐热炸药-二氨基六硝基联苯，代号DIPAM.

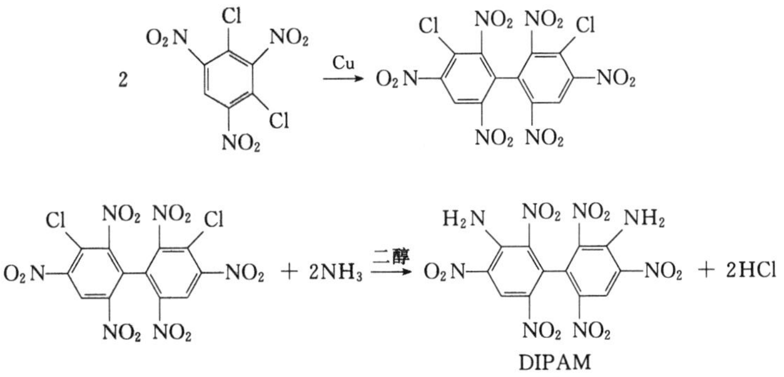

2 Cl, O_2N, NO_2, Cl, NO_2 $\xrightarrow{Cu}$ Cl, NO_2, NO_2, Cl, O_2N, NO_2, NO_2, NO_2

Cl, NO_2, NO_2, Cl, O_2N, NO_2, NO_2, NO_2 + $2NH_3$ $\xrightarrow{二醇}$ H_2N, NO_2, NO_2, NH_2, O_2N, NO_2, NO_2, NO_2 + 2HCl

DIPAM

一个分子氯代苦基氯与两个分子苯胺发生Ullmann缩合反应，生成1,3-二［苯胺基］-2,4,6三硝基苯中间体。该中间体经硝化可生成九硝基化合物，也是一个较好的耐热炸药，代号NONAN：

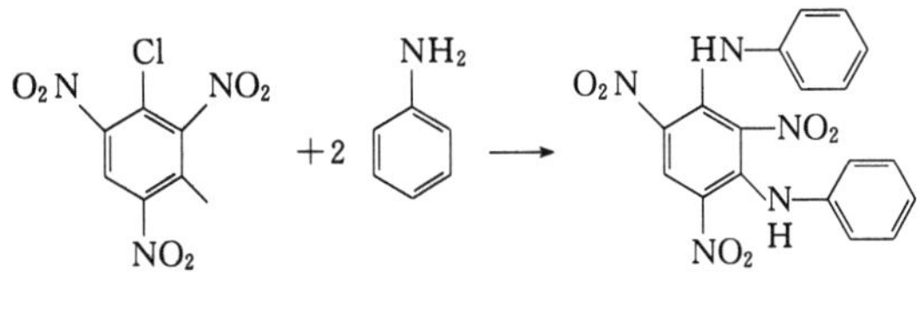

Cl, O_2N, NO_2, NO_2 +2 NH_2 ⟶ HN, O_2N, NO_2, N, H, NO_2

NH_2, O_2N, NO_2, NO_2 $+ 6HNO_3 \xrightarrow{H_2SO_4}$ NO_2, O_2N, NO_2, NH, O_2N, NO_2, NO_2, N—H, NO_2, NO_2, NO_2 $+ 6H_2O$

NONAN

过去制备氯代苦基氯曾经使用过单一硝酸，混酸硝化间二氯苯的方法[1,2]，也有人用斯蒂酚酸经吡啶成盐后氯化的方法[3,4]，前者反应周期长达二十小时，后者工艺处理复杂，成本昂贵，给氯代苦基氯制备造成很大困难。本文使用超酸硝化剂与老工艺相比有其长处，为氯代苦基氯制备提供一条新的工艺路线。

Cl, Cl $+ KNO_3 \xrightarrow{H_2S_2O_7}$ Cl, O_2N, NO_2, NO_2 $+ 3KHSO_4$

2 氯代苦基氯制备最佳工艺条件的单因素研究

氯代苦基氯制备是以发烟硫酸与硝酸钾组成的超酸硝化剂硝化间二氯苯，控制最佳料比，反应温度及反应时间等工艺条件，获得熔点 120℃的产品（粗品）。为此，我们首先研究硝化温度，料比等单因素的影响。

2.1 硝化温度的影响

取 80 毫升 50%的发烟硫酸，35.5 克硝酸钾，硝化温度为 120℃，140℃，160℃，对氯代苦基氯得率影响如图 1。

2.2 不同浓度发烟硫酸的影响

取 35.5 克的硝酸钾，不同浓度的发烟硫酸（以硫酸总量计）80 毫升，硝化温度 130℃，保温时间 3 小时，进行硝化，对氯代苦基氯得率影响如图 2。

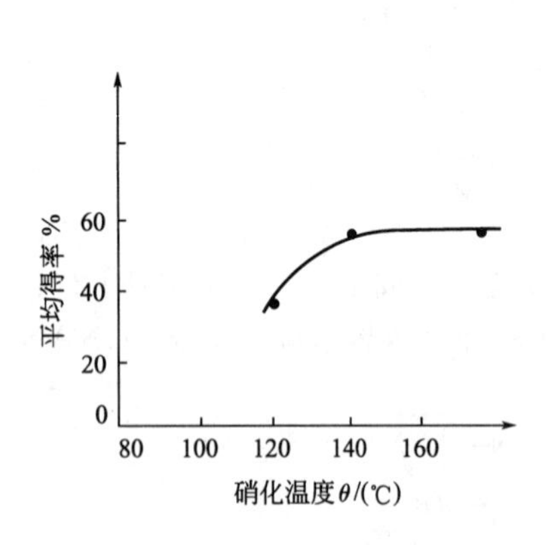

图 1 硝化温度对平均得率的影响

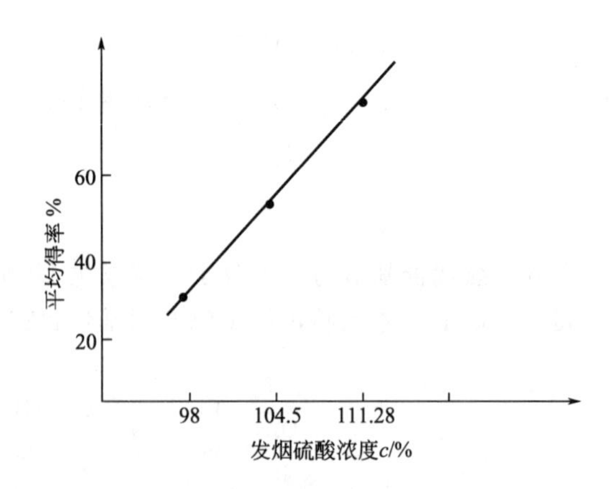

图 2 发烟硫酸浓度对平均得率的影响

2.3 不同硝酸钾用量的影响

取 80 毫升 50%的发烟硫酸，不同用量的硝酸钾，硝化温度 130℃，保温时间 3 小时，进行硝化，对氯代苦基氯得率影响如图 3。

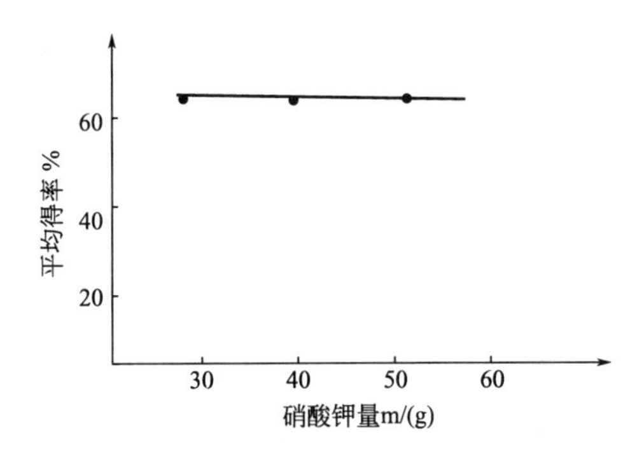

图 3 硝酸钾用量对平均得率的影响

从上述单因素研究发现：硝化温度 130℃，使用 50%的发烟硫酸，硝酸钾用量为 35.5 克时，具有较高得率。

3 氯代苦基氯制备最佳工艺条件的正交实验

在上述单因素影响的研究基础上，为获得最佳组合，我们用正交实验研究它们之间的相互影响，从而获得考虑综合因素后的最佳工艺条件。正交实验数据如下。

表 1 硝化间二氯苯的正交实验

序号	间二氯苯 (g)	50%发烟硫酸的量(ml)	保温温度 (h)	保温时间 (h)	产品熔点 (℃)	产品重量 (g)	得率 (%)
1	15.8	200	160	1	104	21.5	71.0
2	15.8	160	160	3	118	7.5	24.8
3	15.8	120	160	5	116	2.0	6.6
4	15.8	160	140	1	90	14.0	46.7
5	15.8	120	140	3	96	11.0	36.7
6	15.8	200	140	5	110	14.5	48.0
7	15.8	120	120	1	84	6.5	21.5
8	15.8	200	120	3	100	16.5	54.5
9	15.8	160	120	5	88	9.5	31.4

表 2 诸因素对产品得率影响正交实验数据处理表

影响因素 / 数据 / 项目			保温温度(℃)	保温时间(h)	50%发烟硫酸的量(ml)
K160℃	1h	160ml	102.4	139.2	713.5
K140℃	3h	160ml	131.4	116.0	102.9
K120℃	5h	120ml	107.4	86.0	64.8
k160℃	1h	200ml	34.1	46.4	57.8
k140℃	3h	160ml	43.8	38.7	34.3
k120℃	5h	120ml	35.8	28.7	21.6
	极差 R		9.7	17.7	36.2

从表 2 中极差及图 4 中可以看出影响得率的最主要因素是发烟硫酸用量，次要因素是保温温度。

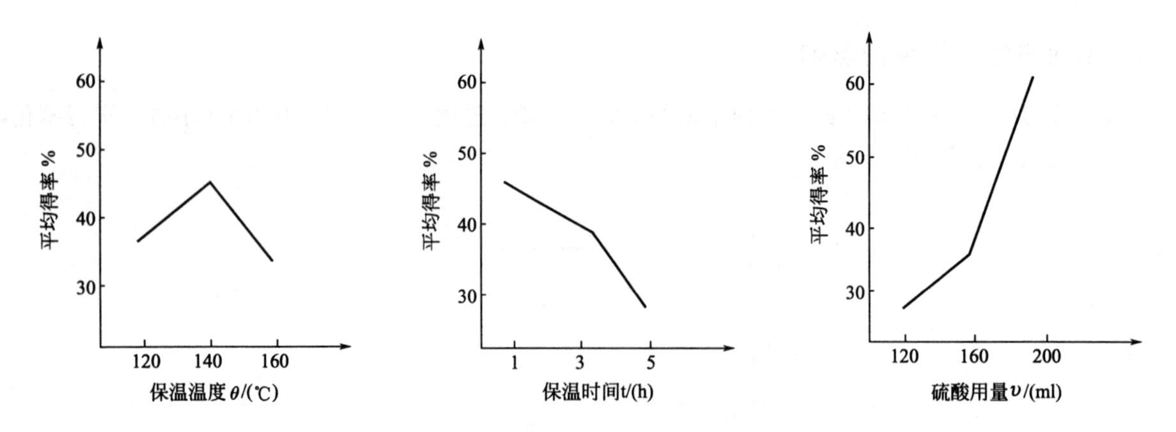

图 4　平均得率与保温温度，时间及硫酸用量关系曲线

表 3　诸因素对产品熔点影响正交实验数据处理表

影响因素 / 数　据 / 项目			保温温度(℃)	保温时间(h)	50%发烟硫酸的量(ml)
K160℃	1h	200ml	338	278	314
K140℃	3h	160ml	296	314	296
K120℃	5h	120ml	272	314	296
k160℃	1h	200ml	113	93	105
k140℃	3h	160ml	99	105	99
k120℃	5h	120ml	91	105	99
	极差 R		22	12	6

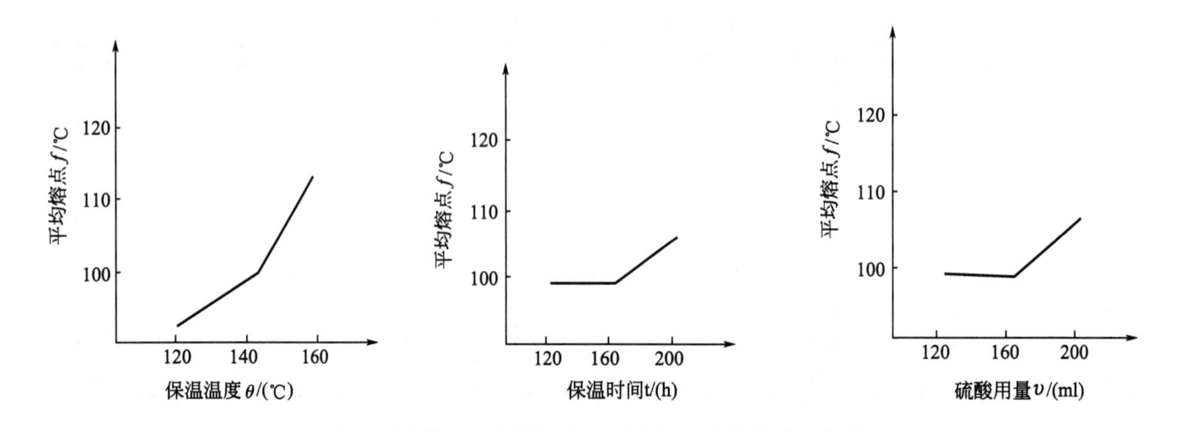

图 5　平均熔点与保温温度，时间及硫酸用量关系曲线

从表 3 中极差及图 5 中可以看出影响熔点的主要因素是保温温度，其次是保温时间。因为保温温度越高，时间越长越有利于其他异构体及杂质的分解，产品越纯，其熔点也越高。

综上所述，既要保证较高得率，又要具有较高熔点，选择保温温度 140℃，保温时间 3 小时，硫酸用量为 200 毫升的工艺条件较为合适。

我们还对发烟硫酸浓度，硝化温度及硝酸钾用量等作了正交实验研究，确定出最佳硝化温度及硝酸钾用量。

4　最佳制备工艺及产品结构鉴定

将 400 毫升 50%发烟硫酸倒入三口瓶中，用油浴升温至 80℃，慢慢加入 71.0 克硝酸钾，加料过程温度不超过 100℃．加料完毕后，将温度升至 130℃，再滴加 31.6 克间二氯苯，此过程不超过 140℃。加完后在 140℃下保温 3 小时。然后冷却至 80℃，将反应混合液倾入冰水中过滤，洗涤及烘干。产品熔点 120℃（粗品），得率为 75%。

上述实验放大五倍，近百次均获得稳定结果。

产品经北京中国科学院化学所进行元素分析，结果如下：

	C	H	O	N	Cl
理论值	25.53	0.35	34.04	14.89	25.18
实测值	25.10	0.43	33.80	14.20	26.47

产品经南京大学现代测试中心测定红外光谱，波谱解析表明波数为 3088cm^{-1}的峰是苯环上$\bar{\gamma}\alpha$-H 伸展振动峰，波数为 1586cm^{-1}的峰和 1425cm^{-1}的峰是苯环$\bar{\gamma}\alpha$-α 骨架振动峰，波数为 1554cm^{-1}和 1365cm^{-1}的峰是芳烃上硝基反对对称振动峰和对称（伸展）振动峰，波数为 902cm^{-1}的峰是苯环上 1,2,3,4,6，五取代峰，波数为 724cm^{-1}的峰是 $\gamma\alpha$-α_1 伸展振动峰。可以认为该产品是氯代苦基氯。

5 结论

（1）采用发烟硫酸-硝酸钾超酸硝化剂制备氯代苦基氯具有工艺操作简单，周期短，成本低等特点。

（2）本工艺中，影响产品平均得率的主要因素是发烟硫酸用量。

（3）采用本工艺条件制备的产品熔点达 120℃，得率不低于 70%。

（4）采用本工艺条件制备的产品，经熔点，元素分析及红外光谱等结构鉴定表明是氯代苦基氯。

参考文献

[1] Sudborough J J，Picton N. JCS，1906，89：589.
[2] Frankland p f，Garner F H. J. SOC. Chem. Ind，1920，29：269.
[3] Warman M，Siels V I. J. Org. Chem，1961，26：2997.
[4] Sharnin G，M. Opz. Xuu.，1968. 3：78.
[5] 金家骏．流体热力学及化学平衡．南京：华东工学院，1979.
[6] 孙荣康，猛炸药化学及工艺学（上册）．北京：国防工业出版社，1981.
[7] Ouoruhko C C. M. Ⅱ Xuu.，1978，51 (3)：685.

Research of the Best Technological Conditions for Preparing Chloropicrylchloride

Lv Chunxu，Deng Aimin，Zheng Mingen

J. of East China Inst. of Techno，No. 1 (1989)，Sum No. 49，PP. 18-23

Abstract: Super acid nitrating agent containning oleum is used for preparing chloropicrylchloride. This is a new technique. The effects of changes in reacting parameters (hitration temperature concentration of oleum and amount of potassium nitrate) are investgated. The best technological conditions are studied with the methods of orthogonal design. The best technique preparing chlorpicrylchloride is given. The structure of the product is identified. Result of elementary analysis and IR spectra show the above product is chloropicrylchloride.

Key Words: Compound，technology，optimization，chloropicrychloride.

（注：此文原载于 华东工学院学报，1989，(1)：18-23）

盐酸法制备三硝基间氯苯胺最佳工艺条件研究

吕春绪，王恩芳

（华东工学院，化工学院，南京.210014）

摘要：三硝基间氯苯胺是精细化工品合成的重要中间体。本文重点讨论盐酸氯化工艺。通过工艺条件单因素影响研究及正交实验，得出盐酸法制备三硝基间氯苯胺最佳工艺。使用该工艺所得产物经红外光谱及元素分析等结构鉴定表明与预期化合物结构相符。

三硝基间氯苯胺是精细化工合成的重要中间体。

分子中同时存在Cl及-NH_2基。通过反应引进化合物中，由于氨基的存在，它可与化合物中存在的硝基形成氢键，如果形成分子内氢键可使其特征密度ρ_0增大；如果形成分子内氢键缩短分子间距可使堆积系数K值升高，这均会使化合物的熔点较大幅度提高。—Cl基可以与有机胺发生Ullmann缩合反应，把芳核引进体系[1,2]。有人利用苦基氯（卤）把苦基（三硝基苯基）引进体系[3-6]，据此，我们通过三硝基间氯苯胺成功地把氨苦基引进分子中，化合物熔点确实升高，热稳定性也有所改善。乙二胺与苦基氯反应：

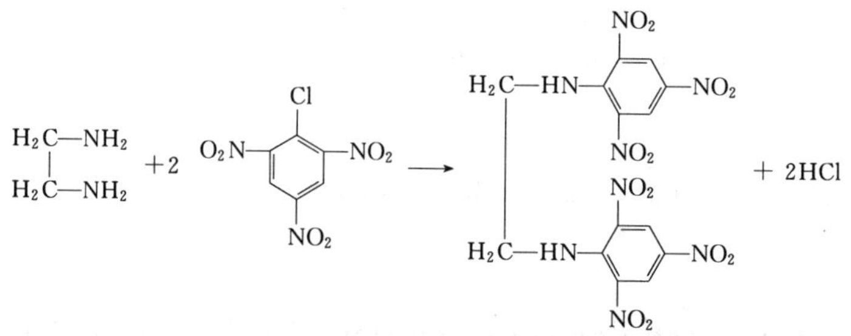

生成的二苦基乙二胺基熔点为231℃。而乙二胺与三硝基间氯苯胺反应：

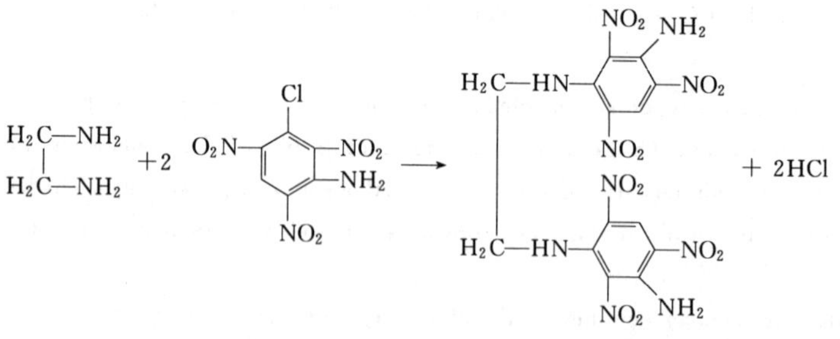

生成的二氨芳基乙二胺其熔点为275℃，熔点上升40多度。二氨苦基乙二胺是个新化合物。

三硝基间氯苯胺还被广泛应用于其它引进氨苦基的精细化工品合成反应中。

利用四硝基苯胺中氨基间位的硝基的不稳定性，采用盐酸氯化工艺是合成三硝基间氯苯胺有较大实际意义的一条工艺路线。

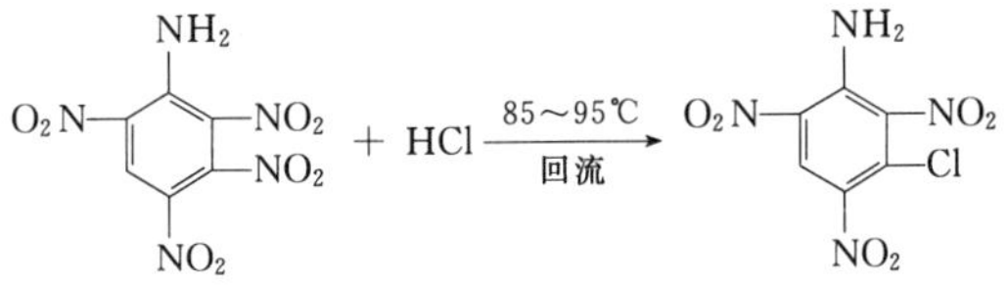

NH_2, O_2N, NO_2, NO_2, NO_2 $+ \ HCl \xrightarrow[\text{回流}]{85\sim95℃}$ NH_2, O_2N, NO_2, Cl, NO_2

实验部分

1. 盐酸法制备三硝基间氯苯胺最佳工艺

在锥形瓶中，加入5克四硝基苯胺（熔点219℃），150毫升浓盐酸，3毫升二甲基苯胺。搅拌使其完全溶解，再在100℃水浴上回流搅拌2小时，冷却过滤、水洗，得到褐色产品，经甲醇洗涤后得黄色晶体，熔点185℃（毛细管法）。

2. 盐酸法制备三硝基间氯苯胺工艺条件单因素实验

表中每组数据为三个平行实验的平均值。

① 反应温度对反应得率及熔点的影响见表1。

表1　反应温度的影响

序号	四硝基苯胺(克)	浓盐酸(毫升)	反应时间(小时)	反应温度(℃)	平均得率(%)	熔点(℃)
1	2.5	75	2	70	基本没反应	196以上
2	2.5	75	2	80	59.8	185
3	2.5	75	2	90	69.4	184
4	2.5	75	2	100	74.1	185

② 反应时间对反应得率及熔点的影响见表2。

表2　反应时间的影响

序号	四硝基苯胺(克)	浓盐酸(毫升)	反应温度(℃)	反应时间(小时)	平均得率(%)	熔点(℃)
1	2.5	75	95	0.5	没反应	210
2	2.5	75	95	1.0	58.8	186
3	2.5	75	95	1.5	69.4	185
4	2.5	75	95	2.0	73.4	186

③ 反应物料比对反应得率及熔点的影响见表3

表3　反应料比的影响

序号	四硝基苯胺(克)	浓盐酸(毫升)	反应温度(℃)	反应时间(小时)	平均得率(%)	熔点(℃)
1	2.5	60	95	2.0	63.6	181
2	2.5	70	95	2.0	68.4	183
3	2.5	80	95	2.0	74.2	185
4	2.5	90	95	2.0	74.1	185

④ 盐酸法制备三硝基间氯苯胺工艺条件正交实验见表4、表5及图1，表中每组数据为三个平行实验的平均值。

⑤ 三硝基间氯苯胺结构鉴定

ⅰ. 熔点（毛细管法）

按上述最佳工艺制备的三硝基间氯苯胺纯品为黄色粉末，熔点184-186℃。与文献报道的产品外观及熔点值185℃相符[7]。

表4　反应条件正交实验数据（四硝基苯胺投料为2.5克）

序号	反应温度(℃)A	反应时间(小时)B	盐酸用量(毫升)C	平均得率(%)
1	(1)80	(1)1.0	(1)60	48
2	(2)90	(2)1.5	(1)60	40
3	(3)100	(3)2.0	(1)60	68
4	(2)90	(1)1.0	(2)75	40
5	(3)100	(2)1.5	(2)75	60
6	(1)80	(3)2.0	(2)75	72
7	(3)100	(1)1.0	(3)90	67.2
8	(1)80	(2)1.5	(3)90	60
9	(2)90	(3)2.0	(3)90	46

表5　反应条件正交实验数据处理

	反应温度(℃)A	反应时间(小时)B	盐酸用量(毫升)C		反应温度(℃)A	反应时间(小时)B	盐酸用量(毫升)C
K1	120	155.2	156	k2	60	53.3	57.3
K2	180	160	172	k3	65.1	60	55.7
K3	195.2	180	167.2	极差R	25.1	8.3	5.3
k1	40	51.7	52				

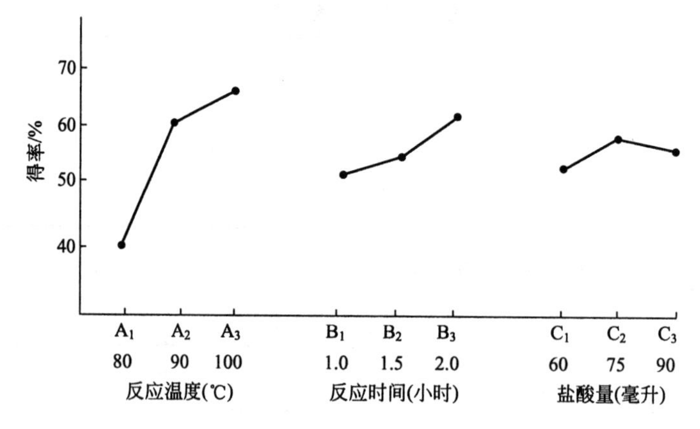

图1　反应条件正交实验效果和因素关系图

ⅱ．红外光谱分析（数据由南京大学现代分析中心提供）

波数 cm^{-1} 1283，1625，1505，为 C-NO_2 振动峰

波数 cm^{-1} 3393 为-NH_2 振动峰

波数 cm^{-1} 3010 为苯环 C-H 振动峰

波数 cm^{-1} 765，725，680 为 C-Cl 振动峰

整个谱图与三硝基间氯苯胺红外标准谱图相符。

ⅲ．元素分析（数据由北京中国科学院化学所提供）

	C	H	O	N	Cl
理论值(%)	27.42	1.14	36.57	21.33	13.50
实测值(%)	27.32	1.11	36.33	20.80	14.43

结果与讨论

① 根据熔点、红外光谱及元素分析等结构鉴定可以确认：该化合物是三硝基间氯苯胺。

② 结合正交实验图及表综合分析可以看出：投入2.5克四硝基苯胺时，以反应温度100℃、反应

时间 2 小时、盐酸量 75 毫升时具有较高的平均得率，即最佳匹配应是 $A_3B_3C_2$。从表的数据还可看出，A 的极差最大 C 的极差最小，三个因素的主次顺序为 A 到 B 到 C。

③ 依据上述三硝基间氯苯胺合成最佳工艺，做了近百次验证实验，大量数据表明：其平均得率可稳定在 75%，熔点在 184-186℃。部分验证实验数据如表 6 所示。

表 6　部分验证实验数据

序号	四硝基苯胺(克)	浓盐酸（毫升）	反应温度(℃)	反应时间(小时)	平均得率(%)	熔点*（℃）
1-1	5	150	100	2	76.2	184.5-186
1-2	5	150	100	2	77.6	183-185
1-3	5	150	100	2	73.4	184-185
1-4	5	150	100	2	75.8	184-185
1-5	5	150	100	2	77.2	184-186
1-6	5	150	100	2	76.2	185-186
1-7	5	150	100	2	75.7	184.5-186
1-8	5	150	100	2	76.8	184-185.5

* 该熔点数据系初熔点及终熔点。

④ 该工艺路线具有工艺简单、产品得率及纯度较高而成本较低、且无污染等特点。

⑤ 影响氯化亲核取代反应因素分析

影响亲核取代反应的因素有亲核试剂、温度、时间及溶剂等。

ⅰ. 在本亲核取代反应中，我们是选用 Cl^- 作亲核试剂，根据有机结构理论，Cl^- 在卤素亲核试剂系列中亲核性是最弱的，据此，在比较强烈的反应条件下进行反应应该是有利的。

ⅱ. 温度越高，反应得率也越高。因为温度高，反应物分子内能增大，分子运动速度加快，亲核试剂与反应物分子之间碰撞频率增加，反应机率增多，在相同时间内，其得率也相对提高。根据阿仑尼乌斯公式，$k=Ae^{-E/RT}$，其中 A、E、B 均为常数，k 为速度常数，显然，温度（T）增高，则速度常数也增大，也就是说，从动力学的角度，反应温度升高对提高速度常数是有利的。

ⅲ. 反应时间越长，反应得率也越高，这是因为反应进行得越完全的缘故。

ⅳ. 在反应过程中，由反应物转变为过渡状态时，凡电荷有所增加时，则反应在强极性溶剂中进行有利；反之，凡电荷有所降低或有所分散时，则反应在弱极性溶剂中进行有利。该反应在形成过波状态时是个电荷分散过程，所以，该反应在极性弱的溶剂中进行较好。实验表明：在浓盐酸介质（偶极矩 1.03D）中比在水介质（偶极矩 1.84D）及乙醇介质（偶极矩 1.70D）中有较高得率。

参考文献

[1] T. D. Tuong Bull. Chem. Soc. Japan 43，1763 (1970).
[2] T. D. Tuong Bull. Chem. Soc. Japan 44，765 (1971).
[3] U. S. 3755321.
[4] J. C. Dacons J. Heter Ocydic Chem. 14，1151 (1977).
[5] U. S. 3414570.
[6] M. D. Coburn Ind. Eng. Chem. Prod. Res. Dev. 25，68 (1986).
[7] E. Yu. Orlova Nitro，Compds. Prod. Zatern. Symp. 441 (1963) Warsaw.

（注：此文原载于　精细化工，1990，7（1）：32-35，42）

乳化剂理论及其选择研究

吕春绪
中国兵器工业民用爆破器材研究所

摘要：文章阐述了乳化炸药中乳化工艺及乳化剂作用的物理化学原理。基于对乳化剂的实验研究，为有效乳化和长贮安定，对乳化剂的选择提出几点看法，特别讨论了混合乳化剂的重要意义。

关键词：乳化炸药，乳化剂，乳化工艺，表面活性理论

1 乳化工艺的物理化学实质

乳化炸药通常由氧化剂、可燃物及其它附加剂所组成，是一种膏状塑性体，它是经乳化工艺制得的。

按物理化学基本原理，乳化工艺是一种液溶胶形成过程．其分散相是无机氧化剂盐的饱和水溶液，连续相是燃料油，依靠乳化剂增加该分散体系的稳定性。它是水分散在油介质中，也称为油包水型体系，以 W/O（water in oil）表示。包覆情况的显微摄影见图 1。

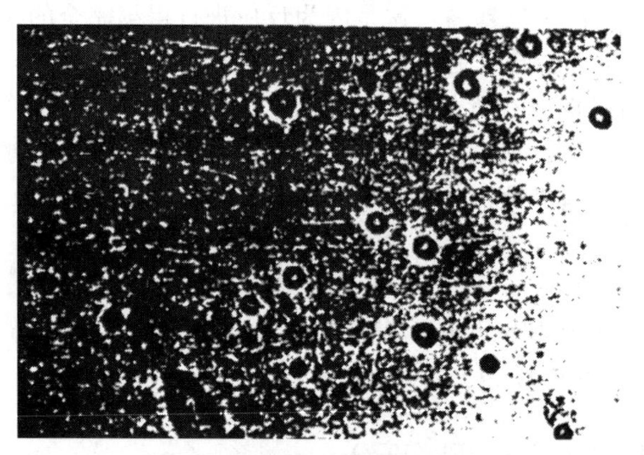

图 1　W/O 型包复显微摄影图

凡能促进乳化和保持乳胶稳定的各种因素，均有利于乳化工艺。当然，配方设计是根本。大量实验表明：石蜡、地蜡、复合蜡、微晶蜡、软化沥青、机油及凡士林等都是较好的油相组分，其用量不超过 5%（质量）。即用尽可能少量的油相，均匀地包覆水相，而成为 W/O 型乳状液。所确定的工艺条件，应是造成乳化的有力措施。

乳化是把整个水相分散在连续相-油中，形成油包水的胶粒。乳状液体系有很大的相界面，界面的大小取决于粒子的分散度。将一颗边长为 1μm 的立方体分散成边长为 1nm 的立方体胶粒，这时胶粒的数目为 10^9 个，其总面积增大了 1000 倍。研究还表明：体系的分散度越大，表面积越大，消耗的功也越大，因而表面能也越大。该表面能近似地可以看作增加分散相表面积时外界所给予的能量。

例如，1g 水成小球时，其面积为 4.83cm^2，表面能为 3.40×10^{-5}J；若将此小球分散成半径为 10^{-7}cm 的细小水滴，可得 2.4×10^{20}个，其总表面积为 3×10^{7}cm^2，表面能约为 200J. 此结果表明，将 1g 水分散成半径为 10^{-7}cm 的细小水滴时，外界要供给 200J 的能量，这 200J 相当于使 1g 水的温度升高 50℃[1]。这个外界能量在工艺上就是由搅拌器设计及转速、乳化器的形状及大小、乳化温度等因素决定的。

乳胶是靠外界能量形成的，比表面剧烈增加，是个热力学非自发过程。即在两相界面上的自由能都有自发减小的趋向。

$$\delta_{w可}=(dG)_{T.P}=\sigma\cdot dA$$

式中：$\delta_{w可}$——以可逆方式形成新表面过程中所消耗的表面功；

$(dG)_{T.P}$——恒温、恒压下自由能增量；

σ——比界面自由能（比表面能），即$\frac{\partial G}{\partial A}$，温度、压力和组成一定条件下，增加一个单位面积时，体系自由能的增加，以 $J\cdot cm^{-2}$表示；

dA——界面积增量

界面能的减少可通过界面（A）的自动缩小，也可通过界面张力（σ）的自动降低来达到。而加入乳化剂有利于后一种作用，使其体系稳定。但已形成的胶粒聚结在一起（破乳）使比表面减少即是热力学自发过程，特别是乳状液分散相粒子较大（其直径在 0.1μm 以上）时，特易聚结，聚结的显微摄影见图 2。

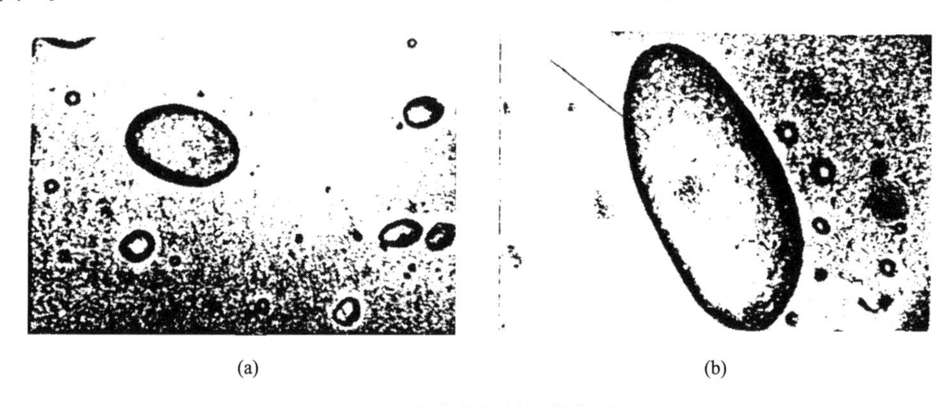

(a) (b)

图 2 乳胶微粒聚结显微摄影图

小珠滴聚结是热力学不可逆过程，大量聚结的最终结果将造成水相及油相分层（破乳），如图 3 所示。

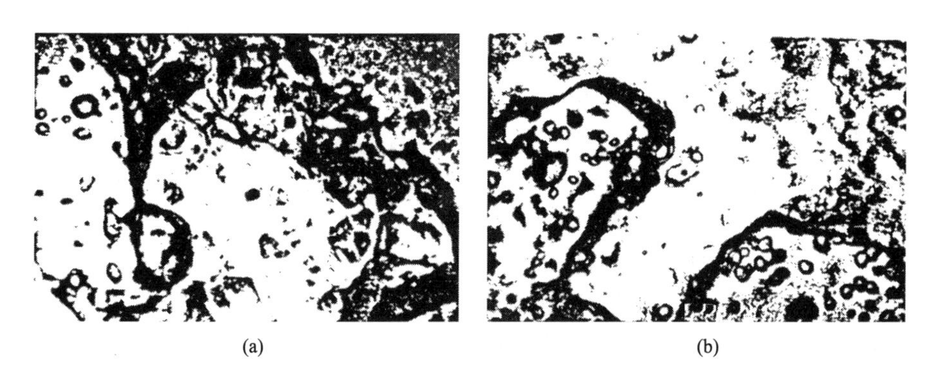

(a) (b)

图 3 乳胶破乳分层显微摄影图

为了防止聚结，保持乳状液稳定以及长期贮存，要靠乳化剂（表面活性物质、电解质、固体粉末等）的作用，特别是表面活性物质，既可防止凝聚，又可降低界面张力，是稳定乳化体系必不可少的。

2 乳化剂作用的物理化学原理

乳化炸药中无论水相还是油相除了主氧化剂及主可燃物外，均有添加剂，如氧平衡调解剂、降低冰点剂、敏化剂、乳化促进剂及乳化剂等。要制得稳定的乳化炸药，其中乳化剂尤为重要。

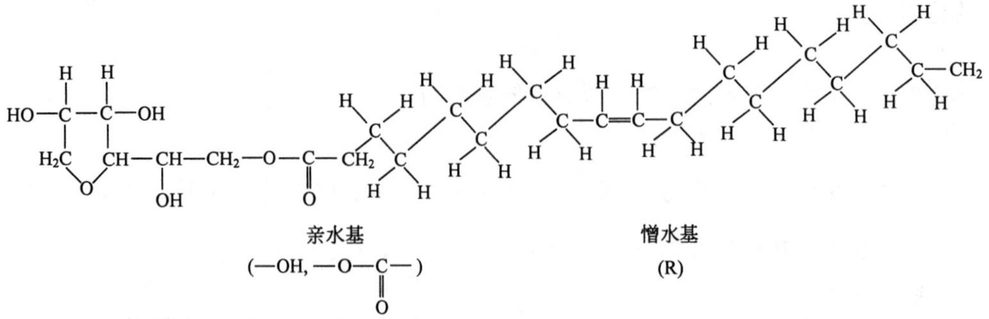

亲水基
(—OH, —O—C(=O)—)

憎水基
(R)

目前，在乳化炸药中得到广泛应用的乳化剂多是非离子型的含羟基的高级脂肪酸酯类，例如司盘-80（Span-80），即失水山梨糖醇单油酸酯，其中—OH 及—COO^- 是极性基团，因而亲水；基本对称的长碳链部分是非极性基团，是憎水的。该乳化剂可以其亲水基去包覆作为氧化剂的硝酸铵水溶液的小液滴，而将其憎水基与连续相燃料油相接触，得到稳定的 W/O 型乳化体系。其作用实质在于：乳化剂结构是由极性基团和非极性基团组成，即两性结构，它们在两相界面浓集，它的极性基团的“头”浸在水中；非极性基团却倾向于“逃”出水溶液。像“尾巴”翘在上面，这种定向排列方式使原表面上分子的不饱和力场得到某种程度上的平衡，且占据了部分表面（界面），使表面自由能大为降低，即降低了界面张力，有利于油相对水相的包覆。

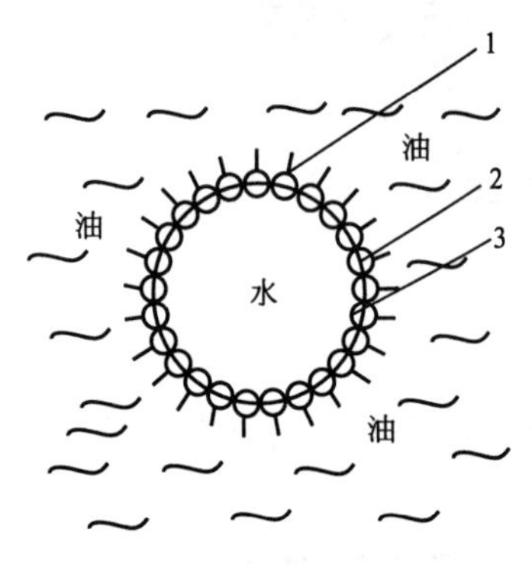

图 4 分散体系示意图
1—亲油基；2—亲水基；3—保护膜

另一方面，它在水相液滴界面上的定向吸附，一头向“水”，一头向油，好像在分散的水相液滴表面上裹了一层保护膜，而且，这种保护膜万一本身局部受到损伤时，也能自动弥补“伤口”（见图 4）。

这层保护膜使分散液滴相互碰撞时难于凝聚，保持乳膜的稳定状态长期存在。

同一乳化剂的不同用量对乳化效果有一定影响，Trable 规则[2]认为同一乳化剂在低浓度时表面张力的降低效应与浓度成正比，但是，达到饱和吸附后，不同链长的直链有机脂肪酸的 $\sqrt{\infty}$（吸附量极值）在定温下是定值，再增加用量也不会降低表面张力。

乳化剂的乳化作用与亲水基的亲水性及憎水基的憎水性有关，为了定量地表示它们的相对大小，人们提出了乳化剂亲憎平衡值 HLB（Hydrophic-Lipophilic Balance）值的概念[2~4]。HLB 值越高，表示亲水性越大；HLB 值越低，表示憎水性越大。HLB 值的大小与乳化剂分子结构有关，有人提出用如下经验公式估算 HLB 值[5~6]。

$$\mathrm{HLB}=\sum(\mathrm{HLB})_{亲水基}-\sum(\mathrm{HLB})_{憎水基}+7$$

式中 $(\mathrm{HLB})_{亲水基}$——各种亲水基对 HLB 值的贡献；

$(\mathrm{HLB})_{憎水基}$——各种憎水基对 HLB 值的贡献。

各种基团均有其对 HLB 值的贡献值。经验计算出某些乳化剂的 HLB 值，例如，失水山梨糖醇单油酸酯的 HLB 值为 4.3，失水山糖梨醇单硬脂酸酯为 4.7，单硬脂酸酯的 HLB 值为 3.8。实验发现乳化剂的 HLB 值为 3～6 时，最适宜作 W/O 型乳化剂。事实上，上述 3 种乳化剂已在乳化炸药中广泛应用。

3 乳化剂及其用最实验研究

乳化剂是乳化炸药组分的核心，它直接影响乳化炸药的性能．目前乳化炸药使用的乳化剂类型很多，除司盘型外，还有硬脂酸单甘油酯、大豆卵磷脂和羊毛脂的衍生物、烷基苯磺酸酯、十甘油十油酸酯、丁二酰亚胺类、聚丙烯酰胺、柠檬酸盐、十八烷基苯磺酸、含有聚丁烯或聚丙烯酚类衍生物等。

对于W/O型乳化剂的种类我们作了大量实验。实验进一步发现对于W/O型乳化剂，乳化剂的HLB值在3～6时乳化效果最佳。司盘型乳化剂乳化效果实验数据见表1。

表1 乳化剂种类对乳化的影响（配方以100g计）

序号	NH_4NO_3	$NaNO_3$	敏化剂	水	司盘85	司盘80	改司盘80	司盘60	司盘40	石蜡	地蜡	凡士林	HLB	物态	挤压出水
1	64	10	12	8	2	—	—	—	—	2	1	1	1.8	坚硬	渗水
2	64	10	12	8	—	2	—	—	—	2	1	1	4.1	软	不易
3	64	10	12	8	—	—	2	—	—	2	1	1	4.3	较软	不易
4	64	10	12	8	—	—	—	2	—	2	1	1	4.7	坚硬	自然渗水
5	64	10	12	8	—	—	—	—	2	2	1	1	6.7	坚硬	自然渗水

研究还指出：一些常温下为固态的乳化剂，例如司盘60，所制备的乳胶在常温下异常表坚硬，自然渗水（破乳），故在乳化剂选择中，以物态的观点，应该选择常温下为液态或半液态的乳化剂，司盘80特别是改性司盘80—混合乳化剂的效果特别突出。

乳化剂的用量也具有重要意义，因为它不仅关系到保护膜的致密程度，即乳胶体的稳定性，也关系到乳胶体的成本高低。实验表明：当乳化剂用量大于0.5%时，就能够顺利地进行乳化，在1.5%～2.0%时比较合适，若贮存时间在120天以上时，以1.8%～2.0%为佳。司盘型乳化剂用量的实验数据见表2。

表2 乳化剂用量对乳化的影响（配方以100g计）

序号	NH_4NO_3	$NaNO_3$	敏化剂	水	司盘80	机油	石蜡	地蜡	乳化时间/s	挤压出水
1	63.5	10	12	8	0	1.5	3	2	不乳化	—
2	63.5	10	12	8	0.5	1.0	3	1	75	易
3	63.5	10	12	8	1	0.5	3	2	60	较易
4	63.5	10	12	8	1.5	0	3	2	55	较不易
5	63.5	10	12	8	2	0	2.5	2	45	较不易

4 乳化剂选择的几点看法

4.1 复合乳化剂的使用有较大的实际意义

大量实验表明，复合乳化剂在乳化及保证长贮安定性方面有重要作用。

被乳化物要求乳化剂具有适宜的HLB值。最常用的油相材料：机油、凡士林、蜡等，要求适宜的HLB值如表3。

表3 被乳化物要求乳化剂适宜的HLB值

被乳化物	要求乳化剂具有的HLB值		被乳化物	要求乳化剂具有的HLB值	
	W/O型乳液	O/W型乳液		W/O型乳液	O/W型乳液
矿物油(重质)	4	10.5	蜂蜡	5	10～16
矿物油(轻质)	4	10	石蜡	4	4
凡士林	4	10.5			

被乳化物的 HLB 值越高，与乳化剂 HLB 值差别越大，两者的亲和力就越差，乳化效果就越不好，相反，会使乳化效果有较大改善，如图 5 所示：

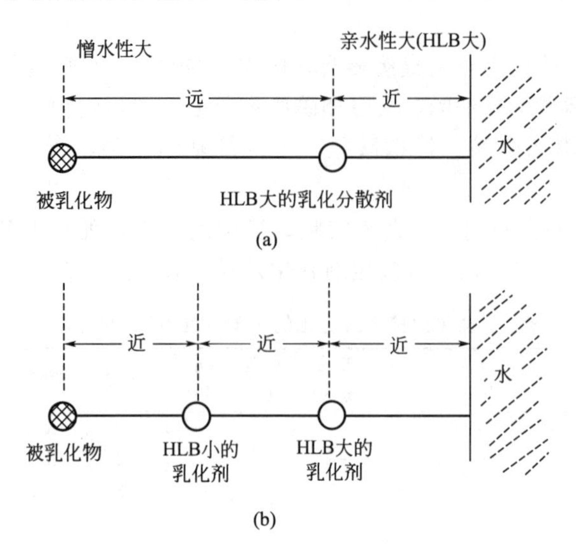

图 5　乳化剂配合使用原理图

(a) 乳化剂溶于水，不能很发乳化；

(b) 用 HLB 值小的乳化剂将两者连接起来，就要稳定乳化

要满足上述要求，只能采用混合（复合）乳化剂。实验表明：复合乳化剂的 HLB 值可由组分中各个乳化剂的 HLB 值按质量平均算出。例如，要得到 HLB=10 的复合乳化剂，可将司盘 60（HLB=4.7）45%，吐温 60（HLB=14.9）55%进行复合，即可得到 HLB=10.3 的复合乳化剂。

4.2　复合乳化剂形成复合界面膜提高乳胶

单一乳化剂不能满足使用要求的主要原因与分子结构有关。以司盘 80 为例，其成分油酸是含一个双键的不饱和酸，在同样条件下，饱和的十八酸分子所占面积为（21～25）$\times10^{-19}m^2$，油酸分子所占面积为（28～55）$\times10^{-19}m^2$，显然，使用不饱和油酸所生成的司盘，在油水两相界面因定向吸附所生成的单分子膜就显得比较“膨松”。而每个油酸分子的双键处电子云分布密集产生了电荷的相斥作用，从而在双键处形成孔腔。所谓乳化炸药破乳就是水分子从油酸分子孔腔中被挤压出所致。采用复合乳化剂后，界面膜上有“复合物”生成，形成所谓的复合界面膜，则此膜的强度增加，紧密堆积就最容易实现，因而液珠就不易凝结，乳酸就更稳定。显然，复合乳化剂中，辅助乳化剂与主乳化剂必须生成“复合物”才能有较好效果。司盘 80 与木糖醇单油酸酯以一定比例混合后，收效显著。实验还发现：在司盘 80 中添加磺酸盐、固体粉末等也有一定效果。显见，选用复合乳化剂是制取长贮安定性较好的乳化炸药的必要条件。

4.3　主乳化剂自身的稳定性是乳化剂选择的关键

目前，乳化炸药普遍使用的主乳化剂司盘 80 或 M201 都是羧酸酯。酯在酸性或碱性介质中易发生水解。因此，乳化炸药的稳定性，最终将归结于乳化剂本身的水解性能。乳化炸药的破乳硬化，多数人认为是由于司盘 80 不同程度水解，失去乳化性能所致。有人研究了介质 pH 值对乳化炸药稳定性的影响、司盘 80 的经值及酸值对乳状液稳定性的影响。都认为中性条件对提高稳定性是有利的。

4.4　主乳化剂分子结构是选择单一乳化剂的重要依据

（1）亲水基处于憎水基末端的乳化剂，具有较好的乳化能力，例如，司盘 80、司盘 60 以及十八

Currently, the only important industrial synthesis of OPP is still the cyclohexanone method, in which a mixture of 2-cyclohexylidene and 2-(1-cyclohexenyl) cyclohexanone is obtained from cyclohexanone via aldol condensation and dehydration (see Figure 1) . After catalytic dehydrogenation of the mixture, OPP is obtained. In this method, the attenuation of catalyst activity has considerable effect on the cost of the product. Our experiments showed that with 2% Pd/C at 320℃, the conversion of the mixture to OPP decreased from the initial rate of 96% to 81% after 6h. With a Cu-Zn-Al catalyst at 360℃, the decrease was from 95% to 62% after 8h. In this article, we will show how the catalyst life can be prolonged.

Pd-zeolite catalysts are stable. Table 1 shows the dehydrogenation results for the synthesis of OPP catalyzed with zeolite Pd/H-ZSM-5 under different conditions. It is obvious that the attenuation rates of the zeolite with Si/Al ratios of 85 : 1 and 200 : 1 are lower than that of the zeolite with a Si/Al ratio of 17 : 1. Because of the strong bonding between palladium and the zeolite surface in high-silicon catalysts, the migration and cementation of metals on the surface and the formation of large particles are slow at these high dehydrogenation temperatures. The high dispersity of the catalyst can be maintained, and its lifetime can be extended.

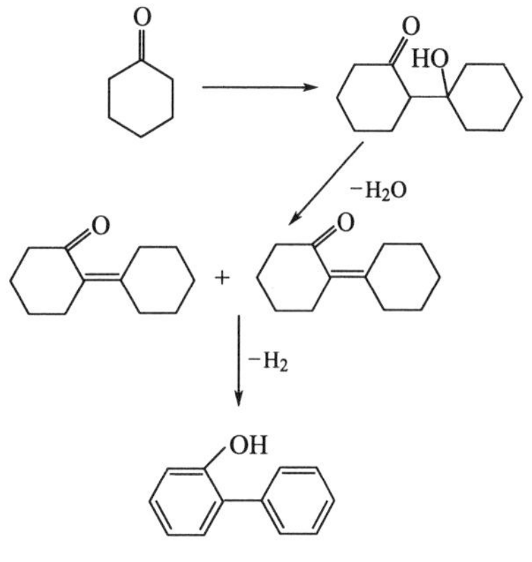

Figure 1 Linear polymers with a variety of bonding schemes are produced by varying the monomers in hydrosilylation polymerization reactions.

The dehydrogenation results for catalysis by Pd/Mg-ZSM-5 are also listed in Table 1. Comparing these results with those for Pd/H-ZSM-5, we conclude that introducing magnesium diminishes the catalyst activity attenuation rates, even at the low Si/Al ratio. The dehydrogenation results were not affected, so the magnesium can be considered an "assistant catalyst" in this reaction.

Surface differences X-ray diffraction shows that no obvious differences exist between the structures of the catalysts with Si/Al ratios of 85 : 1 and 200 : 1 before and after dehydrogenation. But the zeolites with the Si/Al ratio of 17 : 1 are different. The zeolite itself is stable at these reaction temperatures, so the difference is due to the pyrolysis of a mixture of condensation products on the surface of the catalyst.

Table 1 Dehydrogenation results with Pd/H-ZSM-5 and Pd/Mg-ZSM-5

Zeolite Si/Al ratio	Dehydrogenation temperature, ℃	Pd/H-ZSM-5			Pd/Mg-ZSM-5			
		Pd concn on surface, %	OPP yield after 1h, %	OPP yield after 10h, %	Pd concn on surface, %	Mg concn on surface, %	OPP yield after 1h, %	OPP yield after 10h, %
200 : 1	280-290	0.42	86	4	0.37	3.5	90	89
	290-300		89	8			92	90
	330-350		91	9			92	92
85 : 1	280-290	0.54	90	7	0.43	6.7	92	91
	290-300		93	8			94	95
	330-350		94	2			95	94
17 : 1	280-290	0.60	96	9	0.57	11.2	94	88
	290-300		97	4			95	87
	330-350		94	2			94	85

The acid catalytic activity of a synthetic zeolite is directly proportional to its aluminum content, so a catalyst prepared from a zeolite with a low Si/Al ratio has more acidic sites on its surface than does a zeolite with a high Si/Al ratio. A mixture of condensation products from cyclohexanone could form a polymer when catalyzed at such an acidic site. Pyrolysis of this polymer at high temperature changes the surface structure. For this reason, the specific surface area of the catalyst may also be decreased; in this case, catalytic dehydrogenation activity decreased considerably (see Table 2). On the other hand, electron spectroscopy for chemical analysis (ESCA) shows that no significant differences exist in palladium concentration on the surface of the catalyst, so that there is little migration of metal at high temperatures.

Table 2 Specific surface areas of catalysts before and after use at 330-350℃

	Pd/H-ZSM-5			Pd/Mg-ZSM-5		
Si/Al ratio	200:1	85∶1	17∶1	200∶1	85∶1	17∶1
Initial surface area, m^2/g	431	390	364	412	386	325
Surface area after, m^2/g	427	391	307	415	382	341

Thus, we have demonstrated a method for producing OPP that uses a plentiful feedstock and a Pd-supported zeolite catalyst system that maintains its activity over a significantly longer time period than is possible using Pd/C or Pt-V_2O_5/C, with no significant loss of surface area. This extended catalyst lifetime has the potential to reduce the cost of OPP synthesis significantly.

References

[1] Goto, H.; Suzuki, M. Japan Patent 76-131863, 1976.
[2] Iodwin, K. U. K. Patent 1, 480, 326, 1977.

(注：此文原载于 Chemical Innovation, 2001, 31 (10): 37-38)

Studies on Catalytic Side-Chain Oxidation of Nitroaromatics to Aldehydes with Oxygen

Wei Yunyang, Cai Minmin, Lv Chunxu
(Nanjing University of Science and Technology, Nanjing 210094, P. R. China)
SCI: 000185494000011, EI: 03457714108

Abstract: The side-chain oxidation of 2-nitrotoluene in liquid phase in the presence of catalytic amount of manganese sulfate and stoichiometric amount of potassium hydroxide with oxygen was studied. In the most favorable conditions, over 80% of conversion of 2-nitrotoluene and 50% of selectivity to 2-nitrobenzaldehyde was achieved. Effects of the reaction parameters on the conversion of the reactant and the selectivity of the product were examined. These results, together with EPR spectroscopic study, show that a benzyl anion was formed in the early stage of the reaction, which was then converted to the final product via a free radical mechanism.

Key Words: oxidation, substituted benzyl anion, free radical, reaction mechanism, active intermediate

1 Introduction

Aromatic aldehydes, important intermediates for the manufacture of pharmaceuticals and fine chemicals, are prepared on industrial scale via oxidation of aromatic alcohols with chemical oxidizers such as potassium permanganate, sodium bichromate or nitric acid. For example, 2-nitrobenzaldehyde, an important intermediate for the manufacture of nifedipine, is prepared in China via the following route:

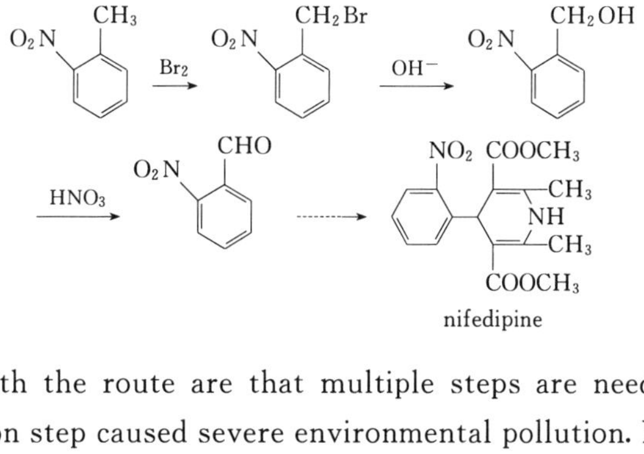

O_2N, CH_3 $\xrightarrow{Br_2}$ O_2N, CH_2Br $\xrightarrow{OH^-}$ O_2N, CH_2OH $\xrightarrow{HNO_3}$ O_2N, CHO $\dashrightarrow$ NO_2, $COOCH_3$, CH_3, NH, CH_3, $COOCH_3$

nifedipine

Problems associated with the route are that multiple steps are needed and the overall yield is low. Moreover, the oxidation step caused severe environmental pollution. In order to solve these problems, several catalytic side-chain oxidation methods are developed in the literature recently. We used the copper-catalyzed air oxidation method developed by Marko and coworkers[1] to oxidize 2-nitrobenzyl alcohol to 2-nitrobenzaldehyde in toluene in the presence of 1, 10-phenanthro-line and hydrazine dicarboxylic acid diethyl ester. The reaction was successful and the yield of recrystallized product was over 70%[2]. But the copper-catalyzed oxidation method still needs multiple steps for the preparation of aromatic aldehydes starting from substituted toluenes.

The simplest and ideal way to prepare aromatic aldehydes would be the direct side-chain oxidation of substituted toluenes to aldehydes with air or oxygen. Most of the reactions are carried out in gas phase, and vanadium pentoxide is usually used as the catalyst[3-6]. The reactions are carried out at high temperatures (usually 300-500℃). Because aldehydes are easier to oxidize than an alkyl group, it is difficult to stop the oxidation at the aldehyde stage. Thus, at a reasonable conversion, selectivity to aldehydes is low, and the main product is usually a substituted benzoic acid. This situation is slightly changed by the use of liquid-phase catalytic side-chain oxidation of substituted toluenes[7,8]. Liquid-phase oxidation, which uses Co or Mn as a catalyst, can be carried out at much lower temperatures and the selectivity to aldehydes can be improved. Sodium bromide or a strong base such as potassium hydroxide is usually used as a catalyst promoter.

In this paper, side-chain oxidation of 2-nitrotoluene in liquid phase in the presence of a catalytic amount of manganese sulfate and a stoichiometric amount of potassium hydroxide with oxygen was studied. In the most favorable conditions, over 80% of conversion of 2-nitrotoluene and 50% of selectivity to 2-nitrobenzaldehyde was achieved. Effects of the reaction parameters on the conversion of the reactant and the selectivity of the product were examined. These results together with electron paramagnetic resonance (EPR) spectroscopic study show that a benzyl anion was formed in the early which was then converted to the final product via a free radical mechanism.

2 Experimental

2.1 Oxidation procedure

To a mixture of 2-nitrotoluene (6.86g, 50mmol) and $MnSO_4 \cdot H_2O$ (0.05g, 0.3mmol) in 50mL of solvent, oxygen (30mL/min) at 0-5℃ was introduced. A solution of KOH (3.3g, 59mmol) in 7mL of methanol was then added to the reaction mixture dropwise. The resultant mixture was stirred at 0-5℃ for 4.5h. KOH was then neutralized with 80% sulfuric acid and the solvent was removed in vacuo. The residue was dissolved in water and acidified with sulfuric acid to pH 2-3. The aqueous solution was then extracted with methylene chloride three times. Removal of methylene chloride gives the oxidation product containing 2-nitrobenzalde-hyde and the unconverted 2-nitrotoluene, which was weighed and analyzed qualitatively by GC-MS (Saturn 2000) and quantitatively by high performance liquid chromatography (HPLC) (Waters 600E chromato-graph, 2487 detector).

2.2 Detection of reactive free radical intermediate by EPR

The reactant and the spin-trapping agent N-tert-butyl-phenylnitrone were added to the sample cell of the EPR spectrometer at 0℃, and the absorption peak was recorded every few minutes. The microwave frequency and power was controlled at 9.4910GHz and 20 to 64mW respectively. Resonance occurred at 3380.75G. $MnSO_4$ shows a strong EPR absorption peak under the reaction conditions. In order to eliminate the interference of Mn on the detection of organic free radicals, $MnSO_4$ was not added in these experiments.

3 Results and discussion

3.1 Effects of KOH concentration on the oxidation

The dependence of reactant conversion (C) and product selectivity (S) on the amount of KOH used (RKOH=o-MNT, the molar ratio of KOH over 2-nitrotoluene) was determined, and results obtained at 0-5℃ for 4.5h is illustrated in figure 1.

Figure 1 shows that both the conversion of 2-nitrotoluene and the selectivity of product increase with the increasing of RKOH/o-MNT until RKOH/o-MNT reaches 1.5. If RKOH/o-MNT is larger than 1.5, the two curves tend to change in different directions, the conversion curve going upward and the selectivity curve going downward. But the upward change of the conversion curve is much slower than the downward change of the selectivity curve.

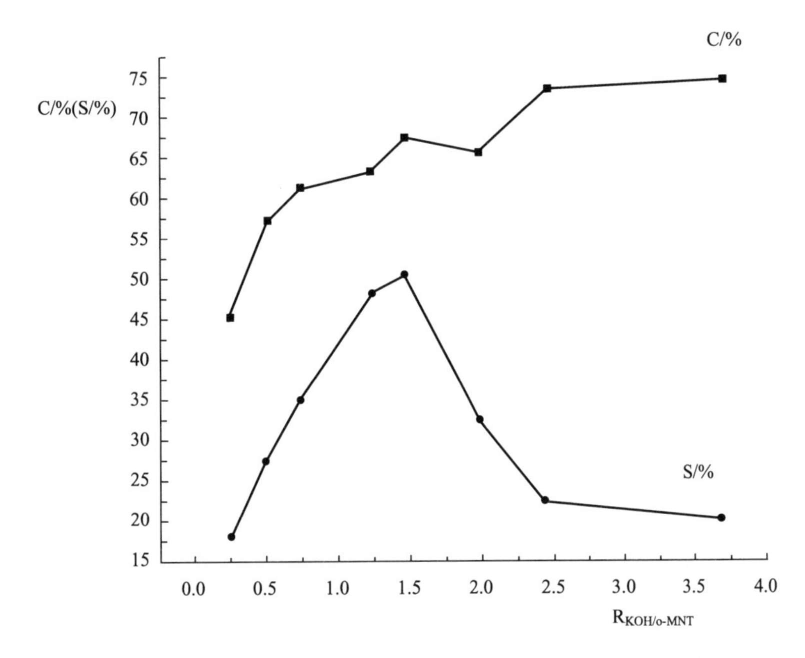

Figure 1 Dependence of reactant conversion (C) and product selectivity (S) on the amount of KOH used (RKOH=o-MNT, the molar ratio of KOH over 2-nitrotoluene).

When RKOH/o-MNT = 1 and the initial concentration of o-MNT (A_0) and KOH (B_0) is 1.0mol · L^{-1}, the oxidation shows second-order kinetics until 6h of reaction in morpholine according to the observed linear relationship between 1=ðA0-Cþ and reaction time t. The second-order constants, the turnover frequencies and the activation energy in the temperature range -5 to 10℃ are listed in table 1.

The above results can be explained by a nitro-stabilized benzyl anion mechanism, i.e. the first and ratedetermining step of the reaction is deprotonation of 2-nitrotoluene to form 2-nitrobenzyl anion. The anion was then converted into a more stable form and was finally transferred to the product via further reactions. If excess amount of KOH is used, side reactions may occur because of which conversion increases and selectivity decreases.

$$C_6H_4(CH_3)(NO_2) + OH^- \underset{}{\overset{-H_2O}{\rightleftharpoons}} C_6H_4(CH_2^-)(NO_2) \longrightarrow C_6H_4(=CH_2)(=N(O)O^-) \dashrightarrow C_6H_4(CHO)(NO_2) + \text{by products}$$

The above hypothesis was further strengthened by the oxidation of 3-nitrotoluene (m-MNT) under the same conditions. The dependence of conversion of m-MNT on the amount of KOH used at 0-5℃ for 4.5h is illustrated in figure 2 together with the conversion of o-MNT under the same conditions for convenience of comparison.

Table 1 Kinetic data for the oxidation of o-MNT in morpholine

T(℃)	$k_2\times10^{-5}$(L $mol^{-1}\cdot s^{-1}$)	Catalyst turnover(h^{-1})	Ea(kJ mol^{-1})
−5	3.56	8	
5	9.26	19	52kJ mol^{-1}
10	12.9	25	

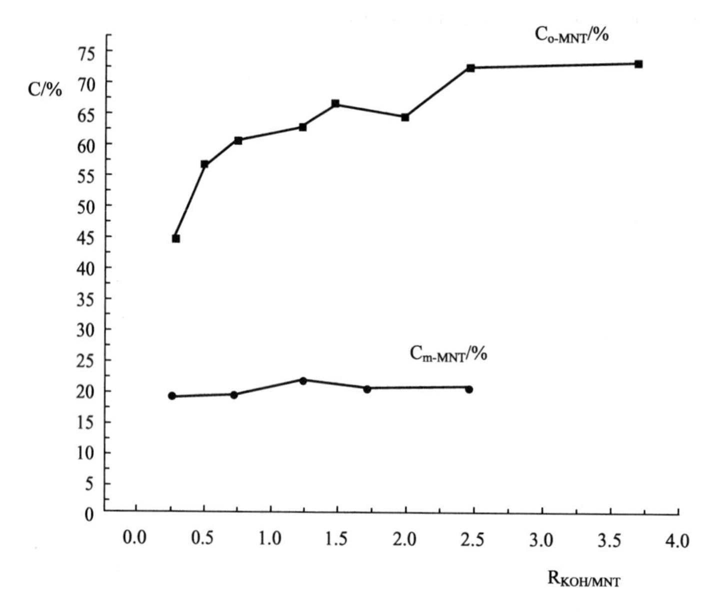

Figure 2 The dependence of reactant conversion (C) on the amount of KOH used.

Table 2 Conversion and selectivity of the reaction in different solvent

Solvent	C(%)	S(%)	Solvent	C(%)	S(%)
Methanol	21	0	Morpholine	63	50
Methanol/benzene(v/v=1/5)	33	0	Propylamine	72	52
2-methoxyethanol	35	3	2-methoxyethylamine	81	50

The curves in figure 2 indicate that the conversion of m-MNT is much lower than the conversion of o-MNT under the same conditions and is almost independent of the amount of KOH used. This evidence supports the hypothesis of benzyl anion formation given above, because a nitro group attached to the meta-position could not stabilize a benzyl anion effectively.

3.2 Effects of solvent on the oxidation

Solvent dependence of the reactant conversion (C) and aldehyde selectivity (S) was also examined and the results are listed in table 2.

It can be seen from table 2 that organic amines with an additional functional group are favorable solvents for the reaction. In protic solvents such as alcohols, the reaction could not take place. These results can again be explained by the hypothesis of benzyl anion formation. Obviously in protic solvents such as alcohols, it is difficult to form a carbanion with metal hydroxides. The basicity of an amine is much stronger, and the formation of a carbanion in an amine solvent would be much easier than in alcohols. Moreover, amines with an additional functional group are good ligands to form complexes with transition metals, which may play a key role in the catalytic cycle.

3.3 Formation of reactive free radical intermediate

Although the mechanism of catalytic side-chain oxidation of aromatics in gas phase has been studied extensively and free radical mechanism is widely accepted, reports about the mechanism of liquid-phase oxidation, especially the direct detection of reactive intermediates is scarce in the literature. The evidence given above confirms the formation of benzyl anion as a reactive intermediate in catalytic side-chain oxidation of aromatics in the presence of a strong base. Interesting phenomena were observed in further studies of the reaction by EPR. In methanol, no free radical signal was observed (figure 3) until 25min of the reaction. In morpholine, a remarkable free radical signal appears within a few minutes of the reaction (figure 4). In 2-methoxyethylamine, a strong free radical signal appeared after a few minutes of the reaction and the splitting is clear (figure 5). These results are in accordance with the experimental data of conversion and selectivity of the reaction in corresponding solvents. Spectrum analysis shows that the free radical is a spintrapping species of 2-nitrobenzyl radical by the spintrapping agent N-tert-butyl-α-phenylnitrone.

$(H_3C)_3C-N(\rightarrow O)=CH(Ph) + H_2\dot{C}-C_6H_4-NO_2 \longrightarrow (H_3C)_3C-N(O^{\cdot})-CH(Ph)-CH_2-C_6H_4-NO_2$

$a_{H\alpha}=3.47G$, $a_{H\beta}=3.47G$, $a_N=11.75G$

Scheme 3

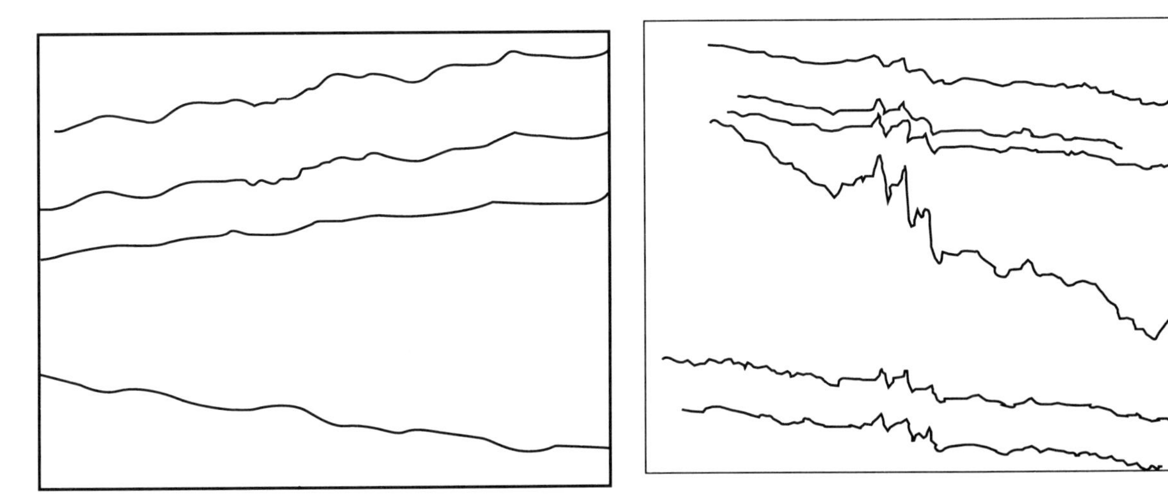

Figure 3 EPR spectrum in methanol. Figure 4 EPR spectrum in morpholine.

The g value for the spin-trapping species was calculated to be 2.0057, which is in good accordance with the ge value of 2.0023 for free radicals. The electronnuclear hyperfine interaction constants

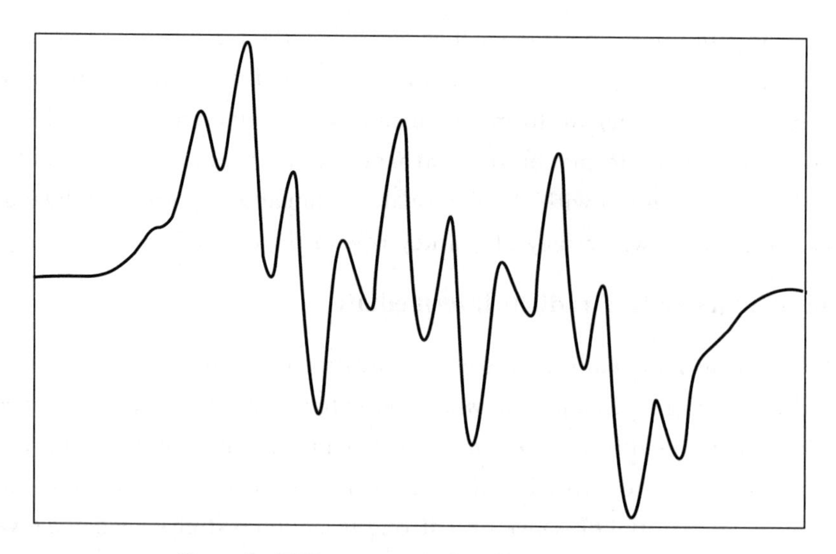

Figure 5 EPR spectrum in 2-methoxyethylamine.

are 11. 75G for a_N, 3. 47G for a_H and 7. 21G for aH respectively. A large a_H is reasonable, considering the electron withdrawing effect of the nitro group.

All the results given above support a conclusion that benzyl anion was the initially formed intermediate in liquid-phase catalytic side-chain oxidation of aromatics in the presence of a strong base, which was then converted to the final product via a free radical mechanism.

References

[1] I. E. Marko, P. R. Giles, M. Tsukazaki et al. , Science 274 (1996) 2044.
[2] Y. Wei, Y. Liu and S. Tian, Chem. React. Eng. Technol. 17 (2) (2001) 124 (Chinese).
[3] J. Zhu, S. Andersson and T. Lars. J. Catal. 126 (1) (1990) 92.
[4] B. Jonson, R. Bernd, R. Larsson et al. J. Chem. Soc. , FaradayTrans. 1 84 (10) (1988) 3547.
[5] Z. Yan, S. Andersson and T. Lars, Appl. Catal. 66 (1) (1990) 149.
[6] H. Zhang, Z. Li and X. Fu, Chinese J. Catal. 9 (4) (1988) 331 (Chinese).
[7] H. V. Borgaonkar, S. R. Raverkar and S. B. Chandalia, Ind. Eng. Chem. Prod. Res. Dev. 23 (3) (1984) 455.
[8] A. Schnatterer and H. Flege, European Patent, 0, 671, 381 A1 (1999).

(注：此文原载于 Catalysis Letters, 2003, 90 (1): 81-84)

1,5-二甲基-2S-(1′-苯基乙基)-2-氮杂-8-氧杂-双环［3.3.0］辛烷-3,7-二酮的合成及其结构表征

胡炳成，吕春绪
（南京理工大学化工学院，南京 210094）
SCI：000221725900024

摘要： 利用 2，3-丁二酮与丙二腈缩合形成双甲基双氰基环并内酰胺，经酸性水解得到双甲基环并内酯，然后和 (S)-(−)-苯基乙基胺反应生成 1,5-二甲基-2S-(1′-苯基乙基)-2-氮杂-8-氧杂-双环［3.3.0］辛烷-3,7-二酮，通过元素分析、红外光谱、核磁共振氢谱和质谱对其结构进行了表征。用 X 射线单晶衍射法测定了目标产物的晶体结构，结果表明，晶体属单斜晶系，$P2_1$ 空间群，$a=0.8777$ (18) nm，$b=1.0102$ (2) nm，$c=0.9009$ (18) nm，$\beta=118.75$ (3)°，$V=0.7003$ (2) nm^3，$D_c=1.296g\cdot cm^{-3}$，$\mu=0.089mm^{-1}$，$F(000)=292$，$Z=2$，$R_1=0.0393$，$wR_2=0.0955$。

关键词： 双甲基双氰基环并内酰胺，双甲基环并内酯，(S)-(−)-苯基乙基胺，X 射线衍射

引言

天然环状四吡咯化合物所具有的优良生物活性愈来愈引起化学家的重视，通过人工合成各种环状四吡咯化合物及其衍生物来模仿天然产物的各种生物活性，是目前有机化学最为活跃的研究领域之一[1-5]。托尼卟吩类化合物就是从海藻微生物中提取的一类具有高抗癌活性的环状四吡咯衍生物[6]，已作为一类新的抗癌药物得到深入研究[7-10]，托尼卟吩及其衍生物的全合成研究也因此受到关注[11-13]。我们利用 2,3-丁二酮和丙二腈为原料，先经缩合、水解制得双甲基环并内酯 4，然后与手性试剂 (S)-(−)-苯基乙基胺反应，制得标题化合物 1,5-二甲基-2-(S)-(1′-苯基乙基)-2-氮杂-8-氧杂-双环［3.3.0］辛烷-3,7-二酮 **5**，并以此作为手性源试剂，利用环并内酰胺-内酯酯环的开环反应来合成托尼卟吩模型及其衍生物[14]。同时，它也可以作为手性源试剂，用来合成分子中含有部分饱和的吡咯烷结构的环状四吡咯化合物，如叶绿素 a 等[15,16]。本文报道了标题化合物的合成情况（合成路

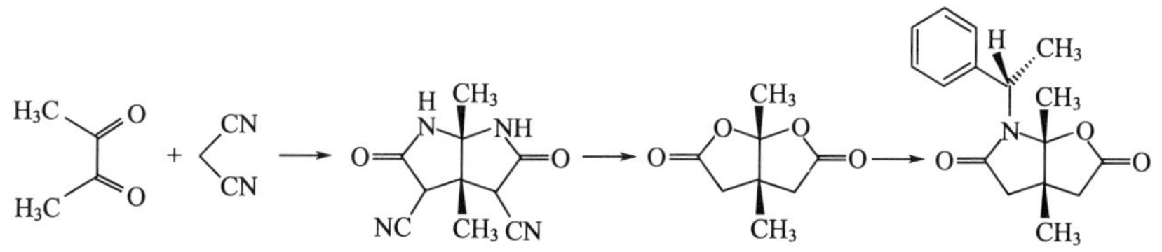

图 1　标题化合物的合成路线

Figure 1　Route of synthesis of the title compound

线如图 1)，并通过 X 射线衍射法测定了其结构。

1 实验部分

1.1 主要仪器与试剂

Cary-50 型紫外-可见分光光度计；Paragon 500 FT-IR 型红外光谱仪，KBr 压片；Bruker DPX-200 型核磁共振仪，溶剂 $CDCl_3$；MAT-8200 型质谱仪；PE-2400 型元素分析仪；Bruker Smart-1000 CCD 型 X 射线衍射仪；Kofler 型显微熔点测定仪，温度计未经校正。

所有药品均由 Aldrich 或 Acros 公司提供。其中，盐酸和氢溴酸的浓度分别为 37%和 48%，二氯甲烷为工业品，重新蒸馏后再使用，无水乙醇在使用前加入氧化钙重新蒸馏，无水氯仿采用氯仿加五氧化二磷蒸馏制得，其余所用试剂均为分析纯；薄层层析（TLC）用硅胶 60 F_{254}（Fluka 公司）厚度为 0.2mm，柱层析用硅胶 32～63μm 60Å（ICN 生物医疗公司）。

1.2 化合物的合成

1.2.1 1,5-二甲基-2,8-二氮杂-3,7-二酮-双环［3.3.0］辛烷-4,6-二腈（3）

将 8.6g（100mmol）丁二酮 1 加入以 2.3g（100mmol）金属钠和 250mL 无水乙醇形成的溶液中，冰浴冷却下边搅拌边慢慢滴加用 100mL 无水乙醇溶解的 13.2g（200mmol）丙二腈 2，控制溶液的温度不超过 0℃。然后在氩气保护、5℃下搅拌 4h，滴加 37%盐酸直至溶液的 PH 值为 1，静置 1h 小时后过滤，加水洗涤至冲洗液为无色，干燥后得到 6.97g（32mmol）无色晶体，为化合物 3，产率 32%。纯度≥98%，m. p. >220℃（分解）；UV-vis（CH_3OH）λ_{max}：209（14300）nm；^{1}HNMR（200 MHz，$CDCl_3$）δ：1.29（s，3H，5-CH_3），1.35（s，3H，1-CH_3），4.62（s，1H，CHCN），4.84（s，1H，CHCN），9.25（s，1H，NH），9.38（s，1H，NH）；IR（KBr）v：3313（s，NH），2989（w，CH），2902（s，CH），2255（s，CN），1750（s，5-lactam），1705（s，5-lactam）cm^{-1}；MS（70eV）m/z（%）：218（M^+，15），217（M^+-H，14），203（M^+-CH_3，20），135（100）；Anal. Calcd for$C_{10}H_{10}N_4O_2$：C 55.04，H 4.62，N 25.68；found C 55.16，H 4.69，N 25.88。

1.2.2 1,5-二甲基-2,8-二氧杂-双环［3.3.0］辛烷-3,7-二酮（4）

在 250mL 热的 48%氢溴酸溶液中加入 4.4g（20mmol）环并内酰胺 3，搅拌回流 45min，经真空浓缩后加入二氯甲烷，有机相经无水硫酸钠干燥后再经旋转蒸发器除去溶剂，所得红色黏稠状的残余物经柱层析[200g 硅胶，$V_{(二氯甲烷)}$：$V_{(甲醇)}$=95：5，R_f=0.29]分离、氯仿/正戊烷重结晶，得到 1.71g（10mmol）无色晶体，为化合物 4，产率 50%。m. p. 128℃；UV-vis（$CHCl_3$）λ_{max}：216（11700）nm；^{1}HNMR（200MHz，$CDCl_3$）δ：1.40（s，3H，5-CH_3），1.71（s，3H，1-CH_3），2.60～2.84（AB-system，4H，2×CH_2）；IR（KBr）v：2993（s，CH），2954（s，CH），1794（s，C=O，lactone），1778（s，C=O，lactone）cm^{-1}；MS（70eV）m/z（%）：171（M^++H，24），126（M^+-CO_2，14），43（100）；Anal. calcd for $C_8H_{10}O_4$：C 56.47，H 5.92；found C 56.24，H 6.07。

1.2.3 1,5-二甲基-2S-(1′-苯基乙基)-2-氮杂-8-氧杂-双环［3.3.0］辛烷-3,7-二酮（5）

室温和氩气保护下将 680mg（4mmol）环并内酯 4、484mg（4mmol）（S)-(－)-苯基乙基胺在 10mL 无水氯仿中搅拌，用 TLC 确定反应的进程，直至反应物完全转化。加入 30mL 饱和氯化钠溶液萃取反应混合物，水相用二氯甲烷洗涤两次。合并有机相，浓缩后得到黄色的油状物质。通过柱层析[15g 硅胶，$V_{(二氯甲烷)}$：$V_{(甲醇)}$=9：1，R_f=0.73]分离、氯仿/正戊烷重结晶，得到 974 mg（3.56 mmol）无色晶体，即为化合物 5，产率 89%. m. p. 224℃；UV-vis（$CHCl_3$）λ_{max}：210（13600），263（10400）nm；^{1}HNMR（200MHz，$CDCl_3$）δ：1.23（s，3H，5-CH_3），1.50（s，3H，1-CH_3），1.82（d，^3J=

7.34Hz，3H,CH-CH_3)，2.35～2.77(A B-system，4H,2×—CH_2—)，4.70(q,^3J=7.06Hz,1H,CH—CH_3)，7.34(m,5H,—C_6H_5)；IR(KBr)v：3005(s,C—H)，2990(s,C—H)，1780(s,C=O,lactone)，1710(s,C=O,lactam)cm^{-1}；MS(70eV)m/z(%)：274(M^++H,12)，273(M^+,55)，161(35)，125(17)，105(100)；Anal. calcd for $C_{16}H_{19}NO_3$：C 70.31，H 7.01，N 5.12；found C 70.04，H 7.26，N 4.93。

1.3 X射线衍射测定

选取尺寸 0.50mm×0.50mm×0.30mm 化合物**5**的无色晶体置于 Bruker Smart-1000CCD 型 X 射线衍射仪上，用经石墨单色器单色化的 Mo Kα 射线（λ=0.071073nm），在室温[(293±2)K]下，以 $w/2\theta$ 扫描方式在 2.58°≤θ≤26.09°范围内共收集 8242 个独立衍射点，其中 2727 个观察点用于结构计算。全部强度数据均应用 Texsan V2.0 程序进行了 Lp 因子校正和经验吸收校正。晶体结构用直接法（SHELXS97 程序）解出，所有非氢原子坐标是在以后的数轮差值 Fourier 合成中陆续确定的，对全部非氢原子的坐标及各向异性热参数进行全矩阵最小二乘修正。分子结构见图 2，晶体数据和结构参数列于表 1，原子坐标参数和热参数见表 2，部分键长和键角列于表 3。

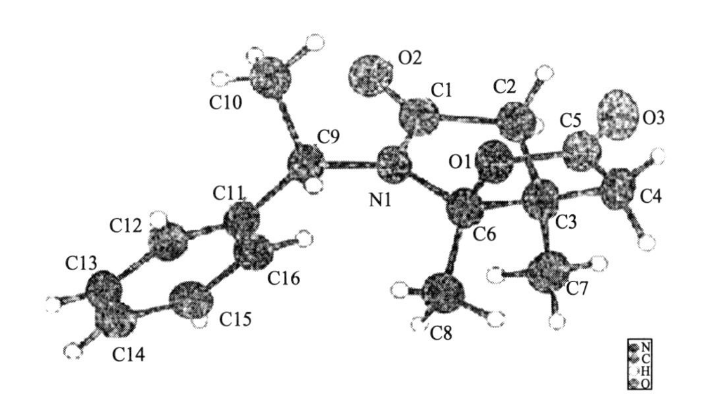

图 2 标题化合物的分子结构图

Figure 2 Molecular structure of the title compound

表 1 晶体数据和结构参数

Table 1 Crystal data and structure refinement for the title compound

Empirical formula	$C_{16}H_{19}NO_3$	Absorption coefficient/mm^{-1}	0.089
Formula weight	273.32	F(000)	292
Temperature/K	293(2)	Crystal size/mm^3	0.50×0.50×0.30
Wavelength/nm	0.071073	θ range for data collection/(°)	2.58 to 26.09
Crystal system	Monoclinic	Index ranges	$-10\leqslant h\leqslant 10, -12\leqslant k\leqslant 12, -11\leqslant l\leqslant 11$
Space group	$P2_1$	Reflections collected	8242
a/nm	0.8777(18)	Independent reflections	2727[R_{int}=0.0451]
b/nm	1.0102(2)	Completeness to theta=26.09°	98.2%
c/nm	0.9009(18)	Max. and min. transmission	0.9737 and 0.9567
α/(°)	90	Refinement method	Full-matrix least-squareson F^2
β/(°)	118.75(3)	Data/restraints/—parameters	2727/1/188
γ/(°)	90	Goodness-of-fit on F^2	1.046
Volume/nm^3	0.7003(2)	Final R indices[$I>2\sigma(I)$]	R_1=0.0393,wR_2=0.0955
Z	2	R indices(all data)	R1=0.0455,$wR2$=0.1024
		Absolute structure parameter	0.0(13)
Density(calculated)g·cm^{-3}	1.296	Largest diff. peak and hol/e·nm^{-3}	158and−154

表 2 原子坐标参数（$\times 10^4$）和热参数（$\times 10 nm^2$）

Table 2 Atomic coordinates ($\times 10^4$) and equivalent isotropic displacement parameters ($\times 10 nm^2$) for the title compound

Atom	x	y	z	U_{eq}^a	Atom	x	y	z	U_{eq}^a
O(1)	1294(2)	1342(2)	6465(2)	42(1)	C(6)	951(3)	1344(2)	7918(2)	36(1)
C(1)	−70(3)	1877(2)	5073(3)	41(2)	C(61)	547(3)	−65(2)	8165(3)	49(1)
O(2)	−49(2)	1945(2)	3743(2)	56(1)	C(7)	4087(3)	1155(2)	10282(3)	39(1)
C(2)	−1441(3)	2324(2)	5497(3)	43(1)	C(71)	5659(3)	1947(3)	10477(3)	55(1)
C(3)	−517(3)	2382(2)	7423(3)	40(1)	C(8)	4372(3)	605(2)	11964(2)	36(1)
C(31)	−1681(3)	2116(3)	8216(3)	48(1)	C(9)	5617(3)	−387(2)	12729(3)	39(1)
C(4)	518(3)	3660(2)	8138(3)	40(1)	C(10)	5980(3)	−930(2)	14277(3)	43(1)
C(5)	2228(3)	3221(2)	9576(3)	39(1)	C(11)	5103(3)	−483(2)	15110(3)	44(1)
O(3)	3288(2)	3912(2)	10731(2)	50(1)	C(12)	3852(3)	498(2)	14360(3)	44(1)
N(1)	2443(2)	1902(2)	9354(2)	36(1)	C(13)	3491(3)	1032(2)	12806(3)	39(1)

a U_{eq} is defined as one third of the trace of the orthogonalized U_{ij} tensor.

表 3 部分键长（nm）和键角（°）

Table 3 Selected bond lengths (nm) and angles (°) for the title compound

O(1)-C(1)	0.1362(3)	O(1)-C(6)	0.1476(2)	C(1)-O(2)	0.1210(3)
C(1)-C(2)	0.1496(3)	C(2)-C(3)	0.1523(3)	C(3)-C(31)	0.1526(3)
C(3)-C(4)	0.1530(3)	C(3)-C(6)	0.1551(3)	C(4)-C(5)	0.1502(3)
C(5)-O(3)	0.1226(3)	C(5)-N(1)	0.1374(3)	N(1)-C(6)	0.1442(2)
N(1)-C(7)	0.1480(3)	C(6)-C(61)	0.1509(3)	C(7)-C(8)	0.1518(3)
C(7)-C(71)	0.1531(3)	C(8)-C(13)	0.1387(3)	C(8)-C(9)	0.1396(3)
C(9)-C(10)	0.1385(3)	C(10)-C(11)	0.1384(3)	C(11)-C(12)	0.1389(3)
C(12)-C(13)	0.1387(3)				
C(1)-O(1)-C(6)	110.62(15)	O(2)-C(1)-O(1)	120.62(19)	O(2)-C(1)-C(2)	129.3(2)
O(1)-C(1)-C(2)	110.08(17)	C(1)-C(2)-C(3)	104.43(17)	C(2)-C(3)-C(31)	114.64(19)
C(2)-C(3)-C(4)	113.30(18)	C(31)-C(3)-C(4)	110.31(18)	C(2)-C(3)-C(6)	102.39(17)
C(31)-C(3)-C(6)	113.38(18)	C(4)-C(3)-C(6)	101.93(16)	C(5)-C(4)-C(3)	105.00(16)
O(3)-C(5)-N(1)	125.04(19)	O(3)-C(5)-C(4)	126.81(19)	N(1)-C(5)-C(4)	108.12(16)
C(5)-N(1)-C(6)	112.41(16)	C(5)-N(1)-C(7)	125.29(17)	C(6)-N(1)-C(7)	121.67(17)
N(1)-C(6)-O(1)	108.18(15)	N(1)-C(6)-C(61)	114.39(17)	O(1)-C(6)-C(61)	106.99(16)
N(1)-C(6)-C(3)	105.27(16)	O(1)-C(6)-C(3)	103.97(15)	C(61)-C(6)-C(3)	117.29(18)
N(1)-C(7)-C(8)	113.49(16)	N(1)-C(7)-C(71)	112.22(18)	C(8)-C(7)-C(71)	111.71(17)
C(13)-C(8)-C(9)	117.62(19)	C(13)-C(8)-C(7)	124.37(18)	C(9)-C(8)-C(7)	118.01(18)
C(10)-C(9)-C(8)	121.80(2)	C(11)-C(10)-C(9)	119.90(2)	C(10)-C(11)-C(12)	118.90(2)
C(13)-C(12)-C(11)	120.80(2)	C(12)-C(13)-C(8)	120.90(2)		

2 结果与讨论

以丁二酮 1 和丙二腈 2 合成环并内酰胺 3 时，由于空间位阻较大，反应较难进行，再加上存在副反应[17]，因此 3 的产率较低，只有 32%。由于未完全反应的反应物及副产物能溶于水，可以通过水洗除去，这样就保证了产物 3 较高的纯度（≥98%）。在合成化合物 5 时，我们还研究了环并内酯 4 与浓氨水的反应，生成环并内酰胺-内酯的产率为 86%，而化合物 5 的产率高达 89%。由于苯基乙基胺比氨的碱性强，有利于反应的进行，从而可以弥补由苯基乙基胺的空间位阻效应带来的不利因素。

2.1 标题化合物的波谱特征

在标题化合物的紫外光谱中，最大吸收波长 210nm 和 263nm 所对应的官能团分别为内酯和苯环。^{1}HNMR 谱图中，化合物在 $\delta=1.23$ 和 1.50 处出现的单峰分别为 C（5）和 C（1）上的甲基氢，由于 C（1）上所连氮及氧等原子的影响，C（1）上的甲基氢化学位移值较大；同样，在苯环及氮原子的作用下，苄甲基氢的化学位移值高达1.82，且在临近质子作用下裂分为二重峰；C（4）和C（6）上的氢核由于化学位移相差不大，造成跃迁能级混合，形成AB高级谱。红外谱图中，1780和1710cm^{-1}左右的强吸收峰表明化合物中内酯和内酰胺结构的存在。质谱和元素分析数据与按化合物分子式计算所得的数据基本一致，从而进一步证实了标题化合物的结构。

2.2 晶体结构分析

表1中的数据表明，标题化合物的晶体为单斜晶系，空间群 $P2_1$，$a=0.8777$（18）nm，$b=1.0102$（2）nm，$c=0.9009$（18）nm，$\beta=118.75$（3）°，$V=0.7003$（2）nm^3，$D_c=1.296g\cdot cm^{-3}$，$\mu=0.089mm^{-1}$，F（000）=292，$Z=2$。最终的偏离因子为 $R_1=0.0393$，$wR_2=0.0955$，$S=1.046$。最终差值电子云密度的最高峰为158$e\cdot nm^{-3}$，最低峰为$-154e\cdot nm^{-3}$。

在标题化合物分子中，N(1)-C(6)-O(1)间的键角为108.18C(2)-C(3)-C(4)间的键角为113.30°，表明内酯环和内酰胺环不在同一平面，而是形成二面角。从C(5)-N(1)-C(6)、C(5)-N(1)-C(7)和C(6)-N(1)-C(7)的键角值可以断定，C(7)位于二面角的外端；同理，结合N(1)-C(7)-C(8)、N(1)-C(7)-C(71)和C(8)-C(7)-C(71)的键角值以及C(7)和C(8)之间的距离(0.1518nm)可以判断出，C(8)也处在二面角的外端。由于键角C(13)-C(8)-C(9)、C(13)-C(8)-C(7)与C(9)-C(8)-C(7)之和为360°，可以得出，C(7)、C(8)、C(9)和C(13)四点处于一个平面上，根据键角C(10)-C(9)-C(8)、C(10)-C(11)-C(12)和C(12)-C(13)-C(8)的值分别为121.8°、118.9°和120.9°，可以断定，C（10）指向二面角里端，苯环平面与内酰胺环平面也呈二面角态势。

参考文献

［1］Jordan，P.；Fromme，P.；Witt，H. T.；Klukas，O.；Saenger，W.；Krauss，N. Three-dimensional structure of cyanobacterial photosystem I at 2.5 Å resolution［J］. Nature，2001，411，909.

［2］Cosnier，S.；Gondran，C.；Wessel，R.；Montforts，F. P.；Wedel，M. J. Poly（pyrrole-metallodeuteroporphyrin）electrodes：towards electrochemical biomimetic devices［J］. Electroanal. Chem. 2000，488，83.

［3］Kraeutler，B.；Matile，P. ChemInform Abstract：Solving the Riddle of Chlorophyll Reakdown［J］. Acc. Chem. Res. 1999，32，35.

［4］Cosnier，S.；Lepellec，A.；Guidetti，B.；Ricolattes，I. J. Enhancement of biosensor sensitivity in aqueous and organic solvents using a combination of poly（pyrrole-ammonium）and poly（pyrrole-lactobionamide）films as host matrices［J］. Electroanal. Chem. 1998，449，165.

［5］Cammidge，A. N.；Oeztuerk，O. Selective Synthesis of meso-Naphthylporphyrins［J］. J. Org. Chem. 2002，67，7457.

［6］Prinsep，M. R.；Caplan，F. R.；Moore，R. E.；Patterson G. M. L.；Smith，C. D. Tolyporphin，a novel multidrug resistance reversing agent from the blue-green alga Tolypothrix nodosa［J］. J. Am. Chem. Soc. 1992，114，385.

［7］Wang，W.-G.；Kishi，Y. Synthesis and Structure of Tolyporphin AO，O-Diacetate［J］. Org. Lett. 1999，1，1129.

［8］Prinsep，M. R.；Patterson，G. M. L.；Larsen，L. K；Smith，C. D. Tolyporphins J and K，two further porphinoid metabolites from the cyanobacterium Tolypothrix nodosa［J］. J. Nat. Prod. 1998，61，1133.

［9］Minehan，T. G.；Kishi，Y.［beta］-selective C-glycosidations：lewis-acid mediated reactions of carbohydrates with silyl ketene acetals［J］. Tetrahedron Lett. 1997，38，6811.

［10］Morliere，P.；Maziere，J. C.；Santus，R.；Smith，C. D.；Prinsep，M. R.；Stobbe，C. C.；Fenning，M. C.；Golberg，J. L.；Chapman，J. D. Tolyporphin：a natural product from cyanobacteria with potent photosensitizing activity against tumor cells in vitro and in vivo［J］. Cancer Res. 1998，58，3571.

［11］Minehan，T. G.；Cook-Blumberg，L.；Kishi，Y.；Prinsep，M. R.；Moore，R. E. Revised structure of Tolyporphin A［J］. Angew. Chem. 1999，111，975.

［12］Prinsep，M. R.；Patterson，G. M. L.；Larsen，L. K.；Smith，C. D. Further tolyporphins from the Blue-Green alga

Tolypothrix nodosa [J] . Tetrahedron，1995，51，10523.

[13] Minehan，T. G. ; Kishi，Y. Angew. Totalsynthese des (+)-Tolyporphin-A-O，O-diacetats mit der vorgeschlagenen StrukturChem [J] . 1999，111，972.

[14] Hartke，K. ; Roeber，H. ; Matusch，R. Zur Reaktion von Biacetyl mit Malononitril unter Alkoholat - Katalyse [J] . Chem. Ber. 1975，108，3256.

[15] Woodward，R. B. Totalsynthese des chlorophylls [J] . Angew. Chem. 1960，72，651.

[16] Montforts，F. P. ; Schwartz，U. M. Ein gezielter Aufbau des Chlorinsystems [J] . Angew. Chem. 1985，97，767.

[17] Eschenmoser，A. ; Wintner，C. E. Natural product synthesis and vitamin B12 [J] . Science. 1977，196，1410.

Synthesis and Structural Characterization of 1,5-dimethyl-2s-(1′-phenylethyl) - 2-aza-8-oxa-bicyclo [3. 3. 0] octane-3,7-dione

Hu Bingcheng，Lv Chunxu

(School of Chemical Engineering，Nanjing University of Science & Technology，Nanjing 210094，China)

Abstract: 1,5-Dimethyl-2S-(1′-phenylethyl)-2-aza-8-oxa-bicyclo [3. 3. 0] octane-3,7-dione was synthesized. Reaction of 2, 3-butandione with malononitrile gave dimethyldicyanobicyclolactam. Under acidic condition the lactam was hydrolyzed to dimethylbicyclolactone，which was converted with (S)-(−)-phenylethyl amine into the target product. The structure was characterized by elemental analysis，IR，^{1}H NMR and MS spectra. The crystal structure was determined by single-crystal X-ray diffraction analysis. The crystal belongs to monoclinic system，space group $P2_1$，unit cell dimensions: a=0.8777 (18) nm，b=1.0102 (2) nm，c=0.9009 (18) nm，β=118.75 (3)°，V=0.7003 (2) nm^3，D_c=1.296g · cm^{-3}，μ=0.089mm^{-1}，F (000) =292，Z =2，R_1=0.0393，wR_2=0.0955.

Keywords: dimethyldicyanobicyclolactam，dimethylbicyclolactone，(S)-(−)-phenyl-ethylamine，X-ray diffraction

(注：此文原载于 有机化学，2004，24 (6)：697-701)

覆炭 γ-Al_2O_3 载 Pt 催化脱氢制备邻苯基苯酚

卫延安，蔡春，吕春绪

（南京理工大学，化工学院，南京，210094）

关键词：覆炭 γ-Al_2O_3，铂催化剂，己烯基环己酮，脱氢

邻苯基苯酚（OPP）是一种重要有机化工产品，用途十分广泛。国外广泛采用以环己酮为原料，经缩合、脱水得到双聚中间体，再经脱氢的制备工艺[1~3]；国内自 90 年代中期开始进行此方面研究[4]，但催化剂的脱氢活性、稳定性及 OPP 的选择性均不很理想，总体收率较低，尚未工业化。在催化剂载体的选取上一般采用金属氧化物如 γ-Al_2O_3 等。载体的性能也对催化剂的活性有着重要影响。炭载体的抗结焦性能良好，比表面积较大。但是，炭载体也有机械强度低、微孔孔径小的缺点[5]。传统的氧化铝载体机械强度高，微孔孔径较大，但其表面酸性较强，易于结焦。苏雅军等[6]曾经探索了 γ-Al_2O_3 载体覆盖炭类物质的方法，所制备的覆炭 γ-Al_2O_3（CCA）具有良好的抗结焦性能。杨继涛等[7]采用炭覆载体制备了 Pt-Sn 催化剂用于甲基环己烷的脱氢，证明该类催化剂具有良好的脱氢活性和稳定性。

本文分别以 γ-Al_2O_3 和覆炭 γ-Al_2O_3(CCA) 为载体，制备了负载金属 Pt 的催化剂，同时加入少量的钾盐作为助催化剂，对其进行了适当改性；采用 CS_2 预中毒的方法，考察了由环己酮二聚物脱氢生成 OPP 的催化剂性能，获得了较为理想的结果。催化剂的稳定性有了明显的改进，OPP 的选择性也有了 明显的提高，总收率达 90%以上，该方法具有较好的工业化前景。

环己酮、36%盐酸、氯铂酸、CS_2、KOH 等均为分析纯试剂；γ-Al_2O_3 为工业品。

PHI-5300 型 X 衍射光电子能谱（ESCA 系统），标准射线管，阳极耙为镁，能量 1240eV，功率 250W，电压 15kV。SP3700 型气相色谱仪，SE-30 弹性石英毛细管柱。进样温度 280℃，检测器温度 290℃，柱温起始为 160℃，保持 1min 后以 20℃/min 升至 260℃。采用 FID 进行检测。

催化剂参照文献[6]方法制备出不同覆炭量的催化剂载体 CCA。1g 氯铂酸溶于水中制成 75mL 的溶液，加 36%HCl 将 pH 调整至 4 左右；将在 350℃焙烧过的 100gγ-Al_2O_3 载体浸渍于上述含 Pt 的水溶液中进行吸附，110℃干燥 1h，置入带加热套的石英玻璃管中，N_2 气保护下，升温至 350℃加热 3h，再换成流量为 25mL/min 的 H_2 气，还原 3h；冷却至室温后取出；催化剂再浸入含有 2g KOH 的 40mL 溶液中吸附，110℃干燥 1h，250℃干燥 2.5h，所得催化剂即可用于脱氢反应。CCA 替代 γ-Al_2O_3 可分别得到 Pt-K_2O/γ-Al_2O_3 和 Pt-K_2O/CCA 这 2 组催化剂。

环己酮二聚物：环己烯基环己酮参照文献[1]方法制备，质量≥99%。

采用自制石英管固定床反应器（内径 1.2cm，催化剂层高度 30cm，玻璃微球形成的蒸发层高度 15cm，外部电加热），将催化剂加热至指定温度，在 30mL/min 的 H_2 气氛下，使用 5μL 微量注射器将计算量 CS_2 分 2 次缓慢加入反应器进料口，由载气将其带入催化剂层中，30min 后，以约 5mL/h 的速度滴加环己酮缩合二聚物，在反应器口收集产物，用气相色谱在线分析反应产物组成。

结果与讨论

选用4种不同覆炭质量的CCA载体，制备负载活性组分Pt的质量分数为0.5%，K_2O的质量分数为1.0%的催化剂进行脱氢实验（表1）。由表1结果可见，CCA中的覆炭量对环己酮二聚物的脱氢活性有一定影响，对OPP选择性的影响较为显著；2# ～3# 催化剂（CCA中的覆炭量在0.5%～3.3%时，OPP的选择性逐渐提高，4# 和5# 催化剂（覆炭量在3.0%～4.6%）时，对OPP的选择性有所降低，因此，覆炭量不应超过3.3%。我们添加了1.0%的K_2O为助催化剂，起到了抑制催化剂表面酸中心的作用[5]，减少了脱氢时的裂解等副反应，降低了催化剂的结焦速度，因而可用较低的覆炭量获得类似的结果。

表1 不同覆炭量的CCA对环己酮二聚物脱氢活性的影响

Table 1 Effect of carbon content in CCA on dehydrogenation of cyclohexenyl-cyclohexanone

Calystyst* No.	ω(C)/% in covered γ-Al_2O_3	Conversion of cyclohexenyl-cyclohexanone/%	Selectivity of OPP/%
1#	0	98.6	92.0
2#	0.5	98.3	92.0
3#	1.5	98.0	94.1
4#	3.3	97.7	95.4
5#	4.6	96.2	93.5

* with ω（Pt）/%=0.2%；ω（K_2O）/%=1.0%；reaction condition：temp. 350℃；H_2 flow rate 50mL/min；and rate 7mL/h；reaction time 10h.

S对Pt的中毒效应早已得到认识[8]，为防止催化剂的失活，通常应避免原料中携带有机硫化合物。King等[9]在以Pt/C为催化剂制备OPP时发现，以金属环烷酸盐为环己酮缩聚催化剂时，缩聚催化剂中含有的少量有机硫杂质，质量分数大约为2×10^{-4}%的硫可以被带入产品二聚物中，所得原料脱氢效果较好。实验结果见表2。

表2 CS_2预中毒对3# 催化剂活性的影响

Table 2 Effect of pre-poisoned CS_2 on the activity of 3# catalyst

Dehydrogenation Time/h	m(CS_2):m(Pt) 0		0.5:100		1.0:100		1.0:50	
	S1	S2	S1	S2	S1	S2	S1	S2
1	99.8	96.5	98.8	95.8	97.8	96.8	97.1	96.6
10	99.8	96.4	98.7	95.6	97.7	96.7	97.0	96.2
20	99.5	93.3	98.3	95.5	97.3	96.3	96.6	95.7
50	99.2	91.4	97.5	94.7	96.6	95.9	96.2	94.1
100	98.3	85.1	97.6	91.6	95.5	94.4	95.1	92.5

Dehydrogenation condition see Table 1. S_1：conversion（%）；S_2：selectivity of OPP（%）

由表2可见，未加入CS_2进行预中毒的催化剂，虽然起始活性较高，但20h后对OPP的选择性明显下降；反应至100h后，OPP的选择性已经低于90%；相比之下，采用微量的CS_2对催化剂预中毒后，催化剂保持了较好的活性和稳定性，OPP的选择性明显提高。

在环己酮二聚物脱氢反应中的主要副反应是氢解反应，加入少量的硫所引起的催化剂的中毒主要是抑制了副反应，提高了OPP的选择性；但当硫的量超过一定值时，催化剂将完全失活。因此，硫的用量必须严格控制才能得到比较理想的结果。在本反应中硫的用量与催化剂中巧的总质量比m(CS_2):m(Pt)=1.0:100为宜。使用CS_2预中毒前后脱氢产品的全分析结果见表3。由表3的结果可以发现，采用催化剂预中毒后可以明显降低氢解产物联苯的含量，由2.6%下降为0.8%，OPP的选择性有较大的提高，这说明CS_2可以使催化剂中毒，虽然降低催化剂的氢解活性，但提高了

OPP 的选择性。

表 3 3# 催化剂脱氢产物组成分析结果

Table 3 Composition of dehydrogenation products obtained on 3# catalyst

m(CS_2)∶m(Pt)	ω(product)/%			
	OH	OH		
0	93.4	2.3	2.6	1.7
1∶100	96.2	1.4	0.8	1.6

Reaction conditions Table 2

采用 XPS 对催化剂预中毒后的结构进行分析发现，Pt 中毒前后 $4d_5$ 的结合能均为 316eV，未发生变化，表明催化剂表面的 Pt 基本保持稳定的 0 价状态；而 S 的 $2p_3$ 电子的结合能由 165eV 升高至 169eV，说明 S 的价态由-2 价向高价态过渡，极有可能是取代 H 与载体中的 O 结合而成为高价态 S，在氢转移过程中，影响到 H 的转移，同时可以抑制-OH 的离去，从而提高了脱氢主产物 OPP 的选择性。

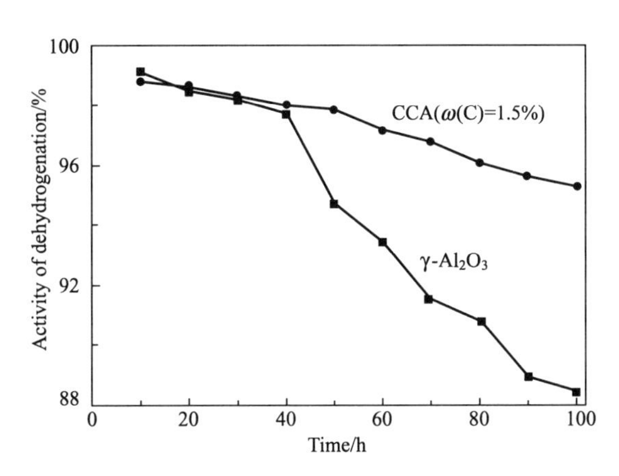

图 1 催化剂脱氢活性与反应时间关系

Fig 1 Dehydrogenation activity of catalysts reaction temperature: 350℃; H_2 flow rate: 60ml/min; LHVS∶0.5h^{-1}; ω (Pt) =0.5%, ω (K_2O) =1%

为比较 Pt/CCA 催化剂与 Pt/γ-Al_2O_3 催化剂的稳定性，我们选取了 3# 催化剂进行长时间的反应，结果见图 1。由图 1 可以看出，3# 催化剂在反应进行到 100h 时，脱氢活性仍能达到 95%以上，活性仅下降了约 5%；而 1# 催化剂在反应 40h 后，活性已开始明显地下降了，至 100h 后仅为 88%左右。由此可见，使用覆炭载体可明显提高催化剂的稳定性。其可能的原因是：一方面覆炭后掩蔽了原来氧化铝载体的酸性中心，降低了催化剂的表面酸性，使催化剂结焦倾向减少；另一方面，覆炭催化剂的 Pt/C 表面 可以具有催化作用，从而使其活性下降较慢，并能维持较高的活性水平[7-10]。

参考文献

[1] CAI Chun（蔡春），Lü Chun-Xu（吕春绪）. Chin J Appl Chem（应用化学）[J]，2001，18（2）：167.

[2] Goto H，Shibamoto N. US 3 997 615 [P]，1976.

[3] Murakami N，Ishiyama S. JP 5 415 119 [P]，1979.

[4] FENG Wang-Yan（冯望烟），CHEN Xu（陈诩），MIU Xue-Ru（缪雪茹）. Chin Petrol Chem（石油化工）[J]，1990，19（4）：238.

[5] CHEN Yang-Guang（陈仰光）. Chin Petrol Chem（石油化工）[J]，1993，22（11）：771.

[6] SU Jun-Ya（苏君雅），YAN Wei（阎炜），YANG Ji-Tao（杨继涛），et al. J Univ Petrol Chin（石油大学学报）

[J]，1995，19 (4)：103.

[7] YANG Ji-Tao (杨继涛)，HAOXue-Song (郝雪松)，SU Jun-Ya (苏君雅)，et al. Chin Petrol Chem (石油化工) [J]，2002，31 (12)：955.

[8] Thomas J M，Lambert R M，Auths (著) . HUANG Zhong-Tao (黄仲涛) Trans (译) . Characterisation of Catalyst (催化剂的表征) [M] . Beijing (北京)：Chemical Industry Press (化学工业出版社)，1987：185.

[9] King I R，Hildon AM. US 4 002 693 [P]，1977.

[10] Elliott C S. US 3 980 716 [P]，1.

Platinum Catalyst Supported on Carbon Covered Alumina for Preparation of O-phenylphenol

Wei Yanan，Cai Chun，Lv Chunxu

(School of Chemical Engineering，Nanjing University of Science and Technology，Nanjing 210094)

Abstract：The platinum catalyst supported on carbon covered γ-Al_2O_3 (CCA) or γ-Al_2O_3 with K_2O as assistant was prepared and used for preparation of o-phenylphenol (OPP) by dehydrogenation of 2-cyclohexenyl-cyclohexanone. The catalyst was pre-poisoned with CS_2 to improve its catalytic activity. The results showed that the catalytic performance of Pt supported on CCA is better than that supported onγ-Al_2O_3. When the cat-alyst consists of[ω(Pt)=0. 5%，ω(K_2O)=1%]/CCA[ω(C)=1. 5%]and m(CS_2)：m (Pt) is 1. 0：100，the selectivity of OPP was more 96% in a reaction condition of 350℃ and flow rate of H_2 at 60mL/min.

Keywords：carbon coveredγ-Al_2O_3，platinum catalyst，cyclohexenyl-cyclohexanone，dehydrogenation

(注：此文原载于 应用化学，2004，21 (8)：859-861)

4,4′-二硝基苯乙烯-2,2′二磺酸合成工艺改进

黄小波，张军科，蔡春，廖伟秋，吕春绪
（南京理工大学，化工学院，江苏，南京.210094）
EI：0045107697

摘要： 以 $MnSO_4$ 为催化剂，用空气氧化 4,4′-硝基甲苯-2,2′-磺酸合成了 4,4′-二硝基二苯乙烯-2,2′-二磺酸，用 UV 跟踪来确定反应终点。较习用工艺——$FeSO_4$ 做催化剂，提高了收率 10%～15%，减少 了 50%废水量，并缩短了反应时间 10h。

关键词： 4,4′-二硝基二苯乙烯-2,2′-二磺酸，4-硝基甲苯-2-磺，合成工艺，改进

引言

4,4′-二硝基二苯乙烯-2,2′-二磺酸是制备 4,4′-二氨基二苯乙烯-2,2′-二磺酸（DSD 酸）的原料。其习用合成工艺是以 NaOH 做碱剂、$FeSO_4$ 做催化剂用空气氧化 4-硝基甲苯-2 磺（p-NTS 酸）。目前，我国各厂家几乎都采用该工艺。但它存在收率低、废水量大、反应时间长等缺点，亟待改进。日本专利[2]采用 $MnSO_4$ 和 Mo、Ba、Fe 等的化合物做混合催化剂，可使反应时间缩短为 12h，收率提高到 80%～82%。前东德专利[3]则以 $MnSO_4$ 为催化剂，采用分步氧化法，收率可达 80%～85%。也有人[4]用 LiOH 做碱剂、Mn（OAc）$_2$ 为催化剂，收率达到 86.5%。

使用非水介质，如液氨、DMSO 或多胺-水混合溶剂等，可以大幅度提高收率且没有废水问题。Bayer 公司对此发表了一系列专利[5,6]。国内也有此类报道[7]。

捷克专利[8,9]介绍，在碱性介中用 NaClO 氧化 NTS 酸，可以用电位计或极谱分析来控制过量氧化，收率可提高到 85%以上。

作者认为：使用非水介质消耗大量溶剂而次氯酸钠腐蚀性强，对设备材质要求较高；LiOH 则价格昂贵。总之，前述方法都不太适合我国国情。而用 $MnSO_4$ 做催化剂的改进工艺则与我国各厂家的现行工艺几乎完全相同改造方便、见效快。因此，作者以 $MnSO_4$ 做催化剂的改进工艺进行了研究。

1 实验

1.1 试剂与仪器

p-NTS 酸制并精制，氢氧化钠、$\omega(H_2SO_4)=98\%$的硫酸为分析纯，$\omega(MnSO_4)=5\%$的 $MnSO_4$ 溶液；VarianFT-80A 核磁共振仪，Bruker Vector 红外光谱仪，Shimadzu UV-1601 紫外-可见-近红外分光光度仪，Shimadzu LC-10A 高效液相色谱仪。

1.2　原理

CH_3　SO_3H　NO_2　p-NTS 酸　$\xrightarrow[NaOH/H_2O, MnSO_4]{[O]}$　SO_3Na　SO_3Na　C═C　H　H　O_2N　NO_2　DNS-Na

1.3　操作

向装有搅拌器、温度计、空气导入管和冷凝管的500mL四口烧瓶中加入20g P-NTS，一定量质量分数5%的$MnSO_4$溶液、水，剧烈搅拌，加热。冒泡通入空气，将14g NaOH溶于50mL水所得溶液滴加入到反应体系中。接着升温至66℃，在此温度下保温反应。用UV-1601分光光度仪跟踪反应至终点。然后用质量分数为98%的硫酸调pH值为7～8，加入食盐。析出DNS-Na浅黄色晶体。自然冷却至室温过滤，用饱和NaCl溶液洗涤至洗液无色，置于105摄氏度下干燥，得到固体19.2g，收率为84.0%。

1.4　产物定性分析

UV：λ_{max}=353nm（文献值[10]：357nm）。

IR（KBr 压片）/cm^{-1}：3443.5，3096.6，1630.3，1581.7，1472.1，1348.9，1145.5，1076.8，1029.7，895.3，818.7，752.1。与Sadder标准红外谱图一致。

^{1}HNMR（D_2O做溶剂）δ：8.07（H^1）、7.94（H^2）、8.28（H^3）、8.62（H^4）。

上述渡谱数据表明，产物即为目标产物DNS-Na。

1.5　产物纯度测定

采用UV-1601紫外-可见-近红外分光光度仪以工作曲线法[4]测得产物纯度为95.51%，由此收率为84.0%；而用Shirnadzu LC-10AD高效液相色谱仪测定，产物纯度为99.42%，符合产品标准[11]。

2　结果与讨论

2.1　物料质量分数的确定

由于该反应的原料、产物在水中的溶解度都不太大，若质量分数过高就会沉淀析出；而质量分数过低则单位反应器体积产量低，废水量大。

表1　物料质量分数对反应的影响

Table 1　The influence of material quality score to reaction

物料质量分数/%	反应现象	物料质量分数/%	反应现象
5	反应正常，但废水量大	11	反应正常，反应时间12-13h
8	反应正常，反应时间7-8h	16	大量泡沫、有沉淀析出，反应难以进行

从表1可看出，较适宜的物料质量分数为8%-11%。

2.2　加料方式的确定

实际操作中，发现向NTS酸溶液加入NaOH的方式比NaOH溶液加入NTS酸的方式有更高的收率。因此本研究采用向NTS酸溶液加入NaOH的加料方式。

表 2　不同加料方式对收率的影响

Table 2　The influenceof yield by different way of feeding

加料方式	收率/%
向 NaOH 溶液加入 NTS 酸	74-78
向 NTS 酸加入 NaOH 溶液	80-85

2.3　反应时间的确定

本实验过程中，采用 UV 跟踪反应的方法来确定反应终点。

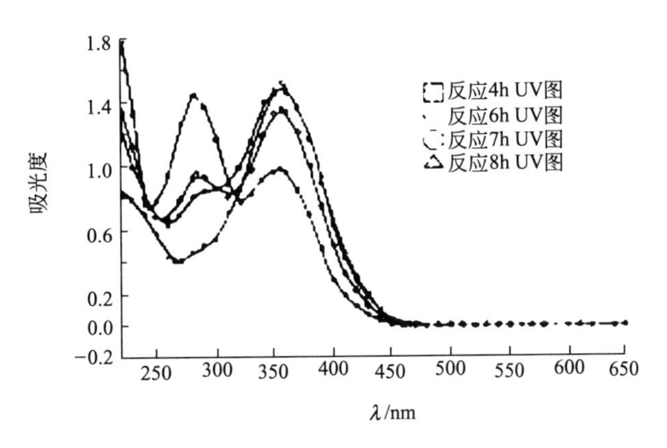

图 1　UV 跟踪图

Fig 1　tracking figure of UV

UV 跟踪图如图 1 所示，其中 λmax（p-NTS）＝277nm，λmax（DNS-Na）＝353nm（文献值[10]分别为：277.8，357nm）。可看出：随着反应时间的增加，原料吸收峰（270-280nm）越来越小，而产物吸收峰（350～360nm）越来越高，8h 后原料峰消失。再结合 HPLC 分析，确定较适宜的反应时间为 8～10h。

2.4　放大实验

在原实验基础上，放大 3 倍原料量进行了多次放大实验，反应收率≥84.5%，这是由于减少了机械损失所致。该结果进一步验证了此改进工艺的可行性。

2.5　改进工艺与习用工艺的比较

表 3　改进工艺与习用工艺的比较

Table 3　The comparison of improved technology and commonly used process

工艺	催化剂	物料质量分数/%	收率/%	t/h	m(产物)：(废水)
习用工艺	$FeSO_4$	5	70-73	18-20	1：22
改进工艺	$MnSO_4$	8-11	≥84	8-10	1：10

从表 3 可看出，改进工艺不仅可提高收率、缩短反应时间，而且还可以减少废水量，具有很好的工业化价值。

3　结论

（1）以 $MnSO_4$ 为催化剂对习用工艺——$FeSO_4$ 做催化剂进行改进，采用空气分步氧化 4-硝基甲苯-2-磺酸合成 4,4′-二硝基二苯乙烯-2,2′-二磺酸是可行的。

（2）采用改进后的工艺可以提高收率10%～15%、缩短反应时间10h。

（3）由于物料质量分数的提高，可减少50%的废水量。

参考文献

[1] 棘克勋．精细有机化工原料及中间体手册［M］．北京：化学工业出版社，1998.

[2] 住友化学工业株式会社．4,4′-Dinitrosfilbene-2,2′-disulfonic acid［P］．JP：38 764，1982-03-03.

[3] Wolter G，Kojtecheff T，Zoelch L. Preparation of 4,4′-dinitrostilbene-2,2′-disulfonic acid［P］．DD：240 200，1986-10-22.

[4] 宋东明，李树德，杨传远．4,4′-二硝基二苯乙烯-2,2′-双磺酸合成研究［J］．染料工业，1995. 32（2）：23-26.

[5] Schnatterer A. Fiege H Preparation of ditrostilbenedisulfonic acid by oxidation of nitrotoluenesulfonic acid in water-solvent mixture［P］．EP：641-773，1995.

[6] Schomaecker R，Waldmann H，Traenckner H J. Preparation of 4,4′-dinitrostilbene-2,2′-disulfonic acid and its salts［P］．EP：684 228，1995.

[7] 李彬，杨薇，才宝荣等．对硝基甲苯邻磺酸溶剂法氧化生产DSD酸的研究［J］．染料工业，1995. 32（6）：23-25.

[8] Kmonickova S，Drabek J，Cermak J，et al. 4,4′-dinitrostilbene-2,2′-disulfonic acid［P］. CS：216 132，1984-06-15.

[9] Kmonickova S，Drabek J，Cemak J，et al. 4,4′-dinitrostilbene-2,2′-disulfonic acid［P］．CS：216 132，1984-06-15.

[10] Barek. J，Danhel I. Spectrophotometric paper chromatograpic analysis of mixtures of 4-dinitrotoluene-2-sulph-onic acid and 4,4′-dinitrostilbene-2,2′-disulfonic acid［J］．Collection Czech Chem Commun，1984，49：2751-2755.

[11] 王琼轩，侯纪蓉，王一兵．世界有机中间体标准［M］．北京：中国环境出版杜，1991.

Improvement on the Synthetic Procedure of 4,4′-Dinitrostilbene-2,2′，-Disulfonic Acid

Huang Xiaobo，Zhang Junke，Cai Chun，Liao Weiqiu，Lv Chunxu

(School of Chemical Engineering，Nanjing University of Science & Technology，Nanjing，210094，China)

Abstract：4,4′-Dinitrostilbene-2,2′-disulfonic acid was synthesized through air oxidation of 4-nitrotoluene-2-sulfonic acid using $MnSO_4$ as catalyst. The termination was Determined by UV tracing. The yield was increased by 10%-15%，the amount of waste water reduced by 50% and the reaction time shortened by 10h in this procedure compared with the conventional procedure using $FeSO_4$ as catslyst.

Key Words：4,4′-dinitrostilbene-2,2′-disulfonic acid，4-nitrotoluene-2-sulfonic acid，synthetic procedure improvement

（注：此文原载于 精细化工，2000，17（3）：175-177）

氧气液相氧化法制备邻硝基苯甲醛

蔡敏敏，魏运洋，蔡春，吕春绪
（南京理工大学化工学院，江苏 南京 210094）
EI：02397110815

摘要： 氧气直接液相氧化邻硝基甲苯一步合成邻硝基苯甲醛，成本低，污染少，腐蚀性小。反应以有机胺为溶剂，在5℃，催化剂加入量0.05g，反应时间4.5h条件下，邻硝基甲苯转化率为41%，邻硝基苯甲醛得率为36%，同时对反应机理也作了初步探讨，认为该反应是碳阴离子反应历程。

关键词： 氧气液相氧化，邻硝基甲苯，邻硝基苯甲醛，碳阴离子

1 前言

邻硝基苯甲醛（以下简称 o-NBA）是重要的医药、染料和农药中间体[1]，主要用于生产抗心绞痛药硝苯吡啶（俗称心痛定），将硝基进一步还原成氨基后，又可合成喹啉环类药物。o-NBA 在工业上应用广泛，需求量大，因此如何高效简捷地氧化硝基甲苯制取相应的硝基芳香醛一直是人们感兴趣的问题，如用无机氧化剂 Cr(Ⅳ)、Co(Ⅲ)、Mn(Ⅲ)、Ce(Ⅳ)氧化取代甲苯制芳香醛常见于报道[2]。但是，应用无机盐氧化剂得率低、选择性差且污染严重。目前国内生产 o-NBA 主要采用三段法，即邻硝基甲苯（以下简称 o-MNT）侧链卤化水解氧化法，此法同样存在步骤较多、污染大、原料成本高、催化剂过氧化物不太稳定等诸多问题。国外从二、三十年代便开始了廉价、易得的氧气直接氧化甲苯或其一元取代物制取相应醛的研究，特别是 Morimoto[3] 等人在六十年代报道了该反应在 Co-Mn-Br-AcOH 系统中醛的产量可大幅提高，此后这方面的研究便吸引了众多学者兴趣。氧气一步氧化制邻硝基苯甲醛，操作简便，成本低，收率理想。受此启发，讨论氧气在碱性条件下液相氧化 o-MNT 一步制得 o-NBA 的反应，考察反应温度、反应时间、催化剂用量对 o-NBA 收率的影响。

2 实验部分

2.1 原料与仪器

邻硝基甲苯为工业品，甲醇、硫酸锰、氢氧化钾、硫酸、二氯甲烷为化学纯，高压氧气钢瓶供气。Saturn2000 气-质联用仪，Waters 600E/2487 高效液相色谱仪。

2.2 邻硝基苯甲醛的合成

一定温度下，往 150mL 四口烧瓶中加入 6.8g（6mL）o-MNT，0.05g 硫酸锰催化剂，和 50mL 有机胺溶剂，在常压下通入纯氧，氧气流量 30mL/mim，将氢氧化钾的甲醇溶液缓慢加入反应体系，

搅拌反应 4.5h 后，蒸出溶剂，加水并酸化，水相用二氯甲烷萃取 3 次，脱色，蒸去溶剂，得产物，称量并分析。

2.3 产品的定性与定量分析

定性分析采用 Saturn2000 气-质联用仪，在保留时间 10.819min 出现一大面积峰，质谱图与标准谱图吻合。定量则采用美国 Waters600E/2487 高压液相色谱仪分析反应产物组成，并计算 o-MNT 的转化率及 o-NBA 的得率。液相色谱分析条件为：

检测波长 254nm　　流动相甲醇/水＝60/40

流速 1mL/min　　Nova-pak　　C_{18} 柱 Φ3.9 mm×150mm

紫外吸收波长由日本岛津 UV-1601 紫外-可见-近红外分光光度计确定。

3 结果与讨论

3.1 反应温度对反应的影响

取－8～－5℃，－4～－1℃，2～5℃，8～10℃ 4 个不同反应温度下的产物 0.05g 左右，配成 0.1mg/mL 甲醇溶液，液相色谱分析其组成，每个样平行分析 2 次，对照各自的标准曲线进一步计算可得 o-MNT 转化率 x 和 o-NBA 的得率 Y 及反应对目标产物的选择性 S。温度对反应的影响可用表 1 表示。

表 1 温度对反应的影响

Table 1 Effect of reaction temperature

T/℃	x,%	Y,%	S,%	T/℃	x,%	Y,%	S,%
－8～－5	14.57	14.29	98.05	2～5	40.97	35.97	87.80
－4～－1	20.35	19.14	94.10	8～10	58.00	25.42	43.83

从表 1 可以看出，随着温度的上升，转化率也随着增加，这是因为随着温度的提高，反应速度加快，在相同的反应时间内消耗的反应物就愈多。表中还显示，在 2～5℃时，转化率有一个突跃，从 20.35%上升到 40.97%。

反应温度同样影响着 o-NBA 的得率，但却与转化率的影响稍有不同。从表 1 可以看出，o-NBA 得率变化是温度对 o-MNT 转化率与反应选择性交叉影响的结果。随着温度的升高，反应对生成 o-NBA 的选择性逐渐降低，o-NBA 的得率先是逐渐增大，但到 8～10℃时，o-NBA 得率反而下降。结合 o-MNT 的转化率可知，虽转化率随温度一直增加，但在较高温度时，反应向深度氧化方向发展，副产物增多，从而影响了 o-NBA 的得率。实验证明理想的反应温度应在 5℃左右。

3.2 反应时间对反应的影响

6mL o-MNT，0.05g 催化剂在 0～5℃下反应 2～7.5h，结果如表 2。

表 2 反应时间对反应的影响

t/h	x,%	Y,%	S,%	t/h	x,%	Y,%	S,%
2.0	18.54	5.80	31.28	6.0	55.75	30.98	55.56
3.0	27.45	18.02	65.64	7.5	60.00	24.63	36.76
4.5	40.97	35.97	87.79				

表 2 说明，o-MNT 转化率随时间延长而增大，而 o-NBA 得率先增后减，且在前 4.5 小时内增加很快，之后平稳下降。这说明随着时间的延长，o-NBA 被继续氧化成酸，另外一个可能的原因是反

应呈多步骤，起始反应相对容易些，而随着转化率的上升，中间产物积聚过多而使得反应不往期待的方向发展，副产物增多。适宜的反应时间应在 4.5h 左右。

3.3 催化剂加入量对反应的影响

6mL o-MNT，50mL 溶剂，在温度 2～5℃、反应时间 4.5h 下，考察了催化剂硫酸锰加入量对反应的影响，结果见表 3。

表 3 催化剂加入量对反应的影响

Table 3 Effect of catalyst content

m/g	x,%	*Y*,%	*S*,%	*m*/g	x,%	*Y*,%	*S*,%
0.0261	40.09	18.90	47.14	0.1078	45.84	16.49	35.97
0.0524	41.85	33.17	79.26				

表 3 可初步说明，催化剂质量的成倍变化对 o-MNT 转化率影响并不大，均在 40%～45%，但对 o-NBA 得率有较大的影响，这进一步表明反应呈多步骤的说法。反应的起始步可能是脱氢，后续才是氧化，而金属催化剂只是作用在反应的后半程。表 3 表明催化剂适宜的加入量在 0.05g 左右，即浓度为 1g/L 左右为宜。过多的催化剂使得副产物增多，选择性下降，反而不理想。

3.4 反应机理的初步探讨

气相条件下氧气氧化甲苯的反应机理已研究得十分透彻，普遍认为是自由基反应，且遵从 Mars-VanKrevelen 反应机理。但是氧气在溶液中氧化甲苯或其衍生物还未有深刻的研究。某些学者基于苄基 C-H 键较弱（解离能 77×4.187kJ/mol），而离解生成的苄基自由基又具有共轭稳定性这一事实，认为甲苯或其衍生物氧化与烯丙基型氧化机理类似，属于苄自由基催化氧化[4]，但笔者认为由于反应在强碱条件下进行，工艺数据又明显表明反应呈多步骤，因此反应可能先生成碳阴离子[5,6]：

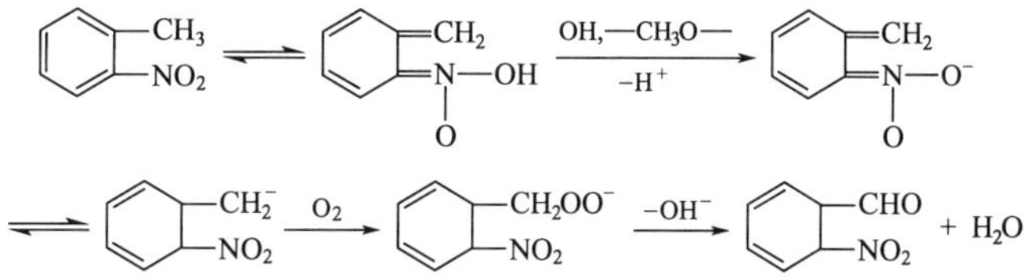

o-MNT 先形成醌式假酸结构，强碱从中夺取一个质子使 o-MNT 形成更易被氧化的邻硝基苯甲基阴离子，此碳阴离子被氧化成过氧化物，进而脱羟成醛。不过目前尚无直接的证据证明反应中碳阴离子的存在，也不能排除碳阴离子经自由基历程生成产物的可能性。

4 结论

a）在催化剂及特定的溶剂作用下，氧气在碱性条件下直接氧化邻硝基甲苯制邻硝基苯甲醛是可行的。在反应时间 5h，氧气流量 30mL/min，温度在 2～5℃，催化剂加入量为 0.05g 左右时，邻硝基甲苯的转化率为 41%，邻硝基苯甲醛的得率可达 36%。

b）初步的机理研究表明，该反应可能是一个碳阴离子反应，进一步的证明有待于研究。

参考文献

[1] 徐克勋．精细有机化工原料与中间体手册 [M]．北京：化学工业出版社，1998，(3)：371-372.

[2] Kei Ohkubo，Shunichi Fuk uzumi. 100% Selective Oxygenation of p-Xylene to p-Tolualdeh yde via Photoinduced Electron Transfer [J]．Organic Letters，2000，2 (23)：3647-3650.

[3] Morimoto T，Ogata Y. Kinetics of The Autoxidation of Toluene Catalyzed by Cobaltic Acetate [J]．J. Chem Soc，

Sect B，1967，62.
[4] 邓景发．催化作用原理导论［M］．长春：吉林科技出版社，1984.479.
[5] Artamkina G A，Grinfel'd A A，Beletaskaya I G. Air Oxidation of in situ Obtained Carban ions in KOH-DM E-Crown system［J］．Tetrahedron Lett，1984，25（43）：4989-4992.
[6] 余远斌，张淑芬，杨锦宗．NaOH对氧气液相氧化硝基甲苯助催化作用的研究［J］．大连理工大学学报，1999，39（1）：56-59.

Preparation of O-nitrobenzaldehyde via Liquid Autoxidation

Cai Minmin，Wei Yunyang，Cai Chun，Lv Chunxu
(Nanjing University of Science and Technology，Nanjing 210094，P. R. China)

Abstract：o-Nitrobenzaldehyde was prepared via liquid phase autoxidation of o-nitrotoluene in oganic amine solvents. The process friendly to environment possesses advantages of low cost and clean production. Under favorable conditions the reactant conversion and the product yield are 41% and 36% respectively. The reaction mechanism was discussed.

Key words：liquid phase autoxidation，o-nitrotoluene，o-nitrobenzaldehyde，carbanions

（注：此文原载于 化学反应工程与工艺，2002，18（1）：23-25）

双亲性席夫碱铜（Ⅱ）配合物的合成及苯甲醇的催化氧化

袁淑军[1,2]，蔡春[1]，吕春绪[1]

（1. 南京理工大学化工学院 南京 210094）

（2. 盐城工学院化工系 盐城 224003）

关键词： 席夫碱，铜配合物，苯甲醇，催化氧化

席夫碱及其金属配合物的种类很多，其中水杨醛席夫碱类配体及其配合物由于其结构类似于卟啉环和酞菁环，有较好的载氧、抗菌和模拟生物酶的催化作用[1-3]，因而它的合成与应用已得到普遍的重视和研究。但大多数研究工作基于配合物与氧的气-固反应、液-固反应。如水杨醛席夫碱的铜配合物的氧化作用已有不少报道，但由于配合物的溶解性差，它的催化反应绝大多数是在无水的有机相中进行的。也有一些水溶性席夫碱配体及配合物的研究报道[4,5]，但水相中的配合物难以催化有机相中底物的反应。虽然有人用外加表面活性剂的方法，利用表面活性剂的增溶、增敏等作用改善了反应的微环境[6,7]，但仍未使其催化作用得到根本的改善。本文合成了双亲媒性席夫碱型配体——双［1-(2-羟基-3-磺酸钠-5-甲基）苯十二酮］缩乙二胺及其铜（Ⅱ）的配合物，并初步探明了该配合物具有较强的催化苯甲醇在稀碱水溶液中的空气氧化作用。这一方法克服了水杨醛席夫碱类配合物必须在某些非质子极性溶剂（如 DMF）中和强碱的作用下才能催化醇类物质的空气氧化反应的弱点[8]。到目前为止，该配体及其金属配合物以及它的催化氧化性能尚未见文献报道。该配体及其铜配合物的合成路线如下：

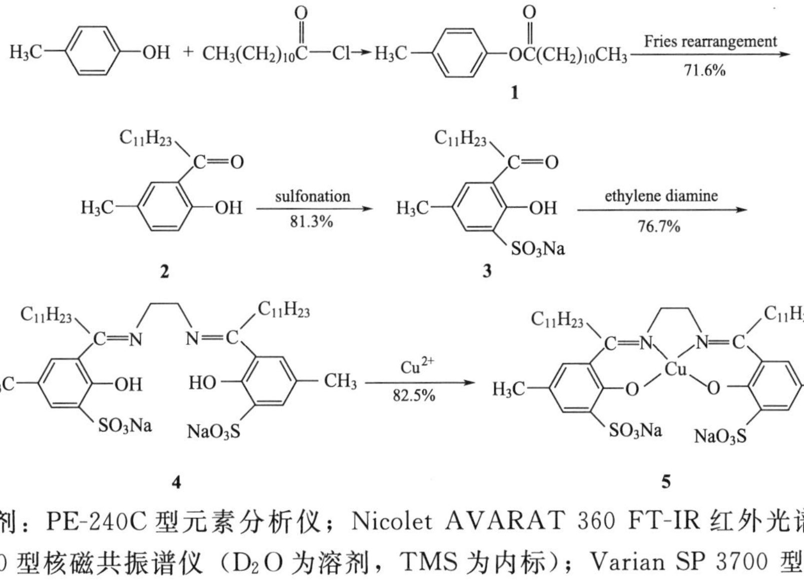

$H_3C-C_6H_4-OH + CH_3(CH_2)_{10}COCl \rightarrow H_3C-C_6H_4-OC(O)(CH_2)_{10}CH_3$ (1) —Fries rearrangement, 71.6%→

2 —sulfonation, 81.3%→ 3 —ethylene diamine, 76.7%→ 4 —Cu^{2+}, 82.5%→ 5

仪器与试剂：PE-240C 型元素分析仪；Nicolet AVARAT 360 FT-IR 红外光谱仪（KBr 压片）；Brukerdmx-300 型核磁共振谱仪（D_2O 为溶剂，TMS 为内标）；Varian SP 3700 型气相色谱仪，FID

检测器，NE-2000 数据处理工作站；BZY-180 型表面张力仪；熔点用 Thiele 熔点管测定。所用试剂均为分析纯。

十二酸对甲酚酯（**1**）：按文献［9］的方法合成，mp 29～31℃。

1-(2-羟基-5-甲基）苯十二酮（**2**）的合成：用无水 $AlCl_3$ 作催化剂对 **1** 进行 Fries 重排，将 29g（0.1mol）1 溶于 80mL CS_2 中，加入 15g $AlCl_3$ 在 45℃下回流 1h 后，蒸去溶剂，升温至 120℃反应 2h 后加 10% HCl 水解，再用 CH_2Cl_2 萃取，在萃取液中加入 0.15mol NaOH 溶液，形成的酚盐固体经抽滤后分离，以 HCl 酸化后用 CH_2Cl_2 萃取，蒸去溶剂后用乙醇重结晶，即得 2。产率 71.6%，mp41～43℃。IR，σ_{max}/cm^{-1}：1643，1612，1511，1204，877，819。元素分析 $C_{19}H_{30}O_2$ 实测值（计算值）/%：C 78.23(78.57)，H 10.47(10.41)。

1-(2-羟基-3-磺酸钠-5-甲基）苯十二酮（**3**）的合成：将 0.05mol **2** 溶于 50mL 二氯乙烷中，在搅拌下和 5℃时滴加 15mL 含 6.5g（0.055mol）氯磺酸的二氯乙烷溶液，反应 1.5h 后，升温到 50℃再反应到无 HCl 放出（约 2h），加入 40mL 水搅拌，静置分出水层，用 $NaHCO_3$ 中和至 pH 值为 6～7，蒸干水后得到淡黄色固体，用 85%乙醇重结晶得纯品，产率 81.3%，100℃下烘干后测熔点，mp>280℃。IR，σ_{max}/cm^{-1}：3200～3500，1641，1250，823。元素分析 $C_{19}H_{29}SO_5Na \cdot 2H_2O$ 实测值（计算值）/%：C 53.01(53.26)，H 7.51(7.76)，S 7.29(7.48)。^{1}HNMR(D_2O)，δ：6.59～7.56 (2H，Ar-H)，2.19(3H，Ar-CH_3)，0.84～1.23[20H，-$(CH_2)_{20}$]。

双亲性配体——双[1-(2-羟基-3-磺酸钠-5-甲基)苯十二酮]缩乙二胺(**4**)的合成：0.05mol 的 3 溶于 60mL 65%的乙醇，加入 0.025mol 乙二胺后，在室温下搅拌 0.5h，慢慢升温至回流状态下反应 2h，冰水冷却后析出淡黄色结晶，用 85%乙醇重结晶，100℃下干燥至恒重，产率 76.7%，mp>260℃(分解)。IR，σ_{max}/cm^{-1}：3200～3500，1612，1250，818。元素分析 $C_{40}H_{62}N_2S_2O_8Na_2 \cdot 2H_2O$ 实测值(计算值)/%：C 56.54(56.85)，H 7.70(7.87)，N 3.47(3.32)，S 7182(7.59)。^{1}HNMR (D_2O)，δ：6.60～7.43(4H，Ar-H)，2.67(6H，Ar-CH_3)，1.95～2.21(4H，=N-CH_2)，0.85～1.23 [40H，-$(CH_2)_{20}$]。

双亲性铜配合物（**5**）的合成：将 0.02mol 配体 **4** 溶于 40mL 50%乙醇中，在搅拌下加入 10mL (0.02mol)$Cu(NO_3)_2$ 的乙醇溶液，在 70℃下保温 1h，稍冷后用 10% NaOH 调节 pH=7，冰水冷却后滤出固体，用乙醇洗涤后真空干燥，产品为墨绿色晶体，产率 82.5%，mp>200℃（分解）。IR，σ_{max}/cm^{-1}：3200～3500，1603，1248，817，540，415。元素分析 $C_{40}H_{60}N_2S_2O_8CuNa_2 \cdot 4H_2O$ 实测值(计算值)/%：C 50.76(50.97)，H 7.21(7.27)，N 3.04(2.97)，S 6.90(6.81)。

按国标 GB/11278—89 临界胶束浓度的测定方法，测定配体 4 和配合物 5 的不同浓度下水溶液的表面张力，作浓度 c(mol/L)-表面张力 γ(mN/m) 图，得出图中拐点处对应的浓度即为临界胶束浓度(CMC)。

苯甲醇催化氧化反应：在三颈瓶中加入 20mL 苯甲醇和 20mL 不同浓度的 NaOH 溶液，在（40±1)℃恒温水浴槽上，加 2g 配合物 5 作催化剂，通空气（流速 50mL/min）反应 5h 后，将反应液用 HCl 酸化至 pH 为 2～3，分出油层，用少量无水 $MgSO_4$ 干燥后，进行气相色谱检测，用色谱数据处理工作站分析各成分含量，得苯甲醛收率。

结果与讨论

双亲性配体及其铜（Ⅱ）配合物的结构分析：在红外光谱中，二者均在 3250～3500cm^{-1}处有较强较宽的吸收峰，这是结晶水的-OH 峰。中间体 **2** 和 **3** 中酚基的 H 因与 $\rangle$C=O 形成氢键而在谱中不出现吸收峰；同样，配体 **4** 中酚-OH 也因与 $\rangle$C=N 形成氢键而不出现吸收峰。配体 **4** 中 1612cm^{-1}处

是 >C=O 在与乙二胺形成席夫碱后成为 >C=N 的吸收峰，而不再有 1641cm^{-1}的 >C=O 吸收。当它与Cu形成配合键后，>C=N 吸收红移了，变为 1603cm^{-1}，同时配合物又比配体多出了酚羟基与铜的 Cu-O（540cm^{-1}）和席夫碱 N 与 Cu 的 Cu-N（415cm^{-1}）2 个峰。说明 Cu 与酚羟基的 O 原子和席夫碱的 N 原同时配位，形成了四齿配合物。另外，1250cm^{-1}左右的宽强峰是-SO_4Na的吸收。在核磁共振谱图中，配体 **4** 和中间体 **3** 在 δ 为 6.6～7.6 处出现内强外弱的四重峰，这符合苯环四取代的特征。δ 为 0.84～1.23 处的多重峰是十二羰基峰，其中 1.23 处的强峰是十二羰基中多个 CH_2 的吸收峰。配体 **4** 比中间体 **3** 又多出了 δ 在 1.95～2.21 处席夫碱结构（$=NCH_2CH_2N=$）中氢核的吸收峰。结合元素分析以及各谱图的分析，说明了合成方法和各产物结构的正确性。

配体和配合物的水溶液浓度与其表面张力的关系见表 1。根据其数据测得配体 **4** 和配合物 **5** 水溶液的临界胶束浓度 CMC，分别为 2.0×10^{-4}和 4.5×10^{-3}mol/L，说明配体和配合物不但可溶于水而且使其水溶液的表面张力大幅度下降（最低值分别为 30.0 和 36.0mN/m），并可形成胶束。实验中发现，配合物 **5** 可在苯甲醇-水中形成均匀的绿色体系，而水杨醛缩乙二胺席夫碱铜配合物[Cu(salen)]在该体系中不能够溶解。

表 1 配体 4 和配合物 5 的水溶液表面张力（20℃，mN/m）

Table 1 Surface tension of aqueous solution of ligand and complex

c(4 or 5)/(mol·L^{-1})	1×10^{-6}	1×10^{-5}	1×10^{-4}	2×10^{-4}	5×10^{-4}	1×10^{-3}	5×10^{-3}	1×10^{-2}
Ligand 4	72.0	65.0	46.0	38.1	34.5	30.0	30.0	30.0
Cu complex 5	72.0	67.0	50.0	44.5	42.0	38.0	36.0	36.0

配合物对苯甲醇催化氧化的活性：考察了配合物 **5** 在不同温度和不同碱液浓度下的催化性能，并与 Cu（salen）进行对比实验。实验中发现 Cu（salen）在苯甲醇-碱水体系中不起催化作用（文献[8]报道该物在 DMF 和 NaOH 中有催化作用），而配合物 **5** 在该体系中有明显的催化活性。在苯甲醇的催化氧化反应中，当 NaOH 质量分数为 5%、10%、15%、20%和 30%时，苯甲醛的产率分别为 40.5%、57.8%、56.6%、54.3%和 48.1%。其催化性能在 40℃时最好，并与碱液浓度有很大关系。可以看出，反应在 40℃和碱液质量分数为 10%时，苯甲醇氧化成苯甲醛的产率最高为 57.8%；质量分数在 10%～20%的 NaOH 范围内产率变化不大，但再提高碱浓度产率会随之降低。可能是因为配合物的稳定性随碱浓度和温度的提高而下降，这从反应液颜色由绿→棕→褐→黑的变化现象也可看出。另外，提高温度和碱液浓度也不利于苯甲醛的稳定。

实验结果表明，该配合物具有一定的双亲媒性、能有效地降低水溶液的表面张力并形成胶束，对苯甲醇在稀碱水溶液中的两相空气氧化生成苯甲醛的反应具有显著的催化作用。

参考文献

[1] Doyle M P，Forbes D C. Chem Rev [J]，1998，98 (2)：911.

[2] CHEN Xin-Bin（陈新斌），ZHU Shen-Jie（朱申杰），GUI Ming-De（桂明德）. Chem J Chin Univ（高等学校化学学报）[J]，2000，21 (7)：1048.

[3] MA Hui-Xuan（马会宣），HU Dao-Dao（胡道道），Fang Yu（房喻）. Chin J Appl Chem（应用化学）[J]，2001，18 (4)：290.

[4] Kevin J Berry，Francisco Moya，Keith S Murry. J Chem Soc，Dalton trans [J]，1982：109.

[5] WANG Jian-Hua（王建华），DU Jun（杜军）. Chin J Appl Chem（应用化学）[J]，1999，16 (3)：114.

[6] TANG Bo（唐波），LIU Yang（刘阳），TANG Xiao-Ling（唐晓玲）. Chem J Chin Univ（高等学校化学学报）[J]，2001，22 (6)：919.

[7] GAO Yue-Ying（高月英），XU Ke-Wei（徐克伟），MAO Yue-Gan（毛岳淦）. Acta Chim Sin（化学学报）[J]，2001，59 (8)：1176.

[8] LIU Wei（刘卫），XI Zu-Wei（奚祖威）. Chin J Catal（催化学报）[J]，1990，11（2）：152.
[9] ZHANG Si-Gui（章思规）Chief-Edr（主编）. Technical Handbook of Organic Fine Chemicals（精细有机化学品技术手册）[M]. Beijing（北京）：Science Press（科学出版社），1992：1109.

Synthesis of an Amphipathic Schiff Base complex with Cu（Ⅱ）and Catalytic Oxidation of Benzyl Alcohol

Yuan Shujun[1,2]，Cai Chun[1]，Lv Chunxu[1]
（1. College of Chemical Engineering，Nanjing University of Science and Technology，Nanjing 210094）
（2. Department of Chemical Engineering，Yancheng Institute of Technology，Yancheng 224003）

Abstract：An amphipathic Schiff base complex with Cu（Ⅱ）was synthesized from p-cresol and dodecanoyl chloride with total yield of 36.8%. The structure and amphipathic property of the ligand and Cu complex have been investigated. The oxidation of benzyl alcohol catalyzed by the complex in alkali solution gave benzaldehyde in yield of 57.8%.

Key words：Schiff base，Cu（Ⅱ）complex，benzyl alcohol，catalytic oxidation

（注：此文原载于 应用化学，2003，20（3）：278-280）

含硼硅酚醛树脂BSP的合成和性能

杜杨[1]，吉法祥[2]，刘祖亮[1]，吕春绪[1]
（1. 南京理工大学化学工程与工艺系；2. 聚合物材料科学与工程系，江苏 南京 210094）
EI：03367625722

摘要：采用氢氧化钡和乙酸锌为催化剂合成了含硼和硅的酚醛树脂BSP，用FT-IR对其结构进行了分析，用TGA和氧指数仪对树脂的耐热性和阻燃性进行了测试。树脂具有明显的高邻位树脂的特点，体系粘度低，合成易于控制；树脂具有良好的耐热性和阻燃性，800℃热失重低于30%，氧指数大于47%。有机硅链段的引入显著降低了树脂的表面能。此外对影响BSP树脂的耐热性和阻燃性的因素进行了研究。

关键词：酚醛树脂，硼酸，有机硅预聚物，耐热性，阻燃性

酚醛树脂是最早的合成树脂，其固化产物不溶不熔，具有良好的电绝缘性和尺寸稳定性，在20世纪的上半叶获得了广泛的应用；后来由于其它合成聚合物的出现，酚醛树脂在某些方面的用途被更易成型的热塑性聚合物所取代；但进入20世纪80年代以后，高新技术产业的崛起为酚醛树脂开辟了新的应用前景，使其在电子、计算机、通讯、航空航天飞行器、生物材料、生化技术和先进复合材料等一系列高技术领域发挥了积极的作用[1]。

在耐热性能方面，矿物填充的酚醛树脂可归入耐热性聚合物的范畴，但一般最高使用温度不超过240℃（用作烧蚀保护材料除外），与现有的一些芳杂环耐温聚合物及元素有机聚合物相比有相一定的差距；但酚醛树脂的优势在于高温下具备良好的刚性保持率和耐蠕变性能，且其原料来源广、合成简便、价格低廉。此外酚醛树脂的耐热性能还可通过化学改性进一步提高，已有文献对其改性方法进行了综述[1]。本文合成了一种新的酚醛树脂BSP，即在酚醛树脂的分子结构中同时引入B-O键及有机硅链段，以期在提高耐热性的同时提高树脂的韧性，并改善树脂的耐水性及储存稳定性。

1 实验部分

1.1 原料

端羟基有机硅预聚物（分子量约1000）的甲苯溶液：自制；苯酚、甲醛水溶液、硼酸、氢氧化钡和乙酸锌：均为化学纯。

1.2 BSP树脂的合成

将计量的苯酚熔化后加入配有温度计、搅拌器、回流冷凝器及真空脱水装置的四口烧瓶中，开动搅拌，加入催化剂氢氧化钡和乙酸锌，搅拌10min左右待催化剂全部溶解，控制体系温度低于60℃，加入37%的甲醛水溶液。缓慢升温，在（65±3)℃反应2h后，加入硼酸，升温至100℃反应，注意观测反应体系的状

态，待体系由浅绿色的透明溶液变混浊时，开始真空脱水；缓慢提高真空度，维持真空度在0.080MPa～0.085MPa快速脱水，当体系温度逐渐升高达到85℃时加入50%端羟基有机硅预聚物的甲苯溶液，加热常压缩合反应，控制反应温度不高于110℃，待甲苯和水的共沸物全部蒸出后，取样测定树脂的凝胶时间，当凝胶时间70s～90s（180℃）时，停止反应，快速将树脂倒出，得浅绿色透明的热固性树脂。

1.3 IR分析

树脂先后分别溶于乙醇和丙酮、静置36h，取上层清液，减压脱除溶剂，与KBr一起研磨、压片，用VECTOR22型傅立叶红外分光光度计测定树脂的红外光谱。

1.4 热失重分析

将树脂在160℃、180℃和200℃氮气气氛下分别固化2h、2h和3h，将所得产物研磨、过60 mesh筛，采用Shimadzu TGA 250型热重分析仪测定热失重曲线；测试条件为：升温速率10℃/min，N_2气氛，气体流速20.00mL/min。

1.5 氧指数的测定

采用热压机在180℃、1MPa下将纯的粉状树脂进行压制、并经打磨制成长80mm、宽（6.5±0.5）mm、厚（3.0±0.5）mm的样条，在HT22型氧指数仪上测定氧指数。

1.6 表面张力的测定

将30%的树脂丙酮溶液均匀涂布于玻载片上，使溶剂在空气中自然挥发，然后在真空烘箱中60℃抽干1h，所得样品用国产JY 282型接触角测定仪分别用蒸馏水和二碘甲烷测定接触角，采用调和平均法[2]计算树脂的表面张力。

2 结果与讨论

2.1 反应原理

采用硼酸、硼砂等含硼化合物进行改性是提高酚醛树脂耐热性的有效途径。硼改性酚醛树脂的方法大致有两种：一种是苯酚与硼酸进行酯化生成苯酚硼酸酯，然后与多聚甲醛缩合[3]；另一种是苯酚与甲醛水溶液先在NaOH催化下反应生成羟甲基苯酚，经脱水，然后与含硼化合物缩合[4]。笔者发现这两种方法在实施工艺上均有难度：反应后期体系的粘度均很高，往往出现凝胶现象，使得反应难于控制；同时改性树脂的储存稳定性能较差，特别是固体粉状树脂，极易团聚结块并形成半固化的产物；此外由于硼酸酯键易于水解，导致最终产品的耐水性较差。这些缺点可能是导致硼改性的酚醛树脂未获得广泛应用的原因。

在BSP树脂的合成中，反应体系的粘度低，反应过程平稳，其原因是采用氢氧化钡和乙酸锌复合催化剂，在苯酚和甲醛反应生成羟甲基苯酚的反应中，生成了较多的邻位结构，导致最终形成的聚合物中直链型分子结构的数目远大于支链型；同时在树脂体系中引入聚硅氧烷链段，以改善树脂的表面性能，提高树脂的耐水性能及储存稳定性。可能的分子结构如下：

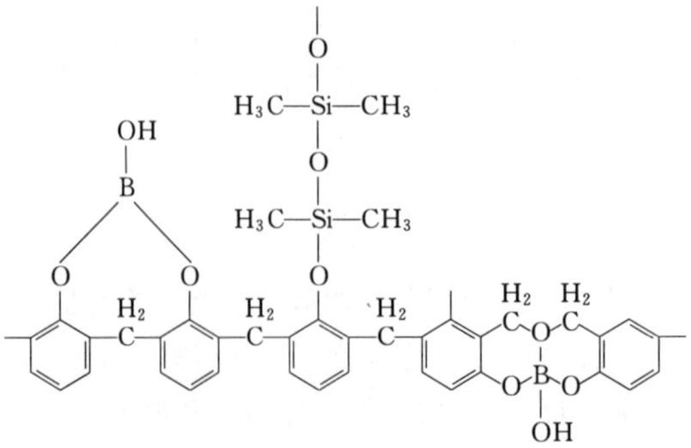

OH B O O H_2 C H_2 C O H_3C—Si—CH_3 O H_3C—Si—CH_3 O H_2 C H_2 H_2 C O C O B O OH

2.2 树脂的红外光谱分析

树脂的红外光谱见 Fig. 1。在指纹区，822cm^{-1}与 755cm^{-1}处的吸收峰的强弱可以表征酚醛树脂苯酚芳环的取代位置。据资料[5]，线形的高邻位的酚醛树脂仅在 750cm^{-1}左右出现一个强的吸收峰，而普通的线形酚醛树脂则在 820cm^{-1}和 750cm^{-1}处显示两个强度几乎相等的吸收峰；本文合成的树脂在 755cm^{-1}处的吸收峰强度远大于 822cm^{-1}处的吸收峰，显示出高邻位树脂的特点，只是由于其热固性树脂，醛与酚的摩尔比大于 1，在反应过程中有少部分甲醛在苯酚的对位发生反应。

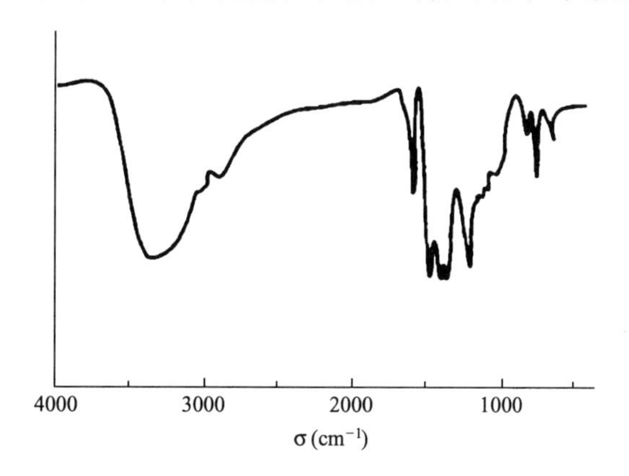

Fig. 1 FT-IR spectra of BSP resin

谱图中其它各吸收峰的归属如下：3380cm^{-1}左右出现的宽吸收带是酚羟基 O—H 键的吸收谱带，1597cm^{-1}、1481cm^{-1}处是苯环骨架的伸缩振动吸收峰，1421cm^{-1}处的吸收峰对应于相邻苯环之间的亚甲基，1385cm^{-1}处的吸收峰归属于硼酸酯键[6]，1222cm^{-1}处是 C—O 键的吸收峰；1200cm^{-1}～1000cm^{-1}之间的振动宽带是二甲基硅氧烷的特征谱带，只是由于树脂中二甲基硅氧烷的含量较少，强度显得较弱。

2.3 BSP 树脂的性能表征

Fig. 2 和 Fig. 3 分别是普通的热固性酚醛树脂（213 型）和 BSP 树脂（苯酚∶甲醛∶硼酸的摩尔比为 1∶1.5∶0.25，有机硅预聚物的量为苯酚质量的 5%）在相同固化、测试条件下所得的热失重曲线；从中可以看出，普通的酚醛树脂失重 5%的温度为 404℃，800℃时的失重率达 75.6%；而 BSP 树脂失重 5%的温度为 514.8℃，而 800℃时的失重率仅为 25%。

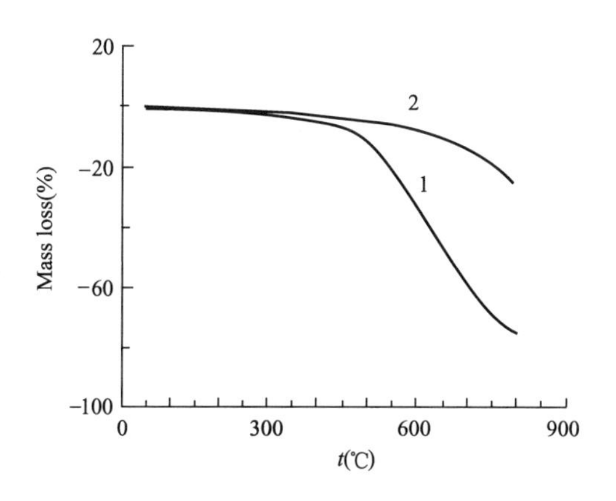

Fig. 2 TGA thermograms of cured commercial 213 phenolic (curve 1) resin and BSP resin (curve 2) BSP composition: n_{PhOH} ∶ n_{CH_2O} ∶ $n_{H_3BO_3}$ = 1∶1.5∶0.25, silicone prepolymer is 5 mass% by PhOH.

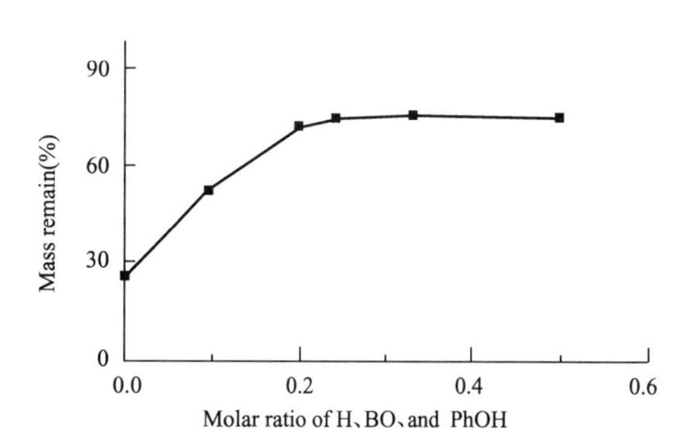

Fig. 3 Dependency of heat-resistance on H_3BO_3 n_{PhOH}∶n_{CH_2O}=1∶1.5, silicone prepolymer is 5 mass% by PhOH.

BSP 树脂的耐热性能（用 800℃时的质量保持率表示）和氧指数随原材料硼酸量的变化关系见 Fig. 4 和 Fig. 5。数据显示随硼酸用量的增加，树脂的耐热性和氧指数值均提高，但当硼酸的量达到 H_3BO_3：PhOH 摩尔比为 0.2 以上时，增加的趋势减缓，尤其是耐热性基本不再增加。树脂的耐热性和氧指数与有机硅预聚物用量的关系如 Fig. 6 和 Fig. 7，由图可见有机硅预聚物对树脂耐热性和氧指数的提高贡献不大，耐热性和氧指数甚至有下降的趋势，这可能是由于有机硅链段的存在妨碍了树脂固化时立体网状结构的形成；但有机硅的引入却使得树脂的表面能显著下降（如 Fig. 7），2%（质量）即可使树脂的表面张力由 43.2×10^{-3}N/m 降低到 25.5×10^{-3}N/m。表面张力的降低有助于提高树脂对增强材料的润湿能力，改善它们之间的界面性能，提高最终材料的力学性能，同时也可改善树脂的储存稳定性。将普通的 213 酚醛树脂和 BSP 分别粉碎、过 200mesh 筛、装满 250mL 广口试剂瓶（自然堆积）。放置一个夏季（6～9 月，南京室温约 28℃～35℃）后，黄色的 213 树脂体积收缩，结为红褐色异常坚硬的块状，经测试，树脂已交联固化、失去了粘接性能；而 BSP 树脂只是轻微结块，块状物疏松且颜色不变、性能也未受到影响。

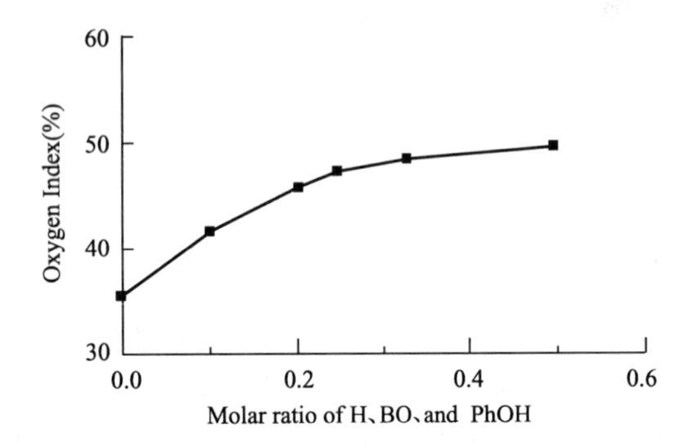

Fig. 4 Dependency of oxygen index on H_3BO_3
n_{PhOH} : n_{CH_2O} = 1 : 1.5,
silicone prepolymer is 5 mass% by PhOH.

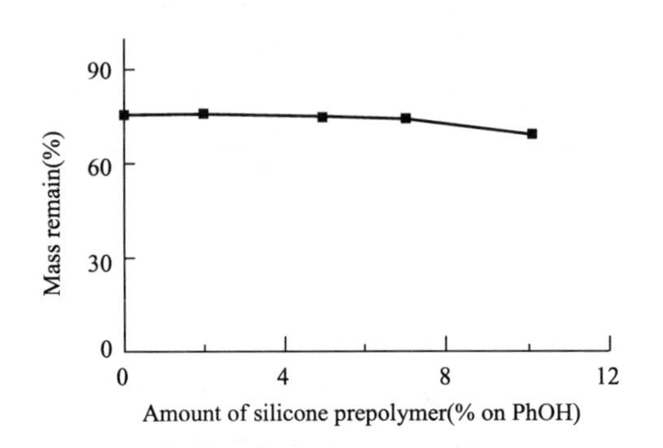

Fig. 5 Effect of amount of silicone prepolymer on heat-resistance n_{PhOH} : n_{CH_2O} : $n_{H_3BO_3}$ = 1 : 1.5 : 0.25

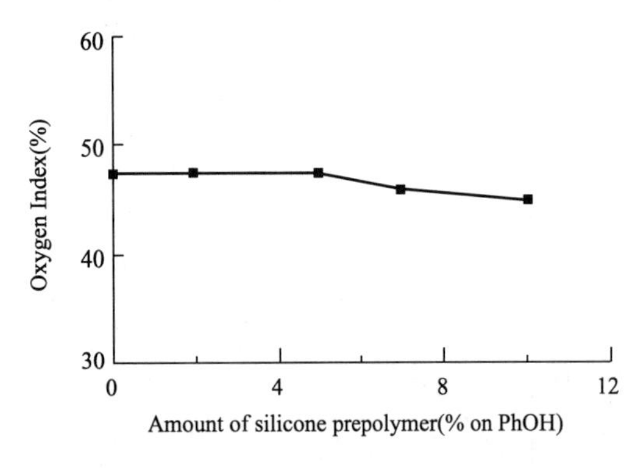

Fig. 6 Effect of amount of silicone prepolymer on oxygen index n_{PhOH} : n_{CH_2O} : $n_{H_3BO_3}$ = 1 : 1.5 : 0.25

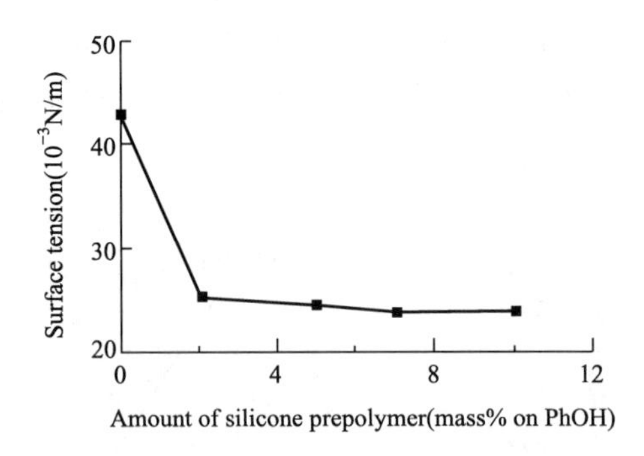

Fig. 7 Effect of silicone prepolymer on surface tension of BSP resin

参考文献

[1] Knop A，Pilato L A. Phenolic Resin. Berlin Heidel-berg，New York Tokyo：Springer-Verlag，1985.

[2] Wu S. 高聚物的界面与粘合 .（Gao juwu De Jiemian Yu Nianhe）. 第一版（First Edition）. 北京：纺织工业出版社（Beijing：Textile Industry Press)，1987.

[3] Hirohata T，Misake T，et al. Zairyo，1989，38（432).

[4] Hoefel H B, Kiessling H J, et al. DE-OS, 2, 436, 360.
[5] Brode G L. Phenolic Resin, in Encyclopedia of Chemical Technology, Vol. 17, John Wiley, New York, 1980.
[6] 贝拉米著．复杂分子的红外光谱（Fuzha Fenzi De Hongwai Guangpu）．第一版（First Edition），北京：科学出版社（Beijing：Science Press），1975.

Synthesis and Properties of Boron-and Silicone-containing Phenolic Resin

Du Yang[1], Ji Faxiang[2], Liu Zuliang[1], Lv Chunxu[1]
(1. Department of Chemical Engng. & Tech.; 2. Department of Polymer Materials Sci. & Engng., Nanjing University of Sci. and Tech., Nanjing 210094, China)

Abstract: Boron-and Silicone-Containing Phenolic Resin (BSPR) was synthesized by using barium hydroxide and zinc acetate composite catalyst. The structure of BSPR was characterized by FT-IR. Its heat-resistant and flame-resistant properties were analyzed using TGA and Oxygen Index (OI) respectively. The BSPR had high o-o′ methylene structure between phenol nuclei in its macromolecule, and the synthesis reaction was readily carried out due to the lower viscosity of the resin. The BSPR exhibited excellent heat-resistance and flame-resistance, mass loss less than 30% at 800℃ and OI larger than 47%. Silicone segment played a significant role on reducing BSPR's surface energy. Furthermore, some effects on the resin's properties were also investigated.

Keywords: phenolic resin, boric acid, silicone prepolymer, heat-resistance, flame-resistane

（注：此文原载于 高分子材料科学与工程，2003，19（4）：44-47）

拉夫替丁的合成

陶锋[1]，吕春绪[2]

（1. 江苏亚邦集团公司医药研究所，江苏常州 213163）

（2. 南京理工大学，江苏南京 210094）

摘要： 2-氨基-4-甲基吡啶经溴代、氧化、还原、再氧化成醛、与乙二醇成缩醛后醚化得 2-[4-(四氢吡喃-2-基氧基)-(2Z)-丁烯基氧基]-4-(1,3-二氧戊环-2-基)吡啶，再将脱 THP 保护基、Gabriel 反应、肼解一锅合成，然后经酰胺化和缩醛水解、与哌啶缩合等反应制得 H_2 受体拮抗剂拉夫替丁，总收率 6.8%。

关键词： 拉夫替丁，H_2 受体拮抗剂，抗溃疡药物，合成

拉夫替丁（lafutidine，1），化学名为 2-[(2-呋喃甲基)亚磺酰基]-N-[(2Z)-4-[[4-(1-哌啶基甲基)-2-吡啶基]氧基]-2-丁烯基]乙酰胺，是由日本 Fujirebio 和 Taiho 公司联合开发的 H_2 受体拮抗剂，2000 年首次在日本上市。本品具有持续的抗分泌作用及潜在的粘膜保护作用，主要用于胃溃疡、十二指肠溃疡及胃炎的治疗[1]。

文献[2]以 2-氨基-4-甲基吡啶为原料，经溴代、氧化、还原、氧化成醛、缩醛化、醚化、脱保护基，再经 Gabriel 反应、肼解、酰胺化、缩醛水解，最后与哌啶缩合得 **1**（图 1）。本文参考文献[3~5]，进行适当改进：改用毒性较小的甲苯替代苯[2]作缩醛化溶剂，无需分离直接与 4-(四氢吡喃-2-基氧基)-(2Z)-丁烯-1-醇反应制得 **6**，收率 80.9%（文献[2]：71.3%）；将脱保护基、Gabriel 反应、肼解等 3 步反应用“一锅法”完成，并改用 DMF 为溶剂直接由 **6** 制得 **7**，简化了工艺，收率由

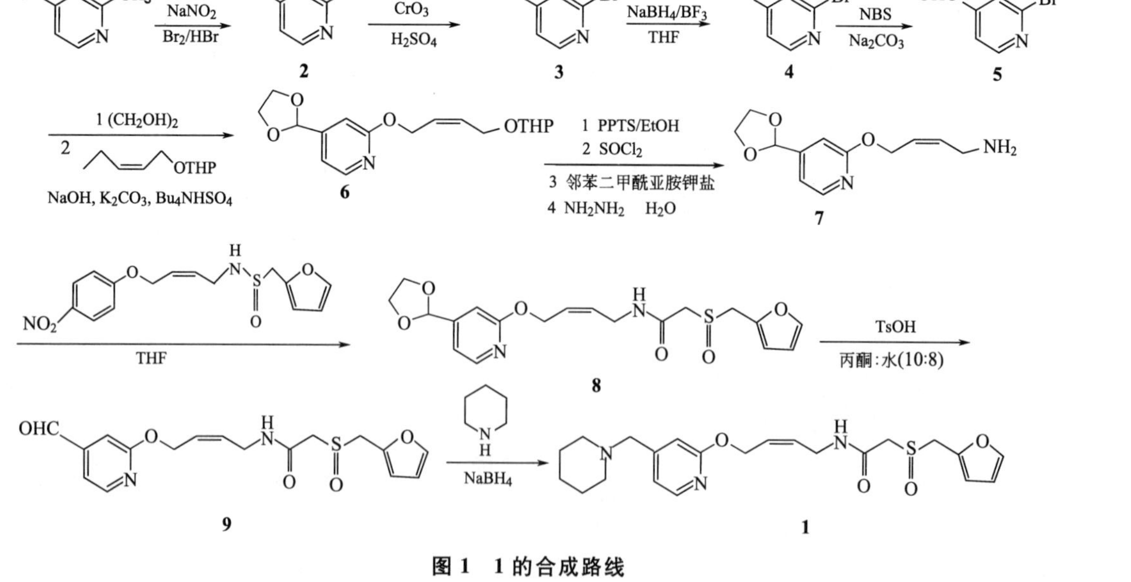

图 1　1 的合成路线

41.2%[2]提高到69.9%；制备**9**时将回流时间延长至48h（文献[2]：24h），并确定最佳溶剂配比为丙酮-水（10∶8），收率59.8%（文献[2]：52%）。改进后的总收率6.8%（文献[2]：总收率3.0%）。

实验部分

熔点用b型管法测定，温度计未校正：Carlo ERBA-1106型元素分析仪；Perkin-Elmer Lambda 2 UV/Vis型光谱仪；Nicolet Impact 410型红外光谱仪；Bruker DRX500型核磁共振仪；VG ZAB-HS型质谱仪。

2-溴-4-吡啶甲醛（5）

按文献[2]操作，由2-氨基-4-甲基吡啶制得白色固体**5**，4步连乘收率47.6%，mp 52.5～54.0℃（文献[2]：4步连乘收率46.4%，mp 52.6～53.5℃）。

2-[4-(四氢吡喃-2-基氧基)-(2Z)-丁烯基氧基]-4-(1,3-二氧戊环-2-基）吡啶（6）

5（18.6g，0.1mol）、乙二醇（12.4g，0.2mol）和对甲苯磺酸（1.9g）溶于甲苯（150mL），回流反应32h，同时用Dean. Stark分水器除水。反应毕加入冰水，用饱和碳酸氢钠溶液调至碱性，静置分层，有机层用水洗，用无水硫酸镁干燥，过滤。滤液中加入4-(四氢吡喃-2-基氧基)-(2Z)-丁烯-1-醇（18.1g，0.1mol）、固体氢氧化钠（13.0g）、碳酸钾（28.0g）和四丁基硫酸氢铵（2.5g），回流反应24h。冷却后用甲苯（200mL）稀释，水洗，用无水硫酸镁干燥，蒸除溶剂后得**6**（26.0g，80.9%）（文献[2]：两步收率71.3%）。

4-[[4-(1,3-二氧戊环-2-基)吡啶-2-基]氧基]-(2Z)-丁烯-1-胺(7)

6（25.7g，80.0mmol）和对甲苯磺酸吡啶盐（3.2g）加至乙醇（600mL）中，55℃搅拌反应18h。反应毕加入饱和碳酸氢钠溶液调至碱性，减压蒸除溶剂，用乙酸乙酯（150mL×3）萃取，有机层水（250mL）洗、干燥，蒸除溶剂后，用氯仿（500mL）溶解残余物。加入三乙胺（12.0g），冰浴冷至5℃以下，滴加氯化亚砜（12.0g，0.1mol），加毕搅拌2h。加入冰水溶解，用饱和碳酸氢钠溶液调至碱性，有机层水洗、干燥，蒸除溶剂后用DMF（500mL）溶解残余物。加入邻苯二甲酰亚胺钾盐（16.7g，90.0mmol）和四丁基硫酸氢铵（2.8g），回流过夜。冷却后过滤，滤液浓缩后用乙酸乙酯溶解残渣，经水洗、干燥后减压蒸除溶剂，得到的固体用乙醇重结晶。加入甲醇（500mL）和80%水合肼（5.2g），回流10h。冷却后过滤，滤液浓缩后静置析晶，过滤、干燥得黄色结晶**7**（13.2g，69.9%），mp 86.5～90.5℃（分解）（文献[2]：3步收率41.2%，mp 88～90℃）。

N-[4-[[4-(1,3-二氧戊环-2-基)吡啶-2-基]氧基]-(2z)-丁烯基]-2-(呋喃-2-基甲基亚磺酰基)乙酰胺(8)

四氢呋喃（300mL）中加入[（呋喃-2-基甲基）亚磺酰基]乙酸对硝基苯酯（13.6g，44.0mmol），搅拌及冰浴下滴加7（10.4g，44.0mmol）的四氢呋喃（100mL）溶液，加毕低温搅拌反应1h，缓慢升至室温，搅拌过夜。浓缩反应液，残余物用乙酸乙酯（300mL）溶解，用1mol/L NaOH溶液（100mL）和水（100mL）洗，干燥后蒸除溶剂得8（16.7g，93.2%）（文献[2]：91%）。

N-[4-[[(4-甲酰基)吡-2-基]氧基]-(2Z)-丁烯基]-2-(呋喃-2-基甲基亚磺酰基)乙酰胺(9)

8（16.7g，41.1mmol）溶于丙酮-水（350mL，10∶8），加入对甲苯磺酸（12.5g）回流反应48h。冷却后用饱和碳酸氢钠溶液调至碱性，浓缩后用乙酸乙酯（100mL×3）萃取剩余物，经水（150mL）洗、无水硫酸镁干燥后蒸除溶剂。残余物用硅胶柱[硅胶120g，流动相为甲醇-氯仿（1∶

100)]，得淡黄色结晶**9**（8.9g，59.8%），mp 68～70℃（文献[3]：52%，mp 67.6～69.9℃）。

拉夫替丁（1）

9(8.5g，23.5mmol）溶于乙醇（200mL)，冰浴下加入哌啶（4.2g，49.3mmol)，搅拌反应 5h，冰浴下加入硼氢化钠(1.2g)，搅拌。反应毕加入乙酸分解未反应的硼氢化钠，浓缩反应液，残余物用乙酸乙酯(150mL)溶解，用 20%乙酸(100mL×3)萃取。水相用乙酸乙酯 30mL×3 洗，用碳酸钾调至碱性，用乙酸乙酯(80mL×3)萃取，经水(50mL×2)洗、无水硫酸镁干燥后蒸除溶剂，剩余物用乙醚-正己烷(3：5)重结晶，得白色结晶 1(4.6g，45.5%)，mp(92.5～94.5℃)(文献［2]：46%，mp 92.7～94.9℃)，纯度 99.2%(HPLC 法)。结构经元素分析 IR、FAB-MS、^{1}HNMR 和^{13}CNMR 等确证。

参考文献

[1] Fujirebio Ts. Lafutidine [J] . Drugs Future，1999，24 (9)：1024-1025.

[2] Hifakawa N，Kashiwaka H，Matsumoto H，et al. Preparation of N-[(pyridyloxy) butenyl]-(furfurylthio) acetamides and analogs as ulcer inhibitors [P] . EP：282077，1988-09-14. (CA1990. 112：20902).

[3] Hirakawa N，Matsumoto H，HosodaA，et al. A novel histamine 2 (H_2) receptor antagonist with gastroprotective activity. Ⅱ. Synthesis and pharmacological evaluation of 2-furfuryl-thio and 2-furfurylsulfinyl acetamide derivatives with heteroaromatic rings [J]. Chem Pharm Bull，1998，46 (4)：616-622.

[4] Ishii A，Nishimura Y，Kondo H，et al. Preparation of N-[(pyridyloxy) alkyl]-or N-[(pyridyloxy) alkenyl]-2-(furfurylsulfinyl) acetamides as histamine H_2 receptor antagonists [P] . JP：5059045，1993-03-09. (CA1993，119：139114).

[5] 闻韧 . 药物合成化学 [M] . 北京：化学工业出版社，1988. 74-77.

Synthesis of Lafutidine

Tao feng[1]，Lv Chunxu[2]

(1. Jiangsu Yabang Group Co. Pharm. R&D. Inst.，Changzhou 213163)

(2. Nanjing University of Science & Technology，Nanjing 210094)

Abstract：Lafutidine，a histamine H_2-receptor antagonist，was synthesized from 2-amino-4-methylpyridine by bromination，oxidation，reduction，oxidation，etherification to give 2-[4-(tetrapyran-2-yloxyl)-(2Z)-buten-oxy]-4-(1，3-dioxolan-2-yl) pyridine which subjected to hydrolysis，chlorination，Gabriel reaction，hydrazinolysis in one pot followed by amidation，hydrolysis and then condensation with piperidine in an overall yield of 6.8 %.

Keywords：lafutidine，histamine H_2-receptor antagonist，anti-ulcer，synthesis

（注：此文原载于 中国医药工业，2004，35 (7)：393-395)

A Polymer Onium Acting as Phase-transfer Catalyst in Halogen-exchange Fluorination Promoted by Microwave

Luo Jun, Lv Chunxu, Cai Chun, Qv Wenchao
(School of Chemical Engineering, Nanjing University of Science & Technology, 210094 Nanjing, PR China)
SCI: 000221704900007

Abstract: A polymerized quaternary ammonium salt polydiallyldimethylammonium chloride, exhibiting high stability to heat and base, was prepared and applied as phase-transfer catalyst (PTC) in halogen-exchange (Halex) fluorination of chloronitrobenzenes to give excellent yields of corresponding fluoronitrobenzenes. Dimethyl sulfoxide was found to be the best solvent when microwave was applied as heating resource.

Keywords: Fluoronitrobenzenes, Polydiallyldimethylammonium chloride, Halogen-exchange fluorination, Microwave, Solvent

1 Introduction

Halogen-exchange (Halex) fluorination is an important method to prepare fluorinated aromatics. To accelerate the reaction between KF and substrates, phase-transfer catalyst (PTC) is often used. But quaternary ammonium salts containing β-H are prone to decompose to certain degree under Halex uorination conditions[1]. At the same time, by-products derived from the decomposition would worsen the reaction greatly. To solve these problems, Yoshida grafted N-(2-ethylhexyl)-4-(N′, N′-dimethyl) aminopyridinium bromide on polystyrene to get a polymer PTC with a relative molecular weight of near 1000g · mol^{-1}[2]. Subsequently, he and co-workers developed divinylbenzene across linked polystyrene supported tetraphenyl-phosphonium bromide and its modified analogue by replacing active hydrogen by methyl group[3]. The two polymers would not decompose obviously even at 210℃. Herein, we report another effcient and stable polymer onium PTC to act as phase-transfer catalyst in halogen-exchange fuorination under the irradiation of microwave.

2 Results and discussion

As we know, effcient and stable PTC should have big relative molecular weight, stable structure and big polarity. So, we think it may be a polymer with cyclic structure. For the sake of convenience for preparation, polydiallyldimethylammonium chloride (**1**) was found to be the best candidate. It was prepared by polymerization of diallyldimethylammonium chloride according to reference[4] (Scheme 1).

NaOH

$(NH_4)_2S_2O_8$

1, KF

MW, solvent

2a,3a: R_1=H, R_2=NO_2
2b,3b: R_1=NO_2, R_2=H
2c,3c: R_1=Cl, R_2=NO_2
2d,3d: R_1=NO_2, R_2=Cl
2e,3e: R_1=NO_2, R_2=NO_2

Scheme 1

As shown in Scheme 1, polymer 1 has a big density of catalytic active component. We reason that it should have high phase-transfer activity. At the same time, the catalytically active part is a five-membered ring, which provide it high stability. And fortunately these were proved by our experimental results.

1 is a very hygroscopic yellowish polymer with an average relative molecular weight of about $2\times10^{-5}g\cdot mol^{-1}$. It would convert to a liquid when exposed to air for only 1min. Furthermore, it is almost unsolvable in some polar aprotic solvents such as Me_2SO, sulfolane, $HCONMe_2$, $MeCONMe_2$ and N-methylpyrrolidone. So, the Halex reaction is a solid (**1**)-solid (KF)-liquid (solution of substrates) triphase procedure. To the best of our knowledge, it is the first time that **1** is used as PTC in Halex reaction. **1** is so hygroscopic that it is usually stored as solution in water; so, it is necessary to to dry it thoroughly before use. At first, its solution is dried at 100℃ in common oven, then grinded to fine powder and further dehydrated at 100℃ in vacuum for 10h, and finally stored in desiccator. To examine its stability to heat, 0.33g of **1** was added in 20ml Me_2SO and refluxed for 6h to find no obvious change. 2.19g of KF was added in this system and refluxed for another 6h to find almost the same phenomenon. So, **1** is very stable to heat and base. When **1** was used as phase-transfer catalyst in the halogen-exchange fluorination of p-chloronitrobenzene (PCNB), only 3h was needed to complete the reaction and 90.1% p-fluoronitrobenzene (PFNB) was obtained. In a control experiment, only 86.8% PFNB yielded after 6h and some tar appeared when cetyltrimethylammnoium bromide was used.

The application of microwave in organic synthesis was found its beginning by Gedye et al.[5] for esterization and have been developing very rapidly from then on. Some authors reported that the fluorination by radioactive elemental fluorine under microwave irradiation could be accelerated tremendously[6-8]. Kidwai et al.[9] got similar results when using KF and $(Me)_4NF$ as fluorinating agent. But they applied only in a very small scale and reacted in sealed tubes. In this work, microwave irradiation was applied to promote Halex fluorination in common scale.

To our surprise, sulfolane and N-methylpyrrolidone, excellent solvents in traditional preparation, did not increase the conversion of PCNB but rather decreased it (Table 1). In fact, the flnal reaction mixture was viscous, filtered to get a saturated solid, washed twice by benzene and changed to brownish. The filtrate was distilled by steam to get nut-brown mother liquids with considerable amount of tar or dispersed black powder. It is known that superheating of liquid is sure to occur when microwave was used as heating resource[10], and so are these solvents according to this work (Table 2). The boiling point increment at the same conditions obviously followed linear plot against dielectric constant with a correlation coefficient of 0.998 (Fig. 1).

Table 1 Preparation of PFNB catalyzed by 1 in 350W microwave

Solvent	Time(h)	Yield(%)	Conversion(%)
Me_2SO	2	92.5	99.7
	1.5	90.3	99.0
$HCONMe_2$	2	6.0	7.2
$MeCONMe_2$	2	31.5	38.2
N-Methylpyrrolidone	2	40.6	61.9
Sulfolane	2	46.2	70.9

Table 2 Superheating phenomenon of some solvents (50 ml) heated by microwave

Solvent	Dielectric constant	bp in oil(℃)	bp in microwave(℃)	bp increment(℃)
$HCONMe_2$	36.71	153	161	8
$MeCONMe_2$	37.78	166	174	8
Me_2SO	48.9	189	199	10
Sulfolane	43.3	287	296	9
N-Methylpyrrolidone	32.0	204	211	7

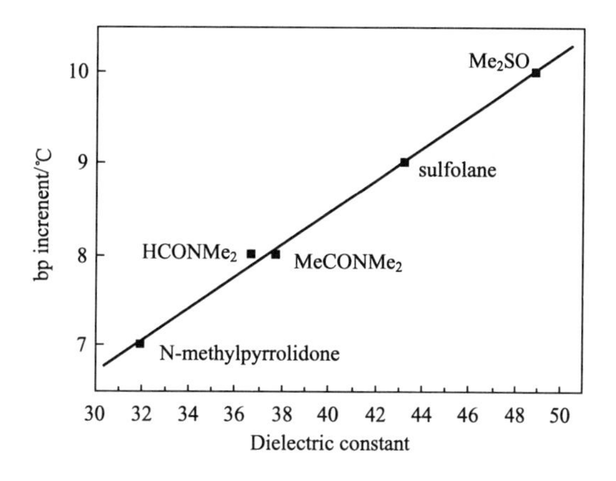

Fig. 1

The superheating phenomenon made excessively high reaction temperature when sulfolane and N-methylpyrrolidone, which would promote the decomposition, were used as solvents. Some evidences were found in blank experiments that sulfolane and N-methylpyrrolidone would darken their color gradually to nut-brown and brown, respectively, when refluxed in microwave field. The former changed to black and obviously some fine black grit appeared at the bottom of the vessel after 2h. Anhydrous fluoride is such a strong base in this system that it can extract proton from solvents, which is a main factor for the formation of a series of by-products[11]. To prove this, 3g of KF was added in 20ml of fresh sulfolane and refluxed in 700W microwave for 2h. A black, viscous and effluvial system was gotten and some one fifth of total mixture solidified to coke-like thing. It was obvious that decomposition was reinforced by KF. It seemed that the decomposition of solvents changed to be the main reaction at elevated boiling point since much lower conversion of substrates were resulted in comparison with Me_2SO (Table 1) . In a word, Me_2SO is the best solvent for microwave-promoted Halex reaction.

In order to identify its catalytic activity, **1** was used to catalyze the preparation of a series of fluoronitrobenzenes from corresponding chloronitrobenzenes (Table 3).

Table 3 Preparation of Ar-F catalyzed by 1

Entry	Substrate(2)	Heating resource	Solvent	Time	Product(3)	Yield(%)
a	O_2N–C₆H₄–Cl	350W(MW)	Me_2SO	2h	O_2N–C₆H₄–F	92.5
b	NO_2, Cl (o-)	200℃(oil bath)	Me_2SO	3h	NO_2, F (o-)	90.1
		300W(MW)	Me_2SO	3h		60.5
c	O_2N, Cl, Cl	200℃(oil bath)	Me_2SO	5h	O_2N, Cl, F	51.1
		350W(MW)	Me_2SO	10min		94.5
d	Cl, NO_2, Cl	200℃(oil bath)	Me_2SO	1h	Cl, NO_2, F	88.6
		350W(MW)	Me_2SO	15min		58.0
		190℃(oil bath)	Me_2SO	30min		41.2
e	O_2N, NO_2, Cl	200W(MW)	Acetonitrile	2h	O_2N, NO_2, F	89.2
		110℃(oil bath)	Acetonitrile	10h		80.5

The results shown in Table 3 illustrated that **1** was an efficient and stable PTC for Halex reaction, and the application of microwave irradiation accelerated the reaction rate for several times and the yields of products were also elevated. At the same time, very small tar and light color of reaction system were observed in experiments. "Non-heat effect" has being presumed to be the reason for the rate enhancement by a great number of authors. And we have also found the kinetic functions of common and microwave-promoted fluorination following different reaction orders in our further research. But, it is beyond the scope of this paper.

3 Conclusion

A stable polymerized quaternary ammonium salt poly-diallyldimethylammonium chloride was developed to act as efficient and stable phase-transfer catalyst in halogen-exchange fluorination to prepare fluoronitrobenzenes with good yields. The application of microwave enhanced the reaction rate several times. Superheating phenomenon showed that sulfolane and N-methylpyrrolidone, excellent solvents for traditional fluorination, were not fit for microwave-promoted fluorination any longer.

4 Experimental

4.1 General

Spray-dried KF was further dried in vacuum at 150℃ for 15h before use. All solvents were subjected to re-distillation and dried over 4A molecular sieve for at least 1 week.

Sanle WHL07S-1 chemistry-oriented microwave oven, designed by the author, was made in Nanjing Sanle Microwave Technology Development Co. Ltd., and the temperature was measured by Reytek® infrared thermometer. Products were inspected by Varian 3700 gas chromatograph, Varian Saturn 2000GC/MS chromatograph and Bomenn MB1543 FT-IR chromatograph (KBr). The IR and

MS spectra of products were identical with authentic standard spectra.

4.2 Preparation of polydiallyldimethylammonium chloride (1)

To a mixture of 30ml of dimethylamine (35.8 wt.%) and 37ml of allyl chloride was added 22.4g of solution of NaOH (40%) at 15℃ with agitation; after 10min, the system was heated to 40℃ and maintained for 6h. The final mixture was filtered and 6ml of water was added, and finally distilled under reduced pressure to get rid of volatile by-products. Consequently, the residue was diluted by water to 48.4g to get a solution of diallyldimethylammonium chloride. At 60℃ in N_2 atmosphere, 33g of solution of $(NH_4)_2S_2O_8$ (4%) in water was added evenly into the above solution within 2h and reacted for further 6h. **1** was recovered by a yield of 92.3%.

4.3 Typical experimental procedure for the preparation of fluoronitrobenzenes

4.3.1 Preparation of 3a-d from 2a-d

To a solution of 25mmol of chloronitrobenzenes (2a-d) in 20ml of Me_2SO was added 0.33g of **1** and 37.5mmol of KF, then heated by microwave with violently stirred for certain time, cooled down to less than 40℃ and filtered; the residue was washed by 2×10ml of benzene, filtrate and lotion were combined to undergo steam distillation. Organic layer was separated and water layer extracted by 3×20ml of benzene. The organic layer and extraction were combined and benzene was removed by simple distillation. Pure products was obtained by rectification under reduced pressure. Traditional heating reaction was all the same except for being heated by oil bath.

4.3.2 Preparation of 2,4-dinitrofluorobenzene (3e) from 2,4-dinitrochlorobenzene (2e)

4.05g of 2,4-dinitrochlorobenzene, 0.37g of **1**, 1.75g of KF and 0.28g of urotropine, acting as polymerizing inhibitor, were added in 20ml acetonitrile, heated by 200W microwave for 2h, then cooled down to room temperature and filtered. The filtrate was examined by GC directly. Traditional reaction was all the same except that the mixture was refluxed in oil bath.

References

[1] J. Clark, D. K. Smith, Chem. Soc. Rev. 28 (1999) 225-231.
[2] Y. Yoshida, Y. Kimura, Tetrahedron Lett. 30 (1989) 7199-7202.
[3] Y. Yoshida, Y. Kimura, M. Tomoi, Chem. Lett. (1990) 769-772.
[4] W. C. Qu, Master's degree dissertation, Nanjing University of Science & Technology, Nanjing, 1998, p. 12.
[5] R. Gedye, F. Smith, K. Westaway et al., Tetrahedron Lett. 27 (1986) 279-282.
[6] O. R. Hwang, S. M. Moerlein, L. Lang, J. Chem. Soc., Chem. Commun. 21 (1987) 1799-1801.
[7] S. Stone-Elander, N. Elander, Appl. Radiat. Isot. 42 (1991) 885-887.
[8] S. Stone-Elander, N. Elander, Appl. Radiat. Isot. 44 (1993) 889-893.
[9] M. Kidwai, P. Sapra, K. Ranjan, Indian J. Chem. 38B (1999) 114-115.
[10] Z. Wang, Research on Behavior of Gas, Liquid and Solid in Microwave Field, PhD Dissertation, Jilin University, Cangchu, 1994, p. 51.
[11] D. V. Adams, J. H. Clark, D. J. Nightingale, Tetrahedron 55 (1999) 7725-7738.

（注：此文原载于 Journal of Fluorine Chemistry, 2004, 125 (5): 701-704）

十六烷基二苯醚二磺酸钠表面化学性质及胶团化作用

许虎君[1,2]，吕春绪[1]，梁金龙[2]

（1. 南京理工大学化工学院，江苏 南京 210094）

（2. 江南大学化工学院，江苏 无锡 214036）

SCI：000233313800009

摘要： 用滴体积法通过表面张力的测定，系统地研究了十六烷基二苯醚二磺酸钠（C_{16}-MADS）在不同温度（298.0K～318.0K）和不同 NaCl 浓度（0～0.50mol・L^{-1}）下的表面活性。结果表明，温度升高使 C_{16}-MADS 溶液的临界胶束浓度（cmc）略有增大，表面极限吸附量（Γ_∞）降低。cmc 随 NaCl 浓度的增大从 1.45×10^{-4}mol・L^{-1}降至 4.10×10^{-5}mol・L^{-1}，但最低表面张力（γcmc）基本不受影响。在 298.0K 与 303.0K 时，NaCl 浓度的增大，Γ_∞ 增大；在 308.0、313.0 与 318.0K 时，NaCl 浓度的增大，出现了 Γ_∞ 从 2.27μmol・m^{-2} 降低至 1.41μmol・m^{-2} 的“反常”现象。胶团形成自由能（ΔG_m^0）随温度和 NaCl 浓度增加负值增大（−63.98～−76.20kJ・mol^{-1}），胶团的形成主要是熵驱动过程。

关键词： 十六烷基二苯醚二磺酸钠，表面张力，临界胶团浓度

十六烷基二苯醚二磺酸钠，简称 C_{16}-MADS 是一种的特殊双联型阴离子表面活性剂。

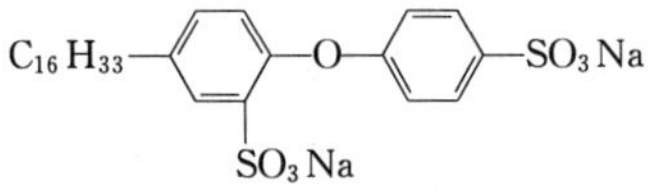

$C_{16}H_{33}$ —O— SO_3Na

SO_3Na

其具有独特的双磺酸盐亲水基，亲水基之间由二苯醚刚性基团联接，分子内产生超共轭效应。因而与传统的表面活性剂相比，有许多优点：优良的水溶性和偶联性，高度分散能力、抗硬水及漂白能力，在强酸、强碱及浓盐溶液中稳定性好[1-2]，已应用于高含量无机盐电解质体系中。但目前无机盐对该产品的作用研究尚未见报道。本文采用表面张力法考察了不同浓度 NaCl 溶液及体系温度下，C^{16}-MADS 溶液表面化学性质及胶团化热力学函数。

1 实验

1.1 主要试剂与仪器

NaCl（AR，上海化学试剂公司，在 500℃烧 5h 后使用）；C_{16}-MADS（自制，高压液相色谱测定及质谱检定，含量 99.8%）；超纯水（无锡新中亚微电子研究所，电导率 7.8×10^{-7}S・cm^{-1}）。

Agilent 1100 高压液相色谱仪；WATERS plat-form ZMD 4000 质谱仪；自制滴体积法表面张力仪。

1.2 样品的处理

C_{16}-MADS样品采用萃取分离法去除未反应的有机杂质，无水乙醇溶解过滤法去除无机盐杂质，在真空烘箱中烘干，经高效液相色谱测定及质谱检测．配制一系列不同浓度的C_{16}-MADS水溶液，分别测定其表面张力（γ），在γ-lgc曲线上无最低点，证明所提纯的样品无杂质。

1.3 表面张力的测定

配制一定浓度系列的C_{16}-MADS溶液，分别在（298.0±0.1）、（303.0 ±0.1）、（308.0±0.1）、（313.0±0.1）、（318.0±0.1）K温度下采用滴体积法测定其表面张力。

2 结果与讨论

2.1 NaCl浓度对C_{16}-MADS溶液表面张力的影响

测定了NaCl浓度在0、0.05、0.10、0.20、0.50mol・L^{-1}及体系温度在298.0、303.0、308.0、313.0、318.0K时C_{16}-MADS溶液的表面张力。图1给出了C_{16}-MADS在298.0K，不同NaCl浓度时的表面张力-浓度对数曲线。

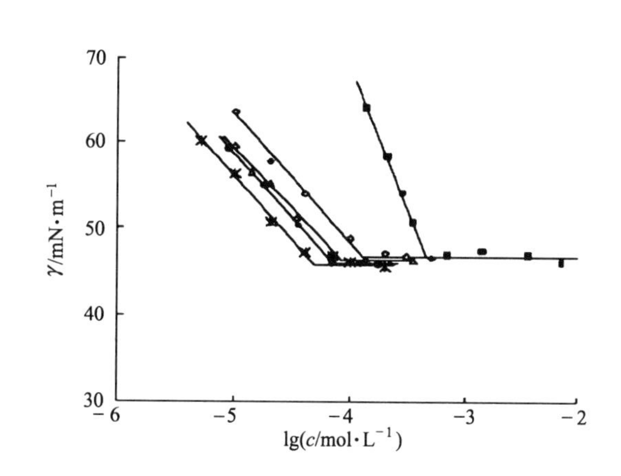

图1 298.0K时NaCl浓度对C_{16}-MADS溶液表面张力的影响

Fig. 1 Effect of NaCl concentration on surface tension of C_{16}-MADS in aqueous solution at 298.0K

C_{NaCl}/mol・L^{-1}：■ 0.00；◇ 0.05；△ 0.10；● 0.20；* 0.50

临界胶团浓度（cmc）和临界胶团浓度下的表面张力（γ_{cmc}）是衡量表面活性剂表面活性的重要参数[3]。对于传统离子型表面活性剂，在其溶液中加入与表面活性剂反离子相同的无机盐时，表面活性得到提高，cmc及γ_{cmc}均降低。这主要是无机盐的加入，部分破坏了水化膜，压缩了离子基团周围的扩散双电层，屏蔽了电荷之间的斥力，使得表面层及胶团中表面活性剂分子排列更为紧密，胶团容易形成[3]。随着NaCl的加入，C_{16}-MADS溶液的cmc明显下降，但由于C_{16}-MADS的两个亲水基之间是由二苯醚刚性基团联接，缺乏柔性弯曲，所以γ_{cmc}基本未变化。

2.2 十六烷基二苯醚二磺酸钠溶液的表面化学性质

根据Gibbs吸附公式，在反离子过量情况下[3-4]

$$-d\gamma=\Gamma sRTd\ln Cs \tag{1}$$

则表面活性剂在溶液表面的极限吸附量（Γ_∞）为

$$\Gamma_\infty=-d\gamma/(RTd\ln Cs)=-d\gamma/(2.303RT\lg Cs) \tag{2}$$

由γ-lgc曲线的斜率可求得极限吸附量（Γ_∞）。

在溶液表面饱和吸附层中表面活性剂分子所占极限平均面积（Amin）为

$$Amin=1/NA\cdot\Gamma_\infty \tag{3}$$

其中Γ_∞为表面活性剂在溶液表面的吸附量，R、NA分别为气体常数和Avogadro常数，T为绝对温标，Cs为溶液中表面活性剂离子的浓度。各体系的表面化学性质计算如表1。

表 1 C_{16}-MADS 溶液的表面化学性质

Table 1 Surface chemical properties of C_{16}-MADS in aqueous solution

C_{NaCl} /mol·L^{-1}	T/K	cmc /mmol·L^{-1}	γcmc /mN·m^{-1}	Γ_∞ /μmol·m^{-2}	Amin /nm^2	cmc /C_{20}	pC_{20}
0.00	298.0	0.451	46.7	2.21	0.75	1.430	3.500
	303.0	0.470	46.3	2.21	0.75	1.542	3.516
	308.0	0.498	45.5	2.18	0.76	1.720	3.542
	313.0	0.542	45.4	2.18	0.76	1.893	3.553
	318.0	0.604	45.4	2.18	0.76	2.115	3.571
0.05	298.0	0.145	46.8	2.27	0.73	2.225	4.192
	303.0	0.150	46.5	2.27	0.73	2.489	4.220
	308.0	0.158	45.0	2.27	0.73	3.021	4.281
	313.0	0.161	45.0	2.24	0.74	3.132	4.289
	318.0	0.166	45.0	2.24	0.74	3.262	4.293
0.10	298.0	0.089	46.3	2.27	0.73	2.566	4.460
	303.0	0.094	45.8	2.27	0.73	2.742	4.465
	308.0	0.100	45.1	2.16	0.77	3.002	4.477
	313.0	0.103	44.9	2.12	0.78	3.176	4.489
	318.0	0.105	44.8	2.10	0.79	3.297	4.497
0.20	298.0	0.062	45.9	2.55	0.66	2.617	4.625
	303.0	0.065	45.8	2.31	0.72	2.890	4.648
	308.0	0.069	45.5	2.08	0.80	3.391	4.691
	313.0	0.073	45.3	1.94	0.86	3.684	4.703
	318.0	0.076	45.2	1.83	0.91	4.061	4.728
0.50	298.0	0.041	45.8	2.58	0.64	2.637	4.808
	303.0	0.044	45.4	2.37	0.70	3.247	4.868
	308.0	0.048	44.8	1.94	0.86	4.250	4.947
	313.0	0.050	45.0	1.60	1.04	5.272	5.023
	318.0	0.053	45.0	1.41	1.12	6.557	5.092

C_{20}: the surfactant molar concentration required to reduce the surface tension of the water by 20 mN·m^{-1}; $pC_{20}=-lgC_{20}$; cmc/C_{20}: ratio measures the tendency of the surfactant to adsorb at the interface relative to its micellization tendency.

由表 1 可知在一定的 NaCl 浓度下，随着体系温度的上升，C_{16}-MADS 溶液的 cmc 逐渐增大。温度升高削弱亲水基团的水合作用，有利于胶团的形成，但温度升高亦引起疏水基周围水的结构破坏又不利于胶团形成，通常情况下温度对疏水基相互作用的影响在较高温度时起主要作用，所以 C_{16}-MADS 溶液的 cmc 随温度上升呈增大趋势。同样，在一定的 NaCl 浓度下，随着体系温度的上升，C_{16}-MADS 溶液表面极限吸附量减少，平均分子面积增大，其主要原因除了极性基团之间的电性排斥外，分子热运动也会随 温度上升而增强，促使其表面极限吸附量有所下降；其次，C_{16}-MADS 离子中的氧原子在水溶液中有与 H^+ 结合而产生质子化的趋势，降低了 C_{16}-MADS 的电负性，但温度升高，这种质子化趋势减弱，C_{16}-MADS 的电负性增强，因而亲水头基间的静电斥力增强，亦导致表面极限吸附量下降。在 298.0K 与 303.0K 时 NaCl 浓度的增加对 C_{16}-MADS 有两方面的作用，一方面改变溶液的离子强度，因而改变表面活性离子的活度。表面吸附作为一种平衡性质，必然随吸附物活度改变而改变；另一方面，由于增加反离子的浓度而有利于反离子与表面活性离子结合，削弱了它们在吸附层中的静电排斥，使吸附分子排列更紧密，导致表面极限吸附量增加。在 308.0、313.0 与 318.0K 的较高温度时，虽然 C_{16}-MADS 溶液具有 298.0K 与 303.0K 时相同的作用，但 NaCl 浓度的增加却使 C_{16}-MADS 的表面极限吸附量出现了下降的“反常”的现。其原因除了分子热运动增强及 C_{16}-MADS 离子的质子化趋势减弱导致其表面极限吸附量下降外，温度/NaCl 复合作用也是一个主要因素。根据 Bjerrum 理论[5]，当 NaCl 浓度增大时，温度升高，溶液中离子水化作用减弱，正负离子之间的距离缩短，且在电负性上，Cl^- 要大于 C_{16}-MADS 离子头基，所以 Na^+ 与 Cl^- 相互间的静电吸引能可以超过热运动能，在溶液中形成一个比较稳定的离子缔合体，即离子对，这样相对减少了 C_{16}-MADS 离子头基的反离子数量，促使原来电负性大于传统阴离子表面活性剂的 C_{16}-MADS 离子头基的电负性及电性斥力均增强，导致表面极限吸附量下降。这也与在实际应用中，该表面活性剂的洗涤能力随洗涤温度升高而降低[6]的状态相吻合。

2.3 十六烷基二苯醚二磺酸钠溶液的胶团化热力学函数的计算

对于 2-1 型离子表面活性剂，当存在大量外加无机盐电解质时，其胶团形成自由能(ΔG_m^0)、焓变(ΔH_m^0)和熵变(ΔS_m^0)可采用热力学函数计算公式[7-8]：

$$\Delta G_m^0 = 3RT\ln ncmc + 2RT\ln 2 \tag{4}$$

$$\Delta H_m^0 = -3RT^2\left(\frac{\alpha \ln cmc}{\alpha p}\right)_p \tag{5}$$

$$\Delta S_m^0 = (\Delta H_m^0 - \Delta G_m^0)/T \tag{6}$$

表 2 的 C16-MADS 水溶液胶团形成的热力学函数值表明，在所有实验温度下，ΔG_m^0 均为负值，说明此过程可以自发进行。ΔH_m^0 称为胶团生成热，它是胶团形成过程中的重要热力学参数[9]，$\Delta H_m^0 < 0$ 说明胶团形成过程为放热过程，这是由于形成胶团时单个 C16-MADS 分子先失去平动能量及由于碳氢链间色散力的相互作用所放的热超过“冰山结构”破坏所需的热。其值较负表示 C16-MADS 分子在溶液中形成胶团的自发趋势较强．在 $C_{NaCl}=0.50\text{mol}\cdot\text{L}^{-1}$ 时 C16-MADS 溶液的驻 ΔH_m^0 较低，所以其 cmc 较小。熵变(ΔS_m^0)反映了形成胶束前后无序性的变化。ΔS_m^0 均为正值，意味着 C16-MADS 分子加入到胶团中这一过程易于进行，伴随着正熵变使分子趋向无序状态，那么 C16-MADS 分子聚集成胶团是如何导致无序数增加呢？在水溶液中，水分子会在 C16-MADS 分子周围形成有序区域，即所谓“冰山结构”，当 C16-MADS 分子形成胶团后，其分子周围“冰山结构”被瓦解，体系无序数增加，使 ΔS_m^0 值为正值。从表 2 中还可以看出，伴随着温度的升高，ΔS_m^0 减小，这是由于当温度增高时，溶液中 C16-MADS 分子聚集成胶团的倾向减弱，促使正熵变的值越来越小。同时。ΔS_m^0 对 ΔH_m^0 的贡献 $-T\cdot\Delta S_m^0$ 为较大的负值，而 ΔH_m^0 的数值的绝对值比相应的 $-T\cdot\Delta S_m^0$ 的绝对值要小一些，因此 C16-MADS 溶液胶团的形成过程主要是熵驱动过程。

表 2　C_{16}-MADS 溶液的胶团化热力学函数

Table 2　Thermodynamic parameters of micellization of C_{16}-MADS

C_{NaCl} /mol·L^{-1}	T/K	ΔG_m^0 /kJ·mol^{-1}	ΔH_m^0 /kJ·mol^{-1}	ΔS_m^0 /kJ·mol^{-1}·K^{-1}	$T\Delta S_m^0$ /kJ·mol^{-1}
0.05	298.0	−63.98	−14.95	0.1645	−49.03
	303.0	−64.79	−15.46	0.1628	−49.33
	308.0	−65.47	−15.97	0.1607	−49.50
	313.0	−66.39	−16.75	0.1586	−49.64
	318.0	−67.20	−17.43	0.1565	−49.77
0.10	298.0	−67.60	−18.27	0.1655	−49.33
	303.0	−68.32	−18.89	0.1631	−49.43
	308.0	−68.99	−19.38	0.1610	−49.61
	313.0	−69.87	−20.11	0.1590	−49.76
	318.0	−70.14	−20.31	0.1567	−49.83
0.20	298.0	−70.30	−20.82	0.1660	−49.48
	303.0	−71.11	−21.53	0.1636	−49.58
	308.0	−71.83	−22.14	0.1613	−49.69
	313.0	−72.56	−22.76	0.1592	−49.80
	318.0	−73.30	−23.21	0.1575	−50.10
0.50	298.0	−73.40	−23.81	0.1664	−49.59
	303.0	−74.06	−24.36	0.1640	−49.70
	308.0	−74.60	−24.74	0.1619	−49.86
	313.0	−75.52	−25.55	0.156	−49.97
	318.0	−76.20	−25.91	0.1581	−50.29

3 结论

（1）在一定的 NaCl 浓度下，随着体系温度的升高，C_{16}-MADS 溶液的 cmc 略有增大，表面极限吸附量降低；cmc 随 NaCl 浓度的增大而明显降低，但最低表面张力（γ_{cmc}）基本不受影响。

（2）随着 NaCl 浓度的增大，在 298.0K 与 303.0K 时，C_{16}-MADS 溶液的表面极限吸附量增大；而在 308.0、313.0 与 318.0K 的较高温度时，C_{16}-MADS 溶液表面极限吸附量呈下降的“反常”的现象。

（3）计算了 C_{16}-MADS 溶液胶团形成过程的热力学函数，表明 C_{16}-MADS 溶液的胶团形成可自发进行，且主要是熵驱动过程。

参考文献

[1] Loughne, T. J.; Quencer, L. B. SCCS, 1992, 68 (1): 24.
[2] Milton, J. R.; Zhu, Z. H.; Hua, X. Y. JAOCS, 1992, 69 (1): 30.
[3] Zhao G. X.; Zhu B. Y. Principles of surfactant action [M]. Beijing: China Light Industry Press, 2003: 226-284.
[4] Yan, Y.; Huang, J. B.; Li, Z. C.; Zhao, X. L.; Ma, J. M. Acta Chimica Sinica, 2002, 60 (7): 1147.
[5] Liu Yun-pu. Physical chemistry. Beijing: China People Education Press, 1979: 132-133.
[6] Li, G. Z.; Sui, H.; Zhu, W. Z. China Surfactant Detergent Cosmetics, 1991, 1: 24.
[7] Zheng, Y. Y.; Zhao, J. X.; Zheng, O.; You, Y.; Qiu, Y. Acta Chimica Sinica, 2001, 59 (5): 690.
[8] Li, X. G.; Zhao, G. X. Acta Phys. 鄄 Chim. Sin., 1995, 11 (5): 450.
[9] Milton, J. R.; Zhu, Z. H.; Gao, T. Journal of Colloid and Interface Science, 1993, 157: 254.

Surfactivity and Micellization of Disodium Hexadecyl Diphenyl Disulfonate in Aqueous Solution

Xu Hujun[1,2], Lv Chunxu[1], Liang Jinlong[2]
([1]School of chemical engineering, Nanjing University of Science and Technology, Nanjing 210094;
[2]School of Chemical Engineering, Southern Yangtze University, Wuxi 214036)

Abstract: By using drop-volume method, the surface tension of disodium hexadecyl diphenyl ether disulfonate (C_{16}-MADS) was measures in different NaCl concentrations (0～0.50mol·L^{-1} NaCl) at different temperatures (298.0～318.0K). The results show that with the increse of temperature, the critical micelle concentration (cmc) of C_{16}-MADS slightly increases, but the maxium surface eccess concentration at the air/water interface (Γ_∞) decreases. The cmc of C_{16}-MADS decreases from 1.45×10^{-4}mol·L^{-1} down to 4.10×10^{-5}mol·L^{-1} as the concerntration of NaCl increases from 0 to 0.50mol·L^{-1}, but the corresponding surface tension at the concentration (γ_{cmc}) shows no variation. As the increase of NaCl concentration, Γ_∞ of C_{16}-MADS increases as the increase of temperature in the range of 298.0～303.0K, but abnormally decreases from 2.27μmol·m^{-2} down to 1.41μmol·m^{-2} in the range of 308.0～318.0K. The micellization free energies (ΔG_m^0) are in the range of −63.98～−76.20kJ·mol^{-1} in the studied ranges of temperature and NaCl concentration. The process of micellization of C_{16}-MADS in aqueous solution is mainly driven by the entropy.

Keywords: Disodium hexadecyl diphenyl ether disulfonate, Surface tension, Critical micelle concentration

（注：此文原载于 物理化学报，2005，21（11）：1240-1243）

N-正辛基硼酸二乙醇胺酯催化合成2-羟基-4-正辛氧基二苯甲酮

李斌栋，吕春绪，郑永勇
（南京理工大学化工学院，南京 210094）

关键词：硼酸酯，合成，羟基正辛氧基二苯甲酮

2-羟基-4-正辛氧基二苯甲酮是一种性能卓越的高效防老化助剂，能吸收240-340nm紫外光，具有延缓泛黄和阻滞物理性能损失的功能，它广泛用于PE、PVC、PP、PS、PC、有机玻璃、丙纶纤维和聚醋酸乙烯酯等方面。而且能为干性酚醛和醇酸清漆类、聚氨酯类、丙烯酸类、环氧类产品及汽车整修漆、粉末涂料、橡胶等制品提供良好的光稳定效果[1-4]。2-羟基-4-正辛氧基二苯甲酮的合成主要有一步法、氯代法和溴代法[5-9]。一步法一般是用三氯甲苯、正辛醇和间苯二酚在四丁基溴化胺和其他催化剂的作用下合成，此种方法主要产品收率仅为70%，分离提纯困难。氯代法主要是利用氯代正辛烷和2,4-二羟基二苯甲酮在环已酮溶剂中用十六烷基三甲基溴化胺作催化剂反应，但是合成产物中的2,4-二正辛氧基二苯甲酮多，提纯分离总收率仅82%。溴代法主要是利用2,4-二羟基二苯甲酮和溴代正辛烷在DMF或环已酮中合成，同氯代法比较存在溴代正辛烷价格高。而且后2种方法都存在回收溶剂困难的缺点。

本文利用2,4-二羟基二苯甲酮及氯代正辛烷直接在N-正辛基硼酸二乙醇胺酯催化条件下合成了2-羟基-4-正辛氧基二苯甲酮。一方面可以省去溶剂，而且提高合成2-羟基-4-正辛氧基二苯甲酮的选择性；另一方面利用溴代正辛烷合成N-正辛基硼酸二乙醇胺酯催化剂在反应中的费用远比溶剂的损耗要低。

采用GC-9790型气相色谱（SE30m×0.53mm×1μm，柱温100-250℃，程序升温，30℃/min）（温岭福立分析仪器厂）、Bomem MB1548型FTIR红外光谱仪（加拿大）、PE-2400型元素分析仪（美国）和H-NMR-300 MHz核磁谱仪（美国）作产物的分析和表征。

2,4-二羟基二苯甲酮、氯代正辛烷、溴代正辛烷均为工业级试剂；二乙醇胺和硼酸均为分析纯试剂。

N-正辛基硼酸二乙醇胺酯的制备：在带有搅拌和分水器的250mL四口烧瓶中加入12.5g硼酸和50g二乙醇胺以及100mL甲苯，回流分水达到理论量10.9mL时结束反应，蒸出甲苯，加入100mL四氢呋喃分离未反应的二乙醇胺，然后减压蒸除残余的四氢呋喃，得到硼酸二乙醇胺酯39.2g，收率为89.9%。经用0.1mol/L盐酸测得其质量分数为99.73%。其FT-IR，σ/cm^{-1}：3412，2934，1635，864。元素分析实测值(理论值)/%：C 44.01(44.24)，H 7.98(8.29)，N 12.87(12.90)。^{1}H-NMR(Aceton)，δ：2.69(—NH)，2.61(—N(CH$_2$)$_2$)，3.59(B(OCH$_2$))，3.82(—N$^+$H$_2$)，2.96(—N$^+$(CH$_2$)$_2$)。

在具有搅拌和回流装置的四口烧瓶中，加入0.1mol溴代正辛烷，0.1mol硼酸二乙醇胺酯（自

制）和无水乙醇 120mL，再加入 0.1mol 无水碳酸钠，加热回流 12h，反应结束，蒸除部分乙醇，离心除去固体，再减压蒸除乙醇得淡黄色透明粘稠 N-正辛基硼酸二乙醇胺酯产品 28.4g，收率 86.1%。经用 0.1mol/L 盐酸滴定测得其质量分数为 99.23%。元素分析实测值(理论值)/%：C58.76(58.36)，H10.14(10.33)，N8.49(8.51)。FT-IR，σ/cm^{-1}：3420，2925，2855，1652，1457，1075，875，720。^{1}H-NMR(Aceton)，δ：0.88($-CH_3$)，1.29($-(CH_2)_6$)，1.47($-NCH_2$)，2.60($-N(CH_2)_2$)，2.80($-N(CH_2CH_2O)_2B=$)，2.98($=B(OCH_2)_2$)，3.56($-N(CH_2CH_2O)_2$)，3.82($-N^+H_2$)。

2-羟基-4-正辛氧基二苯甲酮的合成：在带有搅拌和回流装置的四口烧瓶中，加入 21.4g 2,4-二羟基二苯甲酮、15.7g 氯代正辛烷、11.7g 无水碳酸钠、1g 溴化钾、1.1g N-正辛基硼酸二乙醇胺酯，升温至 155-160℃反应 8h 后取样分析，当 ω（2-羟基-4-正辛氧基二苯甲酮）=98.5%，ω（2,4-二羟基二苯甲酮）<0.3%时，停止反应，冷却至 80℃加水 40mL 和甲苯 10mL 洗涤，然后减压蒸馏得主产品 2-羟基-4-正辛氧基二苯甲酮 30.1g，加 60mL 无水乙醇重结晶得到 27.8g 2-羟基-4-正辛氧基二苯甲酮。

结果与讨论

所制得的 2-羟基-4-正辛氧基二苯甲酮，熔点为 47.2-48.4℃（文献值 47-49℃[1]）；红外光谱分析结果表明，1635cm^{-1}（C=O），1350cm^{-1}（—OH）与标准谱图（Sadtler8476）一致，GC 分析为 99.83%（面积归一法）；元素分析实测值（理论值）/%：C77.36(77.30)，H7.95(7.98)。^{1}H-NMR（Aceton），δ：0.88（$-CH_3$），1.29（$-CH_2-CH_2-CH_2-CH_2-$），1.80（$-CH_2$），2.08($-CH_2$)，4.10($-OCH_2$)，3.39(—OH)。

在合成 2-羟基-4-正辛氧基二苯甲酮的反应中，利用催化剂可以有效地防止副产物的生成和副反应的进行。在比较胺盐系列的反应中选择了 N-正辛基硼酸二乙醇胺酯，一方面它有比较好的表面活性，另一方面它在常温下是液态，在 160-180℃时比较稳定，而此时反应更容易形成均相，对反应更加有利。实验表明，加大催化剂在反应中的比例有利于反应的转化，但增加太多会使成本增加太大，取占反应物质量 5%的比例基本上与占 20%时的结果一致。

表 1　反应物摩尔比对反应的影响

Table 1　The effect of reactant molar ratio on the reaction

n-(2,4-dihydroxybenzophenone):n(octyl chloride)	ω(2,4-dihydroxybenzophenone)/%	ω(2-hydroxy-4-octyloxybenzophenone)/%	ω(2,4-dioctyloxybenzophenone)/%
1.0∶1.0	10.2	89.3	0.1
1.0∶1.05	0.3	98.5	0.8
1.0∶1.1	0.2	98.3	1.1
1.0∶1.2	0	97.6	1.9

Reaction temperature：160℃，reaction time 8h

在反应温度和催化剂添加量一定的条件下，发现氯代正辛烷与 2,4-二羟基二苯甲酮的摩尔比对反应的选择与转化率都有影响。

从表中可以发现，增加氯代正辛烷的摩尔比有利于反应的进行，但是比例太大时，2,4-二羟基二苯甲酮可以完全转化，可 2,4-二正辛氧基二苯甲酮要增加，对合成目标产物不利。

反应温度对副产物 2,4-二正辛氧基二苯甲酮的生成有影响。实验发现反应温度高于 160℃时，转化率增加，反应时间明显缩短但是副产物增多，当反应温度低于 140℃时，产物转化率降低，反应时间加长，反应进行 16h 还很难进行到底，实验发现在 150-160℃时反应 8h 可以降低副产物，并达到较好的转化率。

参考文献

[1] MENG Xue-An（孟学安）. Additive for Plastic and Rubber（塑料和橡胶助剂）[M]. Beijing（北京）：Chemistry Industry Press（化学工业出版社），2002：258.

[2] Mervin G，Andrea R，James P，et al. US 6699298 [P]，2004.

[3] Di Stefano S，Gupta A. Polym Chem [J]，1980，21（2）：179.

[4] Gugamus F. Polym Degrad Stab [J]，1993，19（1）：117.

[5] Allan P，Chakrabarti P，Brown N J，et al. US 4323710 [P]，1982.

[6] Malfrold P，Mulders J. EP 0032275 [P]，1982.

[7] Takeuchi M，Amagai A，Mizuno K，et al. EP 0950950 [P]，1999.

[8] Tanimoto S，Toshimistu A. Bull Inst Chem Res [J]，1992，69（5～6）：560.

[9] FAN Ping（范平），LI Hong-Tu（李鸿图），DENG Ming（邓明），et al. J Liaoning Univ（辽宁大学学报）[J]，1995，22（3）：7.

Synthesis of 2-Hydroxy-4-octyloxybenzophenone Catalyzed by N-octyl-diethanolamine Borate

Li Bindong，Lv Chunxu，Zheng Yongyong

（Nanjing University of Science & Technology，Nanjing 210094，China）

Abstract：2-Hydroxy-4-octyloxybenzophenone was synthesized from 2,4-dihydroxybenzophenone and octyl chloride in the presence of N-octyl-diethanolamine borate，which was synthesized from diethanolamine borate and octyl bromide. When the molar ratio of 2,4-dihydroxybenzophenone to octyl chloride was 1∶1.05，reaction temperature was 155～160 ℃，reaction time was 8 h and m（N-octyl-diethanolamine borate）/m（2,4-(dihydroxybenzophenone)）=0.05，the purity of the alkylated product was 98.5%，and the yield was 92.3%.

Key words：borate，synthesis，hydroxyoctyloxybenzophenone

（注：此文原载于 应用化学，2005，22（5）：572-574）

Synthesis of Metallodeuteroporphyrin-IX-dimethylester and Its First Using in Aerobic Catalytic Oxidation of Cyclohexane

Ma Dengsheng, Hu Bingcheng, Wang Fang, Lv Chunxu
(Nanjing University of science & technology, Nanjing, Jiangsu, 210094, P. R. China)
SCI: 000263627800008; EI: 20090611896518

Abstract: Metallodeuteroporporphyrin-IX-dimethylesters (MDP-IX-DME) of Cu (Ⅱ), Co (Ⅱ), Ni (Ⅱ) and Zn (Ⅱ) were synthesized, characterized and used successfully for the first time to catalyze aerobic oxidation of cyclohexane. The results showed that deuteroporphyrins-IX-dimethylester-cobalt (Ⅱ) (DP-IX-DMECo) had the best performance among the four efficient catalysts. Comparing with the reported tetraphenylporphyrin cobalt (Ⅱ) (CSTPPCo), under the reactionconditions of 150℃ and 0.8MPa, using MDP-IX-DMECo as catalyst led to an exciting result with enhancing cyclohexane conversion from 16.2% to 18.6%, selectivity of cyclohexanol and cyclohexanone from 82% to 84.6% and cyclohexane turnover number from 4×105 to 8.5×105. This was a great progress for hexane catalystic oxidation with air.

Keywords: metallodeuteroporphyrins-IX-dimethylester, cyclohexane, catalytic oxidation, air, synthesis

Introduction

Oxidation of cyclohexane to cyclohexanol and cyclohexanone is a very important industrial process. At the present time, the process is carried out without catalyst or using a cobalt salt catalyst in industry with a poor cyclohexane conversion of 3%~6%. Because of the limited cyclohexane conversion, people have been making a lot of efforts to discover better ways of cyclohexane oxidation to cyclohexanol and cyclohexanone and got some important achievements especially in the way of metalloporphyrin catalystic oxidation of cyclohexnae. These progresses mainly includes: (a) metalloporphyrins catalyzing cyclohexane oxidation with oxidants such as peracids, alkylhydroperoxides, hydrogen peroxide, iodosylbenzene (PhI0), hypochlorite. etc.. Its disadvantages are that metalloporphyrins are decomposed by oxidants, and there must be solvents and phase-transfer reagents in the reaction system; (b) metalloporphyrins catalyzing cyclohexane oxidation with air in the present of reductants[00]. This route consumes the same mole of redutants as products; (c) metalloporphyrins catalyzing cyclohexane oxidation with air in the present of electric current, or the illu-mination of light[0-0]. This production route possesses good selectivity and high conversion, but reaction conditions are very strict. (d) multihalogenmetalloporphyrins catalyzing cyclohexane oxidation directly with air[0-0]. But this way is very difficult to apply to industrial production because of the heavy price of multi-halogen-metalloporphyrins. Recently metallotetraporphyrins catalyzing cyclohexane oxidation with air in the absence of reductants were reported[00]. In the last reported paper[0] the cyclohexane conversion was 16.2% and the turnover number was 4×10^5 with simple cobalt tetraphenylporphyrin as catalyst and cyclohexane conversion was 10 times and turnover number was 27 times of those for unsupported cobalt porphyrin. The catalysts used in all the papers above are metallotetraporphyrins (MTPP) and metallodeuteroporphyrin-IX-dimethylesters (DP-IX-DME) have not been reported to catalyze cyclohexane oxidation or other hydrocabon oxidation.

In this paper, we reported simple deuteroporphyrin-IX-dimethylesters (DP-IX-DME) catalysing cyclohexane oxidation into cyclohexanol and cyclohexanone with air in the absence of any cocatalysts or reductants. We have previously studied the effects of reaction temperature and pressure on metallotetraphenylporphyrin-cobalt (TPPCo) catalyzing cyclohexane oxidation to cyclohexanol and cyclohexanone and applied the two conditions to the metallodeuteroporporphyrin-IX-dimethylesters (MDP-IX-DME) catalyzing cyclohexane oxidation. The optimum reaction temperature was 150℃, and the optimal reaction pressure is 0.8MPa. The catalysis of DP-IX-DMECu, DP-IX-DMECo, DP-IX-DMENi, DP-IX-DMEZn on cyclohexane oxidation with air were studied. Comparing with the reported data[45], the results were exciting with enhancing cyclohexane conversion from 10.48% to 18.6%, the results were exciting with enhancing cyclohexane conversion from 10.48% to 18.6%, the selectivity of cyclohexanol and cyclohexanone from 79.2% to 84.6%, and turnover number from 2.9×10^5 to 8.5×10^5. The cyclohexane conversion and turnover number is 16.9 times and 79 times respectively of those for reported unsupported-tetraphenylporphyrin-cobalt (Ⅱ) (TPPCo).

1 Experimental

1.1 Instruments and reagents

Metallodeuteroporphyrins were characterized by IR spectra with a Perkin-Elmer Model 783 IR spectrophotometer, MS with a finnigan TSQ QUANTUM ULTRA AM mass spectrometer and ^{1}HNMR with a Bruker-300NMR spectrometer in the solvent of $CDCl_3$. The products of cyclohexane catalytic oxidation were determined by GC-MS with a finnigan DSQ QUAN-TUM mass spectrometer and a PEG-2000 column (50m×0.5mm).

$Cu(AC)_2\cdot6H_2O$, $Co(AC)_2\cdot6H_2O$, $Ni(AC_3)_2\cdot6H_2O$, $Zn(AC_3)_2\cdot6H_2O$, $Fe_2SO_4\cdot7H_2O$, resorcinol, ether, methanol, dichloromethane, N, N-dimethylform-amide (DMF) are of AR. Hemin of 94% was purchased from Hainning Hetianlong Bio. Co., LTD. (Zhejiang, China).

1.2 Reaction equipment

The reactor was designed and manufactured by us. Its experimental equipment diagram is given in Fig. 1.

1.3 Synthesis of metallodeuteroporphyrin-IX-dimethylesters (MDP-IX-DME)

1.3.1 Synthesis and characterization of deut-eronporphyrin-IX-dimethylester (DP-IX-DME)

In this paper, MDP-IX-DME was synthesized from hemin **1**, and the general Process is shown as

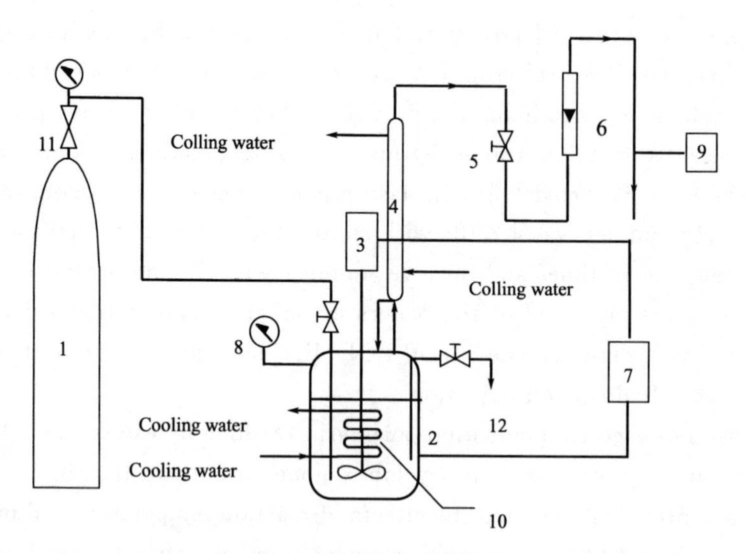

Fig. 1 Experimental equipment diagram

(1) air tank; (2) reactor; (3) agitator; (4) reflux condenser; (5) stainless steel needle valve; (6) gas flowmeter; (7) pressure controls; (8) pressure meter; (9) oxygen meter; (10) cooling coil; (11) reducing valve; (12) sample connection

scheme 1.

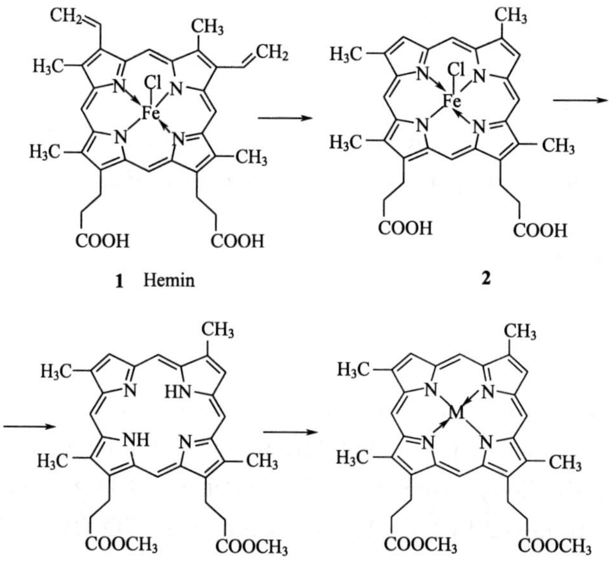

1 Hemin　　2

3　　M=Cu,Co,Ni,Zn

Scheme 1 The synthesis of Metal-deuteroporphyrin-dimethylesters

The synthesis of DP-IX-DME **3** was carried out according to the reference[43], and the yield was 85%. Structure characterization: IR (cm^{-1}, KBr): 3400 (m, -NH stretching vibration), 2900 (w, $—CH_3$, $—CH_2$ stretching vibration), 1730 (s, C=O stretching vibration), 1430 (m), 1360 (m), 1300 (w), 1195 (s), 1160 (s, C-O stretching vibration), 1105 (m), 980 (m), 920 (w), 840 (s), 790 (w), 725 (s), 710 (w), 675 (m), 640 (w). MS (ESI, MS2, Spray voltage: 3500v, Collision energy: 45v, m/z): 539 ($[M+H]^+$, 17%), 466 ($[M+H\text{-}CH_2COOCH_3]^+$, 100%), 465 ($[M\text{-}CH_2COOCH_3]^+$, 29%), 393 ($[M+H\text{-}2CH_2COOCH_3]^+$, 100%), 392 ($[M\text{-}2CH_2COO\ CH_3]^+$, 9%). $^1HNMR(CDCl_3)$: $\delta=-4.02$(s, 2H, NH), 3.26, 3.29, 3.31(2t, $^3J=$

7.5, 4H, CH_2CO_2R), 3.60(s, 6H, 13-, 17-OCH_3), 3.68~3.72(4s, 12H, 2-, 7-, 12-, 18-CH_3), 4.36, 4.39, 4.41(2t, ^3J=7.5, 4H, 13-, 17-$\underline{CH_2}CH_2CO_2R$), 9.05(s, 2H, 3-, 8-H), 9.96~10.03 (4s, 4H, 5-, 10-, 15-, 20-H).

1.3.2 Synthesis of metallodeuteroporphyrin-IX-dimethylesters (MDP-IX-DME)

Adding 0.49g (0.91mmol) DP-IX-DME **3** and 20mL DMF in a flask; adding 1mmol Cu(Ⅱ)/Co(Ⅱ)/Ni(Ⅱ)/Zn(Ⅱ)acetic salts respectively; heating the mixture under reflux for 0.5h with stirring then cooling it to room-temperature; recovering the solvent on a rotatory evaporator (70℃ with a circulating water vaccum pump) then taking out the residual solid and grinding it. Then wash the powder with 100mL deionized water through a Buchner funnel twice. Drying the filter cake to constant weight and getting MDP-IX-DME with the yield more than 98%. Structure characterzation: (a) DP-IX-DMECu: IR (cm^{-1}, KBr): 2914, 2880 (w, -CH_3, -CH_2), 1730 (s, C=O stretching vibration), 1435 (m), 1415 (w), 1385 (w), 1355 (m), 1320 (m), 1305 (w), 1280 (m), 1235 (m), 1202 (s, C-O stretching vibration), 1170 (m), 1140 (m), 1125 (m), 1115 (w), 1055 (w), 970 (w), 955 (w), 930 (w), 890 (m), 840 (s), 825 (s), 745 (m), 680 (w), 670 (w). MS (ESI, MS2, Spray voltage: 3500v, Collision energy: 15v, m/z): 599 ($[M]^+$, 19%), 526 ($[M+H-CH_2COOCH_3]^+$, 100%), 453 ($[M-2CH_2COOCH_3]^+$, 18%). ^{1}HNMR ($CDCl_3$): δ=3.66 (s). Here is only one pick because of para-magnetism of Cu^{2+}. (b) DP-IX-DMECo: MS (ESI, MS2, Spray voltage: 3500v, Collision energy: 16v, m/z): 595 ($[M]^+$, 7%), 522 ($[M-CH_2COOCH_3]^+$, 70%), 449 ($[M-2CH_2COOCH_3]^+$, 100%). ^{1}HNMR: 3.50 (s). (c) DP-IX-DMENi: MS (ESI, MS2, Spray voltage: 3500v, Collision energy: 16v, m/z): 594 ($[M]^+$, 5%), 521 ($[M-CH_2COOCH_3]^+$, 100%), 448 ($[M-2CH_2COOCH_3]^+$, 65%). ^{1}HNMR ($CDCl_3$): δ=3.41, 3.43, 3.45 (2t, ^3J=7.5, 4H, 13-, 17-$CH_2\underline{CH_2}CO_2R$), 3.53 (s, 6H, 13-, 17-$OCH_3$), 3.68~3.82 (4s, 12H, 2-, 7-, 12-18-CH_3), 4.62, 4.64, 4.66 (2t, ^3J=7.4, 4H, 13-, 17-$\underline{CH_2}CH_2CO_2R$), 9.27 (s, 2H, 3-, 8-H), 10.22, 10.41, 10.34, 10.30 (4s, 4H, 5-, 10-, 15-, 20-H); (d) DP-IX-DMEZn: MS (ESI, MS2, Spray voltage: 3500v, Collision energy: 16v, m/z): 600 ($[M]^+$, 42%), 527 ($[M-CH_2COOCH_3]^+$, 100%), 454 ($[M-2CH_2COOCH_3]^+$, 9%).

1.4 Cyclohexane oxidation with air catalyzed by metallodeuteroporphyrin-IX-dimethylesters

Adding 1000mL cyclohexane and 12mg MDP-IX-DME to the reactor, sealing the reactor and heating it to 150℃; inletting compressed air, stirring, and maintaining the reaction system pressure of 0.8MPa by controlling offgas with the needle valve. Recording the concentration of oxygen in offgas by the oxygenmeter on real-time. Stop the reaction when the concentration of oxygen in offgas stops to increase.

2 Results and discussion

2.1 The effects of catalysts on cyclohexane conversion

The products of cyclohexane oxidation catalyzed by MDP-IX-DME were measured by GC-MS. From GC-MS analysis data it could be found that the products were mainly cyclohexanol and cyclohexanone with small amounts of by-products such as acids, esters, polylols, and hyperoxide. Contrast experiment was operated under the same temperature and pressure in the absence of

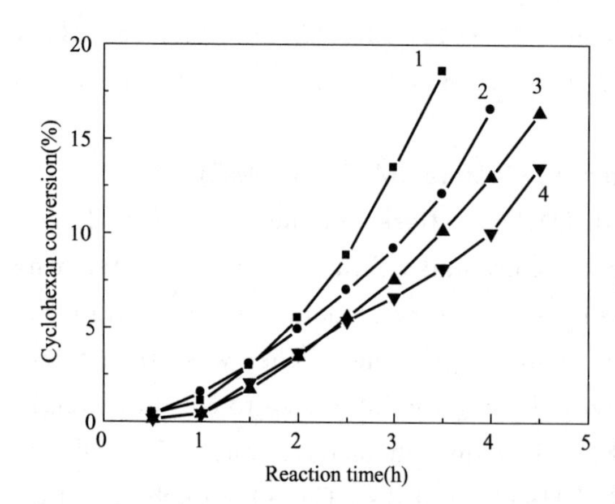

Fig. 2 The effects of catalysts on cyclohexane conversion. C cat. =12ppm, P=0. 8MPa. T=150℃

(1) DP-IX-DMECo; (2) DP-IX-DMECu; (3) DP-IX-DMEZn; (4) DP-IX-DMENi.

MDP-IX-DME, and the result showed that no oxidation of cyclohexane occurred. This indicated that metalloporphyrins acted catalysis during cyclohexane oxidation with air.

The curve of cyclohexane conversion for different MDP-IX-DME with time is shown in Fig. 2. From the Fig. 2 it could be seen that cyclohexane conversions for all catalysts were low within 1h, and then increased rapidly. This indicated that the reaction needs a stimulating phase. The sequence of cyclohexane conversion increasing speed for the four catalysts is DP-IX-DMECo> DP-IX-DMECu > DP-IX-DMEZn > DP-IX-DMENi. The curve for DP-IX-DMECo rose rapidly to 18. 6% within 3. 5h, The curve for DP-IX-DMECu rose to 16. 6% within 4h, the curve for DP-IX-DMEZn rose to 16. 4% within 4. 5h, and the curve for DP-IX-DMENi rose to 13. 52% within 4. 5h. Comparing with CSTPPFe (Ⅲ)[42] and CSTPP-Co[45], all DP-IX-DME in this paper had a higher cyclohexane conversion.

2. 2 The effects of catalysts on cyclohexanol and cyclohexanone selectivity

Fig. 3 shows how cyclohexanol and cyclohexanone selectivity changed with reaction time. It could be seen that cyclohexanol and cyclohexanone selectivity for each case decreased with time from a high point of about 95%. The selectivity dropped slowly to about 90% for a long reaction time from the beginning, but in the last thirty minutes it dropped rapidly to about 80%. For the instance of DP-IX-DMECo, cyclohexanol and cyclohexanone selectivity decreased from 94. 9% at 0. 5h to 90. 2% at 3h and then to 84. 6% at 3. 5h. This reflected that cyclohexanol and cyclohexanone selectivity decreased with time by an accelerated speed. This is because of that as the reaction proceeded, more and more cyclohexanol and cyclohexanone were generated and further oxidated to adipic acid and its esters. This point was proved by GC-MS analysis data.

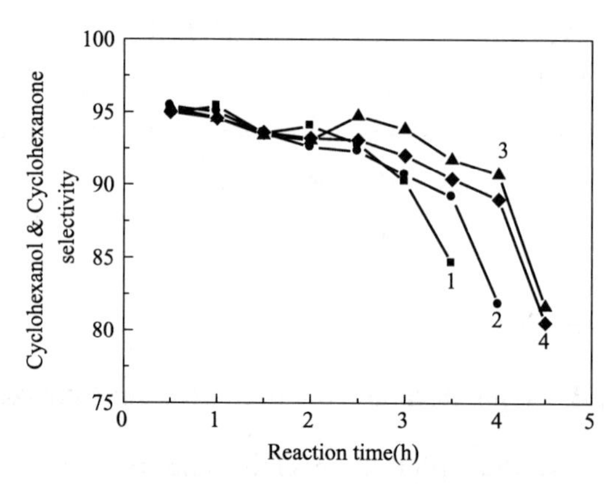

Fig. 3 The effects of catalysts on cyclohexanol and cyclohexanone selectivity. $C_{cat.}$ =12ppm, P=0. 8MPa. T=150℃ (1) DP-IX-DMECo; (2) DP-IX-DMECu; (3) DP-IX-DMEZn; (4) DP-IX-DMENi.

Among the four MDP-IX-DME, DP-IX-DMECo had the rapidest decreasing speed of cyclohexanol and cyclohexanone selectivity, so did DP-IX-DMECu, DP-IX-DMEZn and DP-IX-DMENi. However, in-consideration of the highest cyclohexane conversion, DP-IX-DMECo is still better than the others.

2. 3 The effects of catalysts on cyclohexanol and cyclohexanone turnover number

Cyclohexane turnover number is an important data to value the catalyst. It is expressed by the following equation:

Turnovernumber $= M_{cyclohexane\ converted}/M_{catalyst} \times 100\%$

Fig. 4 shows the curve of cyclohexane turnover number changing with time for different MDP-IX-DME. As shown in the Fig. 4, cyclohexane turnover number for all catalysts increased rapidly with time. Among the four cases, the cyclohexane turnover number for DP-IX-DMECo increased most rapidly followed by those for DP-IX-DMECu, DP-IX-DMEZn, DP-IX-DMENi consequently. The cyclohexane turnover number for DP-IX-DMECo was up to the maximum value of 85417 within 3h. Comparing with the cyclohexane turnover number for the oxidation of cyclohexane with air catalyzed by CSTPPCo (9.3×10^4 at 1.5h, after that it increased a little)[45], the lowest value for MDP-IX-DME was 9.5×10^4 (DP-IX-DME-Ni, 1.5h) but it increased constantly with reaction time and rose up to 6.2×10^5 at 4.5h.

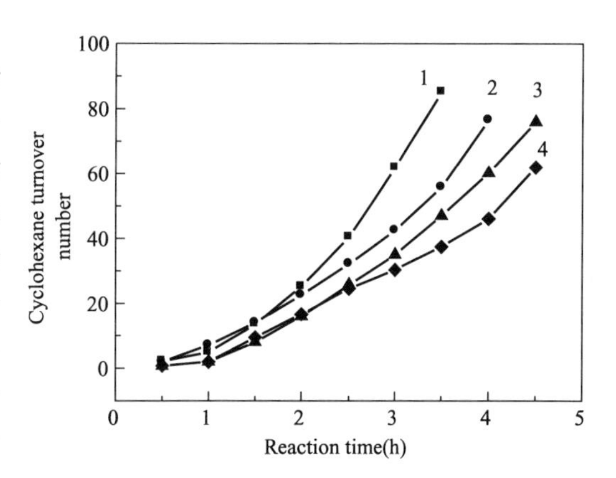

Fig. 4 The effects of catalysts on cyclohexane turnover number $C_{cat.}=12ppm$, $P=0.8MPa$. $T=150℃$

(1) DP-IX-DMECo; (2) DP-IX-DMECu; (3) DP-IX-DMEZn; (4) DP-IX-DMENi.

3 Conclusions

From the analysis above, we can draw two main conclusions: the first is that metallodeuteroporphyrin-IX-dimethylester is a better catalyst for aerobic cyclohexane to cyclohexanol and cyclohexanone than metallotetraphenylporphyrin. Metallodeuteropor-phyrin-IX-dimethylester can be easily and cheaply prepared by hemin. The second is that among the four metal-catalysts, DP-IX-DMECo have the best catalystic activity which obtained maximal cyclohexane conversion and turnover number in the least reaction time.

References

[1] A. K. Suresh, M. M. Sharma, T. Sridhar, Ind. Eng. Chem. Res. 39 (2000) 3958.
[2] B. Meunier, Chem. Rev. 92 (1992) 1411.
[3] Y. Iamamoto, Y. M. Idemori, S. Nakagaki, J. Mol. Catal. 99 (1995) 187.
[4] F. G. Doro, J. R. L. Smith, J. Mol. Catal. A: Chem. 164 (2000), 97-108.
[5] E. Porhiel, A. Bondon, Eur. J. Inorg. Chem. 5 (2000) 1097-1105.
[6] X. B. Zhang, C. Ch. Guo, J. B. Xu, J. Mol. Catal. A: Chem. 154 (2000) 31-38.
[7] C. -C. Guo, M. -D. Gui, S. -J. Zhu, Chin. J. Org. Chem. 14 (1994) 163.
[8] C. -C. Guo, Z. -P. Li, Chem. J. Chin. Univ. 18 (1997) 242.
[9] V. W. Bowny, K. U. Ingold, Biochem. Biophy. Res. Commun. 113 (1991) 5699.
[10] J. Haber, R. Iwanejko, T. M. Lodnicka, J. Mol. Catat. 55 (1989) 268-275.
[11] B. Meunier, Chem. Rev. 92 (1992) 1411-1456.
[12] P. Hoffmann, G. Labat, A. Robert, B. Meunier, Tetrahedron Lett. 31 (1990) 1991-1994.
[13] M. J. Nappa, C. A. Tolman, Inorg. Chem. 24 (1985) 4711-4719.
[14] J. T. Groves, W. J. Kruper, T. E. Nemo, R. S. Meyers, J. Mol. Catal. 7 (1980) 169-177.
[15] J. T. Groves, T. E. Nemo, R. S. Meyers, J. Am. Chem. Soc. 101 (1979) 1032-1033.
[16] P. Battioni, J. -P. Lallier, L. Barloy, D. Mansuy, J. Chem. Soc. , Chem. Commun. (1989) 1149-1151.
[17] Melo, M. Moraes, O. R. Nascimento, C. M. C. Prado, J. Mol. Catal. A: Chem. 109 (1996) 189-200.
[18] D. Mansuy, J. F. Bartoli, M. Momenteau, Tetrahedron Lett. 23 (1982) 2781-2784.

[19] C.-C. Guo，G. Huang，Z.-P. Li，J.-X. Song，J. Mol. Catal. A：Chem. 170 (2001) 43-49.
[20] C.-C. Guo，H.-P. Li，J.-B. Xu，J. Catal. 185 (1999) 345-351.
[21] A. N. de Sousa，M. E. M. D. de Carvalho，Y. M. Idemori，J. Mol. Catal. A：Chem. 169 (2001) 1-10.
[22] B. Meunier，W. Guilmet，M. E. de Carvalho，R. Poiblanc，J. Am. Chem. Soc. 106 (1984) 6668-6676.
[23] J. P. Collman，H. Tanaka，R. T. Hembre，J. I. Brauman，J. Am. Chem. Soc. 112 (1990) 3689-3690.
[24] L. M. Slaughter，J. P. Collman，T. A. Eberspacher，J. I. Brauman，Inorg. Chem. 43 (2004) 5198-5204.
[25] P. Battioni，J. P. Renaud，J. E. Bartoli，M. Reina-Artiles，M. Fort，D. Mansuy，J. Am. Chem. Soc. 110 (1988) 8462-8470.
[26] T. G. Traylor，S. Tsuchiya，Y. S. Byun，C. Kim，J. Am. Chem. Soc. 115 (1993) 2775-2781.
[27] F. S. Vinhado，C. M. C. Prado-Manso，H. C. Sacco，Y. Iamamoto，J. Mol. Catal. A：Chem. 174 (2001) 279-288.
[28] S. Campestrini，U. Tonellato，J. Mol. Catal. A：Chem. 171 (2001) 37-42.
[29] C. Poriel，Y. Ferrand，P. Le Maux，J. Rault-Berthelot，G. Simonneaux，Tetrahedron Lett. 44 (2003)：1759-1761.
[30] E. I. Karasevich，Y. K. Karasevich，Kinet. Catal. 41 (2000) 492.
[31] E. I. Karasevich，Y. K. Karasevich，Kinet. Catal. 41 (2000) 485.
[32] D. Mansuy，M. Fontecave，J. Chem. Soc.，Chem. Commun. (1983) 253.
[33] L. N. Ji，M. Liu，A. K. Hsieh，J. Mol. Catal. A 70 (1991) 247.
[34] P. Leduc，P. Battioni，J. F. Bartoili，Tetrahedron Lett. 29 (1988) 205.
[35] Maldotti，A. R. Bartocci，E. Polo，J. Chem. Soc.，Chem. Commun. (1991) 1487.
[36] K. Teramura，T. Tanaka，T. Yamamoto，J. Mol. Catal. A：Chem. 165 (2001) 299.
[37] K. Teramura，T. Tanaka，M. Kani，T. Hosokawa，T. Funabiki J. Mol. Catal. A：Chem. 165 (2001) 299 208 (2004) 299-305.
[38] J. E. Lyons，P. E. Ellis，H. K. Myers and R. W. Wagner，J. Catal. 141 (1995) 311-315.
[39] J. E. Lyons，P. E. Ellis Jr.，J. Catal. 155 (1995) 59-73.
[40] M. W. Grinstaff，M. G. Hill，Science 264 (1994) 1311.
[41] J. F. Bartoli，P. Battioni，W. R. Defoor，J. Chem. Soc.，Chem. Commun. (1994) 23.
[42] C.-C. Guo，M.-F. Chu，Q. Liu，Y. Liu，D.-C. Guo，X.-Q. Liu，Appl. Catal. A：General 246 (2003) 303-309.
[43] D.-S. Ma，B.-C. Hu，B. Cao，C. X. Lu，Chin. J. Appl. Chem. 9 (2006) 961-964.
[44] C.-C. Guo，G. Huang，X.-B. Zhang，D.-C. Guo，Appl. Catal. A：General 247 (2003) 261-267.
[45] C.-C. Guo，M.-F. Chu，Q. Liu，Y. Liu，D.-C. Guo，X.-Q. Liu，Appl. Catal. A：General 246 (2003) 303-309.

(注：此文原载于 Catalysis Communications，2009，10 (6)：781-783)

吡咯及二氢吡咯类化合物的合成研究进展

蔡超君，胡炳成，吕春绪
（南京理工大学化工学院 南京 210094）
SCI：000232416200028

摘要： 吡咯衍生物单体是一类重要的五元氮杂环化合物，用途非常广泛。根据母环氧化状态的不同，吡咯衍生物单体可分为吡咯、二氢吡咯以及四氢吡咯等三类化合物，该文综述了它们的合成方法及其合成研究进展情况。

关键词： 二氢吡咯，四氢吡咯，吡咯衍生物，合成

前言

吡咯衍生物单体是一类重要的五元氮杂环化合物，作为精细化工产品的重要中间体，在医药[1,2]、食品、农药、日用化学品[3]、涂料、纺织、印染[4]、造纸、感光材料、高分子材料等领域[5,6]有着广泛的用途。当前，天然环状四吡咯化合物的全合成研究是有机化学领域的一大热点[7]，其中一个非常重要的环节就是各种取代吡咯单体合成子的构筑，这些吡咯单体合成子获得的难易程度往往是决定反应路线是否合理的关键因素[8-10]。虽然人们开发出许多种吡咯衍生物的化学合成方法[11,12]，但是至今却鲜有文献较系统和全面地对其研究进展进行过报道。本文根据吡咯母环氧化状态的不同，将吡咯单体衍生物分为吡咯、二氢吡咯以及四氢吡咯等三类化合物，并在此基础上，对各类吡咯化合物的合成方法及其目前的进展情况作一综述。

1　吡咯环化合物的合成及进展

吡咯母环完全不饱和的吡咯单体主要用于卟啉类化合物的全合成[13-19]中。由于对它的合成主要有环化反应和环转化反应两类[20-24]，因而本文在此基础上又以其反应物类型的不同并结合其在各类四吡咯化合物合成中的地位对它们做进一步的分类说明。

1.1　环化反应进展

1.1.1　Knorr 合成

1884 年 Knorr 提出由 β-二羰基化合物与 α-氨基酮合成 3,5-二甲基-2,4-二乙酯基吡咯，从此以后大量的吡咯二酯通过此方法大规模地合成。许多其它的吡咯，尤其是用于卟啉合成中的吡咯，也是利用这一方法制备的，只不过改变不同的取代基而已。但是人们还是在此方法的基础上提出了许多新方法。例如，由 Knorr 合成方法人们尝试用 β-二羰基化合物与 α-氨基酸或 α-氨基腈发生反应，结果得到了 3 位羟基或氨基取代的吡咯衍生物，这是 Knorr 合成所不能得到的化合物。因此人们提出了许多类似于 Knorr 合成的方法，而且它们也在以后许多吡咯化合物单体的合成中得到了应用，如 1990 年

R. B. Woodward 等[25]在叶绿素 a 的全合成一文中，报道了叶绿素 a，卟啉和二氢卟酚[26]的各吡咯单体 A，B，C，D 的合成都是在以后的类似于 Knorr 合成反应的基础上得到的。我们在托尼卟吩的其中两个合成子的合成中也以 Knorr 反应为基础，经过溴化、甲酰化、脱羧等步骤生成我们所需要的托尼卟吩化合物的 C 环和 D 环。此法具有合成路线简捷，反应选择性高，产物收率高，有较高实用价值等优点。

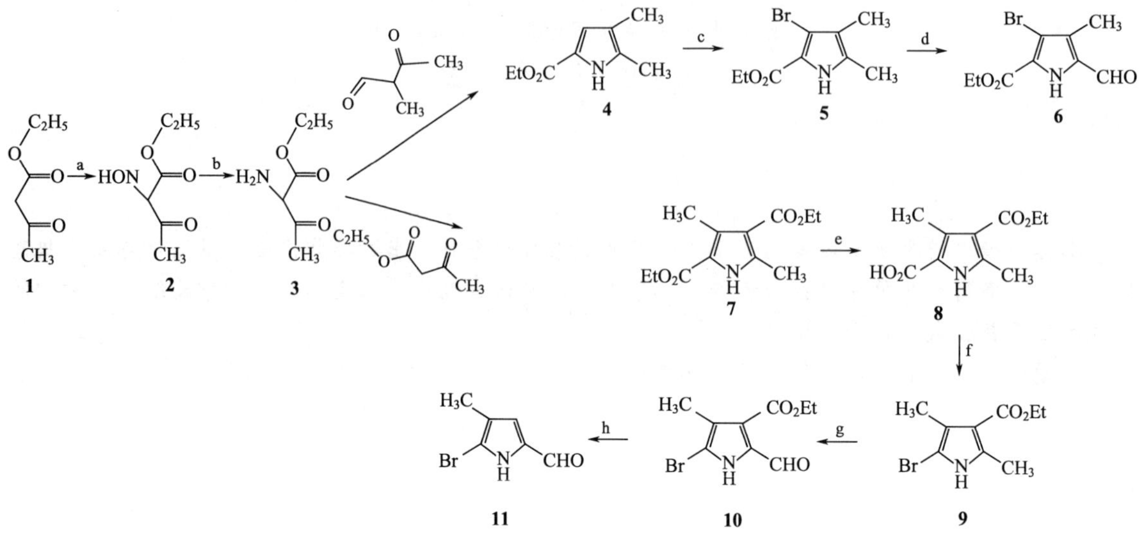

图 1 托尼卟吩中 C 环和 D 环的合成

Figure 1 Synthesis of Building Block for Ring C and Ring D of Tolyporphin

(a) $NaNO_2$，HOAc.（b）Zn，HOAc.（c）Br_2，HOAc.（d）SO_2Cl_2.（e）KOH.（f）Br_2.（g）Ce $(NH_4)_2$ $(NO_3)_6$，HOAc.（h）KOH

1.1.2 Paal-Knorr 合成

由于当吡咯环上有四个取代基时，它就具有抗菌、抗滤过性病原体、抗痉挛、抗氧化等药效，而通过 Paal-Knorr 合成能生成这种类型的吡咯化合物，其中当胺参与反应时所生成的是 N 上含有取代基的吡咯化合物。但由于这类环化合成的反应条件要求比较高（如反应中需通过加入沸腾的乙酸来延长反应时间等），并且对于不对称的 1,4-二羰基化合物较难生成所需的产物，针对上述情况 Minetto 等[27]提出了通过微波辐射促进 Paal-Knorr 反应的新方法，并且通过实验证明通过此方法能够解决以前人们所遇到的问题。其反应过程如下：

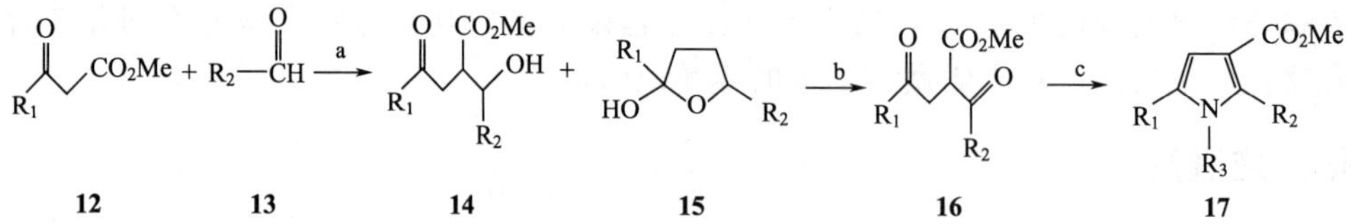

图 2 微波辐射条件下的 Paal-Knorr 合成

Figure 2 Microwave-Assisted Paal-Knorr Reaction

(a) Et_2Zn，CH_2I_2.（b）PCC.（c）R_3-NH_2；MW

1.1.3 烯胺参与的缩合反应

不对称的偶姻、酮酯类化合物以及氨或胺反应生成吡咯化合物的这类反应称为 Feist 合成。由于 2,3-二芳基类吡咯用 Feist 合成方法能得到接近 50%的产率，因而在有芳基取代的吡咯合成上应用较广泛，但研究者们不满足于此，为了克服其产率不高以及反应时间较长等缺点，他们在反应中以二价钐化物[28]为还原剂做了一系列反应，最后由实验结果证明二价钐化物能促进反应，提高吡咯类化合物的产率。

1.1.4 醛与硝基烷烃的反应

以前八乙基卟啉的合成中多是以2-乙酯基-3,4-二乙基 5-甲基吡咯为原料，而这种吡咯化合物则由乙酸丙酯与2,4-戊二酮通过 Knorr 合成反应而得，但由于反应的原料制备较为困难并且吡咯环上5位的甲基取代对后面的反应有较大影响，所以 Barton[29]、Ono[30]以及 Sessler[31]等人提出以3,4-二取代-吡咯-2-酯为反应原料，在此他们用醛与硝基乙烷在 DBU 催化下发生加成反应，而后再与乙酸酐发生酰化反应最后得到化合物**18**，化合物**18**与异氰基乙酸丁酯通过 Michael 加成以及成环反应而最后生成吡咯衍生物。为了后面反应的需要可以再通过碘化反应得到碘代吡咯。碘代吡咯在线性以及环状四吡咯化合物中发挥了至关重要的作用[32,33]。不过通过这种方法所得到的碘化产物**20**较少，因而最近 Rapoport 等[34]认为可以由苄酯通过氧化降解以及脱羧、碘化等反应得到的碘化产物**20**，虽然其路线相对较长，但其产量却能达到前者的近十倍。

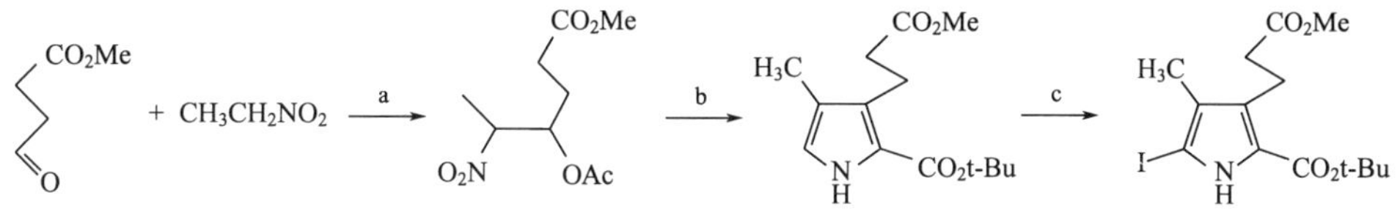

CO_2Me + $CH_3CH_2NO_2$ →(a) →(b) →(c)

18 **19** **20** **21** **22**

图3 醛与硝基乙烷的反应

Figure 3 The Reaction of Aldehyde and Nitroethane

(a) DBU；Ac_2O. (b) $CNCH_2CO_2$t-Bu；t-BuTMG. (c) NIS.

1.1.5 异腈的环化反应

由于异腈的环化反应能直接制备2或2，5位未取代的吡咯，因而成为了人们关注的热点。简单异腈 XCH_2NC，用碱脱质子后产生的负离子和不饱和亲电基团反应，形成中间体，该中间体通过5-endo-dig 过程，得到2位未取代的吡咯化合物。其中甲苯磺酰基甲基异氰（TOSMIC）的应用范围最广泛，因为它所需要的反应条件温和，而且甲苯磺酰基在合环后的芳化步骤中常常离去。近来有文献报道，用 TOSMIC 和市场上大量出购的芳烯烃只需经过一步反应就能得到3-芳基或3,4-二芳基吡咯，大大缩短了其合成路线[35]。

1.2 环转化反应进展

吡咯化合物的形成可以通过其它的环状化合物转化而成，其中比较常见的是通过五元环或六元环转化，下面我们就这两种类型各举一例加以说明。

对于五元环转换反应的例子，工业上的吡咯已经由呋喃与氨通过气相反应而制备，并且有大量文献及专利对其反应条件进行了描述。同样1位取代的吡咯衍生物也可以由呋喃、糠醛以及乙酰呋喃与伯胺反应得到。后来人们又进一步发现：在酸性条件下糠醛开环后与芳胺反应生成斯吞郝盐类，这类物质再闭环反应而得到吡咯化合物或3-羟基吡啶鎓盐。

在叶吩化合物（Bonellin）一合成子 D 环的合成中曾有人以六元环化合物环己烯为原料经过四步反应而得到吡咯化合物[36]。

同样，我们在叶绿素 a 的 D 环构造中，也以六元环化合物环己烯为原料首先在分子中引入三丁基锡甲基，然后再经过开环、缩合等几步反应最后得到我们的产物。其中哈格曼乙酯17是以乙酰乙酸甲酯和甲醛为原料合成的，在实验中我们发现在这一步骤中如果采用 DBU 为催化剂，用比理论量多10%的多聚甲醛代替甲醛，那么反应收率能从40%～48%提高到58%[37]。

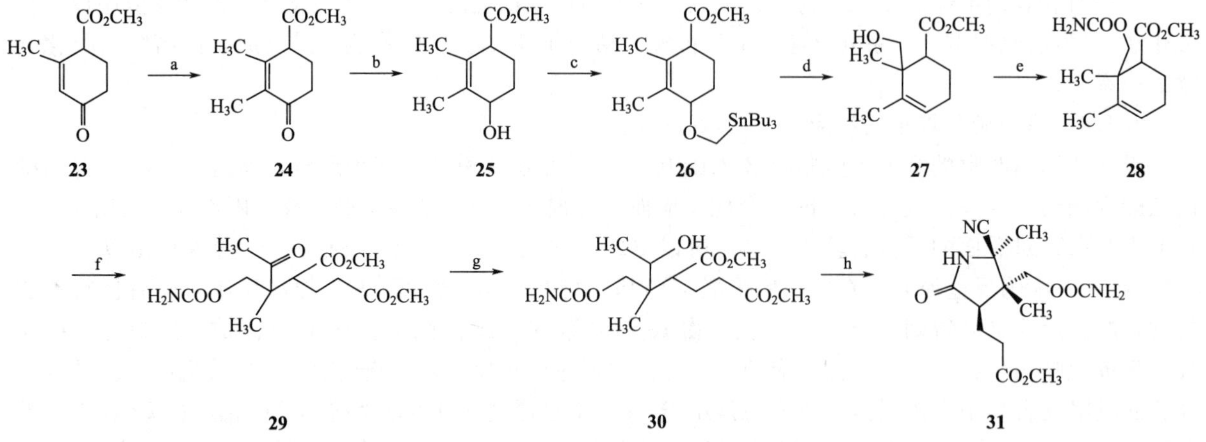

图 4 叶绿素 a 中 D 环的合成

Figure 4 Synthesis of Building Block for Ring D of Chlorophyll a

(a) Ethyether, NaH, ICH_3. (b) $LiAl[(CH_3)_3CO]_3H$, Ethylether; NaH_2PO_4.
(c) Bu_3SnCH_2I, KH, THF, DMPU; Chromatography. (d) LiBu, NH_4Cl. (e) BzlNHCOCl. (f) H_2O_2
(g) Zn, AcOH; (h) NaOEt, NH_3, KCN, MeOH

2 二氢吡咯环化合物的合成进展

在大分子化合物的全合成中，二氢吡咯化合物主要用作半饱和的卟吩等环状四吡咯化合物及植物色素、藻青素、藻红素等线性四吡咯化合物中的单体[38-43]。其中应用较多的主要有两类：二氢吡咯酮和 4,5-二氢化-3*H* 化合物，同样我们按照其反应物的不同进行分类说明。

2.1 二羰基化合物反应生成二氢吡咯酮

二氢吡咯酮经常成为许多四吡咯化合物中的合成子，由于其分子结构中含部分饱和的吡咯结构，可以使其吸光度得到提高从而改善四吡咯化合物的许多性能，所以它是研究者们较为关注的一类化合物。Robert[44]等在植物胆汁色素的 A 环的合成中由二羰基化合物通过加成转化为相应的亚硫酸酯羟腈至最后生成植物胆汁色素的 A 环（3,4-烷基取代的二氢吡咯酮），虽然在此过程中避免了大量使用无水 HCN 并且也使反应在羟腈还原转化阶段变得更为容易，但由于在反应中需引入的氰化物对人体及环境都造成伤害，于是我们在光敏剂托尼卟吩化合物全合成研究[45,46]中，对于其中的一合成子（3-甲基-2-吡咯酮）改用了其它方法：先由 Knorr 缩合生成完全不饱和吡咯化合物，然后再通过氧化及脱羧得到所需要的产物。此法不但不需以氰化物为反应原料，而且具有反应条件不苛刻，反应过程中副产物较少，产率较高等优点。

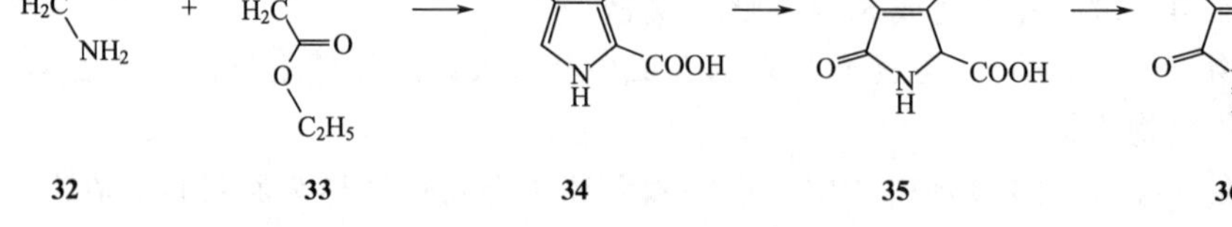

图 5 托尼卟吩 B 环的合成

Figure 5 Synthesis of Building Block for Ring B of tolyporphin

(a) $NaNO_2$, AcOH, Zn, H_2O. (b) H_2O_2. (c) NaOH, KI.

2.2 羰基化合物与氨基丙醛缩二甲醇化合物反应生成二氢吡咯酮

Jocabi[47]等在植物色素和藻青素等线性四吡咯化合物的合成中，二氢吡咯酮的制备是通过羰基化合物与二甲缩醛氨基丙酮经过几步反应生成二氢吡咯酮，其中此羰基化合物是由市场售价较便宜的丁内酯在对氯苯硒阴离子作用下开环得到。并且从图中可以看到吡咯酮的N原子上的取代基以及烷链上的对氯苯硒阴离子都可以在后期处理中将产物与TFA反应而脱去。这种方法由于其反应原料便宜，因而得到大量地推广。

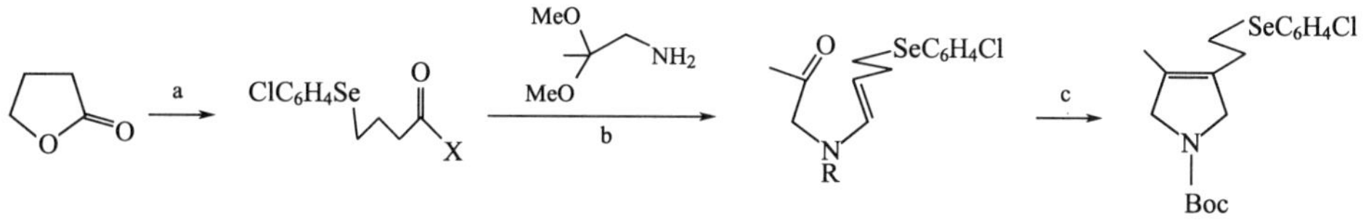

图6 吡咯酮的合成

Figure 6 Synthesis of pyrrolinone

(a) p-Cl-C_6H_4-Se^-；$(COCl)_2$. (b) NEt_3；H_3O^+. (c) $(BOC)_2O$；t-BuOK

2.3 环戊酮转化为二氢吡咯化合物

叶吩、细菌叶吩、咕啉等氢化卟啉化合物在自然界的生物过程中扮有十分重要的角色，尤其叶吩和细菌叶吩在癌症治疗的动力学疗法中发挥着重要作用。然而咕啉环的合成一直是广大科研工作者遇到的一大难题。据Johann Mulzer等[48]在维生素B_{12}的A-B-半咕啉环的合成一文中对A、B环的合成路线做了一新的尝试，即A、B环都来自于同一四氢吡咯烯酮，此四吡咯烯酮由乳酸酯通过一系列的反应转化而成。虽然取得了一定的成果，但人们并未停止对其中各吡咯单体合成的更进一步完善。4,5-二氢化-3H吡咯化合物作为钴啉氨酸的A环，Jacobi等大大地优化了其合成路线，在咕啉环的合成史上取得了又一重大飞跃。例如它与Eschenmoser等人在钴啉胺酸的全合成中合成D环所需要的22步反应相比较，4,5-二氢化-3H吡咯的合成从市场可购的2,3-二甲基-2-环戊酮出发只需11步即可[49]。

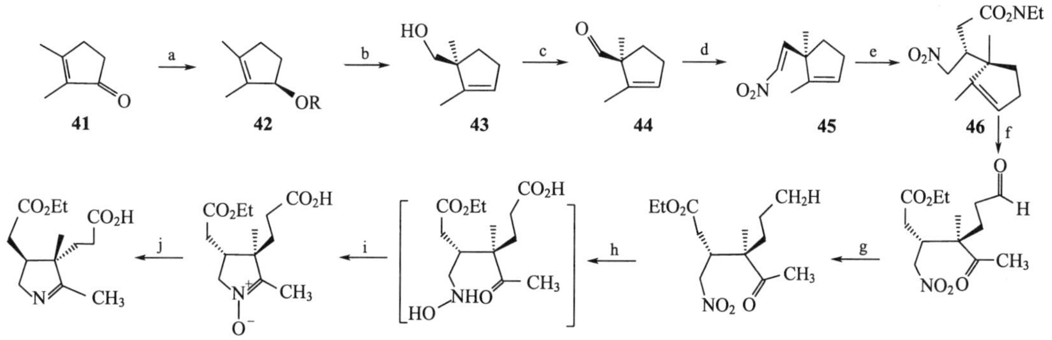

图7 钴啉胺酸的D环的合成

Figure 7 Synthesis of Building Block for Ring D of Cobyric Acid

(a) (S)-CBSB-butyloxazaborolidinecatalyst, 0.6equivBH_3.THF. (b) n-BuLi, THF. (c) SO_3.Py, DMSO, NEt_3, CH_2Cl_2. (d) 0.025mol NaOH, CH_3NO_2, CTACl. (e) NEt_3, MsCl, ethyl acetate. (f) ethyl acetate, LiHMDS, HMPA, THF. (g) O_3, PPh_3. (h) $NaClO_2$, 2, 3-dimethylbutene, t-BuOH, KH_2PO_4. (i) HCO-NH_4^+, Pd/C. (j) $TiCl_3$, THF, water, NaOAc.

3　四氢吡咯环化合物的合成进展

四氢吡咯化合物在卟吩、异菌卟吩以及咕啉等四吡咯化合物的环框架的构造中特别在咕啉环的合成上发挥着重要作用[50-52]，另外吡咯烷酮在医药、食品、日用化学品、涂料、高分子聚合等领域，在纺织、印染、造纸、感光材料、农畜牧业等方面也有许多用途。

3.1　烯酸酯反应生成四氢吡咯酮

Eschenmoser 等[53]曾由炔醇出发通过反应生成丙二酸乙酯最后得到四氢吡咯烯酮。由于其原料炔醇在市场上有大量出售，因而此方法一直受到人们的欢迎。而后来 Jacobi 等[54-56]人提出的由炔酸转化为二氢吡咯化合物的方法实际上与 Eschenmoser 等提出的由炔醇转化为四氢吡咯酮的方法是相同的机理。

同样以前托尼卟吩一合成子四氢吡咯酮的制备也一直都采用 Eschenmoser 等在可啉合成中提出的合成方法：以 2-(1,1-二甲基)-丙炔基丙二酸二乙酯为反应物，经过 8～10 步反应来实现，虽然此方法由于原材料易得而被得到广泛应用，但该法由于所用反应物分子中含有炔基，其合成存在有一定的难度，使整个合成路线缺乏实用价值。鉴于此，我们选用了较易获取的戊烯酸酯和羟基内酯等为原料来合成四氢吡咯酮[57]，实验表明其合成路线简捷，反应选择性较高，具有较高的实用价值。

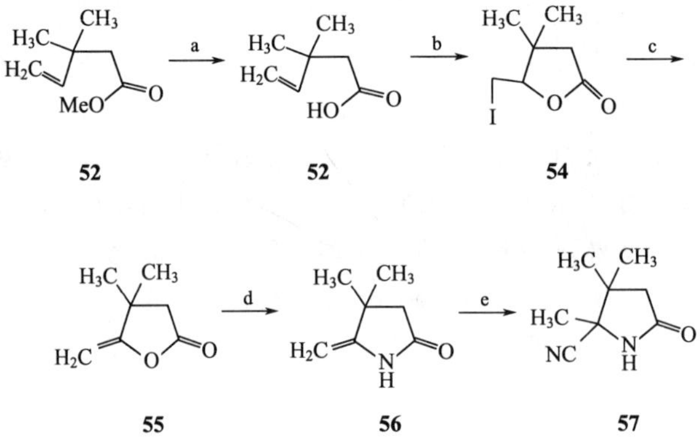

52　　52　　54

55　　56　　57

图 8　吡咯酮的合成

Figure 8　Synthesis of pyrrolinones

(a) KOH，H_2O/CH_3OH. (b) $NaHCO_3$，I_2，KI，H_2O. (c) DBU，Benzene.
(d) NH_3；Toluol. (e) KCN，$KHCO_3$，H_2O

3.2　环并内酰胺-内酯开环生成四氢吡咯酮

我们在探索托尼卟吩结构模型 A 环结构单元的合成方法时应用了此方法，实验证明其合成路线设计新颖且反应收率较高[58]。它是以 2,3-丁二酮和丙二腈为原料，经缩合、水解、氨化、氰基取代开环以及重氮甲烷酯化等反应合成出产物四氢吡咯酮

3.3　炔氨反应生成四氢吡咯酮

以前 Eschenmoser 等首先通过二酮胺环化然后再脱水转化为四氢吡咯烯酮。由于在最后脱水过程中要求温度在 150℃以上而导致四氢吡咯烯酮的分解，并且反应物二酮胺的合成也较为困难，于是 Jacobi 等在 Eschenmoser 的基础上提出了由炔氨以 n-Bu_4NF 为催化剂直接转化为四氢吡咯烯酮的方案并取得了较为满意的效果[59]。

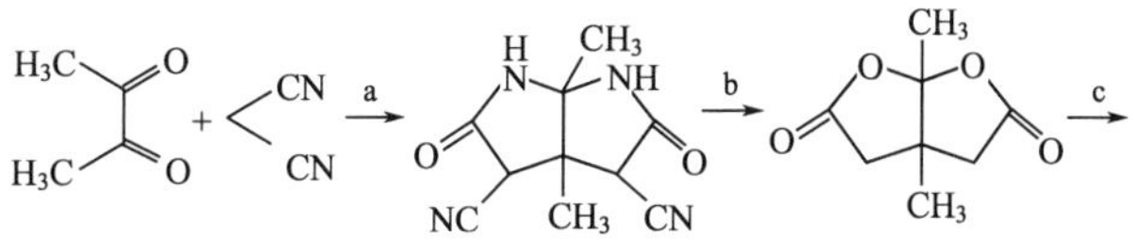

58 59 60 61

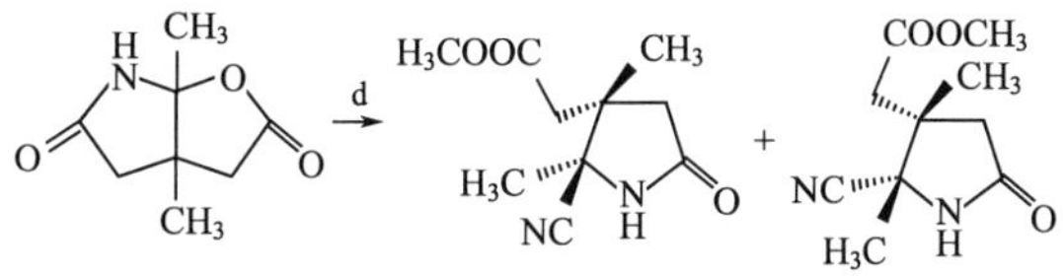

62 63 64

图 9 四氢吡咯酮的合成

Figure 9 Synthesis of tetrahydropyrrolinone

(a) EtOH，NaOEt；37% HCl.（b）48% HBr.（c）concentrated watery NH_3.

(d) KCN，MeOH；CH_2N_2，MeOH.

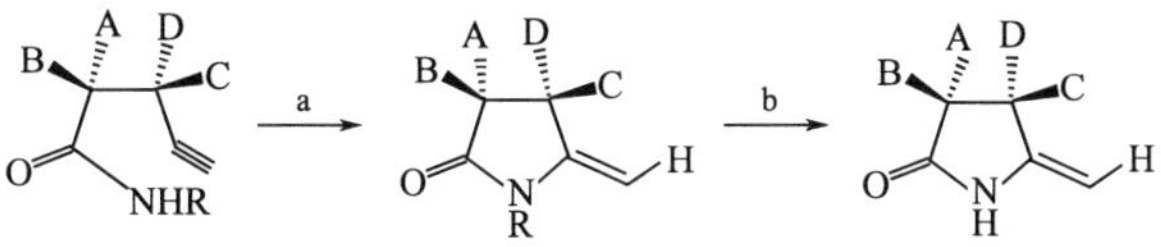

65 66 67

(1) A-D=H.（2）A，B=Me；C，D=H.（3）A，B=H；C，D=Me.

(4) A=H；B，C，D=Me.（5）A，C=H；B，D=Me

图 10 四氢吡咯酮化合物的合成

Figure 10 Synthesis of tetrahydropyrrolinone compound

(a) n-Bu_4NF.（b）Na，NH_3.

3.4 二乙酯氨化环合

由于2-硫代-5-氧代-四氢吡咯是维他命 B_{12} 等四吡咯化合物合成中两环连接时其合成子所经常要转化的形式。而我们经常是先通过生成二氢吡咯二酮然后再转化为2-硫代-5-氧代-四氢吡咯的方法来达到目的，其中二氢吡咯二酮也被统称为环B亚氨化合物，且人们对二氢吡咯二酮合成方法也日益走向了成熟。首先人们从腈类化合物着手经过酰化等反应合成二氢吡咯二酮，但由于产品得率较低而被放弃。接着人们又开始进一步的研究，以羧酸为反应原料逐步反应得到产物，虽然与上次相比取得了较大的进步但仍然存在着许多问题，如反应物和产物较难分离等。最后经过大量实验研究确定了从二乙酯出发经过四步反应最后得到产品[60]。

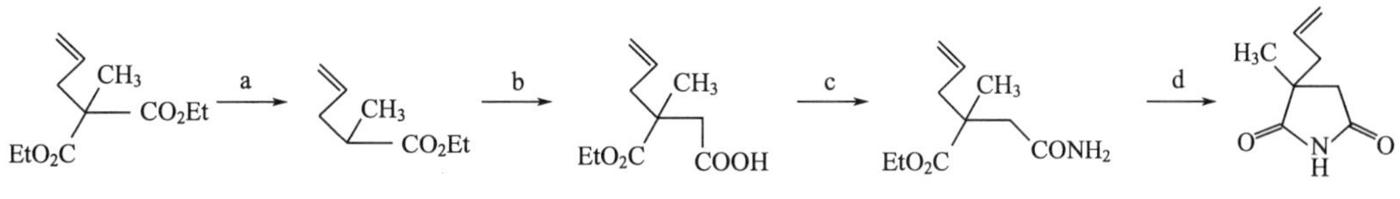

68 69 70 71 72

图 11 吡咯二酮的合成

Figure 11 Synthesis of pyrrolidione

(a) KOH.（b）XCH_2COOH.（c）$IBCF/NH_3$.（d）n-Bu_4NF.

4 展望

随着现代实验条件的改善以及光谱技术等检测手段的不断完善，自然界中更多的含吡咯环的大分子化合物被发现，其在医药、生物、食品等领域的用途也越来越多地被人们意识，同时含取代基的各种吡咯单环化合物在人类生活中也扮演着非常重要的角色，因而它无论是作为中间体还是作为其本身，其合成方法的研究都将成为研究者们所关注的一个重点。另外随着人们对天然产物性能的进一步改良，对模拟天然产物活性的多吡咯化合物的制备等都不断地迫切需要人们合成含各种取代基的吡咯衍生物，并使之工业化也是广大合成工作者们努力的方向。所以随着对各吡咯衍生物合成研究的不断深入，必将推动天然产物化学、材料化学、药物化学以及能源技术等的向前发展，同时也将不断丰富有机合成方法学的内容。

参考文献

[1] Ortega, H. G.; Crusats, J.; Feliz, M.; Ribo, J. M. Modulation of the porphyrin inner proton exchange rates by the steric effects of bridge substitution [J]. J. Org. Chem. 2002, 67, 4170.

[2] Yu, S.; Saenz, J.; Srirangam, J. K. Facile synthesis of N-aryl pyrroles via Cu (Ⅱ)-mediated cross coupling of electron deficient pyrroles and arylboronic acids [J]. J. Org. Chem. 2002, 67, 1699.

[3] Bullington, J. L.; Wolff, R. R.; Jackson, P. F. Regioselective preparation of 2-substituted 3, 4-diaryl pyrroles: A concise total synthesis of ningalin B [J]. J. Org. Chem. 2002, 67, 9439.

[4] Lee, D.; Swager, T. M. Defining space around conducting polymers: Reversible protonic doping of a canopied polypyrrole [J]. J. Am. Chem. Soc. 2003, 125, 6870.

[5] Azioune, A.; Ben Slimane, A.; Ait Hamou, L.; Pleuvy, A.; Chehimi, M. M.; Perruchot, C.; Armes, S. P. Synthesis and characterization of active ester-functionalized polypyrrole-silica nanoparticles: application to the covalent attachment of proteins [J]. Langmuir 2004, 20, 3350.

[6] Shenoy, S. L.; Cohen, D.; Erkey, C.; Weiss, R. A. A solvent-free process for preparing conductive elastomers by an in situ polymerization of pyrrole [J]. Ind. Eng. Chem. Res. 2002, 41, 1484.

[7] 胡炳成，吕春绪，刘祖亮. 天然环状四吡咯化合物的合成研究进展 [J]. 有机化学，2004, 24, 270.

[8] Neya, S.; Funasaki, N. meso-Tetra (tert-butyl) porphyrin as a precursor of porphine [J]. Tetrahedron Lett. 2002, 43, 1057.

[9] Naik, R.; Joshi, P.; Kaiwar S. P.; Deshpande R. K. Facile synthesis of meso-substituted dipyrromethanes and porphyrins using cation exchange resins [J]. Tetrahedron 2003, 59, 2207.

[10] Cammidge A. N.; Ozturk, O. Selective Synthesis of m eso-Naphthylporphyrins [J]. J. Org. Chem. 2002, 67, 7457.

[11] Santra, S.; Kumaresan, D.; Agarwal N.; Ravikanth N. cis-Pyridyl core-modified porphyrins for the synthesis of cationic water-soluble porphyrins and unsymmetrical non-covalent porphyrin arrays [J]. Tetrahedron, 2003, 59, 2353.

[12] Li, Z.; Xia, C. G. Epoxidation of olefins catalyzed by manganese (Ⅲ) porphyrin in a room temperature ionic liquid [J]. Tetrahedron Lett. 2003, 44, 2069.

[13] Taniguchi, S.; Hasegawa, H.; Yanagiya, S.; Tabeta, Yusuke S. The first isolation of unsubstituted porphyrinogen and unsubstituted 21-oxaporphyrinogen by the '3+1' approach from 2, 5-bis (hydroxymethyl) pyrrole and tripyrrane derivatives [J]. Tetrahedron 2001, 57, 2103.

[14] Fox, S.; Hudson R.; Boyle, R. W. Use of orthoesters in the synthesis of meso-substituted porphyrins [J]. Tetrahedron Lett. 2003, 44, 1183.

[15] Takase, M.; Ismael, R.; Murakami, R.; Ikeda, M.; Kim, D.; Shinmori, H.; Furuta, H.; Osuka, A. Efficient synthesis of benzene-centered cyclic porphyrin hexamers [J]. Tetrahedron Lett. 2002, 43, 5157.

[16] Shinso, M.; Fukuda, Y.; Shishido, K. The efficient entry into the tricyclic core of halichlorine [J]. Tetrahed-

ron Lett. 2000, 41, 929.
[17] Ravikamth, M. Tetrahedron Lett. 2000, 41, 3709.
[18] Lintuluoto, J. M.; borovkov, V. V.; Inoue, Y. Tetrahedron Lett. 2000, 41, 4781.
[19] Santander, C. P.; Scott, A. I. Tetrahedron Lett. 2002, 43, 6967.
[20] Gryko, D. T.; Jadach, K. A Simple and Versatile One-Pot Synthesis of m eso-Substituted trans-A2B-Corroles [J]. J. Org. Chem. 2001, 66, 4267.
[21] Alvarez, A.; Guzman, A.; Ruiz, A.; Velarde, E. Synthesis of 3-arylpyrroles and 3-pyrrolylacetylenes by palladium-catalyzed coupling reactions [J]. J. Org. Chem. 1992, 57, 1653.
[22] Bishop, J. E.; Connell, J. F.; Rapoport, H. The reaction of thioimides with phosphorus ylides [J]. J. Org. Chem. 1991, 56, 5079.
[23] Jacobi, P. A.; Brielmann, H. L.; Hauck, S. I. ynthesis of cyclic enamides by intramolecular cyclization of acetylenic amides [J]. Tetrahedron Lett. 1995, 36, 1193.
[24] Gilchrist, T. L. Heterocyclic Chemistry, Vol 1, Ed.: Lanzhou University Press, p. 159.
[25] Woodard, R. B.; Ayer, W. A.; Beaton, J. M.; Bickelhaupt, F.; Boonnett, R.; Buchschacher, P.; Closs, G. L.; Dutler, H.; Hannah, J.; Hauck, F. P.; Ito, S.; Langemann, A.; Legoff, E.; Leimgrubet, W.; Lwowski, W.; Sauer, J.; Valenta, Z.; Volz, H. Tetrahedron 1990, 46, 7599.
[26] Shenoy, S. L.; Cohen, D.; Erkey, C.; Weiss, R. A.; Ind. Eng. Chem. Res. 2002, 41, 1484.
[27] Minetto, G.; Raveglia, L. F.; Taddei, M. Organic Lett. 2004, 6, 389.
[28] Farcas, S.; Namy, J. L. Tetrahedron 2001, 57, 4881.
[29] Barton, D. H. R.; Kerbagore, J.; Zard, S. Tetrahedron 1990, 46, 7587.
[30] Ono, N.; Kawamura, H.; Bougauchi, M. Tetrahedron 1990, 46, 7483.
[31] Sessler, J. L.; Mozattari, A.; Johnson, M. Org. Synth. 1991, 10, 68.
[32] Tang, J. S.; Verkade, J. G. J. Org. Chem. 1994, 59, 7793.
[33] Jacobi, P. A.; Guo, J. S.; Rajeswari, S.; Zheng, W. J. J. Org. Chem. 1997, 62, 2907.
[34] Bishop, J. E.; O'Connell, J. F.; Rapoport, H. J. Org. Chem. 1991, 56, 5079.
[35] Smith, N. D.; Huang, D. H.; Cosford N. D. P. Org. Lett. 2002, 4, 3537.
[36] Montforts, F. P.; Gerlach, B.; Hoeper, F. Chem. Rev. 1994, 94, 327.
[37] Hu, B. -C; Lv, C. -X; Liu, Z. -L. Chin. J. Appl. Chem. 2003, 20 (10), 1012.
[38] Zhao, K. -H.; Ran, Y.; Li, M.; Sun, Y. -N.; Zhou, M.; Storf, M.; Kupka, M.; Bohm, S.; Bubenzer, C.; Scheer, H. Biochemistry 2004, 43, 11576.
[39] Tu, S. -L.; Gunn, A.; Toney, M. D.; Britt, R. D.; Lagarias, J. C. J. Am. Chem. Soc. 2004, 126, 8682.
[40] Tu, B.; Ghosh, B.; Lightner, D. A. J. Org. Chem. 2003, 68, 8950.
[41] Taniguchi, M.; Ra, D.; Mo, G.; Balasubramanian, T.; Lindsey, J. S. J. Org. Chem. 2001, 66, 7342.
[42] Kretzer, R. M.; Ghiladi, R. A.; Lebeau, E. L.; Liang, H.-C.; Karlin, K. D. Inorg. Chem. 2003; 42, 3016.
[43] Krayushkin, M. M.; Yarovenko, V. N.; Semenov, S. L.; Zavarzin, I. V.; Ignatenko, A. V.; Martynkin, A. Y.; Uzhinov, B. M. Org. Lett. 2002, 4.
[44] Schoenleber, R. W.; Kim, Y.; Rapoport, H. J. Am. Chem. Soc. 1984, 106, 2645.
[45] Hu, B.-C. Ph. D. Thesis, Nanjing University of Science and Technology, Nanjing, 2004 (in Chinese).
[46] Hu, B.-C.; Lv, C.-X. Chin. J. Appl. Chem. 2004, 21 (10).
[47] Jacobi, P. A.; DeSimone, R. W.; Ghosh, I.; Guo, J.-S.; Leung, S. H.; Pippin, D. J. Org. Chem. 2000, 65, 8478.
[48] Mulzer, J.; List, B.; Bats, J. W. J. Am. Chem. Soc. 1997, 119, 5512.
[49] Mulzer, J.; Riether, D. Org. Lett. 2000, 2, 3139.
[50] Jacobi, P. A.; Lanz, S.; Ghosh, I.; Leung, S. H.; Lolwer, F.; Pippin, D. Org. Lett. 2001, 3, 831.
[51] Stevens, R. V.; Chang, J. H.; Lapalme, R.; Schow, S.; Schlageter, M. G.; Shapiro, R.; Weller, H. N. J. Am. Chem. Soc. 1983, 105, 7719.
[52] Stolowich, N. J.; Wang, J.; onathan Spencer, J. B.; atricio Santander, P. J.; Roessner, C. A.; Scott, A.

I. J. Am. Chem. Soc. 1996, 118, 1657.
[53] Montforts, F. P. ; Schwartz, U. M. liebigs Ann. Chem. 1985, 1228.
[54] Jacobi, P. A. ; Li, Y. Org. Lett. 2003, 5, 701.
[55] Jacobi, P. A. ; Liu H. J. Org. Chem. 1999, 64, 1778.
[56] Jacobi, P. A. ; Liu H. J. Org. Chem. 1999, 121, 1958.
[57] Hu, B.-C. ; Lv, C.-X. ; Liu, Z.-L. Hua Xue Tong Bao 2004, 67 (4), w028.
[58] Hu, B.-C. ; Lv, C.-X. ; Liu, Z.-L. Chin. J. Org. Chem. 2004, 24, 697.
[59] Jacobi, P. A. ; Brielmann, H. L. ; Hauck, S. I. J. Org. Chem. 1996, 61, 5013.
[60] Battersby, A. R. ; Westwood, S. W. J. Chem. Soc. Perkin Tran. 1. 1987, 50, 1679.

Recent progress in the study on Synthesis of Pyrrole and Dihydropyrrole Compounds

Cai Chaojun, Hu Bingcheng, Lv Chunxu

(School of Chemical Engineering, Nanjing University of Science and Technology, Nanjing 210094)

Abstract: In this paper the various methods of forming pyrrole, dihydropyrrole and pyrrolidine compounds were reviewed.

Keywords: dihydropyrrole, pyrrolidine, pyrrole derivative, synthesis

(注：此文原载于 有机化学，2005，25 (10)：1311-1317)

化学还原新体系制备负载 Ni-B 非晶态催化剂及其催化加氢性能

孙昱，吴秋洁，李斌栋，吕春绪，邹祺
南京理工大学化工学院，南京，210094

摘要： 采用水合肼和硼氢化钾为共还原剂，适当比例的甲醇、乙醇和水的混合物为溶剂，制备得到负载催化剂 P1。经过 TEM、XRD、SAED 表征结果表明，该催化剂是一种纳米级非晶态合金。将 P1 催化剂用于邻氯硝基苯加氢反应中，并与常规方法所制备的非晶态催化剂 P2 的加氢性能进行了比较，P1 催化剂的活性和选择性略大于 P2 催化剂，但制备 P1 催化剂所用的硼氢化钾的量仅相当于制备 P2 催化剂的 50%。讨论了水合肼、硼氢化钾、pH 值和溶剂对催化剂性能的影响。

关键词： 共还原剂，非晶态合金，催化加氢

非晶态合金催化剂因其特殊的物理化学性质及其在众多催化反应中表现出的优异活性和选择性而愈来愈受到人们的重视[1-3]。负载 Ni-B 非晶态合金具有比表面积大、催化活性高和热稳定性好的优点而备受青睐。目前，负载 Ni-B 非晶态合金的主要制备方法为浸渍还原法，其基本过程是在冰浴条件下，把硼氢化钾水溶液滴加到浸渍好的催化剂前驱体上，此方法操作复杂、均匀搅拌困难、催化剂制备的可重现性差，难于大规模工业化生产。况且，在还原催化剂前驱体时，相当一部分硼氢化钾被无效水解[4]，导致催化剂制备成本上升。本文采用硼氢化钾和水合肼为共还原剂，以适当比例的甲醇、乙醇和微量的水为溶剂制备非晶态催化剂，并将其用于邻氯硝基苯的催化加氢。研究发现，与常规方法所制备的非晶态催化剂相比，该法制得的催化剂催化活性高、催化剂可重现性好，并且具有操作简单、催化剂制备成本低、易于大规模生产的特点。

1 实验部分

1.1 试剂与仪器

$NiCl_2 \cdot 6H_2O$、KBH_4、CH_3CH_2OH、CH_3OH 等为分析纯试剂，$NH_2NH_2 \cdot H_2O$（85%）为化学纯，邻氯硝基苯为工业纯，硅胶为青岛海洋化工厂提供的工业品。

采用 JA1100 型电感耦合等离子直读光谱仪（ICP）分析负载催化剂中组分；德国 Brucker 公司 D8 转靶 X 射线衍射仪（XRD）分析催化剂的结构；JEOL 2011 型透射电镜（TEM）观察催化剂的形貌；同时利用选区电子衍射（SAED）确定晶相结构。用福立 GC9790 型气相色谱仪（SE 30m×0.53mm×1μm，FID 检测器）分析加氢产物各组分的含量，分析条件为：N_2 气为载气，柱温 80℃，峰面积归一法确定含量。

1.2　催化剂制备

在烧杯中加入 2gSiO_2（BET 比表面积为 198m^2/g）和所需要的 $NiCl_2$ 水溶液（理论负载量为 20%），浸渍过夜，343K 水浴搅拌干燥，373K 烘 6h，在 473K 下焙烧 4h 得催化剂前驱体。在置于 60℃水浴中的四口烧瓶中依次加入 13mL 甲醇、2mL 乙醇、3mL 水合肼（$n(NH_2NH_2 \cdot H_2O)/n(Ni^{2+})=9.1$）、0.05gNaOH、0.4mL 水和 0.4g 硼氢化钾（$n(KBH_4)/n(Ni^{2+})=1.25$），搅拌溶解。称取一定量烘干后的催化剂前驱体迅速加入到四口烧瓶中还原，搅拌一定时间后得到的 Ni-B/SiO_2 样品用去离子水反复洗涤，再用无水乙醇洗涤 3 次，保存在无水乙醇中备用。所制备的催化剂记为 P1。

常法制备负载型催化剂按照文献[5]方法进行，将处理好的催化剂前驱体置于烧杯中，用 1.25 倍于载体体积的一定浓度的 KBH_4 溶液（$n(KBH_4)/n(Ni^{2+})=2.5$）逐滴加入，反应迅速进行，放出大量气体，载体逐渐变黑。待无气体产生时，用去离子水反复洗涤，再用无水乙醇洗涤 3 次，保存在无水乙醇中备用。所制备的催化剂记为 P2。

1.3　催化剂活性评价

在 500mL 的高压釜中依次加入湿催化剂 2g、邻氯硝基苯 2g 和无水乙醇 100mL，通入 H_2 气 4 次以除去空气，开启搅拌器，搅拌速度 1000r/min，将高压釜缓慢加热到 363K，通入 H_2 至 1.0MPa。通过观察和记录高压釜内压强随时间的变化监测加氢反应进程，反应 0.5h，抽出釜内溶液，过滤出催化剂，用气相色谱仪分析滤液。

2　结果与讨论

2.1　P1 催化剂表征结果

ICP 分析 P1 型催化剂得出 Ni-B 催化剂中的 B 与 Ni 的质量比为 11∶89，Ni 的负载量为 18%（质量分数）。XRD 分析结果见图 1，图 1 中显示，在 2θ 为 22°有 1 个衍射峰，此峰为无定型的 SiO_2，位于 2θ 为 45°附近还有 1 个弥散峰，为 Ni 基非晶态合金的特征衍射峰[6]。选区电子衍射（SAED）表征见图 2。图 2 中显示，新鲜样品的电子衍射谱主要是非晶衍射漫散环图，进一步证明了催化剂的非晶态结构[7]。催化剂样品的 TEM 照片见图 3。从图 3 中可看出，新鲜样品为球形颗粒状，分散较好，颗粒直径均在 50nm 左右，无团聚现象。

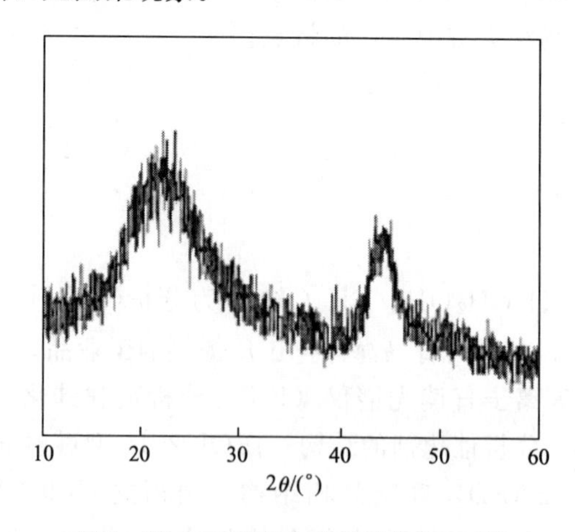

图 1　Ni-B 非晶态合金样品的 XRD 谱

Fig. 1　XRD pattern of Ni-B amorphous alloy sample

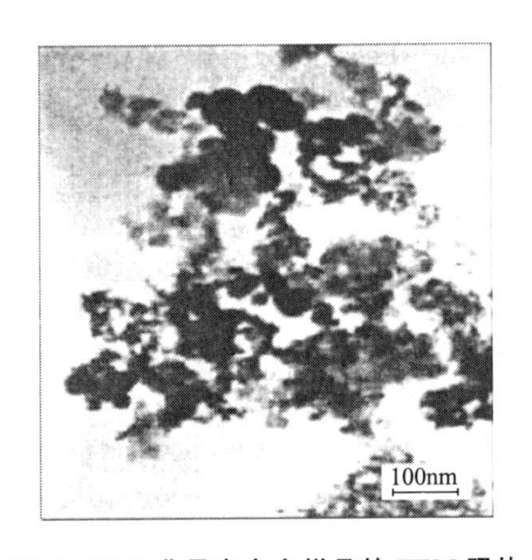

图2 Ni-B非晶态合金样品的TEM照片

Fig. 2 TEM image of Ni-B amorphous alloy sample

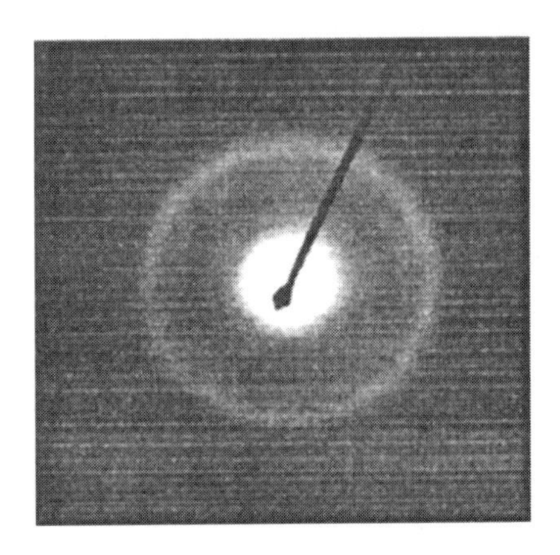

图3 Ni-B非晶态合金样品的SAED照片

Fig. 3 SAED image of Ni-B amorphous alloy sample

2.2 催化剂活性评价

催化剂P1和P2催化加氢邻氯硝基苯制备邻氯苯胺的结果见表1。

表1 催化剂和催化加氢邻氯硝基苯制备邻氯苯胺

Table 1 Hydrogenation of *o*-chloronitrobenzene to *o*-chloroaniline over catalysts P1 and P2

Catalyst	Selectivity of *o*-CAN/%	Conversion of *o*-CNB/%
P1	94.3	100
P2	93.9	98.7

从表1可以知道，P1催化剂的活性稍大于P2催化剂，选择性大致相当，说明P1与P2的性质基本相似，但在制备P1催化剂时硼氢化钾的量相比于镍的量的摩尔比为1.25，仅相当于P2催化剂所用的硼氢化钾量的50%。研究[4]表明，在水溶液中制备非晶态催化剂硼氢化钾和镍的摩尔比必须大于1.8～1.9，否则金属离子难以还原完全，因为相当一部分硼氢化钾被水解，由于硼氢化钾的价格较高，导致催化剂制备成本上升。在水合肼、硼氢化钾共还原剂体系中，水合肼起到了还原剂的作用，文献[8,9]表明，水合肼在还原制备纳米级镍催化剂时是一种优良的还原剂。

2.3 催化剂制备因素分析

2.3.1 水合肼的添加量对催化剂性能的影响

在其它反应条件与制备P1时一致的情况下，讨论水合肼的添加量对催化剂性能的影响，结果见图4。图中可见，随着水合肼含量的增加，催化剂的活性和选择性均增加，当$n(NH_2NH_2 \cdot H_2O)/n(Ni^{2+})=9.1$时，邻氯硝基苯的转化率和邻氯苯胺的选择性分别达到100%和94.30%。再增加水合肼的量时，催化剂的活性不再增大，选择性略微下降。这可能与形成的催化剂粒子大小有关。文献[10]报道用水合肼还原负载Ni^{2+}形成负载纳米镍催化剂时，增加水合肼的量，会使镍的成核速率增大，当成核速率远远大于生长速率时，催化剂颗粒来不及长大，形成的催化剂颗粒也就越小，当$n(NH_2NH_2 \cdot H_2O)/n(Ni^{2+})>10$时，催化剂的颗粒度达到一常数，不再随水合肼的用量增加而变小。本文实验结果表明，当水合肼的量达到$n(NH_2NH_2 \cdot H_2O)/n(Ni^{2+})=9.1$时，可能催化剂的颗粒达到最小值，因而活性达到最高，再多增加水合肼的量无益。

2.3.2 硼氢化钾的量对催化剂性能的影响

在其它反应条件与制备P1时一致的情况下，讨论还原体系中硼氢化钾的量对催化剂性能的影响，结果见图5。图中可见，随着硼氢化钾量的增加，催化剂的活性增加，邻氯苯胺的选择性呈上升趋势，当$n(KBH_4)/n(Ni^{2+})=1.25$时，催化剂的活性和选择性均达最佳。继续增加硼氢化钾的量，催化剂活性保持不变，但催化剂的选择性有所下降。硼氢化钾在还原 Ni^{2+} 时能够产生一部分硼，硼在催化剂形成非晶态过程中起到重要作用，还能够供电子给金属 Ni，使催化剂的活性和选择性增大。

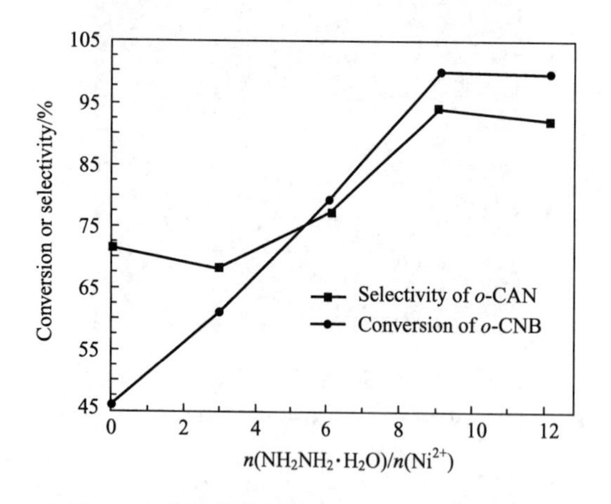

图 4　水合肼的添加量对催化剂性能的影响

Fig. 4　Influence of hydrazine hydrate amount on catalytic performance

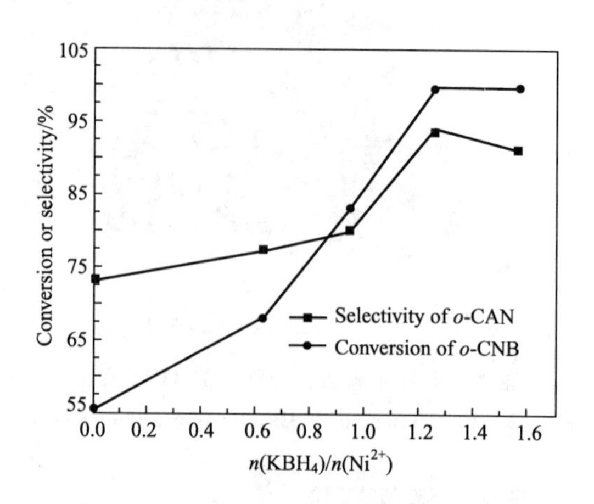

图 5　硼氢化钾的量对催化剂性能的影响

Fig. 5　Influence of potassium borohydride amount on catalytic performance

2.3.3　氢氧化钠的用量对催化剂性能的影响

在其它反应条件与制备 P1 时一致的情况下，讨论还原体系中氢氧化钠的用量对催化剂性能的影响，结果见表 2。

表 2　氢氧化钠的用量对催化剂性能的影响

Table 2　Influence of NaOH amount on catalytic performance

No.	NaOH/g	Selectivity of o-CAN/%	Conversion of o-CNB/%
1	0.03	91.2	86.5
2	0.05	94.3	100
3	0.1	80.6	70

从表 2 可以看出，催化剂的活性和邻氯苯胺的选择性随氢氧化钠的增加呈先增加后减少的趋势，在 NaOH 为 0.05g 情况下，邻氯硝基苯已全部转化，邻氯苯胺表现出最高的选择性 94.3%。原因可能与催化剂前驱体的还原度有关，水合肼在碱性的条件下，体现出更负的化学电位，具有更强的还原能力[11]，促使催化剂前驱体还原完全。从硼氢化钾还原剂来看，碱性增加可以防止硼氢化钾水解。但是，NaOH 量过多时，Ni^{2+} 容易与 OH^- 产生 $Ni(OH)_2$，不利于水合肼与其配位而影响其还原，并且，由于硼氢化钾可还原的游离的 Ni^{2+} 离子减少，造成还原度的下降，使催化剂的活性降低。

2.3.4　溶剂配比对所制备催化剂性能的影响

常规化学还原法制备非晶态合金都是在水溶液中进行，以往的研究[9]也尝试在非水溶液中还原金属离子来制得非晶态合金，结果是在 95%的乙醇中制备的催化剂活性小于在水溶液中制备的催化剂。溶剂效应对制备的催化剂活性影响极大，本文溶剂对制备催化剂的活性影响结果见表 3。

从表 3 看出，催化剂在碱度、还原剂的用量都相似的情况下，溶剂的不同所得的催化剂活性也不一样，在甲醇、水体系中，水含量的增加导致催化剂活性的减小。在甲醇、乙醇、水体系中，乙醇量的增加会使催化剂活性减少，催化剂的活性和选择性在甲醇为 13mL、乙醇 2mL、水 0.4mL 时达到最高。在乙醇、水体系中，催化剂的活性都较低。原因可能与溶剂的溶剂化能力、镍离子在不同溶剂中的电子状态及游离情况有关。从还原与溶剂化的角度来看，在溶液中镍盐被还原的先决条件是镍盐必须被溶剂化形成自由离子，水的介电常数最大，对于镍盐的溶剂化能力最强，甲醇次之，乙醇的介

表 3 溶剂配比对所制备催化剂性能的影响

Table 3 Influence of solvent proportion on catalytic performance

No.	CH_3OH/mL	C_2H_5OH/mL	H_2O/mL	Selectivity of *o*-CAN/%	Conversion of *o*-CNB/%
1	13	0	2	92.5	95.4
2	15	0	0.4	88.4	100
3	13	2	0.4	94.3	100
4	10	5	0.4	94.1	98.1
5	5	10	0.4	95.0	66.4
6	0	12	3	87.8	57.5
7	0	3	12	90.0	55.4

电常数最小，对镍盐的溶剂化能力最弱，所以在水含量多的溶剂中，镍盐迅速被溶剂化，迅速溶剂化的后果是镍离子很快进入溶剂中，而还原速度可能低于溶剂化速度，导致一部分镍从催化剂载体表面脱附，使催化剂的有效活性位减少，进而催化剂的活性减少。在乙醇含量大的溶剂中，溶剂的溶剂化能力减小，在载体表面已溶剂化的镍离子被还原，但是无法再深入还原内层的镍盐，使实际利用的镍减少，催化剂的活性降低。由此可以设想，催化剂的还原与脱附是一个动态的平衡过程，负载有效镍的量取决于还原速度和镍盐的溶剂化速度，镍盐的溶剂化速度若大于还原速度，镍容易从催化剂载体表面脱附，使负载的镍减少，催化剂的活性减少；镍盐的溶剂化速度小于还原速度则使有效镍减少，只有使溶剂化速度和还原速度达到平衡时，才能高效地形成负载镍催化剂。

参考文献

[1] Zu Y. Liaw B J，Chiang S J. Appl Catal A：General [J]，2005，284：97.

[2] Wang M. Li H X. Wu Y D，et al. Mater Lett [J]，2003，57：2954.

[3] WANG Wen-Jing（王文静），YAN Xin-Huan（严新焕），XU Dan-Qian（许丹倩），et al. J Chin Catal（催化学报）[J]，2004，(5)：369.

[4] Chen Y. Catal Today [J]，1998，44：3.

[5] MA Ai-Zeng（马爱增）. J Chin Catal（催化学报）[J]，1999，(6)：603.

[6] WANG You-Zhen（王友臻），QIAO Ming-Hua（乔明华），HU Hua-Rong（胡华荣）. Acta Chim Sin（化学学报）[J]，2004，62 (14)：1349.

[7] ZHANG Xiao-Xin（张晓昕），MA Ai-Zeng（马爱增），MU Xu-Hong（慕旭宏）. Acta Phys-chim Sin（物理化学学报）[J]，2000，16 (2)：180.

[8] Du Y，Chen H L，Chen R Z，et al. Appl Catal A：General [J]，2004，277：259.

[9] Boudjahem A G，Monteverdi S，Mercy M，et al. J Catal [J]，2004，221：325.

[10] Abdel-Ghani Boudjahem，Serge Monteverdi，Michel Mercy，et al. Langmuir [J]，2004，20：208.

[11] ZHENG Hua-Gui（郑化桂），ZENG Jing-Hui（曾京辉），YU Hua-Ming（余华明），et al. J Univ Sci Tech China（中国科学技术大学学报）[J]，1999，29 (06)：722.

[12] Chen Y Z，Chen Y C. Appl Catal A：General [J]，1994，115：45.

Preparation of Supported Ni-B Amorphous Alloy Catalyst via New Chemical Reduction System and Its Catalytic Hydrogenation Properties

Sun Yu，Wu Qiujie，Li Bindong，Lv Chunxu，Zou Qi

School of Chemistry，Nanjing University of Science and Technology，Nanjing 210094

Chinese Journal of Applied Chemistry，2006，23 (9)，1047-1051

Abstract: The supported catalyst P1 was prepared with hydrazine hydrate and potassium borohydride as co-reducers in a mixed solvent consisting of methanol，ethanol and water in a proper proportion. The catalyst was a kind of nanosized amorphous alloy characterized by TEM, XRD and SAED. Catalyst P1 was used to catalyze the hydrogenation of *o*-chloronitrobenzene, and compared with the amorphous catalyst P2 prepared by the common method. As it is shown in the results, the activity and selectivity of P1 was only half of that in the preparation of P2. The effects of hydrazine hydrate, potassium borohydride, pH and solvent on the properties of the catalyst were also discussed.

Keywords: *co*-conducer, amorphous alloy, catalytic hydrogenation

（注：此文原载于 应用化学，2006，23（9），1047-1051）

Kinetics Approach for Hydrogenation of Nitrobenzene to *p*-Aminophenol in a Four-Phase System

Li Guangxue[1,2], Ma Jianhua[2], Peng Xinhua[1], Lv Chunxu[1]
(1 Department of Fine Chemicals, Nanjing University of Science and Technology, Nanjing 210094, China)
(2 Institute of Chemical Engineering, Anhui University of Science and Technology, Huainan 232001, China)
SCI: 000288465000012

Abstract: The kinetics of the catalytic hydrogenation of nitrobenzene (NB) on a Pt/carbon catalyst in an acid medium to p-aminophenol (PAP), with aniline (AN) as a byproduct, was investigated in a batch reactor. By the elimination of the external and internal diffusion effects by good mixing, use of a superfine catalyst and the maintenance of a constant nitrobenzene concentration, the exponential macro-kinetics with respect to hydrogen pressure and reaction temperature are reported.

$$r_{PAP}=k_0\exp\left(-\frac{E_{PAP}}{RT}\right)p_H^{\alpha}=1.912\times10^5\exp\left(-\frac{2.874}{T}\right)p_H^{0.3251}$$

$$r_{AN}=k_0\exp\left(-\frac{E_{AN}}{RT}\right)p_H^{\beta}=7.083\times10^{33}\exp\left(-\frac{28.186}{T}\right)p_H^{2.8342}$$

$$-r_{NB}\approx r_{PAP}+r_{AN}=k_0\exp\left(-\frac{E_{PAP}}{RT}\right)p_H^{\alpha}+k_0\exp\left(-\frac{E_{AN}}{RT}\right)p_H^{\beta}$$

$$=1.912\times10^5\exp\left(-\frac{2.874}{RT}\right)p_H^{0.3251}+7.083\times10^{33}\exp\left(-\frac{28.186}{RT}\right)p_H^{2.8342}$$

In the kinetic equation, the reaction orders α, β for p-aminophenol (APA) and aniline (AN) formation are 0.3251 and 2.8342, respectively, while the activation energies are EPAP 23.89kJ/mol and EAN 234.34kJ/mol, respectively. The reaction results correspond well with experimental data.

Key Words: Multi-phase flow, Catalysis, Reaction kinetics, Nitrobenzene, p-Aminophenol, Aniline.

Introduction

p-Aminophenol is an important intermediate for synthesizing dyestuffs, antioxidants, photographic developers, urethanes, agrochemicals and pharmaceuticals. In 1874, Baeyer first prepared *p*-aminophenol(PAP) by reducing *p*-nitrosophenol with metallic tin in an acid medium[1]. Conventionally, *p*-aminophenol is prepared from chlorobenzene *via* the *p*-nitrochlorobenzene intermediate using iron powder (the Bechamp process) or by using molecular hydrogen reduction technology.

It is noted that Henke disclosed a more convenient process for the manufacture of *p*-aminophenol, which was by direct hydrogenation of nitrobenzene (NB) in the presence of Pt/C catalyst and a mineral acid[2]. Nitrobenzene reduction mechanism established by Haber, from his work on the electrochemical reduction of nitrobenzene, has been widely accepted[3]. The initial step was hydrogenolysis of the NO_2 functional group to form the key intermediate nitrosobenzene, followed by conversion to hydroxylamine, azoxy-, azo-, hydrazo-*etc*. (Scheme-I).

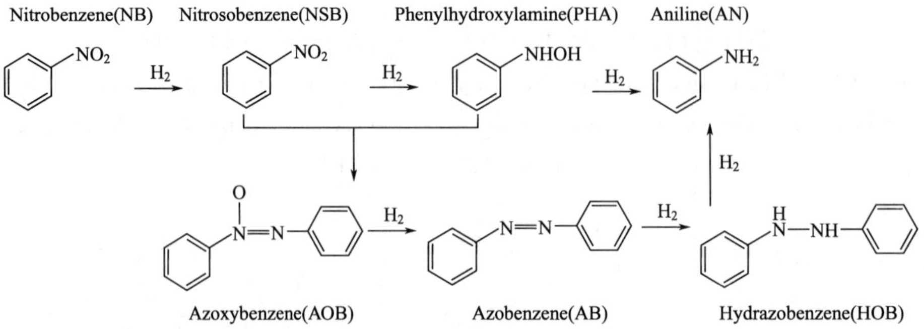

Nitrobenzene(NB)　Nitrosobenzene(NSB)　Phenylhydroxylamine(PHA)　Aniline(AN)

Azoxybenzene(AOB)　Azobenzene(AB)　Hydrazobenzene(HOB)

Scheme-I: Reaction pathway involved in nitrobenzene hydrogenation

Rylander and Ternary postulated that the hydrogenation of nitrobenzene to N-phenyl hydroxylamine (PHA) was followed by *in situ* a Bamberger rearrangement to *p*-aminophenol in the presence of an acid and the formation of aniline was the main competing side reaction in this process (Scheme-Ⅱ)[4,5].

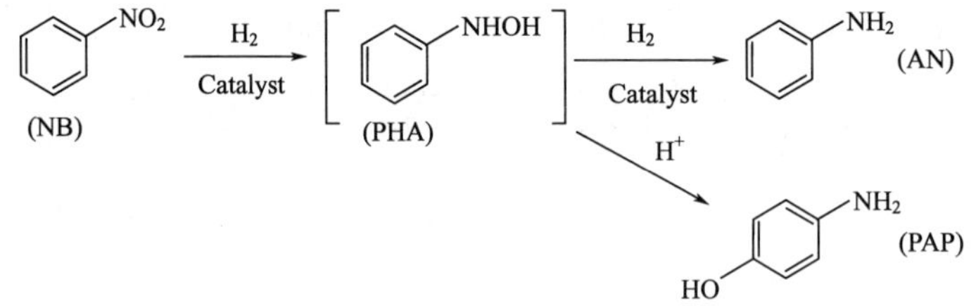

(NB)　(PHA)　(AN)　(PAP)

Scheme-Ⅱ: Hydrogenation reaction of nitrobenzene to *p*-aminophenol

The methodology concerning a one-pot preparation of *p*-aminophenol from nitrobenzene has received much attention because of the environmental economy and commercial potential compared with the *p*-nitrochlorobenzene route, especially the reaction behaviour with various metal catalysts such as Pt[1,4,6-11], Pd[1,12], Ni[13] and sulfides of Mo and W[1,14]. To enhance the efficiency and selectivity of the hydrogenation to *p*-aminophenol, various surfactants, such as octadecyl trimethyl ammonium chloride, dimethyldodecylamine sulfate, trimethylamine chloride, dimethyl alkylamine oxide[7,12,15-17] and some non-ionic surfactants[8], have been investigated. The addition of morpholine, phosphorous acid[18] and dimethyl sulfoxide[4,6,10-11] favoured the formation of N-phenyl hydroxylamine. However, N-phenyl hydroxylamine separation was undesirable due to its carcinogenic and explosive properties[19]. The influences of catalysts and reaction conditions on the reaction activity and selectivity were also addressed in the references cited.

However, the rapid kinetics and exothermicity of nitrobenzene hydrogenation makes a detailed kinetic discussion difficult. Juang *et al*.[20] found that the rate-determining step for the catalytic hydrogenation of nitrobenzene to *p*-aminophenol was the process involving hydrogen atom adsorption on to the metal surface. A reaction kinetics model was set as reported by Juang *et al*.[20].

$$-\frac{d[PhNO_2]}{dt}=k[H_2]^{\frac{1}{2}} \tag{1}$$

Rode *et al*[21] suggested a kinetic equation as in formula (2), giving an overall reaction rate in a four-phase system reactor with mass transfer effect. R_A, which is present on both sides of the equation, is a coupling or implicit function. Notations are given in the reference.

$$R_A=wk_1+wk_2\left[\frac{A_{NB}^{*}-R_A/K_A a_B}{1+K_A(A_{NB}^{*}-R_A/K_L a_B)}\right] \tag{2}$$

An exponential rate equation is rather convenient for auto-control chemistry, although a hyperbolic type equation from the Langmuir-Hinshelwood model can be obtained *via* complex deduction and computation. Therefore, the aim of this work is to discuss the macro kinetics of both *p*-aminophenol and aniline formation without any mass transfer effects on the reaction rate. An exponential expression for the reaction kinetics was found to fit the experimental data and the kinetic parameters have been obtained. This result is of importance as a guide for the engineering challenge of scaling up the reaction and characterizing the reactor.

Experimental

Nitrobenzene, H_2SO_4 (98%), dodecyl trimethyl ammonium bromide and dimethyl sulfoxide were obtained commercially with chemical purity 99% or over. Standard *p*-aminophenol was purchased from Sinopharm Chemicals Reagent Co. Ltd. Hydrogen with 99.999% purity (GB/T7445-95) was supplied by Nanjing Special Gas Factory. Nitrogen, with more than 99.5% purity (GB/T7445-95) from a commercial source was directly used from the compressed gas cylinder. 2-5% Pt/C catalyst was prepared by impregnation of charcoal (U-101, Tangshan United Carbon Technology Co. Ltd., China) with chloroplatinic acid solution followed by formaldehyde reduction in basic conditions.

A 0.25L Autoclave Hastelloy was supplied by Dalian Tongda Reactor Factory and a 500L autoclave reactor obtained from Hangzhou Yuanzheng Chemical Engineering Equipment Co. Ltd., China.

Procedure: Two reactors were used for the hydrogenation experiment, one reactor of 0.25L capacity, the other a 500L autoclave reactor with gas-inducing turbine. The reactor was connected to a hydrogen reservoir held at a pressure higher than that of the reactor through a constant pressure regulator. Hydrogen was supplied from this reservoir to the reactor through a non-return valve. The gas consumed during the course of the reaction was monitored by the pressure drop in the reservoir vessel, measured using a transducer. Initial data were obtained at different temperatures, catalyst loading, agitation speed and partial pressure of hydrogen in a stirred high-pressure slurry reactor.

For a typical hydrogenation experiment the reactor was filled with nitrogen gas and charged with nitrobenzene, 2-5% Pt/C catalyst, 14-18% H_2SO_4 solution, quaternary ammonium salt and dimethyl sulfoxide. The contents were flushed with hydrogen while stirring. After the desired temperature was attained, the system was pressurized with hydrogen to a defined level. The course of the reaction was monitored by the observed pressure drop in the reservoir vessel as a function of time. Samples were analyzed at intervals during the reaction. Upon completion, the solid catalyst was separated by filtration. The filtrate was kept at temperature of 30℃ and adjusted to pH 5.0-5.5and then extracted with toluene in order to remove aniline, unconverted nitrobenzene and traces of impurities. The pH of the aqueous layer after extraction was adjusted to 7 by the addition of aqueous ammonia leading to the complete precipitation of *p*-aminophenol.

Results and discussion

Kinetics model fundamentals: The hydrogenation reaction of nitrobenzene to *p*-aminophenol was conducted in a four-phase system, *i. e.* gas phase-H_2, inorganic phase-dilute sulfuric acid (which dissolves the aminophenol and aniline to form the corresponding salt), organic phase-nitrobenzene, solid phase-powdery Pt/C catalyst (which is distributed in the nitrobenzene). As the catalytic hydrogenation proceeded, the proportion of the nitrobenzene phase in the system decreased (for a defined amount of nitrobenzene) and the ratio of acid phase increased, with the subsequent formation of *p*-aminophenol, aniline and water. Accordingly, the reaction rate in each phase can be expressed in moles per unit mass of catalyst per unit time, *i. e.*

$$-r_{NB}=-\frac{dn_{NB}}{wdt},r_{PAP}=\frac{dn_{PAP}}{wdt},r_{AN}=\frac{dn_{AN}}{wdt}$$

The reaction was influenced by many factors, such as surfactant, catalyst activity, reaction temperature and pressure, agitation strength and cooling capacity *etc.* In order to improve the reaction rate, a special device was employed to provide vigorous agitation and eliminate external diffusion resistance. It reduced the size of the H_2 gas bubbles and nitrobenzene and acid liquid droplets and mixed them well with the solid catalyst. The external diffusion resistance existed in two forms; one was the diffusion of gaseous H_2 into the solid catalyst *via* liquid nitrobenzene, the other was the diffusion of products (N-phenyl hydroxylamine, aniline and water) on catalyst surface to the acid phase *via* the nitrobenzene phase. Accordingly, the catalyst was finely grounded (to millimicron particle size) which was sufficient to eliminate the internal surface diffusion resistance to H_2, nitrobenzene and intermediates or some formations existed in catalyst channels. Depending on the conditions, the macro-kinetics of the hydrogenation reaction at the active site of the catalyst can follow several reasonable hypotheses; (a) the reaction system could be pseudo-homogeneous; (b) the formation of *p*-aminophenol and aniline from nitrobenzene could be a parallel reaction; (c) the nitrobenzene concentration in the system could be constant, giving

$$C_{NBO}=C_{NB}=\frac{\rho_{NB}}{M_{NB}}=9.786\text{kmol/m}^3.$$

Therefore, the differential equation for the complex. reaction (Scheme-Ⅱ) could be expressed as follows:

$$\begin{cases} -r_{NB}=r_{PAP}+r_{AN} \\ r_{PAP}=\dfrac{dn_{PAP}}{wdt}=k_{PAP}C_{NB}^{\alpha'}p_H^{\alpha}=kp_H^{\alpha} \\ r_{AN}=\dfrac{dn_{AN}}{wdt}=k_{AN}C_{NB}^{\beta'}p_H^{\beta}=k'p_H^{\beta} \end{cases}$$

The boundary conditions for the formulae above are:

$$t=0,n_{BNO}\neq 0,n_{AN}=0,n_{PAP}=0;t=t,n_{NBO}=n_{NB}+n_{AN}+n_{PAP}$$

Estimation of the reaction order for the exponential macro-kinetics model: Combining the finite difference expressions to give the reaction rate,

$$r_{PAP}=\frac{1}{w}\frac{dn_{PAP}}{dt}=\frac{1}{w}\frac{\Delta n_{PAP}}{\Delta t}$$

$$\text{and } r_{AN}=\frac{1}{w}\frac{dn_{AN}}{dt}=\frac{1}{w}\frac{\Delta n_{AN}}{\Delta t}$$

and logarithm for the exponential kinetics model,

$$\ln r_{PAP}=\alpha \ln P_H+\ln k$$

$$\text{and } \ln r_{AN}=\beta \ln P_H+\ln k$$

Linear regression analysis using the Least Squares method or the linear equation in Excel 2003 for the experimental data at 90°C (Fig. 1) gave the following formular,

$$\alpha=\frac{n\sum(\ln r_{PAP}\ln p_H)-\sum\ln r_{PAP}\sum\ln p_H}{n\sum(\ln p_H)^2-(\sum\ln p_H)^2}=0.3251$$

$$\text{and}\quad \ln k=\frac{\sum\ln r_{PAP}-\alpha\sum\ln p_H}{n}=0.376$$

$\beta=2.8342$ and $\ln k'=-33.444$ were obtained in the same way. The rate equation effectively explains why low hydrogen pressure facilitated the p-aminophenol yield and a high hydrogen pressure accelerated aniline formation. Moreover, the selectivity of p-aminophenol with respect to aniline will decrease with increasing the hydrogen pressure.

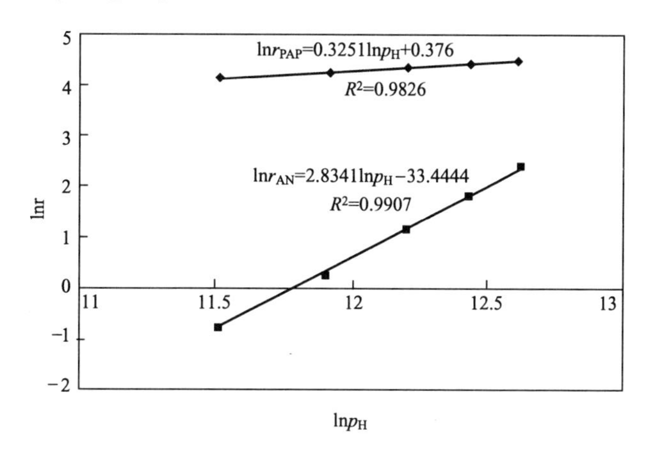

Fig. 1 ln$r$-lnp_H linear relation

Estimation of the Arrhenius rate constant: Since the reaction speed increases with increasing temperature and the rate of formation of aniline exceeds that of p-aminophenol, the activation energy of aniline must be higher than that of p-aminophenol. According to the Arrhenius equation $k=k_0\exp\left(-\frac{E}{RT}\right)$ or its linearized form $\ln k=-\frac{E}{RT}+\ln k_0$, the linear regression analysis results using the Least Squares method for the experimental data or the linear equation using Excel 2003 for the experimental data from Fig. 2 could be

$$\frac{E_{PAP}}{R}=\frac{\sum\frac{1}{T}\sum\ln k-n\sum\left(\frac{1}{T}\ln k\right)}{n\sum\frac{1}{T^2}-\left(\sum\frac{1}{T}\right)^2}=2.847$$

so the activation energy of the p-aminophenol formation was $E_{PAP}=23.89$kJ/mol and the rate constant k_0 for the p-aminophenol formation was 191185.6, giving the equation

$$\ln k_0=\frac{\sum\ln k+\frac{E_{PAP}}{R}\sum\frac{1}{T}}{n}=12.161$$

Similarly, since $\frac{E_{AN}}{R}=28.186$, the activation energy of aniline formation is $E_{AN}=234.34$kJ/mol and $k_0'=7.083\times10^{33}$.

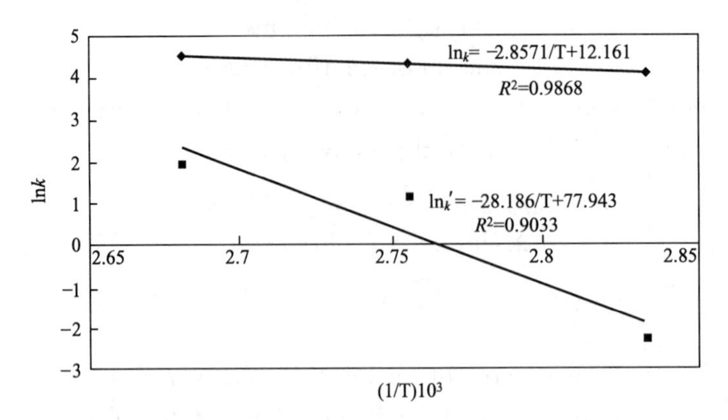

Fig. 2 lnk-1/T linear relation

The results effectively explain a high *p*-aminophenol yield at low temperature as well as the corresponding aniline formation at high temperature. So, the selectivity of *p*-aminophenol with respect to aniline can be improved by employing a low reaction temperature. For *p*-aminophenol formation, the rate equation is

$$r_{PAP}=k_0\exp\left(-\frac{E_{PAP}}{RT}\right)P_H^{\alpha}=1.912\times10^5\exp\left(-\frac{2.874}{T}\right)P_H^{0.3251} \tag{3}$$

and for aniline, $$r_{AN}=k_0'\left(-\frac{E_{AN}}{RT}\right)P_H^{\beta}=7.083\times10^{33}\exp\left(-\frac{28.186}{T}\right)P_H^{2.8342} \tag{4}$$

Since conversion of most of the nitrobenzene gives *p*-aminophenol and aniline, the disappearance rate of nitrobenzene was approximately equal to the sum of the formation rates of *p*-aminophenol and aniline, *i. e.*,

$$\begin{aligned}-r_{NB}\approx r_{PAP}+r_{AN}&=k_0\exp\left(-\frac{E_{PAP}}{RT}\right)P_H^{\alpha}+k_0\left(-\frac{E_{AN}}{RT}\right)P_H^{\beta}\\&=1.912\times10^5\exp\left(-\frac{2.874}{T}\right)P_H^{0.3251}+7.083\times10^{33}\exp\left(-\frac{28.186}{T}\right)P_H^{2.8342}\end{aligned} \tag{5}$$

The rate of the hydrogenation and the selectivity for *p*-aminophenol or aniline, are significantly influenced by the catalyst, surfactant and mixing so that their influences are difficult to quantify. The catalyst is affected by the pore diameter and distribution, type of group at the surface of the activated carbon support and the method used for loading Pt. A catalyst with zero activity was found in present experiment using almost the same preparation method as one with high activity. The type and amount of the surfactant influenced the mass transport of H_2 from the gas phase to the nitrobenzene phase then to the catalyst surface and influenced the mass transport of N-phenyl hydroxylamine from the catalyst surface to the nitrobenzene phase then to the acid phase. As the mixing mode influences the mass transport in the four phase reaction system, scale-up effects may be significant.

The equation showed an excellent agreement with further experimental data, but an induced period of product formation was clearly observed. This was because of the initial hydrogen diffusion from the gas phase to the nitrobenzene phase and Pt/C catalyst active sites, after sequential adsorption, surface reaction and desorption, then product diffusion from the nitrobenzene phase to an acid phase. The induction period lasted about 10-30min.

In this reaction system, considering the presence of a constant nitrobenzene concentration and a continuous decrease of the volume of the nitrobenzene phase as the hydrogenation processed (when some inert organic solvent such as benzene was added to the system), the reaction order for nitrobenzene was deduced by varying the nitrobenzene concentration.

The reaction temperature and pressure affected product selectivity. Although both factors could improve reaction rate and enhance the conversion of nitrobenzene, the selectivity for *p*-aminophenol formation could become poor because of faster aniline formation. The selectivity can be quantitatively expressed according to the following equation

$$S_{PAP}=\frac{r_{PAP}}{r_{AN}}=\frac{kP_H^{\alpha}}{k'P_H^{\beta}}=\frac{k}{k'}P_H^{\alpha-\beta}=2.699\times10^{-27}\exp\left(\frac{25.312}{T}\right)P_H^{-2.5091} \tag{6}$$

When the temperature is increased, the disappearance rate of the nitrobenzene and the rate of *p*-aminophenol and aniline formation all increased, but aniline increases more than *p*-aminophenol, so low temperature favoured the selectivity to *p*-aminophenol and the activation energies are $E_{PAP}<E_{AN}$.

When the hydrogen pressure was increased, the disappearance rate of nitrobenzene and the rate of *p*-aminophenol and aniline formation, all increased, but aniline increased more than *p*-aminophenol, so low hydrogen pressure favoured the selectivity to *p*-aminophenol and the order for the reaction to *p*-aminophenol and aniline is $\alpha<\beta$.

The phenomenon of the induction period of about 20min can be explained by the time required for hydrogen diffusion from the gas phase to the nitrobenzene phase then to the Pt/C catalyst active sites, followed by adsorption of the reactive species, surface reaction (s), desorption of the products N-phenyl hydroxylamine or aniline, then the product diffusion from the nitrobenzene phase to the acid phase.

Conclusion

This first report on the macro-kinetics of the hydrogenation of nitrobenzene to *p*-aminophenol in a four-phase reactor provides an evaluation of a novel reaction sequence for *p*-aminophenol preparation. A rate equation is proposed depending on an exponential function of hydrogen pressure and temperature to quantitatively describe a selected result in the competitive formation of *p*-aminophenol and aniline.

Notation

r_{PAP}: rate of hydrogenation for *p*-aminophenol (PAP) ($kmol/m^3 \cdot s$)

r_{AN}: rate of hydrogenation for aniline (AN) ($kmol/m^3 \cdot s$)

$-r_{NB}$: overall rate of hydrogenation for nitrobenzene (NB) ($kmol/m^3 \cdot s$)

dn_{NB}: differential mole of nitrobenzene (kmol)

dn_{PAP}: differential mole of *p*-aminophenol (kmol)

dn_{AN}: differential mole of aniline (kmol)

w: catalyst loading (kg/m^3)

dt: differential time (s)

E_{PAP}: activation energy for *p*-aminophenol formation (kJ/mol)

E_{AN} : activation energy for aniline formation (kJ/mol)

C_{NB0} : concentration for nitrobenzene at t=0 ($kmol/m^3$)

C_{NB} : concentration for nitrobenzene ($kmol/m^3$)

ρ_{NB} : density for nitrobenzene (kg/m^3)

M_{NB} : mole weight (kg/kmol)

k_{PAP} : rate constant for *p*-aminophenol (kmol/kg · s)

k_{AN} : rate constant for aniline (kmol/kg · s)

k_0 : specific rate constant for *p*-aminophenol

k_0' : specific rate constant for aniline

p_H : hydrogen pressure

α : order for *p*-aminophenol formation

β: order for aniline formation

R : gas content, 8.314 kJ/kmol K

T : reaction temperature (K)

s_{PAP} : reaction selectivity for *p*-aminophenol to aniline

Acknowledgements

The authors wish to acknowledge sincerely the financial support from the High Technology Program from the development and reform commission of Anhui Province (Project No. 2005-huainan-4) and the Natural Science Key Research Program from Education Bureau of AnhuiProvince. (Project No. 2002kj283zd).

References

[1] S. Mitchell, Kirk-Othmer Encyclopaedia of Chemical Technology, Wiley, New York, edn. 4, p. 481, 580 (1992).

[2] C. O. Henke and J. V. Vaughen, Reduction of Aryl Nitro Compounds, US Patent 2198249 (1940).

[3] F. Haber, *Zeit. Electrochem.*, 4, 506 (1898).

[4] P. N. Rylander, I. M. Karpenko and G. R. Pond, Process for Preparing *para*-Aminophenol, US Patent 3715397 (1973).

[5] A. L. Ternary, Contemporary Organic Chemistry, WB Saunders Co., Philadelphia, p. 661 (1976).

[6] P. N. Rylander, I. M. Karpenko and G. R Pond, Selective Hydrogenation of Nitroaromatics to the Corresponding N-Arylhydroxylamine, US Patent 3694509 (1972).

[7] L. Spiegler, Manufacture of Aromatic *para*-Hydroxyamines, US Patent 2765342 (1956).

[8] B. B. Brown and A. E. F75ededck, Aminophenol Production, US Patent 3535382 (1970).

[9] T. J. Dunn, Method for Preparing *p*-Aminophenol, US Patent 4264529 (1981).

[10] D. C. Caskey and D. W. Chapman, Process for Preparing *p*-Aminophenol and Alkyl Substituted *p*-Aminophenol, US Patent 4415753 (1983).

[11] D. C. Caskey and D. W. Chapman, Process for Preparing *p*-Aminophenol and Alkyl Substituted *p*-Aminophenol, US Patent 4571437 (1986).

[12] R. G. Benner, Process for Preparing Aminophenol, US Patent 3383416 (1968).

[13] C. V. Rode, M. J. Vaidya, R. Jaganathan and R. V. Chaudhari, Single Step Hydrogenation of Nitrobenzene to *p*-Aminophenol, US Patent 6403833 (2002).

[14] N. P. Greco, Hydrogenation of Nitrobenzene to *p*-Aminophenol, US Patent 3953509 (1976).

[15] S. S. Sathe, Process for Preparing *p*-Aminophenol in the Presence of Dimethyldodecylamine Sulfate, US Patent 4176138 (1979).

[16] E. L. Derrenbacker, Process for the Selective Preparation of *p*-Aminophenol from Nitrobenzene, US Patent 4307249 (1981).

[17] D. C. Miller, Surfactant Improvement for *para*-Aminophenol Process, US Patent 5312991 (1994).

[18] J. R. Kosak, Catalysis Organic Reactions, Marcel Dekker, New York (1988).

[19] M. Studer and P. Baumeister, Process for the Catalytic Hydrogeneration of Aromatic Nitro Compounds, US Patent 6096924 (2000).

[20] T. M. Juang, J. C. Hwang, H. O. Ho and C. Y. Chen, *J. Chin. Chem. Soc.* 2, 135 (1988).

[21] C. V. Rode, M. J. Vaidya, R. Jaganathan and R. V. Chaudhari, Chem. Eng. Sci., 4, 1299 (2001).

（注：此文原载于 Asian Journal of Chemistry，2011，23 (3)，1001-1005）

2-氨基-6-烷氧基嘌呤的合成

陆鸿飞，陆明，吕春绪，章丽娟
（南京理工大学化工学院，江苏南京 210094）

摘要： 2-氨基-6-烷氧基嘌呤是重要的核苷类药物的中间体。以鸟嘌呤为原料，通过酰化制得 N^2·9-二乙酰鸟嘌呤（Ⅰ），收率 95.8%。氯化水解得到 2-氨基-6-氯嘌呤（Ⅱ），收率 75.8%。通过 2-氨基-6-氯嘌呤（Ⅱ）与醇钠反应得到 2-氨基-6-甲氧基嘌呤（Ⅲa)，收率 96.5%；2-氨基-6-乙氧基嘌呤（Ⅲb)，收率 79.6%；2-氨基-6-异丙氧基嘌呤（Ⅲc)，收率 73.2%；2-氨基-6-三氟乙氧基嘌呤（Ⅲd)，收率 49.8%。用元素分析，核磁共振，质谱，红外光谱对所得产品进行了分析，证明所得产品为目标产品。重点研究了 2-氨基-6-三氟乙氧基嘌呤（Ⅲd）的合成，其合成的最佳工艺条件为：n(Ⅱ)∶n($NaOCH_2CF_3$)＝1∶5，用二甲亚砜（DMSO）作为反应溶剂，其用量为 m(DMSO)∶m(Ⅱ)＝13∶1，用碘化亚铜（CuI）为催化剂，其用量为 n(CuI)∶n(Ⅱ)＝2∶1，反应温度为 160℃。

关键词： 嘌呤衍生物，烷氧基化

嘌呤衍生物是一类很重要的抗病毒药物中间体，对癌症、艾滋病、白血病等有特殊功效[1,2]。近年来的研究结果显示，在嘌呤环的 6 位引入某些取代基团将使嘌呤化合物具有很好的生物活性[3-5]，其中烷氧基就是这些活性基团之一。在嘌呤的 6 号位引入烷氧基的常用方法是将 6-氯嘌呤或者 6-羟基嘌呤首先与吡啶反应生成吡啶盐，然后将吡啶盐在相应的醇中反应合成烷氧基嘌呤[6]，这种方法存在两个问题：一是忽略了吡啶的毒性，二是反应时间长达 60h，成本较高。

本文通过 6-氯嘌呤与醇钠反应合成目标产物，反应过程中无有毒物质产生，收率和反应选择性都很高，降低了合成成本。在嘌呤环引入烷氧基的同时，重点研究了在嘌呤环引入三氟乙氧基。三氟乙氧基是近年来一种很流行的活性基团，将三氟乙氧基引入母环能够增强药效并减少对生物体的副作用。

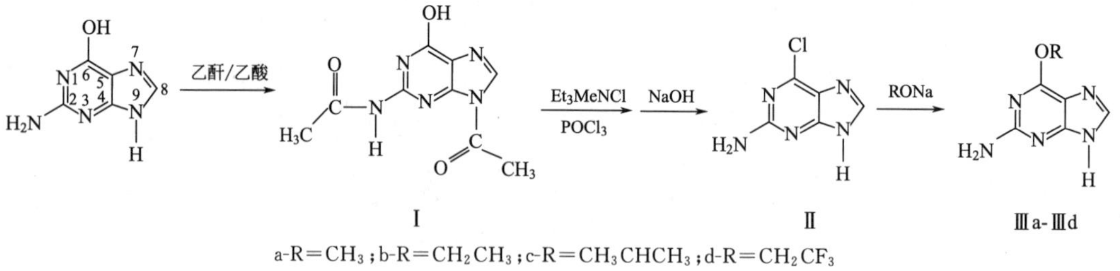

乙酐/乙酸　Et_3MeNCl / $POCl_3$　NaOH　RONa

Ⅰ　Ⅱ　Ⅲa-Ⅲd

a-R＝CH_3；b-R＝CH_2CH_3；c-R＝CH_3CHCH_3；d-R＝CH_2CF_3

1 实验

1.1 试剂与仪器

鸟嘌呤为工业品（常州康丽制药有限公司，质量分数大于 99.5%），异丙醇钠和三氟乙醇钠为实验室自制，其他原料为 CP。Yansco 显微熔点仪，温度计未校正；Brucker-Avance DPX 300MHz 核磁共振仪；PE-2400 元素分析仪；Finningan 公司 Ultra Am TSQ quntium 型高分辨质谱仪。

1.2 N^2·9-二乙酰基鸟嘌呤（Ⅰ）的合成[7]

在三口烧瓶中加入 5g 鸟嘌呤，30mL 乙酐，60mL 乙酸。油浴加热至 135℃，搅拌下回流。反应至反应液澄清结束，趁热过滤，滤液冷却，减压蒸出乙酸和未反应的乙酐，回收套用，析出固体，过滤，将滤饼用去离子水洗至中性，干燥，将样品、乙酸和活性炭加热搅拌，趁热过滤，滤液冷却析出固体，过滤，烘干，得白色晶体Ⅰ，收率 95.8%，熔点>250℃。IR（KBr，γ/cm^{-1}）：3200，1700，1530。^{1}HNMR（DMSO-d_6，300MHz）：2.16～2.50（m，6H，-CH_3），8.10（br，2H，-NH-和=CH-），11.56（s，^{1}H，-OH）。MS（EI），m/Z：234.10（M-），192.1204，150.0764；元素分析，实测值（理论值）/%：ω(C) =45.62（45.69），ω(H) =3.67（3.83），ω(N) =29.12（29.79）。

1.3 2-氨基-6-氯嘌呤（Ⅱ）的合成[7]

在三口烧瓶中加入 4.68g Ⅰ，50mL1，2-二氯乙烷，0.005g 甲基三乙基氯化铵。滴加 9mL 三氯氧磷，回流下搅拌反应 3h，冷却过滤，滤饼用二甲亚砜（DMSO）精制，得白色晶体Ⅱ，收率 75.8%，熔点>250℃。IR（KBr，γ/cm^{-1}）：822（Cl—），1636、1292，630（NH_2-）;^{1}HNMR（DMSO-d_6，300MHz）：6.75（m，3H，—NH_2 和—NH—），8.01（m，1H，—CH—）；MS（EI），m/Z：169.13（M-），151.20，135.13，116.84；元素分析，实测值（理论值）/%：ω(C)=35.53(35.42)，ω(H)=2.41(2.38)，ω(N)=41.25(41.30)。

1.4 2-氨基-6-甲氧基嘌呤（Ⅲa）的合成

在三口烧瓶中加入 3.4gⅡ，5.4g 甲醇钠，20mL 甲醇。加热至回流，反应 10h，TCL 跟踪原料反应完。蒸出甲醇至反应体系成为黏稠物，加水约 30mL 至全部溶解，加 20mL 甲苯搅拌 30min，静置，过滤，水相用稀盐酸调节滤液 pH 至中性，析出白色固体，过滤，滤饼烘干。得白色固体Ⅲa，收率 96.5%。^{1}HNMR（DMSO δ）：3.98（s，3H，—OCH_3），6.81（s，1H，-N=CH-N），8.13（s，3H，-NH_2 和-NH-）；MS（EI），m/Z：166.02（M+1，35），149.01（100），134.02（17）；元素分析，实测值（理论值）%：ω（C）=43.52（43.64），ω（H）=4.23（4.27），ω（N）=42.51（42.40）。

1.5 2-氨基-6-乙氧基嘌呤的（Ⅲb）合成

在三口烧瓶中加入 3.4gⅡ，6.8g 乙醇钠，20mL 乙醇。加热至回流，反应 10h，TLC 跟踪原料反应完。蒸出乙醇至反应体系成为黏稠物，加水约 30mL 至全部溶解，加 20mL 甲苯搅拌 30min，静置，过滤，水相用稀盐酸调节滤液 pH 至中性，析出白色固体，过滤，滤饼烘干。得白色固体Ⅲb，收率 79.6%。^{1}HNMR（DMSO δ）：1.45～1.51（m，3H，CH_3），4.46（m，2H，-O-CH_2-），7.12（s，^{1}H，-N=CH-N），8.13（s，3H，-NH_2 和-NH-）；MS（EI），m/Z：180.07（M+1，52），163.06（9），152.03（100），135.03（98）；元素分析，实测值（理论值）%：ω（C）= 46.95（46.92），ω(H)=5.04（5.06），ω(N)=39.12（39.09）。

1.6 2-氨基-6-异丙氧基嘌呤（Ⅲc）的合成

冰水浴冷却，氮气保护下在 30mL 干燥的异丙醇中加入 2.3g 金属钠，缓慢放出氢气。金属钠完全反应后加入 3.4gⅡ，加热至回流，反应 10h，TLC 跟踪原料反应完。蒸出异丙醇至反应体系成为黏稠物，加水约 30mL 至全部溶解。加 20mL 甲苯搅拌 30min，静置，过滤，水相用稀盐酸调节滤液

pH 至中性，析出白色固体，过滤，滤饼烘干。得白色固体Ⅲc，收率 73.2%。^{1}HNMR（DMSO δ）：1.62～1.65（d，6H，CH_3），4.76（m，1H，-O-CH-），7.13（s，1H，-N＝CH-N），8.13（s，3H，$-NH_2$ 和-NH-）；MS（EI），m/Z：194.00（M＋1，24），151.95（100）；元素分析，实测值（理论值）%：ω(C)＝49.69（49.73），ω(H)＝5.71（5.74），ω(N)＝36.29（36.25）。

1.7　2-氨基-6-三氟乙氧基嘌呤（Ⅲd）的合成

冰水浴冷却、氮气保护下，在 10mL 干燥的异丙醇中加入 2.3g 金属钠，缓慢放出氢气。金属钠完全反应后加入 3.5gⅡ，加干燥的 DMSO 40mL，160℃反应 10h，TLC 跟踪原料反应完。蒸出溶剂至反应体系成为黏稠物，冷却后加水约 30mL 至全部溶解。加 20mL 甲苯搅拌 30min，静置，过滤，水相用稀盐酸调节滤液 pH 至中性，析出白色固体，过滤，滤饼烘干。得白色固体Ⅲd，收率 49.8%。^{1}HNMR（DMSO δ）：3.91～3.96（m，2H，$-O-CH_2-$），7.99（s，lH，-N＝CH-N），8.13（s，3H，$-NH_2$ 和-NH-）；MS（EI），m/Z：233.97（M＋1 100），217.03（43）；元素分析，实测值（理论值）%：ω(C)＝36.09（36.06），ω(H)＝2.61（2.59），ω（N)＝30.09（30.04）。

2　结果与讨论

本文重点研究了 2-氨基-6-三氟乙氧基嘌呤（Ⅲd）的合成工艺。

2.1　溶剂种类对Ⅲd 合成的影响

由于三氟乙氧基团的特殊性质，反应需要在极性非质子溶剂中进行，本文考察了 6 种极性非质子溶剂中的三氟乙氧基化反应，N,N-二甲基甲酰胺（DMF）、N,N-二甲基乙酰胺（DMAc）、二甲亚砜（DMSO）、2-吡咯烷酮（NMP）、六甲基磷酰三胺（HMPA）和环丁砜。结果见表 1。

表 1　溶剂种类对Ⅲd 合成的影响

Table 1　Effect of different solvent on the yield of Ⅲd

	溶剂种类					
	DMF	DMAc	DMSO	NMP	HMPA	环丁砜
介电常数(25℃)	36.71	37.78	48.9	32.0	29.6	43.3
偶极矩/(10^{-2}C·m)(25℃)	12.88	12.41	13.34	13.64	13.34	16.04
产品收率/%	26.8	25.3	49.8	12.1	37.9	56.9

溶剂的介电常数和偶极距反映了溶剂极性的大小，极性越大的溶剂越容易增强反应的活性。表中所列几种极性非质子溶剂中，环丁砜极性最大，在此反应中产品的收率也最高，达到 56.9%，HMPA 的极性和 DMF 极性相似，但由于 HMPA 的稳定性好于 DMF，因此产品收率也高于 DMF。DMSO 作为反应溶剂时产品的收率仅低于环丁砜，但是环丁砜的价格远高于 DMSO，因此，使用 DMSO 做反应溶剂。

2.2　溶剂用量对Ⅲd 合成的影响

用 DMSO 作为溶剂，反应温度为 160℃，Ⅱ为 3.4g（20mmol），原料投料比为：n(Ⅱ)：$n(NaOCH_2CF_3)$＝1∶5，n(CuI)∶n(Ⅱ)＝2∶1。选择不同的 DMSO 用量，TLC 跟踪反应至无原料点为反应终点。结果见表 2。溶剂用量从 20mL 上升到 40mL，Ⅲd 的收率也从 19.3%上升到 49.8%，溶剂用量对产品收率的影响明显。溶剂从 40mL 升到 50mL 时，Ⅲd 的收率几乎无变化，因此，选择溶剂 DMSO 用量为 40mL。溶剂 DMSO 与原料Ⅱ的质量比为：m(DMSO)∶m(Ⅱ)＝13∶1。

表 2 溶剂用量对Ⅲd 合成的影响

Table 2 Effect of solvent dosage on the yield of Ⅲd

	溶剂用量/mL			
	20	30	40	50
产品收率/%	19.3	34.7	49.8	50.4

2.3 反应温度对Ⅲd 合成的影响

用 DMSO 作为溶剂，m(DMSO)∶m(Ⅱ)=13∶1，Ⅱ为 3.4g(20mmol)，原料投料比为：n(Ⅱ)∶n($NaOCH_2CF_3$)=1∶5，n(CuI)∶n(Ⅱ)=2∶1。选择不同的反应温度，考察其对Ⅲd 合成的影响。TLC 跟踪反应至无原料点为反应终点。结果见表 3。

表 3 反应温度对Ⅲd 合成的影响

Table 3 Effect of reaction temperature on the yield of Ⅲd

	温度/℃				
	100	120	140	160	180
产品收率/%	未反应	11.9	35.8	49.8	DMSO 分解

温度低于 100℃时，反应无法进行，随着反应温度的提高，产品的收率由 120℃时的 11.9%上升到 160℃时的 49.8%。但反应温度进一步提高，溶剂的稳定性受到了很大影响，当反应温度达到 180℃时，DMSO 分解得非常厉害，放出难闻的甲烷与二氧化硫气味。这些分解出的气体在工业生产中会造成很大安全隐患，因此，选择反应温度为 160℃。

2.4 催化剂种类对Ⅲd 合成的影响

用 DMSO 作为溶剂，m(DMSO)∶m(Ⅱ)=13∶1，Ⅱ为 3.4g(20mmol)，原料投料比为：n(Ⅱ)∶n($NaOCH_2CF_3$)=1∶5，考察不加催化剂（Bald）和 CuCl、CuI_2、CuI 三种催化剂对Ⅲd 合成的影响，结果见表 4。

表 4 催化剂种类对Ⅲd 合成的影响

Table 4 Effect of different catalyst on the yield of Ⅲd

	催化剂种类			
	Bald	CuCl	CuI_2	CuI
产品收率/%	21.0	21.3	34.9	49.8

加入 CuCl 和不加催化剂时的实验结果基本一样，由此可知，反应过程中起催化作用的不是亚铜离子和氯离子。碘化亚铜的催化效果明显优于碘化铜的催化效果。由表 4 不难推导出以下催化机理。由于 CuI 中碘离子的强亲核性，发生卤素交换反应，首先生成碘取代嘌呤，同时，碘离子的离去效应也很强，碘离子离去时，嘌呤环上形成碳正离子，亚铜离子的存在保证了碳正离子的稳定性，使碳正离子能够顺利地与三氟乙氧基阴离子结合而生成产品。

Cl
N N
H_2N N N H
CuI
⊖ I ⊕
N N
H_2N N N H
⊕
N N
H_2N N N H
⊖ OCH_2CF_3
OCH_2CF_3
N N
H_2N N N H

2.5 催化剂用量对Ⅲd 合成的影响

用 DMSO 作为溶剂，m(DMSO)∶m(Ⅱ)=13∶1，Ⅱ为 3.4g（20mmol），原料投料比为：n(Ⅱ)∶

$n(NaOCH_2CF_3)$=1∶5，考察了 CuI 用量对反应的影响，结果见表 5。

表 5 催化剂用量对Ⅲd 合成的影响

Table 5 Effect of catalyst dosage on the yield of Ⅲd

	n(CuI)∶n(Ⅱ)		
	1∶1	2∶1	3∶1
产品收率/%	36.5	49.8	50.1

催化剂用量对反应的影响比较大，当 n(CuI)∶n(Ⅱ)=1∶1 时，产品的收率只有 36.5%，而当 n(CuI)∶n(Ⅱ)=2∶1 时，产品的收率可达 49.8%，催化剂用量进一步提高，产品的收率提高不明显。因此，选择 n(CuI)∶n(Ⅱ)=2∶1。

2.6 原料配比对Ⅲd 合成的影响

用 DMSO 作为溶剂，m(DMSO)∶m(Ⅱ)=13∶1，Ⅱ为 3.4g（20mmol），催化剂 n（CuI）∶n（Ⅱ）=2∶1 时，考察了不同原料配比对Ⅲd 合成的影响，结果见表 6。

表 6 不同原料配比对Ⅲd 收率的影响

Table 6 Effect of ratio of raw materials on the yield of Ⅲd

	n(Ⅱ)∶$n(NaOCH_2CF_3)$		
	1∶3	1∶5	1∶10
产品收率/%	40.5	49.8	49.9

当 n（Ⅱ）∶n（$NaOCH_2CF_3$）从 1∶3 下降到 1∶5 时，产品的收率从 40.5%上升到 49.8%，此时三氟乙醇钠用量的增加，能明显提高产品的收率，继续提高三氟乙醇钠的用量，即使当 n(Ⅱ)∶$n(NaOCH_2CF_3)$ 为 1∶10 时，产品的收率也几乎没有什么变化。因此，选择 n(Ⅱ)∶$n(NaOCH_2CF_3)$=1∶5。

2.7 水分、惰性气体保护对合成Ⅲd 的影响

该反应要求在无水环境且有惰性气体保护的条件下在密闭装置中进行，所用试剂在使用前需进行充分干燥，水分的存在会改变整个反应体系中的电荷分布，使产品的收率很低甚至无产品生成。

3 结论

以鸟嘌呤为原料，通过酰化反应制得 N^2·9-二乙酰鸟嘌呤，收率 95.8%。氯化水解得到 2-氨基-6-氯嘌呤（Ⅱ），收率 75.8%。通过烷氧基化反应制得 2-氨基-6-甲氧基嘌呤（Ⅲa）、2-氨基-6-乙氧基嘌呤（Ⅲb）、2-氨基-6-异丙氧基嘌呤（Ⅲc）和 2-氨基-6-三氟乙氧基嘌呤（Ⅲd），其最高收率分别为 96.5%，79.6%，73.2%和 49.8%。合成 2-氨基-6-三氟乙氧基嘌呤（Ⅲd）的最佳工艺条件为：n(Ⅱ)∶$n(NaOCH_2CF_3)$=1∶5，用二甲亚砜（DMSO）作反应溶剂，其用量为 m(DMSO)∶m(Ⅱ)=13∶1，用碘化亚铜（CuI）为催化剂，其用量为 n(CuI)∶n(Ⅱ)=2∶1,反应温度为 160℃。

参考文献

[1] Van A A，Petros M，Piet A. Antiviral activity of C-alkylated pruine nucleosides obtained by crose-coupling with tetralkyltin reagents [J] . J Med chem，1993，36：2938-2942.

[2] Estep K G，Josef K A. Synthesis and structure-activity relantionships of 6-heterocyclic-substituted purine as inactivation modiflers of cardiac sodium channels [J] . J Med chem，1995，38：2582-2595.

[3] Ian R Hardcastle，Christine E Arris，Johanne B，et al. N^2-Substituded O^6-cyclohexylmethyl-guanine derivatives：potent inhibitors of cyclin-dependent kinases 1 and 2 [J] . J Med chem，2004，47：3710-3722.

[4] Narender P，Suyerl B，Padmanava P，et al. Synthesis and reactions of 2-chloro-and 2-tosyl-oxy-2′-deoxyinosine derivatives ［J］. J Org Chem，2005，70：7188-7195.

[5] Ashleigh E G，Christine E Arris，Johanne B，et al. Probing the ATP ribose-binding doman of cyclin-dependent kinases 1 and 2 with O^6-Substituted guanine derivatives ［J］. J Med chem，2002，45：3381-3393.

[6] 林紫云，朱莉亚，梁晓天. 新法合成 6-烷氧基和 6-氨基鸟嘌呤核苷衍生物 ［J］. 有机化学，2000，20（4）：510-513.

[7] 陆鸿飞，陆明，吕春绪，等. 2-氨基-6-卤代嘌呤的合成 ［J］. 精细化工，2004，23（4）：386-392.

Synthesis of 2-Amino-6-Alkoxypurine

Lu Hongfei，Lu Ming，Lv Chunxu，Zhang Lijuan

（School of Chemical Engineering，Nanjing University of Science and Technology，Nanjing 210094，Jiangsu，China）

Fine Chemicals，2006，23（11），1104-1107

Abstract：2-Animo-6-alkoxypurine is an important intermediate of nucleotide. From guanine，through acylation，chloridization and hydrolysis，2-amino-6-chloropurine（Ⅱ，75.8% yield）was obtained. Reaction of Ⅱ with a series of sodium alkoxides gave 2-amino-6-methoxypurine（Ⅲa，96.5% yield），2-amino-6-ethoxypurine（Ⅲb，79.6% yield），2-amino-6-isopropoxypurine（Ⅲc，73.2% yield）and 2-amino-6-triflouroethoxypurine（Ⅲd，49.8% yield）. The products were characterized by elementary analysis，^{1}HNMR，MS and IR. Ⅲd was studied emphatically，and the best reaction conditions were：n(Ⅱ)∶n($NaOCH_2CF_3$)＝1∶5，DMSO as solvent，m(DMSO)∶m(Ⅱ)＝13∶1，CuI as catalyst，n(CuI)∶n(Ⅱ)＝2∶1 and reaction temperature 160℃.

Key words：purine derivatives；alkoxy reaction

（注：此文原载于 精细化工，2006，23（11）：1104-1107）

Selective Oxidation of *p*-Chlorotoluene Catalyzed by Co/Mn/Br in Acetic Acid-Water Medium

Hu Anjun, Lv Chunxu, Li Bindong, Huo Ting
(Department of Chemical Engineering, Nanjing University of Science and Technology, Nanjing, 210094, China)
SCI: 000288465000012

Liquid-phase selective oxidation of *p*-chlorotoluene catalyzed by Co/Mn/Br with molecular oxygen under atmospheric pressure was studied. Acetic acid-water was used as the reaction medium in place of the commonly used pure acetic acid, which inhibited the further oxidation of *p*-chlorobenzaldehyde and enhanced the selectivity. Moreover, it is advantageous to use acetic acid-water as the reaction medium because water is a product of the reaction and the separation of the acetic acid-water mixture thus formed is a difficult and uneconomic process. The effects of the initial water concentration, Co/Mn and Br/(Co+Mn) mole ratios, amount of the catalyst, *p*-chlorotoluene/solvent volume ratio, reaction temperature, reaction time, and oxygen flow rate were investigated. The optimum reaction conditions were as follows: initial water concentration of the solvent 10 wt%, Co/Mn mole ratio 0.67, Br/(Co+Mn) mole ratio 0.4, amount of the catalyst accounting for 4 wt% of the substrate, *p*-chlorotoluene/solvent volume ratio 1.5, reaction temperature 106, reaction time 10h, and the oxygen flow rate 10mL/min. Under these conditions, a 14.3% yield of *p*-chlorobenzaldehyde was obtained at 19.7% conversion of *p*-chlorotoluene with 72.4% selectivity.

1 Introduction

p-Chlorobenzaldehyde is an important intermediate in the production of dyes, agricultural chemicals, and pharmaceuticals, commercially manufactured mainly by chlorination of *p*-chlorotoluene to form *p*-chlorobenzal chloride which is then subjected to acid hydrolysis, a seriously polluting process[1]. Consequently, the development of effective and selective catalytic systems for the oxidation of *p*-chlorotoluene by molecular oxygen to *p*-chlorobenzaldehyde is being rejuvenated. However, selective oxidation of *p*-chlorotoluene is much more complicated because the aldehyde itself is more susceptible to further oxidation to carboxylic acid and the higher the conversion of the substrate, the lower the selectivity of aldehyde will be. Hence, oxidation of *p*-chlorotoluene with high selectivity still remains a chemical challenge.

In recent years, the vapor phase oxidation of methylbenzenes has attracted great interest and special attention has been paid to the improvement of solid catalysts of these vapor phase oxidations, especially the vanadium-containing catalysts[2-16]. Despite the advantageous handling of reactants and products, the vapor phase oxidation has such disadvantages as high reaction temperature and the formation of a large amount of carbon dioxide, contributing to global warming. Further, the conversion has to be kept low to attain a high selectivity of aldehydes, and the low concentration of substrates in the feed mixture poses the problem of recovery.

Co/Mn/Br catalyst is a highly efficient catalyst for the oxidation of methylbenzenes in acetic acid (Figure 1), being widely used in the production of aromatic acids[17]. Many improved methods of this catalytic system have been investigated to convert methylbenzenes to the corresponding aldehydes selectively, but most of them are on the selective oxidation of toluene or substrates with electron-donating groups, *p*-methoxytoluene, for example[18-21]. Moreover, the recovery and the reuse of the commonly used acetic acid is still a problem in that water is a product of the reaction and the separation of the acetic acid-water mixture thus formed with distillation or extraction suffers from many drawbacks: distillation is generally uneconomic because of the high costs and the high energy consumption, and extraction is limited by phase separation and distribution of the components[22-30]. As far as we know, water may have some positive effects on the selective oxidation. Being a strong deactivator for the production of aromatic acids, water may inhibit the further oxidation of *p*-chlorobenzaldehyde and thus increase the selectivity of the reaction[17]. *p*-Chlorobenzylic bromide, a kind of inactive form of bromide, rapidly forming in metal/bromide autoxidation (Figure 1, step Ⅶ), can be largely eliminated because water greatly increases its rate of solvolysis[22]. Furthermore, the acetic acid-water mixed solvent can be readily recovered and reused. Therefore, in this paper, we report an improved process for selective oxidation of *p*-chlorotoluene in acetic acid-water medium using a conventional Co/Mn/Br catalyst system. In view of the recovery and reuse of the substrate that are necessary in the industry, a higher selectivity will lead to a higher yield of *p*-chlorobenzaldehyde at a certain amount of substrate. Subsequently, the main aim of the present work is to explore the optimized reaction conditions for high selectivity at relatively high conversion.

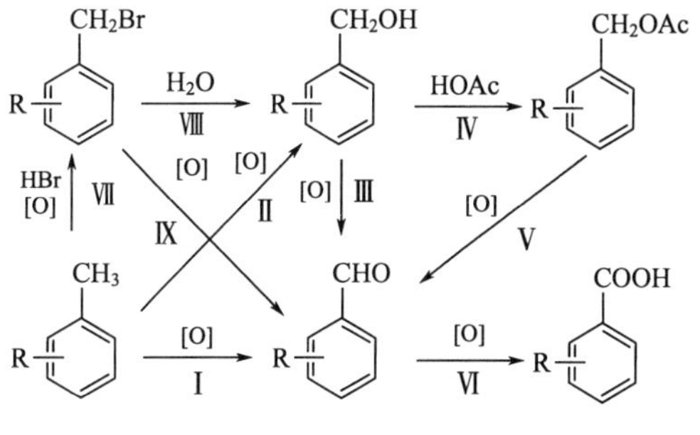

Figure 1 Reaction network during the oxidation of methylbenzenes with a Co/Mn/Br catalyst[22].

2 Experimental section

2.1 Instruments and reagents.

Gas chromatography (GC) analysis was performed on an Agilent GC-6820 equipped with a 30m×0.32mm×0.5μm HP-Innowax capillary column and a flame ionization detector. Mass spectroscopy (MS) spectra were measured on a Trace Ultra-trace DSQ GC-MS spectrometer and a liquid chromatography MS (LC-MS) spectrometer. *p*-Chlorotoluene was purified and analyzed by GC to ensure the absence of auto-oxidation products and impurities before use. Water was distilled. Pure oxygen was used as the source of molecular oxygen. Other reagents were all analytical grade and were used as received.

2.2 *p*-Chlorotoluene oxidation with molecular oxygen.

A composite mixture of a catalyst comprising $Co(OAc)_2 \cdot 4H_2O$, $Mn(OAc)_2 \cdot 4H_2O$, and KBr in specified quantities were added in the acetic acid-water mixture of defined initial water concentration and known amounts of substrate in a reaction flask. While stirring, the reaction system was heated to the required temperature. As soon as the temperature was reached, oxygen was introduced from the bottom of the reaction mixture at a desired flow rate. After a specified time, the reaction mixture was cooled to room temperature and filtered. A gray mixture of black manganese dioxide with the white *p*-chlorobenzoic acid and filtrate **a** were obtained. The gray mixture was washed with 1 mol/L NaOH, and black manganese dioxide and filtrate **b** were gained. The acetic acid-water mixed solvent was recollected with vacuum distillation of filtrate **a**; a certain amount of water was added to the remainder and the mixture attained was separated. The water phase containing catalyst was evaporated to dryness to recover the catalyst and the organic phase was neutralized by 0. 1mol/L NaOH and separated. Water phase **c** and organic phase **d** were acquired. Organic phase **d** was analyzed by GC using chlorobenzene as the internal standard. Combining water phase **c** with filtrate **b**, the solution formed was neutralized by 1 mol/L HCl, and *p*-chlorobenzoic acid precipitated. After filtration and drying, *p*-chlorobenzoic acid was obtained.

3 Results and discussion

3.1 Effect of the initial water concentration.

p-Chlorotoluene conversion at different initial water concentrations is plotted in Figure 2a as a function of reaction time. From Figure 2a, it is clear that *p*-chlorotoluene conversion increases with reaction time but at a given reaction time the conversion decreases with increasing initial water concentration; it decreases slightly when the initial water concentration is lower than 10 wt%, a further increase of the initial water concentration, however, causing a noticeable decreasing effect. Actually, when the initial water concentration is 20 wt%, the catalytic activity of the present system is so low that, after 14h of reaction, the conversion is still lower than 5%. Water results in the shift of the equilibrium toward the reverse reaction, and the higher the water concentration, the more remarkable the trend is; at high water concentrations, a soluble form of manganese (Ⅳ) can form which subsequently reacts with water to form the highly insoluble manganese (Ⅳ) dioxide[31], causing the reduction of the catalytically active form of manganese. Loss of the catalytically active form of manganese results in further reduction of the activity of the composite catalyst because there is a synergistic interaction between Co and Mn[17]. Increasing the water concentration also decreases the solubility of *p*-chlorotoluene. All these factors mentioned above lead to the decrease of the conversion with increasing water concentration especially when the initial water concentration is higher than 10 wt%.

p-Chlorobenzaldehyde selectivity at different initial water concentrations is plotted in Figure 2b as a function of reaction time (when the initial water concentration is 20wt%, the selectivity changes with reaction time being omitted due to the low conversion). As can be seen from Figure 2b, because *p*-chlorobenzaldehyde is an intermediate of the consecutive reactions (Figure 1), for each acetic acid-water mixed solvent with different initial water concentrations, an increase in the reaction time provides a continuous increase in the selectivity, and after a maximal selectivity, a decrease in the selectivity is observed. Though there is a continuous decrease of the selectivity with reaction time in the range of 2-14h when pure acetic acid was used, it could be inferred that the same trend of selectivity exists and the maximal selectivity occurs before 2h of reaction. Besides, after 4 h of reaction, when the initial water concentration is lower than 10 wt %, with increasing initial water concentration, because of the strong deactivation effect of water on reaction Ⅵ (Figure 1), the selectivity of *p*-chlorobenzaldehyde increases at a given reaction time as expected. Further increase of the initial water concentration, however, results in a decrease in selectivity, and this effect could be explained if we notice that when the initial water concentration is higher than 10 wt%, there is an obvious decrease of catalytic activity and, consequently, it is difficult for the *p*-chlorobenzyl alcohol and the *p*-chlorobenzyl acetate to be further oxidized to *p*-chlorobenzaldehyde (reactions Ⅲ and Ⅴ, Figure 1). Moreover, an excessive amount of water may have a simultaneous inhibition effect on reaction Ⅰ (Figure 1). Taking the *p*-chlorotoluene conversion and *p*-chlorobenzaldehyde selectivity into account, the desirable initial water concentration could be 10 wt%.

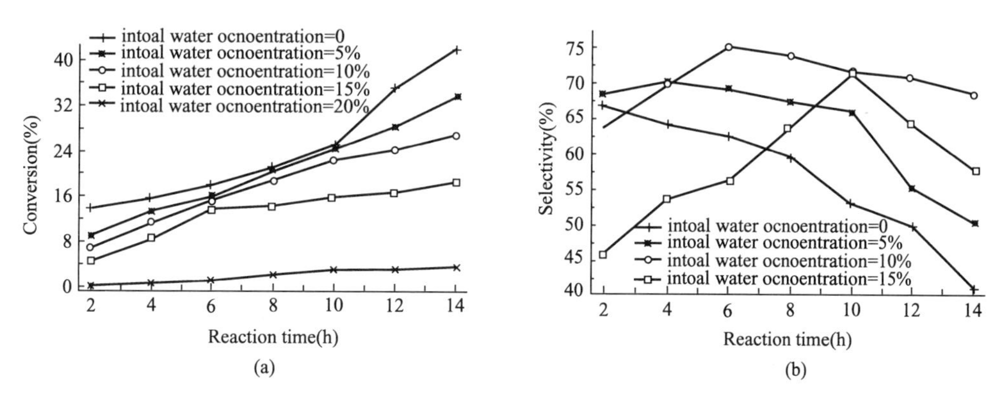

Figure 2 (a) *p*-Chlorotoluene conversion at different initial water concentrations as a function of reaction time. (b) *p*-Chlorobenzaldehyde selectivity at different initial water concentrations as a function of reaction time. Reaction conditions are as follows: substrate 30mL, acetic acid-water mixed solvent 20mL with different initial water concentrations, Co 0.75g, Mn 1.08g, Br 0.38g, and oxygen flow rate 30mL/min.

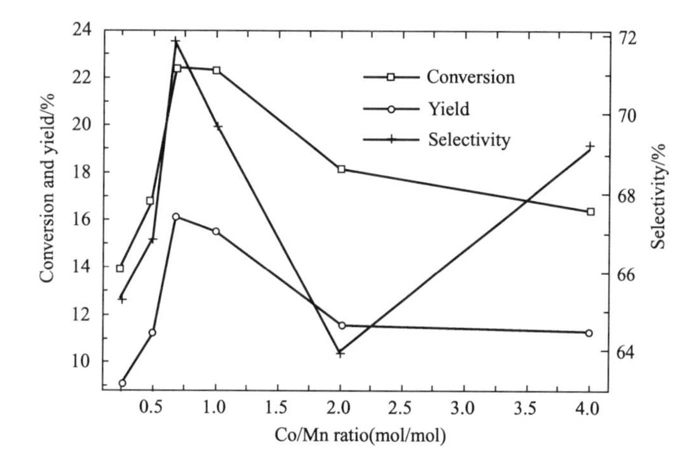

Figure 3 Effect of Co/Mn mole ratio on the oxidation of *p*-chlorotoluene. Reaction conditions are as follows: substrate 30mL, HAc-H_2O 20mL, initial water concentration of the solvent 10 wt%, Br 0.38g, Co+Mn 7.4mmol with different Co/Mn mole ratios, oxygen flow rate 30mL/min, and refluxing for 10h.

3.2 Effect of the Co/Mn mole ratio.

Figure 3 indicates the effect of the Co/Mn mole ratio on the oxidation of *p*-chlorotoluene. Changing the Co/Mn mole ratio from 0.25 to 0.67, with an increase of the Co/Mn mole ratio, sharp increases of conversion, selectivity, and yield take place until they peak at 22.5%, 71.8%, and 16.1%, respectively, when the Co/Mn mole ratio is 0.67. When the Co/Mn mole ratio is in the range of 0.67-2.0, decreases of conversion, selectivity, and yield occur. When the Co/Mn ratio is higher than 2.0, the decrease of conversion continues, while there is a slow increase of selectivity, and accompanying this fall of conversion and rise of selectivity, the yield levels off after a decrease. Theoretically, the synergistic interaction between Co and Mn reaches the highest level when the Co/Mn mole ratio is 1.0. In the present reaction system, however, partial Mn forms manganese (IV) dioxide. Therefore, better catalytic activity could be obtained when the amount of Mn exceeded that of Co slightly.

3.3 Effect of the Br/(Co+Mn) mole ratio.

It is necessary to use bromide to reduce the induction period of the reaction. [17] The presence of bromide also increases the oxidation rate due to its role of electron transfer and free radical initiation. [17] Moreover, a reduction or elimination of Mn (Ⅳ) formation has been observed with the addition of bromide. [31] However, an excessive amount of bromide may result in the formation of metal-multibromides and the inactive form of organic bromide. [17] It is reported that while metal-monobromides are probably responsible for the selective oxidation of C-H bonds in methylbenzenes, [18] metal-multibromides are active for the further oxidation of aldehydes. [18,32] Therefore, the Br/(Co+Mn) mole ratio is a crucial factor and its effect on the oxidation is explored (Figure 4). When the Br/(Co+Mn) mole ratio is lower than 0.4, due to the advantageous effects of bromide mentioned above, the conversion and yield all increase with the increase of the Br/(Co+Mn) mole ratio. When the Br/(Co+Mn) mole ratio is higher than 0.4, consumption of bromide owing to the formation of organic bromide and metal-multibromides leads to a decrease of the conversion. Formation of metal-multibromides also leads to a sharp decrease of the selectivity.

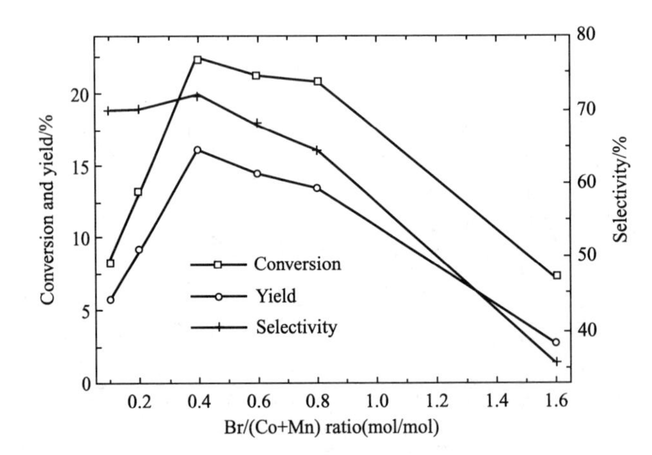

Figure 4 Effect of Br/(Co+Mn) mole ratio on the oxidation of *p*-chlorotoluene. Reaction conditions are as follows: substrate 30mL, HAc-H_2O 20mL, initial water concentration of the solvent 10 wt%, Co 0.75g, Mn 1.08g, different amounts of KBr with the change of the Br/(Co+Mn) mole ratio, oxygen flow rate 30mL/min, and refluxing for 10h.

3.4 Effect of the amount of catalyst.

The effect of the amount of catalyst on the oxidation of *p*-chlorotoluene is indicated in Figure 5. Conversion increases significantly with the increase of the amount of catalyst from 1 to 4 wt%. But, conversion decreases with a further increase of the amount of the catalyst. The same trend is seen for the yield while the amount of catalyst does not have a strong effect on the selectivity. It is bewildering that the effect of the amount of catalyst on the conversion in the present catalytic system is not the same as the effect in the traditional catalytic reaction. This behavior can be identified with the so-called "catalyst inhibitor conversion" phenomenon widely existing in the auto-oxidations catalyzed by transition metal salts and some biomimetic catalytic systems. [33-34]

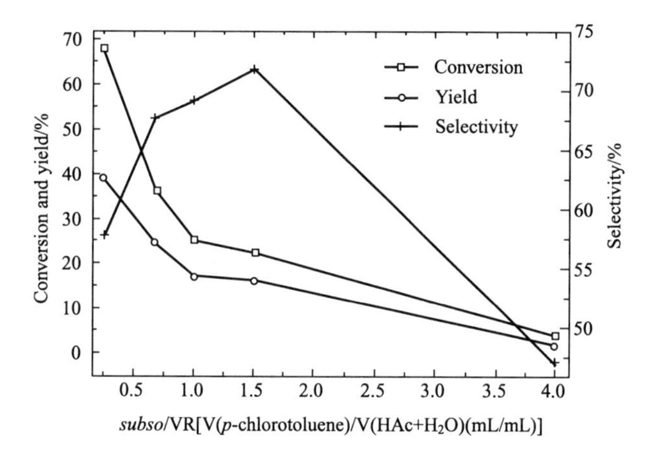

Figure 5 Effect of catalyst amount on the oxidation of p-chlorotoluene. Reaction conditions are as follows: substrate 30mL, HAc-H_2O 20mL, initial water concentration of the solvent 10 wt%, different catalyst amounts with the same Co/Mn/Br mole ratio 1/1. 5/1, oxygen flow rate 30mL/min, and refluxing for 10h.

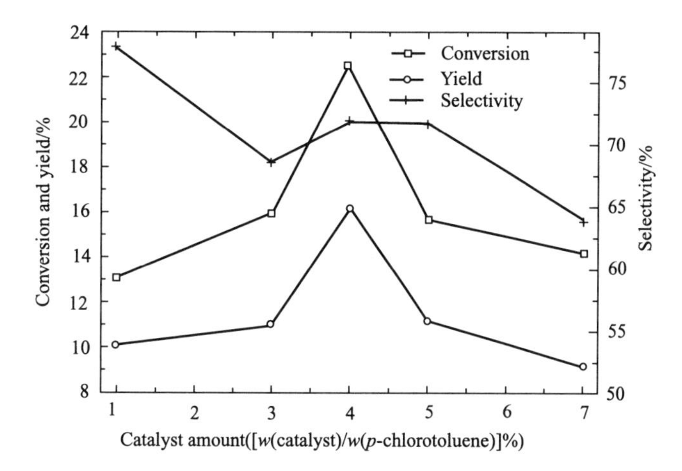

Figure 6 Effect of p-chlorotoluene/solvent volume ratio on the oxidation of p-chlorotoluene. Reaction conditions are as follows: substrate+solvent 50mL with different subsolVRs, initial water concentration of the solvent 10 wt%, Co 0. 75g, Mn 1. 08g, Br 0. 38g, oxygen flow rate 30mL/min, and refluxing for 10h.

3. 5 Effect of the substrate/solvent volume ratio.

The effect of the substrate/solvent volume ratio (subsolVR) on the oxidation of p-chlorotoluene is shown in Figure 6. There is a noticeable increase of selectivity followed by a fast decline with the increase of the subsolVR. The conversion and the yield decrease with the increasing subsolVR. When the subsolVR is 1. 5, the selectivity reaches the highest value; actually, the amount of p-chlorotoluene converted at this subsolVR almost equals the amount of p-chlorotoluene converted when the conversion is the highest (subsolVR) 0. 25) . Apparently, the less of the solvent, the more beneficial the process is. To obtain the highest selectivity and save solvent, it is advantageous to fix the substrate/solvent volume ratio at 1.5.

3.6 Effect of the temperature.

Varying the temperature from 60 to 106℃, the effect of the temperature on oxidation is studied (Figure 7). It is found that the oxidation is highly sensitive to the temperature. When the temperature is below 70℃, the substrate is scarcely converted. When the temperature increases from 80 to 106℃, the selectivity slightly decreases while the conversion increases considerably. Therefore, it is suitable to control the temperature at 106℃.

3.7 Effect of the reaction time.

The changes of conversion and selectivities of different products of the oxidation with time are shown in Figure 8 The selectivity of *p*-chlorobenzaldehyde increases quickly, and the selectivity of *p*-chlorobenzyl acetate and *p*-chlorobenzoic acid rises slowly in the initial 6h. In the same period of time, the selectivity of *p*-chlorobenzyl alcohol decreases rapidly due to the consecutive reactions Ⅲ and Ⅳ (Figure 1). With the further increase of time, the aldehyde concentration in the reaction system is enhanced constantly, leading to its severe further oxidation to an acid, which can be seen in the decrease of the selectivity of aldehyde and the increase of the selectivity of acid.

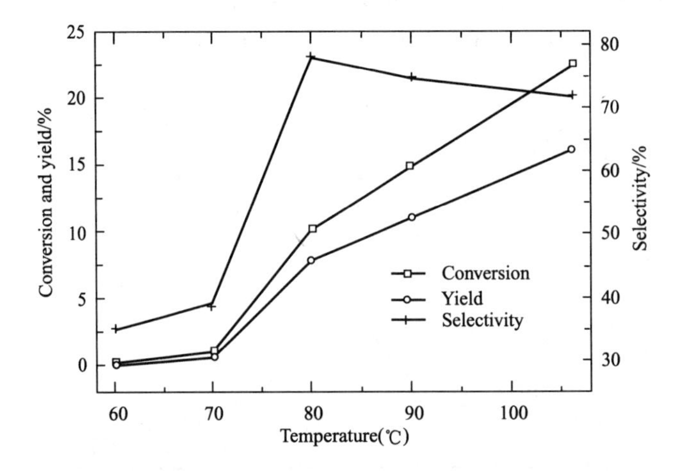

Figure 7 Effect of temperature on the oxidation of *p*-chlorotoluene. Reaction conditions are as follows: substrate 30mL, HAc-H_2O 20mL, initial water concentration of the solvent 10wt %, Co 0.75g, Mn 1.08g, Br 0.38g, oxygen flow rate 30mL/min, reaction time 10h, and different temperatures.

3.8 Effect of the oxygen flow rate.

The dependence of conversion, selectivity, and yield on the oxygen flow rate was investigated in the range of 10-90mL/min. The oxygen flow rate does not have a strong effect on the conversion and yield. When the flow rate is lower than 30mL/min, there is no significant change in the selectivity. With a further increase of the oxygen flow rate, the selectivity decreases rapidly with the enhanced deep oxidation of aldehyde to acid. It is obvious that a minor oxygen flow rate is beneficial for the selective oxidation of *p*-chlorotoluene. Therefore, the oxygen flow rate can be controlled at 10mL/min, at which a 14.3% yield of *p*-chlorobenzaldehyde was obtained at a 19.7% conversion of *p*-chlorotolu-

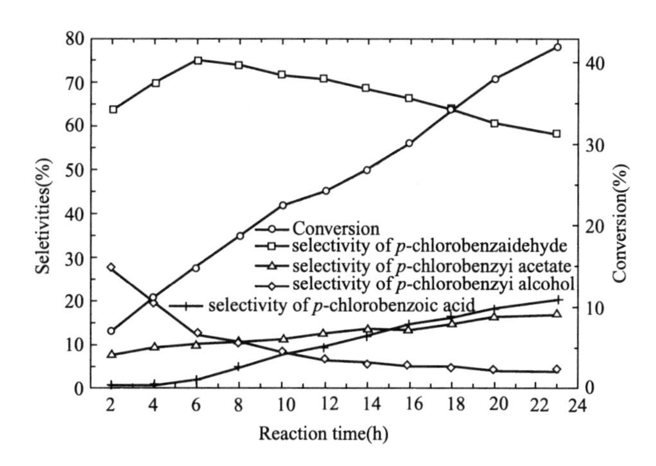

Figure 8 Effect of reaction time on the oxidation of *p*-chlorotoluene. Reaction conditions are as follows: substrate 30mL, HAc-H_2O 20mL, initial water concentration of the solvent 10 wt%, Co 0.75g, Mn 1.08g, Br 0.38g, oxygen flow rate 30mL/min, and refluxing for different periods of time

ene with 72.4% selectivity (see Figure 9).

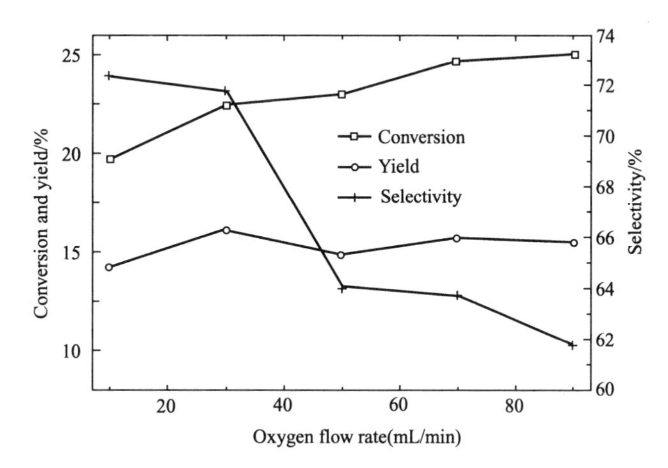

Figure 9 Effect of oxygen flow rate on the oxidation of *p*-chlorotoluene. Reaction conditions are as follows: substrate 30mL, HAc-H_2O 20mL, initial water concentration of the solvent 10 wt%, Co 0.75g, Mn 1.08g, Br 0.38g, and refluxing for 10h at different oxygen flow rates

4 Conclusions

Liquid-phase selective oxidation of *p*-chlorotoluene catalyzed by Co/Mn/Br with molecular oxygen in acetic acid-water medium was studied. The changes of *p*-chlorotoluene conversion and *p*-chlorobenzaldehyde selectivity with reaction time at different initial water concentrations were studied. The results showed that, with the development of the reaction, the conver-sion increases constantly; the selectivity increases first, and after the maximal selectivity occurs, it decreases and the maximal selectivity occurs at a different reaction time with different initial water concentrations. When the initial water concentration is lower than 10 wt%, with increasing initial water concentration, the conversion decreases slightly at a given reaction time; as for the selectivity, after 4h of reaction, it increases at a given reaction time as expected. Further increase of the initial water concentration, however, causes noticeable decrease of the conversion and the selectivity. Due to the strong synergistic interaction between Co and Mn and the strong effect of Br on the oxidation, the Co/Mn and Br/(Co+Mn) mole ratios are optimized. When the Co/Mn mole ratio is 0.67, i.e., when the amount of Mn exceeds that of Co slightly, and the Br/(Co+Mn) mole ratio is 0.4, better catalytic activity could be obtained. The amount of catalyst does not have a strong effect on the selectivity, while it has a strong effect on the conversion, so it is advisable to keep it in the proper range to acquire better catalytic activity. As for the p-chlorotoluene/solvent volume ratio, it is advantageous to fix it at 1.5. The effects of reaction temperature, reaction time, and oxygen flow rate were also investigated. To sum up, the improved process for the selective oxidation of *p*-chlorotoluene has such merits as high selectivity, easy recovery, and reuse of solvent, which are of great importance from an industrial point of view.

Acknowledgment

This work was supported by the Science and Technology Development Project Foundation of Jiangsu Province (No. BE2004045).

Reference

[1] Yoo, J. S. Gas-phase oxygen oxidation of alkylaromatics over CVD Fe/Mo/borosilicate molecular sieve VI. Effects of para-substituents in toluene derivatives. *Appl. Catal.*, *A*: *Gen.* 1996, 135, 261-271.

[2] Gunduz, G.; Akpolat, O. Catalytic vapor-phase oxidation of toluene to benzaldehyde. Ind. Eng. Chem. Res. 1990, 29, 45-48.

[3] Elguezabal, A. A.; Corberan, V. C. Selective oxidation of toluene on $V_2O_5/TiO_2/SiO_2$ catalysts modified with Te, Al, Mg, and K_2SO_4. Catal. Today 1996, 32, 265-272.

[4] Ponzi, M.; Duschatzky, C.; Carrascull, A.; Ponzi, E. Obtaining benzaldehyde via promoted V_2O_5 catalysts. Appl. Catal., A: Gen. 1998, 169, 373-379.

[5] Konietzni, F.; Zanthoff, H. W.; Mailer, W. F. The role of active oxygen in the AMM-VxSi-catalysed selective oxidation of toluene. J. Catal. 1999, 188, 154-164.

[6] Bru¨ckner, A. A new approach to study the gas-phase oxidation of toluene: probing active sites in vanadia-based catalysts under working conditions. Appl. Catal., A: Gen. 2000, 200, 287-297.

[7] Bulushev, D. A.; Kiwi-Minsker, L.; Renken, A. Vanadia/titania catalysts for gas-phase partial toluene oxidation spectroscopic characteriza-tion and transient kinetics study. Catal. Today 2000, 57, 231-239.

[8] Bulushev, D. A.; Kiwi-Minsker, L.; Renken, A. Transient kinetics of toluene partial oxidation over V/Ti oxide catalysts. Catal. Today 2000, 61, 271-277.

[9] Bulushev, D. A.; Kiwi-Minsker, L.; Zaikovskii, V. I.; Lapina, O. B.; Ivanov, A. A.; Reshetnikov, S. I.; Renken, A. Effect of potassium doping on the structural and catalytic properties of V/Ti-oxide in selective toluene oxidation. Appl. Catal., A: Gen. 2000, 202, 243-250.

[10] Besselmann, S.; Loffler, E.; Muler, M. On the role of monomeric vanadyl species in toluene adsorption and oxidation on V_2O_5/TiO_2 catalysts: a Raman and in situ DRIFTS study. J. Mol. Catal. A: Chem. 2000, 162, 401-411.

[11] Bentrup, U.; Bruckner, A.; Martin, A.; Lucke, B. Selective oxida-tion of p-substituted toluenes to the corresponding benzaldehyde over $(VO)_2P_2O_7$: an in situ FTIR and EPR study. J. Mol. Catal. A: Chem. 2000, 162, 391-399.

[12] Bulushev, D. A.; Reshetnikov, S. I.; Kiwi-Minsker, L.; Renken, A. Deactivation kinetics of V/Ti-oxide in toluene partial oxidation. Appl. Catal., A: Gen. 2001, 220, 31-39.

[13] Larrondo, S.; Barbaro, A.; Irigoyen, B.; Amadeo, N. Oxidation of toluene to benzaldehyde over VSb0.8Ti0.2O4 effect of the operating condi-tions. Catal. Today 2001, 64, 179-187.

[14] Martin, A.; Bentrup, U.; Wolf, G. U. The effect of alkali metal promotion on vanadium-containing catalysts in the vapour phase oxidation of methyl aromatics to the corresponding aldehydes. Appl. Catal., A: Gen. 002, 227,

131-142.

[15] Kiwi-Minsker, L.; Bulushev, D. A.; Rainone, F.; Renken, A. Implication of the acid-base properties of V/Ti-oxide catalyst in toluene partial oxidation. J. Mol. Catal. A: Chem. 2002, 184, 223-235.

[16] Chen, M.; Zhou, R. X.; Zheng, X. M. Partial oxidation of p-tert butyl toluene to p-tert butyl benzaldehyde. Appl. Catal., A: Gen. 2003, 242, 329-334.

[17] Partenhermer, W. Methodology and scope of metal/bromide au-toxidation of hydrocarbons. Catal. Today 1995, 23, 69-158.

[18] Yin, G. C.; Cao, G. Y.; Fan, S. H.; Hai, X. D. Selective oxidation of m-phenoxytoluene to m-phenoxybenzaldehyde with methanol as an additive in acetic acid. Appl. Catal., A: Gen. 1999, 185, 277-281.

（注：此文原载于 Industrial & Engineering Chemistry Research，2006，45，5688-5692）

A Polymer Imidazole Salt as Phase-Transfer Catalyst in Halex Fluorination Irradiated by Microwave

Liang Zhengyong, Lv Chunxu, Luo Jun, Li Bindong
School of Chemical Engineering, Nanjing University of Science and Technology, Nanjing 210094, China
SCI: 000247423400004

Abstract: A new imidazole polymer salt was synthesized in order to develop a high efficieney phase-transfer catalyst for multi-phase reactions. The polymer salt was prepared easily by co-polymerization of 1-1′-(1, 4-butamethylene) bis (imidazole) and 1, 2-dibromoethane, and has the properties of excellent chemical and thermal stability and high ionic conductivity. It was applied as phase-transfer catalyst in the fluorination of chloronitrobenzenes under the irradiation of microwave and gave excellent yields of corresponding fluoronitrobenzenes. In addition, the enhanced mechanism of microwave was studied and found "non-thermal" effect was a great factor.

Key words: Microwave chemistry; Polymer imidazole salt; Halogen-exchange fluorination; Fluoronitrobenzenes; Non-thermal effect

1 Introduction

Halogen-exchange (halex) fluorination is an important method to prepare fluorinated compounds. To accelerate the reaction between KF and substrates, phase transfer-catalyst (PTC) is often used. But quaternary ammoniumsalts containing β-H are prone to take place Hofmann Elimination during halex fluorination. At the same time, by-products derived from the decomposition would worsen the reaction greatly. To improve PTC's catalytic activity under the condition of high temperature, quaternary ammonium or phosphonium salts are grafted on polymer carrier and form a new kind of phase-transfer catalyst named polymer PTC. Yoshida grafted N-(2-ethylhexyl)-4-(N, N′-dimethyl) aminopyridinium bromide on polystyrene to get a polymer PTC (noted: Cat. A)[1]. Subsquently, he and his co-workers developed divinylbenzene across linked polystyrene supported tetraphenylphosphonium bromide and its modifled analogue by replacing active hydrogen by methyl group (Note: Cat. B and Cat. C)[2]. The three polymers would not decompose obviously even over 210℃. Luo and his co-workers also reported an effective and stable polydiallyldimethylammonium chloride (PDMDAAC) as PTC was used in hales fluorination[3]. Herein, we describe another stable and effective polyalkylimidazolimu bromide 1 (Scheme 1) acts as PTC in halex fluorination under the irradiation of microwave.

2 Results and discussion

As we know, efflcient and stable PTC should have big relative molecular weight, stable structure and big polarity. As shown in Scheme 1, polymer **1** has a big density of catalytic component, so it should have high phase-transfer activity. In fact, polymer imidazole salts have received much attention because of their remarkable properties including catalytic activity, chemical and thermal stability and high ionic conductivity[4]. In addition, polymer imidazole salts have several beneflcial features for phase transfer catalyst: (1) reactivity of the monomer molecule can be controlled easily by modiflcation of the alkyl chain and reaction temperature, which makes it possible to control PTC's polymerization degree and density of catalytic active component,

N⁀N—H —(1. K; 2. $Br(CH_2)_4Br$)→ N⁀N—$(CH_2)_4$—N⁀N —($Br(CH_2)_4Br$)→ [Br^- N⁺⁀N—$(CH_2)_4$—N⁀N⁺(BrH^-)—CH_2CH_2]$_n$ **1**

2a: 4-Cl
2b: 2-Cl
2c: 2,4-Cl_2
2d: 3,4-Cl_2
2e: 2,5-Cl_2
2f: 4-Cl,3-NO_2

Cl_n-C$_6$H-NO_2 (**2a~2f**) —(SD-KF; PTC,MW)→ F_m-C$_6$H-NO_2-Cl_{n-m} (**3a~3f**)

3a: 4-F
3b: 2-F
3c: 2,4-F_2
3d: 3-Cl,4-F
3e: 5-Cl,2-F
3f: 4-F,3-NO_2

Scheme 1

(2) addition of polymerization initiator is not required, which makes polymerization manageable. For the reasons above, polymer **1** is very suitable for halex fluorination on high temperature. Its decomposition temperature determined by TGA is about 340℃. The TGA data of polymer **1** was demonstrated in Fig. 1. Catalytic capability of different polymer catalysts on preparation of 4-fluoronitrobenzene (PFNB) via halex fluorination from 4-chloronitrobenzene (PCNB) with the conventional heating was tested. The results were shown in Table 1. From Table **1**, it could be found that polymer **1** had excellent catalytic effect than the other polymer catalyst for the sake of its higher density of catalytic active center. [bmim] Br was unstable owing to elimination or substitution processes under the condition of high temperature and KF[5]. While polymer **1** was more stable due to its greater steric hindrance effect.

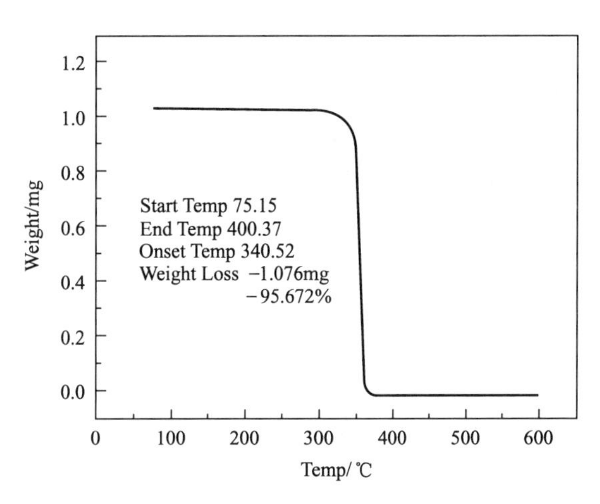

Fig. 1 TGA data of polymer 1.

Reaction rate can be accelerated remarkably in some reactions under the irradiation of microwave, which was attributed to "thermal effect" and "non-thermal effect" of microwave[6]. The reason why "thermal effect" can accelerate reaction rate is known as overheat. So, it cannot change kinetics under the conditions of same reactant, catalyst and product when compared to conventional heating while "non-thermal effect" can change kinetic feature, reduce activation energy and even change reaction order[7]. The promoting effect of "non-thermal effect" is possibly attributed to an enhancement in the polarity of the system from the ground state towards the transition state[8]. It is consistent with the extension of anionic charge distributing in transition state.

Table 1　Comparison of different polymer catalysts in halogen-exchange fluorination

Catalyst	Mw	PCNB (mmol)	KF (mmol)	Solvent	Temperature (℃)	Time (h)	Yield of PFNB(%)	Reference
Polymer 1(0.5g)	28,400[a]	50	75	Me_2SO(10mL)	192	2	92.8	
Cat. A(2.43g)	970	25	37.5	$TMSO_2$(15g)	180	5	87.0	[1]
Cat. B(4.5g)	892	50	75	$TMSO_2$(30g)	180	4	72.0	[2]
Cat. C(4.2g)	840	50	75	$TMSO_2$(30g)	180	5	83.0	[2]
PDMDAAC(0.33g)	200,000	25	37.5	Me_2SO(20mL)	185	3	90.1	[3]
[bmim]Br(0.5g)	219	50	75	Me_2SO(10mL)	192	2	78.4	

a　Mw=28，400；Mn=24，483，D=1.160.

Table 2　Fluorination of different substrates in Me_2SO under irradiation of microwave

Entry	Reactant(**2a-2f**) (substance $C_6H_5NO_2$)	KF (equiv.)	Power (W)	Temperature (℃)[a]	Time (h)	Product(**3a-3f**) (substance $C_6H_5NO_2$)	Yield (%)[b]
1	4-Cl	1.5	300	202	1.5	4-F	94.2
2	2-Cl	1.5	300	203	2	2-F	87.8
3	2,4-Cl_2	3	250	203	2	2,4-F_2	70.2
4	3,4-Cl_2	1.5	250	206	0.5	3-Cl,4-F	87.4
5	2,5-Cl_2	1.5	250	204	1	5-Cl,2F	84.4
6[c]	4-Cl,3-NO_2	1.5	200	90	4	4-F,3-NO_2	88.3

a　Determined by Reytek infrared thermometer.

b　Determined by GC.

c　Reacted in 15 mL CH_3CN.

Owing to the defect of our microwave equipment without temperature controller, we could not study "non-thermal effect" of microwave through comparing the activation energy (E_a) between microwave irradiation and conventional heating. But based on the theoretical premise that microwave can change kinetics feature, "non-thermal effect" can be revealed from the change of kinetics curve under microwave and convention heating conditions.

The kinetics experiments were carried based on considerations below: (1) to eliminate the influence of diffusion, all reactions were carried at the same stirring speed. (2) To eliminate the influence of PTC's decomposition on fluorination reaction rate during the process of reaction, all reactions were carried without PTC. The kinetics curve on condition of microwave irradiation and conventional heating were shown in Figs. 2 and 3. From Fig. 2, we could find the reaction followed pseudo-first order kinetics based on the C (PCNB) under the conventional heating because the concentration of the F^- was essentially constant due to the excess of insoluble KF present throughout the reaction. To our surprise, the relationship of ln C (PCNB) and reaction time was not linear any more on condition of microwave irradiation (Fig. 2). The reaction followed pseudo-second order kinetics based on the C^2 (PCNB) (see Fig. 3). Different substrates were halex fluorinated in Me_2SO under the irradiation of microwave. The results were shown in Table 2. From Table 2, it could be found that all reactions were performed well under the irradiation of microwave. The reason why 3, 4-dichloronitrobenzene could be promoted best is that the "non-thermal effect" of microwave was most signiflcant in the fluorination. To our surprise, the fluorination reaction rate of 2,4-dichloronitrobenzene could be accelerated slightly. The reasonmay be the limitation ofmass transfer between inorganic solid phase and organic liquid phase besides of reaction activity of substrate, which is also the reason why more and more scientists devoted themselves to research and development of high efflciency phase-transfer catalyst.

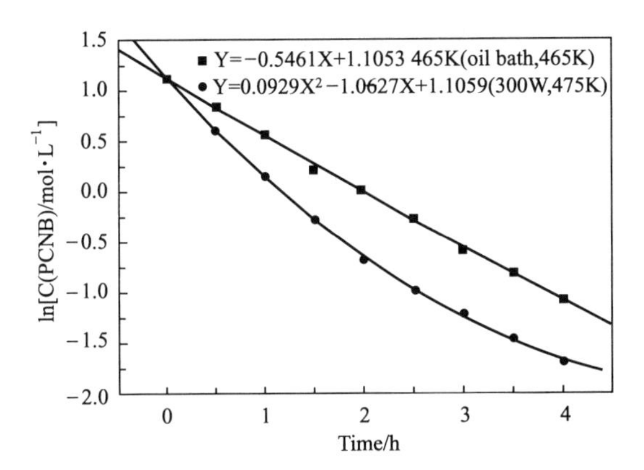

Fig. 2 First-order kinetic plot for fluorination of PCNB in Me_2SO_4 under the irradiation of microwave.

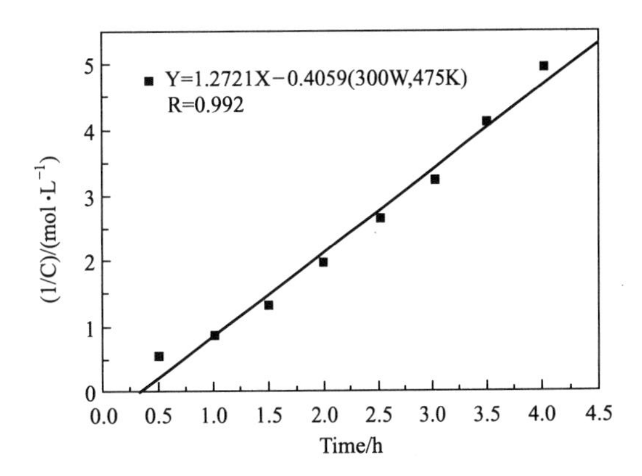

Fig. 3 Second-order kinetic plot for fluorination of PCNB in Me_2SO under the irradiation of microwave.

3 Conclusion

Polymer **1** is a new kind of polymer phase transfer catalyst, which has the superiority of convenience for preparation, good thermal stability and high catalytic activity. So it is very suitable for halogen-exchange fluorination at high temperature. Some fluorinitrobenzene derivatives were prepared via halogen-exchange fluorination catalyzed by polymer **1** from corresponding chloronitrobenzene derivatives under the irradiation of microwave. The yields of products under the optimum reaction conditions were in the range of 70. 2-94. 2%, and the reaction time can be shortened more or less when compared with conventional heating. The catalytic effect is better than other kind of polymer phase transfer catalysts. In addition, "non-thermal effect" of microwave was found in fluorination through comparing macroscopic kinetics between conventional heating and microwave heating.

4 Experimental

Thermal gravimetric analyses data were obtained by SHIMADZU DTA-50 thermal analyzer. Molecule weight was recorded on AGILENT 1100 gel permeation chromatograph. Reactions were carried in Sanle WHL07S-1 chemistry-oriented microwave oven made in Nanjing Sanle Microwave Technology and Development Co. Ltd. Yields were determined by Agilent 4890 gas chromatograph.

4.1 Synthesis of polyalkylimidazolium bromide

1-1′-(1,4-Butamethylene) bis (imidazole) was prepared by a coupling reaction of 1mol 1,4-dibromobutane with 2mol imidazole potassium salt in THF according to procedures described in reference[9]. A 0. 1mol 1,2-dibromoethane was added dropwise to a stirred mixture of 0. 1mol 1-1′-(1,4-butamethylene) bis (imidazole) and 50mL toluene, then the mixture was heated and maintained at 110℃ for 24h. Cooled to room temperature and flltered, yellow crude product could be obtained. The crude product was wished several times with acetic ether and chloroform successively and dried in vacuum at 100℃ for 36h before use.

4.2 Preparation of fluorinated nitrobenzene derivates

Seventy-five millimole or 150mmol of KF, 0.5g of polyalkylimidazolium bromide 1, 10mL of Me_2SO and 50mmol of substrate were placed in an open vessel equipped with a mechanical stirrer and water condenser, then heated by microwave with violently stirred and maintained at reflux temperature for certain time. After cooled down to temperature and filtered, a sample of crude product mixture was analyzed by gas chromatography. Conversion and yield were determined by the internal standard method.

References

[1] Y. Yoshida, Y. Kimura, Tetrahedron Lett. 30 (1989) 7199-7202.
[2] Y. Yoshida, Y. Kimura, M. Tomoi, Chem. Lett. (1990) 769-772.
[3] J. Luo, C. X. Lv, C. Cai, et al. J. Fluorine Chem. 125 (2004) 701-704.
[4] K. Suzuki, M. Yamaguchi, S. Hotta, et al. J. Photochem. Photobiol. A: Chem. 164 (2004) 81-85.
[5] C. B. Murray, G. Sandford, S. R. Korn, J. Fluorine Chem. 123 (2003) 81.
[6] D. Stnerga, P. Gainard, Tetrahedron. 52 (1996) 5505-5510.
[7] B. Dayal, K. Rao, G. Salen, Steroids. 60 (1995) 453-457.
[8] S. Marque, H. Snoussi, A. Loupy, et al. J. Fluorine Chem. 125 (2004) 1847-1851.
[9] S. K. Sharma, M. Tandon, J. W. Lown, J. Org. Chem. 65 (2000) 1102-1107.

(注：此文原载于 Journal of Fluorine Chemistry, 2007, 128 (6): 608-611)

Self-assembly Microstructures of Amphiphilic Polyborate in Aqueous Solutions

Wang Haiying[1], Chai Liyuan[1], Hu Anjun[2], Lv Chunxu[2]*, Li Bingdong[2]
(1. School of Metallurgical Science and Technology, Central South University, Changsha 410083, P. R. China)
(2. School of chemical engineering, Nanjing University of Science and Technology, Nanjing 210094, P. R. China)
SCI: 000267415600030, EI: 200992312109276

Abstract: A new type of amphiphilic homo-polyborate was synthesized with the surface tension (γ_{CMC}) in deionized water 31.2mN · m^{-1} (pH=7, 25℃), which is similar to low-molecule surfactants. Self-assembly behaviors of the polyborate in aqueous solutions were investigated by TEM. Self-aggregation morphologies of the polyborate can be controlled by the concentration of NaCl. With the increase of the concentration of NaCl, micromorphologies from the amphiphilic polyborate transformed from spherical polymeric vesicles to treelike or even vertebra-like morphologies. It was also found that, the polymeric vesicles show a strong stability against ethanol addition, even when the concentration of ethanol increases to a volume ratio of 33%, they still maintain good micromorphologies. A possible mechanism of vesicles formation was proposed based on packing parameter theory.

Keywords: Self-assembly, Vesicles, Spontaneous, Polyborate

1 Introduction

Amphiphilic molecules in solution tend to self-assembly to produce diverse microstructures such as micelle, vesicle, which have attracted considerable attention over the past decades because of their scientific interests and potential applications in drug delivery, biomembrane mimic, nanoreactors[1-6]. Among the various morphologies, the closed bilayer membranes, polymeric vesicles (also called polymersomes), attract much attention due to their more stability and higher membrane toughness than regular low-molecule lipidsomes[3-6]. Up to now, polymeric vesicles have been prepared generally from a variety of some amphiphilic block copolymers containing covalently linked hydrophobic block and hydrophilic block. Vesicles morphologies formed with block copolymers can be controlled in aqueous solution depending on the fraction of hydrophobic/hydrophilic block, as well as solvent properties, such as the composition, pH, or ionic strength[7,8]. A variety of amphiphilic block copolymers have been synthesized to explore the interesting phenomenon, for example, poly (styrene-*b*-4-vinylpridine) (PS_{80}-*b*-$P4VP_{110}$)[9], polymethylene-*b*-poly (dimethyl siloxane)-b-polymethylene (PM-*b*-PDMS-*b*-PM)[10], ethyl cellulose-graft-polystyrene (EC-g-PS)[11]. In general, input of external energy during these vesicles preparation is required, so that several conventional methods have been developed, such as ultrasonic, dialysis, extrusion, etc[12-16].

During the past ten years, growing interests have been paid on investigating the spontaneous formation of amphiphilie's vesicles due to their easy process and quite stability when compared with "conventional" vesicles[16]. By far, the most spontaneous vesicle-forming systems reported in the literatures are based on aqueous mixtures of oppositely charged amphiphiles[17]. Recently, copolymers with di/triblock of PEO/PPO[18], PEO/PBO[19,20] and some complexes of block ionomers and surfactants[21] were also found to possess the properties of spontaneous vesicles formation, and great attentions have been paid on these. In fact, until now, few about self-assembling character of homopolymers in aqueous solution have been reported. As a consequence, spontaneous vesicles made from homopolymers have been rather scarce.

In the present work, a novel amphiphilic linear homopolymer surfactant containing boron was synthesized. The prepared polyborate consists of a hydrophobic C_{12} segment which is attached to a hydrocarbon chain by a hydrophilic link containing a borate unit(the polyborate **4a**, see Scheme 1). The target amphiphilic polyborate has an excellent surface activity in aqueous solution similar to low-molecule surfactants, and shows the ability to spontaneously self assembly to form polymeric vesicles in aqueous solutions. And by transmission electron microscopy (TEM), the transition between the assembled geometries triggered by different concentration of NaCl and ethanol was also observed. The interesting phenomenon is that the morphologies of the vesicles changes spherical vesicles to tree-like and vertebra-like structure, which, to our knowledge, has not been reported in the literatures.

2 Experiment section

2.1 Materials and process

The polyborate **4a** and its monomer **4b** were synthesized according to the following procedures. N, N-dihydroxyethyl dodecylamine **1** and triethyl borate **2** were dissolved in benzene. The mixture was stirring for 1h at 60℃ under reduced pressure to remove the produced ethanol by $ZnCl_2$ powder. After N-hydroxymethyl acrylamide **3** and azoisobutyronitrile (AIBN) (molar ratio: [1]/[2]/[3]/[AIBN] =1 : 1 : 1:0.006)added, the mixture was kept for additional 1.5h, and then was maintained at 75℃ for 2h under normal pressure without stirring. The obtained polymer, isolated by precipitation into acetone, followed by filtration to remove unreacted material and oligopolymers, is white or light-yellow powder with the yield 78%. With phenothiazine to replace AIBN, we obtained viscous liquid monomer **4b** in the similar way (see Scheme 1). Molecular weight of **4a**, measured with SEC apparatus (using Plgel MIXED-C 79911GP-MXC column and calibrated with polystyrene standards, the effluent is DMF, Agilent 1100), is 28400 $g \cdot mol^{-1}$ with polydiversity index Mw/Mn 1.16. The HN-MR and FT-IR analysis of **4a** and **4b** was reported in our previous study[22]. Other reagents used are all analytical grades and were used as received.

2.2 Surface activity

Surface tension in deionized water at 25℃ (pH=7) was determined by surface tensionmetry

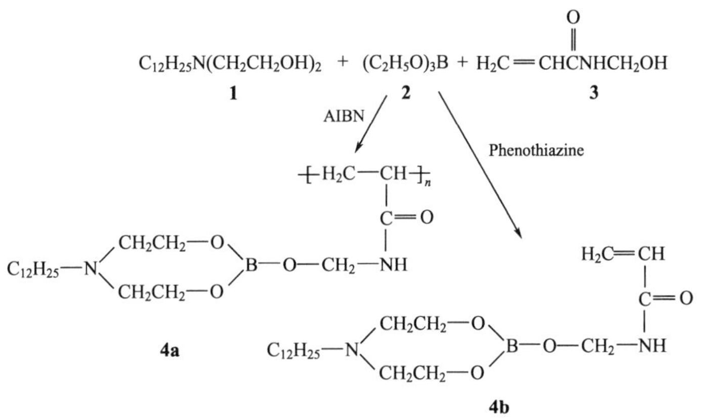

$C_{12}H_{25}N(CH_2CH_2OH)_2$ + $(C_2H_5O)_3B$ + $H_2C{=}CHC(O)NHCH_2OH$

1 2 3

AIBN

Phenothiazine

4a

4b

Sheme 1 Synthesis of polyborate and its monomer

(Du Nouy ring, JYW-200) at 25±0.2℃. The measurement was calibrated with respect to the surface tension of the pure water (γ_{CMC}=72.0mN·m^{-1}).

2.3 Transmission electron microscopy (TEM)

A droplet of wt. 0.1% aqueous samples in solution was placed onto a carbon-coated copper grid. Filter paper was applied to suck away the excess liquid. The specimens were examined on a JEOL JEM-2100 electron microscope at room temperature.

3 Results and discussion

3.1 Surface activity

In order to examine the surface activity of the polyborate, surface tensions were measured in deionized water. The plot of surface tension versus concentration of the polyborate **4a** and its monomer **4b** is shown in Figure 1. The breakpoint of the plot generally determines the CMC (Critical Micelle Concentration) value of the surfactant, and the surface tension values vary slightly after CMC or remain constant afterwards. The surface tension values of the polyborate and its monomer in aqueous solutions are listed in Table 1.

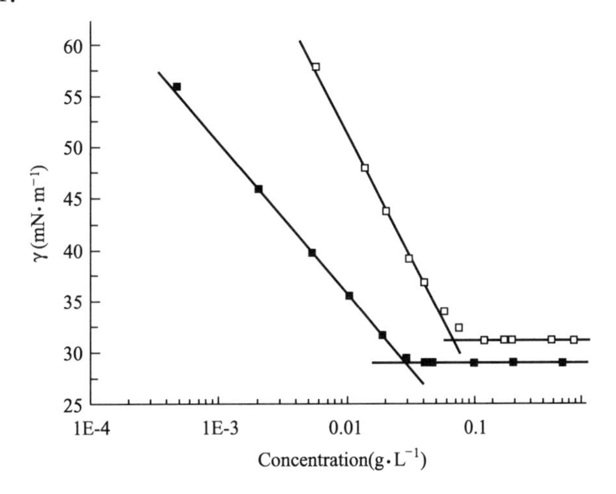

Figure 1 surface tension (γ) plots of polyborate 4a (□) and monomer 4b (■) in deionized water (pH=7.0, 25℃)

As can be see from Table 1, the CMC of polyborate **4a** in water was 68.4mg・L^{-1}, and the surface tension γ_{CMC} was 31.2mN・m^{-1} (pH=7), while the CMC of monomer **4b** was about 28.3 mg・L^{-1} and the surface tension γ_{CMC} 29.2mN・m^{-1}. The γ_{CMC} values suggest that the surface activity of the polyborate is as good as the monomer. The subsequent measurements also showed the NaCl addition has no significant influence on the surface tension of the polyborate **4a** and its monomer **4b** due to their nonionic properties.

Table 1 Values of CMC, γ_{CMC}, the maximum interfacial excess concentration Γ_m, the minimum area per molecule A, and the packing parameters P of 4a and 4b in aqueous solution.

	solution	CMC /mg・L^{-1}	γ_{CMC} /mN・m^{-1}	Γ_m /μmol・m^{-2}	A/nm^2	P
Polyborate, **4a**	water	68.4	31.2	4.85	0.343	0.604
	0.1 M NaCl	67.0	31.1	4.78	0.348	0.596
	0.5 M NaCl	64.0	30.9	4.84	0.344	0.602
Monomer, **4b**	water	28.3	29.2	2.66	0.625	0.332
	0.1 M NaCl	27.7	29.2	2.60	0.640	0.324
	0.5 M NaCl	26.5	29.0	2.67	0.623	0.334

3.2 Self-assembly property

The borate samples were dissolved in freshly distilled water. After the solution was kept at room temperature for 24h, self-assembly morphology of the polyborate and monomer was investigated by TEM directly. As indicated by TEM measurements, no vesicles and other aggregates from monomer **4b** in aqueous solutions were observed, whereas, after addition of NaCl (0.5M), thread-like micelle morphology appears (shown in Figure 2a) in this solution.

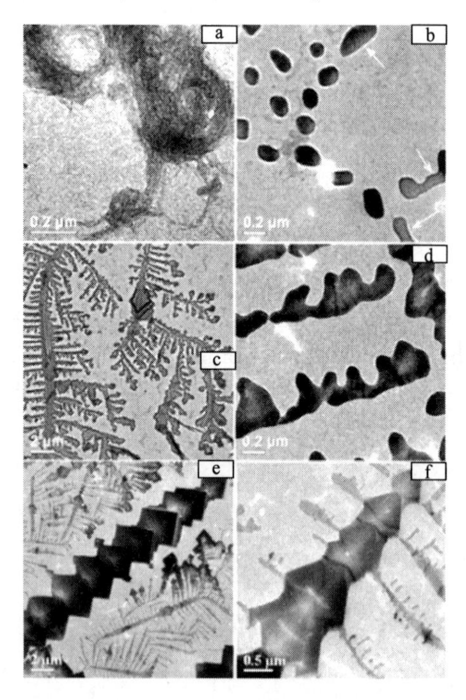

Figure 2 TEM image of 4b micelles in (a) 0.5M NaCl, and TEM image of 4a aggregates in (b) 0.1M NaCl, in (c, d) 0.5M NaCl, in (e, f) 0.9M NaCl. Vesicles formation coexists with fusion and fission (arrows)

In previous work[22], we reported the polyborate **4a** has a tendency to produce spherical polymeric vesicles spontaneously in pure water with diameters of ~20nm, though they were associated seriously. When NaCl concentration in aqueous solution increasing to 0.1M, **4a** exhibited spontaneous formation of ordered polymersomes with uniform diameters (150~250nm, see Figure 2b) without any normal mechanical process, such as extrude, dialysis, sonication. Self-assembly microstuctures generally are dynamic aggregates. Upon higher ionic strength, the molecules polar group will undergo "dehydration"[19]. The caused increasingly hydrophobic will reduce their exchange kinetics[17] and drive polyborate molecules to rearrange to form more stable bilayer vesicles. It should be noted that fusion or fission phenomena among these polymersomes were observed, just as the areas designated by white arrows in Figure 2b.

Further experiments showed self-assembly morphology of **4a** in aqueous solution changed greatly with the increase of NaCl concentration. In 0.5M NaCl, polymersomes from **4a** were treelike aggregate structure (Figure 2c and enlarged 2d). A bilayer is composed of two monolayers. The bilayer curvature free energy, E, can be given by Helfrich-type model[23]:

$$E/A=\frac{1}{2}K[(c+c_0)^2+(c-c_i)^2] \tag{1}$$

Where A is per unit area for bilayers, K the bilayer bending modulus, c the vesicle curvature ($c=1/R$, R is the vesicle radius), c_o the outer monolayer spontaneous curvature, and c_i the inner monolayer spontaneous curvature. For single component bilayers, c_i necessarily equals c_o. Then we can imagine the free energy E/A will be reduced by decreasing c. When $c=0$, the bilayers has the lowest E/A and become "flat" with zero spontaneous curvature. High degree of fusion among vesicles raised by increasing NaCl concentration enable polymeric bilayers broken and lower their curvature and then form a long-strip treelike microstructures. As NaCl concentration in aqueous solution increased to 0.9 M, further fusion of these aggregates was strengthened to make polymeric bilayers curvature low enough to produce flat bilayer structures leading to the formation of a vertebra-like morphology composed of many rectangular aggregates (Figure 2e and 2f).

It is well known that ethanol leads to the destruction of surfactant aggregates in that it weakens the hydrophobic property of surfactants[24]. Effects of ethanol addition on morphologies of **4a** aggregates in NaCl aqueous solution were analyzed by TEM, which were shown in Figure 3. When the volume of ethanol reaches 33% of the total solution volume ($V_{ethanol}$ is equal to 33%), the polyborate **4a** was still found to be able to form spherical polymersomes spontaneously in 0.1 M NaCl (Figure 3a). It indicates that the formed vesicular aggregates from the polyborate **4a** have a strong resistance to large concentration of ethanol as a result of high toughness and low mobility of polymer bilayers[3,25], which exhibit potential applications in drug-release or nanoreactors. However, compared Figure 3a with 2b, we can see that ethanol has a great impact on the self-assembly of the polyborate **4a** evidently because the polymersomes sizes were decreased after adding ethanol. At higher NaCl concentration ($V_{ethanol}=33\%$), **4a** aggregates morphology changes greatly: in 0.5M NaCl, the aggregates transformed from tree-like morphology into tubular morphology while in 0.9M NaCl the vertebra-like morphology was also impaired greatly (Figure 3c). It is interesting that a number of spherical polymersomes appeared again and aligned in a highly ordered array with increasing ethanol concentration in 0.9M NaCl, though the polydiversity increased (Figure 3d). This implied there should exist some interactions among them. These phenomena also indicate that the complicated microsturctures of tree-like and vertebra-like morphologies (in Figure 2c, 2e) originate from the self-assembly of the polyborate and its fusion process. As a whole, aggregates morphology of **4a** in aqueous solution could be expected to be adjusted by addition of NaCl and ethanol.

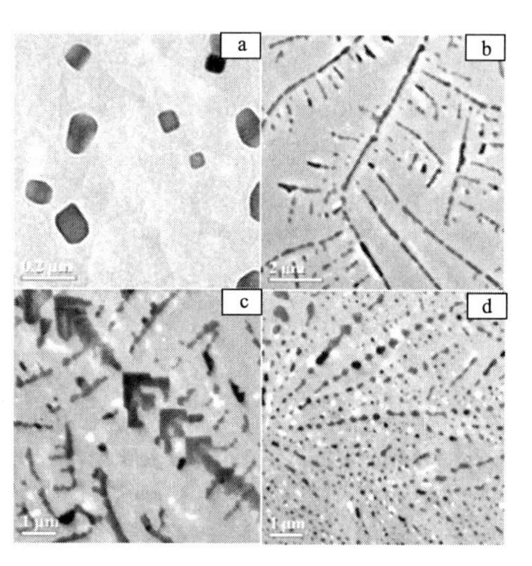

Figure 3 TEM images of 4a at $V_{ethanol}=33\%$ in (a) 0.1M, (b) 0.5M and (c) 0.9M NaCl; and TEM image at $V_{ethanol}=$ 66% in (d) 0.5M NaCl.

3.3 Mechanism of vesicles formation

The concept of molecular packing parameter introduced by Israelachvili, Mitchell and Ninham has been widely applied to explain self-assembly phenomenon of amphiphilic molecules[26-28]. The molecular packing parameter is defined as $P=v_c/a_0 l_c$, where v_c is the volume of the hydrophobic tail of amphiphilic molecules, l_c is the length of the hydrophobic tail, and a_0 is the equilibrium head group area of the molecule at the aggregate surface. Generally, v_c and l_c can be calculated by using Eq. (2) and Eq. (3)[29] respectively, and a_0 can be obtained using Gibbs Equation, Eq. (4) and Eq. (5)[30].

$$v_c=(27.4+29.6n)\times 10^{-3} nm^3 \tag{2}$$

$$l_c=(0.15+0.1265n) nm \tag{3}$$

$$\Gamma_m=-\left(\frac{1}{2.303RT}\right)\left(\frac{\partial\gamma}{\partial \log C}\right)T \tag{4}$$

$$A=\frac{1}{N_A\Gamma_m} \tag{5}$$

Where n is carbon number of hydrophobic chain, Γ_m is the maximum interfacial excess concentration in moles per square meter, A is the minimum area per molecule, $\partial\gamma/\partial\log C$ is the slope of the surface tension of amphiphilic molecules, and N_A is Avogadro's constant. The maximum interfacial excess concentration is obtained from the slope of the surface tension versus the logarithm of the concentration. Packing parameter P can be calculated by approximating the a_0 by the A obtained from the surface tension.

The well-known relation between the molecular packing parameter and the aggregate shape are as follows: $0\leqslant P\leqslant 1/3$ corresponds to sphere micelles (cone-shaped amiphiphilies molecules), $1/3<P\leqslant 1/2$ corresponds to rod/thread-like micelles (truncated cone), $1/2<P\leqslant 1$ to bilayer or vesicle (larger truncated cone).

Vesicle formation of the polyborate **4a** could be explained by packing parameter theory. Here, the polyborate **4a** was considered as monomer **4b** congeries formed in some ways, in other words, we regarded the polyborate **4a** solution as a analogic monomer solution to carry out the calculations. Then, depending on l_c, v_c and a_0 of single molecules (segment) by measurement of surface tension isotherms of the amphiphiles in aqueous solution, we calculated the packing parameter ($P=v_c/a_0 l_c$), which are listed in Table 1.

The addition of NaCl almost does not influence the values of P, Γ_m and A (Table 1). Nevertheless, the A of polyborate **4a** at interfaces in both water and NaCl aqueous solution are much lower than monomer **4b**. This would be attributed to the covalent link between monomer segments along the polyborate hydrocarbon chain. As a result, p values of the polyborate **4a** increase significantly to about 0.6 in contrast to about 0.32 of the monomer **4b** in both solutions. Thus, according to the packing parameter theory, the monomer **4b** molecules would show cone-shape in aqueous solution, and it is likely that no bilayer aggregates or vesicles but only thread-like micelles would appear theoretically, as was confirmed by TEM. High concentration of NaCl is beneficial to self-assembly of monomer **4b** consequently. On the other hand, since P values of the polyborate **4a**$>1/2$, monomer segment of **4a** main chain

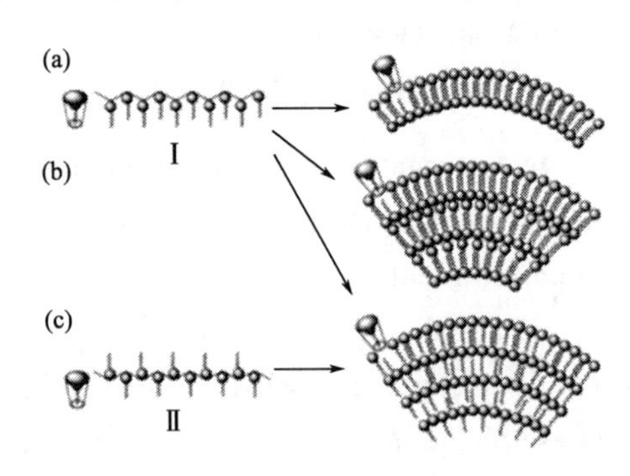

Scheme 2 (a) Unilamellar vesicles formation by structure I of 4a, (b) Multilamellar vesicles formation by structure I of 4a, and (c) Multilamellar vesicles formation combined by structure I and II of 4a.

in both water and NaCl solution would be larger truncated-cone shape and is consequently apt to form bilayers or vesicles. Generally the walls in such aggregates are formed with the hydrophobic chain of amphiphilic polymers[20]. A possible mechanism of **4a** vesicles formation in aqueous solution is illustrated by Scheme 2. The hydrophobic chains of **4a** are expected to be distributed at the same side (see structure I in Scheme 2) or contrary side (see structure II in Scheme 2) along the polymer backbones. As a result, the macromolecules possessing these structures would be located, according to the scheme 2, to construct the outer and internal layer of asymmetry bilayers respectively with a non-zero spontaneous curvature, and then was assembled with each other to form unilamellar or multilamellar vesicles. The increased membrane thickness leads to low permeability and high stability of the polymersomes. After NaCl addition, the dehydration effect and fusion process will make vesicles more stable or lower the bilayers curvature to form the treelike or vertebra-like morphologies.

Conclusions

In summary, a novel amiphiphilic polyborate in aqueous solution was synthesized, which has 28400g · mol^{-1} molecule-weight and high surface activity close to its monomer. It is more interesting that TEM measurement indicates that the obtained polyborate could spontaneously self-assembly to form spherical polymersomes of low polydiversity. And the transition to tree-like and even vertebra-like structures in aqueous solutions was found with increasing NaCl concentration. Ethanol generally is liable to disrupt amphiphilies microstructures, however, polyborate at $V_{ethanol}\%=33\%$ still can form stable polymersomes in 0. 1M NaCl. The packing parameter theory is helpful to give an illustration of the formation mechanism of the polyborate vesicles. After ethanol addition, the tree-like and vertebra-like structures from polybotate were found to be disrupted to form tubular structure or make spherical polymersomes reemerged again in solution. The self-assembly morphologies from polyborate are tunable by addition of NaCl and ethanol, which may provide some possibilities for drug-release vehicles and nanoreactor.

Acknowledgments

Financial support was given by National Key Technology R&D Program (No: 2007BAC25B01), Key Project of NSFC (No: 50830301), and Key Project of Chinese Ministry of Education of China (No: 308019).

References

[1] You LC, Schlaad H. J Am Chem Soc 2006; 128: 13336.
[2] Gao JP, Wei YH, Li BY, Han YC. Polymer 2008; 49: 2354.
[3] Opsteen JA, Cornelissen JJLM, Hest, JCM. Pure Appl Chem 2004; 76: 1309.
[4] Zhang WD, Zhang W, Zhou NC, Cheng ZP, Zhu J, Zhu XL. Polymer 2008; 49: 4569.
[5] Discher DE, Eisenberg A. Science 2002; 297: 96.
[6] Walther A, Goldmann AS, Yelamanchili RS, Drechsler M, Schmalz H, Eisenberg A, Müller AHE. Macromolecules 2008; 41 (9): 3254.
[7] Strandman S, Zarembo A, Darinskii AA, Loflund B, Butcher SJ, Tenhu H. Polymer 2007; 48: 7008.
[8] Lefevre N, Fustin C, Varshney SK, Gohy JF. Polymer 2007; 48: 2306.
[9] Gao LC, Shi LQ, Zhang WQ, An YL, Liu Z, et al. Macromolecules 2005; 38: 4548.

[10] Wang JD, Horton JH, Liu GJ, Lee SY, Shea KJ. Polymer 2007; 48: 4123.
[11] Liu WY, Liu RG, Li YX, Kang HL, Shen D, Wu M, Huang Y. Polymer 2009; 50: 211.
[12] Engberts JB, Kevelam J. Current opinion in Colloid & Science 1996; 1: 779.
[13] Wang W, McConaghy AM, Uchegbu IF. Langmuir 2001; 17 (3): 631.
[14] Li XL, Ji J, Shen JC. Macromol Rapid Commun 2006; 27: 214.
[15] Liu Y, Zhao DY, Ma RJ, Xiong DA, An YL, Shi LQ. Polymer 2009; 50: 855.
[16] Segota S, Tezak D. Advances in Colloid and Interface Science 2006; 121: 51.
[17] Svenson S. Current opinion in Colloid & Science 2004; 9: 201-222.
[18] Bryskhe K, Jansson J, Topgaard D, Schillen K, Olsson U. J. Phy. Chem. B 2004; 108: 9710.
[19] Harris JK, Rose GD, Bruening ML. Langmuir, 2002; 18: 5337.
[20] Norman AI, Ho DL, Lee JH, Karim A. J Phys Chem B 2006; 110: 62.
[21] Kabanov AV, Bronich TK, KAbanov VA, Yu K, Eisenberg A. J Am Chem Sci 1998; 120: 9941.
[22] Wang HY, Li BD, Hu AJ, Lu CX. Chem J Chin Univ 2007; 28: 358.
[23] Walker SA, Zasadzinski JA. Langmuir 1997; 13: 5076.
[24] Bauer A, Kopschutz C, Stolzenburg M, Forster S, Mayer C. Journal of Membrane Science 2006; 284: 1.
[25] Zhu JT, Jiang Y, Liang HJ, Jiang W. J Phys Chem B 2005; 109 (18): 8619.
[26] Israelachvili JN, Mitchell DJ, Ninham BW. Biochim Biophys Acta 1977; 470: 185.
[27] Engberts JBFN, Kevelam J. Current Opinion in Colloid & Interface Science 1996; 1: 779.
[28] Nagarajan R. Langmuir 2002; 18 (1): 31.
[29] Zhao GX, Zhu BY. China Light Industry Press: Beijing, Principles of surfactant action. 1st ed; 2003 (chapter 4).
[30] Sanz MA, Granizo N, Gradzielski M, Rodrigo MM, Valiente M. Colloid Polym Sci 2005; 283: 646.

(注：此文原载于 Polymer，2009，50 (13)：2976-2980)

盐酸曲唑酮的合成

薛叙明[1]，贺新[1]，刘长春[1]，吕春绪[2]

（1. 常州工程职业技术学院，江苏常州 213164）

（2. 南京理工大学化工学院，江苏南京 210094）

摘要： 间氯苯胺与二乙醇胺和氢溴酸反应得到1-(3-氯苯基) 哌嗪二氢溴酸盐，再经烷基化得1-(3-氯苯基)-4-(3-氯丙基) 哌嗪单盐酸盐 (4)；另由邻氯吡啶与氨基脲盐酸盐反应制得1，2，4-三唑并［4,3-α］吡啶-3 (2H)-酮 (7)。4和7经缩合、成盐得到盐酸曲唑酮，总收率为30.4%（以间氯苯胺计）。

关键词： 盐酸曲唑酮，抗抑郁药，合成

盐酸曲唑酮（Trazodone hydrochloride 1)，化学名为2-[3-[4-(3-氯苯基)-1-哌嗪基]丙基]-1,2,4-三唑[4,3-α]吡啶-3(2H)-酮盐酸盐，由意大利 Angelini公司开发，现已在多国上市。本品为特异性5-羟色胺再摄取抑制剂，具有α_1肾上腺素能拮抗作用与抗组胺作用。1与苯丙胺类药物不同，对中枢神经系统无兴奋作用，临床主要用于治疗抑郁症和伴随抑郁症状的焦虑症以及药物依赖者戒断后的情绪障碍[1-3]。

本研究参考相关文献[4-9]，确定了1的合成路线（图1)，并进行了工艺改进。首先以氢溴酸代替盐酸[4]，与二乙醇胺反应制得双-(2-溴乙基）胺氢溴酸盐，再与间氯苯胺（2）反应制得重要中间体1-(3-氯苯基）哌嗪二氢溴酸盐（3)，收率为61.4%。3与1-溴-3-氯丙烷制的关键中间体1-(3-氯苯基)-4-(3-氯丙基）哌嗪单盐酸盐（4)，反应液用36%盐酸调至pH=5即可除去反应副产物1，3-双［4-(3-氯苯基）哌嗪基］丙烷（5)，操作简便。另用邻氯吡啶（6）与氨基脲盐酸盐一步反应得1,2,4-三唑并［4,3-α］吡啶-3（2H)-酮（7）反应温度由155～160℃降至128～130℃，反应完毕后冷却析晶，再经过滤、水洗即得产品，革除了文献[5]中繁琐的萃取操作，收率由58%提高至70%。本研究以氢氧化钠代替文献[6,7]中的氢化钠或碳酸钾作为缚酸剂，使4和7在无水乙醇中缩合得曲唑酮；最后成盐酸盐时，用35%氯化氢乙醇溶液（w/v）代替氯化氢气体[5]制得1。改进后的工艺总收率以邻氯吡啶计达49.6%（文献[5]：39.1%，以2计为30.4%，操作简便，适合工业化生产。

实验部分

YRT-3型熔点仪；Agilent1100型色谱仪［色谱柱C18柱（4.6mm×250mm，5μm)，流动相甲醇-0.01mol/L磷酸二氢铵缓冲液（80∶20)，检测波长248nm，流速1.0ml/min，柱温25℃］；Elementar Vario EL Ⅲ型元素分析仪；Abatar360FT-IR型红外光谱仪（KBr压片)；AVANCE AV-300型核磁共振仪。

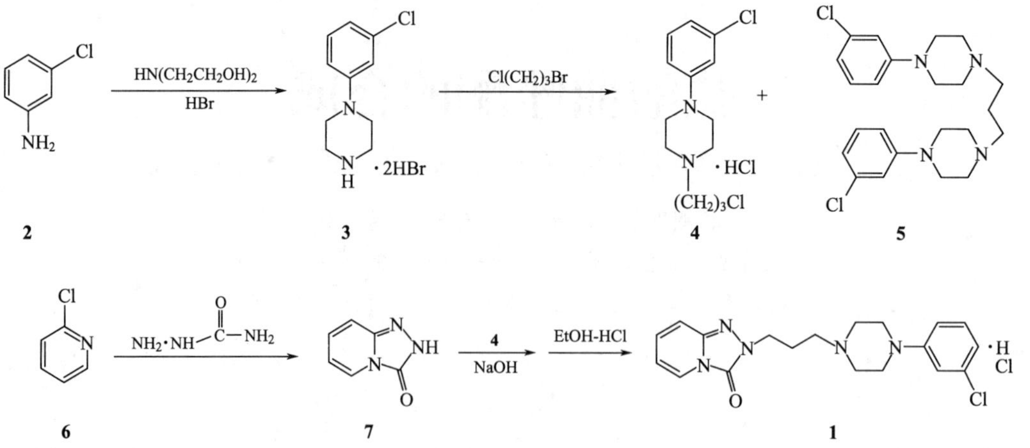

图1 1的合成路线

1-(3-氯苯基)哌嗪二氢溴酸盐(3)

二乙醇胺(110g,1.05mol)和48%氢溴酸(500g,2.96mol)加至500ml四口烧瓶中,搅拌下升温至115～130℃,常压蒸馏10h后减压蒸馏1h。残留物冷却至50℃以下,缓慢滴加2(127.5g,1mol),滴毕升温至150℃反应5h。冷却至78℃以下,滴加95%乙醇(200ml),继续冷却到5℃以下,保温搅拌2h,过滤,滤饼烘干,得类白色晶体3(220g,61.4%),mp>250℃(文献[4]):收率49.1%。纯度>99.0%(HPLC归一化法)。元素分析($C_{10}H_{15}Br_2ClN_2$)实测值(理论值,%):C33.52(33.50),H4.57(4.22),N7.74(7.81)。

1-(3-氯苯基)-4-(3-氯丙基)哌嗪单盐酸盐(4)

丙酮(140ml)、1-溴-3-氯丙烷(110g,0.7mol)、水(76ml),3(179g,0.5mol)加至1L四口烧瓶中,搅拌下于25℃滴加30%氢氧化钠水溶液(105g,0.79mol),滴毕保温反应24h。静置分层,水层用丙酮(120ml×2)提取,合并有机相,减压蒸馏除丙酮,残留物中加入36%盐酸(约8ml)调至pH=5,副产物5析出,滤除。滤液中加入36%盐酸(约5ml)调至pH=3,过滤,得类白色固体4(108g,70%),mp199～200.5℃(文献:收率72.6%,mp195～197℃)。纯度>99.0%(HPLC归一化法)。

2,4-三唑并[4,3-α]吡啶-3(2H)-酮(7)

乙二醇单乙醚(200ml)、6(80g,0.7mol)和氨基脲盐酸盐(153g,1.37mol)加至500ml四口烧瓶中,加热回流1h至全溶,稍冷后滴加浓硫酸(3g,0.03mol),滴毕于128～130℃反应24h。冷却至25℃,析出晶体,过滤,滤饼用水洗至中性,干燥,得浅黄色晶体7(66g,70%),mp 226.5～228.5℃(文献:收率58%,mp229℃)。纯度>99.0%(HPLC面积归一化法)。

盐酸曲唑酮(1)

无水乙醇(600ml)、7(67.5g,0.5mol)和4(155g,0.5mol)加至1L四口烧瓶中,搅拌下加入固体氢氧化钠(41.3g,1.03mol),加热至回流反应30h。趁热过滤,于60℃左右在滤液中滴加35%氯化氢无水乙醇溶液(w/v)(50g,0.21mol),冷却至30～40℃,有晶体析出,继续冷却至5℃,过滤,滤饼用无水乙醇(60ml)洗涤,得类白色固体1(180g,85%),纯度>98.5%(HPLC归一化法)。粗品用6倍量无水乙醇(1.08L)重结晶,得白色晶体1(150g,70.8%),mp222.5～224.5℃(文献:收率67.5%,mp222～225℃)。纯度>99.0%(HPLC外标法)。IR(KBr)v

(cm^{-1})：3394，3069，2943，2853，1704，1640，1595，943，883，743。HNMR（$CDCl_3$）δ：2.52～2.62（m，2H，CH_2），2.93～3.03（m，2H，CH_2），3.15～3.22（m，2H，CH），3.61～3.77（m，8H，哌嗪环），4.14～4.18（t，J＝6Hz，2H，CH_2），6.50～6.55（m，1H，ArH），6.78（dd，J_1＝8.1Hz，J_2＝2.1Hz，1H，ArH），6.88～6.93（m，2H，ArH），7.09～7.22（m，3H，ArH），7.74（d，J＝6.9Hz，1H，ArH）。

参考文献

[1] James SP，Mendelson WB. The use of trazodone as a hypnotic：a critical review [J]. J Clin Psychiatry，2004，65 (6)：752-755.

[2] Odagaki Y，Toyoshima R，Yamauchi T. Trazodone and its active metabolite m-chlorophenylpiperazine as partial agonists at 5-HT_{1A} receptors assessed by [35S] GTPgammaSbinding [J]. J Psychopharm，2005，19 (3)：235-241.

[3] Gorman JM. Treating generalized anxiety disorder [J]. J Clin Psychiatry，2003，64 (Suppl 2)：24-29.

[4] 董新荣，杨建奎，李霞等．盐酸奈法唑酮的合成 [J]．化学研究，2006，17 (4)：48-50.

[5] 陈国良，谈冬生，徐丽英等．盐酸曲唑酮的合成 [J]．中国药物化学杂志，2002，12 (2)：92-93.

[6] Galiendo G，Di Carlo R，Greco G，et al. Systhesis ang biological activity of benzotriazole derivatives structurally related to trazodone [J]. Eur J Med Chem，1995，30 (1)：77-84.

[7] Budai Z，Mezei T，Reiter J，et al. Preparation of substituted 1，2,4-triazolo [4,3-a] pyridine-3 (2H)-one and its salts[P]. HU：50817，1990-03-28. (CA 1990，113：231382).

[8] Temple DL，Lobeck WG. Phenoxyethyl-1，2,4-triazol-3-one antidepressants：US，4338317 [P]. 1982-07-06. (CA 1982，97：182432).

[9] Lei B，Tam TF，Karimian K，et al. Methods of manufacture of nefazodone：US，5900485 [P]. 1999-05-04. (CA 1999，130：311820).

Synthesis of Trazodone Hydrochloride

Xue Xuming[1]，He Xin[1]，Liu Changchun[1]，Lv Chunxu[2]

(1. Changzhou Institute of Engineering Technology，Changzhou 213164；

2. College of Chemical Engineering，Nanjing University of Science & Technology，Nanjing 210094)

Abstract：Trazodone hydrochloride was synthesized from m-chloroaniline with an overall yield of 30.4% (based on m-chloroaniline) by cyclocondensation with diethanolamine and hydrobromic acid to give 1-(3-chloro-phenyl) piperazine dihydrobromide，which was reacted with 1-bromo-3-chloropropane，followed by condensation with 1，2,4-triazolo [4,3-α] pyridine-3 (2H)-one.

Key words：Trazodone hydrochloride，antidepressant；synthesis

（注：此文原载于 中国医药工业杂志，2008，39 (11)：809-810）

A New PEG-1000-Based Dicationic Ionic Liquid Exhibiting Thermoregulated Biphasic Behavior with Toluene and Its Application in One-Pot Synthesis of Benzopyrans

Zhi Huizhen, Lv Chunxu, Zhang Qiang and Jun Luo
(School of Chemical Engineering, Nanjing University of Science and Technology, Nanjing, 210094, China)
SCI: 000266003700019

Abstract: A new recyclable temperature-dependant phase-separation system comprised of PEG-1000-based dicationic acidic ionic liquid and toluene was developed and used in one-pot three-component condensation to prepare benzopyrans in excellent yields (up to 93%).

4H-Benzopyrans are important compounds due to their biological and pharmacological activities. Some methods for the synthesis of 5-oxo-5,6,7,8-tetrahydro-4H-benzo [b] pyrans have been reported and the conventional synthesis involves a three-component condensation of dimedone with aromatic aldehyde and malononitrile (or cyanoacetates)[1]. Other effective methods include the use of microwave, ultrasonic irradiation and ionic liquid[2]. However, these procedures have some drawbacks each, such as involving high reaction temperature, use of expensive catalyst or low yield. Benzopyrans have also been attracting our interest recently due to their importance.

Ionic liquids (ILs), as environmental friendly reaction media or catalysts, have attracted increasing attention[3]. ILs have many excellent advantages such as negligible volatility, excellent thermal stability, remarkable solubility, and a variety of available structures. Among the numerous ILs developed, recoverable ones have been hot research targets owing to their obvious advantages. Unfortunately, most ILs show a degree of solubility in commonly used organic solvents, which causes many diffculties for product separation and recovery of the IL, especially in homogeneous catalytic reaction systems. On the contrary, for the IL-liquid biphasic catalysis, substrate and catalyst transfer is less effcient though the separation is relatively simple. In view of both the advantages and disadvantages of homogeneous and heterogeneous catalytic reactions, some excellent work has been done to improve catalyst recovery; multiphase systems, such as phase-transfer catalysis[4], thermoregulated phase-transfer catalysis[5], and liquid-liquid biphasic catalysis[6], have been studied. Some researchers have also reported some novel temperature-dependent IL-liquid biphasic catalytic systems. van den Broeke and co-workers reported a fluorous ionic liquid exhibiting temperature-dependent phase behavior with toluene[7]. Jiang and co-workers reported polyether-based ionic liquids demonstrating such behavior with toluene and heptane[8]. Wang and co-workers reported very recently new temperature-controlled heteropolyanion-based acidic ILs acting as reaction-induced self-separation catalysts for esterification[9]. There are also other results about such ILs reported recently[10,11]. However, these ILs inevitably showed some drawbacks such as low recovery ratio, high cost or difficulty of synthesis, etc.

We now report a new temperature-dependent biphasic system, including recoverable novel PEG-1000-based dicationic acidic ionic liquid 4 (PEG_{1000}-DAIL, Scheme 1) and toluene, and its application in the synthesis of 5-oxo-5,6,7,8-tetra-hydro-4H-benzo [b] pyrans by a three-component condensation.

Scheme 1 Synthesis of PEG_{1000}-DAIL

PEG_{1000}-DAIL is a hygroscopic viscous brown liquid (ρ=1.21g · mL^{-1}, 25℃), and stable to air and heat. Treatment of it under vacuum at 100℃ for 72h resulted in no obvious loss of weight. No distinct phase transition enthalpy could be indicated according to DSC determination. The solubility properties of PEG_{1000}-DAIL in some traditional solvents were investigated and the results are shown in Table 1. It can be seen from Table 1 that PEG_{1000}-DAIL is soluble in water and very soluble in polar organic solvents (entries 1-3,5-9). However, it is nearly insoluble in toluene at low temperature, but miscible at high temperature. The phase behavior of PEG_{1000}-DAIL with toluene was then investigated (Fig. 1) which showed that this system is very similar to a fluorous system[7], i. e., PEG_{1000}-DAIL exhibits a temperature-dependent phase behavior with toluene. For equivalent volumes of PEG_{1000}-DAIL and toluene, the phase transformation temperature is about 80℃.

The PEG_{1000}-DAIL/toluene temperature-dependent biphasic catalytic system was then applied in the synthesis of benzo-pyrans by three-component condensation of malononitrile, 5,5-dimethyl-1,3-cyclohexanedione and 4-nitrobenzaldehyde, and the reaction procedure is demonstrated in Fig. 2

Table 1 Solubility of the PEG_{1000}-DAIL in some solvents[a]

	T/℃				
Solvent	25	40	60	80	100
H_2O	s	s	s	s	
EtOH	s	s	s		
CH_2Cl_2	s				
PhMe	i	i	i	s	s
Et_2O	i	i			
Acetone	i	I			
Benzene	i	i	i	i	
EtOAc	i	i	I		
Petroleum ether	i	i	i	i	

[a] s: soluble; i: insoluble

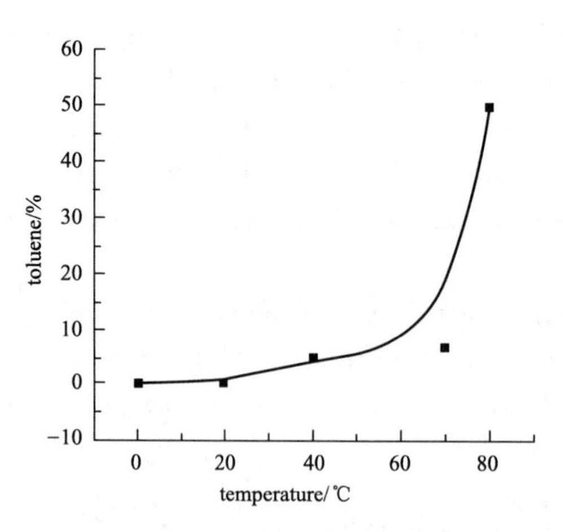

Fig. 1 Relationship between temperature and volume of toluene in PEG_{1000}-DAIL (2mL) /toluene system

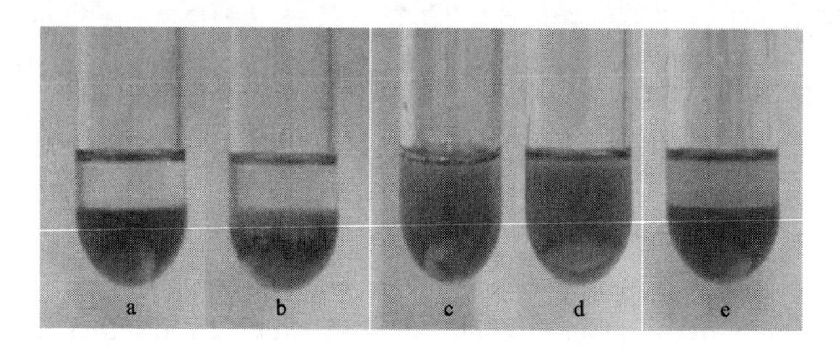

Fig. 2 Reaction procedure in PEG_{1000}-DAIL/toluene. a) PEG_{1000}-DAIL/toluene at room temperature; b) after addition of substrates; c) homogeneous phase at 80℃; d) phase-separation along with cooling; e) complete phase-separation after cooling to room temperature.

Four reaction media were compared and the results were listed in Table 2. It was found that almost no product was obtained in toluene only without any catalyst (entry 1, Table 2). Low yield also observed when catalysed by conc. H_2SO_4 (entry 2, Table 2) may be owing to too strong acidity. When PEG_{1000}-DAIL was used as both solvent and catalyst, the yield would be improved to 78% (entry 3, Table 2). In comparison, the PEG_{1000}-DAIL/toluene system gave the best result (93% yield, entry 4, Table 2) due to lower viscosity.

Table 2 Comparison of one-pot condensation of 4-nitrobenzaldehyde, malononitrile, 5, 5-dimethyl-1, 3-cyclohexanedione in different media

Entry	Solvent	T/℃	t/h	Yield/%
1	PhMe	80	3	trace
2	H_2SO_4	80	1	46
3	PEG_{1000}-DAIL	80	1	78
4	PEG_{1000}-DAIL/PhMe	80	1	93

When the final reaction mixture was cooled to room temperature, the upper layer of toluene, containing product, was removed by decantation, fresh subtrates and toluene were then recharged to the residual PEG_{1000}-DAIL and the mixture was heated to react once again. The procedure was repeated 10 times and the results indicated that PEG_{1000}-DAIL could be recycled without apparent loss of catalytic activity, Fig. 3, and only 7.5% loss of weight was observed after 10 times recycling.

When different aromatic aldehydes were used as substrates along with dimedone and malononitrile, the product benzo-pyrans could also be obtained in high isolated yields (82-93%) after reaction for 1h at 80℃ (entries 1-13, Table 3). However, higher temperature (90℃) and longer time (1.5h) were required for the less active substrate ethyl cyanoacetate (entries 14-16, Table 3).

In conclusion, the PEG_{1000}-DAIL/toluene system exhibited temperature-dependent phase behavior. 5-Oxo-5,6,7,8-tetra-hydro-4H-benzo [b] pyrans could be efficiently synthesized in excellent yields in this system by one-pot three-component condensation reaction of aromatic aldehydes, malononitrile (or ethyl cyanoacetate) and 5, 5-dimethyl-1, 3-cyclohexanedione. PEG_{1000}-DAIL could be effciently recovered by simple decantation after reaction without any apparent loss of catalytic activity and little loss of weight even after 10 times recycling. The PEG_{1000}-DAIL/toluene system has several advantages: (1) PEG_{1000}-DAIL is a strong Brønsted acid and shows superior catalytic activity; (2) PEG_{1000}-DAIL can be separated by simple decantation without apparent loss of catalytic activity and little loss of weight; (3) this catalytic system has a wide range of applications for different substrates and the products can be obtained conveniently and in excellent yield and purity. In a word, PEG_{1000}-DAIL/toluene system is an excellent recyclable catalytic reaction media.

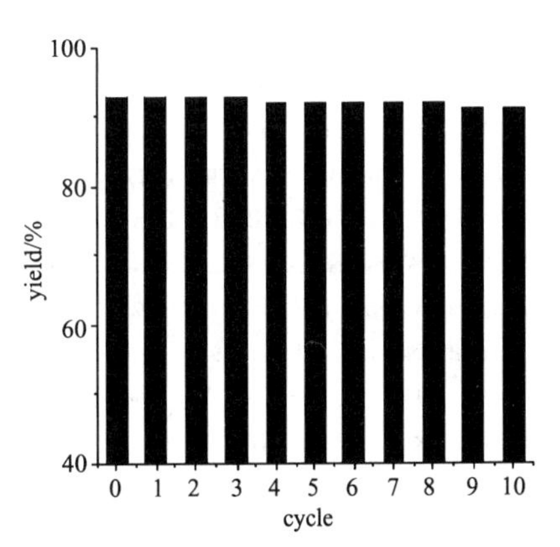

Fig. 3 Repeating reactions using recovered PEG_{1000}-DAIL

This project was financially supported by Doctoral Fund of Ministry of Education of China for New Teacher (No. 200802881029).

Notes and references

Procedure for snynthesis of PEG_{1000}-DAIL: Imidazole (0.5 mol) and sodium ethoxide (0.5 mol) were dissolved in ethanol (50 mL), stirred at 70 ℃ for 8 h, then ethanol was removed by rotatory evaporation under reduced pressure, the residue was washed with ether three times and dried in vacuum to give intermediate 1, ^1H NMR(D_2O)δ 6.84(s,2H,2×CH), 7.44(s,2H,2×CH). Intermediate 2 was prepared by reported method.[12] H NMR($CDCl_3$)δ 3.61-3.71(m,95.7H, $(OCH_2CH_2)_n$). Then compound 2 (0.25 mmol) and 1 (0.5 mol) were charged into ethanol (50 mL) and stirred at reflux for 10 h, then the mixture was filtered and the liquid was washed with ether several times. And the liquid was evaporated under reduced pressure to give 3 as a colorless oil. ^1H NMR($CDCl_3$,TMS)δ 3.45-3.66(m,91.2H,$(OCH_2CH_2)_n$), 3.72(4H,2×CH_2), 4.22(4H,2×NCH_2), 7.27(s,2H,2×CH), 7.35(s,2H,2×CH), 8.59(s,2H,2×CH). 3 was dissolved in ethanol (30mL). 1,3-propane sultone (1,3-PS) was added dropwise into the solution in 30min, and the mixture was stirred at reflux for 8 h, then H_2SO_4 was added dropwise in 30 min. The final solution was stirred for 8 h at 50 ℃ and was evaporated under reduced pressure to give PEG_{1000}-DAIL 4 as viscous brown liquid, ^1H NMR(D_2O)δ 2.15(t,4H, J=7Hz,2×CH_2), 2.77(m,4H,2×CH_2), 3.45-3.66(m,90.3H,$(OCH_2CH_2)_n$), 3.74(4H,2×CH_2), 4.21-4.30(8H, 4×NCH_2), 7.41(s,4H,4×CH), 8.71(s,2H,2×CH); ^{13}C NMR(D_2O)δ 25.2, 47.7, 47.9, 49.1, 68.6, 68.7, 69.6, 122.3, 123.1, 136.0; IR (cm^{-1}): 3849, 3400, 3020, 2883, 1732, 1591, 1456, 1431, 1377, 1122, 1109, 972, 883, 798, 677, 577; ESI-MS: 666.4, $M^{++}/2$, n=21, 688.4, $M^{++}/2$, n=22, 260.9 (62).

General procedure for condensation reaction: Aromatic aldehydes (1 mmol), 5,5-dimethyl-1,3-cyclohexanedione (1 mmol), malononotrile (or ethyl cyanoacetate) (1 mmol), PEG_{1000}-DAIL (2 mL), toluene (2 mL) were added into a tube reactor. The mixture was heated at 80℃ (90 ℃ for ethyl cyanoacetate) with vigorous stirring for 1 h (1.5 h for ethyl cyanoacetate). the mixture was cooled to 20 ℃ and the upper phase was separated by decantation. After recrystallizing in toluene, the solid products were washed once with water and dried. PEG_{1000}-DAIL was reused without any treatment.

[1] (a) G. Kaupp, M. R. Naimi-Jamal, J. Schmeyers, Tetrahedron, 2003, **59**, 3753; (b) S. Balalaie, M. Sheikh-Ahrmadi, M. Bararjanian, Catal. Commun., 2007, **8**, 1724; (c) T. S. Jin, A. Q. Wang, A. X. Wang, J. S. Zhang, T. S. Li,

Synlett., 2004, 871; (d) I. Devi, P. J. Bhuyan, Tetrahedron Lett., 2004, **45**, 8625; (e) S. Balalaie, M. Bararjanian, A. M. Amani, B. Movassagh, Synlett., 2006, 263.

[2] (a) S. J. Tu, Y. Gao, C. Guo, D. Shi, Z. Lu, Synth. Commun., 2002, **32**, 2137; (b) J. T. Li, W. Z. Xu, L. C. Yang, T. S. Li, Synth. Commun., 2004, **34**, 4565; (c) S. Abdolmohammadi, S. Balalaie, Tetrahedron Lett., 2007, **48**, 3299; (d) K. Bourahla, A. Derdour, M. Rahmouni, F. Carreaux, J. P. Bazureau, Tetrahedron Lett., 2007, **48**, 5785; (e) Q. P. Ding, J. Wu, Org. Lett., 2007, **9**, 4959; (f) B. C. Ranu, R. Jana, S. Sowmiah, J. Org. Chem., 2007, **72**, 3152; (g) J. J. Ma, J. C. Li, R. X. Tang, X. Zhou, Q. H. Wu, C. Wang, M. M. Zhang, Q. Li, Chin. J. Org. Chem., 2007, **27**, 640; (h) X. S. Fan, X. Y. Hu, X. Y. Zhang, J. J. Wang, Aust. J. Chem., 2004, **57**, 1067; (i) B. C. Ranu, R. Jana, S. Sowmiah. J. Org. Chem., 2007, **72**, 3152.

[3] (a) P. Wasserscheid, W. Keim, Angew. Chem. Int. Ed., 2000, **39**, 3772; (b) J. Dupont, R. F. de Souza, P. A. Z. Suarez, Chem. Rev., 2002, **102**, 3667; (c) R. Sheldon, Chem. Commun., 2001, 2399; (d) M. Poliakof, J. M. Fitzpartrick, Science, 2002, **297**, 807; (e) L. A. Blanchard, K. R. Hancu, E. J. Beckman, J. F. Brennecke, Nature, 1999, **399**, 28; (f) R. D. Rogers, K. R. seddon, Science, 2003, **302**, 792; (g) T. L. Greaves, C. J. Drummond, Chem. Rev., 2008, **108**, 206; (h) N. Peter, T. Ben, V. H. Kristof, V. M. Luc, K. Binnemans, Crystal growth Design, 2008, 8, 1353; (i) V. I. Parvulescu, C. Hardacre, Chem. Rev. 2007, **107**, 2615; (j) S. -G. Lee, Chem. Commun., 2006, 1049; (h) Z. F. Fei, T. J. Geldbach, D. B. Zhao, P. J. Dyson. Chem. Eur. J., 2006, 12, 2122.

[4] (a) T. Ooi, K. Maruoka. Angew. Chem. Int. Ed., 2007, **46**, 4222; (b) Z. Xi, N. Zhou, Y. Sun, K. Li. Science, 2001, **292**, 1139; (c) V. K. Dioumaev, R. M. Bullock. Nature, 2000, **424**, 530.

[5] Y. H. Wang, X. W. Wu, F. Cheng, Z. L. Jin, J. Mol. Catal. A: Chem., 2003, **195**, 133.

[6] S. Sunitha, S. Kanjilal, P. S. Reddy, B. N. Rachapudi. Tetrahedron Lett., 2007, **48**, 6962.

[7] J. van den Broeke, F. Winter, B. J. Deelman, G. van Koten, Org. Lett. 2002, **4**, 3851-3854.

[8] L. Wei, J. Y. Jiang, Y. H. Wang, Z. L. Jin, J. Mol. Catal. A: Chem., 2004, **221**, 47.

[9] Y. Leng, J. Wang, D. R. Zhu, X. Q. Ren, H. Q. Ge, L. Shen. Angew. Chem. Int. Ed., 2009, **48**, 168.

[10] (a) J. M. Crosthwaite, S. N. V. K. Aki, E. J. Maginn, J. F. Brennecke, J. Phys. Chem. B, 2004, **108**, 5113; (b) A. M. Scurto, S. N. V. K. Aki, J. F. Brennecke, J. Am. Chem. Soc., 2002, **124**, 10276; (c) C. S. Consorti, G. L. P. Aydos, G. Ebeling, J. Dupont, Org. Lett., 2008, **10**, 237; (d) A. Riisager, R. Fehrmann, R. W. Berg, R. van Hal, P. Wasserscheid, Phys. Chem. Chem. Phys., 2005, **7**, 3052; (e) Y. Y. He, T. P. Lodge, Chem. Commun., 2007, 2732; (f) N. Karodia, S. Guise, C. Newlands, J. A. Andersen, Chem. Commun., 1998, **21**, 2341; (g) P. J. Dyson, D. J. Ellis, T. Welton, Can. J. Chem., 2001, **79**, 705.

[11] (a) Z. F. Bai, Yiyong He, Timothy P. Lodge. Langmuir, 2008, **24**, 5284; (b) Z. Zeng, B. S. Phillips, J. -C. Xiao, J. M. Shreeve. Chem. Mater., 2008, **20**, 2719; (c) B. Breure, S. B. Bottini, G. -J. Witkamp, C. J. Peters J. Phys. Chem. B, 2007, **111**, 14265; (d) A. Kumar, P. L. Dubin, M. J. Hernon, Y. J. Li, W. Jaeger. J. Phys. Chem. B, 2007, **111**, 8468; (e) U. Domańska, Z.? oflek-Tryznowska, M. Królikowski. J. Chem. Eng. Data, 2007, **52**, 1872; (f) J. Ortega, R. Vreekamp, E. Marrero, E. Penco. J. Chem. Eng. Data, 2007, **52**, 2269; (g) P. Nockemann, K. Binnemans, B. Thijs, T. N. Parac-Vogt, K. Merz, A. -V. Mudring, P. C. Menon, R. N. Rajesh, G. Cordoyiannis, J. Thoen, J. Leys, C. Glorieux. J. Phys. Chem. B, 2009, **113**, 1429; (h) G. D. Yadav, B. G. Motirale. Ind. Eng. Chem. Res., 2008, **47**, 9081.

[12] H. W. Gibson, N. Yamaguchi, J. W. Jones, J. Am. Chem. Soc., 2003, **125**, 3522.

(注：此文原载于 Chemical Communication，2009，2878-2880)

Influence of Solvent on the Structures of two One-dimensional Cobalt (Ⅱ) Coordination Polymers with Tetra-chloroterephthalate

Fang Yongqin, Lu Ming, Lv Chunxu

(School of Chemistry and Chemical Engineering, Nanjing University of Science and Technology, Nanjing 210093, People's Republic of China, and The Research Institute of Fine Chemical Engineering, Jiangsu Polytechnic University, Changzhou 213164, People's Republic of China)

SCI: 000263071500012, EI: 20090811911849

The title cobalt coordination polymers, catena-poly[[[aquatripyridinecobalt(Ⅱ)]-μ-tetra-chloroterephthalato]pyridine solvate], $\{[Co(C_8Cl_4O_4)(C_5H_5N)_3(H_2O)]\cdot C_5H_5N\}_n$, (I), and catena-poly[[[diaquadipyridinecobalt (Ⅱ)]-μ-tetra-chloroterephthalato] 1, 4-dioxane trihydrate], $\{[Co(C_8Cl_4O_4)(C_5H_5N)_2(H_2O)_2]\cdot C_4H_8O_2\cdot 3H_2O\}_n$, (Ⅱ), have been prepared with tetrachloroterephthalic acid (H_2BDC-Cl_4) under different solvent media. Both complexes form infinite cobalt (Ⅱ)-tetrachloroterephthalate polymeric chains. In (I), two independent Co^{II} ions are six-coordinated through N_3O_3 donorsets in slightly distorted octahedral geometries provided by two carboxylate and three pyridine ligands, and one water molecule. The structure of (Ⅱ) contains two independent Co^{II} atoms, each lying on a twofold axis, which adopt a tetragonally distorted N_2O_4 octahedral geometry *via* two carboxylate groups, two pyridine ligands and two water molecules. The different stoichiometry of coordinated and solvent guest molecules leads to different two-dimensional supramolecular networks, with (I) utilizing C-H$\cdots\pi$ and weak π-π inter-actions and (Ⅱ) utilizing mainly conventional hydrogen bonding.

Comment

The rapid development in crystal engineering of coordination polymers has afforded access to crystalline solid-state materials with potential applications in the areas of catalysis, magnetism, gas absorption, nonlinear optics and drug delivery (Férey, 2008; Wang *et al.*, 2008). To date, the aromatic bridging ligand benzene-1,4-dicarboxylate (BDC) has been widely utilized to fabricate coordination polymers based on paddlewheel units with robust networks and attractive properties as conventional porous materials (Li *et al.*, 1999; Serre *et al.*, 2002; Xiao *et al.*, 2005; Du *et al.*, 2007). Derivatives of BDC with electron-donating or-accepting substituents, such as 2,3,5,6-tetramethylbenzene-1,4-dicarboxylate (TBDC) and 2,3,5,6-tetrafluorobenzene-1,4-dicarboxylate (BDC-F_4), have also received considerable attention in the design of porous materials with an efficient hydrogen-storage capacity (Kitaura *et al.*, 2004; Chun *et al.*, 2005). Few investigations of solvent effects have been carried out on crystalline compounds using substituted BDC building blocks (Chen *et al.*, 2008). In the latter report, solvent-regulated assemblies of manganese (Ⅱ) coordination polymers with 2,3,5,6-tetrachlorobenzene-1,4-dicarboxylic acid ($H_2BDC\text{-}Cl_4$) exhibit different binding modes and dimensionality in the final coordination networks with variable auxiliary solvent co-ligands. As part of our investigation of the coordination behavior of $H_2BDC\text{-}Cl_4$ and divalent transition metal ions, we report the solvent-directed assembly of two Co^{II} coordination polymers, $\{[Co(BDC\text{-}Cl_4)(C5H_5N)_3\text{-}(H_2O)]\cdot C_5H_5N\}_n$, (I), and $\{[Co(BDC\text{-}Cl_4)(C_5H_5N)_2(H_2O)_2]\cdot C_4H_8O_2\cdot 3H_2O\}_n$, (Ⅱ).

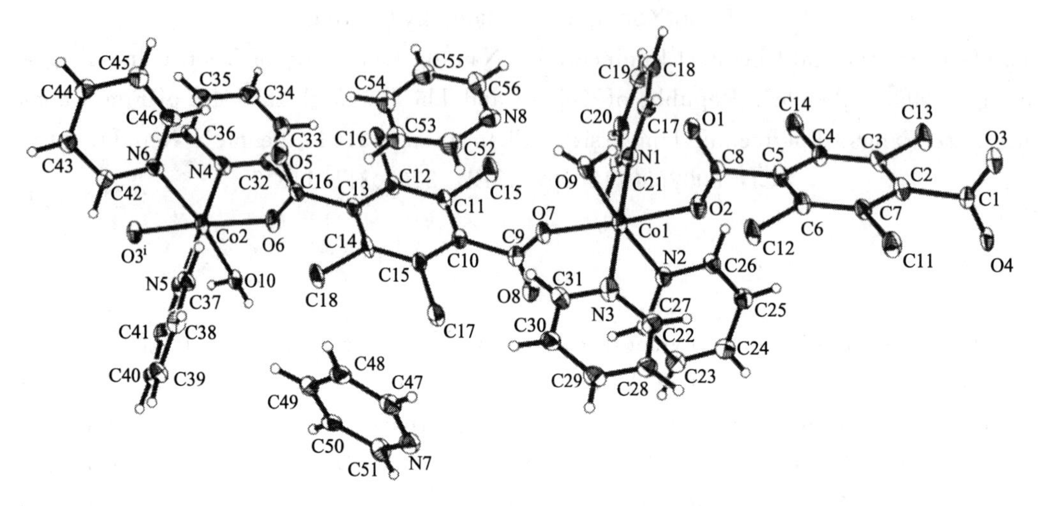

Figure 1　The structure of (I), with displacement ellipsoids drawn at the 30% probability level. [Symmetry code: (i) x-1, y-1, z+1.]

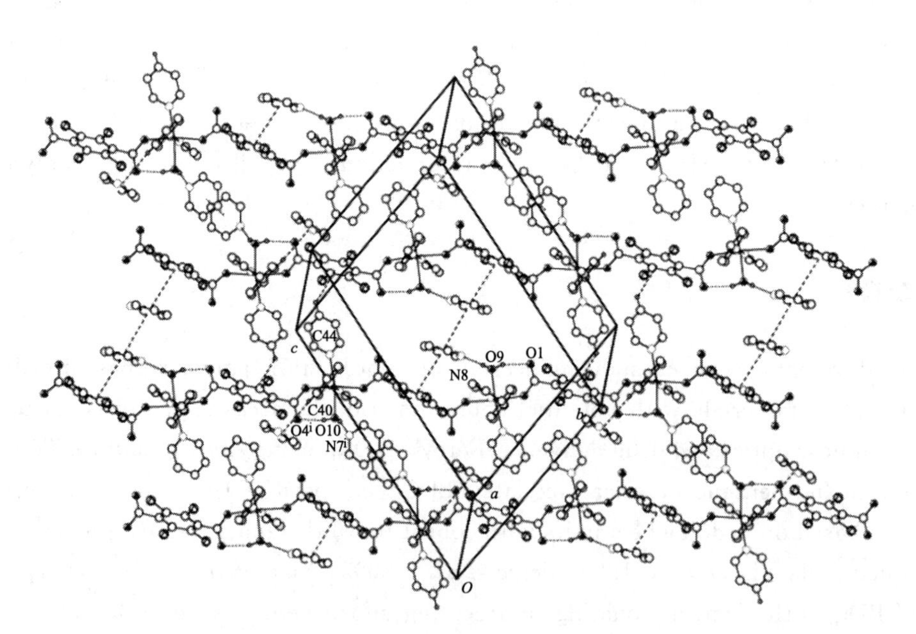

Figure 2　A view of part of the two-dimensional supramolecular network of (I), approximately normal to the *bc* plane. Hydrogen bonds and aromatic interactions are shown as dashed lines. [Symmetry codes: (i) x-1, y-1, z+1; (ii)-x+1, -y, -z+1.]

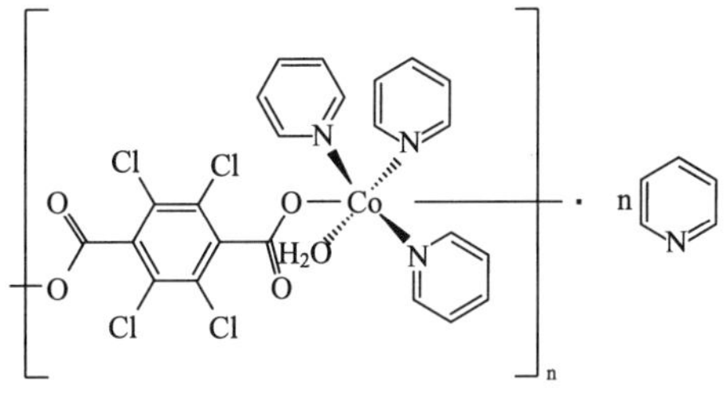

(I) 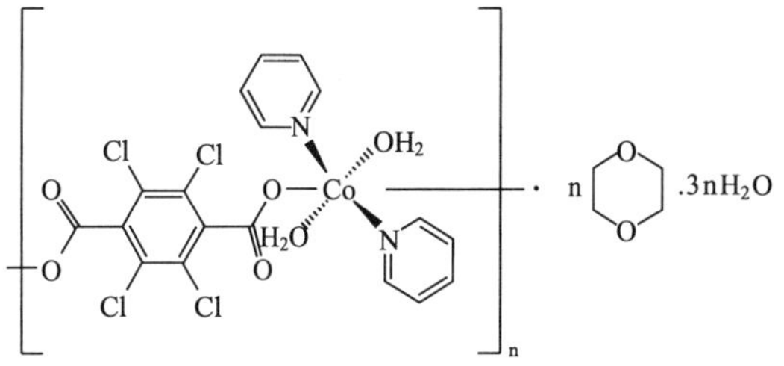 (II)

The asymmetric unit of (I) contains two Co^{II} atoms, two BDC-Cl_4 ligands, six coordinated pyridine molecules and two water molecules, along with two pyridine molecules of crystallization (Fig. 1 and Table 1). The two Co^{II} atoms exhibit similar slightly distorted octahedral coordination geometries, with the N_3O_3 coordination environment comprising two BDC-Cl_4 anions, one aqua ligand and three pyridine molecules. In the Mn^{II}-BDC-Cl_4 polymer crystallized from a similar pyridine-water medium, the Mn^{II} ion is six-coordinated through two BDC-Cl_4 anions, two aqua ligands and two pyridine molecules (Chen *et al.*, 2008). Within the anionic unit, the rotation angles (φ_{rot}) of the tetrachlorinated benzene rings relative to the carboxylate groups are in the range 81.0 (2)-86.2 (1)°, similar to the corresponding values for other polychlorinated carboxylate compounds (Maspoch *et al.*, 2004).

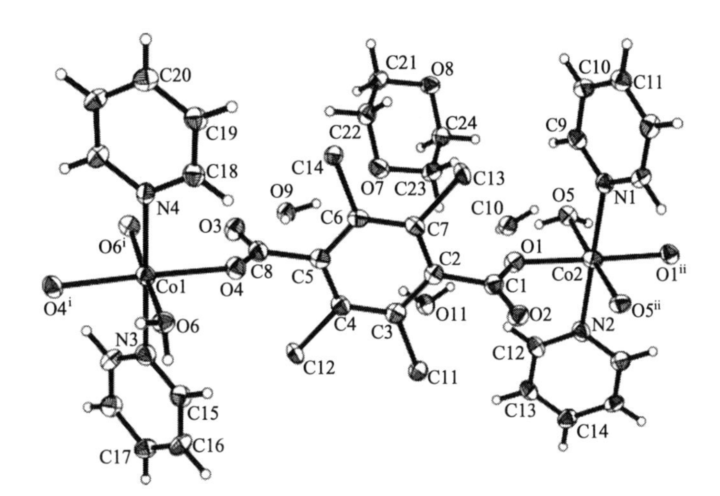

Figure 3 The structure of (Ⅱ), with displacement ellipsoids drawn at the 30% probability level.

[**Symmetry codes: (i) $-x+\frac{1}{2}$, $-y+2$, z; (ii) $-x+\frac{3}{2}$, $-y+1$, z.**]

Each BDC-Cl_4 ligand adopts a bis-monodentate coordination mode, bridging adjacent Co^{II} centers into an infinite linear chain with decorated pyridine molecules, which runs along the [111] direction, as illustrated in Fig. 2.

There exist strong intramolecular O-H···O interactions (entries 2 and 3 in Table 2) between the carboxylate O atoms and the aqua ligands, which further stabilize the one-dimensional polymeric chain motif. Intermolecular O-H···N bonds (entries 1 and 4 in Table 2) join the uncoordinated pyridine molecules and the coordination chain together. Because there are a large number of coordinated and uncoordinated pyridine molecules in the structure, multiple weak π-π interactions are noted between the benzene and pyridine rings and between the pyridine rings (*e. g.* C10-C15, N7/C47-C51 and N8/C52-C56), with centroid-to-centroid distances in the range 3.508 (2)-3.641 (2) Å; these link the one-di-

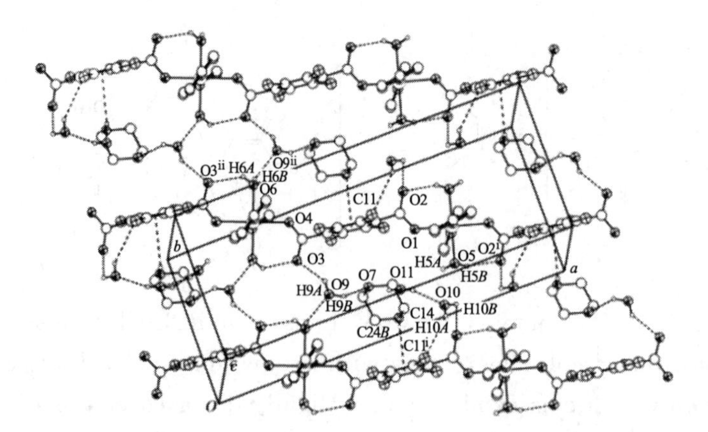

Figure 4 A view of part of the two-dimensional hydrogen-bonding network in (Ⅱ), approximately normal to the ab plane. O-H···O and C-H···π interactions are shown as dashed lines.

[Symmetry codes: (i) $-x+\frac{3}{2}$, $-y+1$, z; (ii) $-x+\frac{1}{2}$, $-y+2$, z; (iii) x, y+1, z; (iv) x, y-1, z.]

mensional hydrogen-bonded chains into a two-dimensional supramolecular network (Fig. 2). The supramolecular architecture is further consolidated by multiple C-H···π interactions between CH groups and the centroids of the N3/C27-C31 pyridine and C2-C7 benzene rings (Fig. 2 and Table 2).

The crystal structure of complex (Ⅱ) also exhibits a one-dimensional coordination motif. In the asymmetric unit, two crystallographically independent centers, Co1 and Co2, lie on twofold rotation axes at ($\frac{1}{4}$, 1, y) and ($\frac{3}{4}$, $\frac{1}{2}$, y), respectively. Each Co^{II} center is coordinated by two carboxylate groups in a monodentate mode from two symmetry-related BDC-Cl_4 ligands, and by two pyridine and two aqua ligands (Table 3). The asymmetric unit also contains one dioxane and three water molecules as guests. The N_2O_4 coordination geometry, provided by two carboxylate O atoms, two pyridyl N atoms and two water molecules, can be described as an elongated octahedron with two pyridyl N atoms occupying the axial sites. Neighboring Co1 and Co2 atoms are bridged by the BDC-Cl_4 dianion in a *trans* fashion (Fig. 3). Propagation of this arrangement generates a one-dimensional linear array, which is similar to that of complex (I) but with fewer coordinated pyridine molecules and more water molecules. Similar to (I), the two carboxylate groups within each BDC-Cl_4 group make φ_{rot} dihedral angles of 81.6 (1) and 87.7 (1)° with the central benzene plane. Analogous intramolecularO-H···O bonds (entries 1 and 2 in Table 4) between uncoordinated carboxylate O atoms and aqua ligands consolidate the one-dimensional polymeric chain. Analysis of the crystal packing shows the existence of two main types of conventional hydrogen-bonding interactions (O-H···O and O-H···Cl; Table 4), which connect the polymeric chains, solvent water molecules and dioxane molecules to form a two-dimensional network parallel to the *ab* plane (Fig. 4). Notably, one hydrogen-bonded pattern (entries 2-4 in Table 4), denoted as R_6^6 (12) (Etter, 1990), is observed in this layer array. Additionally, the dioxane solvent molecule has a C-H···π interaction with the benzene ring of the BDC-Cl_4 anion (Table 4). However, in contrast to (I), adjacent layers have no signiflcant aromatic ring-based stacking inter-actions.

Experimental

To a solution of $CoCl_2 \cdot 6H_2O$ (23.8mg, 0.1mmol) in a mixture of pyridine and water (1∶1

v/v, 10ml) was added a solution of $H_2BDC\text{-}Cl_4$ (30.4mg, 0.1mmol) in water (5ml). After stirring for 30min, the reaction mixture was filtered and left to stand at room temperature. After 3d, pink block-shaped crystals of (I) suitable for X-ray diffraction were obtained by slow evaporation of the filtrate in a yield of 78% (54.2mg, based on $H_2BDC\text{-}Cl_4$). Analysis found: C 48.42, H 3.18, N 8.13%; calculated for $C_{56}H_{44}Cl_8Co_2N_8O_{10}$: C 48.37, H 3.19, N 8.06%. For the preparation of (Ⅱ), an aqueous solution (5ml) of $CoCl_2 \cdot 6H_2O$ (23.8 mg, 0.1mmol) was added to a dioxane and water solution (1∶1 v/v, 10ml) of $H_2BDC\text{-}Cl_4$ (30.4 mg, 0.1mmol) with stirring for 30 min. Pyridine (0.5ml) was then added and allowed to diffuse into the mixture at room temperature. After 7 d, pink block-shaped crystals of (Ⅱ) suitable for X-ray diffraction were collected by slow evaporation of the filtrate in a yield of 67% (46.4 mg, based on $H_2BDC\text{-}Cl_4$). Analysis found: C 37.73, H 3.94, N 4.09%; calculated for $C_{22}H_{28}Cl_4CoN_2O_{11}$: C 37.90, H 4.05, N 4.02%.

Compound (I)

Crystal data

$[Co(C_8Cl_4O_4)(C_5H_5N)_3\text{-}(H_2O)] \cdot C_5H_5N$

$M_r = 695.23$

Triclinic, $P\bar{1}$

$a = 8.7128(10)$ Å

$b = 17.8408(17)$ Å

$c = 21.227(2)$ Å

$\alpha = 67.879(1)°$

$\beta = 79.839(2)°$

$\gamma = 84.962(1)°$

$V = 3007.9(5)$ Å^3

$Z = 4$

Mo Kα radiation

$\mu = 0.97$ mm^{-1}

$T = 291(2)$ K

0.30×0.26×0.24mm

Data collection

Bruker SMART APEXII CCD diffractometer

Absorption correction: multi-scan (*SADABS*; Sheldrick, 2003)

$T_{min} = 0.76$, $T_{max} = 0.79$

23660 measured reflections

11691 independent reflections

8460 reflections with $I > 2\sigma(I)$

$R_{int} = 0.027$

Refinement

$R[F^2 > 2\sigma(F^2)] = 0.051$

$wR(F^2) = 0.109$

$S = 1.01$

11691 refiections

757 parameters

H-atom parameters constrained

$\Delta\rho_{max}=0.27e\mathring{A}^{-3}$

$\Delta\rho_{min}=-0.62e\mathring{A}^{-3}$

Table 1　Selected bond lengths (Å) for (I).

Co1-O7	2.0619	Co2-O6	2.066	Co1-N3	2.155	Co2-N5	2.167
Co1-O9	2.122	Co2-O3[i]	2.127	Co1-N1	2.158	Co2-N4	2.170
Co1-O2	2.126	Co2-O10	2.144	Co1-N2	2.171	Co2-N6	2.190

Symmetry code: (i) x-1; y-1; z+1.

Table 2　Hydrogen-bond geometry (Å,°) for (I).

Cg1 and Cg2 denote the centroids of the N3/C27-C31 and C2-C7 rings, respectively.

D-H···A	D-H	H···A	D···A	D-H···A
O9-H9A···N8	0.85	1.91	2.718(3)	159
O9-H9B···O1	0.85	1.97	2.768(3)	157
O10-H10B···O4[i]	0.85	1.85	2.686(3)	166
O10-H10A···N7[ii]	0.85	1.98	2.814(3)	167
C40-H40···Cg1[ii]	0.93	2.99	3.813(4)	149
C44-H44···Cg2[iii]	0.93	2.94	3.638(4)	133

Symmetry codes: (i) x-1; y-1; z+1; (ii)-x+1, y, -z+1; (iii)-x+1, -y+1, -z+1.

Compound (Ⅱ)

Crystal data

$[Co(C_8Cl_4O_4)(C_5H_5N)_2(H_2O)_2]\cdot C_4H_8O_2\cdot 3H_2O$

$M_r=697.19$

Orthorhombic, *Pcca*

a=21.4257(19)Å

b=8.6567(8)Å

c=33.537(3)Å

$V=6220.3(10)\mathring{A}^3$

Z=8

Mo Kα radiation

$\mu=0.95mm^{-1}$

$T=291(2)K$

0.28×0.24×0.22mm

Data collection

Bruker SMART APEXII CCD diffractometer

Absorption correction: multi-scan (*SADABS*; Sheldrick, 2003)

$T_{min}=0.77$, $T_{max}=0.81$

46466 measured reflections

6133 independent reflections

4296 reflections with $I>2\sigma(I)$

$R_{int}=0.066$

Refinement

$R[F^2>2(F^2)]=0.043$

$wR(F^2)=0.077$

$S=1.02$
6133 reflections
366 parameters
H-atom parameters constrained
$\Delta\rho_{max}=0.48e\text{Å}^{-3}$
$\Delta\rho_{min}=-0.42e\text{Å}^{-3}$

Table 3 Selected bond lengths (Å) for (Ⅱ).

Co1-O6^{i}	2.0971	Co2-O5^{ii}	2.1043	Co1-O4i	2.1469	Co2-O1	2.1302
Co1-O6	2.0971	Co2-O5	2.1043	Co1-O4	2.1469	Co2-N1	2.172
Co1-N4	2.146	Co2-O1^{ii}	2.1302	Co1-N3	2.171	Co2-N2	2.177

Symmetry codes: (i) $-x+\frac{1}{2}$; $-y+2$; z; (ii) $-x+\frac{3}{2}$; $-y+1$; z.

H atoms bonded to C atoms were positioned geometrically (C-H=0.93 and 0.97Å for pyridine and dioxane H atoms, respectively) and included in the refinement in the riding-model approximation. All water H atoms were located in difference maps and then subsequently geometrically optimized. U_{iso} (H) values for all H atoms were set at $1.2U_{eq}$ (C, O). The electron-density electron-density trough in (I) (-0.62eÅ^{3}) is 0.66Å from atom Co1.

Table 4 Hydrogen-bond geometry (Å,°) for (Ⅱ). *Cg*1 denotes the centroid of the C2-C7 ring.

D-H···A	D-H	H···A	D···A	D-H···A
O5-H5B···O2^{ii}	0.85	1.97	2.657(3)	137
O6-H6A···O3^{i}	0.85	2.24	2.684(2)	113
O6-H6B···O9^{iii}	0.85	2.10	2.629(2)	120
O9-H9A···O3	0.85	2.09	2.829(2)	145
O9-H9B···O7	0.85	2.03	2.726(3)	138
O10-H10A···Cl1^{iv}	0.85	2.71	3.3714(19)	136
O11-H11A···O10	0.85	2.37	3.199(3)	164
C24-H24B···Cg1^{iv}	0.97	2.97	3.715(3)	135

Symmetry codes: (i) $-x+\frac{1}{2}$; $-y+2$; z; (ii) $-x+\frac{3}{2}$; $-y+1$; z; (iii) x; y+1; z; (iv) x; y−1; z.

For both compounds, data collection: *APEX2* (Bruker, 2003); cell refinement: *APEX2* and *SAINT* (Bruker, 2001); data reduction: *SAINT*; program (s) used to solve structure: *SHELXTL* (Sheldrick, 2008); program (s) used to refine structure: *SHELXTL*; molecular graphics: *SHELXTL* and *DIAMOND* (Brandenburg, 2005); software used to prepare material for publication: *SHELXTL*.

The authors gratefully acknowledge financial support from Jiangsu Polytechnic University.

Supplementary data for this paper are available from the IUCr electronic archives (Reference: GA3116). Services for accessing these data are described at the back of the journal.

References

[1] Brandenburg, K. (2005). *DIAMOND*. Version 3.0d. Crystal Impact GbR, Bonn, Germany.
[2] Bruker (2001). *SAINT* and *SHELXTL*. Bruker AXS Inc., Madison, Wisconsin, USA.
[3] Bruker (2003). *APEX2*. Bruker AXS Inc., Madison, Wisconsin, USA.
[4] Chen, S. C., Zhang, Z. -H., Huang, K. -L., Chen, Q., He, M. -Y., Cui, A. J., Li, C., Liu, Q. & Du, M. (2008). *Cryst. Growth Des*. 8, 3437-3445.
[5] Chun, H., Dybtsev, D. N., Kim, H. & Kim, K. (2005). *Chem. Eur. J*. 11, 3521-3529.
[6] Du, M., Jiang, X. -J. & Zhao, X. J. (2007). *Inorg. Chem*. 46, 3984-3995.

[7] Etter, M. C. (1990). *Acc. Chem. Res.* 23, 120-126.
[8] Férey, G. (2008). *Chem. Soc. Rev.* 37, 191-214.
[9] Kitaura, R., Iwahori, F., Matsuda, R., Kitagawa, S., Kubota, Y., Takata, M. & Kobayashi, T. C. (2004). *Inorg. Chem.* 43, 6522-6524.
[10] Li, H., Eddaoudi, M., O'Keeffe, M. & Yaghi, O. M. (1999). *Nature* (*London*), 402, 276-279.
[11] Maspoch, D., Domingo, N., Ruiz-Molina, D., Wurst, K., Tejada, J., Rovira, C. & Veciana, J. (2004). *J. Am. Chem. Soc.* 126, 730-731.
[12] Serre, C., Millange, F., Thouvenot, C., Nogue's, M., Marsolier, G., Louer, D. & Fe'rey, G. (2002). *J. Am. Chem. Soc.* 124, 13519-13526.
[13] Sheldrick, G. M. (2003). *SADABS*. Version 2. 10. University of Go¨ttingen, Germany.
[14] Sheldrick, G. M. (2008). *Acta Cryst*. A64, 112-122.
[15] Wang, B., Cote, A. P., Furukawa, H., O'Keeffe, M. & Yaghi, O. M. (2008). *Nature* (*London*), 453, 207-211.
[16] Xiao, D. R., Wang, E. -B., An, H. Y., Su, Z. M., Li, Y. G., Gao, L., Sun, C. Y. & Xu, L. (2005). *Chem. Eur. J.* 11, 6673-6686.

(注：此文原载于 Acta Crystallographica Section C, 2009, 65 (2): 86-90)

$[Hmim]_3PW_{12}O_{40}$: A High-efficient and Green Catalyst for the Acetalization of Carbonyl Compounds

Dai Yan[a], Li Bindong[a], Quan Hengdao[b], Lv Chunxu[a]

[a]School of Chemical Engineering, Nanjing University of Science and Technology, Nanjing 210094, China[b] National Institute of Advanced Industrial Science and Technology, Ibaraki 3058565, Japan

SCI: 000278442300011

Abstract: $[Hmim]_3PW_{12}O_{40}$ was developed and used in the acetalization of carbonyl compounds in excellent yields. The ionic liquidheteropoly acid hybrid compound and reaction medium formed temperature-dependent phase-separation system with the ease of product as well as catalyst separation. The catalyst was recycled more than 10 times without any apparent loss of catalytic activity.

Key words: $[Hmim]_3PW_{12}O_{40}$, Phosphotungstic acid, Carbonyl compounds, Acetalization, Ionic liquid

Acetalization plays an important role in organic synthesis for the protection of carbonyl groups[1], and usually is achieved by treatment of carbonyl compounds with an alcohol or diol in the presence of Brønsted or Lewis acids[2]. However, those procedures own some drawbacks such as many by-products and serious environmental pollution. Besides that, the separation of the products from the catalytic system is also difficult.

Recently, heteropoly acids (HPAs) especially Keggin type have attracted increasing interest due to their high acidity, low toxicity and tunable redox properties[3]. Unfortunately, most HPAs exhibit some difficulties for product separation in homogeneous catalytic systems. The deposition of HPAs on porous solid supports has been considered as an effective means for improving properties and obtaining better performance in many potential heterogeneous catalytic applications[4]. However, HPAs supported on carrier by physisorption still show a degree of solubility in the reaction medium, which causes low recovery ratio and difficulty of separation. Thus, to develop novel HPAs and their salts is a crucial task at present.

Ionic liquids (ILs) absorbed more and more interests of researchers and successfully used in acetalization reaction due to low corrosion tendencies, good thermal stability and reusage[5]. In view of the advantages of ILs and HPAs, the ionic liquid based hybrid molecular materials have been studied[6]. Some researchers have already reported IL-solid catalytic systems. However, these ILs inevitably showed some drawbacks such as low recovery ratio, low yield or difficulty of synthesis. In addition, they were used as electrochemical rather than as catalysts. Herein, we report a Keggin phosphotungstate anion based Brønsted acidic ionic salt $[Hmim]_3PW_{12}O_{40}$ could be used as a higheffi-cient, reusable and environmental-friendly catalyst for the acetalization of carbonyl compounds (Scheme 1).

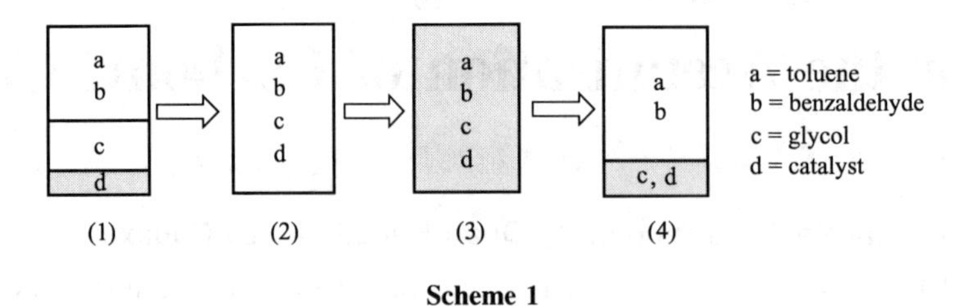

Scheme 1

1 Results and discussion

$[Hmim]_3PW_{12}O_{40}$ was obtained from phosphotungstic acid and N-methyl imidazole with 95% yield at room temperature[7]. It is white solid with high melting point (>300℃), which is absolutely different from conventional ionic liquids. $[Hmim]_3PW_{12}O_{40}$ can be dissolved in dimethyl sulfoxide, however, it does not dissolve readily in other standard organic solvents and water at room tempera-ture. The unique property is coincident with the requirements of a solid catalyst.

To study the catalytic activity of $[Hmim]_3PW_{12}O_{40}$, the acetalization of benzaldehyde and glycol was chosen as reaction model. The reaction was performed smoothly and 2-phenyl-1, 3-dioxolane was obtained with 97% yield[8]. It may be mentioned here that the catalyst and reaction medium formed temperature-dependent phase-separation system in the acetalization of benzaldehyde (Fig. 1) . The outstanding feature of the catalyst is that it scarcely dissolves in standard organic solvents and water at room temperature, but can dissolve in glycol at high temperature due to the formation of H-bonding between $[Hmim]_3PW_{12}O_{40}$ and glycol. Thus, the reaction system was heterogeneous at room tem-perature and then became homogeneous at reflux during the reaction. As the consumption of glycol, the system became heterogeneous and induced a spontaneous self-separation of the catalyst. The unu-sual behavior was also observed in the acetalization of other carbonyls and glycol.

$$\text{N-methylimidazole} + H_3PW_{12}O_{40} \xrightarrow{\text{rt, 24h}} [\text{HmimH}^+]_3 PW_{12}O_{40}^{3-}$$

$$R^1COR^2 \xrightarrow[{[Hmim]_3PW_{12}O_{40}}]{ROH} R^1R^2C(OR)_2 \qquad R^1COR^2 \xrightarrow[{[Hmim]_3PW_{12}O_{40}}]{HOCH_2CH_2OH} \text{2-}R^1\text{-2-}R^2\text{-1,3-dioxolane}$$

Fig. 1 Illustration of the acetalization of benzaldehyde with glycol using $[Hmim]_3PW_{12}O_{40}$ as catalyst:

(1) heterogeneous mixture before the reaction at room temperature;

(2) homogeneous mixture during the reaction at reflux;

(3) near completion of the reaction; and (4) at the end of the reaction at room temperature.

Because glycol was one of the reactants, the catalyst layer was recycle used without any further purification. Performance of recycling used $[Hmim]_3PW_{12}O_{40}$ was investigated in the reaction of ben-zaldehyde with glycol. As shown in Table 1, $[Hmim]_3PW_{12}O_{40}$ could be easily recycled and reused 10 times without obvious loss of activity. Moreover, the recovery yield of the catalyst was obtained more than 93% after reused 10times.

Table 1 The recycle of $[Hmim]_3PW_{12}O_{40}$.

Cycles	1	2	3	4	5	6	7	8	9	10
Yields[a]	97	98	97	96	95	94	93	93	92	92

a Chemical yields of the reaction.

In order to prove the merits of the catalyst in the present work, we compared the catalytic activity of $[Hmim]_3PW_{12}O_{40}$ with other catalysts reported previously, and the data were listed in Table 2. Although inorganic salts and Brønsted acids could catalyze the acetalization reaction, the yields were dissatisfactory (Table 2, entries 1-4). Moreover, some of these catalysts are not stable in air. In cases of I_2, $H_4SiW_{12}O_{40}$, $H_3PW_{12}O_{40}$/Pan and $H_3PW_{12}O_{40}$, due to their shortcomings of difficult recovery and too many by-products, they could not be considered as economic and green catalysts even though some of them displayed good catalytic capacity. As for [TMPSA] HSO_4,

Table 2 Comparison of the effect of catalysts in acetalization of benzaldehyde with glycol.

Entry	Catalyst	Yield(%)	Entry	Catalyst	Yield(%)
1	$FeCl_3 \cdot 6H_2O$	58[2g]	7	$H_3PW_{12}O_{40}$/PAn	76[2i]
2	H_2NSO_3H	43[2a]	8	IL-1000/TsOH	92[2j]
3	$Bi(OTf)_3 \cdot 4H_2O$	73[2c]	9	[TMPSA]HSO_4	77[2k]
4	Al-SBA-15	65[2f]	10	[PSebim]HSO_4	72[2d]
5	I_2	92[2h]	11	$H_3PW_{12}O_{40}$	68
6	$H_4SiW_{12}O_{40}$	84[2g]	12	$[Hmim]_3PW_{12}O_{40}$	97

Reaction conditions: benzaldehyde (50mmol), glycol (75mmol), catalyst (0.44mmol), toluene (8mL), reflux for 2h.

Table 3 Acetalization of carbonyl compounds.

Entry	Carbonyl compounds R^1	R^2	Alcohol	Time(h)	Yield(%)[a]
1[b]	C_6H_5—	H—	Methanol	2	82
2[b]	C_6H_5—	H—	Ethanol	2	79
3[b]	C_6H_5—	H—	Butan-1-ol	3	81
4[c]	C_6H_5—	H—	Glycol	2	97
5[c]	C_2H_5—	H—	Glycol	2.5	99
6[c]	C_3H_7—	H—	Glycol	2.5	96
7[c]	iso-C_4H_9—	H—	Glycol	3	93
8[c]	4-OH—C_6H_5—	H—	Glycol	4	78
9[c]	3-OH—C_6H_5—	H—	Glycol	4	85
10[c]	2-OH—C_6H_5—	H—	Glycol	6	62
11[c]	4-NO_2—C_6H_5—	H—	Glycol	1.5	97
12[c]	3-NO_2—C_6H_5—	H—	Glycol	1.5	96
13[c]	2-NO_2—C_6H_5—	H—	Glycol	4	88
14[c]	4-Cl—C_6H_5—	H—	Glycol	1.5	95
15[c]	3-Cl—C_6H_5—	H—	Glycol	1.5	94
16[c]	C_6H_5—	CH_3—	Glycol	5	89
17[c]	—C_4H_6—	—CH_2—	Glycol	2	98
18[c]	C_2H_5—	CH_3—	Glycol	3.5	90

Reaction conditions: carbonyl compound (50mmol), alcohol ([b]150mmol; [c]75mmol), toluene (8mL), $[Hmim]_3PW_{12}O_{40}$ (0.44mmol), reflux. [a] Detected by GC/MS. Yields were b ased on the starting c arbonyl compounds.

[PSebim] HSO_4 and IL-1000/TsOH, they gave the desired products in moderate to excellent yields, however, these catalysts are expensive and require special efforts for preparation. $[Hmim]_3PW_{12}O_{40}$ could be easily recycled and reused 10 times without obvious loss of activity and weight. Thus, in view of excellent catalytic capacity, low cost, outstanding stability and ready reutilization, $[Hmim]_3PW_{12}O_{40}$ was chosen as an effective catalyst for acetalization reaction of benzaldehyde and glycol.

In order to evaluate the generality of the process, we extended experiments to a series of alcohols and carbonyl compounds. The reaction proceeded smoothly. Corresponding acetals and ketals were obtained in high yields in most cases (Table 3) . Due to its unique structure with two active sites, glycol usually reacts more actively than monohydroxy alcohol. The yields of acetalization of glycol and benzaldehydes vary depending upon the types and positions of substitutions. The results showed that the electron-donating groups decrease the yield, whereas electron-withdrawing groups increase it. Due to steric factors, o-hydroxybenzaldehyde reacted more slowly than p-hydroxybenzaldehyde. Because alkyl group provides more electron density and steric crowding, acetophenone has a lower reactivity than benzaldehyde.

2 Conclusion

In conclusion, $[Hmim]_3PW_{12}O_{40}$ was facilely prepared and used as a high-efficient catalyst for the acetalization of carbonyl compounds. $[Hmim]_3PW_{12}O_{40}$ has several advantages: (1) it and reaction medium can form temperature-dependent phase-separation system with the ease of product as well as catalyst separation; (2) it was recycled more than 10 times without any apparent loss of weight and catalytic activity; (3) it has a wide range of application for different substrates and the products can be obtained conveniently in good to high yields. In a word, $[Hmim]_3PW_{12}O_{40}$ is an excellent reusable catalyst.

Acknowledgments

We are grateful for the financial support from Nanjing University of Science and Technology.

References

[1] T. W. Greene, P. G. M. Wuts, Protective Groups in Organic Synthesis, Wiley, New York, 1991.

[2] (a) B. Wang, Y. Gu, G. Song, et al. J. Mol. Catal. A: Chem. 233 (2005) 121.

(b) X. Y. Shen, H. F. Jiang, Y. C. Wang, Chin. J. Org. Chem. 28 (2008) 782;

(c) N. M. Leonard, M. C. Oswald, D. A. Freiberg, et al. J. Org. Chem. 67 (2002) 5202;

(d) Y. Wang, D. Jiang, L. Dai, Catal. Commun. 9 (2008) 2475;

(e) B. T. Gregg, K. C. Golden, J. F. Quiin, Tetrahedron 64 (2008) 3287;

(f) P. Srivastava, R. Srivastava, Catal. Commun. 9 (2008) 645;

(g) J. H. Dong, S. L. Hu, S. J. Yang, Chin. Appl. Chem. Ind. 34 (2005) 611;

(h) F. Gao, S. J. Yang, Chin. Appl. Chem. Ind. 34 (2005) 28;

(i) S. J. Yang, Y. J. Zhang, X. X. Du, et al. Rare Metals 27 (2008) 89;

(j) H. Z. Zhi, J. Luo, W. Ma, et al. Chem. J. Chin. Univ. 29 (2008) 2007;

(k) D. Fang, K. Gong, Q. Shi, et al. Catal. Commun. 8 (2007) 1463.

[3] (a) M. Tajbakhsh, J. Mol. Catal. A: Chem. 287 (2008) 5.

(b) A. S. Dias, I. V. Kozhevnikov, J. Mol. Catal. A: Chem. 305 (2009) 104;

(c) P. S. Kishore, M. M. Heravi, S. Sadjadi, et al. Catal. Commun. 9 (2008) 504.

[4] (a) N. Lucas, A. Bordoloi, A. P. Amrute, et al. Appl. Catal. A 352 (2009) 74.

(b) J. H. Sepu'lveda, J. C. Yori, C. R. Vera, Catal. Appl. A 288 (2005) 18;
(c) D. P. Sawant, J. Justus, V. V. Balasubramanian, et al. Chem. Eur. J. 14 (2008) 3200;
(d) S. Minakata, M. Komatsu, Chem. Rev. 109 (2009) 711.
[5] (a) H. Zhi, C. Lu¨, Q. Zhang, et al. Chem. Commun. 20 (2009) 2878.
(b) J. Dupont, R. F. Souza, P. A. Z. Suarez, Chem. Rev. 102 (2002) 3667;
(c) T. Welton, Chem. Rev. 99 (1999) 2071.
[6] (a) Y. Leng, J. Wang, D. Zhu, Angew. Chem. Int. Ed. 48 (2009) 168.
(b) K. Wang, W. W. Zhang, S. Yin, et al. Chin. J. Appl. Chem. 26 (2009) 32;
(c) A. B. Bourlinos, K. Raman, R. Herrera, et al. J. Am. Chem. Soc. 126 (2004) 15358;
(d) D. Chen, Q. Zhang, G. Wang, et al. Electrochem. Commun. 9 (2007) 2755;
(e) J. H. Shi, G. Pan, Chin. J. Catal. 29 (2008) 629.
[7] Phosphotungstic acid (5mmol) and deionizedwater (15mL) was added drop wisely into N-methyl imidazole (15mmol) in 30min with constant stirring. Then the materials were stirred at room temperature for 24h. Finally, the produced white solid was flltrated, washed and dried (95% yield). Mp>300℃; IR (KBr): n (cm^{-1}) 3330, 3159, 2871, 1585, 1548, 1438, 1384, 1330, 1151, 1080, 979, 893, 808, 740, 622, 592; ^{1}H NMR (300 MHz, DMSO-d$_6$, TMS): d 3.85 (s, 3H, CH$_3$), 7.60 (s, 1H, CH), 7.65 (s, 1H, CH), 8.92 (s, 3H, CH); Anal. Calcd. for $C_{12}H_{21}N_6PW_{12}O_{40}$: C 4.61, H 0.68, N 2.69, P 0.97, W 68.76; found C 4.59, H 0.71, N 2.68, P 0.98, W 68.73; ESI-MS (energy: 50v, m/z, z = 3): 958.3 (48%), 701.5 (100%), 470.6 (41%).
[8] Representative procedure: the mixture of benzaldehyde (50mmol), glycol (75mmol), toluene (8mL) and [Hmim]$_3$PW$_{12}$O$_{40}$ (0.44mmol) was stirred at 80℃ for certain time. Water formed in the process was removed by azeotropic distillation using toluene as water-taken reagent. After the reaction, the organic layer was separated by decantation, and the product was detected by GC-MS (GC yield: 97%). The catalyst was reused without any treatment.

（注：此文原载于 Chin. Chem. Lett, 2010, 21 (6): 678-681)

A Mild and Efficient Method for Nucleophilic Aromatic Fluorination using Tetrabutylammonium Fluoride as Fluorinating Reagent

Hu Yufeng, Luo Jun, Lv Chunxu
(School of Chemical Engineering, Nanjing University of Science & Technology, Nanjing 210094, China)
SCI: 000275396700007

Abstract: Anhydrous tetrabutylammonium fluoride ($TBAF_{anh}$.) has been found to be a highly efficient fluorinating reagent for nucleophilic aromatic fluorinations such as fluorodenitration or halogen exchange (Halex) reaction. The products were formed in high to excellent yields under surprisingly mild reaction conditions and no phenol or ether side-products were detected in these reactions.

Keywords: TBAF anh., Fluoroaromatics, Fluorodenitration, Halex

Fluoroaromatics have applications in many sectors, including the pharmaceutical and agrochemical industries due to the unusual properties of the C-F bond[1-4]. Because of the industrial importance of fluoroaromatics, mild and selective methods for their preparation are desirable, thus an expansive set of nucleophilic reagents has been developed to replace various C—X functional groups with C—F via nucleophilic aromatic fluorination. Alkali metal fluoride such as potassium fluoride is the traditional reagent for nucleophilic aromatic fluorination. However, its limited solubility in organic solvent and low nucleophilicity require generally vigorous conditions[5]. CsF being both more reactive and more expensive and LiF and NaF being completely unreactive. Thus, a number of metal fluoride reagents such as KF/18-crown-6, "freeze-dried" KF and "spray-dried" KF have been reported to solve these problems[6]. However, these above-mentioned processes are generally less efficient than those using tetraalkylammonium fluorides[7-10]. $TBAF_{anh}$. was found to be an efflcient fluorinating reagent for the nucleophilic fluorination[11-14]. $TBAF_{anh}$. is a soluble, highly nucleophilic fluoride-ion source. It can be dissolved in dimethyl sulfoxide (DMSO), CH_3CN and THF, and it can fluorinate primary alkyl halides and tosylates rapidly at low temperature (−35℃) for 5min in THF[12]. The rapid rate observed for S_N2 reactions that feature $TBAF_{anh}$. prompted us to investigate nucleophilic aromatic substitution (S_NAr) reactions with this reagent. Below we describe variants of fluorodenitration and Halex processes for aromatic fluorination. The results show that $TBAF_{anh}$. permits these reactions to be performed under surprisingly mild conditions.

1 Experimental

A mixture of $TBAF_{anh}$. (30mmol), dried THF (10mL), nitroaromatics or chloroaromatics (25mmol) were stirred for certain time at room temperature or refluxed under nitrogen atmosphere. The reaction was trapped by GC. Side-products were obtained by flash column chromatography. 4,4′-Dinitrodiphenyl ether: ^{1}HNMR (500MHz, $CDCl_3$, δ ppm): 7.16~7.18 (d, 4×CH, J=7.5Hz), 8.29~8.30 (d, 4×CH, J=7.5Hz); ^{13}CNMR (500MHz, $CDCl_3$, δ ppm): 160.71, 144.19, 126.19, 119.31; Melting point: 145℃~146℃; MS (m/z): 260, 138, 123, 95, 77; IR (KBr): 3099.91, 1582.87, 1464.64, 1321.93, 1308.05, 1281.75, 1146.28, 1088.96cm^{-1}. p-Nitrophenyl methyl sulfide: ^{1}HNMR (500MHz, $CDCl_3$, δ ppm): 2.58 (s, CH_3), 7.31~7.32 (d, 2×CH, J=8.9Hz), 8.15~8.17 (d, 2×CH, J=8.9Hz); ^{13}CNMR (500MHz, $CDCl_3$, δ ppm): 148.92, 144.74, 124.99, 123.85, 14.82; Melting point: 68℃ ~ 69℃; MS (m/z): 169, 139, 111, 108, 77, 45; IR (KBr): 1530.34, 1359.32, 1334.26, 745.42cm^{-1}.

2 Results and discussion

The usual reagents employed to effect fluorodenitration or Halex are spray-dried KF (SD-KF); a high-boiling-point polar aprotic solvent, such as sulfolane; and a phase-transfer catalyst to improve the solubility of the fluoride ion. Prolonged and vigorous heating is often required. Such harsh conditions preclude syntheses in which the late introduction of fluorine substituents is desired. In contrast, many aromatic compounds undergo fluorodenitration or Halex reactions at room temperature in THF upon exposure to $TBAF_{anh}$. (Scheme 1) . Some examples are summarized in Table 1.

R_1=CHO, NO_2, CN; R_2=NO_2, Cl

Scheme 1 Fluorodenitration and Halex reactions of nitro-and chloroaromatic compounds by $TBAF_{anh}$.
(a) $TBAF_{anh.}$, THF, in RT.

As the data in Table 1 indicate, aromatics with electron-withdrawing groups at the 2-or 4-position (entries 2-4,6,7 and 9) underwent smooth denitration to afford the corresponding fluoroaromatics after a few hours either at room temperature or at reflux. While aromatics with electron-withdrawing groups at the 3-position required prolong the reaction time to drive completion (entries 1 and 8) . A comparison of the meta-ortho-and para-dinitrobenzene examples shows fluorination of meta-dinitrobenzene requires 40h, whereas ortho-and para-dinitrobenzene are exhaustively fluorinated in 1.5h under identical conditions. Nitroaromatic substrates bearing more strongly electron-withdrawing (cyano) substituents are also fluorinated rapidly. It is noteworthy that 2-and 4-nitrobenzonitriles are fluorinated quantitatively within 1h. In addition, $TBAF_{anh}$. performs fluorinations even on relatively weakly activated arenes, for example, nitrobenzaldehydes and chlorobenzaldehydes, fluorodenitration and Halex fluorinations of 2-and 4-nitrobenzaldehydes and 2-and 4-chlorobenzaldehydes require heating at reflux for ten or twenty hours to give more than 80% yields. Adams and Clark reported that NO_2^- is superior to chloride as a leaving group in S_NAr[15], as the data in Table 1 confirmed the conclusion (entries 4, 6, 7, 9, 10 and 12-14). Entries 8 and 15 were chosen as examples of meta-substituent, which are of interest due to the difficulty of forming such fluoroaromatics via Halex fluorination, thus entries 1 and 8 are established industrial methods of forming such fluoroaromatics.

Table 1 Fluorodenitration and Halex reactions of various nitro-and chloroaromatic compounds using $TBAF^{a}_{anh}$.

Entry	Substrate	Product	T(℃)	Time(h)	Yield(%)
1	O_2N-⌬-NO_2	F-⌬-NO_2	RT	1.5	98.5
2	NO_2 O_2N-⌬	NO_2 F-⌬	RT	40	85.3
3	NO_2 ⌬ NO_2	F ⌬ NO_2	RT	1.5	97.4
4	O_2N-⌬-CHO	F-⌬-CHO	Reflux	10	82.1
5	CHO O_2N-⌬	CHO F-⌬	Reflux	48	<5
6	NO_2 ⌬ CHO	F ⌬ CHO	Reflux	10	80.6
7	NO_2 ⌬ CN	F ⌬ CN	RT	1	97.3
8	CN O_2N-⌬	CN F-⌬	RT	36	94.6
9	O_2N-⌬-CN	F-⌬-CN	RT	1	98.2
10	Cl-⌬-CHO	F-⌬-CHO	Reflux	20	79.7
11	CHO Cl-⌬	CHO F-⌬	Reflux	48	trace
12	Cl ⌬ CHO	F ⌬ CHO	Reflux	20	77.3
13	Cl ⌬ CN	F ⌬ CN	RT	30	95.8
14	Cl-⌬-CN	F-⌬-CN	RT	40	92.5
15	CN Cl-⌬	CN F-⌬	Reflux	48	trace

a Yields based on GC area, corrected by the presence of an internal standard, and biphenyl as standard.

As the data in Table 1 indicate, aromatics with electron-withdrawing groups at the 2-or 4-position (entries 2-4, 6, 7 and 9) underwent smooth denitration to afford the corresponding fluoroaromatics after a few hours either at room temperature or at reflux. While aromatics with electron-withdrawing groups at the 3-position required prolong the reaction time to drive completion (entries 1 and 8). A comparison of the meta-ortho-and para-dinitrobenzene examples shows fluorination of meta-dinitrobenzene requires 40h, whereas ortho-and para-dinitrobenzene are exhaustively fluorinated in 1.5h under identical conditions. Nitroaromatic substrates bearing more strongly electron-withdrawing (cyano) substituents are also fluorinated rapidly. It is noteworthy that 2-and 4-nitrobenzonitriles are fluorinated quantitatively within 1h. In addition, $TBAF_{anh}$. performs fluorinations even on relatively weakly activated arenes, for example, nitrobenzaldehydes and chlorobenzaldehydes, fluorodenitration and Halex fluorinations of 2-and 4-nitrobenzaldehydes and 2-and 4-chlorobenzaldehydes require heating at reflux for ten or twenty hours to give more than 80% yields. Adams and Clark reported that NO_2^- is superior to chloride as a leaving group in S_NAr[15], as the data in Table 1 confirmed the conclusion (entries 4, 6, 7, 9, 10 and 12-14). Entries 8 and 15 were chosen as examples of meta-substituent, which are of interest due to the difficulty of forming such fluoroaromatics via Halex fluorination, thus entries 1 and 8 are established industrial methods of forming such fluoroaromatics.

In order to compare the reaction efficiency for nucleophilic aromatic fluorination of various fluorinating reagents, we chose the conversion of p-dinitrobenzene (PDNB) to p-fluoronitrobenzene (PFNB). Representative fluorinations are shown in Table 2.

Table 2 Fluorodenitration reactions of PDNB using various fluorinating reagents.

Entry	Fluorinating reagent	Solvent	T(℃)	Time(h)	Yield(%)[a]
1	SD-KF(4 equiv)	15mL DMSO	189	1	82.7
2	CsF(4 equiv)	15mL DMSO	120	1.5	91.4
3	$TMAF_{anh}$. (1.4 equiv)	10mL DMF	80	1	95.6
4	$TBAF_{anh}$. (1.2 equiv)	15mL THF	RT	1.5	98.5

a Yields calculated by GC area, corrected by the presence of an internal standard, and biphenyl as standard.

As the data in Table 2 indicate, the activity of KF is the weakest and it needs very high temperature to make the reaction proceed successfully. Attempted displacement of 25mmol of PDNB by 100mmol of SD-KF in the presence of 2.5mmol of Ph_4PBr, 1.25mmol of TEG-Me_2 and 25mmol of PDC in 15mL of DMSO at 189℃ for 1h gave only 82.7% of PFNB. In the mean time, because of weak stability of NO_2^-, there will result some by-products, such as phenols and ethers (we had isolated the 4,4′-dinitrodiphenyl ether from the reaction mixture). Generally, side reaction can be controlled effectively by adding NO_2^- trapping agent, such as phthaloyl dichloride (PDC)[16]. The high basicity of the fluoride can also result in the formation of by-products arising from reaction with the solvent. p-Nitrophenyl methyl sulfide incorporated products (derived from DMSO) were also isolated. In contrast, the activity of CsF is stronger than that of KF. So it can make the reaction temperature decrease evidently. $TMAF_{anh}$. was observed to be more efficient than SD-KF and CsF but was less efficient than $TBAF_{anh}$. $TMAF_{anh}$. gave 95.6% yield of PFNB after 1h at 80℃ in DMF. No phenols or ethers were detected in either reaction and, unlike the method of fluorodenitration with SD-KF no brown fumes were observed. Whereas $TBAF_{anh}$. permits the reaction to be performed under surprisingly mild reaction conditions. $TBAF_{anh}$. gave 98.5% yield of PFNB after 1.5h at only RT. No phenols or ethers were detected in reaction. Tetraalkylammonium fluorides are known to be a particularly effective fluorodenitration reagent partly because it forms a stable nitrite which is less likely to back-attack the fluoro product leading to a loss in yield and the formation of phenol and ether side-products. This may indicate that the displaced NO_2^- can exist in this system with not very high temperature as effectively stable R_4NNO_2[10].

In conclusion, $TBAF_{anh}$. is a highly efflcient fluorinating reagent for selective nucleophilic aromatic fluorination of a wide range of nitroaromatics and chloroaromatics. In addition, high yields of the products, surprisingly mild reaction conditions, little side reactions and easy work-up are worthy advantages of this method.

Acknowledgment

We are grateful to the National Basic Research (973) Program of China (No. 613740101).

References

[1] Y. C. Wu, C. Wu, X. M. Zou, et al. Chin. Chem. Lett. 16 (2005) 1293.
[2] X. J. Song, X. H. Tan, Y. G. Wang, Chin. Chem. Lett. 17 (2006) 1443.
[3] C. Z. Zhang, J. L. Lei, S. M. Cai, et al. Chin. Chem. Lett. 13 (2002) 666.
[4] D. J. Adams, J. H. Clark, D. J. Nightingale, Tetrahedron 55 (1999) 7725.
[5] Y. Sasson, S. Negussie, M. Royz, et al. Chem. Commun. (1996) 297.
[6] D. W. Kim, C. E. Song, D. Y. Chi, J. Am. Chem. Soc. 124 (2002) 10278.
[7] D. J. Adams, J. H. Clark, L. B. Hansen, et al. J. Fluorine Chem. 92 (1998) 123.
[8] J. H. Clark, D. Wails, J. Fluorine Chem. 70 (1995) 201.
[9] D. J. Adams, J. H. Clark, H. McFarland, J. Fluorine Chem. 92 (1998) 127.
[10] N. Boechat, J. H. Clark, J. Chem. Soc., Chem. Commun. 11 (1993) 921.
[11] S. D. Kuduk, R. M. DiPardo, M. G. Bock, Org. Lett. 7 (2005) 577.
[12] H. R. Sun, S. G. DiMagno, Angew. Chem. Int. Ed. 45 (2006) 2720.
[13] H. R. Sun, S. G. DiMagno, J. Am. Chem. Soc. 127 (2005) 2050.
[14] W. X. Zhang, X. L. Hou, Chin. Chem. Lett. 17 (2006) 1037.
[15] D. J. Adams, J. H. Clark, Chem. Soc. Rev. 28 (1999) 225.
[16] H. Suzuki, N. Yazawa, Y. Yoshida, et al. Bull. Chem. Soc. Jpn. 63 (1990) 2010.

(注：此文原载于 Chinese Chemical Letters, 2010, 21 (2): 151-154)

Synthesis of Chalcones via Claisen-Schmidt Reaction Catalyzed by Sulfonic Acid-Functional Ionic Liquids

Qian Hua[1], Liu Dabin[1], Lv Chunxu[2]

(1. School of Chemical Engineering, Nanjing University of Science and Technology, Nanjing 210094, PR. China)

(2. Jiangsu Pharmaceutical Intermediate Research Center, Nanjing 210094, PR. China)

SCI: 000286027900080, EI: 20110313594441, ISPT: 00002286027900080

Abstract: Four recyclable Sulfonic acid-functional ionic liquids were prepared and used as dual catalyst and solvent for the synthesis of chalcones by Claisen-Schmidt condensation between acetophenone and benzaldehyde. All the four ionic liquids proved to be very active, leading to 85-94% yield of chalcone in the presence of 20% ionic liquids. The chalcones can simply be separated from mixtures by simple decantation. After removal of the water, the ionic liquids can be readily recovered and reused successfully without any significant loss of catalytic activity, which are potential substitutes for conventional homogenous/heterogeneous acidic catalysts.

1 Introduction

Chalcones and its derivatives are attracted increasing attention due to numerous pharmacological applications. They are main precursors for the biosynthesis of flavonoids and exhibit various biological activities such as anti-cancer, anti-inflammatory and anti-hyperglycemic agents[1-3]. Traditionally, chalcones could be obtained via the Claisen-Schmidt condensation carried out in acidic or basic media under homogeneous conditions with many drawbacks such as catalyst recovery and waste disposal problems[4]. As a potential alternative, heterogeneous catalysts have also been used for the Claisen-Schmidt condensation, including Lewis acids[5,6]. Brønsted acids[7], solid acids[8-10], solid bases[11-13]. However, most of them still require expensive toxic solvents to facilitate the heat and mass transfer of the liquid phase in reaction systems.

Due to particular properties, ionic liquids (ILs) have been emerged as a powerful alternative to conventional molecular organic solvents, and becoming a greener alternative to volatile organic solvents. Some common ILs have been tried in Claisen-Schmidt condensation, and show a significant enhancement in the yield of chalcone[14-17]. However, most of them are liposoluble and products need to be extracted with extra organic solvents, and ILs are difficult to recovery[16]. Furthermore, some ILs are poisonous, which are harmful for the environment[17]. We are especially interested in developing the potential use of efficient and clean ILs catalysts. Novel Sulfonic acid-functional ILs are prepared through a simple and atom-economicneutralization reaction. In comparison to imidazolium-based ILs, the Sulfonic acid-functional ILs have a relatively lower cost, lower toxicity, and higher catalytic activity[18]. We tried a series of acid-ILs in the nitration of DAPT (3,7-diacetyl-1,3,5,7-tetraazacyclo-[3.3.1]-octane) to form the most powerful military explosives HMX (1,3,5,7-tetranitro-1,3,5,7-tetraazacyclooctane), and showed great success[19,20]. In continuation of our work in studying acid-catalyzed reactions in ILs, we report an efficient and environmentally friendly solvent-free method for the synthesis of chalcone using Sulfonic acid-functional ILs as dual catalyst and solvent, which is a potential substitute for conventional homogenous/heterogeneous acidic catalysts.

2 Experimental section

2.1 Materials and reagents

1,4-Butanesultone and trifluoroacetic acid were of reagent grade and obtained from J&K Co. Ltd. All others chemicals were also from commercial sources and of reagent grade. Dichloromethane were dried over 4Å molecular sieve before use.

2.2 Synthesis of SO_3H-functionals ILs

In a typical experiment, a mixture of Triethylamine (0.1mol) and 1,4-butanesultone (0.1mol) was charged into a 250mL round-bottom flask. The reaction was proceed at 40℃ for 24h and vigorous stirred. After filtrating, the solid product was washed with ethyl ether (3×10mL) and dried under vacuum (70℃, 0.1MPa). Then, zwitterions (0.1mol) was dissolved in 50mL de-ionized water, 0.1mol acid (nitric acid, sulphuric acid, 4-toluene sulfonic acid, trifluoroacetic acid) was added dropwise with vigorously stirring and the reaction mixture stirred for an additional 6h at 60℃. Extra water was removed by rotary evaporation (70℃, 0.1MPa), and SO_3H-functional ILs were obtained with the total yield of 68%.

2.3 Characterization of ILs

The synthesized ILs were identified by ^{1}HNMR and ^{13}CNMR spectral data. Its relative acidity was conducted by UV-visible spectrophotometer with 4-nitroaniline used as indicator.

[TEBSA][NO_3] ^{1}HNMR(300MHz, D_2O): δ 1.03-1.08(t, 9H, J=6.7 Hz), 1.60-1.63(m, 2H), 2.74-2.78(t, 2H, J=6.6 Hz), 2.98-3.11(m, 8H). ^{13}C NMR(75.5 MHz, D_2O): δ 58.87, 54.23, 51.40, 23.07, 21.12, 7.11.

[TEBSA][HSO_4] ^{1}HNMR(300MHz, D_2O): δ 1.08-1.11(m, 9H, J=7.2 Hz), 1.64(t, 2H), 2.77-2.81(t, 2H, J=7.2Hz), 3.00-3.14(m, 8H). ^{13}C NMR(75.5 MHz, D_2O): δ 56.38, 52.68, 50.17, 21.80, 19.84, 6.98.

[TEBSA][CF_3COO] ^{1}HNMR(300MHz, D_2O): δ 1.03-1.07(t, 9H, J=7.0 Hz), 1.61(m, 4H), 2.74-2.79(t, 2H, J=7.5 Hz), 2.97-3.10 (m, 8H). ^{13}C NMR (75.5 MHz, D_2O): δ 165.5, 117.4, 59.07, 54.83, 52.67, 24.15, 22.90, 7.86.

[TEBSA][pTSO] ^{1}HNMR(300MHz, D_2O): δ 1.05-1.08(t, 9H, J=6.3 Hz), 1.64(m, 4H), 2.23(s, 3H), 2.80-2.82(t, 2H, J=6.6 Hz), 3.01-3.11(m, 8H), 7.19-7.22(d, 2H, J=7.8 Hz), 7.51-7.54(d, 2H, J=7.7 Hz). ^{13}CNMR(75.5 MHz, D_2O): δ 143.6, 142.4, 129.5, 124.1,

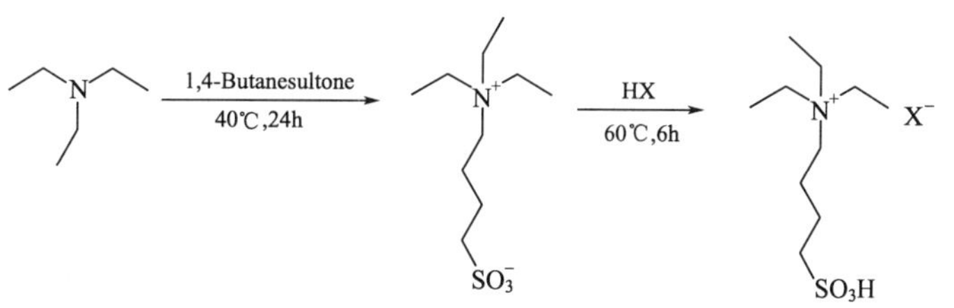

1,4-Butanesultone 40℃,24h HX 60℃,6h

TEBSA [TEBSA][X]

Scheme 1 The Synthesis Process of [TEBSA] [X] (X=NO_3^-, HSO_4^-, *p*-TsO$^-$, CF_3COO^-)

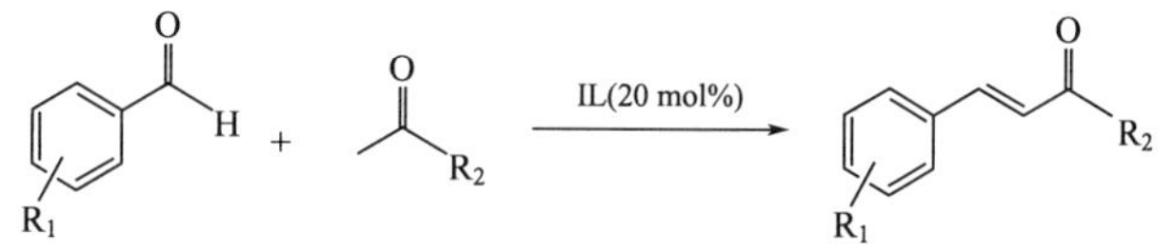

IL(20 mol%)

1 2 3

Scheme 2 Claisen-Schmidt Condensation Catalyzed by Ionic Liquids (ILs)

58.69, 52.01, 51.83, 22.48, 20.51, 18.57, 7.55.

2.4 Reaction prodedure.

A mixture of benzaldehyde (20mmol), acetophenone (20mmol) and IL (4mmol) was charged into a magnetically stirred three-necked-bottom flask equipped with a condenser system. The mixture was heated up to 120℃ and vigorously stirred under N_2 atmosphere for 4-6h. After the reaction was completed (TLC detected), the upper organic phase was separated from IL by decantation and chalcone was obtained. The dissolved organic species in IL could be extracted with ethyl acetate (3×5mL), and the extractant was combined with the upper organic phase and then washed with saturated solution of Na_2CO_3 (5mL) and water (3×5mL), dried over Na_2SO_4 and analyzed by GC. IL was recovered from the aqueous solution by removing water at 70℃ under vacuum (0.1MPa) for 5h and reused without significantly lost of its reactivity.

3 Results and discussion

The measurement of the Brønsted acidic scale of these SO_3H-functional ILs was conducted on an UV-240 spectrophotometer with a basic indicator according to the literature previously reported.[21] With the addition of acidic ILs, the absorbance of the unprotonated form of the basic indicator decreased, so the[I]/[IH$^+$]ratio could be calculated from the measured absorbance differences, and then the Hammett function, H_0, is calculated by using eq. 1. This value can be regarded as the relative acidity of the ILs.

$$H_0 = pK(\mathrm{I})_{aq} + \log([\mathrm{I}]/[\mathrm{IH}^+]) \quad (1)$$

Where pK $(I)_{aq}$ is the pK_a value of 4-nitroaniline, $pK(I)_{aq}=0.99$,[I]and[IH$^+$]are, respectively the molar concentrations of the unprotonated and protonated forms of 4-nitroaniline.

The results of acidities of four ILs are shown in Fig. 1. The maximal absorbance of the unprotonated form of 4-nitroaniline was observed at 350 nm in CH_2Cl_2 (Fig. 1. spectra a). We could determine the[I]/[IH$^+$]ratio by measuring the absorbance when each IL was added (spectra b-e), and then the Hammett function (H_0) is calculated (Table 1).

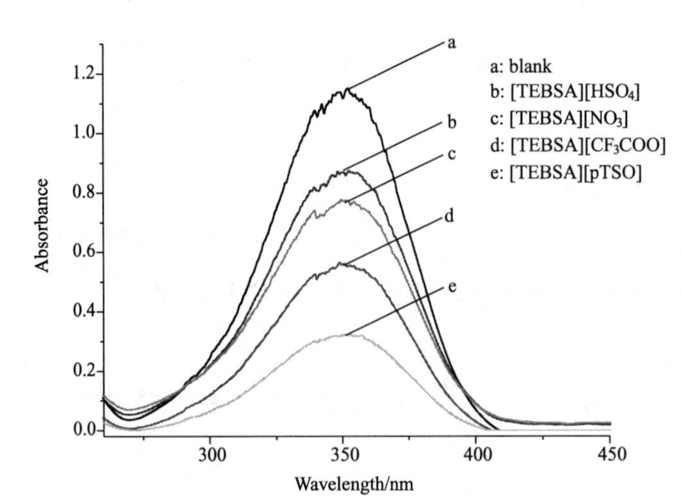

Figure 1 Absorption spectra of 4-nitroaniline in various ionic liquids (ILs) in dichloromethane.

Table 1 Hammett Function (H_0) Values of Ionic Liquids (ILs) in Dichloromethane at Room Temperature

ionic liquid	absorbance	[I](%)	[IH+]	Hammett function, H_0
blank	1.025	100.0	0	
[TEBSA][pTSO]	0.871	73.6	26.4	1.44
[TEBSA][CF_3COO]	0.782	68.1	31.9	1.32
[TEBSA][NO_3]	0.568	52.8	47.2	1.03
[TEBSA][HSO_4]	0.323	19.9	80.1	0.38

Ultimately, we obtained the acidity order of four SO_3H-functional ILs: [TEBSA][HSO_4]>[TEBSA][NO_3]>[TEBSA][CF_3COO]>[TEBSA][pTSO]. It is clear that acidities of ILs depend on the nature of anion, which will effectively affect their catalytic activities.

For the beginning of this study, benzaldehyde and acetone were employed to compare the catalytic performance of different SO_3H-functional ILs.

Table 2 Claisen-Schmidt Condensation Catalyzed by Different SO_3H-Functional Ionic Liquids (ILs)[a]

entry	ionic liquid	yield(%)[b]	entry	ionic liquid	yield(%)[b]
1	none	<5	4	[TEBSA][NO_3]	91
2	[TEBSA][pTSO]	85	5	[TEBSA][HSO_4]	94
3	[TEBSA][CF_3COO]	88			

a Reaction conditions: a mixture of benzaldehyde (10mmol), acetone (10mmol), and IL (2mmol) was vigorously stirred for 4h at 120℃ under a N_2 atmosphere. b GC yield, based on benzaldehyde.

As shown in Table 2, <5% yield of chalcone was detected by GC in the absence of ILs (entry 1), which indicated that the catalyst should be absolutely necessary for the Claisen-Schmidt reaction. All the four ILs proved to be very active, leading to 85-94% yield of chalcone in the presence of 20% ILs (entries 2-16). The best catalyzed IL is[TEBSA][HSO_4], with the yield of chalcone 94%. In addition, the immiscibility of the resulted chalcone with the IL facilitates the separation in work-up procedure.

The Claisen-Schmidt condensation reactions of other substituted benzaldehydes and acetophenones in the presence of [TEBSA][HSO_4]were accomplished under the optimized reaction conditions described above and the results are presented in Table 3.

Under the optimal reaction conditions, the condensation reactions between other substituted benzaldehydes and ketones in the presence of[TEBSA][HSO_4]were conducted and the results are summarized in Table 3. All benzaldehydes and ketones with different structural features were found to be suitable substrates. The condensation accomplished smoothly with reasonable to good yields ranged from 84% to 96%. Furthermore, it was found that electronic effect on the reactions was very limited. Benzaldehydes with electron-withdrawing groups such as nitro group only gave a little higher yield than other electro-donating groups such as methoxy group (entries 1-4) . The decline of yield between acetophenone and benzaldehydes may due to the steric hindrance of phenyl (entries 7-10).

Table 3 Claisen-Schmidt Condensation Catalyzed by [TEBSA] [HSO_4]

Entry	Compounds	R_1	R_2	Time(h)	Yield(%)
1	3a	H	$-CH_3$	4.0	94
2	3b	4-NO_2	$-CH_3$	3.5	96
3	3c	4-Cl	$-CH_3$	4.5	94
4	3d	4-OCH_3	$-CH_3$	4.5	91
5	3e	3-OCH_3	$-CH_3$	4.0	92
6	3f	2-OCH_3	$-CH_3$	4.0	94
7	3g	H		4.5	86
8	3h	4-NO_2		4.0	89
9	3i	4-Cl		5.0	85
10	3j	4-OCH_3		5.0	84

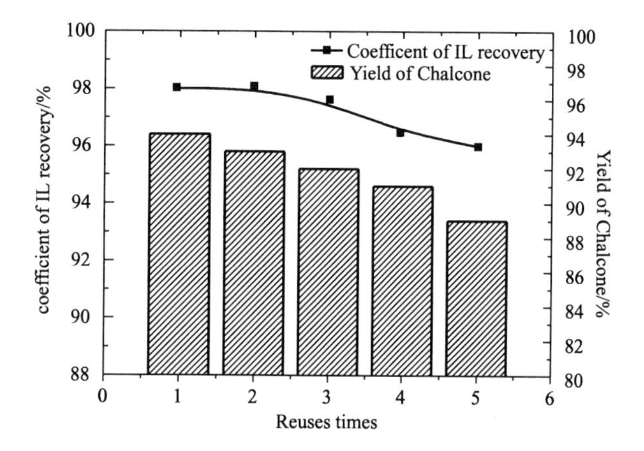

Figure 2 Recycling and reuse of [TEBSA] [HSO_4] in the Claisen-Schmidt reaction.

Ultimately, the recyclability of ILs is central to their utilities. In order to investigate the reusable properties of these Brønsted acidic ILs, recycle experiments of[TEBSA][HSO_4]were tested. After reaction, IL was recovered from the aqueous solution by removing water at 70℃ under vacuum (0.1MPa) for 5h, IL was reused for the next cycle. As shown in Fig. 2, [TEBSA][HSO_4]can be reused at least five times without appreciable decrease in yield. This indicates that these acidic ILs as dual catalysts and solvents for Claisen-Schmidit condensation were reusable. It is worthy of noting, when the IL lost its activity after repeated use, it could be easy recovered by washed with dichloromethane in water, then removed water at reduced pressure and catalyzed new reaction.

4 Conclusions

Four readily available and economic task-specific ILs have been used as recyclable catalysts for the Claisen-Schmidt condensation of benzaldehydes and acetophenones to synthesize chalcones. The merit of this methodology is that it is simple, efficient and eco-friendly. Chalcone as a separate phase can be conveniently decanted out from the ILs. The ILs can be reused after heat treatment under vacuum (70℃, 0.1MPa). The present study shows that this process is potentially interesting as an example of homogenous/heterogeneous acidic catalysis.

Acknowledgement

This work has been supported by the 8th Explosive & Propellants Funding of the General Reserve Department (hzy08020305-5). We also thank Science Development Fund of Nanjing University of Science and Technology (2010GJPY008 and 2010ZYTS020) for financial assistance.

References

[1] Xia, Y.; Yang, Z. Y.; Bastow, K. F.; Nakanishi, Y.; Lee, K. H. Antitumou agents. Part 202: novel 2′-amino chalcones: design, synthesis and biological evaluation. Bioorg. Med. Chem. Lett. 2000, 10, 699.

[2] Hsieh, H. K.; Tsao, L. T.; Wang, J. P. Lin, C. N. Synthesis and anti-inflammatory effect of chalcones. J. Pharm. Pharmacol. 2000, 52, 163.

[3] Satyanarayana, M.; Tiwari, P.; Tripathi, B. K.; Srivastava, A. K.; Pratap, R. Synthesis and antihyperglycemic activity of chalcone based aryloxypropanolamines. Bioorg. Med. Chem. 2004, 12, 883.

[4] Dhar, D. N. Chemistry of Chalcones and Related Compounds; Wiley: NewYork, 1981.

[5] Iranpoor, N.; Kazemi, F.; Iranpoor, N.; Kazemi, E. $RuCl_3$ catalyses aldol condensations of aldehydes and ketones. Tetrahedron 1998, 54, 9475.

[6] Narender, T.; Reddy, K. P. A simple and highly efficient method for the synthesis of chalcones by using borontrifluoride-etherate. Tetrahedron Lett. 2007, 48, 3177.

[7] Szell, T.; Sohar, I. New nitsochalcones. Can. J. Chem. 1969, 47, 1254.

[8] Drexler, M. T.; Amiridis, J. The Effect of Solvents on the Heterogeneous Synthesis of Flavanone over MgO. J. Catal. 2003, 214, 136.

[9] Choudary, B. M.; Ranganath, K. V.; Yadav, J.; Kantam, M. L. Synthesis of flavanones using nanocrystalline MgO. Tetrahedron Lett. 2005, 46, 1369.

[10] Saravanamurugan, S.; Palanichamy, M.; Arabindoo, B.; Murugesan, V. Liquid phase reaction of 2-hydroxyacetophenone and benzaldehyde over ZSM-5 catalysts. J. Mol. Catal. A: Chem. 2004, 218, 101.

[11] Daskiewicz, J. B.; Comte, G.; Barron, D.; Pietro, A. D.; Thomasson, F. Organolithium mediated synthesis of prenylchalcones as potential inhibitors of chemoresistance. Tetrahedron Lett. 1999, 40, 7095.

[12] Sebti, S.; Solhy, A.; Smahi, A.; Kossir, A.; Oumimoun, H. Dramatic activity enhancement of natural phosphate catalyst by lithium nitrate. An efficient synthesis of chalcones. Catal. Commun. 2002, 3, 335.

[13] Wang, X.; Cheng, S. Solvent-free synthesis of flavanones over aminopropyl-functionalized SBA-15. Catal. Commun. 2006, 7, 689.

[14] Yang, S. D.; Wu, L. Y.; Yan, Z. Y.; Pan, Z. L.; Liang, Y. M. A novel ionic liquid supported organocatalyst of pyrrolidine amide: Synthesis and catalyzed Claisen-Schmidt reaction. J. Mol. Catal. A: Chem. 2007, 268, 107.

[15] Dong, F.; Jian, C.; Hao, F. Z.; Kai, G.; Liang, L. Z. Synthesis of chalcones via Claisen-Schmidt condensa-

tion reaction catalyzed by acyclic acidic ionic liquids. Catal. Commun 2008, 9, 1924.

[16] Pilar, F.; Antonio, G. H, L. Assessment of the suitability of imidazolium ionic liquids as reaction medium for base-catalysed reactions Case of Knoevenagel and Claisen-Schmidt reaction. J. Mol. Catal. A: Chem. 2004, 214, 137.

[17] Shen, J. H.; Wang, H.; Liu, H. C.; Sun, Y.; Liu, Z. M. Brønsted acidic ionic liquids as dual catalyst and solvent for environmentally friendly synthesis of chalcone. J. Mol. Catal. A: Chem. 2008, 280, 24.

[18] Qi, X. F.; Cheng, G. B.; Lv, C. X. Selective nitration of chlorobenzene on super-acidic metal oxides. Cent. Eur. J. Energ. Mater. 2007, 4, 105.

[19] Qian, H.; Ye, Z. W.; Lv, C. X. Ultrasonically promoted nitrolysis of DAPT to HMX in ionic liquid. Ultrason. Sonochem. 2008, 15, 326.

[20] Qian, H. Dintrogen pentoxide and its use in nitration. Nanjing: Nanjing University of Science and Technology, 2008.

[21] Thomazeau, C.; Bourbigou, H.; Luts, S.; Gillber, B. Determination of an Acidic Scale in Room Temperature Ionic Liquids. J. Am. Chem. Soc. 2003, 125, 5264.

（注：此文原载于 Industrial& Engineering Chemistry Research，2011，50 (2)：1146-1149）

Insight into Acid Driven Formation of Spiro-[oxindole] xanthenes from Isatin and Phenols[#]

Hongwei Yang[§], Khuloud Takrouri[§], Michael Chorev[§ ¶*]

[§] Laboratory for Translational Research, Harvard Medical School, One Kendall Square, Building 600, Cambridge, MA 02139 and [¶]Department of Medicine, Brigham and Women's Hospital, 75 Francis Street, Boston, MA 02115.

SCI: 000306705800002

[#] *Dedicated by Dr. Hongwei Yang to Professor Chunxu Lu upon his retirement.*

Abstract: One-pot synthesis of symmetrical spiro [oxindole] xanthenes **2** was achieved by reacting excess of *p*-substituted phenols with isatin in the presence of *p*-TsOH in refluxing *p*-xylene. The intermediary 3,3-bis (2-hydroxy-5-substitutedphenyl) oxindoles **1** undergo ring forming dehydration generating the targeted spiro compounds **2**. Presence of acid liable groups such as methoxy and tert-butyl on the phenol leads to product heterogeneity.

Key words: Spiro [oxindole] xanthenes; One-pot synthesis; 3,3-diaryloxindoles; Demethylation.

1 Introduction

Spiroxanthene are known for their diverse spectroscopic and biological activities. Of interest are compounds such as spiro [isobenzofuran-1 (3*H*), 9′-(9*H*) xanthene]-3-one that are thermochromic, [1]polyspiro (fluorene-9,9′-xanthene) that are light emitters,[2] bis (spiroxanthene)-type compounds that act as donors in dynamic redox systems,[3] 9,9′-spirobixanthene-1,1′-diol used in Rh-catalyzed asymmetric hydrogenation reactions,[4] and important dyes such as fluoresceine and rhodamine, which are derivatives of 3*H*-spiro [isobenzofuran-1,9′-xanthen]-3-ones[5] used as fluorescent dyes for labeling proteins,[6] riboses[7] and cell monitoring processes.[8] At the same time, interesting biological activities were reported for compounds such spiro-xanthene-9′,2-[1,3,4] thiadiazole derivatives as potential analgesic and anti-inflammatory agents,[9] spiro [9,10-dihydroanthracene]-9,3′-pyrrolidine and derivatives thereof as selective serotonin receptor antagonists,[10] rhodamine-based probes for real-time imaging of phagocytosis[11] and xanthene-spiropiperidines as opiates and centrally-active drugs.[12,13]

Spirooxindole alkaloids represent an important group of natural products characterized by interesting biological activities including anti-cancer, inhibition of neuronal transmision and antimicrobial activity.[14-17] However, only limited number of studies reported the preparation of spiro [3*H*]-indole-

3,9′-[9*H*] xanthen]-2 (1*H*)-ones[18,19] that can also serve as a precursor for the more familiar spiroacridines such as the spiro [9*H*-acridine-9,3′-[3*H*] indol]-2′ (1*H*)-ones.[19]

While numerous reports described the preparation of spiro [fluorene-xanthenes][2,20,21] and spiro-hetrocyclic xanthenes such as fluoran,[5,22,23] spiro [piperidine-xanthene],[13,24] spiro [pyrrolidine-xanthene],[24] and spiro [thiadiazole-xanthene],[9] only few were dedicated to the preparation of spirooxindole-xanthenes via thermal condensation of isatin derivatives with simple phenols in presence of concentrated sulphuric acid[19] or dilute aqueous *p*-TsOH.[25] Recently, during our medicinal chemistry effort to optimize our lead oxindole, 3-(5-*tert*-butyl-2-hydroxyphenyl)-3-phenylindolin-2-one, as an inhibitor of cap-dependent translation,[26,27] we observed the formation of novel symmetrical and non-symmetrical spiro [oxindole] xanthenes. Herein, we report the one-pot synthesis of this class of compounds and discuss the mechanistic aspect involved in this reaction.

2 Result and Discussion

The formation of spiro [oxindole] xanthenes in the reaction of isatin with *p*-alkyl-and *p*-methoxyphenols in the presence of strong acid is preceded by the formation of intermediary 3,3-[di-2-hydroxylphenyl]-and 3,3-[di-2-methoxyphenyl] oxindoles. Synthesis of symmetrical 3,3-diaryloxindoles was previously reported.[28,29] In these methods isatins were reacted with phenol, ethoxycarbonate-, sulfate-, alkyl-and halide-substituted aromatics in the presence of a strong acid. Protolytic activation of the 3-oxo in isatin is followed by an electrophilic attack on the *para*-position of the substituted phenyl leading to high yields of the symmetrical 3,3-diaryloxindoles. Under similar conditions, *para*-alkyl-substituted phenols generate 3,3-bis (2-hydroxy-5-alkylphenyl) indoline-2-ones which are prone to eliminate H_2O and undergo ring closure forming spiro [oxindole] xanthenes. Herein, we focused on the reaction of these *p*-substituted phenols with isatin and studied in detail their conversion to the corresponding spiro [oxindole] xanthenes.

2.1 Reaction of Isatin with p-Alkylphenols

Acid catalyzed double electrophilic aromatic substitution of the electronically rich *p*-alkylphenols by isatin in refluxing DCE led to a fast formation of 3,3-bis (2-hydroxy-5-alkylphenyl) oxindoles **1** followed by a slow disappearance that was accompanied by slow accumulation of the corresponding 2′, 7′-dialkylspiro [indoline-3,9′-xanthen]-2-ones (Scheme 1 and Table 1). We hypothesize that in the presence of strong acid these 3,3-diaryloxindoles transform into the corresponding spiro compounds **2**. For example, 1.5 hrs after addition of *p*-TsOH to a mixture of isatin and p-tert-butylphenol we observe 81% conversion to the 3,3-disubstituted oxindole **1a** and no detectable presence of the corresponding spiro compound (Table 1, Entry 2). However, 18 hrs after reaction initiation we see decrease in the amount of **1a** to 67% and the appearance of 11% of the 2′,7′-di-*tert*-buthylspiro [indoline-3,9′-xanthen]-2-one **2a**. This process continues and at the end of 120 hrs there are only 6% of **1a** left while the spiro compound **2a** increased to 60% (Table 1, Entry 5). Similar change in the distribution of reaction products was observed also for other *p*-alkylphenols. Initially, we observe a build-up of symmetrical 3,3-disubstituted oxindoles **1** to reach conversion of >75%. Progressive transformation of **1** to spiro [oxindole] xanthene **2** results in a slow build-up of the latter reaching levels of 30-60% (Table 1).

Scheme 1: Reaction conditions for the synthesis of 3, 3-bis (2-hydroxy-5-alkylphenyl) oxindoles

1 and their in-situ conversion into 2′, 7′-dialkylspiro [indoline-3,9′-xanthen]-2-ones **2.**

p-TsOH(5 eq)
DCE,85℃

2eq **1eq** **1** **2**

a: R=*t*-Bu c: R=*i*-Pr
b: R=*n*-Bu d: R=*c*-pentyl

TABLE 1 Time dependence of the distribution of products in the reaction of isatin with different p-alkylphenols in DCE in the presence of p-TsOH[a]

Entry	R	Reaction Time	Product Yield [b] (%)	
			1	2
1	a-*t*-Bu	15 min	63	0
2		1. 5h	81	0
3		18h	67	11
4		66h	29	40
5		120h	6	60
6	b-*n*-Bu	19 h	83	17
7		48h	64	36
8		68h	54	46
9	c-*i*-Pr	19 h	79	11
10		48h	61	25
11		68h	53	32
12	d-*c*-pentyl	19 h	76	24
13		48h	53	46
14		68h	50	50

a Reaction conditions: *p*-alkylphenol/isatin/*p*-TsOH, 2eq: 1eq: 5eq. b Product yield was determined by analytical RP-HPLC.

Confirmation that the intermediary 3,3-bis (2-hydroxy-5-alkylphenyl) oxindoles **1** undergoes in-situ cyclization into the corresponding spiro [oxindole] xanthenes **2** was achieved when a solution of isolated **1a** was refluxed in DCE in the presence of 5 equivalents of *p*-TsOH. After 132 hours we observed~70% conversion of the oxindole **1a** into the spiro compound **2a** (not shown) . Nevertheless, for preparative purposes the cyclization step in DCE at 85℃ is too slow and requires optimization to achieve faster conversion rates of the 3,3-disubstituted oxindoles **1** into the spiro compounds **2** and improvement of the total yields of the latter.

2. 2 Reaction of Isatin with *p*-Methoxyphenol

We chose the reaction between isatin and *p*-methoxyphenol as a model reaction for optimization purposes of the one-pot synthesis of spiro [oxindole] xanthenes and screened different reaction conditions (Scheme 2 and Table 2) . These included different solvents, acids and reaction temperatures. The choice of *p*-methoxyphenol was aimed to have an aromatic system that is highly activated toward electrophilic aromatic substitution leading to an accelerate conversion of the intermediary 3,3-diaryloxindole **1** into the corresponding spiro compound **2.** Evidently, we got more than we asked for in terms of side reactions and multiplicity of reaction products. Importantly, the results provided useful insight to the benefits and shortcomings of the one-pot synthesis of spiro [oxindole] xanthenes **2.**

Scheme 2: Model reaction used to optimize conditions for one-pot synthesis of spiro [oxindole] xanthenes 2e-I-Ⅲ from isatin and *p*-methoxyphenol.

OH H₃CO + O O N H → HO OCH₃ OCH₃ OH O N H **1e** + OH H₃CO OCH₃ OH O N H **1f** +

H₃CO O O N H OCH₃ **2e-I** + H₃CO O O N H OH **2e-Ⅱ** + HO O O N H OH **2e-Ⅲ**

TABLE 2 Screening Conditions for One-pot Synthesis of Spiro [oxindole] xanthenes (2e-Ⅰ-Ⅲ) from Isatin and *p*-Methoxyphenol[a]

Entry	Catalyst	Solvent	Temp.	Reaction Time	Product Yield %[b]				
					1e	1f	2e-Ⅰ	2e-Ⅱ	2e-Ⅲ
1	*p*-TsOH(5 eq.)	DCE	reflux	12h	55	8	15	21	0
			reflux	48h	38	7	31	24	0
			reflux	72h	22	7	43	28	0
2	1)$ZnCl_2$;2)12 N HCl	neat	140℃	2h	59	10	0	0	0
3	$MeSO_3H$(5 eq.)	DCM	reflux	30h	74	18	5	3	0
4	TfOH(5eq.)	DCM	RT	30h	73	19	3	0.7	0
5	TfOH(5eq.)	DCM	reflux	30h	74	17	2	0.5	0
6	*p*-TsOH(5eq.)	H_2O	reflux	72h	14	0	0	29	0
7	H_2SO_4(1eq.)	neat	140℃	3h	9	6	40	32	13
8	Eaton's(4ml/mmol)	neat	140℃	14h	9	0	26	43	16
9	$MeSO_3H$(4eq.)	neat	140℃	14h	0	0	0	14	82
10	*p*-TsOH(5eq.)	*p*-xylene	140℃	4h	0	3	34	50	13

a Reaction conditions: 4-methoxyphenol /isatin, 2eq: 1eq) . b determined by RP-HPLC.

The reaction of *p*-methoxyphenol with isatin in refluxing DCE and in the presence of 5 eq. of *p*-TsOH, the same reaction conditions described in Scheme 1, followed the same time-dependent product distribution reported in Table 1. We observed a time-dependent decline in the amount of the rapidly formed 3,3-bis (2-hydroxy-5-methoxyphenyl) oxindole (**1e**), decreasing from 55% at 12h to 22% at 72 h (Table 2, Entry 1), and the concomitant increase in the formation of the corresponding spiro [oxindole] xanthene **2e-Ⅰ**, increasing from 15% to 43%. Accompanying the predominant symmetrical 3, 3-disubstituted oxindole **1e** we observed the formation of a small amount (7-8%) of the non-symmetrical 3-(5-hydroxy-2-methoxyphenyl)-3-(2-hydroxy-5-methoxyphenyl) oxindole (**1f**) whose concentration did not change during the 72h we monitored the reaction. Apparently, **1f** does not undergo cyclization to form a spiro compound. The slow time-dependent increase in the concentration of

2e-Ⅱ (21% and 28% after 12h and 72h, respectively) suggest that the formation of the non-symmetrical spiro [oxindole] xanthene **2e-Ⅱ** does not originate from **1f** but rather from the acid-driven and time-dependent demethylation of one of the methoxy groups on **2e-Ⅰ**. Notably, the corresponding non-symmetrical 3, 3-diaryloxindole was absent in the reaction of isatin with *p*-alkylphenols (Scheme 1 and Table 1). This absence can be related to: 1. Greater activation of the *p*-methoxy-vs. *p*-alkylphenol toward aromatic electrophilic substitution; 2. Smaller difference between the -ortho Hammett substitution constants σ of OH and OMe as compared to the OH and Me substituents; [30] and 3. Smaller steric hindrance imposed on the *ortho* position by the methoxy as compared to alkyls that are greater than methyl.

In an attempt to shorten the reaction time and improve the yield of spiro [oxindole] xanthenes **2** we screened a host of different reaction conditions using different acids, temperatures and solvents. Decent formation of 3, 3-diaryl-oxindoles **1** but no formation of spiro compounds **2** was observed following the use of $ZnCl_2$ and concentrated HCl (Table 2, Entry 2), which were successfully used Tseng et al. to generate moderate yield of a derivative of spiro (fluorene-9, 9′) xanthene from 2, 7-dibromo-fluoren-9-one and resorcinol. [2] Reacting isatin with *p*-methoxyphenol in DCM in the presence of triflic acid at RT or under reflux or in the presence of $MeSO_3H$ under reflux (Table 2, Entries 3-5) led to a fast formation of 3, 3-diaryl-oxindoles **1** with yields of ~92% but the subsequent conversion to the spiro [oxindole] xanthenes **2**, even after extended reaction times (30h), was very sluggish. Refluxing a mixture of isatin and *p*-methoxyphenol in water in the presence of *p*-TsOH, conditions which were previously used successfully by Ahadi et al. to prepare derivatives of spiro [indoline-3, 9′-xanthene] trione from isatin and dimedone, [31] yielded after 3 days a mixture of only 14% of **1e** and 29% of **2e-Ⅱ** (Table 2, Entry 6).

Rewardingly, we have identified several reaction conditions that led to good yields of spiro [oxindole] xanthenes **2** (85-97%) and a high consumption of the intermediary precursor 3, 3-diaryloxindoles **1** (85-100%) (Table 2, Entries 7-10). Heating isatin and *p*-methoxyphenol in neat acids such as sulfuric acid[5,19] and Eaton's reagent[32] yields 85% of the spiro compounds but also left significant amounts of the precursor 3, 3-diaryl-oxindoles **1** (15% and 9%, respectively) (Table 2, Entries 7 and 8). The same reaction in neat $MeSO_3H$[20] resulted in complete disappearance of the intermediary precursor 3, 3-diaryl-oxindoles **1e** and the formation of 96% of demethylated spiro compounds (14% of **2e-Ⅱ** and 82% of **2e-Ⅲ**) (Table 2, Entry 9).

Indeed, monitoring the reaction of isatin and *p*-methoxyphenol in refluxing *p*-xylene in the presence of 5 eq of *p*-TsOH revealed a fast generation of the oxindole **1e** reaching ~70% after 5 min of reaction and complete disappearance after ~2hrs(Figure 1 and Table 2, cf. Entry 1 with Entry 10). These changes are accompanied by a build-up of the parent spiro 2′, 7′-dimethoxyspiro [indoline-3, 9′-xanthen]-2-one (**2e-Ⅰ**) and the non-symmetrical mono-methylated 2′-hydroxy-7′-methoxyspiro [indoline-3, 9′-xanthen]-2-one (**2e-Ⅱ**) that peak at 1 hr reaching ~ 46% and ~ 37%, respectively. Moreover, a slow and consistent build-up of the symmetrical non-methylated 2′-hydroxy-7′-methoxyspiro [indoline-3, 9′-xanthen]-2-one (**2e-Ⅲ**) on the expense of **2e-Ⅰ** and **2e-Ⅱ** is observed throughout the reaction reaching ~50% after 44hr. Importantly, in refluxing *p*-xylene we observe an almost quantitative conversion of isatin to the different spiro forms that peaks at ~96% after 2. 5 hrs and remains at this level throughout the rest of the reaction (Figure 1). The reaction of isatin and *p*-methoxyphenol to generate the different forms of spiro compounds **2** is much faster in *p*-xylene than in DCE (cf. Figures 1 and Table 2, Entry 1) where it takes 72 hrs to reach ~71% conversion of

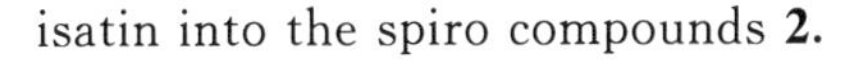
isatin into the spiro compounds **2**.

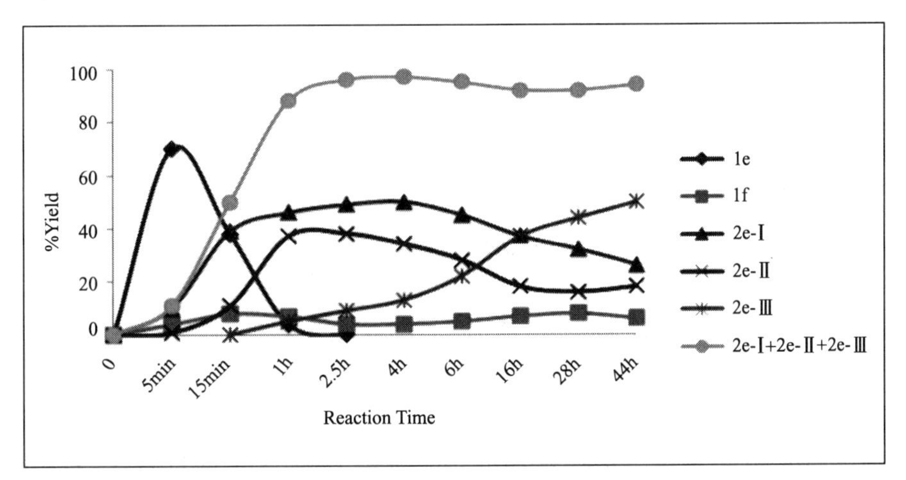

Figure 1 Kinetic profile of the reaction of *p*-methoxyphenol (2 eq) and isatin (1 eq) carried out in refluxing *p*-xylene in the presence of *p*-TsOH (5 eq) (Scheme 2). The time course marked by-●- was the total conversion into spiro compounds 2e-Ⅰ+2e-Ⅱ+2e-Ⅲ. Note the non-linear time scale.

2.3 Reaction of Isatin with p-tert-Butylphenol in Refluxing p-Xylene

Our interest in 3-(5-*tert*-butyl-2-hydroxyphenyl)-3-phenylindolin-2-one as lead anti-cancer agent[26] led us to choose *p-tert*-butylphenol as representative of the above mentioned *p*-alkylphenols, for studying the course of reaction in the one-pot synthesis of spiro (oxindole) xanthenes **2** in the optimized reaction conditions consisting of refluxing *p*-xylene in the presence of 5eq. *p*-TsOH (Scheme 3 and Figure 2). The formation of the symmetrical oxindole **1a** reached the maximum of ~88% after ~1 min and was almost fully consumed after 2 hrs. The conversion to spiro compounds **2** peaked at 72% after ~1h and remained at this level throughout the rest of the reaction. During the extended reaction time (45h) we observed a change in the composition of the mixture of spiro compounds **2.** The continuous build-up of parental symmetrical spiro **2a-Ⅰ** reached peak concentration of ~ 60% at 1hr. Subsequent loss of spiro **2a-Ⅰ** was accompanied by increased formation of non-symmetrical mono-*tert*-butylated spiro **2a-Ⅱ** and symmetrical di-hydroxy spiro **2a-Ⅲ**, reaching concentrations of ~30% each after extended reaction times (~33h). Evidently, in the presence of strong Brønsted acid and elevated temperature (140℃) we observed gradual removal of the acid labile *tert*-butyl group, a side-reaction not observed in refluxing DCE at 85℃ (Table 1, Entries 1-5). De-*tert*-butylation under acidic conditions was previously reported for systems such as 4, 4'-di-*tert*-butyl-3, 3'-dimethyl-2, 2-dihydroxybiphenyl,[33] 2-methoxy-5-*tert*-butylbenzophenones, [34] and 4-*tert*-butylcalix [4] arene.[35]

Scheme 3: One-pot synthesis of spiro [oxindole] xanthenes **2a-Ⅰ-Ⅲ** in the reaction of isatin and *p-tert*-butylphenol in refluxing *p*-xylene in the presence of 5eq. *p*-TsOH.

p-TSOH(5eq)
p-xylene,140℃

1eq **2eq** **1a** **2a-Ⅰ** **2a-Ⅱ** **2a-Ⅲ**

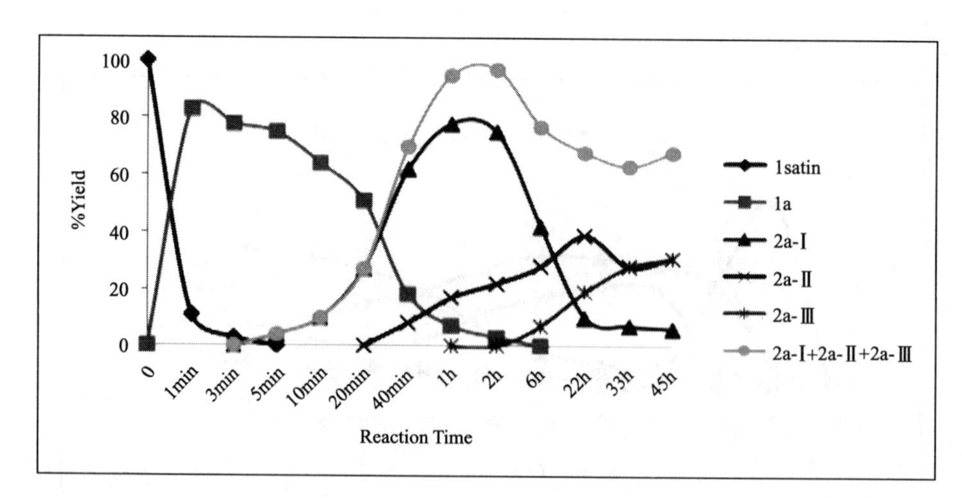

Figure 2 Kinetic profile of the reaction of *p-tert*-butylphenol (2 eq) with isatin (1 eq) in refluxing *p*-xylene, in the presence of *p*-TsOH (5 eq), forming 3, 3-bis (5-*tert*-butyl-2-hydroxyphenyl) oxindole (1a) that transformed into the respective di-*tert*-butyl-, mono-*tert*-butyl, and non-alkylated-spiro compounds 2a-Ⅰ, 2a-Ⅱ and 2a-Ⅲ. The time course marked by-●-is the total conversion into spiro compounds 2a-Ⅰ+2a-Ⅱ+2a-Ⅲ. Note the non-linear time scale.

2.4 Reaction Mechanism

Our observations suggest that in the acid catalyzed reaction of isatin with phenols formation of the 3, 3-diaryl-oxindoles **1** is kinetically-controlled reaction while the formation of the spiro [oxindole] xanthenes **2** is a thermodynamically-controlled one.

In support of the above, we treated symmetrical oxindole **1a** in refluxing DCE or *p*-xylene in presence of 5eq. of *p*-TsOH and observed its transformation into spiro [oxindole] xanthene **2a-I** (80% in 132 h and 100% after 2 h, respectively) (data not shown). Taken together, in both solvents the formation of the spiro [oxindole] xanthenes **2** required the in situ formation of 3, 3-disubstituted-oxindoles **1** as a precursor. The higher reaction temperature in *p*-xylene led to accelerated reaction rates and higher yields of spiro [oxindole] xanthenes **2** in shorter reaction times.

The absence of detectable levels of mono-methylated and non-methylated oxindoles, 3-(2, 5-dihydroxyphenyl)-3-(2-hydroxy-5-methoxyphenyl)-and 3, 3-bis (2, 5-dihydroxyphenyl)-oxindoles, respectively, in the reaction mixtures of *p*-methoxyphenol with isatin under different reaction conditions (Table 2 and Scheme 2), together with our observation that the stable low level (~8%) of the full methylated non-symmetrical 3-(5-hydroxy-2-methoxyphenyl)-3-(2-hydroxy-5-methoxyphenyl) oxindole (**1f**), suggest that the parent dimethylated spiro **2e-Ⅰ** is the only precursor for the demethylated spiro forms **2e-Ⅱ** and **2e-Ⅲ**. This conclusion is supported by the observed time-dependent conversion of the oxindole **1e** in refluxing DCE in the presence of *p*-TsOH into all three spiro compounds **2e-Ⅰ-Ⅲ** (Scheme 4 and Figure 3). We therefore propose a slow acid catalyzed sequential demethylation of the parent spiro **2e-Ⅰ** as a minor side-reaction leading to the formation of a mixture of spiro compounds. Importantly, we did not detect formation of demthylated oxindoles and we assume that their contribution to the formation of the different forms of demethylated spiro compounds **2** is negligible.

Scheme 4: Direct conversion of 3, 3-diaryloxindole **1e** into the dimethoxy-, mono-methoxy-and dihydroxy-spiro compounds **2e-Ⅰ**, **2e-Ⅱ** and **2e-Ⅲ**, respectively.

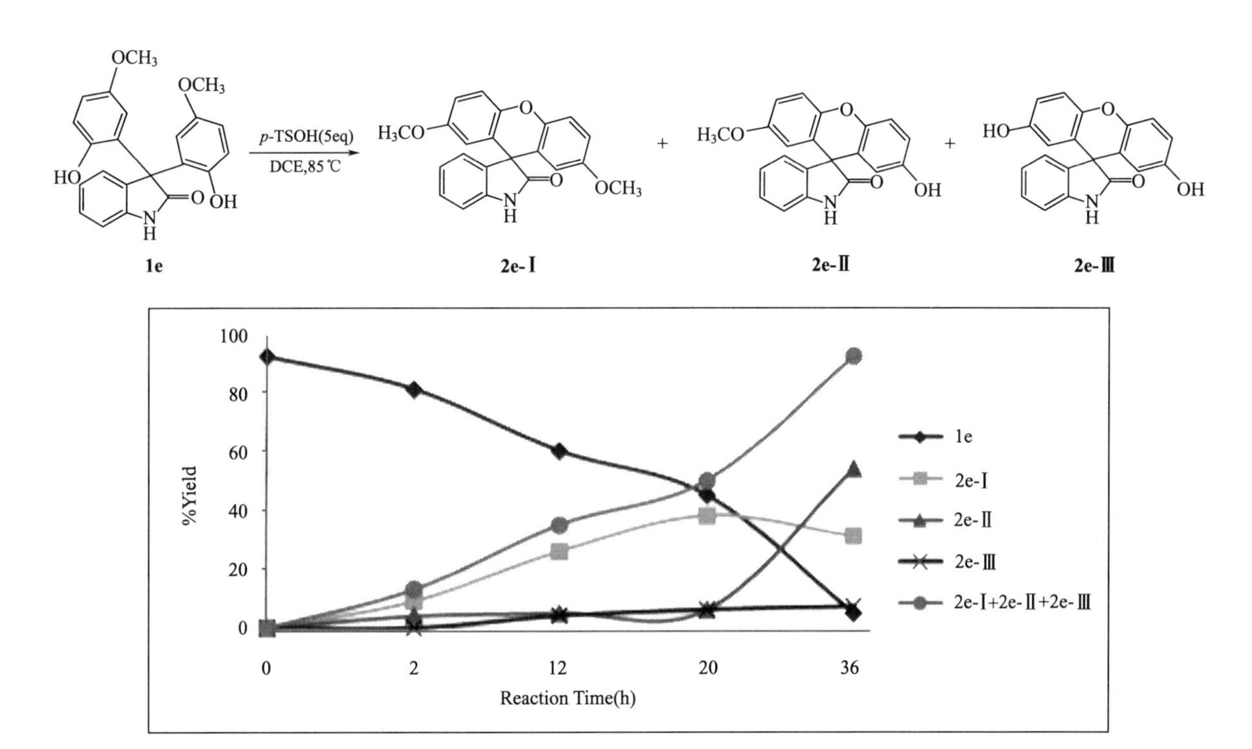

Figure 3: Kinetic profile of the direct conversion of 1e in refluxing DCE, in presence of *p*-TsOH (5 eq), to the respective dimethoxy-, mono-methoxy-, and dihydroxy-spiro compounds 2e-Ⅰ, 2e-Ⅱ and 2e-Ⅲ. The time course marked by-●-is the total conversion into spiro compounds 2e-Ⅰ +2e-Ⅱ +2e-

Ⅲ. Note the non-linear time scale.

Further confirmation for the in-situ demethylation of the parent spiro compound was achieved by treatment of the parent symmetrical spiro compound **2e-Ⅰ** by *p*-TsOH, in refluxing *p*-xylene. As anticipated, we observed a time-dependent demethylation generating the mono-and non-methylated spiro compounds **2e-Ⅱ** and **2e-Ⅲ**, respectively (Figure 1S). Demethylation of methoxyphenyl under similar acidic conditions was previously reported in systems such as 2, 2′-dimethoxybiphenyl, 3, 4-dimethoxyphenyl butanoic acid, nor-bostricoidin, and 2-methoxybenzophenone.[33,34,36,37]

Importantly, the same reaction pathway described above for the formation of spiro compounds **2** from *p*-methoxyphenol and isatin (Scheme 2) was confirmed also for the reaction of *p-tert*-butylphenol and isatin (Scheme 3). Exposing 3, 3-bis (5-*tert*-butyl-2-hydroxyphenyl) oxindole (**1a**) to *p*-TsOH in refluxing *p*-xylene led to a time-dependent conversion into the parent spiro [oxindole] xanthenes **2a-Ⅰ** reaching ~58% after 1 h (Figure 4). Slow de-*tert*-butylation converted spiro **2a-Ⅰ** into the 2′-*tert*-butylspiro [indoline-3, 9′-xanthen]-2-one (**2a-Ⅱ**) and spiro [indoline-3, 9′-xanthen]-2-one (**2a-Ⅲ**). After 2.5 h we observed complete consumption of the oxindole **1a** and the presence of~51% of the spiro **2a-Ⅰ**, ~22% of spiro **2a-Ⅱ** and noticeable amounts of **2a-Ⅲ** (<1%). Moreover, after 22 h the total amount of spiro compounds did not change (~70%) but there was a reduction in the amount of **2a-Ⅰ** (~17%) and an increase in **2a-Ⅱ** and **2a-Ⅲ** (~40% and ~15%, respectively). Evidently, extention of reaction time after the complete consumption of the oxindole **1a** results in dealkylation that increased the formation of de-*tert*-butylated spiro compounds **2a-Ⅱ** and **2a-Ⅲ** on the expense of the parent spiro **2a-Ⅰ**.

Based on the findings described above the plausible mechanism of the acid catalyzed transfor-

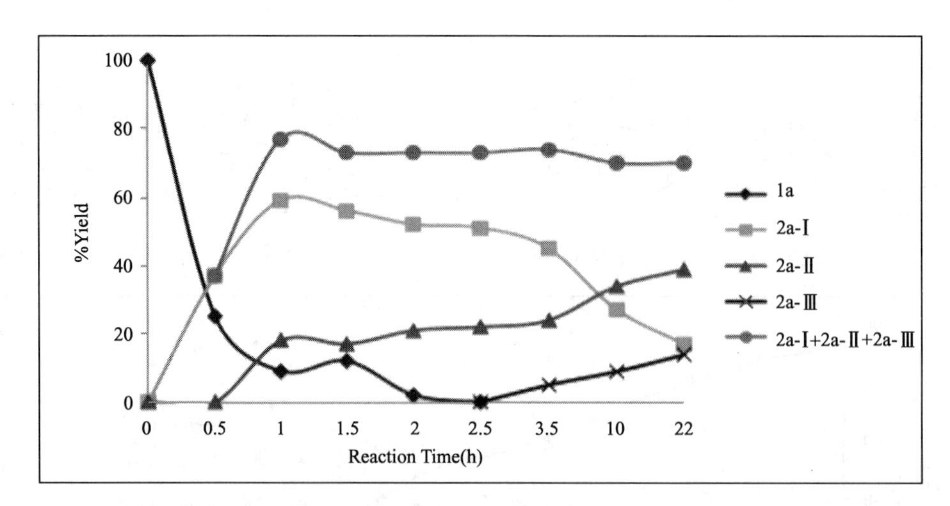

Figure 4 Kinetic profile of the formation of di-*tert*-butyl-, mono-*tert*-butyl-, and non-substituted spiro [oxindole] xanthenes, 2a-Ⅰ, 2a-Ⅱ, and 2a-Ⅲ, respectively, from the isolated 3, 3-bis (5-*tert*-butyl-2-hydroxyphenyl) oxindole (1a) in refluxing *p*-xylene in the presence of *p*-TsOH (5 eq). The time course marked by-●-is the total conversion into spiro compounds 2a-Ⅰ+2a-Ⅱ+2a-Ⅲ. Note the non-linear time scale.

mation of 3, 3-di (*o*-hydroxyaryl) oxindoles **1** to the spiro [oxindole] xanthenes **2** is summarized in Scheme 5. Protonation of one of the *ortho*-hydroxyls on 3, 3-bis (2-hydroxy-5-methoxyphenyl) oxindole (**1e**) results in the loss of water and the formation of the parent symmetrical 2′, 7′-dimethoxyspiro [indoline-3, 9′-xanthen]-2-one (**2e-Ⅰ**). Subsequently, stepwise demethylation leads initially to the formation of the mono-methylated non-symmetrical 2′-hydroxy-7′-methoxyspiro [indoline-3, 9′-xanthen]-2-one (**2e-Ⅱ**) and is followed by the generation of the symmetrical non-methylated 2′, 7′-dihydroxyspiro [indoline-3, 9′-xanthen]-2-one (**2e-Ⅲ**). Similarly, stepwise de-tert-butylation of the parent spiro **2a-Ⅰ** leads to the formation of partial and full de-*tert*-butylated spiro compounds **2a-Ⅱ** and **2a-Ⅲ** (Figure 4). This reaction pathway is compatible with our observations and explains the time-dependent composition of products in the reaction mixture. Importantly, these observations provide sound guidelines for optimizing the yields of either the parent spiro or its derived partially or fully dealkylated spiro forms.

Scheme 5: Plausible mechanism for the formation of the spiro [oxindole] xanthenes **2** from the 3, 3-bis (2-hydroxy-5-methoxyphenyl) oxindole (**1e**) under strong Brønsted acid conditions.

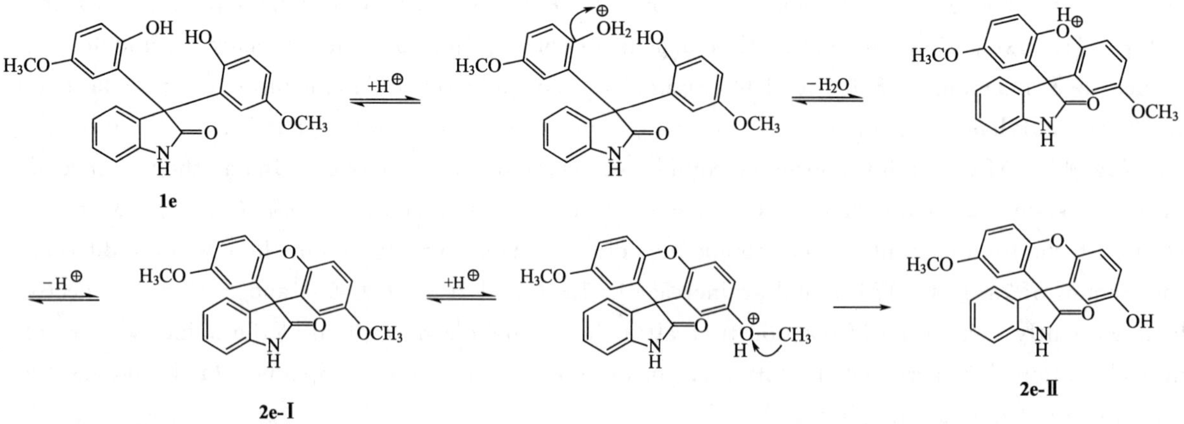

1e $\xrightarrow{+H^{\oplus}}$ $\xrightarrow{-H_2O}$ $\xrightarrow{-H^{\oplus}}$ **2e-Ⅰ** $\xrightarrow{+H^{\oplus}}$ $\longrightarrow$ **2e-Ⅱ**

2e-Ⅲ

2.5 Preparative Applications

The mechanistic insight described above led to the optimization that was applied to the preparation of spiro [oxindole] xanthenes **2** listed in Table 3. The 2-fold molar excess of phenol and the adjustment of the reaction times to maximize the yield of the parent spiro product (Scheme 6) resulted in a wide range of isolated yields (Table 3).

Scheme 6: Preparative one-pot synthesis of spiro [oxindole] xanthenes **2** under optimized conditions.

p-TsOH(5 eq), *p*-xylene, 140 ℃

4eq **1eq**

1 **2**

1a: R_1=*t*-Bu
1b: R_1=*n*-Bu
1c: R_1=*i*-Pr
1d: R_1=*c*-pentyl
1g: R_1=Me
1n: R_1=Ph
1e: R_1=OMe
1i: R_1=O-*n*-Bu
1k: R_1=OPh

1f: R_1=OMe
1j: R_1=O-*n*-Bu
1m: R_1=OPh

2a-Ⅰ: R_1=*t*-Bu
2b: R_1=*n*-Bu
2c: R_1=*i*-Pr
2d: R_1=*c*-pentyl
2g: R_1=Me
2n: R_1=Ph
2e-Ⅰ: R_1=OMe
2i-Ⅰ: R_1=O-*n*-Bu
2k: R_1=OPh
2e-Ⅲ: R_1=OH

2a-Ⅱ: R_1=*t*-Bu, R_2=H
2e-Ⅱ: R_1=OMe, R_2=OH
2i-Ⅱ: R_1=O-*n*-Bu, R_2=OH

2e-Ⅲ: R_2=H
2e-Ⅲ: R_2=OH
2e-Ⅲ: R_2=OH

TABLE 3 One-pot Synthesis of Spiro [oxindole] xanthenes 2 from Isatin and Phenols (p-R-C_6H_4-OH) Following Method B[a]

Entry	*p*-Substituted Phenol	Reaction Time(h)	Isolated Yield(%) of Parent Spiro **2**	Total Isolated Yield of All Forms of Spiro Products[b]
1	R_1=*t*-Bu	1.5	**2a-Ⅰ**: 35	53[c]
2	R_1=*n*-Bu	5	**2b**: 90	90
3	R_1=*i*-Pr	5	**2c**: 63	63
4	R_1=*c*-pentyl	2	**2d**: 60	60
5	R_1=Ph	9	**2n**: 40	40[d]
6	R_1=Me	4	**2g**: 54	54[e]
7	R_1=OH	1	**2e-Ⅲ**: 96	96
8	R_1=OMe	4	**2e-Ⅰ**: 38	85[f]
9	R_1=O-*n*-Bu	1.5	**2i-Ⅰ**: 25	61[g]
10	R_1=OPh	8	**2k**: 42	42[h]

a Reaction conditions: *p*-alkylphenol: isatin: *p*-TsOH=4 eq : 1 eq : 5 eq, the reaction was carried out in refluxing *p*-xylene; b Isolated yields. c Isolated also 9% of each **2a-Ⅱ** and **2a-Ⅲ**. d Isolated also 22% of **1n**. e Isolated also 9% of **1g**. f Isolated also 35% of **2e-Ⅱ** and 12% of **2e-Ⅲ**. g Isolated also 36% of **2i-Ⅱ**, 19% of **1i** and 12% of **1j** (Figure 2S). h Isolated also 10 % of **1k** and 13% of **1m**.

The excellent yields (90% and 96%) obtained for the reaction of *p-n*-butylphenol and *p*-hydroxyphenol (Table 3, Entries 2 and 7, respectively) with isatin can be attributed to the unfavorable dealkylation of the former phenol and the high reactivity and complete symmetry of the latter one. Moderate yields (54-63%) were observed for *p-i*-propyl-, *p-c*-pentyl-and *p*-methylphenol (Table 3, Entries 3, 4 and 6, respectively). The yields of the parent spiro compounds obtained from the *p-c*-pentyl-and *p*-methylphenol (**2d** and **2g**, respectively), as measured by RP-HPLC in the reaction mixtures were 13 and 16% higher than the isolated yields, respectively. In addition, to maximize the yield of **2g** the reaction was quenched prior to complete consumption of the intermediary oxindole **1g** (Table 3, Entry 6). Low yields (25-42%) were obtained for *p-tert*-butyl-, *p*-phenyl-, *p*-methoxy-, *p-n*-butoxy-and *p*-phenoxyphenol (Table 3, Entries 1, 5 and 8-10). Also, in most of these cases the yields, as measured by RP-HPLC in the reaction mixtures, were higher than the isolated yields (60, 46, 33 and 51% for **2a-Ⅰ**, **2e-Ⅰ**, **2i-Ⅰ** and **2k**, respectively). Moreover, these low isolated yields could be attributed to the partial conversion that led to the presence of residual symmetrical oxindoles in the reaction mixtures (22% of **1n**, 19% of **1i**, and 10% of **1k**, Table 3, Entries 5, 9 and 10, respectively) as well as to the formation of non-symmetrical oxindoles (12% of **1j** and 13% **1m**, Table 3, Entries 9 and 10, respectively) that cannot undergo cyclization to form the corresponding spiro compounds. Importantly, in some cases dealkylation of the parent spiro compounds compromised the isolated yield significantly (18% of **2a-Ⅱ** + **2a-Ⅲ**, 35% of **2e-Ⅱ** and 12% **2e-Ⅲ**, and 36% of **2i-Ⅱ**, Table 3, Entries 1, 8 and 9, respectively).

3 Conclusions

The synthesis of spiro (oxindole) xanthenes **2** from *p*-substituted-phenols and isatin in presence of *p*-TsOH was studied in detail to provide insight about the reaction pathway and the accompanying side-reactions. The set of phenols chosen to develop the optimized one-pot synthesis is very challenging. They include acid labile substituents which are susceptible to dealkylation thus giving rise to a heterogenous mixture of spiro (oxindole) xanthenes. Nevertheless, we were able to develop a practical one-pot synthesis of an important class of compounds. Future efforts will be targeted to the expansion of the diversity of substituted isatins and the replacement of the oxygen in the xanthenes moiety with other hetero atoms.

4 Experimental Section

4.1 General Procedures

Method A

p-TsOH · H_2O (1.9 g, 10mmol) was added to a solution of alkylphenol (5.0mmol) and isatin (294 mg, 2 mmol) in 1, 2-dichloroethane (15 mL). The reaction mixture was stirred under reflux for 72 h. The crude residues obtained after the removal of the solvent under vacuum was purified by flash column chromatography (FCC).

Method B

A solution of alkylphenol (8.0mmol) and isatin (294 mg, 2.0mmol) was stirred under reflux in *p*-xylene (5 mL). Then, *p*-TsOH · H_2O (1.9 g, 10mmol) was added and the reaction was refluxed for the specified time (Table 3) . The crude residue obtained after removal of solvent under vacuum was separated on FCC.

4.2 Reaction of Isatin with 4-Isopropylphenol

Method A: The reaction was carried out with 4-isopropylphenol (680 mg, 5.0mmol) . The crude was purified by FCC on silica gel (SNAP Cartridge, KP-Sil, 100 g) employing variable gradient of EtOAc in cyclohexane (5-35%, flow-rate 40mL/min) to afford two products **1c** and **2c**:

3, 3-Bis (2-Hydroxy-5-Isopropylphenyl) Indolin-2-One (1c): 435 mg of **1c** was obtained as a white solid (yield 54%); mp = 240-241℃; RP-HPLC on a C18 Xbridge column (4.6 × 100mm, 1mL/min), t_R=5.78min, purity of 99.6%, employing a linear gradient system of acetonitrile-water: 70%-100% B in A for 15min. Where A is 0.1% AcOH in water and B is 0.1% AcOH in acetonitrile; ^{1}H NMR ($CDCl_3$, 400MHz) in ppm: δ 1.13 (m, 12H), 2.74 (m, 2H), 6.25 (d, J= 7.2Hz, 1H), 6.42 (t, J=7.2Hz, 1H), 7.64-6.83 (m, 4H), 6.82 (d, J=7.2Hz, 1H), 7.01 (d, J=8.4Hz, 2H), 7.19-7.21 (m, 1H), 9.14 (s, 1H); ^{13}C NMR ($CDCl_3$, 100MHz) in ppm: δ 14.4, 21.3, 24.5, 33.4, 60.7, 111.4, 116.7, 119.6, 122.8, 125.8, 126.7, 127.9, 128.0, 128.6, 131.6, 140.8, 141.5, 150.8, 155.2, 185.5; HRMS (ESI) calcd for $C_{26}H_{28}NO_3$: 402.19909, found: m/z=402.20609 [M+H]$^+$.

2′, 7′-Diisopropylspiro [Indoline-3, 9′-Xanthen]-2-One (2c): 190 mg of **2c** was obtained as a white solid (yield 25%); mp = 338-339℃; RP-HPLC on a C18 Xbridge column (4.6 × 100mm, 1mL/min), t_R=5.77min, purity of 99.9%, employing a linear gradient system of acetonitrile-water: 70%-100% B in A for 15 min. Where A is 0.1% AcOH in water and B is 0.1% AcOH in acetonitrile; ^{1}H NMR (DMSO-$d6$, 400MHz) in ppm: δ 1.01 (d, J = 6.8Hz, 12H), 3.40 (q, J = 6.8Hz, 2H), 6.38 (d, J = 1.6Hz, 2H), 6.91 (d, J = 7.2Hz, 1H), 6.97 (t, J = 8.0Hz, 1H), 7.04 (d, J = 8.0Hz, 1H), 7.13 (d, J = 8.4Hz, 2H), 7.18 (dd, J = 8.4Hz and J = 1.6Hz, 2H), 7.30 (t, J = 7.2Hz, 1H), 10.60 (s, 1H); ^{13}C NMR (DMSO-$d6$, 100MHz) in ppm: δ 19.2, 24.5, 27.1, 33.3, 53.6, 56.7, 110.7, 117.3, 121.4, 123.5, 125.1, 125.6, 127.4, 129.7, 136.4, 143.1, 144.1, 149.6, 179.4; HRMS (ESI) calcd for $C_{26}H_{26}NO_2$: 384.18853, found: m/z=384.19662 [M+H]$^+$.

Method B: The reaction was carried out with 4-isopropylphenol (1.08 g, 8.0mmol) and refluxed for 5h. The crude was purified by FCC on silica gel (SNAP Cartridge, KP-Sil, 100 g) employing variable gradient of EtOAc in cyclohexane (5-35%, flow-rate 40mL/min) to afford 480 mg of **2c** as a white solid (yield 63%; Table 3, entry 3) .

4.3 Reaction of Isatin with 4-tert-Butylphenol

Method A: The reaction was carried out with 4-*tert*-butylphenol (750 mg, 5.0mmol) . The crude was purified by FCC on silica gel (SNAP Cartridge, KP-Sil, 100 g) employing variable gradient of EtOAc in cyclohexane (5-25%, flow-rate 40mL/min) to afford two products **1a** and **2a-Ⅰ**:

3, 3-Bis (5-*tert*-Butyl-2-Hydroxyphenyl) Indolin-2-One (1a): 154 mg of **1a** was obtained as a white solid (yield 18%); mp = 168-170℃; RP-HPLC on a C18 Xbridge column (4.6 × 100mm, 1mL/min), t_R=10.40min, purity of 99.5%, employing a linear gradient system of acetonitrile-water: 50%-100% B in A for 15 min. Where A is 0.1% AcOH in water and B is 0.1% AcOH in acetonitrile; ^{1}H NMR (CD_3OD, 400MHz) in ppm: δ 1.12 (s, 9H), 1.14 (s, 9H), 6.61 (m, 1H), 6.79-6.85 (m, 3H), 6.99-7.10 (m, 4H), 7.35-7.29 (m, 2H); ^{13}C NMR (CD_3OD, 100MHz) in ppm: δ 26.8, 30.7, 61.1, 105.0, 110.1, 114.7, 121.8, 125.0, 126.1, 126.4, 128.1, 132.5, 142.1, 185.1; HRMS (ESI) calcd for $C_{28}H_{32}NO_3$: 430.23039, found: m/z=430.23790 $[M+H]^+$.

2′, 7′-Di-*tert*-Butylspiro [Indoline-3, 9′-Xanthen]-2-One (2a-Ⅰ): 297 mg of **2a-Ⅰ** was obtained as a white solid (yield 36%); mp=234-236℃; RP-HPLC on a C18 Xbridge column (4.6×100mm, 1mL/min), t_R=7.82min, purity of 99.8%, employing a linear gradient system of acetonitrile-water: 70%-100 % B in A for 15 min. Where A is 0.1% AcOH in water and B is 0.1% AcOH in acetonitrile; ^{1}H NMR ($CDCl_3$, 400MHz) in ppm: δ 1.12 (s, 18H), 6.62 (d, J=1.6Hz, 2H), 6.98 (d, J=8.4Hz, 1H), 7.04-7.06 (m, 2H), 7.15 (d, J=9.2Hz, 2H), 7.29-7.32 (m, 3H), 8.75 (s, 1H); ^{13}C NMR ($CDCl_3$, 100MHz) in ppm: δ 31.5, 34.5, 54.3, 110.4, 116.8, 120.6, 123.7, 125.8, 126.5, 129.1, 135.4, 141.7, 146.3, 149.7, 180.6; HRMS (ESI) calcd for $C_{28}H_{30}NO_2$: 412.21983, found: m/z=412.22715 $[M+H]^+$.

Method B: 4-*tert*-butylphenol (1.2 g, 8.0mmol) was used, and the reaction was refluxed for 1.5 h. The crude was purified by FCC on silica gel (SNAP Cartridge, KP-Sil, 100 g) employing manually controlled gradient of EtOAc in cyclohexane (5-25%, flow-rate 40mL/min) to afford mixture of **2a-Ⅰ**, **2a-Ⅱ** and **2a-Ⅲ**. The mixture was seperated by Preparative HPLC with a C8 Xterra OBD column (5μM, 19×150mm) employing a gradient of A (15%-60% in 15min followed by 60%-100% in 20min, flow-rate 20mL/min) in B, Where A is 0.1% TFA in water and B is 0.1% TFA in acetonitrile. The purified components are reported according to their order of elution.

Spiro [Indoline-3, 9′-Xanthen]-2-One (2a-Ⅲ): 54 mg of white solid was obtained (yield 9%, Prep HPLC retention time 13.40); mp=361-363℃; RP-HPLC on a C18 Xbridge column (4.6×100mm, 1mL/min), t_R=4.51min, purity of 98.3%, employing a linear gradient system of acetonitrile-water: 50%-100% B in A for 15 min. Where A is 0.1% AcOH in water and B is 0.1% AcOH in acetonitrile; ^{1}H NMR (DMSO-d_6, 500MHz) in ppm: δ 6.60 (dd, J=1.5Hz and 7.5Hz, 2H), 6.94 (d, 7.5Hz, 1H), 6.98-7.02 (m, 3H), 7.04 (d, J=8.0Hz, 1H), 7.23 (dd, J=1.0Hz and 8.0Hz, 2H), 7.31 (t, J=7.5Hz, 3H), 10.66 (s, 1H); ^{13}C NMR (DMSO-d_6, 125MHz) in ppm: δ 53.3, 110.7, 117.5, 121.8, 123.6, 124.5, 125.7, 127.9, 129.8, 136.6, 143.1, 151.2, 179.3. HRMS (ESI) calcd for $C_{20}H_{14}NO_2$: 300.09463, found: m/z=300.10250 $[M+H]^+$.

2′-*tert*-Butylspiro [Indoline-3, 9′-Xanthen]-2-One (2a-Ⅱ): 64 mg of white solid was obtained (yield 9%, Prep HPLC retention time 17.42 min); mp=273-274℃; RP-HPLC on a C18 Xbridge column (4.6×100mm, 1mL/min), t_R=7.96min, purity of 98.8%, employing a linear gradient system of acetonitrile-water: 50%-100% B in A for 15 min. Where A is 0.1% AcOH in water and B is 0.1% AcOH in acetonitrile; ^{1}H NMR ($CDCl_3$, 400MHz) in ppm: δ 1.14 (s, 6H), 1.23 (s, 3H), 6.64 (d, J=2.4Hz, 1H), 6.68 (d, J=8.0Hz, 1H), 6.89-6.99 (m, 2H), 7.03-7.08 (m, 2H), 7.18-7.35 (m, 5H); ^{13}C NMR ($CDCl_3$, 100MHz) in ppm: δ 31.5, 34.9, 54.2, 110.8, 116.9, 117.4, 121.2, 122.9, 123.9, 126.0, 126.7, 127.6, 129.4, 135.7, 139.3, 141.6, 146.6, 149.5, 151.8, 181.3. HRMS (ESI) calcd for $C_{24}H_{22}NO_2$: 356.15723, found: m/z=356.16517 $[M+H]^+$.

288 mg of **2a-Ⅰ** was obtained as a white solid (yield 35%; retention time 20.46 min; Table 3, entry 1).

4.4 Reaction of Isatin with 4-Butylphenol

Method A: The reaction was carried out with 4-butylphenol (750 mg, 5.0mmol). The product was purified by flash chromatography with reversed phase column employing a variable gradient of ace tonitrile in H_2O (50%-100%, flow-rate 40mL/min) to afford two products **1b** and **2b**:

3, 3-Bis (5-Butyl-2-Hydroxyphenyl) Indolin-2-One (1b): 388 mg of white solid was obtained (yield 45%); mp=100-102℃; RP-HPLC on a C18 Xbridge column (4.6×100mm, 1mL/min), t_R=7.27min, purity of 99.9%, employing a linear gradient system of acetonitrile-water: 70%-100% B in A for 15 min. Where A is 0.1% AcOH in water and B is 0.1% AcOH in acetonitrile; ^1H NMR ($CDCl_3$, 400MHz) in ppm: δ 0.90 (t, J=7.6Hz, 6H), 1.29 (m, 4H), 1.46 (m, 4H), 2.48 (t, J=7.6Hz, 4H), 6.30 (d, J=7.6Hz, 1H), 6.46 (t, J=7.6Hz, 1H), 6.61-6.64 (m, 2H), 6.72-6.76 (m, 2H), 6.82 (d, J=7.6Hz, 1H), 6.95 (d, J=8.0Hz, 1H), 7.00 (d, J=8.0Hz, 1H), 7.16 (d, J=8.0Hz, 1H), 9.14 (s, 1H); ^{13}C NMR ($CDCl_3$, 100MHz) in ppm: δ 14.2, 22.3, 33.8, 35.0, 60.7, 111.4, 116.8, 119.7, 122.9, 123.2, 125.8, 126.4, 128.6, 129.1, 129.7, 130.2, 131.6, 134.7, 135.5, 140.7, 150.7, 155.2, 185.6; HRMS (ESI) calcd for $C_{28}H_{32}NO_3$: 430.23039, found: m/z=430.23724 $[M+H]^+$.

2′, 7′-Dibutylspiro [Indoline-3, 9′-Xanthen]-2-One (2b): 190 mg of white solid was obtained (yield 23%); mp=138-140℃; RP-HPLC on a C18 Xbridge column (4.6×100mm, 1mL/min), t_R=9.17min, purity of 99.7%, employing a linear gradient system of acetonitrile-water: 70%-100% B in A for 15 min. Where A is 0.1% AcOH in water and B is 0.1% AcOH in acetonitrile; ^1H NMR ($CDCl_3$, 400MHz) in ppm: δ 0.83 (t, J=7.6Hz, 6H), 1.21 (m, 4H), 1.39 (m, 4H), 2.42 (t, J=7.6Hz, 4H), 6.45 (d, J=2.0Hz, 2H), 6.98 (d, J=8.0Hz, 1H), 7.03-7.06 (m, 2H), 7.09 (dd, J=8.0Hz and 2.0Hz, 2H), 7.15 (d, J=8.0Hz, 2H), 7.25-7.29 (m, 1H), 9.40 (s, 1H); ^{13}C NMR ($CDCl_3$, 100MHz) in ppm: δ 14.1, 22.5, 33.8, 35.1, 53.9, 110.7, 117.2, 120.6, 123.8, 125.7, 127.1, 129.0, 136.1, 138.1, 141.7, 149.7, 181.3. HRMS (ESI) calcd for $C_{28}H_{30}NO_2$: 412.21983, found: m/z=412.22788 $[M+H]^+$.

Method B: The reaction was carried out with 4-butylphenol (1.2 g, 8.0mmol) and refluxed for 5 h. The crude was purified by FCC on silica gel (SNAP Cartridge, KP-Sil, 100 g) employing manually control gradient of EtOAc in cyclohexane (0-25%, flow-rate 40mL/min) to afford 740 mg of **2b** as a white solid (yield 90%; Table 3, entry 2).

4.5 Reaction of Isatin with 4-Isopropylphenol

Method A: The reaction was carried out with 4-isopropylphenol (680 mg, 5.0mmol). The crude was purified by FCC on silica gel (SNAP Cartridge, KP-Sil, 100 g) employing variable gradient of EtOAc in cyclohexane (5-35%, flow-rate 40mL/min) to afford two products **1c** and **2c**:

3, 3-Bis (2-Hydroxy-5-Isopropylphenyl) Indolin-2-One (1c): 435 mg of **1c** was obtained as a white solid (yield 54%); mp=240-241℃; RP-HPLC on a C18 Xbridge column (4.6×100mm, 1mL/min), t_R=5.78min, purity of 99.6%, employing a linear gradient system of acetonitrile-water: 70%-100% B in A for 15 min. Where A is 0.1% AcOH in water and B is 0.1% AcOH in acetonitrile; ^1H NMR ($CDCl_3$, 400MHz) in ppm: δ 1.13 (m, 12H), 2.74 (m, 2H), 6.25 (d, J=7.2Hz, 1H), 6.42 (t, J=7.2Hz, 1H), 7.64-6.83 (m, 4H), 6.82 (d, J=7.2Hz, 1H), 7.01 (d, J=8.4Hz, 2H), 7.19-7.21 (m, 1H), 9.14 (s, 1H); ^{13}C NMR ($CDCl_3$, 100MHz) in ppm: δ 14.4, 21.3, 24.5, 33.4, 60.7, 111.4, 116.7, 119.6, 122.8, 125.8, 126.7, 127.9, 128.0, 128.6, 131.6, 140.8, 141.5, 150.8, 155.2, 185.5; HRMS (ESI) calcd for $C_{26}H_{28}NO_3$: 402.19909, found: m/z=402.20609 $[M+H]^+$.

2′, 7′-Diisopropylspiro [Indoline-3, 9′-Xanthen]-2-One (2c): 190 mg of **2c** was obtained as a white solid (yield 25%); mp=338-339℃; RP-HPLC on a C18 Xbridge column (4.6×100mm, 1mL/min), t_R=5.77min, purity of 99.9%, employing a linear gradient system of acetonitrile-water: 70%-100% B in A for 15 min. Where A is 0.1% AcOH in water and B is 0.1% AcOH in acetonitrile; ^{1}H NMR (DMSO-*d*6, 400MHz) in ppm: δ 1.01 (d, J=6.8Hz, 12H), 3.40 (q, J=6.8Hz, 2H), 6.38 (d, J=1.6Hz, 2H), 6.91 (d, J=7.2Hz, 1H), 6.97 (t, J=8.0Hz, 1H), 7.04 (d, J=8.0Hz, 1H), 7.13 (d, J=8.4Hz, 2H), 7.18 (dd, J=8.4 Hz and J=1.6Hz, 2H), 7.30 (t, J=7.2Hz, 1H), 10.60 (s, 1H); ^{13}C NMR (DMSO-*d*6, 100MHz) in ppm: δ 19.2, 24.5, 27.1, 33.3, 53.6, 56.7, 110.7, 117.3, 121.4, 123.5, 125.1, 125.6, 127.4, 129.7, 136.4, 143.1, 144.1, 149.6, 179.4; HRMS (ESI) calcd for $C_{26}H_{26}NO_2$: 384.18853, found: m/z=384.19662 [M+H]$^+$.

Method B: The reaction was carried out with 4-isopropylphenol (1.08 g, 8.0mmol) and refluxed for 5h. The crude was purified by FCC on silica gel (SNAP Cartridge, KP-Sil, 100 g) employing variable gradient of EtOAc in cyclohexane (5-35%, flow-rate 40mL/min) to afford 480 mg of **2c** as a white solid (yield 63%; Table 3, entry 3).

4.6 Reaction of Isatin with 4-Cyclopentyl phenol

Method A: The reaction was carried out with 4-cyclopentylphenol (810 mg, 5.0mmol). The crude was purified by FCC on silica gel (SNAP Cartridge, KP-Sil, 100 g) employing variable gradient of EtOAc in cyclohexane (5-35%, flow-rate 40mL/min) to afford two products **1d** and **2d**:

3, 3-Bis (5-Cyclopentyl-2-Hydroxyphenyl) Indolin-2-One (1d): 315 mg of **1d** was obtained as a white solid (yield 35%); mp=244-246℃; RP-HPLC on a C18 Xbridge column (4.6×100mm, 1mL/min), t_R=8.19min, purity of 99.7%, employing a linear gradient system of acetonitrile-water: 70%-100% B in A for 15 min. Where A is 0.1% AcOH in water and B is 0.1% AcOH in acetonitrile; ^{1}H NMR ($CDCl_3$, 400MHz) in ppm: δ 1.44 (s, 4H), 1.57-1.69 (m, 8H), 1.95 (s, 4H), 2.81-2.86 (m, 2H), 6.29 (d, J=7.6Hz, 1H), 6.46 (t, J=7.6Hz, 1H), 6.61-6.65 (m, 2H), 6.71-6.77 (m, 2H), 6.82 (d, J=7.2Hz, 1H), 6.97-7.01 (m, 2H), 7.18 (d, J=7.2Hz, 1H), 9.05 (s, 1H); ^{13}C NMR ($CDCl_3$, 100MHz) in ppm: δ 25.4, 34.8, 45.2, 60.8, 111.3, 116.7, 119.5, 122.8, 125.6, 126.5, 127.3, 128.2, 128.6, 131.6, 138.4, 139.2, 140.7, 150.7, 155.1, 185.4; HRMS (ESI) calcd for $C_{30}H_{32}NO_3$: 454.23039, found: m/z=454.23767 [M+H]$^+$.

2′, 7′-Dicyclopentylspiro [Indoline-3, 9′-Xanthen]-2-One (2d): 235 mg of **2d** was obtained as a white solid (yield 27%); mp=256-258℃; RP-HPLC on a C18 Xbridge column (4.6×100mm, 1mL/min), t_R=9.59min, purity of 99.9%, employing a linear gradient system of acetonitrile-water: 70%-100 % B in A for 15 min. Where A is 0.1% AcOH in water and B is 0.1% AcOH in acetonitrile; ^{1}H NMR ($CDCl_3$, 400MHz) in ppm: δ 1.35-1.41 (m, 4H), 1.52-1.60 (m, 4H), 1.63-1.69 (m, 4H), 1.87-1.92 (m, 4H), 2.75-2.79 (m, 2H), 6.47 (s, 2H), 6.98-7.05 (m, 3H), 7.15 (d, J=0.8Hz, 4H), 7.26-7.30 (m, 1H), 8.85 (s, 1H); ^{13}C NMR ($CDCl_3$, 100MHz) in ppm: δ 25.4, 34.7, 45.4, 53.9, 110.5, 117.2, 120.5, 123.8, 125.9, 127.7, 129.0, 135.8, 141.6, 149.8, 180.7. HRMS (ESI) calcd for $C_{30}H_{30}NO_2$: 436.21983, found: m/z=436.22651 [M+H]$^+$.

Method B: The reaction was carried out with 4-cyclopentylphenol (1.3 g, 8.0mmol) and refluxed for 2h. The crude was purified by FCC on silica gel (SNAP Cartridge, KP-Sil, 100 g) employing variable gradient of EtOAc in cyclohexane (5-35%, flow-rate 40mL/min) to afford 523 mg

2d as a white solid (yield 60%; Table 3, entry 4) .

4.7 Reaction of Isatin with p-Cresol

Method B: The reaction was carried out with *p*-cresol (864 mg, 8.0mmol) and refluxed for 4 h. The crude was purified by FCC on silica gel (SNAP Cartridge, KP-Sil, 100 g) employing manually control gradient of EtOAc in cyclohexane (0-40%, flow-rate 40mL/min) to afford two products **1g** and **2g** (Table 3, entry 6) .

2′, 7′-Dimethylspiro [Indoline-3, 9′-Xanthen]-2-One (2g): 355 mg of **2g** was obtained as a white solid (yield 54%); mp=386-388℃; RP-HPLC on a C18 Xbridge column (4.6×100mm, 1mL/min), t_R=9.72min, purity of 99.2%, employing a linear gradient system of acetonitrile-water: 30%-100% B in A for 15 min. Where A is 0.1% TFA in water and B is 0.1% TFA in acetonitrile; ^{1}H NMR ($CDCl_3$, 400MHz) in ppm: δ 2.10 (s, 6H), 6.35 (s, 2H), 6.90 (d, J = 7.2Hz, 1H), 6.98 (t, J=7.2Hz, 1H), 7.03 (d, J=7.6Hz, 1H), 7.10 (s, 4H), 7.30 (d, J=7.6Hz, 1H), 10.63 (s, 1H); ^{13}C NMR ($CDCl_3$, 100MHz) in ppm: δ 20.9, 53.3, 110.8, 117.2, 121.3, 123.6, 125.7, 127.6, 129.7, 130.5, 133.1, 136.8, 142.9, 149.2, 179.4; HRMS (ESI) calcd for $C_{22}H_{18}NO_2$: 328.12593 , found: m/z=328.13320 $[M+H]^+$.

3, 3-Bis (2-Hydroxy-5-Methylphenyl) Indolin-2-One (1g): 60 mg of **1g** was obtained as a white solid (yield 9%); mp = 355-358℃; RP-HPLC on a C18 Xbridge column (4.6×100mm, 1mL/min), t_R = 8.64min, purity of 90.2%, employing a linear gradient system of acetonitrile-water: 30%-100% B in A for 15 min. Where A is 0.1% TFA in water and B is 0.1% TFA in acetonitrile; ^{1}H NMR (DMSO-d_6, 500MHz) in ppm: δ 2.06 (s, 6H), 6.50 (s, 1H), 6.55 (s, 2H), 6.76-6.87 (m, 3H), 6.93 (d, J=7.5Hz, 1H), 6.98-7.03 (m, 2H), 7.23 (td, J=7.5and 1.0Hz, 1H), 9.31 (s, 1H), 9.75 (s, 1H), 10.94 (s, 1H); ^{13}C NMR (DMSO-d_6, 125MHz) in ppm: δ 21.2, 60.7, 110.8, 116.0, 119.6, 122.6, 124.5, 126.9, 127.6, 128.8, 129.6, 130.0, 130.7, 132.6, 142.5, 153.0 155.6, 184.2; HRMS (ESI) calcd for $C_{22}H_{20}NO_3$: 346.13649, found: m/z=346.14412 $[M+H]^+$.

4.8 Reaction of Isatin with 4-Phenylphenol

Method B: The reaction was carried out with 4-phenylphenol (8.0mmol, 1.36g) and refluxed for 9 h. The crude was purified by FCC on silica gel (SNAP Cartridge, KP-Sil, 100g) employing manually control gradient of EtOAc in cyclohexane (0-40%, flow-rate 40mL/min) to afford two products **1n** and **2n** (Table 3, entry 5) .

2′, 7′-Diphenylspiro [Indoline-3, 9′-Xanthen]-2-One (2n): 360mg of white solid was obtained (yield 40%); mp 311℃ decomposed ; RP-HPLC (C18) t_R=23.2min, purity of 91%; 1H NMR (DMSO-d_6, 400MHz) in ppm: δ 6.76 (d, J = 2.4Hz, 2H), 7.01-7.10 (m, 3H), 7.26-7.40 (m, 13H), 7.63 (dd, J=8.4and 2.5Hz, 2H); ^{13}C NMR (DMSO-d_6, 100MHz) in ppm: δ 53.8, 110.9, 118.3, 122.2, 125.6, 125.9, 126.9, 128.5, 129.7, 130.2, 136.7, 139.7, 143.3, 150.9, 179.2; HRMS (ESI) calcd for $C_{32}H_{22}NO_2$: 452.15723, Found: m/z=452.16454 $[M+H]^+$.

3, 3-Bis (4-Hydroxybiphenyl-3-yl) Indolin-2-One (1n): 209 mg of white solid was obtained (yield 22%); mp 191-192℃; RP-HPLC (C18) t_R=25.6min, purity of 92%; ^{1}H NMR (CD_3OD, 500MHz) in ppm: δ 6.81-6.85 (m, 3H), 6.99 (t, J = 8.0Hz, 2H), 7.08 (d, J = 7.0Hz, 1H), 7.14 (t, J = 7.5Hz, 1H); 7.20-7.28 (m, 5H), 7.33-7.38 (m, 4H), 7.41 (d, J =

8.0Hz, 2H), 7.46 (d, J=9.0Hz, 2H); ^{13}C NMR (CD_3OD, 125MHz) in ppm: δ 64.2, 110.0, 115.5, 116.8, 122.1, 126.1, 126.3, 127.3, 127.9, 128.1, 128.7, 131.9, 132.5, 133.3, 137.9, 140.4, 141.1, 142.0, 155.5, 157.1, 182.5; HRMS (ESI) calcd for $C_{32}H_{24}NO_3$: 470.16779, Found: m/z=470.17567 $[M+H]^+$.

4.9 Reaction of Isatin with Hydroquinone

Method B: The reaction was carried out with hydroquinone (8.0mmol, 880 mg) and refluxed for 1h. The crude was purified by FCC on silica gel (SNAP Cartridge, KP-Sil, 100 g) employing manually control gradient of EtOAc in cyclohexane (10-45%, flow-rate 40mL/min) to afford 638 mg of **2e-Ⅲ** as a white solid (yield 96%, Table 3, entry 7).

4.10 Reaction of Isatin with *p*-Methoxyphenol

p-TsOH · H_2O (1.9 g, 10mmol) was added to a refluxing and stirred solution of 4-methoxyphenol (1.08 g, 8.0mmol) and isatin (294 mg, 2.0mmol) in *p*-xylene (5 mL) and the reaction was allowed to proceed for 20 min. The crude residue obtained after removal of solvent under vacuum was separated on a silica gel column (SNAP Cartidge, KP-Sil, 100 g) employing a manually controlled gradient of EtOAc in cyclohexane (0-50%, flow-rate 40mL/min). The purified components are reported according their order of elution. Analytical RP-HPLC was carried out on a C18 Xbridge column (5μM, 4.6×100mm, 1mL/min), employing a linear gradient of 0-60% B in A for 18 min followed by 60%-100% B in A for 12min, where A is 0.1% AcOH in H_2O and B is 0.1% AcOH in acetonitrile.

2′, 7′-Dimethoxyspiro [Indoline-3, 9′-Xanthen]-2-One (2e-Ⅰ): Evaporating under vacuum yielded 130 mg of white solid (yield 18%); mp 302-303℃; RP-HPLC (C18) t_R=18.71min, purity of 99.4%; ^1H NMR (DMSO-d_6, 400MHz) in ppm: δ 3.55 (s, 6H), 6.1 (d, J=3.2Hz, 2H), 6.54 (s, 1H), 6.90-7.05 (m, 5H), 7.16 (d, J=8.8Hz, 1H), 7.32 (d, J=8.0Hz, 1H), 10.65 (s, 1H); ^{13}C NMR (DMSO-d_6, 100MHz) in ppm: δ 54.0, 56.0, 110.8, 112.3, 115.2, 116.3, 118.3, 121.9, 123.6, 125.7, 130.0, 135.7, 143.1, 145.6, 150.4, 155.6, 178.9; HRMS (ESI) calcd for $C_{22}H_{18}NO_4$: 360.11576, Found: m/z=360.12285 $[M+H]^+$.

3, 3-Bis (2-Hydroxy-5-Methoxyphenyl) Indolin-2-One (1e) and 3-(5-Hydroxy-2-Methoxyphenyl)-3-(2-Hydroxy-5-Methoxyphenyl) Indolin-2-One (1f): A fraction containing a mixture of **1e** and **1f** obtained from separation described above was further separated by a subsequent flash chromatography on a reverse phase column (SNAP Cartidge, KP-C18-HS, 120 g) employing a linear gradient of acetonitrile (20%-40%, flow-rate 40mL/min for 1h) in H_2O. The purified components are reported according their elution time.

3-(5-Hydroxy-2-Methoxyphenyl)-3-(2-Hydroxy-5-Methoxyphenyl) Indolin-2-One (1f): Evaporating under vacuum yielded 70 mg of white solid (yield 9%); mp=322-333℃; RP-HPLC (C18): t_R=15.16min, purity of 99.1%); ^1H NMR (DMSO-d_6, 400MHz) in ppm: δ 3.35 (s, 3H), 3.54 (s, 3H), 6.26 (s, 1H), 6.31 (s, 1H), 6.61 (d, J=8.4Hz, 1H), 6.77 (d, J=8.4Hz, 1H), 6.82-6.84 (m, 3H), 6.95-7.02 (m, 2H), 7.25 (t, J=7.2Hz, 1H), 8.96 (s, 1H), 9.55 (s, 1H), 11.02 (s, 1H); ^{13}C NMR (DMSO-d_6, 100MHz) in ppm: δ 55.8, 57.1, 60.5, 110.9, 114.4, 115.2, 116.2, 120.4, 122.7, 125.4, 126.7, 129.6, 132.2, 142.3, 150.2, 151.5, 152.0, 153.1, 183.8; HRMS (ESI) calcd for $C_{22}H_{20}NO_5$: 378.12632, found: m/z=378.13354 $[M+H]^+$.

3, 3-Bis (2-Hydroxy-5-Methoxyphenyl) Indolin-2-One (1e): Evaporating under vacuum yielded 275 mg of white solid (yield 36%); mp=242-244℃; RP-HPLC (C18): t_R=16.54min, purity of 96.9%; ^{1}H NMR (DMSO-d_6, 400MHz) in ppm: δ 3.53 (s, 6H), 6.28 (s, 1H), 6.34 (s, 1H), 6.62 (m, 2H), 6.84-6.89 (m, 3H), 6.94-7.01 (m, 2H), 7.24 (t, J=7.6Hz, 1H), 9.16 (s, 1H), 9.60 (s, 1H), 11.01 (s, 1H); ^{13}C NMR (DMSO-d_6, 100MHz) in ppm: δ 55.6, 55.8, 60.7, 110.8, 114.3, 116.1, 122.6, 126.8, 129.0, 132.2, 142.5, 151.5, 183.6; HRMS (ESI) calcd for $C_{22}H_{20}NO_5$: 378.12632; found: m/z=378.13322 [M+H]$^+$.

2′-Hydroxy-7′-Methoxyspiro [Indoline-3, 9′-Xanthen]-2-One (2e-Ⅱ): Evaporating under vacuum yielded 180 mg of white solid (yield 26%); mp=260-261℃; RP-HPLC (C18): t_R=15.80min, purity of 95.9%; ^{1}H NMR (DMSO-d_6, 400MHz) in ppm: δ 3.54 (s, 3H), 5.98 (t, J=3.6Hz, 2H), 6.68 (dd, J=8.8Hz and 2.8Hz, 1H), 6.88-7.04 (m, 5H), 7.13 (d, J=9.2Hz, 1H), 7.29 (t, J=8.0Hz, 1H), 9.14 (s, 1H), 10.63 (s, 1H); ^{13}C NMR (DMSO-d_6, 100MHz) in ppm: δ 53.9, 56.0, 110.7, 112.8, 115.1, 117.1, 118.2, 121.8, 123.6, 125.7, 129.8, 136.2, 143.1, 144.3, 145.7, 153.6, 155.4, 179.2. HRMS (ESI) calcd for $C_{21}H_{16}NO_4$: 346.10011; found: m/z=346.10710 [M+H]$^+$.

2′, 7′-Dihydroxyspiro [Indoline-3, 9′-Xanthen]-2-One (2e-Ⅲ): Evaporating under vacuum yielded 35 mg of white solid (yield 5%); mp=339℃ decopose; RP-HPLC (C18): t_R=12.90min, purity of 98.2%; 1H NMR (CD_3OD, 400MHz) in ppm: δ 5.98 (d, J=3.2Hz, 2H), 6.67 (dd, J=2.8 Hz and 8.8Hz, 2H), 6.94-7.05 (m, 5H), 7.28 (t, J=8.0Hz, 1H); ^{13}C NMR (DMSO-d_6, 100MHz) in ppm: δ 56.7, 110.5, 112.8, 117.0, 118.1, 121.4, 123.6, 125.7, 129.6, 136.6, 143.0, 144.5, 153.4, 179.4; m/z = 332.1 (M + H)$^+$. HRMS (ESI) calcd for $C_{20}H_{14}NO_4$: 332.08446; found: 332.09206 [M+H]$^+$.

4.11 Reaction of Isatin with p-Butoxyphenol

p-TsOH · H_2O (1.9g, 10.0mmol) was added to a refluxing and stirred solution of 4-butoxyphenol (1.33g, 8.0mmol) and isatin (294mg, 2.0mmol) in p-xylene (5 mL) and the reaction was allowed to proceed for 2h. The crude residue obtained after removal of solvent under vacuum was separated on a silica gel column (SNAP Cartridge, KP-Sil, 100g) employing a manually controlled gradient of EtOAc in cyclohexane (0-35%, flow-rate 40mL/min). The purified components are reported according their order of elution. Analytical RP-HPLC was carried out on a C18 Xbridge column (5μM, 4.6×100mm, 1mL/min), employing a linear gradient system of 0-100% B in A for 30min, where A is 0.1% TFA in H_2O and B is 0.1% TFA in acetonitrile (Table 3, entry 9).

2′, 7′-Dibutoxyspiro [Indoline-3, 9′-Xanthen]-2-One (2i-Ⅰ): 220 mg of white solid was obtained (yield 25%); mp 221-223℃; RP-HPLC (C18) t_R=26.41min, purity of 97.2%; ^{1}H NMR (DMSO-d_6, 500MHz) in ppm: δ 0.81-0.84 (m, 6H), 1.28-1.32 (m, 4H), 1.53-1.55 (m, 4H), 3.71-3.77 (m, 4H), 5.98-5.99 (m, 2H), 6.89-6.95 (m, 3H), 7.00-7.05 (m, 2H), 7.14 (d, J = 8.4Hz, 2H), 7.31-7.34 (m, 1H), 10.63 (s, 1H); ^{13}C NMR (DMSO-d_6, 125MHz) in ppm: δ 14.3, 19.3, 31.3, 68.1, 110.8, 112.9, 115.5, 118.2, 121.8, 123.6, 125.7, 130.0, 135.8, 143.1, 145.1, 155.0, 164.3, 179.0; HRMS (ESI) calcd for $C_{28}H_{30}NO_4$: 444.20966, Found: m/z=444.21696 [M+H]$^+$.

3-(5-Butoxy-2-Hydroxyphenyl)-3-(2-Butoxy-5-Hydroxyphenyl) Indolin-2-One (1j): 110 mg of white solid was obtained (yield 12%); mp 234-235℃; RP-HPLC (C18) t_R=21.85min, purity of 97.4%; 1H NMR (CD_3OD, 500MHz) in ppm: δ 0.86-0.93 (m, 6H), 1.18-1.25 (m, 2H),

1.37-1.41 (m, 2H), 1.47-1.51 (m, 2H), 1.59-1.63 (m, 2H), 3.72-3.78 (m, 4H), 6.40-6.42 (m, 2H), 6.61 (d, J=2.5Hz, 1H), 6.65 (dd, J=9.0 and 2.5Hz, 1H), 6.72-6.73 (m, 1H), 6.81-6.86 (m, 2H), 6.99 (d, J=7.5Hz, 1H), 7.06 (td, J=7.5and 0.5Hz, 1H), 7.29 (td, J=7.5and 1.0Hz, 1H); ^{13}C NMR (CD_3OD, 125MHz) in ppm: δ 13.0, 19.0, 30.9, 31.3, 68.1, 68.7, 110.2, 113.2, 114.8, 115.6, 116.1, 119.9, 122.2, 125.3, 126.2, 128.3, 132.4, 150.0, 151.0, 153.1, 184.4; HRMS (ESI) calcd for $C_{28}H_{32}NO_5$: 462.22022, Found: m/z=462.22787 [M+H]$^+$.

3, 3-Bis (5-Butoxy-2-Hydroxyphenyl) Indolin-2-One (1i): 180 mg of white solid was obtained (yield 19%); mp: 96-97℃; RP-HPLC (C18) t_R=24.0min, purity of 92%; ^{1}H NMR (CD_3OD, 500MHz) in ppm: δ 0.88-0.94 (m, 6H), 1.35-1.39 (m, 4H), 1.57-1.63 (m, 4H), 3.72-3.74 (m, 4H), 6.42 (s, 1H), 6.46 (s, 1H), 6.60-6.66 (m, 2H), 6.80 (d, J=2.0Hz, 1H), 6.85 (d, J=8.0Hz, 2H), 6.96-7.02 (m, 2H), 7.24 (t, J=8.0Hz, 1H); ^{13}C NMR (CD_3OD, 125MHz) in ppm: δ 13.1, 19.1, 31.3, 60.7, 64.3, 68.2, 110.2, 114.4, 115.7, 116.3, 119.7, 122.2, 126.3, 128.3, 132.0, 142.1, 148.8, 150.9, 152.4, 153.0, 184.6; HRMS (ESI) calcd for $C_{28}H_{32}NO_5$: 462.22022, Found: m/z=462.22878 [M+H]$^+$.

2′-Butoxy-7′-Hydroxyspiro [Indoline-3, 9′-Xanthen]-2-One (2i-Ⅱ): 280 mg of white solid was obtained (yield 36%); mp 187℃ decoposed; RP-HPLC (C18) t_R = 20.2min, purity of 94%; ^{1}HNMR (CD_3OD, 500MHz) in ppm: δ 0.87 (t, J=7.5Hz, 3H), 1.32-1.39 (m, 2H), 1.55-1.61 (m, 2H), 3.68-3.74 (m, 2H), 6.09 (d, J=2.5Hz, 1H), 6.11 (d, J=2.5Hz, 1H), 6.73 (dd, J=8.5and 2.5Hz, 1H), 6.82 (dd, J=8.5and 2.5Hz, 1H), 6.95 (d, J=7.5Hz, 1H), 7.01-7.07 (m, 4H), 7.30 (td, J=8.0 and 1.0Hz, 1H); ^{13}C NMR (CD_3OD, 125MHz) in ppm: δ 13.0, 19.0, 31.2, 54.2, 67.9, 110.2, 112.1, 115.6, 116.4, 117.6, 120.8, 123.5, 125.3, 129.1, 136.4, 142.2, 145.0, 153.0, 155.0, 181.0; HRMS (ESI) calcd for $C_{24}H_{22}NO_4$: 388.14706, Found: m/z=388.15455 [M+H]$^+$.

4.12 Reaction of Isatin with p-Phenoxyphenol

p-TsOH · H_2O (1.9 g, 10.0mmol) was added to a refluxing and stirred solution of 4-phenoxyphenol (8.0mmol, 1.48g) and isatin (294mg, 2.0mmol) in *p*-xylene (5mL) and the reaction was allowed to proceed for 8 h. The crude residue obtained after removal of solvent under vacuum was separated on a silica gel column (SNAP Cartridge, KP-Sil, 100 g) employing a manually controlled gradient of EtOAc in cyclohexane (0-40%, flow-rate 40mL/min). The purified components are reported according their order of elution. Analytical RP-HPLC was carried out on a C18 Xbridge column (5μM, 4.6×100mm, 1mL/min), employing a linear gradient system of 0-100% B in A for 30min, where A is 0.1% TFA in H_2O and B is 0.1% TFA in acetonitrile (Table 3, entry 10).

2′, 7′-Diphenoxyspiro [Indoline-3, 9′-Xanthen]-2-One (2k): 410 mg of white solid was obtained (yield 42%); mp 277-278℃; RP-HPLC (C18) t_R = 25.38min, purity of 99%; ^{1}H NMR ($CDCl_3$, 500MHz) in ppm: δ 6.44 (d, J=2.5Hz, 2H), 6.87 (d, J=8.0Hz, 4H), 6.91-6.95 (m, 3H), 7.00-7.10 (m, 4H), 7.20-7.28 (m, 7H), 8.12 (s, 1H); ^{13}C NMR ($CDCl_3$, 125MHz) in ppm: δ 54.0, 110.6, 118.0, 118.6, 120.6, 121.7, 123.1, 124.2, 125.9, 129.6, 129.9, 134.7, 141.3, 147.8, 152.3, 157.8, 179.4; HRMS (ESI) calcd for $C_{32}H_{22}NO_4$: 484.14706, Found: m/z=484.15512 [M+H]$^+$.

3-(5-Hydroxy-2-Phenoxyphenyl)-3-(2-Hydroxy-5-Phenoxyphenyl) Indolin-2-One (1m): 130 mg of white solid was obtained (yield 13%); mp 135-136℃; RP-HPLC (C18) t_R=18.52min, purity of

94%; ^{1}H NMR (CD_3OD, 500MHz) in ppm: δ 6.55 (d, J=2.5Hz, 1H), 6.73-6.85 (m, 9H), 6.93-7.03 (m, 3H), 7.11-7.26 (m, 5H), 7.35 (d, J=8.0Hz, 1H), 7.65 (d, J=8.0Hz, 1H); ^{13}C NMR (CD_3OD, 125MHz) in ppm: δ 60.4, 109.8, 116.0, 116.6, 117.3, 118.6, 119.4, 119.6, 121.0, 122.2, 123.7, 125.7, 128.4, 129.9, 132.9, 135.5, 137.6, 142.1, 148.9, 151.8, 153.8, 158.6, 180.7; HRMS (ESI) calcd for $C_{32}H_{24}NO_5$: 502.15762, Found: m/z=502.16528 [M+H]$^+$.

3, 3-Bis (2-Hydroxy-5-Phenoxyphenyl) Indolin-2-One (1k): 100 mg of white solid was obtained (yield 10%); mp 133-135℃; RP-HPLC (C18) t_R=22.8min, purity of 96%; ^{1}H NMR (CD_3OD, 500MHz) in ppm: δ 6.70-6.78 (m, 12H), 6.94 (t, J=7.0Hz, 2H), 7.07-7.09 (m, 5H), 7.16 (t, J=7.5Hz, 1H); ^{13}C NMR (CD_3OD, 125MHz) in ppm: δ 62.1, 110.3, 116.1, 116.8, 121.1, 122.6, 126.0, 128.3, 129.6, 134.3, 135.6, 141.0, 148.9, 153.8, 158.4, 180.9; HRMS (ESI) calcd for $C_{32}H_{24}NO_5$: 502.15762, Found: m/z=502.16549 [M+H]$^+$.

Acknowledgment is made to Egenix, Inc. for supporting this research.

Supporting Information Available:

General procedures and individual one-pot syntheses of spiro [oxindole] xanthenes derived from the reaction of isatin with the following phenols: 4-*tert*-butylphenol, 4-*n*-butylphenol, 4-*i*-propylphenol, 4-cyclopropylphenol, *p*-cresol, 4-phenylphenol, hydroquinone, 4-butoxyphenol, and 4-phenoxyphenol. It also has figures of the time course profiles for the demethylation of 2′, 7′-dimethoxyspiro [indoline-3, 9′-xanthen]-2-one (**2e-Ⅰ**) (Figure 1S) and the acid catalyzed reaction of 4-*n*-butoxyphenol with isatin (Figure 2S). Copies of ^{1}H-and ^{13}C-NMR and HR-MS are also provided.

References

[1] Chiga, K.; Harada, Y. JP 2010001339 A 20100107 AN 2010: 15181.
[2] Tseng, Y. -H.; Shih, P. -I.; Chien, C. -H.; Dixit, A.; Shu, C. -F. *Macromolecules* 2005, 38, 10055-10060.
[3] Suzuki, T.; Tanaka, S.; Higuchi, H.; Kawai, H.; Fujiwara, K. *Tetrahedron. Lett.* 2004, 45, 8563-8567.
[4] Wu, S.; Zhang, W.; Zhang, Z.; Zhang, X. *Org. Lett.* 2004, 6, 3565-3567.
[5] McCullough, J.; Daggett, K. *J. Chem. Educ.* 2007, 84, 1799-1802.
[6] Ojemyr, L.; Sanden, T.; Widengren, J.; Brzezinski, P. *Biochemistry* 2009, 48, 2173-2179.
[7] Conibear, P. B.; Jeffreys, D. S.; Seehra, C. K.; Eaton, R. J.; Bagshaw, C. R. *Biochemistry* 1996, 35, 2299-2308.
[8] Goncalves, M. S. *Chem. Rev.* 2009, 109, 190-212.
[9] Hafez, H. N.; Hegab, M. I.; Ahmed-Farag, I. S.; el-Gazzar, A. B. *Bioorg. Med. Chem. Lett.* 2008, 18, 4538-4543.
[10] Glennon, R.; Westkaemper, R. In *U. S. Pat. Appl. Publ.* 2003, p 21 pp.
[11] Kenmoku, S.; Urano, Y.; Kojima, H.; Nagano, T. *J. Am. Chem. Soc.* 2007, 129, 7313-7318.
[12] Buckett, W. R.; Crossland, N. J.; Galt, R. H.; Matusiak, Z.; Pearce, R. J.; Shaw, J. S.; Turnbull, M. J. *Br. J. Pharmacol.* 1977, 61, 146P-147P.
[13] Galt, R. H.; Horbury, J.; Matusiak, Z. S.; Pearce, R. J.; Shaw, J. S. *J. Med. Chem.* 1989, 32, 2357-2362.
[14] Abdel-Rahman, A.; Keshk, E.; Hannab, M.; El-Bady, S. *Bioorg. Med. Chem.* 2004, 12, 2483-2488.
[15] Galliford, C.; Scheidt, K. *Angew. Chem. Int. Ed. Engl.* 2007, 46, 8748-8758.
[16] Trost, B.; Brennan, M. *Synthesis*, 2009, 3003-3025.
[17] Lim, K. -H.; Sim, K. -M.; Tan, G. -H.; Kam, T. -S. *Phytochemistry* 2009, 70, 1182-1186.
[18] Wexler, H.; Barboiu, V. *Rev. Roum. Chim.* 1976, 21, 127-138.

[19] Joshi, K. C. J., R.; Dandia, A.; Garg, S.; Ahmed, N. *J. Heterocyclic. Chem.* 1989, 26, 1799-1802.
[20] Xie, L.; Liu, F.; Tang, C.; Hou, X.; Hua, Y.; Fan, Q.; Huang, W. *Org. Lett.* 2006, 8, 2787-2790.
[21] Liu, F.; Xie, L. H.; Tang, C.; Liang, J.; Chen, Q. Q.; Peng, B.; Wei, W.; Cao, Y.; Huang, W. *Org. Lett.* 2009, 11, 3850-3853.
[22] Patel, S.; Patel, M.; Patel, R. *J. Iran. Chem. Soc.* 2005, 2, 220-225.
[23] Corrie, J.; Craik, J.; Munasinghe, V. *Bioconjugate Chem.* 1998, 9 160-167.
[24] Quintas, D.; Garcl a, A.; Dominguez, D. *Tetrahedron Lett.* 2003, 44, 9291-9294.
[25] Bazgir, A.; Tisseh, Z.; Mirzaei, P. *Tetrahedron Lett.* 2008, 49, 5165-5168.
[26] Natarajan, A.; Fan, Y. H.; Chen, H.; Guo, Y.; Iyasere, J.; Harbinski, F.; Christ, W. J.; Aktas, H.; Halperin, J. A. *J. Med. Chem.* 2004, 47, 1882-1885.
[27] Natarajan, A.; Guo, Y.; Harbinski, F.; Fan, Y. H.; Chen, H.; Luus, L.; Diercks, J.; Aktas, H.; Chorev, M.; Halperin, J. A. *J. Med. Chem.* 2004, 47, 4979-4982.
[28] Klumpp, D.; Yeung, K.; Prakash, G.; Olah, G. *Org. Chem.* 1998, 63, 4481-4484.
[29] Garrido, F.; Ibanez, J.; Gonalons, E.; Giraldez, A. *Eur. J. Med. Chem.* 1975, 10, 143-146.
[30] Charon, M. *Can. J. Chem.* 1960, 48, 2493-2499.
[31] Ahadi, S.; Khavasi, H.; Bazgir, A. *Chem. Pharm. Bull.* 2008 56, 1328-1330.
[32] Chou, C. -H.; Shu, C. -F. *Macromolecules*, 2002, 35, 9673-9677.
[33] Yamato, T.; Hideshima, C.; Parkash, G.; Olah, G. *J. Org. Chem.* 1991, 56, 3192-3194.
[34] Kulka, M. *J. Am. Chem. Soc.* 1954, 76, 5469-5471.
[35] Aleksiuk, O.; Biali, S. E. *Tetrahedron Lett.* 1993, 34, 4857-4860.
[36] Opitz, A.; Roemer, E.; Hass, W.; Gorls, H.; Werner, W.; Grafe, U. *Tetrahedron*, 2000, 56, 5147-5155.
[37] Kemperman, G.; Roeters, T.; Hilberink, P. *Eur. J. Org. Chem.* 2003, 1681-1686.

(注：此文原载于 Current Organic Chemistry，2012，16：1581-1593).

附　　录

一、吕春绪教授被SCI、EI收录的论文

截止2012年6月，吕春绪教授在国内外核心期刊以上及重要会议上共发表论文560篇．其中SCI检索论文48篇，EI检索论文123篇。

（一）SCI检索论文

1. Peng X H，Chen T Y，Lü C X，Sun R K. Synthesis of New Polynitrostilbenes. Organic Preparations and Procedures International，1995，27（4）：475.（SCI收录号：A1995RK24600005）.
2. 彭新华，吕春绪．皂土催化剂上芳烃的硝酸酯硝化反应的区域选择性研究．有机化学，2000，20（4）：570.（SCI收录号：000088890900024）.
3. Peng X H，Suzuki H，Lü C X. Zeolite-Assisted Nitration of Neat Toluene and Chlorobenzene with a Nitrogen Dioxide/mole Cular Oxygen System. Remarkable Enhancement of Para-Selectivity. Tetrahedron Letters. 2001，42（26）：4357.（SCI收录号：000169311800020）.
4. Cai C，Lü C X. Dehydrogenation with Pd/ZSM-5 for o-Phenyl Phenol Synthesis. Chemical Innovation. 2001，31（10）：37.（SCI收录号：000171543800008）.
5. 曹阳，吕春绪，吕早生，蔡春．硝酰阳离子和二氧化氮分子的弯曲变形研究．物理化学学报．2002，18（6）：527（SCI收录号：000176546800011）.
6. 周新利，李燕，刘祖亮，朱长江，王俊德，吕春绪．FTIR光谱遥测红外药剂的燃烧温度．光谱学与光谱分析．2002，22（5）：764.（SCI收录号：000179099400018）.
7. Wei Y Y，Cai M M，Lü C X. Studies on Catalytic Side-Chain Oxidation of Nitro Aromatics to Aldehydes with Oxygen. Catalysis Letters. 2003，90（1-2）：81.（SCI收录号：000185494000011）.
8. 周新利，李燕，刘祖亮，朱长江，王俊德，吕春绪．利用光谱分析法测定炸药的燃烧瞬间爆炸温度．光谱学与光谱分析．2003，23（5）：982.（SCI收录号：000186376200041）.
9. 胡炳成，吕春绪，刘祖亮．天然环状四吡咯化合物的合成研究进展．有机化学．2004，24（3）：270.（SCI收录号：000220150600004）.
10. 胡炳成，吕春绪．1，5-二甲基-2S-(1′-苯基乙基)-2-氮杂-8-氧杂-双环［3.3.0］辛烷-3,7-二酮的合成及其结构表征．有机化学．2004，24（6）：697.（SCI收录号：000221725900024）.
11. 蔡春，吕春绪．利用heck反应合成1-羟基-4-(3-吡啶基）丁-2-酮．有机化学．2004，24（6）：673.（SCI收录号：000221725900018）.
12. Luo J，Lü C X，Cai C，Qu W C. A polymer Onium Acting as Phase-Transfer Catalyst in Halogen-Exchange Fluorination Promoted by Microwave. Journal of Fluorine Chemistry. 2004，125（5）：701.（SCI收录号：000221704900007）.
13. Zhang Y X，Liu D B，Lü C X. Preparation and Characterization of Reticular Nano-HMX. Propellants Explosives Pyrotechnics. 2005，30（6）：438.（SCI收录号：000234438200009）.
14. 蔡超君，胡炳成，吕春绪．吡咯及二氢吡咯类化合物的合成研究进展．有机化学．2005，25（10）：1311.（SCI收录号：000232416200028）.
15. 许虎君，吕春绪，叶志文．十六烷基二苯醚二磺酸钠表面化学性质及胶团化作用．物理化学学报．2005，21（11）：1240.（SCI收录号：000233313800009）.
16. Li G X，Liu X S，Peng X H，Lü C X. Bis（1-methyl-1H-imidazole-kappa N-3）Dithiocyanato-Copper（Ⅱ）．Acta Crystallogra-phica Section E-Structure Reports Online. 2006，62（2）：259.（SCI收录号：000235033700037）.
17. 胡炳成，吕春绪，蔡超君．2-氰基-2，3，3-三甲基-5-硫代-吡咯烷的合成新法．有机化学．2006，26（2）：219.

(SCI 收录号：000235412300012).

18. 许虎君，王中才，刘晓亚，吕春绪．酯基 Gemini 型季铵盐表面活性剂与 SDS 的相互作用．物理化学学报．2006，22（4）：414.（SCI 收录号：000237040000005).
19. 李广学，彭新华，吕春绪．含苯并五元环方酸衍生物的二阶非线形光学性质的理论研究．有机化学．2006，26（6）：839.（SCI 收录号：000238300000014).
20. Hu A J，Lü C X，Li B D，Huo T. Selective Oxidation of p-Chlorotoluene Catalyzed by Co/Mn/Br in Acetic Acidwater Medium. Industrial & Engineering Chemistry Research. 2006，45（16）：5688.（SCI 收录号：000239278500032).
21. 户安军，吕春绪，李斌栋，霍婷．对氯甲苯选择性氧化新催化体系研究．有机化学．2006，26（8）：1083.（SCI 收录号：000239795000009).
22. 王海鹰，李斌栋，户安军，吕春绪．可聚硼酸酯表面活性剂的表面化学性质及与 LAS 相互作用．物理化学学报. 2007，23（02）：253.（SCI 收录号：000244264100022).
23. 王海鹰，李斌栋，户安军，吕春绪．聚硼酸酯表面活性剂囊泡自发形成．高等学校化学学报．2007，28（2）：358.（SCI 收录号：000244413000040).
24. 户安军，吕春绪，李斌栋．甲苯类化合物的选择性氧化．化学进展．2007，19（2-3）：292.（SCI 收录号：000245139600013).
25. Liang Z Y，Lv C X，Luo J，Li Bin Dong. A polymer imidazole salt as phase-transfer catalyst in halex fluorination irradiated by microwave. Journal of Fluorine Chemistry. 2007，128（6）：608.（SCI 收录号：000247423400004).
26. Hu A J，Lv C X，Wang H Y，Li B D. Selective oxidation of p-chlorotoluene with Co $(OAc)_2/MnSO_4/KBr$ in acetic acid-water medium. Catalysis Communications. 2007，8（8）：1279.（SCI 收录号：000248075700021).
27. Qian H，Ye Z W，Lv C X. An efficient and facile synthesis of hexanitrohexaazaisowurtzitane（HNIW）. Letters in Organic Chemistry. 2007，4（7）：482.（SCI 收录号：000250492200006).
28. Li Y Q，Lu M，Lv C X，Zhu S J，Liu Q F. Facile synthesis of 9-(N-amino-4-fluoromethyl-3-pyrrolidinyl)-2-fluoro-6-aminopurine from trans-4-hydroxy-L-proline. Chinese Chemical Letters. 2007，18（11）：1313.（SCI 收录号：000251475500003).
29. Lu M，Lv C X. A computer model for formulation of ANFO explosives. Science and Technology of Energetic Materials. 2007，68（4）：117.（SCI 收录号：000255861000004).
30. Qi X F，Cheng G B，Lv C X，Qian D S. Nitration of simple aromatics with NO_2 under air atmosphere in the presence of novel Bronsted acidic ionic liquids. Synthetic Communications. 2008，38（4）：537.（SCI 收录号：000253042300007).
31. Hua Q，Ye Z W，Lv C X. Ultrasonically promoted nitrolysis of DAPT to HMX in ionic liquid. Ultrasontics Sonochemistry. 2008，15（4）：326.（SCI 收录号：000254443900012).
32. Qian H，Lv C X. Ye Z W. Synthesis of CL-20 by clean nitrating agent dinitrogen pentoxide. Journal of the Indian Chemical Society. 2008，85（4）：434.（SCI 收录号：000256416100014).
33. 职慧珍，罗军，马伟，吕春绪．新型 PEG 双子温控离子液体中的缩醛反应．高等学校化学学报．2008，29（10）：2007.（SCI 收录号：000260453500020).
34. Cheng G B，Duan X L，Qi X F，Lv C X. Nitration of aromatic compounds with NO_2/air catalyzed by sulfonic acid-functionalized ionic liquids. Catalysis Communications. 2008，10（2）：201.（SCI 收录号：000260988200018).
35. Ma D S，Hu B H，Lv C X. Selective aerobic oxidation of cyclohexane catalyzed by metallodeuteroporphyrin-IX-dimethylester. Catalysis Communications. 2009，10（6）：781.（SCI 收录号：000263627800008).
36. Zhi H Z，Luo J，Feng G A，Lv C X. An efficient method to synthesize HMX by nitrolysis of DPT with N_2O_5 and a novel ionic liquid. Chinese Chemical Leters. 2009，20（4）：379.（SCI 收录号：000265281600001).
37. Zhi H Z，Lv C X，Zhang Q，Luo J. A new PEG-1000-based dicationic ionic liquid exhibiting temperature-dependent phase behavior with toluene and its application in one-pot synthesis of benzopyrans. Chemical Communications. 2009，(20)：2878.（SCI 收录号：000266003700019).
38. Wang H Y，Chai L Y，Hu A J，Lv C X，Li B D. Self-assembly microstructures of amphiphilic polyborate in aque-

ous solutions. Polymer. 2009，50（13）：2976.（SCI收录号：000267415600030）.

39. Fang Yongqin，Lu Ming，Lv Chunxu. Influence of solvent on the structure of two one-dimensional cobalt（Ⅱ）coordination. Polymers with tetra-chloro-terephthalate. 2009，65（2）：86.（SCI收录号：000263071500012）.

40. 王海鹰，立元，吕春绪．聚（2-丙烯酰胺甲基-6-十二烷基硼酸二乙醇胺酯）与十二烷基苯磺酸钠混合溶液的表面活性．物理化学．2010，26（1）：73.（SCI收录号：000273849300013）.

41. Dai Y，Li B D，Quan H D，Lv C X. $CeCl_3$ center dot $7H_2O$ as an efficient catalyst for one-pot synthesis of beta-amino ketones by three-component Mannich reaction. Chinese Chemical Letters. 2010，21（1）：31.（SCI收录号：000273898200008）.

42. Hu Y F，Luo J，Lv C X. A mild and efficient method for nucleophilic aromatic fluorination using tetrabutylammonium fluoride as fluorinating reagent. Chinese Chemical Letters. 2010，21（2）：151.（SCI收录号：000275396700007）.

43. Dai Y，Li B D，Quan H D，Lv C X. $[Hmim]_3PW_{12}O_{40}$：A high-efficient and green catalyst for the acetalization of carbonyl compounds. Chinese Chemical Letters. 2010，21（6）：678.（SCI收录号：000278442300011）.

44. Qian H，Liu D B，Lv C X. Synthesis of Chalcones via Claisen-Schmidt Reaction Catalyzed by Sulfonic Acid-Functional Ionic Liquids. Industrial & Engineering Cheimistry Research. 2011，50（2）：1146.（SCI收录号：000286027900080）.

45. Li G X，Ma J H，Peng X H，Lv C X. Kinetics Approach for Hydrogenation of Nitrobenzene to p-Aminophenol in a Four-Phase System. Asian Journal of Chemistry. 2011，23（3）：1001.（SCI收录号：000288465000012）.

46. Qian H，Liu D B，Lv C X. An Efficient Synthesis of 1,3,5,7-tetranitro-1,3,5,7-tetraazacyclooctane（HMX）by Ultrasonic Irradiation in Ionic Liquid. Letters in Organic Chemistry. 2011，8（3）：184.（SCI收录号：000288985000006）.

47. Yang Hongwei，Qi Xiufang，Wen Liang，Lv Chunxu，Cheng Guangbin. Regioselective Mononitration of Aromatic Compounds Using Keggin Heteropolyacid Anion Based Brønsted Acidic Ionic Salts. Industrial & Engineering Chemistry Research. 2011，50（19）：11440.（SCI收录号：000295237000058）.

48. Yang Hongwei，Takrouri Khuloud，Chorev Michael. Insight into Acid Driven Formation of Spiro-[oxindole] xanthenes from Isatin and Phenols. Currentoranic Chemistry. 2012，16（13）：1851.

（二）EI检索论文

1. 蔡春，罗桂琴，吕春绪．甲苯的硝酸-离子交换树脂硝化及动力学研究．华东工学院学报．1991，（4）：40.（EI收录号：92021219993）.

2. Lü C X，Lü Z S，Cai C. Investigation on Condensation Reaction and Its Kinetics for 3,3′-Dicholoro-4，6-Dinitro Phenylamine. Proceeding of the 17th International Pyrotechnics Seminar Combined with the 2nd Beijing International Symposium on Pyrotechnics and Explosives（ISPE）. 1991，1：205.（EI收录号：92011202622）.

3. 金铁柱，吕春绪．乙酰苯胺选择性硝化及影响因素分析．含能材料．1994，2（4）：26.（EI收录号：96043124174）.

4. 陆明，孙荣康，吕春绪，惠君明．六硝基芪合成新方法．南京理工大学学报．1994，（5）：8.（EI收录号：95012534891）.

5. 陆明，孙荣康，吕春绪．2，5，7，9-四硝基-2，5，7，9-四杂双环[4，3，0]-壬-8-酮的合成．南京理工大学学报. 1995，19（4）：321.（EI收录号：95122955922）.

6. Cai C，Lü C X. Modification of NaZSM-5 Zeolite and Its Application in Mononiyration of Toluene. Proceedings of the Third ISPE China Ordnance Society，Beijing. 1995：95.（EI收录号：96043144864）.

7. Chen T Y，Liu Z L，Lü C X. Research on Expanded Ammonium Nitrate Explosive. Proceedings of the Third ISPE China Ordnance Society，Beijing. 1995：99.（EI收录号：96043144865）.

8. Liu Z L，Lü C X，Lu M. Research and Application of Expanded Ammonium Nitrate. Proceedings of the Third ISPE China Ordnance Society，Beijing. 1995：220.（EI收录号：96043144889）.

9. Lu M，Wei Y Y，Lü C X. Synthesis of Some Azacyclic Compounds. Proceedings of the Third ISPE China Ordnance Society，Beijing. 1995：231.（EI收录号：96043144891）.

10. Peng X H，Lü C X，Chen T Y. Regioselective Nitration of Aromatic Hytrocarbons with Solid Acid as Cataly-

sis. Proceedings of the Third ISPE China Ordnance Society, Beijing. 1995: 237. (EI 收录号: 96043144893).
11. 金铁柱，魏运洋，奚美虹，吕春绪．十二烷基苯硝化选择性测定．南京理工大学学报．1995，19 (3): 243. (EI 收录号: 95082806769).
12. Lü C X, Gu J L, Peng X H. Research on the Selective Nitration of Acetanilide and Its Kinetics. Proceedings of the Third ISPE China Ordnance Society, Beijing. 1995: 225. (EI 收录号: 96043144890).
13. 吕春绪．膨化硝酸铵防结块性能的研究．南京理工大学学报．1995，19 (5): 464. (EI 收录号: 95122955958).
14. 蔡春，吕春绪，金序兰．改性沸石及其在甲苯一段硝化中的应用．南京理工大学学报．1995，19 (4): 321. (EI 收录号: 95122955921).
15. 陆明，吕春绪，魏运洋．2，5，7，9-四硝基-2，5，7，9-四氮杂双环 [4，3，0] 一壬-8-酮的合成与表征．南京理工大学学报．1995，19 (4): 325. (EI 收录号: 95122955922).
16. 蔡春，吕春绪．改性沸石对甲苯一段硝化的影响．含能材料．1995，3 (4): 22. (EI 收录号: 96043124204).
17. 彭新华，吕春绪．Menke 条件下固体酸催化剂上甲苯的区域选择性硝化．含能材料．1995，3 (4): 27. (EI 收录号: 96043124205).
18. 蔡春，吕春绪．金属硝酸盐-醋酐硝化反应机理研究．华东工学院学报．1996，17 (4): 312. (EI 收录号: 93061626525).
19. 吕春绪，顾建良．乙酰苯胺选择性硝化最佳工艺条件研究．南京理工大学学报．1996，22 (2): 113. (EI 收录号: 96063205092).
20. 彭新华，吕春绪．固体酸催化剂和载体上芳烃的选择性硝化研究．南京理工大学学报．1996，20 (2): 117. (EI 收录号: 96063205093).
21. 陈天云，吕春绪，叶志文．改性硝酸铵性能研究．含能材料．1996，4 (4): 169. (EI 收录号: 97033563554).
22. Wei Y Y, Lü C X. Research on Synthesis Technology of TNAD. Proceedings of Beijing Autumn ISPE (1996). 1996: 122. (EI 收录号: 96123479049).
23. 惠君明，刘祖亮，吕春绪．硝酸铵敏化技术．爆炸与冲击．1996，16 (2): 178. (EI 收录号: 96063224845).
24. 吕春绪，蔡春．苯酚两相硝化反应及其机理研究．兵工学报．1996，17 (4): 312. (EI 收录号: 98084353667).
25. 叶志文，吕春绪，刘祖亮．复合表面活性剂在硝酸铵溶液中胶束热力学性质研究．爆破器材．1996，25 (6): 1. (EI 收录号: 97023543144).
26. 顾建良，吕春绪．乙酰苯胺混酸硝化速率常数的测定．含能材料．1996，4 (3): 132. (EI 收录号: 96123472858).
27. 吕春绪，顾建良．乙酰苯胺选择性硝化反应速度研究．兵工学报．1998，19 (1): 28. (EI 收录号: 98054183172).
28. 吕春绪，刘祖亮，叶志文．硝酸铵膨化机理研究．火炸药学报．1998，21 (2): 16. (EI 收录号: 98074307477).
29. 陆明，吕春绪，刘祖亮．岩石膨化硝铵炸药生产工艺和质量控制．火炸药学报．1998，21 (4): 16. (EI 收录号: 98104415491).
30. 陈天云，吕春绪．表面活性剂和添加剂降低硝酸铵吸湿性研究．火炸药学报．1998，21 (4): 19. (EI 收录号: 98104415492).
31. 叶志文，刘祖亮，吕春绪．表面活性剂改善膨化硝酸铵晶变的研究．爆破器材．1998，27 (4): 8. (EI 收录号: 98104400196).
32. 陈天云，吕春绪，刘祖亮．煤矿许用膨化硝铵炸药的研究．爆破器材．1998，27 (6). 8. (EI 收录号: 99034593976).
33. 陆明，吕春绪，惠君明．提高 HNS 合成收率的研究．含能材料．1998，6 (2): 78. (EI 收录号: 98094388223).
34. 蔡春，吕春绪．提高甲苯一段硝化中对位产物收率的研究．现代化工．1998，18 (3): 31. (EI 收录号: 98044170560).
35. 陆明，刘祖亮，吕春绪．膨化硝铵炸药用量配方研究．爆破器材．1998，27 (5): 10. (EI 收录号: 98124493231).
36. Lü C X. Messages from Foreign Colleagues. Explosion. 1999，9 (3): 212. (EI 收录号: 00045114107).
37. 蔡春，吕春绪．甲苯及其硝基衍生物在火焰离子化检测器上重量校正因子的测定．含能材料．1999，7 (1): 28. (EI 收录号: 99054689411).

38. 陆明，吕春绪．硝酸铵膨化过程中心结晶参数研究．含能材料．1999，7 (2)：79.（EI 收录号：99084731755）．
39. 陈天云，吕春绪．改性铵油炸药研究．火炸药学报．1999，22 (2)：34.（EI 收录号：99084764463）．
40. Lü C X. Research and Development Powder Industrial in China. Propellants，Explosives，Pyrotechnics. 1999，24 (1)：27.（EI 收录号：99064703937）．
41. 蔡春，吕春绪．1，4，6，9-四硝基-1，4，6，9-四氮杂双环［4，4，0］癸烷的合成新方法．火炸药学报．1999，22 (1)：17.（EI 收录号：99034593995）．
42. Lu M，Lü C X，Liu Z L，Hu M M. Study on the Expansion Processing of Expanded Ammonium Nitrate. Marine Geology. 1999，(3)：15.（EI 收录号：99104834008）．
43. 叶志文，刘祖亮，吕春绪．某些表面活性剂在硝酸铵水溶液中的表面行为研究．含能材料．2000，8 (1)：31.（EI 收录号：00055168019）．
44. 胡炳成，刘祖亮，吕春绪．影响膨化硝铵性能的因素．爆破器材．2000，29 (2)：16.（EI 收录号：40055179596）．
45. 陆明，吕春绪．膨化硝铵震源药柱配方的优化设计．兵工学报．2000，21 (3)：217.（EI 收录号：00095329795）．
46. 黄小波，蔡春，吕春绪．4,4′-二硝基二苯乙烯-2,2′-二磺酸合成工艺改进．精细化工．2000，17 (3)：175.（EI 收录号：0045107697）．
47. 黄小波，张军科，蔡春，廖伟秋，吕春绪．DSD 酸合成进展．精细化工．2000，17 (4)：245.（EI 收录号：00055160936）．
48. 叶志文，吕春绪．膨化硝铵固体表面自由能的计算及分析．爆破器材．2000，29 (5)：1.（EI 收录号：00115422115）．
49. 程广斌，吕春绪，彭新华．TiO_2-柱撑黏土催化剂上氯苯的选择性硝化．精细化工．2001，18 (7)：426.（EI 收录号：01386651107）．
50. 吕早生，吕春绪，蔡春．紫外光谱在 2，4－间二硝基间二氯苯与间氯苯胺缩合反应动力学中的应用．武汉科技大学学报．2001，24 (1)：153.（EI 收录号：01466725083）．
51. 吕春绪，陆明，陈天云，叶志文．膨化硝酸铵自敏化理论的微气泡研究．兵工学报．2001，22 (4)：485-488.（EI 收录号：02026825571）．
52. 吕春绪．膨化硝酸铵晶体特性研究．兵工学报．2002，23 (3)：316-319.（EI 收录号：0242143949）．
53. 叶志文，吕春绪，刘祖亮．膨化硝酸铵微观结构研究．爆破器材．2002，31 (1)：7.（EI 收录号：02146910164）．
54. 陆明，吕春绪，刘祖亮．硝酸铵的膨化机理研究．兵工学报．2002，23 (1)：30.（EI 收录号：02407124316）．
55. 陆明，吕春绪，刘祖亮．低爆速膨化硝铵炸药的实验研究．爆破器材．2002，31 (2)：1.（EI 收录号：02397116963）．
56. 陆明，吕春绪．绿色化学与工业炸药．爆破器材．2002，31 (4)：13.（EI 收录号：03027311573）．
57. 叶志文，吕春绪，刘祖亮．一种改性硝酸铵晶体性质．兵工学报．2002，23 (3)：316.（EI 收录号：02427143949）．
58. 周新利，刘祖亮，刘三虎，陆明，吕春绪．粉状硝铵炸药用木粉的比表面积和孔径分布测定．爆破器材．2002，31 (3)：1.（EI 收录号：02407127620）．
59. 罗军，蔡春，吕春绪．微波促进聚乙二醇催化卤素交换氟化反应．精细化工．2002，19 (10)：593.（EI 收录号：02487246151）．
60. 蔡敏敏，魏运洋，蔡春，吕春绪．氧气液相氧化法制备邻硝基苯甲醛．化学反应工程与工艺．2002，18 (1)：23.（EI 收录号：02397110815）．
61. 周新利，李燕，刘祖亮，朱长江，王俊德，吕春绪．FTIR 光谱遥测红外药剂的燃烧温度．光谱学与光谱分析．2002，22 (5)：764.（EI 收录号：03027311705）．
62. 罗军，蔡春，吕春绪．卤素交换氟化合成含氟芳香族化合物研究进展．现代化工．2002，22：43.（EI 收录号：02427142322）．
63. 程广斌，吕春绪．SO_4^{2-}/WO_3-ZrO_2 催化剂上氯苯的区域选择性合成研究．含能材料．2002，10 (4)：167.（EI 收录号：03097382215）．
64. 高大元，吕春绪，董海山，陈天云，叶志文．工业炸药的底面输出波形研究．爆破器材．2002，31 (6)：1.（EI 收录号：03087370416）．

65. 程广斌，吕春绪，彭新华．固体酸催化剂作用下选择性硝化研究氯苯的区域选择性研究．火炸药学报．2002，25（1）：61.（EI收录号：01526777521）.

66. 袁淑军，吕春绪，蔡春．一种新席夫碱型表面活性剂的合成及其表面活性．精细化工．2002，19（11）：626.（EI收录号：2527294805）.

67. 吕早生，吕春绪．用Ip作为N2O4-O3硝化取代芳烃的定位判据．武汉科技大学学报．2002，25（4）：352.（EI收录号：4238195444）.

68. 蔡敏敏，魏运洋，吕春绪．氧气液相氧化法一元取代甲苯制备芳香醛．火炸药学报．2002，25（1）：59.（EI收录号：01526777520）.

69. 陈厚和，马惠华，刘大斌，吕春绪．纳米RDX的制备技术与机械感度．弹道学报．2003，15（3）：11.（EI收录号：04358330326）.

70. 刘桂涛，吕春绪，曲红霞．超细RDX爆速和做功能力的研究与测试．爆破器材．2003，32（3）：1.（EI收录号：03327585042）.

71. Lu M，Lü C X，Liu Z L. The Preparation，Characteristic and Application of Self-Sensitizable Ammonium Nitrate. Proceedings of the 2003 International Autumn Seminar on Propellants，Explosives and Pyrotechnics Guilin. China October 15-18. 2003：358.（EI收录号：04208166492）.

72. 陆明，吕春绪．乳化炸药配方设计的数学方法．爆破器材．2003，32（4）：1.（EI收录号：03427682749）.

73. 高大元，许敏，董海山，李宝涛，吕春绪，陈天云．TATB及其杂质的热分解研究．爆破器材．2003，32（5）：5.（EI收录号：03507782018）.

74. Wei Y Y，Cai M M，Lü C X. Studies on Catalytic Side-Chain Oxidation of Nitroaromatics to Aldehydes with Oxygen. Catalysis Letters. 2003，90（1-2）：81.（EI收录号：03457714108）.

75. Gao D Y，Xu R，Dong H S，Li B T，Lü C X. Study on Thermal Stability of TCTNB and TCDNB. Proceedings of the 2003 International Autumn Seminar on Propellants，Explosives and Pyrotechnics Guilin. China October 15-18. 2003：198.（EI收录号：04208166460）.

76. Lü C X，Liu Z L，Lu M，Chen T Y. A Study on Self-Sensitization Theory of Expanded Ammonium Nitrate Explosive. Proceedings of the 2003 International Autumn Seminar on Propellants. Explosives and Pyrotechnics Guilin. China October 15-19. 2003：724.（EI收录号：04208166552）.

77. 卫延安，蔡春，吕春绪．催化剂预中毒在合成邻苯基苯酚的应用．精细化工．2003，20（8）：466.（EI收录号：03387643252）.

78. 杜杨，金发香，刘祖亮，吕春绪．含硼硅酚醛树脂的合成与性质．高分子材料科学工程．2003，19（4）：44.（EI收录号：03367625722）.

79. 陆明，吕春绪．氧平衡对粉状硝铵炸药爆炸性能影响的数学计算方法．兵工学报．2004，25（2）：225.（EI收录号：04298268556）.

80. 卫延安，蔡春，吕春绪．用于合成邻苯基苯酚的新型覆炭脱氢催化剂．石油化工．2004，33（2）：109.（EI收录号：04138091600）.

81. 程广斌，侍春明，彭新华，吕春绪．钼磷酸催化下甲苯的选择性硝化．含能材料．2004，12（2）：110.（EI收录号：04288262086）.

82. 刘祖亮，周新利，吕春绪．膨化硝铵绝热分解的加速量热法研究．南京理工大学学报（自然科学版）．2004，28（6）：643.（EI收录号：05098866044）.

83. Lü C X，Liu Z L，Hu B C. Research on Safety Properties of Expanded Ammonium Nitrate. International Symposium on Safety Science and Technology，Shanghai. 2004，（4）：1604.（EI收录号：05058814382）.

84. 曾贵玉，李海波，曹虎，吕春绪．纳米RDX的撞击感度与分析．含能材料．2004，2（suppl）：102.（EI收录号：05038798102）.

85. Lü C X. Theoretical Research of Expanded AN Explosive. Explosion. 2005，15（3）：107.（EI收录号：06039643004）.

86. Zhang Y X，Lü C X，Liu D B，Guo L W. Preparation of Microcrystals of Organic Compounds with PolarGroups and Inorganic Salts by Reprecipitation. Japanese Journal of Applied Physics，Part 1：Regular Papers and Short Notes and Review Papers. 2005，44（7A）：5319.（EI收录号：06069679538）.

87. Zhang Y X, Liu D B, Lü C X. Preparation and Characterization of Reticular Nano-HMX. Propellants, Explosives, Pyrotechnics. 2005, 30 (6): 438. (EI收录号: 06019628366).

88. 蔡春，吕春绪．五氧化二氮硝解合成1，4，6，9-四硝基-1，4，6，9-四氮杂双环［4，4，0］癸烷．火炸药学报. 2005，28（2）：50.（EI收录号：064610245212）.

89. 高大元，董海山，李宝涛，吕春绪．TATB TCTNB和TCDNB的爆轰性能．火炸药学报．2005，28（2）：68.（EI收录号：064610245218）.

90. 许虎君，吕春绪，叶志文．一类绿色表面活性剂合成与性能．南京理工大学学报．2005，29（1）：62.（EI收录号：05179069608）.

91. 周新利，胡炳成，刘祖亮，吕春绪．耐冻膨化硝铵炸药的制备．含能材料．2005，13（1）：49.（EI收录号：05149026034）.

92. 袁淑军，蔡春，吕春绪．含席夫碱结构的表面活性剂与铜（Ⅱ）配合物的稳定性及抗菌性．南京理工大学学报. 2005，29（5）：598.（EI收录号：06029637474）.

93. 卫延安，蔡春，吕春绪．焙烧处理对H2SM-5负载金属催化剂脱氢活性的影响．南京理工大学学报．2005，29（3）：320.（EI收录号：05349312328）.

94. 张永旭，吕春绪，刘大斌．重结晶法制备纳米RDX. 火炸药学报．2005，28（1）：49.（EI收录号：064610245238）.

95. 周新利，刘祖亮，吕春绪．有机酸类固体可燃剂改进膨化硝铵炸药流药性研究．火炸药学报．2005，28（1）：58.（EI收录号：064610245241）.

96. Hu A J, Lü C X, Huo T. Selective Oxidation of p-Chlorotoluene Catalyzed by Co/Mn/Br in Acetic Acid-Water Medium. Industrial & Engineering Chemistry Research. 2006, 45 (16): 5688. (EI收录号: 063710105852).

97. 梁政勇，徐珍，李斌栋，吕春绪．对氯苯甲醛卤素交换氟化新工艺研究．现代化工．2006，26（suppl）：209.（EI收录号：063810123967）.

98. 钱华，叶志文，吕春绪．HZSM-5催化下N_2O_5对氯苯的选择性硝化研究．含能材料．2007，15（1）：56.（EI收录号：20071310515182）.

99. 王海鹰，李斌栋，户安军，吕春绪．聚硼酸酯表面活性剂囊泡自发形成．高等学校化学学报．2007，28（2）：358.（EI收录号：20071810580611）.

100. Hu An-Jun, Lü, Chun-Xu, Wang Hai-Ying, Li Bin-Dong. Selective oxidation of p-chlorotoluene with Co $(OAc)_2$/MnSO4/KBr in acetic acid-water medium. Catalysis Communications. 2007, 8 (8): 1279. (EI收录号：20072210627125).

101. 方东，施群荣，巩凯，刘祖亮，吕春绪．离子液体催化甲苯绿色硝化反应研究．含能材料．2007，15（2）：122.（EI收录号：20072210631425）.

102. 戴燕，胡炳成，吕春绪．乙二胺合成ZSM-5沸石分子筛的结构表征及其催化性能．南京理工大学学报（自然科学版）．2007，31（2）：252.（EI收录号：20072510662795）.

103. 职慧珍，罗军，吕春绪，户安军．离子液体在两相催化反应中的应用．现代化工．2007，27（8）：67.（EI收录号：20073710812418）.

104. 梁政勇，李斌栋，吕春绪，冯超．2,6-二氯苯甲醛卤素交换氟化动力学研究．化学工程．2007，35（8）：33.（EI收录号：20074010839386）.

105. 曾贵玉，周建华，吕春绪，黄辉．无机物对硝酸铵相转变的影响．含能材料．2007，15（4）：400.（EI收录号：20074110859964）.

106. Lu Ming, Lv Chunxu. A computer model for formulation of ANFO explosives. Science and Technology of Energetic Materials. 2007, 68 (4): 117. (EI收录号: 20074610922487).

107. 王海鹰，柴立元，吕春绪．可聚合表面活性剂的应用研究．现代化工．2007，27（12）：67.（EI收录号：20080311034043）.

108. 刘丽荣，吕春绪，张晓波．磷酸二氢钾催化下甲苯的选择性硝化．含能材料．2008，16（1）：63.（EI收录号：20081311171188）.

109. 叶方青，曾贵玉，吕春绪，黄辉．聚合物对硝酸铵相转变的影响．含能材料．2008，16（1）：77.（EI收录号：20081311171192）.

110. 方东，施群荣，巩凯，刘祖亮，吕春绪．芳香族化合物绿色硝化反应研究进展．含能材料．2008，16（1）：103.（EI 收录号：20081311171199）.
111. 职慧珍，罗军，马伟，吕春绪．PEG 型酸性温控离子液体中芳香酸和醇的酯化反应．高等学校化学学报．2008，29（4）：772.（EI 收录号：20082111273487）.
112. 刘丽荣，吕春绪，张晓波．硅钨酸催化下甲苯的选择性硝化．含能材料．2008，16（3）：341.（EI 收录号：20083111422507）.
113. 齐秀芳，程广斌，吕春绪．酸性离子液体存在下甲苯的硝酸硝化（Ⅱ）．含能材料．2008，16（4）：398.（EI 收录号：20084211645146）.
114. Cheng，Guangbin，Duan，Xuelei，Qi，Xiufang，Lu，Chunxu. Nitration of aromatic compounds with NO2/air catalyzed by sulfonic acid-functionalized ionic liquids. Catalysis Communications. 2008，10（2）：201.（EI 收录号：20084311655396）.
115. 曾贵玉，吕春绪，黄辉．不同状态膨化硝酸铵的自敏化结构研究．兵工学报．2008，29（12）：1417.（EI 收录号：20090611897062）.
116. 职慧珍，罗军，马伟，吕春绪．新型 PEG 双子温控离子液体中的缩醛反应．高等学校化学学报．2008，29（10）：2007.（EI 收录号：20084711718466）.
117. 曾贵玉，黄辉，高大元，吕春绪．添加剂对 ANFO 雷管起爆感度及作功能力的影响．爆炸与冲击．2008，28（5）：467.（EI 收录号：20091011942387）.
118. Ma，Dengsheng，Hu，Bingcheng，Lu，Chunxu. Selective aerobic oxidation of cyclohexane catalyzed by metallo-deuteroporphyrin-IX-dimethylester. Catalysis Communications. 2009，10（6）：781.（EI 收录号：20090611896518）.
119. Fang，Yongqin，Lu，Ming，Lu，Chun-Xu. Influence of solvent on the structures of two one-dimensional cobalt（Ⅱ）coordination polymers with tetra-chloro-terephthalate. Acta Crystallographica Section C：Crystal Structure Communications. 2009，65（2）：86.（EI 收录号：20090811911849）.
120. Wang Haiying，Chai Liyuan，Hu Anjun，Lü Chunxu，Li Bingdong. Self-assembly microstructures of amphiphilic polyborate in aqueous solutions. Polymer. 2009，50（13）：2976.（EI 收录号：20092312109276）.
121. 曾贵玉，高大元，黄辉，吕春绪．有机玻璃法测试硝铵炸药的爆压．兵工学报．2009，30（5）：541.（EI 收录号：20092812176996）.
122. 马晓明，李斌栋，吕春绪，陆明．无氯 TATB 的合成及其热分解动力学．火炸药学报．2009，32（6）：24.（EI 收录号：20101412821149）.
123. Qian Hua，Liu Dabin，Lv Chunxu. Synthesis of chalcones via Claisen-Schmidt reaction catalyzed by sulfonic acid-functional ionic liquids. Industrial and Engineering Chemistry Research. 2011，50（2）：1146.（EI 收录号：20110313594441）.

二、吕春绪教授主要著作及专利

1. 吕春绪 著．表面活性理论与技术．南京：江苏科技出版社，1991.
2. 吕春绪 著．硝化理论．南京：江苏科技出版社，1993.
3. 吕春绪 等编著．工业炸药．北京：兵器工业出版社，1994.
4. 吕春绪 著．耐热硝基芳烃化学．北京：兵器工业出版社，2000.
5. 吕春绪 著．膨化硝铵炸药．北京：兵器工业出版社，2001.
6. 吕春绪 等著．工业炸药理论．北京：兵器工业出版社，2003.
7. 吕春绪 著．硝酰阳离子理论．北京：兵器工业出版社，2006.
8. Lv Chun xu，et al. The Theory of Industrial Explosive. Beijing：The Publishing House of Ordnance Industry，2007.
9. 吕春绪 著．膨化硝酸铵自敏化理论．北京：兵器工业出版社，2008.
10. 吕春绪 等著．药物中间体化学．北京：化学工业出版社，2008.
11. 吕春绪 主编．有机中间体制备．第三版．北京：化学工业出版社，2009.
12. 吕春绪 等著．炸药的绿色制造．北京：国防工业出版社，2010.

13. 吕春绪，惠君明，胡刚．耐寒高爆速液体混合炸药．1985. 专利号：85102869.
14. 吕春绪，惠君明，张正才，王成，白淑芳．工业低爆速混合炸药及制法．1988. 专利号：89105369.7.
15. 孙荣康，吕春绪，惠君明，罗桂琴，白淑芳．六硝基芪合成新方法．1989. 专利号：89105951.2.
16. 孙荣康，吕春绪，惠君明，罗桂琴，白淑芳．六硝基芪—Ⅱ的合成新方法．1990. 专利号：90106206.5.
17. 吕春绪，刘祖亮，陈天云，陆明，叶志文，胡炳成，王依林，惠君明．煤矿无梯硝铵炸药．1997. 专利号：97107202.7.
18. 陆明，秦卫国，刘祖亮，吕春绪，胡炳成，陈天云，叶志文，王依林，惠君明．震源药柱及其制法．1997. 专利号：97107103.9.
19. 刘祖亮，盛君杰，吕春绪，孔少义，陈天云，江志华，陆明，陈建强，惠君明，曹洁明，王依林，丁福来．膨化硝铵炸药连续混合装置．1997. 专利号：97236298.3.
20. 吕春绪，刘祖亮，王依林，陆明，陈天云，叶志文，胡炳成，惠君明．硝酸铵膨化剂．2000. 专利号：00112322.X.
21. 吕春绪，蔡春，黄小波．4,4′-二氨基二苯乙烯-2,2′-二磺酸合成工艺．2000. 专利号：00112135.9.
22. 陆明，吕春绪，刘祖亮，陈天云，叶志文，杜扬，胡炳成．高性能粉状铵油炸药及其制法．2003. 专利号：03112696.0.
23. 程广斌，吕春绪，张书，李斌栋，罗军，芮益民．1，2-丙二胺的合成方法．2007. 专利号：200710134091.8.
24. 叶志文，吕春绪．可聚合乳化剂丙烯酰氧基 Span80 及其制备方法、用途．2007. 专利号：200710135564.6.
25. 曹端林，郑志花，崔建兰，门吉英，吕春绪，陆明．间甲酚生产中的含盐废水处理方法．2007. 专利号：200710139539.5.
26. 曹端林，崔建兰，郑志花，赵林秀，吕春绪，陆明．一种间甲酚的制备工艺．2007. 专利号：200710139542.7.
27. 吕春绪，程广斌，刁春莉．硝基苯甲醛的合成方法．2007. 专利号：200710024363.9.
28. 程广斌，吕春绪，孙海龙．间羟基苯乙胺的合成方法．2007. 专利号：200710024886.3.
29. 胡炳成，刘祖亮，吕春绪，周维友．金属次卟啉化合物的制备及其应用方法．2007. 专利号：201010180790.8.
30. 胡炳成，刘祖亮，吕春绪，周维友．金属次卟啉化合物的制备及其应用方法．2007. 专利号：200710024885.9.
31. 吕春绪，叶方青，李斌栋，曾贵玉，周新利，杜扬．聚合物改性相稳定硝酸铵的制备方法．2007. 专利号：200710192273.0.
32. 周新利，吕春绪，刘祖亮．增敏型膨化铵油炸药及其制备方法．2008. 专利号：200810123716.5.
33. 李斌栋，吕春绪，梁政勇．微波卤素交换氟化制备氟代化合物的方法．2008. 专利号：200810123715.0.
34. 李斌栋，吕春绪，崔小刚．直接水解间甲苯胺制备间甲酚的方法．2008. 专利号：200810123714.6.
35. 叶志文，吕春绪．间硝基甲苯气相加氢制备间甲苯胺用催化剂及其制备方法和再生方法．2008. 专利号：200810234622.5.
36. 罗军，吕春绪，李斌栋，刘西．Balz-Schiemann 反应中副产物三氟化硼的回收利用方法．2008. 专利号：200810234321.2.
37. 李斌栋，吕春绪，许昕．氟代苯甲醛的氧化制备方法．2008. 专利号：200810236174.2.
38. 李斌栋，吕春绪，孙昱．催化加氢制备氯代苯胺的方法．2008. 专利号：200810236175.7.
39. 李斌栋，戴燕，罗军，吕春绪，李晶晶．双酚 AF 的制备方法．2008. 专利号：200810243703.1.
40. 李斌栋，戴燕，李晶晶，吕春绪，罗军．一步制备双酚 AF 的方法．2008. 专利号：200810243702.7.
41. 罗军，职慧珍，吕春绪，张强．聚乙二醇桥链的中性双阳离子型咪唑类离子液体一锅煮制备方法．2008. 专利号：200810243203.8.
42. 罗军，职慧珍，吕春绪，张强．一类具有氟两相性质的酸性功能化离子液体及其用途．2009. 专利号：200910028133.9.
43. 罗军，职慧珍，吕春绪，张强．一类具有氟两相性质的酸性功能化离子液体的制备方法．2009. 专利号：200910028134.3.
44. 罗军，职慧珍，吕春绪，张强．一类具有氟两相性质双子离子液体的制备方法．2009. 专利号：200910028135.8.
45. 罗军，职慧珍，吕春绪，张强．一类具有氟两相性质双子离子液体以及利用该离子液体制得的催化剂及其用途. 2009. 专利号：200910028136.2.

三、吕春绪教授获奖情况

序号	成果(项目)名称	获奖情况					主要合作者
		奖别(国家、省、部)名称	等级	排名	年份	证书号码	
01	3021炸药研究	江苏省科技进步奖	三等	1	1978		惠君明
02	SJY高爆速液体炸药	国家科技进步奖	三等	2	1985	85-KF5-3-161-2	吴腾芳
03	猛炸药化学及工艺学(下册)	兵总(部级)优秀教材奖	二等	2	1987		刘荣海 王文博
04	六硝基芪新工艺研究	兵总(部级)科技进步奖	二等	3	1991	91-BJ06-2-35-03	孙荣康
05	六硝基芪合成新方法	国家发明奖	三等	2	1995	95-KFD05-3-10-2	孙荣康、陆明
06	HF粉状岩石铵梯油炸药研究	江苏省科技进步奖	三等	3	1995	J3-1-3	郑礼木
07	工业炸药	兵总(部级)优秀教材奖	二等	1	1995	95-2-016-1	刘祖亮 倪欧琪
08	岩石膨化硝铵炸药	兵总(部级)科技进步奖	特等	1	1997	97-BM06-0-001-1	刘祖亮
09	岩石膨化硝铵炸药连续生产工艺	江苏省科技进步奖	三等	3	1998	J3-10-03	盛君杰
10	岩石膨化硝铵炸药	国家科技进步奖	二等	1	1998	30-2-003-01	刘祖亮
11	结晶梯恩梯工艺研究	江苏省科技进步奖	三等	2	1998	98-BJ06-3-045-2	惠君明
12	膨化硝铵炸药理论及其应用	国防科学技术奖	二等	1	2003	2003STFJ2191-1	刘祖亮 陆明
13	膨化硝酸铵及膨化硝酸铵炸药技术推广	中国民爆行业协会科学技术奖	一等	1	2008	2008EMTA-1-01	刘祖亮 陆明
14	自敏化技术及其在工业炸药中的应用	部技术发明奖	二等	2	2010	2009-142	刘祖亮
15	药物中间体化学	部科技进步奖	三等	1	2010	2010-JBA0429-3-1	胡炳成 叶志文
16	膨化硝铵炸药自敏化理论研究	中国民爆行业协会科学技术奖	特别奖	1	2010	2010EMTA-T-01	叶志文 胡炳成

四、吕春绪教授培养的学生

1. 吕春绪教授培养的硕士生

编号	姓名	入学时间	论文名称	工作单位
1	邓爱民	1985.09	1,3-二[3′-氯苯氨基]-2,4,6-三硝苯基的合成及硝化研究	沈阳理工大学研究生学院
2	王恩芳	1985.09	二氨基苦基乙烯二胺的合成	常州市范群干燥设备有限公司
3	吕早生	1989.09	一种新型高能耐热炸药的合成	武汉科技大学化学工程与技术学院
4	陆明	1987.09	两步合成六硝基芪的工艺研究	南京理工大学化工学院302教研室
5	蔡春	1988.09	硕博连读	南京理工大学化工学院302教研室
6	丁云	1988.09	表面活性剂改善硝酸铵及其炸药性能的研究	日本
7	顾建良	1990.09	乙酰苯胺和对甲氧基乙酰苯胺的选择性硝化研究	昆山
8	胡炳成	1991.09	表面活性剂在硝酸铵改性中的应用研究	南京理工大学化工学院302教研室

续表

编号	姓名	入学时间	论文名称	工作单位
9	金铁柱	1992.09	正十二烷基苯的合成与选择性硝化	南京东松国际贸易有限公司
10	叶志文	1993.09	硝酸铵膨化机理的研究	南京理工大学化工学院302教研室
11	李志平	1993.09	高能量密度材料的合成研究	中国人民大学化学系
12	董国勋	1994.09	控制氧化机理及应用	上海
13	戴晖	1995.09	甲苯的混酸选择性硝化研究	美国
14	蔡敏敏	1996.09	硕博连读	美国
15	黄小波	1997.09	DSD酸合成新工艺	宁波武盛化学有限公司
16	罗军	1998.09	硕博连读	南京理工大学化工学院302教研室
17	罗世平	1995.09	论甲基纤维素系列产品的应用与发展-兼论泸州化工厂甲基纤维素系列产品的发展战略	中国北方化工集团公司
18	李司南	1999.09	亚甲基类化合物合成2,4,6,8-四硝基-2,4,6,8-四氮杂双环[3,3,0]辛烷的探索和研究	美国
19	亓希国	1999.01	城市建筑物拆除控制爆破的几个安全管理问题	公安部治安管理局
20	曹奇	1999.01	民用爆破器材安全管理	江苏省公安厅
21	商健	1999.01	民用爆炸物品管理系统	南京市公安局
22	曹平	1991.01	中国税务人才研究	南京市国家税务局征管科技处
23	赵国强	1999.01	膨化硝铵炸药连续生产线工艺研究	广西柳州威奇化工有限公司
24	李斌栋	1999.09	催化氢化合成DSD酸工艺研究	南京理工大学化工学院302教研室
25	刘桂涛	1999.09	超细(微米或亚微米)RXD炸药应用特性研究	南京理工大学泰州办公室主任
26	高荣	2000.09	三苯基膦合成新工艺研究	句容
27	迟波	2001.09	季铵盐Gemini表面活性剂的合成及表面活性的研究	南京工业大学食品与轻工学院
28	户安军	2001.09	iNOS抑制剂的设计与合成及iNOS抑制剂活性测定	苏州诺华制药科技有限公司
29	叶方青	2001.09	氟代脱硝合成芳香族氟化物的研究	南京公安局特警支队二大队
30	高棉(缅甸)	2001.09	Research on Polymer Bonded Explosives	缅甸
31	梁政勇	2002.09	螯合性表面活性剂的合成及性能研究	郑州大学化工与能源学院
32	钱华	2003.09	阴离子二聚表面活性剂的合成及表征	南京理工大学民爆中心
33	郑永勇	2003.09	一种新型含硼表面活性剂的合成与性能研究	上海医药工业研究院
34	吴秋洁	2004.09	Ni-B非晶态合金催化加氢制备氟代苯胺的研究	南京理工大学民爆中心
35	邹琪	2004.09	单脂肪酸甘油的合成与应用	华东理工大学实验13楼318室
36	陈祥	2004.09	阴离子表面活性剂脂肪酸甲酯磺酸盐的合成与性能研究	南京紫高科技实业有限公司
37	霍婷	2004.09	控制氧化合成对氯苯甲醛的研究	西安瑞联近代电子材料有限责任公司科研一部
38	徐珍	2004.09	卤素交换氟化制备氟代苯甲醛研究	武汉科技大学化工学院
39	刘登红	2004.09	托尼卟吩单体2-溴-5-甲酰基-3-甲基吡咯的合成与工艺研究	淮安
40	方亚辉	2005.09	控制氧化合成苯甲醛的研究	上海应用技术学院
41	冯超	2005.09	七氟醚的合成研究	新加坡
42	翟羽佳	2005.09	可聚合硼酸酯表面活性剂的合成及性能研究	中共河南省委统战部党外知识分子工作处
43	陈丽丽	2005.09	两种光稳定剂的合成研究	四川省成都市温江区长安路233号
44	孙卫芳	2005.09	间甲苯胺合成间氟苯甲醛的研究	中化上海有限公司上海市黄浦区河南南路33号1909室
45	田璇	2005.09	间苯二酚低污染合成工艺研究	江苏圣泰科合成化学有限公司
46	李霞	2005.09	固体酸催化甲苯的区域选择性硝化研究	南京市红宝丽股份有限公司研究院
47	林卫娜	2005.09	一种共聚高分子硼酸酯表面活性剂的合成与性能研究	南京市红宝丽股份有限公司
48	刘红梅	2005.09	氧化合成六氟丙酮的工艺研究	中国江苏国际经济技术合作公司
49	陈怡欣	2006.09	3-脱氧-1,2-O-异丙叉-D-呋喃木糖的合成	南通大学地理学科学院
50	季世春	2006.09	2,6-二甲基吗啉及其衍生物的合成和工艺研究	南京海陵药业
51	崔小刚	2006.09	直接水解间甲苯胺制备甲苯酚工艺研究	江苏恒瑞医药股份有限公司
52	唐大林	2006.09	间苯二酚衍生物的合成和工艺研究	上海药明康德新药开发有限公司

续表

编号	姓名	入学时间	论文名称	工作单位
53	施春燕	2006.09	吉西他滨中间体的合成	常熟华益化工有限公司
54	许昕	2006.09	间氟甲苯控制氧化合成间氟苯甲醛	安徽六国化工股份有限公司研发部
55	刘东胜	2006.09	相稳定硝酸铵相转变影响因素研究	南京理工大学环境学院
56	谯娟	2007.09	聚合物相稳定硝酸铵的研究	西安近代化学研究所
57	李晶晶	2007.09	合成双酚AF的新工艺研究	上海胜邦质量检测有限公司
58	李鹏	2007.09	橡胶防老剂4020合成工艺研究	广东省公安消防总队中山市公安消防支队
59	何永利	2007.09	4-氨基二苯胺合成工艺研究	上海药明康德新药开发有限公司
60	顾苗	2007.09	甘油合成环氧氯丙烷的工艺研究	上海药明康德新药开发有限公司
61	马小明	2007.09	芳烃的N_2O_5清洁催化硝化及其机理研究	南京理工大学
62	沈青燕	2008.09	曼尼希法合成二聚的苯并三唑类紫外吸收剂	江苏汇鸿国际集团土产进出口股份有限公司
63	杜彬斌	2008.09	2,6-二羟基甲苯合成工艺研究	常州华日新材有限公司
64	卞成明	2008.09	圆筒式离心萃取器的程序设计及其在BuNENA后处理中的应用	北京理工大学
65	祝迎花	2008.09	苯并三唑类紫外吸收剂的合成工艺研究	合肥天鹅制冷科技有限公司
66	尤婷	2008.09	聚丙烯酸钾相稳定硝酸铵的研究	南京纺织品进出口股份有限公司
67	仲崇超	2009.09	西司他丁中间体的合成及其工艺研究	南京理工大学
68	余玉静	2009.09	四氢罂粟碱甲酸盐及其拆分研究	南京理工大学
69	谢丽	2009.09	聚异丁烯丁二酸醇胺酯的合成、性质及应用	南京理工大学
70	周遂	2009.09	地塞米松磷酸钠工艺合成研究	南京理工大学
71	郭晓晶	2009.09	聚异丁烯丁二酸糖醇酯的合成及应用研究	南京理工大学
72	薛叙明	2003.09	访问学者	常州工程技术学院化工系

2. 吕春绪教授培养的博士及博士后

编号	姓名	入学时间	论文名称	工作单位
1	蔡春	1990.09	芳香族化合物的控制硝化及机理研究	南京理工大学化工学院302教研室
2	彭新华	1993.09	固体催化剂和载体上芳烃的选择性硝化反应研究	南京理工大学化工学院305教研室
3	董国勋	1996.03	硝酸铵膨化机理研究	上海
4	陆明	1996.09	膨化硝铵炸药研究	南京理工大学化工学院302教研室
5	赵省向	1997.02	EAKR分子间炸药研究	兵器工业第二零四研究所
6	陈天云	1997.09	硝酸铵的自敏化及其应用研究	合肥工业大学化学工程学院
7	吕早生	1998.09	N_2O_4-O_3硝化芳香化合物的宏观动力学及机理的研究	武汉科技大学化学工程与技术学院
8	蔡敏敏	1998.09	氧气/空气液相氧化一元取代甲苯成芳香醛工艺及机理研究	美国
9	程广斌	1999.03	固体酸催化剂上芳香族化合物区域选择性硝化反应研究	南京理工大学化工学院305教研室
10	罗军	1999.09	微波促进卤素交换氟化反应研究	南京理工大学化工学院302教研室
11	黄龙祥	1999.09	药物中间体合成及应用研究	常州梓瑞化工有限公司
12	曹阳	2000.03	硝酰阳离子的结构、光谱与硝化反应中的电子转移	江苏悦华药业有限公司
13	胡炳成	2000.03	吡咯烷硫酮衍生物的合成及结构表征	南京理工大学化工学院302教研室
14	高大元	2000.09	混合炸药爆轰与安全性能实验与理论研究	中国工程物理研究院化工材料研究所
15	周新利	2000.09	膨化硝酸铵自敏化理论及其炸药的物理性能和改性研究	南京理工大学化工学院302教研室
16	叶志文	2000.09	阳离子型Gemini表面活性剂的合成、性质及应用研究	南京理工大学化工学院302教研室
17	陶锋	2000.09	氟两相技术在替米沙坦中间体合成中的应用研究	常州亚邦制药有限公司
18	杜杨	2000.09	硝酸铵相稳定高聚物合成及应用研究	南京理工大学化工学院302教研室
19	袁淑军	2001.02	新型金属络合性表面活性剂的合成、性质及应用研究	江苏省盐城工学院纺织服装学院

续表

编号	姓名	入学时间	论文名称	工作单位
20	卫延安	2001.02	合成邻苯基苯酚的催化剂的制备及性能研究	南京理工大学化工学院
21	张永旭	2001.02	含能材料纳米粉体的制备和性能研究	上海广茂达光艺科技股份有限公司
22	方永勤	2002.02	氟代芳香醛的制备及仿生催化剂的应用	常州大学设计研究院
23	刘学民	2002.02	新型表面活性剂合成技术研究	江南大学化学与材料工程学院
24	张所信	2002.02	药物中间体合成	连云港淮海工学院化学工程学院
25	马登生	2002.02	卟啉衍生物的合成及应用研究	西南科技大学国防科技学院
26	许虎君	2002.02	烷基二苯醚二磺酸钠的合成、聚集体系及应用性能研究	江南大学化学与材料工程学院
27	李斌栋	2002.09	含硼表面活性剂的合成、性质及应用研究	南京理工大学化工学院302教研室
28	刘桂涛	2003.02	含能材料及其理论研究	南京理工大学泰州学院办公室
29	李广学	2003.02	硝基苯Pt/C催化加氢制备对氨基苯酚的宏观动力学研究	安徽理工大学化工学院
30	孙昱	2003.09	氯代硝基苯选择性加氢制备氯代苯胺的研究	武汉青山区武汉科技大学化工学院
31	王海鹰	2003.09	新型可聚合硼酸酯表面活性剂及其聚合物的合成、性能研究	湖南中南大学冶金科学与工程学院
32	沈健	2004.02	生物医用高分子材料的研制及其基础研究	江苏省教育厅
33	陆鸿飞	2004.02	嘌呤2,6位取代衍生物的合成	江苏科技大学生化学院
34	齐秀芳	2004.02	Bronsted酸性离子液体存在下的硝化(解)反应研究	西南科技大学国防科技学院
35	蔡超君	2004.02	托尼卟吩模型中各结构单元的合成及表征	凯瑞斯德(苏州)有限公司
36	刘丽荣	2004.09	固体酸和酸性离子液体催化甲苯选择性硝化反应的研究	连云港淮海工学院化学工程学院
37	梅震华	2004.09	工业炸药自敏化理论研究	南京市公安局治安支队
38	王艺	2005.02	环氧氯丙烷理论与合成技术研究	美国
39	梁政勇	2005.02	卤素交换氟化反应技术研究	郑州大学化工与能源学院
40	户安军	2005.02	甲苯类化合物的选择性氧化	苏州诺华制药科技有限公司
41	曾贵玉	2005.02	炸药微观结构对性能的影响研究	中国工程物理研究院化工材料研究所
42	钱华	2005.09	五氧化二氮在硝化反应中的应用研究	南京理工大学化工学院民爆中心
43	职慧珍	2005.09	具有温控两相性质的聚乙二醇型双子离子液体及其在有机合成中的应用	南京师范大学化学与材料科学学院
44	叶方青	2006.02	相稳定硝酸铵的理论与技术研究	南京市公安局特警支队二大队
45	胡玉锋	2006.02	氯代脱硝法合成含氟芳香族化合物的研究	南京理工大学化工学院
46	戴燕	2006.09	芳基六氟异丙醇的合成及其应用研究	中国林业科学研究院林产化学工业研究所
47	唐双凌	2006.01	硝酸铵的钝感化改性研究	南京理工大学化工学院
48	夏明珠	2006.01	膦酰基羧酸的合成与性能	南京理工大学化工学院
49	马晓明	2007.01	芳烃的N_2O_5清洁催化硝化及其机理研究	南京理工大学化工学院
50	何志勇	2007.01	N_2O_5硝解制备HMX及机理研究	南京台硝化工有限公司
51	陈厚和(博士后)	2006	纳米黑索金制备技术与炸药性能研究	南京理工大学化工学院303教研室
52	杨红伟(博士后)	2008	新型转译起始的抑制抗癌试剂研究及螺旋化合物的合成	南京理工大学化工学院305教研室
53	朱晨光(博士后)	2008	微泡雾形成与遮蔽机理研究	南京理工大学化工学院303教研室